HEINEMANN CHEMISTRY 1

6TH EDITION

COORDINATING AUTHOR
Melissa MacEoin

AUTHORS
Drew Chan
Chris Commons
Penny Commons
Lanna Derry
Elizabeth Freer
Elissa Huddart
Louise Lennard
Mick Moylan
Pat O'Shea
Bob Ross
Kellie Vanderkruk

VCE UNITS 1 AND 2 • 2023–2027

Pearson Australia
(a division of Pearson Australia Group Pty Ltd)
459–471 Church Street
Level 1, Building B
Richmond, Victoria 3121
www.pearson.com.au

First published 2023 by Pearson Australia
2026 2025 2024 2023
10 9 8 7 6 5 4 3 2 1

VCE Heinemann Project Leads: Bryonie Scott, Misal Belvedere, Malcolm Parsons
Senior Development Editor: Fiona Cooke
Development Editor: Leanne Peters
Schools Programme Manager: Michelle Thomas
Senior Production Editor: Casey McGrath
Series Designer: Anne Donald
Rights & Permissions Editor: Amirah Fatin Binte Mohamed Sapi'ee
Publishing Services Analyst: Jit-Pin Chong
Illustrators: Bruce Rankin, DiacriTech, Anne Donald, QBS
Editor: Maryanne Park
Proofreader: Margaret Trudgeon
Indexer: Penelope Goodes

Printed in Malaysia (CTP-VVP)

National Library of Australia Cataloguing-in-Publication entry

A catalogue record for this book is available from the National Library of Australia

ISBN 978 0 6557 0006 7

Pearson Australia Group Pty Ltd ABN 40 004 245 943

Disclaimer
The selection of internet addresses (URLs) provided for this book was valid at the time of publication and was chosen as being appropriate for use as a secondary education research tool. However, due to the dynamic nature of the internet, some addresses may have changed, may have ceased to exist since publication, or may inadvertently link to sites with content that could be considered offensive or inappropriate. While the authors and publisher regret any inconvenience this may cause readers, no responsibility for any such changes or unforeseeable errors can be accepted by either the authors or the publisher.

Indigenous Australians
We respectfully acknowledge the traditional custodians of the lands upon which the many schools throughout Australia are located. We acknowledge traditional Indigenous Knowledge systems are founded upon a logic that developed from Indigenous experience with the natural environment over thousands of years. This is in contrast to Western science, where the production of knowledge often takes place within specific disciplines. As a result, there are many cases where such disciplinary knowledge does not reflect Indigenous worldviews, and can be considered offensive to Indigenous peoples.

Some of the images used in *Heinemann Chemistry 1* 6th edition might have associations with deceased Indigenous Australians. Please be aware that these images might cause sadness or distress in Aboriginal or Torres Strait Islander communities.

Practical activities
All practical activities, including the illustrations, are provided as a guide only and the accuracy of such information cannot be guaranteed. Teachers must assess the appropriateness of an activity and take into account the experience of their students and facilities available. Additionally, all practical activities should be trialled before they are attempted with students and a risk assessment must be completed. All care should be taken and appropriate personal protective clothing and equipment should be worn when carrying out any practical activity. Although all practical activities have been written with safety in mind, Pearson Australia and the authors do not accept any responsibility for the information contained in or relating to the practical activities, and are not liable for loss and/or injury arising from or sustained as a result of conducting any of the practical activities described in this book.

HEINEMANN CHEMISTRY 1

6TH EDITION

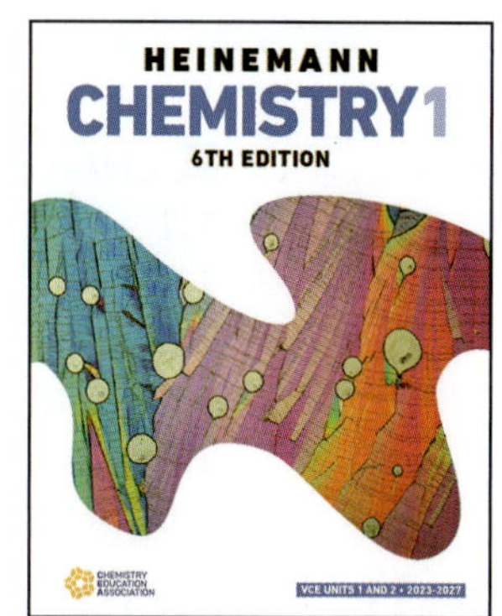

Writing and development team

We are grateful to the following people for their time and expertise in contributing to the **Heinemann Chemistry 1** project.

Melissa MacEoin
Chemistry teacher
Coordinating author and reviewer

Doug Bail
Education consultant
SPARKlab developer

Reuben Bolt
Deputy Vice-Chancellor First Nations Leadership
Reviewer

Erin Bruns
Chemistry teacher
Author and answer checker

Drew Chan
Chemistry teacher, Head of Chemistry
Author

Chris Commons
Retired chemistry teacher and honorary chemistry researcher
Author and reviewer

Penny Commons
Honorary chemistry lecturer and tutor
Author

Lanna Derry
Chemistry teacher, Head of Science
Coordinating author (Year 12) and reviewer

Elizabeth Freer
Chemistry teacher
Author

Renee Hill
Chemistry teacher, Head of Chemistry
Author

Elissa Huddart
Chemistry teacher, Head of Science
Author

Carissa Kelly
Chemistry teacher
Answer checker

Jacqueline Kerr
Chemistry teacher
Author

Rowan Kidd
Chemistry teacher
Author

Louise Lennard
Chemistry teacher
Author

Tanneale Marshall
Laboratory technician
Reviewer

Mick Moylan
Chemistry lecturer
Author

Pat O'Shea
Chemistry teacher
Author

Bob Ross
Retired chemistry teacher
Author

Sophie Selby-Pham
Scientist
Answer checker

Karen Smith
Laboratory technician
Reviewer

Kellie Vanderkruk
Environmental scientist and former chemistry lecturer
Author

Nic Volkmann
Chemistry teacher and edupreneur
VOLK video developer

Gregory White
Scientist
Answer checker

Laurence Wooding
Laboratory technician
Reviewer

The Publisher wishes to thank and acknowledge Warwick Clarke and Geoff Quinton for their contribution in creating the original content for the 5th edition and their longstanding dedicated work with Pearson and Heinemann.

The Chemistry Education Association was formed in 1977 by a group of chemistry teachers from secondary and tertiary institutions. It aims to promote the teaching and learning of chemistry in schools. The CEA has established a tradition of providing professional development opportunities for teachers and up-to-date text and electronic materials, together with other resources supporting both students and teachers. The CEA offers scholarships and bursaries to students and teachers to further their interest in chemistry, assists the Science Teachers Association of Victoria (STAV) to present the VCE Chemistry Conference and sponsors prizes for the STAV Science Talent Search.

Contents

Unit 1 How can the diversity of materials be explained?

■ AREA OF STUDY 1

How do the chemical structures of materials explain their properties and reactions?

AREA OF STUDY 2

How are materials quantified and classified?

AREA OF STUDY 3

Heinemann Chemistry 1 6th edition includes a comprehensive set of resources to support Area of Study 3 via your Pearson Places bookshelf.

Unit 2 How do chemical reactions shape the natural world?

AREA OF STUDY 1

How do chemicals interact with water?

AREA OF STUDY 2

How are chemicals measured and analysed?

AREA OF STUDY 3

Heinemann Chemistry 1 6th edition includes a comprehensive set of resources to support Area of Study 3 via your Pearson Places bookshelf.

How to use this book

Heinemann Chemistry 1 6th edition

Heinemann Chemistry 1 6th edition has been written to the new VCE Chemistry Study Design 2023–2027. The book covers Units 1 and 2. Explore how to use this book below.

Case study

Case studies place chemistry in an applied situation or relevant context. Text and artwork refer to the nature and practice of chemistry, applications of chemistry and associated issues, and the historical development of chemical concepts and ideas.

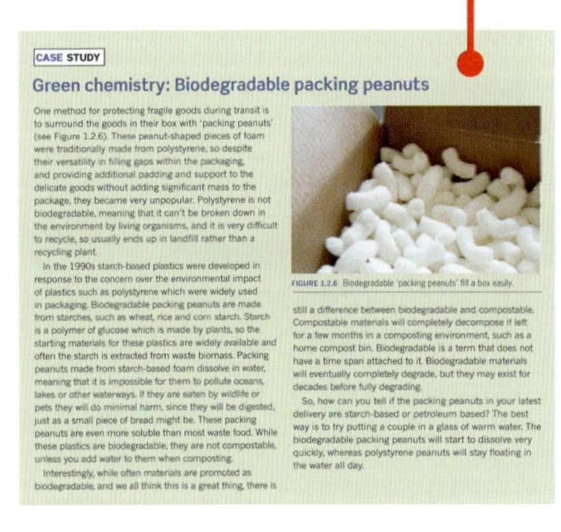
CASE STUDY

Green chemistry: Biodegradable packing peanuts

One method for protecting fragile goods during transit is to surround the goods in their box with 'packing peanuts' (see Figure 1.2.6). These peanut-shaped pieces of foam were traditionally made from polystyrene, so despite their versatility in filling gaps within the packaging, and providing additional padding and support to the delicate goods without adding significant mass to the package, they became very unpopular. Polystyrene is not biodegradable, meaning that it can't be broken down in the environment by living organisms, and it is very difficult to recycle, so usually ends up in landfill rather than a recycling plant.

In the 1990s starch-based plastics were developed in response to the concern over the environmental impact of plastics such as polystyrene which were widely used in packaging. Biodegradable packing peanuts are made from starches, such as wheat, rice and corn starch. Starch is a polymer of glucose which is made by plants, so the starting materials for these plastics are widely available and often the starch is extracted from waste biomass. Packing peanuts made from starch-based foam dissolve in water, meaning that it is impossible for them to pollute oceans, lakes or other waterways. If they are eaten by wildlife or pets they will do minimal harm, since they will be digested, just as a small piece of bread might be. These packing peanuts are even more soluble than most waste food. While these plastics are biodegradable, they are not compostable, unless you add water to them when composting.

Interestingly, while often materials are promoted as biodegradable, and we all think this is a great thing, there is still a difference between biodegradable and compostable. Compostable materials will completely decompose if left for a few months in a composting environment, such as a home compost bin. Biodegradable is a term that does not have a time span attached to it. Biodegradable materials will eventually completely degrade, but they may exist for decades before fully degrading.

So, how can you tell if the packing peanuts in your latest delivery are starch-based or petroleum based? The best way is to try putting a couple in a glass of warm water. The biodegradable packing peanuts will start to dissolve very quickly, whereas polystyrene peanuts will stay floating in the water all day.

FIGURE 1.2.6 Biodegradable 'packing peanuts' fill a box easily.

Case study: Analysis

These case studies include real-world data that can be analysed and evaluated.

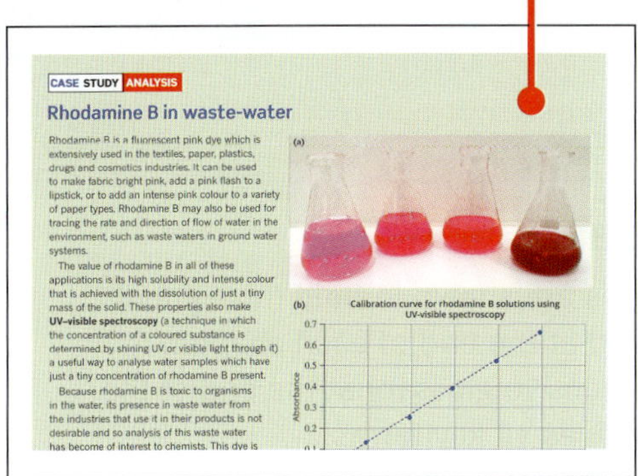
CASE STUDY ANALYSIS

Rhodamine B in waste-water

Rhodamine B is a fluorescent pink dye which is extensively used in the textiles, paper, plastics, drugs and cosmetics industries. It can be used to make fabric bright pink, add a pink flash to a lipstick, or to add an intense pink colour to a variety of paper types. Rhodamine B may also be used for tracing the rate and direction of flow of water in the environment, such as waste waters in ground water systems.

The value of rhodamine B in all of these applications is its high solubility and intense colour that is achieved with the dissolution of just a tiny mass of the solid. These properties also make **UV–visible spectroscopy** (a technique in which the concentration of a coloured substance is determined by shining UV or visible light through it) a useful way to analyse water samples which have just a tiny concentration of rhodamine B present.

Because rhodamine B is toxic to organisms in the water, its presence in waste water from the industries that use it in their products is not desirable and so analysis of this waste water has become of interest to chemists. This dye is

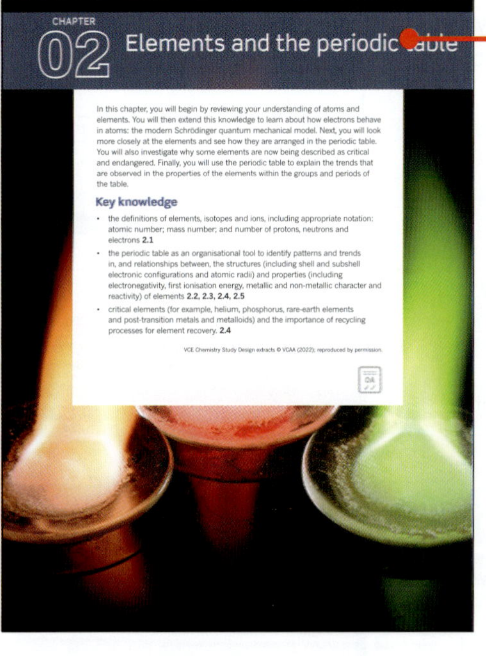
CHAPTER 02 Elements and the periodic table

In this chapter, you will begin by reviewing your understanding of atoms and elements. You will then extend this knowledge to learn about how electrons behave in atoms: the modern Schrödinger quantum mechanical model. Next, you will look more closely at the elements and see how they are arranged in the periodic table. You will also investigate why some elements are now being described as critical and endangered. Finally, you will use the periodic table to explain the trends that are observed in the properties of the elements within the groups and periods of the table.

Key knowledge

- the definitions of elements, isotopes and ions, including appropriate notation: atomic number; mass number; and number of protons, neutrons and electrons **2.1**
- the periodic table as an organisational tool to identify patterns and trends in, and relationships between, the structures (including shell and subshell electronic configurations and atomic radii) and properties (including electronegativity, first ionisation energy, metallic and non-metallic character and reactivity) of elements **2.2, 2.3, 2.4, 2.5**
- critical elements (for example, helium, phosphorus, rare-earth elements and post-transition metals and metalloids) and the importance of recycling processes for element recovery **2.4**

VCE Chemistry Study Design extracts © VCAA (2022); reproduced by permission.

Chapter opener

Chapter opening pages link the study design to the chapter content. Key knowledge addressed in the chapter is clearly listed. To help you find where each outcome is covered in the chapter, the relevant section numbers are written in bold.

Highlight

Highlight boxes focus on important information such as key definitions and summary points.

ChemFile

ChemFiles include interesting information and real-world examples.

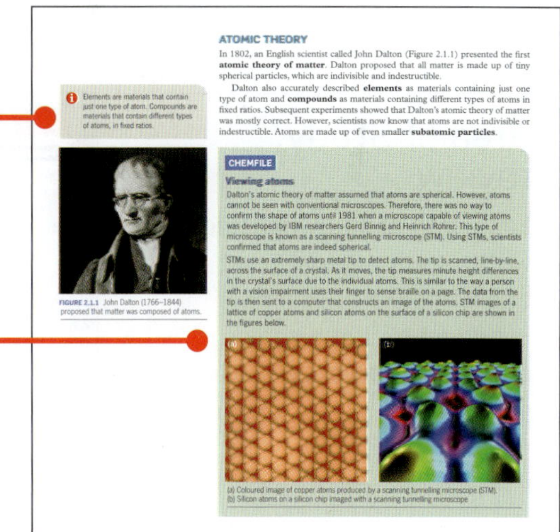
ATOMIC THEORY

In 1802, an English scientist called John Dalton (Figure 2.1.1) presented the first **atomic theory of matter**. Dalton proposed that all matter is made up of tiny spherical particles, which are indivisible and indestructible.

Dalton also accurately described **elements** as materials containing just one type of atom and **compounds** as materials containing different types of atoms in fixed ratios. Subsequent experiments showed that Dalton's atomic theory of matter was mostly correct. However, scientists now know that atoms are not indivisible or indestructible. Atoms are made up of even smaller **subatomic particles**.

Elements are materials that contain just one type of atom. Compounds are materials that contain different types of atoms, in fixed ratios.

FIGURE 2.1.1 John Dalton (1766–1844) proposed that matter was composed of atoms.

CHEMFILE

Viewing atoms

Dalton's atomic theory of matter assumed that atoms are spherical. However, atoms cannot be seen with conventional microscopes. Therefore, there was no way to confirm the shape of atoms until 1981 when a microscope capable of viewing atoms was developed by IBM researchers Gerd Binnig and Heinrich Rohrer. This type of microscope is known as a scanning tunnelling microscope (STM). Using STMs, scientists confirmed that atoms are indeed spherical.

STMs use an extremely sharp metal tip to detect atoms. The tip is scanned, line-by-line, across the surface of a crystal. As it moves, the tip measures minute height differences in the crystal's surface due to the individual atoms. This is similar to the way a person with a vision impairment uses their finger to sense braille on a page. The data from the tip is then sent to a computer that constructs an image of the atoms. STM images of a lattice of copper atoms and silicon atoms on the surface of a silicon chip are shown in the figures below.

(a) Coloured image of copper atoms produced by a scanning tunnelling microscope (STM).
(b) Silicon atoms on a silicon chip imaged with a scanning tunnelling microscope

Icons

This icon is used to alert you to engage with auto-corrected questions through Pearson Places.

These icons indicate when it is the best time to engage with a worksheet (WS), a practical activity (PA) or exam questions (EQ) in the *Heinemann Chemistry 1 Skills and Assessment* book.

Worked example

Worked examples are set out in steps that show thinking and working. Each Worked example is followed by a Worked example: Try yourself that allows you to test your understanding.

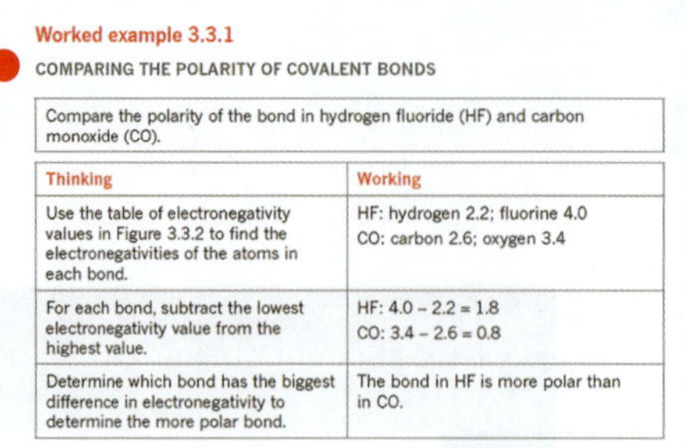

Worked example 3.3.1

COMPARING THE POLARITY OF COVALENT BONDS

Compare the polarity of the bond in hydrogen fluoride (HF) and carbon monoxide (CO).

Thinking	Working
Use the table of electronegativity values in Figure 3.3.2 to find the electronegativities of the atoms in each bond.	HF: hydrogen 2.2; fluorine 4.0 CO: carbon 2.6; oxygen 3.4
For each bond, subtract the lowest electronegativity value from the highest value.	HF: 4.0 − 2.2 = 1.8 CO: 3.4 − 2.6 = 0.8
Determine which bond has the biggest difference in electronegativity to determine the more polar bond.	The bond in HF is more polar than in CO.

Chapter review

Each chapter concludes with a list of key terms and questions that test your understanding of the key knowledge covered in the chapter.

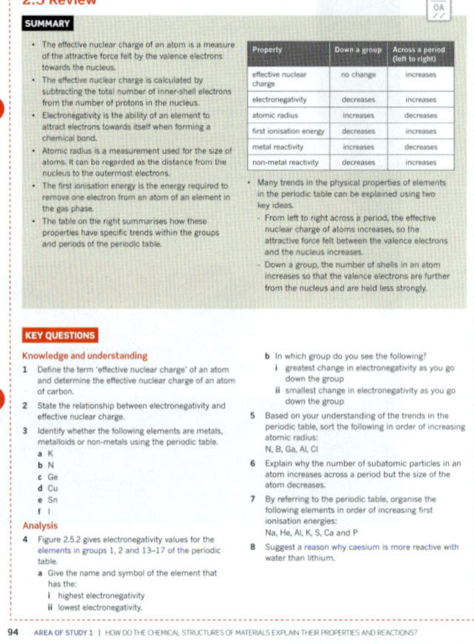

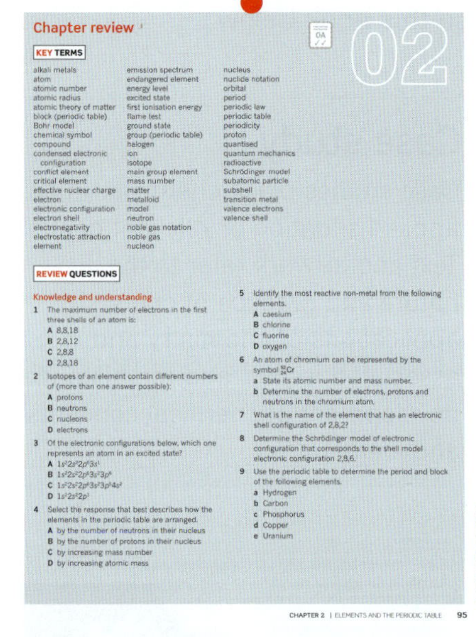

Section summary

Each section includes a summary to help you consolidate key points and concepts.

Section review

Each section concludes with questions that test your ability to recall, explain and apply key concepts.

Area of Study review

Each area of study concludes with a comprehensive set of exam-style questions, including multiple choice and short answer, to support you in your exam preparation.

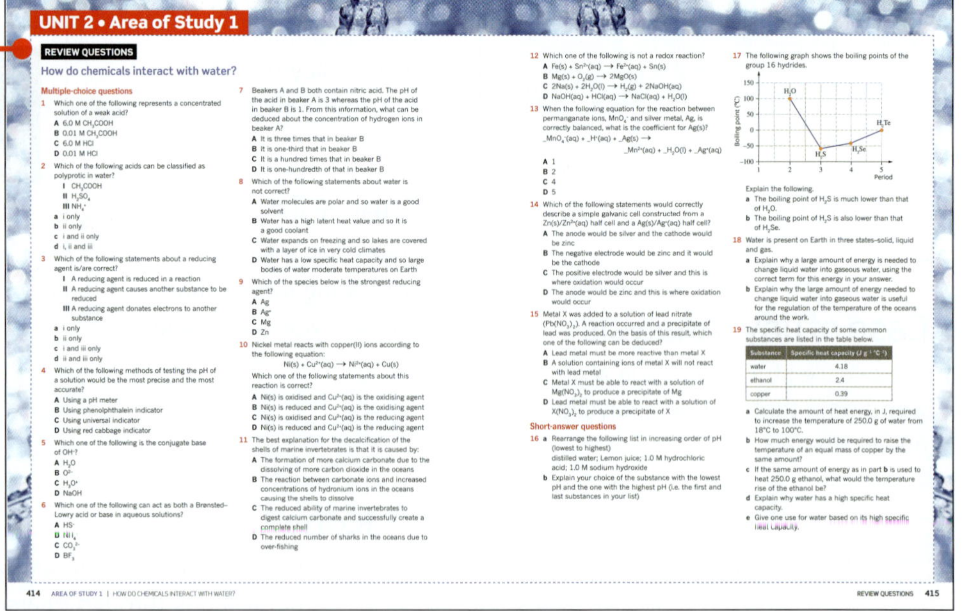

Answers

Comprehensive answers for all section review, chapter review and Area of Study review questions are provided via the *Heinemann Chemistry 1 6th edition eBook + Assessment.*

Glossary

Key terms are shown in **bold** throughout and listed at the end of each chapter. A comprehensive glossary at the end of the book defines all key terms.

250
200

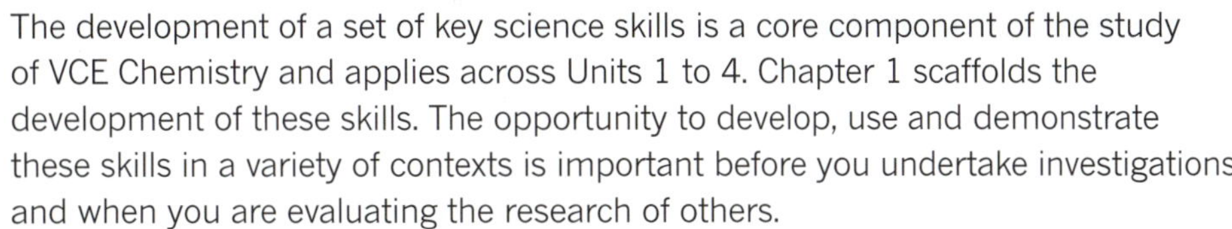

CHAPTER 01 Scientific investigation

The development of a set of key science skills is a core component of the study of VCE Chemistry and applies across Units 1 to 4. Chapter 1 scaffolds the development of these skills. The opportunity to develop, use and demonstrate these skills in a variety of contexts is important before you undertake investigations and when you are evaluating the research of others.

Although this chapter can be read as a whole, it is best to refer to it and use it when the need arises as you work through other chapters. For example, you may need a refresher on what is required to be included in a conclusion. This chapter also contains useful checklists to help you plan investigations, graph results and complete aspects of a scientific report, such as that required for Area of Study 3 in Unit 2. Similarly, when performing a practical investigation, refer to this chapter to make sure that you collect data properly and that your data is of high quality.

Key science skills

Develop aims and questions, formulate hypotheses and make predictions

- identify, research and construct aims and questions for investigation **1.1, 1.2**
- identify independent, dependent and controlled variables in controlled experiments **1.2**
- formulate hypotheses to focus investigations **1.2**
- predict possible outcomes of investigations **1.2**

Plan and conduct investigations

- determine appropriate investigation methodology: case study; classification and identification; controlled experiment; fieldwork; literature review; modelling; product, process or system development; simulation **1.1**
- design and conduct investigations; select and use methods appropriate to the selected investigation methodology, including consideration of sampling technique and size, equipment and procedures, taking into account potential sources of error and causes of uncertainty; determine the type and amount of qualitative and/or quantitative data to be generated or collated **1.1, 1.2, 1.3**
- work independently and collaboratively as appropriate and within identified research constraints, adapting or extending processes as required and recording such modifications in a logbook **1.3**

Comply with safety and ethical guidelines

- demonstrate safe laboratory practices when planning and conducting investigations by using risk assessments that are informed by safety data sheets (SDS), and accounting for risks **1.2**
- apply relevant occupational health and safety guidelines while undertaking practical investigations **1.2**
- demonstrate ethical conduct when undertaking and reporting investigations **1.2**

Generate, collate and record data

- systematically generate and record primary data, and collate secondary data, appropriate to the investigation, including use of databases and reputable online data sources **1.3**
- record and summarise both qualitative and quantitative data, including use of a logbook as an authentication of generated or collated data **1.3**
- organise and present data in useful and meaningful ways, including schematic diagrams, flow charts, tables, bar charts, line graphs and calibration curves **1.3, 1.4**

Analyse and evaluate data and investigation methods

- process quantitative data using appropriate mathematical relationships and units, including calculations of ratios, percentages, percentage change and mean **1.4**
- use appropriate numbers of significant figures in calculations **1.3, 1.4**
- plot graphs involving two variables that show linear and non-linear relationships **1.4**
- identify and analyse experimental data qualitatively, handling, where appropriate, concepts of: accuracy, precision, repeatability, reproducibility, resolution, and validity of measurements; and errors (random and systematic) **1.3**
- identify outliers, and contradictory, provisional or incomplete data **1.4**
- repeat experiments to evaluate the precision of data **1.3**
- evaluate investigation methods and suggest ways to improve precision, and to reduce the likelihood of errors **1.3, 1.5**

Construct evidence-based arguments and draw conclusions

- distinguish between opinion, anecdote and evidence, and scientific and non-scientific ideas **1.1, 1.2**
- evaluate data to determine the degree to which the evidence supports the aim of the investigation, and make recommendations, as appropriate, for modifying or extending the investigation **1.5**
- evaluate data to determine the degree to which the evidence supports or refutes the initial prediction or hypothesis **1.5**
- use reasoning to construct scientific arguments, and to draw and justify conclusions consistent with evidence and relevant to the question under investigation **1.5**
- identify, describe and explain the limitations of conclusions, including identification of further evidence required **1.5**
- discuss the implications of research findings and proposals **1.5**

Analyse, evaluate and communicate scientific ideas

- use appropriate chemical terminology, representations and conventions, including standard abbreviations, graphing conventions, algebraic equations, units of measurement and significant figures **1.3, 1.4, 1.5, 1.6**
- discuss relevant chemical information, ideas, concepts, theories and models and the connections between them **1.1, 1.2, 1.3, 1.4, 1.5**
- analyse and explain how models and theories are used to organise and understand observed phenomena and concepts related to chemistry, identifying limitations of selected models/theories **1.1, 1.2, 1.3, 1.4**
- critically evaluate and interpret a range of scientific and media texts (including journal articles, mass media communications and opinions in the public domain), processes, claims and conclusions related to chemistry by considering the quality of available evidence **1.2, 1.5, 1.6**
- apply sustainability concepts (green chemistry principles, development goals and the transition from a linear towards a circular economy) to analyse and evaluate responses to chemistry-based scenarios, case studies, issues and challenges **1.2**
- identify and explain when judgments or decisions associated with chemistry-related issues may be based on sociocultural, economic, political, legal and/or ethical factors, and not solely on scientific evidence **1.1**
- use clear, coherent and concise expression to communicate to specific audiences and for specific purposes in appropriate scientific genres, including scientific reports and posters **1.6**
- acknowledge sources of information and assistance, and use standard scientific referencing conventions **1.6**

OA ✓✓

1.1 The nature of scientific investigations

Chemistry is the study of matter, how it behaves and interacts with other matter. As scientists, chemists extend their understanding using the scientific method, which involves investigations that are carefully designed, conducted and reported. Well-designed research is based on a sound knowledge of what is already understood about a subject, as well as careful preparation, measurements and observations.

OBSERVATIONS

When you make **observations** during a scientific investigation, you use your senses and a wide variety of instruments and laboratory techniques. **Qualitative** observations provide information about what is present. For example, the colour of a flame in a flame test, as shown in Figure 1.1.1, can indicate the metal ions present, and the distance moved by a component in chromatography can help identify the sample. In contrast, **quantitative** observations provide numerical information to answer questions such as how many, how much and how often; for example, by taking a reading on an electronic balance, or measuring the initial and final volumes on a burette. Quantitative observations are accompanied by relevant units such as grams (g) or millilitres (mL) (Figure 1.1.2).

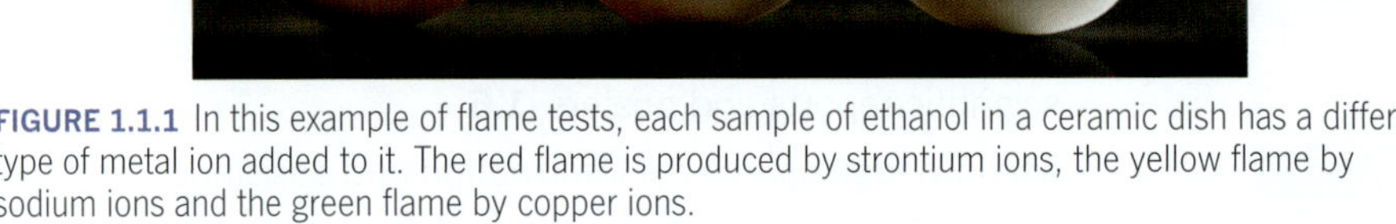

FIGURE 1.1.1 In this example of flame tests, each sample of ethanol in a ceramic dish has a different type of metal ion added to it. The red flame is produced by strontium ions, the yellow flame by sodium ions and the green flame by copper ions.

FIGURE 1.1.2 Masses are an example of quantitative observations and they are measured using an electronic balance.

● *You will now be able to answer key question 4.*

Interpreting observations

A student might observe that a solution of copper(II) chloride, when sprayed into a flame, creates a green flame. They may also observe that a solution of copper(II) nitrate also creates a green flame. However, a solution of lithium chloride creates a red flame.

How observations are interpreted depends on past experience and knowledge, but to an enquiring mind the above observation could provoke further questions, such as:

- Do all copper(II) compounds create a green flame?
- Does the concentration of the solution sprayed into the flame change the intensity of the flame colour?
- Are there any metal compounds that do not change the colour of the flame when they are sprayed into it?

Many of these questions cannot be answered through observation alone, but they can be answered through scientific investigations. Good scientists have acute powers of observation and enquiring minds, they rely on evidence and trends in evidence and they make the most of chance opportunities. They gather and record their observations carefully, so that they can be referred to in the future.

THE SCIENTIFIC METHOD

Scientists observe, consider what is already known by consulting the work of other scientists, and then they ask questions.

Scientific inquiry is not a linear process. Figure 1.1.3 illustrates the **scientific method**, in which a research question is investigated by forming a **hypothesis** (a prediction based on scientific reasoning that can be tested experimentally) and then testing it. Scientists will not necessarily complete these steps in the order shown below, and some steps may need to be repeated or altered to more accurately address the **research question** that they are trying to answer.

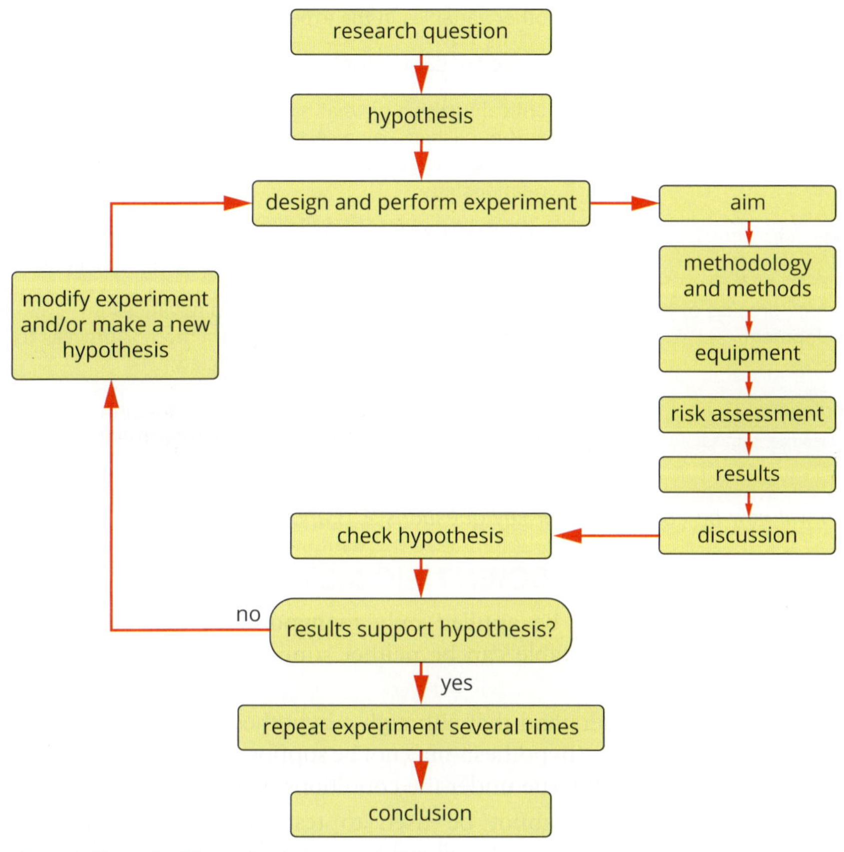

FIGURE 1.1.3 The scientific method

Scientific investigation methodologies

Scientists test their ideas about their research question using various investigation **methodologies**. The methodology is a brief description of the general approach taken to investigate the research question and the reasons why this approach is taken. Using the evidence gained from those investigations, scientists suggest possible explanations for the things they have observed.

Practical investigations involve direct experiences or hands-on activities. Suitable methodologies for a practical investigation would be controlled experiments, simulations, fieldwork or modelling. In comparison, a research investigation could follow the methodology of a literature review or could start with a case study.

The different approaches that you could use for investigations are outlined in Table 1.1.1.

The methodology is the general approach used to investigate the research question. The method (procedure) is the set of specific steps that are taken to collect data during an investigation.

TABLE 1.1.1 Some scientific investigation methodologies

Type of methodology	Explanation
case study	investigation of a real or hypothetical situation, such as an activity, event or problem, often involving analysis of data within a real-world context
classification and identification	using features or properties to classify or identify a substance
controlled experiment	experimental investigation that involves formulating a hypothesis and testing the effect of an independent variable on the dependent variable while controlling all other variables in the experiment
fieldwork	collecting data outside the laboratory
literature review	critical analysis of what has already been investigated and published, using secondary data from other people's investigations to explain events or propose new ideas or relationships
modelling	using models as representations of objects, systems or processes to aid understanding or make predictions
product, process or system development	using scientific understanding and advances in technology to design a new tool, method or process to meet the demands or needs of society
simulation	using mathematical models or computer simulations to test hypotheses or conduct virtual experiments

- *You will now be able to answer key questions 2, 3 and 5.*

LIMITATIONS OF THE SCIENTIFIC METHOD

The scientific method can be applied only to hypotheses that can be tested. A hypothesis that is not testable can be neither supported nor disproved by the scientific method.

It is important to understand that although a hypothesis may be supported by experimental data, the same hypothesis may not be supported in all circumstances—it has only been found to be true under the conditions that have been tested.

The scientific method cannot be used to test morality or ethics. These judgements belong to the fields of philosophy, history, politics and law. Science can, however, provide valuable information that people can consider when making these judgements. For example, science can be used to predict the environmental consequences of pollution and the medical consequences of chemical weapons. However, it is not designed to make value or moral judgements about either.

NON-SCIENTIFIC FACTORS THAT INFLUENCE SCIENTIFIC INVESTIGATIONS

When you investigate questions or issues related to applications of the principles of chemistry in society, it is important to be able to distinguish between factors that are scientific and those that are non-scientific. Often an investigation into a chemical issue must consider non-scientific information, including sociocultural, economic, political, legal or ethical factors.

For example, a discussion on the issue of climate change must involve scientific data, such as the concentration of carbon dioxide in the atmosphere, but it must also consider political factors, such as governmental support of the United Nations Paris Agreement (2015) on climate change.

Non-scientific factors may be classified under more than one category.

Sociocultural factors

Sociocultural factors are those related to individuals, communities, cultures and society. Questions these may raise include:

- Who is directly involved?
- Who will benefit?
- Who might be negatively affected?

Economic factors

Economic factors are those related to costs and resources. Examples might include:

- Who will pay for the research?
- Who will pay for the development, production and application?
- Will this cause a loss of profit for another stakeholder?
- Will there be any costs involved for the government?

Political factors

Political factors are those related to government or public affairs. Examples might include:

- What is the relevant government policy?
- Is there a difference in opinion between political parties?

Legal factors

Legal factors are those connected to law (legislation) or rules. Examples might include:

- What state legislation covers this area?
- What federal legislation covers this area?

Ethical factors

Ethical factors are those related to moral principles, and the need to determine what is right and what is wrong. Examples might include:

- Is one group of people advantaged over another by the desired outcome?
- Does the desired outcome prevent anyone from meeting their basic needs?
- Has the investigation been reported honestly, or have some results been omitted because they didn't support the desired outcome?

Table 1.1.2 gives examples of non-scientific factors that may have been considered during the rapid development of COVID-19 vaccines in 2020–21.

TABLE 1.1.2 Examples of non-scientific factors affecting the development of COVID-19 vaccines

Sociocultural	Economic	Political	Legal	Ethical
Why is vaccination necessary?	Who will pay for the research, development and production?	What agreements have been made for the supply of various vaccines?	Have the vaccines been approved for use in Australia by the Therapeutic Goods Administration?	Are there vulnerable groups who should be prioritised for vaccination?

- *You will now be able to answer key questions 1 and 6.*

Non-scientific factors, such as sociocultural, economic, political, legal or ethical factors, may influence a chemical issue and should be considered as part of some investigations.

1.1 Review

SUMMARY

- Examples of useful methodologies for an investigation include case studies, classification and identification, a literature review, modelling, simulations and controlled experiments.
- Well-designed experiments are based on a sound knowledge of what is already understood or known, and on careful observation.
- The scientific method is an accepted procedure for conducting investigations.
- A hypothesis is a possible explanation for a set of observations that can be used to make predictions, which can then be tested experimentally.
- Controlled experiments allow us to examine one factor at a time: they are a commonly used methodology for testing hypotheses.
- Science helps us to understand a situation or phenomenon. It is used in conjunction with other considerations, such as sociocultural, economic, political, legal and ethical factors.

KEY QUESTIONS

Knowledge and understanding

1 List three non-scientific factors that might be associated with an investigation.

2 The scientific method is a multistep process. Which two of the following are important parts of the scientific method?

A observations made by eye and with instrumentation

B subjective decisions based on data collected

C careful manipulation of results to fit your ideas

D the use of prior knowledge to help objectively interpret new data

3 Distinguish between the terms 'methodology' and 'method'.

4 Distinguish between 'qualitative' and 'quantitative' observations.

Analysis

5 Rewrite this table in your notebook, matching each type of investigation to the appropriate methodology.

Type of methodology	Type of investigation
case study	using secondary sources to find information about recycling of metals from second-hand computers
simulation	designing an experiment with an independent and dependent variables, and keeping everything else constant
controlled experiment	using a computer program to look at rotating three-dimensional models of molecules
literature review	in-depth study about a particular chemical process

6 Copy this table into your workbook. Identify whether the specific factor or resource related to the mining of copper in Australia would be scientific or non-scientific (identifying the category of non-scientific factor).

Factor	Scientific	Non-scientific (sociocultural, economic, political, legal and ethical)
geological assessment of the chosen area		
needs of the residents for employment		
local government regulations		
opinions of local residents about mining		
assessment of nearby waterways for potential pollution of ground water		

1.2 Planning investigations

Taking the time to carefully plan and design an investigation before you begin will help you maintain a clear and concise focus throughout. In this section you will learn about some of the key steps to take when planning an investigation:

- choosing a topic
- developing and refining your investigation (i.e. determining your research question, hypothesis, aim, methodology and methods, and variables)
- modifying an existing investigation
- complying with safety guidelines
- applying ethical principles
- sourcing and evaluating information.

CHOOSING A TOPIC

When you choose a topic, consider the following:

- Do you find the research topic interesting?
- Is there background information on your topic that relates to your course?
- Does your school laboratory have the resources for you to perform the investigation?
- Can you collect clear, measurable data?

Several practical research areas are suggested in Table 1.2.1. You will learn more about useful research techniques for topics such as these later in this section.

TABLE 1.2.1 Potential research areas that could be investigated

Materials (Unit 1 investigations)	Chemical reactions (Unit 2 investigations)
endangered elements in the periodic table	analysis of water from local waterways for salt content
producing and using 'greener' polymers	effect of pH or temperature on the rusting of iron nails
the chemistry of Aboriginal and Torres Strait Islander Peoples' art practices	acid–base titrations of household products, such as vinegar
the sustainability of a commercial product or material	identify the process needed to develop a UV-stable natural indicator

DEVELOPING AND REFINING YOUR INVESTIGATION

When you begin a scientific investigation, you first have to develop and evaluate a research question, decide on an appropriate investigation methodology and method, determine the associated variables, formulate a hypothesis and define the aim. Each of these can be refined as the planning of your investigation continues.

Determining your research question

Once you have selected a research topic, you need to design a research question on which your investigation will be based. The research question must be focused on and limited to ideas that are within your abilities to investigate, or the resources and equipment you have available.

For example, for an experimental investigation, you might form the research question: 'How does the solubility of ionic compounds change as the temperature increases?' You will be able to investigate this research question using the ionic compounds that you have at school, such as potassium nitrate and copper(II) sulfate.

In a **controlled experiment**, the question should refer to the relationship between the independent variable and dependent variable. For example: How does temperature affect the solubility of potassium nitrate in water? Other examples of research questions are shown in Table 1.2.2 on page 11.

Once you have decided upon a topic or idea of interest, the first thing you need to do is conduct a search of the relevant literature; that is, you must read scientific reports and other articles on the topic to find out what is already known, and what is not known or not yet agreed upon. The literature also gives you important information for the introduction to your report and ideas for experimental methods. Use this information to generate potential research questions. Figure 1.2.1 on the following page outlines steps for developing ideas for a proposed research question.

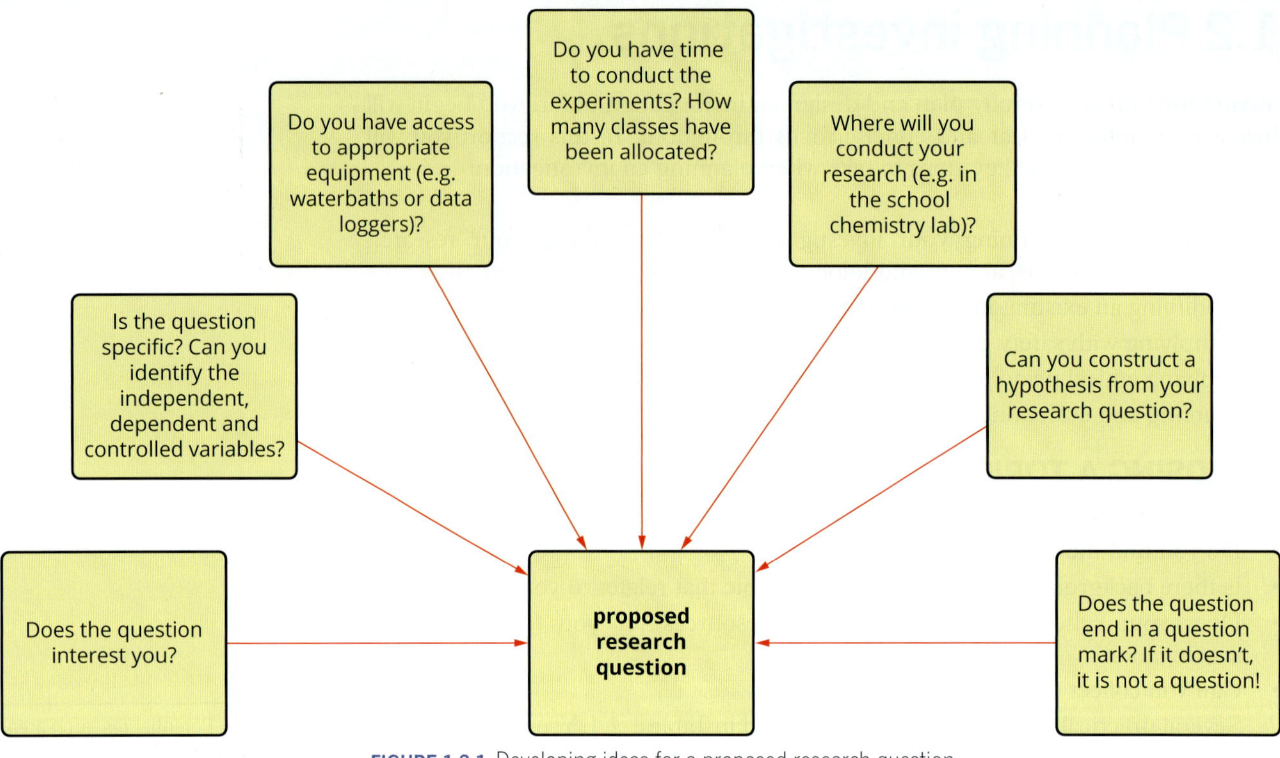

FIGURE 1.2.1 Developing ideas for a proposed research question

Your research question might be too broad or too vague to investigate effectively. Consider using the following checklist for the research question:

- ❑ Relevance—make sure your question is related to your chosen topic.
- ❑ Clarity and measurability—make sure your prediction can be framed as a clear hypothesis that can be measured.
- ❑ Knowledge and skills—make sure you have a level of knowledge and laboratory skills that will allow you to explore the question. Keep the question simple and achievable.
- ❑ Advice—seek advice from your teacher on your question. Their experience may lead them to consider aspects of the question that you have not thought about.

Making predictions and constructing a hypothesis

A hypothesis is a prediction based on scientific reasoning that can be tested experimentally. Carefully designed experiments are conducted to determine whether the predictions made in a hypothesis are accurate or not. If the results of an experiment do not fall within an acceptable range, the hypothesis is rejected. If the predictions are found to be accurate, the hypothesis is supported. If, after many different experiments, one hypothesis is supported by all the results obtained so far and is considered to have been verified using the scientific method, then this explanation can be given the status of a **theory** or **principle**.

In a practical investigation, a hypothesis defines a proposed relationship between two variables and takes the form of cause and effect.

For example, an investigation is made into the effect of pH of an acidic solution on the mass of a seashell submerged in the solution for two days. A potential hypothesis could be: If the pH of the solution decreases from 5 to 3, then the mass of the seashell will decrease more in the lower pH solution because a lower pH corresponds to a greater H^+ concentration, and the calcium carbonate in the seashell reacts with H^+ ions.

Determining the aim

The **aim** is a statement describing in detail what will be investigated. For example: 'To determine how the pH of an acidic solution affects the mass loss of a seashell that is left in the solution for two days'.

Selecting an appropriate methodology and method

When you plan a scientific investigation, you will need to think about the best way to address the research question. The scientific investigation methodology (Table 1.1.1) and the method/s (also known as procedure/s) selected will depend on the aim of the investigation and the research question.

Factors to consider when selecting an appropriate methodology include:

- Do you have access to a laboratory, materials and chemicals (for a controlled experiment or product/process/system development)?
- Do you have school or library access (for a literature review or case study)?
- Do you have computer access (e.g. international databases for classification and identification or for access to simulations)?

The **method** is the set of specific steps that are taken to collect data during the investigation. You will learn more about determining your method/s in Section 1.3.

Defining your variables

The **variables** are the factors that change during your experiment. An experiment or investigation determines the relationship between variables.

There are three categories of variable:

- The **independent variable** is the variable that is controlled by the researcher.
- The **dependent variable** is the variable that may change in response to a change in the independent variable. This is the variable that you will measure or observe.
- **Controlled variables** are all the variables that must be kept constant during the investigation.

A **valid** experiment should have only one independent variable. If it were to have more than one, you would not be sure which independent variable was responsible for the changes observed in the dependent variable. Table 1.2.2 gives examples of research questions and potential independent and dependent variables.

TABLE 1.2.2 Examples of research questions and corresponding independent and dependent variables

Research question	Independent variable	Dependent variable
How does the volume of water affect the temperature at which a saturated solution of potassium nitrate is formed?	volume of water, e.g. 5 mL, 10 mL, 15 mL, 20 mL	temperature of the solution at which crystals start to form
How does the price per 100 mL of vinegar relate to the concentration of ethanoic acid in the vinegar?	brand of vinegar and its price, e.g. Brand A, 50c per 100 mL; Brand B, $1.02 per 100 mL; Brand C, $3.50 per 100 mL	concentration of ethanoic acid in mol L^{-1}
How does the concentration of $Cu(NO_3)_2$ solution affect the mass of copper that is displaced by zinc?	concentration of $Cu(NO_3)_2$ solution, e.g. 0.01 M, 0.1 M, 0.5 M	mass of solid copper formed in the solution after a given period of time

Linking the planning parts of the investigation together

Your research question, hypothesis, aim and variables should all link together. Table 1.2.3 on the following page provides an example of a research question, variables, potential hypothesis and aim that link together.

CHEMFILE

Can a hypothesis be absolutely proven?

The simple answer to this question is 'no'. While scientists can do experiments that test a hypothesis, there is always the possibility that a different experiment or even a repeat of an experiment will contradict the hypothesis, demonstrating that it is not always true.

If scientists perform a large number of different experiments to test a hypothesis and each experiment supports the hypothesis, the hypothesis can be said to be proven beyond reasonable doubt.

The inability to absolutely prove a hypothesis is of more importance in some fields of science than others. For example, researchers in biology, medicine and environmental science will usually perform large numbers of experiments to ensure the validity of a hypothesis. A new drug for the treatment of malaria will undergo large numbers of laboratory tests, followed by thousands of trials with human volunteers, before it is considered effective and safe for use. On the other hand, experiments in chemistry are usually highly reproducible.

When writing a research question for a controlled experiment, include the independent and dependent variables. For example, what is the effect of [*the independent variable*] on [*the dependent variable*]?

TABLE 1.2.3 Example of research question, variables, potential hypothesis and aim

Research question	How does the pH of an acidic solution affect the mass of a seashell submerged in the solution for two days?
Independent variable	the pH of the solution
Dependent variable/s	the mass of the seashell after two days
Controlled variables	the acid used to make the solution, the volume of acidic solution, the temperature of the solution, the size and shape of the seashell
Potential hypothesis	If the pH of the solution decreases from 5 to 3 then the mass of the seashell will decrease more in the lower pH solution because a lower pH corresponds to a greater H^+ concentration, and the calcium carbonate in the seashell reacts with H^+ ions.
Aim	To determine how the pH of an acidic solution affects the mass loss of a seashell that is left in the solution for two days.

● *You will now be able to answer key questions 3, 4, 5 and 7.*

MODIFYING AN EXISTING INVESTIGATION

When you are designing a research question for a controlled experiment, it is often easiest to modify an investigation that you have already conducted in class. For example, you could use an existing method and choose a different independent variable (which might change the dependent variable). Table 1.2.4 shows possible changes that could be made to a practical investigation that you have already undertaken in class.

TABLE 1.2.4 Examples of how to change an existing practical investigation into a student-designed investigation

Existing practical investigation	Potential student experiment
investigate displacement reactions of metals	How does the reactivity of a metal such as magnesium, zinc, aluminium and iron, affect the rate at which it displaces copper from copper(II) sulfate?
acidity of soda water and the effect of heating	How does the pH of soda water change as the temperature of the water is increased?
making natural indicators from flowers	How does the colour of the flowers used to make natural indicators affect the colours that the indicator turns in acidic and basic solutions?

The topics listed in Table 1.2.4 are only suggestions. Figure 1.2.2 shows how you can modify an existing research question.

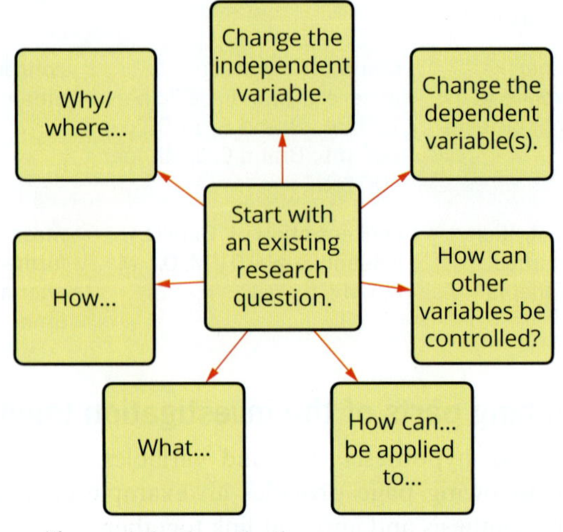

FIGURE 1.2.2 How to modify an existing investigation

COMPLYING WITH SAFETY GUIDELINES

Everything we do involves some risk. A **risk assessment** is performed for a controlled experiment to identify, assess and control hazards. Always identify the risks and control them to keep everyone safe.

To identify risks, think about:

- the activity you will be carrying out
- the equipment or chemicals you will be using or producing. For example, when hydrochloric acid reacts with sodium thiosulfate, the toxic gas, sulfur dioxide, SO_2, is produced, so this reaction must be conducted in a fume cupboard.

Figure 1.2.3 shows a flow chart of how to consider and assess the risks involved in a practical investigation.

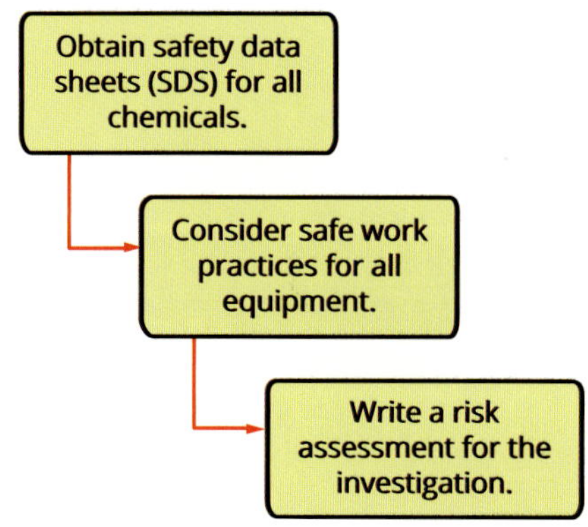

FIGURE 1.2.3 Steps involved in identifying risks

Ethical and safety considerations must be the highest priority at all times during a practical investigation.

Occupational health and safety

Occupational health and safety refers to all the measures that employers need to provide to ensure their employees are safe at work. Schools must also ensure that equipment and processes used in school laboratories are safe for all students, teachers and technical staff.

Chemical codes

The chemicals at school or at the hardware store have warning symbols on their labels. These symbols are a chemical code indicating the nature of the contents (Table 1.2.5). From 1 January 2017, the Globally Harmonised System of Classification and Labelling of Chemicals (GHS) pictograms were introduced into Australia. Some of the pictograms that you might see denote chemicals that are corrosive, pose a health hazard or are flammable.

TABLE 1.2.5 GHS pictograms used as warning symbols on chemical labels

GHS pictogram	Use	GHS pictogram	Use	GHS pictogram	Use
	flammable liquids, solids and gases; including self-heating and self-igniting substances		oxidising liquids, solids and gases, may cause or intensify fire		explosion, blast or projection hazard
	corrosive chemicals; may cause severe skin and eye damage and may be corrosive to metals		gases under pressure		fatal or toxic if swallowed, inhaled or in contact with skin
	low level toxicity; this includes respiratory, skin and eye irritation, skin sensitisers and chemicals harmful if swallowed, inhaled or in contact with skin		hazardous to aquatic life and the environment		chronic health hazards; this includes aspiratory and respiratory hazards, carcinogenicity, mutagenicity and reproductive toxicity

Safety data sheets

Each chemical substance has an accompanying document called a **safety data sheet (SDS)** (Figure 1.2.4). An SDS contains important safety and first aid information about each chemical you commonly use in the laboratory. If the products of a reaction are toxic to the environment, you must pour your waste into a special container (not down the sink). This is something to discuss with your teacher or the laboratory technician.

1. IDENTIFICATION OF THE MATERIAL AND SUPPLIER

Product Name: HYDROCHLORIC ACID – 20% OR GREATER

2. HAZARDS IDENTIFICATION

Classified as Dangers Goods by the criteria of the Australian Dangerous Goods Code (ADG Code) for Transport by Road and Rail; DANGEROUS GOODS.

This material is hazardous according to Safe Work Australia; HAZARDOUS SUBSTANCE.

Classification of the substance or mixture:
Corrosive to Metals - Category 1
Skin Corrosion - Sub-category 1B
Eye Damage - Category 1
Specific target organ toxicity (single exposure) - Category 3

SIGNAL WORD: DANGER

Hazard Statement(s):
H290 May be corrosive to metals.
H314 Causes severe skin burns and eye damage.
H335 May cause respiratory irritation.

Precautionary Statement(s):

Prevention:
P234 Keep only in original container.
P260 Do not breath mist / vapours / spray.
P264 Wash hands thoroughly after handling.
P271 Use only outdoors or in a well-ventilated area.
P280 Wear protective gloves / protective clothing / eye protection / face protection.

FIGURE 1.2.4 Extract of a safety data sheet (SDS) for concentrated hydrochloric acid

Protective equipment

Everyone who works in a laboratory should wear items that help keep them safe. This is called personal protective equipment (PPE) (Figure 1.2.5) and includes:

- safety glasses
- closed-toed shoes
- disposable gloves when handling chemicals
- an apron or a lab-coat if there is risk of damage to clothes
- a fume cupboard that should be used when toxic or corrosive gases are being handled or produced
- ear protection if there is risk to hearing.

FIGURE 1.2.5 It is important to wear appropriate personal protective equipment as identified in a risk assessment.

● *You will now be able to answer key questions 1, 2 and 6.*

APPLYING ETHICAL PRINCIPLES

Ethics are a set of moral principles by which your actions can be judged as right or wrong. Every society or group of people has its own principles or rules of conduct.

Applying ethical principles in chemistry means:

- using integrity and honesty when recording and reporting the outcomes of your investigation and when using other people's data (such as in a literature review).
  - This means you should always record your data in a bound logbook, rather than on scraps of paper, using the correct number of significant figures/ decimal places for the reported value (see Section 1.3).
  - You should NEVER make up raw or processed data and you should record all data that you gathered in the investigation, even if it is not used in later calculations.
- recognising the importance of social, economic and political values when forming conclusions using scientific understanding (see Section 1.1)
- acknowledging the work of others by including in-text citations and details in a list of references (see Section 1.6).

Green chemistry

Green chemistry is an approach to chemistry that aims to design products and processes that efficiently use renewable raw materials, and minimise hazardous effects on human health and the environment. This approach is one of the key concepts related to **sustainability** in chemistry. The goals of green chemistry are described in terms of twelve principles (summarised in Table 1.2.6).

Since they were developed in 1991, interest in applying the twelve principles of green chemistry to chemical processes across the world has continued to increase. Of these, seven principles are particularly relevant to VCE Chemistry. The replacement of petroleum-sourced plastics such as polystyrene, with plant-sourced **biodegradable** plastics, which can break down in the environment over a number of years (see the case study Green chemistry: Biodegradable packing peanuts), is just one recent example of how green chemistry is being applied to solving real-world problems. In the school laboratory, we can apply the principles of green chemistry by using minimum quantities of chemicals in experiments, thus reducing the quantity of product that must be disposed of after the experiment is finished.

Many of the twelve principles of green chemistry can be employed or considered by VCE Chemistry students.

TABLE 1.2.6 Seven of the twelve principles of green chemistry that are most relevant to VCE Chemistry

Principle	Description
Waste prevention	It is better to prevent waste than to treat it or clean it up after it has been produced.
Atom economy	Chemical processes should be designed to maximise incorporation of all reactant materials used in the process into the final product.
Designing safer chemicals	Chemical products should be designed to achieve their intended function while minimising toxicity.
Design for energy efficiency	Chemical processes should be designed for maximum energy efficiency and with minimal negative environmental and economic impacts.
Use of renewable feedstocks	Raw materials or feedstocks should be made from renewable (mainly plant-based) materials, rather than from fossil fuels whenever practicable.
Catalysis	Catalysts should be selected to generate the same desired product(s) with less waste and using less energy and reagents in chemical reactions.
Design for degradation	Chemical products should be designed so that at the end of their use they break down into harmless degradation products and do not persist in the environment.

● *You will now be able to answer key question 8.*

CASE STUDY

Green chemistry: Biodegradable packing peanuts

One method for protecting fragile goods during transit is to surround the goods in their box with 'packing peanuts' (see Figure 1.2.6). These peanut-shaped pieces of foam were traditionally made from polystyrene, so despite their versatility in filling gaps within the packaging, and providing additional padding and support to the delicate goods without adding significant mass to the package, they became very unpopular. Polystyrene is not biodegradable, meaning that it can't be broken down in the environment by living organisms, and it is very difficult to recycle, so usually ends up in landfill rather than a recycling plant.

In the 1990s starch-based plastics were developed in response to the concern over the environmental impact of plastics such as polystyrene, which were widely used in packaging. Biodegradable packing peanuts are made from starches, such as wheat, rice and corn starch. Starch is a polymer of glucose which is made by plants, so the starting materials for these plastics are widely available and often the starch is extracted from waste biomass. Packing peanuts made from starch-based foam dissolve in water, meaning that it is impossible for them to pollute oceans, lakes or other waterways. If they are eaten by wildlife or pets they will do minimal harm, since they will be digested, just as a small piece of bread might be. These packing peanuts are even more soluble than most waste food. While these plastics are biodegradable, they are not compostable, unless you add water to them when composting.

Interestingly, while often materials are promoted as biodegradable, and we all think this is a great thing, there is still a difference between biodegradable and compostable. Compostable materials will completely decompose if left for a few months in a composting environment, such as a home compost bin. Biodegradable is a term that does not have a time span attached to it. Biodegradable materials will eventually completely degrade, but they may exist for decades before fully degrading.

FIGURE 1.2.6 Biodegradable 'packing peanuts' fill a box easily.

So, how can you tell if the packing peanuts in your latest delivery are starch-based or petroleum-based? The best way is to try putting a couple in a glass of warm water. The biodegradable packing peanuts will start to dissolve very quickly, whereas polystyrene peanuts will stay floating in the water all day.

SOURCING INFORMATION

When you are sourcing information during your search of the literature, researching experimental methods or investigating a broader issue, consider whether that information is from primary or secondary sources. You should also consider the advantages and disadvantages of using resources such as books or the internet.

Primary and secondary sources

Primary and secondary sources provide valuable information for research.

Primary sources of information are created by a person directly involved in an investigation. Examples of primary sources are results from research and **peer-reviewed** scientific articles. Peer-reviewed journal articles (articles that have been reviewed by other experts in the field) are likely to be up to date and reliable. Examples of **secondary sources** of information include textbooks, newspaper articles and websites that present a synthesis, review or interpretation of primary sources. All sources of information may have a **bias** (a focus on only one part or one direction of the data or evidence), so you need to determine if they are reliable sources of information. You will learn about assessing the quality of data in Section 1.3. Table 1.2.7 compares characteristics and examples of primary sources and secondary sources.

TABLE 1.2.7 Summary of primary and secondary sources

	Characteristics	Examples
Primary sources	• first-hand records of events or experiences • written at the time the event happened • original documents	• results of experiments • scientific journal or magazine articles • reports of scientific discoveries • photographs, specimens, maps and artefacts • interviews with experts • websites (if they meet the criteria above)
Secondary sources	• interpretations of primary sources • written by people who did not see or experience the event • use information from original documents but rework it	• textbooks • biographies • newspaper articles • magazine articles • radio and television documentaries • websites that interpret the scientific work of others • podcasts

Primary sources of information are created by a person directly involved in an investigation. Secondary sources of information are a synthesis, review or interpretation of primary sources.

Using books and the internet

Peer-reviewed scientific journals are the best sources of information, but you are unlikely to have access to many of these, and much of the information is difficult to interpret if you are not an expert in the field.

As books, magazines and internet searches will be your most commonly used resources for information, you should be aware of their limitations (Table 1.2.8).

TABLE 1.2.8 Advantages and disadvantages of book and internet resources

	Book resources	Internet resources
Advantages	• written by experts • authoritative information • reviewed to ensure information is accurate • logical, organised layout • content is relevant to the topic • contain a table of contents and index to help you find relevant information	• quick and easy to access • allow access to hard-to-find information • allow access to a vast amount of information from around the world • up-to-date information • may be interactive and use animations to enhance understanding
Disadvantages	• may not have been published recently • can only be used by one person at a time	• time-consuming looking for relevant information • search engines may not display the most useful sites • cannot always tell how up-to-date information is • difficult to tell if information is accurate • hard to tell who has responsibility for authorship and if they are biased • information may not be well ordered • only a small proportion of sites are educational

CRITICALLY EVALUATING INFORMATION

Not all sources of information are credible. **Critical thinking** involves asking questions when evaluating the content and its origin, including:

- Who created this message? What are the qualifications, expertise, reputation, and **affiliation** (who they work for or are associated with) of the author/s?
- Why was the information written?
- Where was the information published?
- When was the information published?
- How often has the information been referred to by other researchers?
- Are the conclusions supported by data or evidence?
- What is implied?
- What has been omitted?

Peer review

Peer review is a process in which other researchers who work in the same field review your work and provide feedback about your methodology and whether your conclusions are justified. Scientists are expected to publish their findings in peer-reviewed journals. Some examples of peer-reviewed chemistry journals are:

- *Australian Journal of Chemistry*
- *Journal of the American Chemical Society*
- *Nature Chemistry*
- *Chemical Reviews*
- *Progress in Polymer Science*
- *Green Chemistry*

Because scientific journals are peer-reviewed, this gives them more credibility than other sources.

Evaluating books and journals

Your textbook is an excellent starting point for reliable information. Other information should be consistent with it. Articles published in newspapers and magazines often present findings of new research, which may or may not be confirmed later, so be careful not to treat such sources of information as established fact. Peer-reviewed journal articles are likely to be up to date and reliable.

Evaluating websites

Remember that anyone can publish anything on the internet, so it is important to evaluate the credibility, currency and content of online information. To evaluate online information, follow this checklist.

- ❑ Credibility—consider who the author is, their qualifications and expertise. Check for their contact information and for a trusted abbreviation in the web address, such as .gov or .edu. Websites using .com may have a bias towards selling a product (although this product could be a reputable science magazine or journal), and .org sites might have a bias towards one point of view (although these sites can be a good starting point for general information).
- ❑ Currency—check the date the information you are using was last revised.
- ❑ Content—consider whether the information presented is fact or opinion. Check for properly referenced sources, and compare information to other reputable sources, including books and science journals.

Table 1.2.9 gives an example of how sources might be compiled during an investigation.

TABLE 1.2.9 Techniques used by Indigenous Australians to remove toxins from food

Type of bush food	Notes	Source	Reliability of source
bush tomato *Solanum centrale*	The fruits were 'probably the most important of all the Central Australian plant foods' (citing Latz 1995). Purplish-green colour when unripe. Traditionally eaten raw when greenish-white to yellow-brown in colour. Dried on plant, then shaken from plant, rubbed into sand, followed by pounding or grinding into a paste with water, formed into balls or cakes (citing Peterson 1979).	Hegarty, M.P., Hegarty, E.E. & Wills, R. (2001). Food Safety of Australian Plant Bushfoods: Publication No 01/28, RIRDC, Barton, ACT.	Credible. Published by Rural Industries Research and Development Corporation (RIRDC), review article with credible primary sources.
Macadamia robusta (toxic type of macadamia nut)	Nuts treated by soaking and baking to remove toxins.	Hegarty, M.P., Hegarty, E.E. & Wills, R. (2001). Food Safety of Australian Plant Bushfoods: Publication No 01/28, RIRDC, Barton, ACT.	Credible. Published by Rural Industries Research and Development Corporation (RIRDC), review article with credible primary sources.

Recognising evidence compared to opinion and anecdote

An important aspect of the scientific method is to make conclusions based on data and evidence.

Data that is authoritative has been published by a credible source—for example, the VCAA Chemistry Databook.

Data is **objective** (free of personal bias) and may be used as evidence, whereas opinion and anecdote are **subjective** (influenced by personal views). An anecdote is a story that might not be typical or representative of a pattern. Table 1.2.10 compares evidence versus opinion and anecdote.

TABLE 1.2.10 Comparison of evidence versus opinion and anecdote

Data and evidence	Opinion and anecdote
objective	subjective
measured, unbiased, replicable, systematic, representative	personal, individual story, non-systematic, not necessarily representative or part of a pattern
example: Scientific consensus among the majority of experts is that greenhouse gases are contributing to climate change.	example: Today is a cold day, therefore global warming is not true.

Identifying scientific versus non-scientific ideas

Beware of publications or sources of information that are presented as science but that are not scientifically valid. Non-scientific ideas and writing can be identified by:

- a lack of data or evidence
- bias—only part of the data or evidence is considered (usually the data supporting the claim)
- poorly collected data or evidence; for example, basing data or evidence on a group that is too small or not representative of the whole
- invalid conclusions (that is, not supported by evidence)
- lack of objectivity—appealing to emotion rather than presenting facts and evidence impartially.

1.2 Review

SUMMARY

- The first step in planning an investigation is to identify the topic.
- A research question is a statement that broadly defines what is being investigated.
- A hypothesis:
  - is a statement that can be tested and is based on previous knowledge and evidence or observations, and addresses the research question
  - often takes the form of a proposed relationship between two or more variables in a cause-and-effect relationship
  - must be testable; that is, able to be supported (verified) or not supported (falsified) by investigation.
- The aim of an investigation is a statement describing in detail what will be investigated.
- The scientific investigation methodology and method/s depend on the aim of the investigation and the research question.
- The three types of variables are:
  - independent—the variable that is selected and manipulated by the researcher
  - dependent—a variable that may change in response to a change in the independent variable, and is measured or observed.
  - controlled variables—variables that are kept constant during the investigation because they may affect the dependent variable.
- Ethical and safety considerations must be the highest priority at all times during a practical investigation.
- Many of the twelve principles of green chemistry can be employed or considered by VCE Chemistry students.
- Primary sources of information are created by a person directly involved in an investigation. Secondary sources of information are a synthesis, review or interpretation of primary sources.
- Use critical thinking to evaluate sources of information by asking questions such as:
  - Who created this message?
  - Why was the information written?
  - When was the information published?
  - Are the conclusions supported by data or evidence?
  - What is implied?
  - What has been omitted?

KEY QUESTIONS

Knowledge and understanding

1 Explain what is meant by each of the following chemical codes.

a

b

c

2 Describe the purpose of a risk assessment.

3 Define the following terms:

a research question
b hypothesis
c aim
d independent variable
e dependent variable
f controlled variables

4 Identify whether each of the following statements is a theory, a hypothesis or an observation.

a Spraying copper(II) chloride solution into a flame causes the flame to turn green
b Particles in a gas move rapidly and randomly
c If 1.0 M sodium hydroxide solution is gradually added to 1.0 M hydrochloric acid, the pH of the solution will increase because sodium hydroxide neutralises hydrochloric acid.

Analysis

5 A student conducted an experiment to investigate if increasing the concentration of the acid would increase the volume of hydrogen gas produced in the reaction between hydrochloric acid and magnesium ribbon. The student knew that the equation for the reaction is:

$$Mg(s) + 2HCl(aq) \longrightarrow MgCl_2(aq) + H_2(g)$$

Write a possible hypothesis for this experiment.

6 Complete the following risk assessment table by inserting an appropriate safety measure for each identified hazard.

Hazard	Safety measure
flammable liquid	
respiratory irritant	
corrosive solution	
contamination of wastewater with organic compounds	
toxic solid	

7 A pair of students developed the following hypothesis: 'For the precipitation reaction between silver nitrate and sodium chloride, as the concentration of sodium chloride solution increases, the mass of silver chloride precipitated will increase.'

a Identify the independent variable.

b Identify the dependent variable.

c Suggest possible controlled variables.

8 Thin-layer chromatography is a separation technique used to observe how many different components are present in a mixture. For a chromatography experiment designed to observe the pigments present in a green extract from spinach leaves, a student found two methods which were similar in procedure, but used different chemicals. Method A used 10 mL acetone as the solvent for extracting the green material, and a mixture of 2 mL ethanol and 2 mL water as the mobile phase for the experiment. Method B used 10 mL acetone as the solvent for extracting the green material, and a mixture of 2 mL acetone and 2 mL cyclohexane as the mobile phase for the experiment.

a Using the internet, source and consult an SDS for each of acetone, ethanol and cyclohexane. Using the SDS as your source of information, record the toxicity and flammability of each liquid in a table.

b State which method would be most likely to be considered as a 'green chemistry' method and explain your answer.

1.3 Data collection and quality

In this section you will learn about data collection in a controlled experiment. You will learn how to record both quantitative and qualitative data. You will also learn about the various factors that contribute to data quality, and the importance of controlled experiments in producing valid results.

KEEPING A LOGBOOK

Throughout Units 1 and 2, and particularly during your practical investigation for Unit 2 Area of Study 3, you must keep a **logbook**. This is generally thought of as a bound book that includes every detail of your research (Figure 1.3.1). The reason for keeping a logbook, rather than recording information on random pieces of paper, is primarily to stay organised and to prevent loss of details. The best place for your logbook to be kept is in the laboratory, so that it is always accessible when you are doing your experimental research. Keeping track of all observations, qualitative and quantitative raw data, modifications to the method and your plans to analyse the data will ensure that you don't forget what you did, or spend extra time repeating work because you have lost your results.

The following checklist will help you remember what to include in your logbook:

- ❑ your ideas when planning the research
- ❑ tables ready for raw qualitative and quantitative data entry
- ❑ notes about what data you will collect and how you intend to analyse it. Refer to Section 1.4 on page 33 for further details
- ❑ records of all materials, methods, details of the experiment and raw data
- ❑ all notes, sketches, photographs and results
- ❑ records of any incidents or errors that may influence the results.

Make sure to date all entries in your logbook. Figure 1.3.1 shows a sample page of a student logbook.

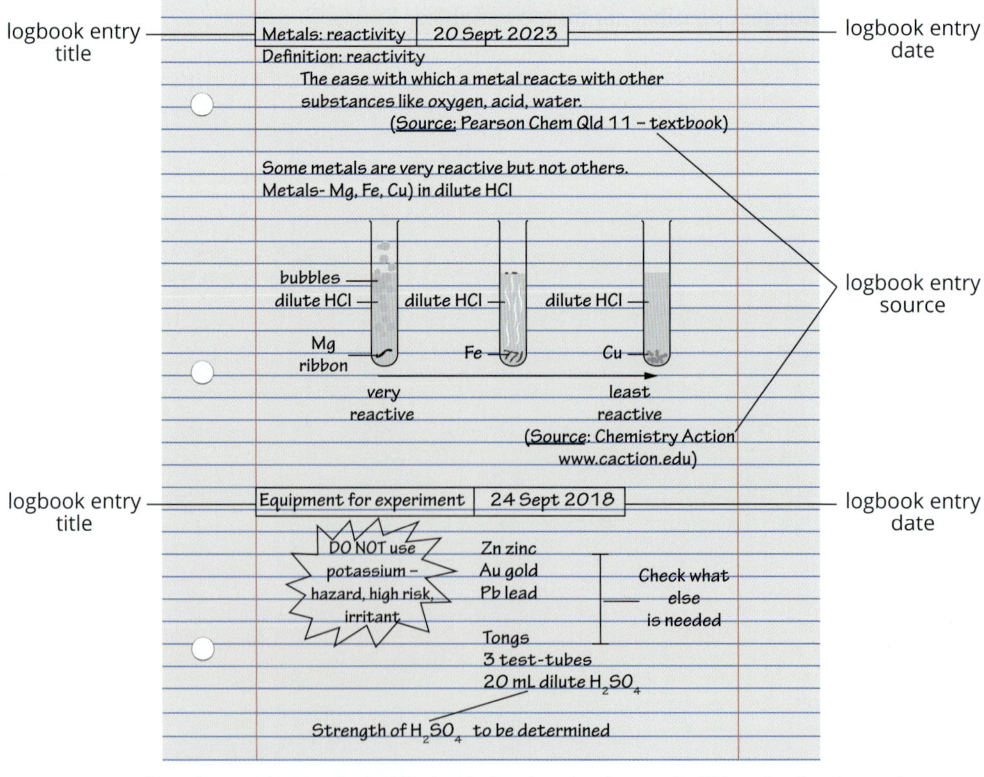

FIGURE 1.3.1 Sample page from a student logbook showing development of ideas in planning and conducting a practical investigation

● *You will now be able to answer key question 3.*

CHEMFILE

When entries in a logbook are legally vital

When a forensic scientist working in a laboratory is given samples from a crime scene, or from a suspected criminal or victim, they are required to keep careful records of the tests they do, as well as the results they obtain. Well-established, peer-reviewed, reliable methods must be used, since their results may determine whether the accused is given a criminal conviction.

If a blood alcohol test result is high enough for the driver to lose their licence, they may also lose their livelihood if they drive as part of their work.

It could be months or years that pass between when the tests have been done and the case finally comes to court. The forensic scientist's records must be clear and unambiguous, so that when the scientist is questioned in the witness box about the validity of their tests, they can answer with confidence.

A forensic scientist collects evidence, wearing gloves and full protective covering to prevent contamination of the sample.

Determining the method

The method for a controlled scientific experiment must consider how the independent variable will be changed, how the dependent variable will be measured and determine the controlled variables that will not change. Any decisions to repeat measurements and change to the next variation (for example, a different concentration) of the independent variable should be included in the method.

Ensure that you have a detailed step-by-step method to follow, which other people could follow if they were to replicate your experiment. Sometimes a diagram of your equipment set-up is a valuable addition to your method.

Modifying the method

The method may need modifying as the investigation is carried out. If this is the case:

- record everything
- note any difficulties you encountered and the ways they were overcome. What were the failures and successes?

If the expected data is not obtained, don't worry. Every test carried out can contribute to the understanding of the investigation as a whole, no matter how much of a disaster it may appear at first. Do your best to critically evaluate the data and identify the limitations of the investigation. Based on this information you should be able to propose further investigations.

Record any difficulties you encounter in the investigation and any changes that you make to the method as a result of the difficulties.

DATA COLLECTION

Before you begin your experiment, you should construct data collection tables in your logbook so that you can fill in the data as you conduct the experiment.

The measurements or observations that you collect during your investigation are your **primary data**. This primary data is also known as **raw data**. When you start to do calculations with that raw data, even an average, then the data is known as **processed data**. Processed data is raw data that has been organised, altered or analysed to produce meaningful information (see Section 1.4, page 33).

Raw and processed data

The data you record in your logbook during the investigation is raw data. This data often needs to be processed or analysed before it can be presented.

Raw data that should be recorded includes:

- all quantitative data (measurements)
- all qualitative observations and other notes.

The raw data may be presented in the form of:

- tables of results
- diagrams and/or photographs of results
- graphs from datalogging equipment.

Processed data may include:

- graphs generated from raw data
- calculations in which raw data is used to determine other quantities
- graphs of processed data showing trends in that data.

Record all raw and processed data in tables in your logbook. Photographs of observations are also useful.

Raw data is the data you measure or observe during the course of the investigation. Processed data is data obtained by applying a calculation or formula to raw data.

● *You will now be able to answer key question 7.*

Qualitative and quantitative data

Qualitative data can be observed and relates to a type or category, such as colour, or states (such as gas, liquid or solid).

Measured numeric values are **quantitative data**. Quantitative data should be accompanied by a relevant unit such as mass (g) or time (s). Numeric data can be discrete or continuous:

- **Discrete data** are values that can be counted or measured, but which can only have certain values. An example is the number of chemistry students in a class or the number of hydrogen atoms in a molecule of water.
- **Continuous data** may be any number value within a given range that can be measured. Examples are time, temperature, mass, pH and concentration.

Determining whether a solution is acidic or basic can be qualitative or quantitative. Figure 1.3.2 shows qualitative and quantitative tests for acidity. Litmus paper gives a qualitative observation indicating the nature of the solution (Figure 1.3.2a), whereas a calibrated pH probe or pH meter gives a quantitative measurement indicating the nature of the solution (Figure 1.3.2b).

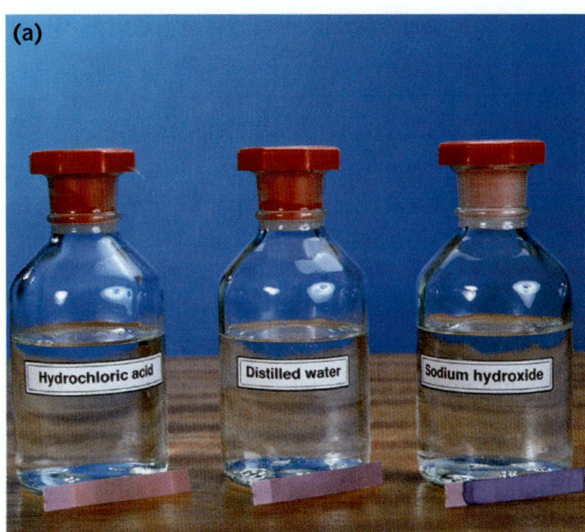

FIGURE 1.3.2 (a) Litmus paper can be used to indicate the acidity of a solution. If the paper turns blue, the solution is basic; if the paper turns pink, the solution is acidic. (b) A pH probe measures pH quantitatively. If the pH < 7 at 25°C, the solution is acidic; if the pH > 7 at 25°C, the solution is basic.

Recording qualitative raw data

Examples of qualitative data that might be observed in an investigation, such as the one being performed in Figure 1.3.3:

- disappearance of a substance
- appearance of a solid (precipitate)
- production of a gas
- colour change (e.g. from colourless to purple)
- temperature change (e.g. increase or decrease without using a thermometer).

When recording qualitative data, focus on the observation. For example, when magnesium is added to hydrochloric acid, bubbles are observed, indicating that a gas is produced. However, don't assume that the gas is hydrogen unless you confirm this; for example, using the 'pop' test.

FIGURE 1.3.3 Qualitative observations include the disappearance of a reactant, the formation of bubbles and the increased heat of a test tube.

Recording quantitative raw data

Have your logbook ready, with prepared data collection tables, before you begin your experiment. Accurately record the correct number of significant figures and decimal places for measurements. For example, burettes should be read to two decimal places. Mass measurements should be recorded to the number of decimal places shown on the balance; for example, 2.034 g.

● *You will now be able to answer key question 4.*

Presenting raw data in tables

Tables organise data into rows and columns, and they vary in complexity according to the nature of your data. Tables can be used to organise raw data and processed data, and to summarise results. Results from all trials should be listed in tables of raw data.

Figure 1.3.4 shows an example of how to present quantitative raw data in a table. Note that the table has the following features:

- a descriptive title
- column headings (including the unit)
- the independent variables (distance between the electrodes) placed in the left column
- the dependent variable(s) placed in the right column(s).

The table is numbered and its title accurately describes what is in the table.

Table 1 The effect of pH on the mass of a seashell submerged in the acidic solution

pH	Trial number	Initial mass (g)	Final mass (g)
6	1	2.054	1.833
	2	2.309	2.089
	3	2.443	2.224
5	1	2.332	2.082
	2	2.056	1.811
	3	2.764	2.512
4	1	2.003	1.723
	2	2.115	1.837
	3	2.367	2.090
3	1	2.508	2.198
	2	2.600	2.292
	3	2.845	2.531

All headings include units and accurately describe the data set that is in the column.

Raw data from which the dependent variable can be calculated is in right-hand column.

All raw data in the same column is quoted to the same accuracy.

The independent variable is in the left-hand column.

Each row shows a different trial; in this case 3 replicates for each pH value.

FIGURE 1.3.4 Features of a good raw data table

Units

Each column in a table of quantitative data should include relevant units. Commonly, these will be SI (International System of Units) units. Examples of commonly used units are shown in Table 1.3.1.

TABLE 1.3.1 Examples of units for various quantities

Quantity	Possible units
mass	g, mg, kg
volume	mL, L
pH	no units
concentration	mol L^{-1} (M), g L^{-1}, %(m/m), %(v/v), %(m/v)
temperature	°C

DATA QUALITY

The results of your data analysis will only be as good as the quality of the data. You should consider this when collecting primary data in your investigations, and also when you evaluate the quality of secondary data from other sources.

Accuracy and precision

If repeated measurements of the same quantity give values that are in close agreement, the measurement is said to be **precise**. To obtain precise results you must minimise random errors. The **precision** of a range of measurements can be compared. For example, a measured mass of 3.1 g is less precise than a measured mass of 3.109 g.

If the average of a set of measurements of a quantity is very close to the **true value**, the value that would be obtained under perfect conditions, then the measurement is said to be **accurate**. To obtain accurate results you must minimise systematic errors. Precision gives no indication of how close the measurements are to the true value and is therefore a separate consideration to **accuracy**. Figure 1.3.5 shows the difference between accuracy and precision.

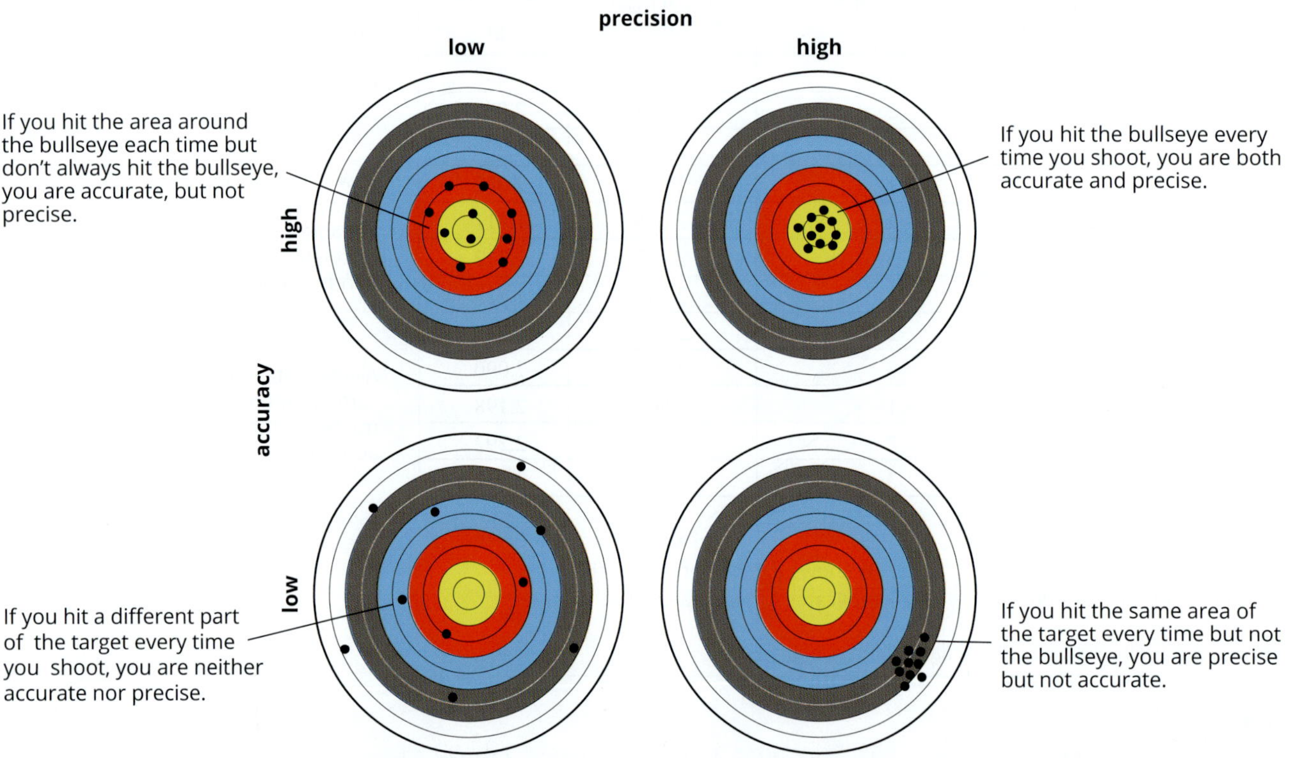

FIGURE 1.3.5 An example of a target illustrates the difference between accuracy and precision.

If the true value is known, then the difference between the true value and a measured value is called the measurement error. In many cases because the true value is unknown, the **measurement error** will also be unknown.

Significant figures

When reporting measurements in raw data tables, you should report a measurement that is as accurate as possible within the limitations of the equipment being used. In doing so, you will be correctly stating the number of **significant figures**. The significant figures indicate the precision of the instrument used and indicate a range within which the value can be considered to exist. For example, a measurement of 56.0 g has three significant figures and indicates that the value is somewhere between 55.95 g and 56.04 g, whereas a measurement of 56 g, which has two significant figures, indicates that the value is somewhere between 55.5 g and 56.4 g.

When giving an answer to a calculation, it is important to take care with the number of significant figures that you use. An answer cannot be more precise than the data or the measuring device used to calculate it. For example, if an electronic balance that measures to the nearest 0.01 g shows that an object has a mass of 1.17 g, then the mass should be recorded exactly as 1.17 g (not as 1.170 g). The number 56 has two significant figures. Recording to three significant figures (e.g. 56.0 g or 55.8 g) would not be scientifically ethical (see Section 1.2, page 15). If this mass of 56 g is used to calculate another value, it would also not be ethical to give an answer that has more than two significant figures. Similarly, it is not correct to cut off decimal places if the equipment is more precise than you require.

Working out the number of significant figures

When counting significant figures, the following rules apply:

- All non-zero digits count as significant figures.
- Zeros may or may not be significant, depending on whether they are part of the measured value or are simply used to position a decimal point.
- Whole numbers written without a decimal point will have the same number of significant figures as the number of digits, with the assumption that the decimal point occurs at the end of the number; for example, 500 has three significant figures. Therefore, a stated volume of '500 mL' will be considered as having three significant figures.

Significant figures are only applied to measurements, and calculations involving measurements (see Section 1.4). They do not apply to quantities that are inherently integers or fractions (for example, a stoichiometric ratio such as 2 or ½ mole), defined quantities (for example, 1 metre equals 100 centimetres), or conversion factors (multiplying by 100 to get a percentage or adding 273 to convert °C to K). For example, the accuracy of a temperature reading of 12°C cannot be increased from two significant figures to three significant figures by converting to kelvin (285 K).

● *You will now be able to answer key question 1.*

Types of errors

Most practical investigations have errors associated with them. The main errors that you will encounter are described below.

Systematic errors

A **systematic error** produces a constant bias in a measurement that cannot be eliminated by repeating the measurement. Systematic errors that affect a practical investigation could include:

- a pH meter that has not been calibrated, so is consistently recording pH values that are greater than they should be
- an unsuitable indicator being used in a **titration** (a precise form of volumetric analysis)
- heat being lost to the surroundings due to lack of insulation during an experiment in which the temperature is being measured.
- a person reading the scale on a measuring cylinder or burette with a constant **parallax error**. In this case, the parallax error occurs in reading a volume where the **meniscus**, the curved upper surface of the liquid, is not at eye level. This is shown in Figure 1.3.6.

Parallax error can also occur with **analogue** (non-digital) meters, such as voltmeters and ammeters, and any instrument in which a line, or needle has to be compared to a fixed scale.

Whatever the cause, the resulting error is in the same direction for every measurement and the average will either be too high or too low as a result. Repeating an experiment will not remove systematic error.

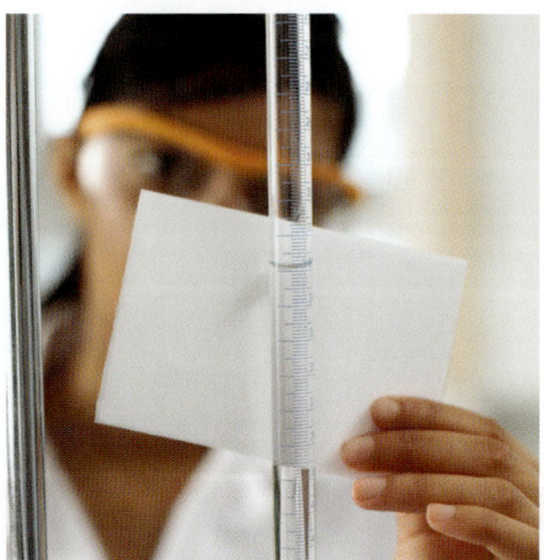

FIGURE 1.3.6 It is important to read the bottom of the meniscus at eye level to avoid parallax error. This student is showing how you can use a piece of white card (or a tile) to improve the contrast between the solution and the scale. She is holding the white card to make it easier for her partner to read the meniscus.

Random errors

Random errors follow no regular pattern. The measurement is sometimes too large and sometimes too small. Random errors in a practical investigation could include:

- error in estimating the second decimal place in the volume read from a burette
- temperature variation within a solution that has not been mixed consistently
- error in estimating the colour of an indicator.

The effects of random errors can be reduced by taking multiple measurements of the same quantity, then calculating a **mean**, or **average**. For example, when you repeat an experiment in which a solid is produced in three separate trials, the three masses of the solid should be added together and divided by three.

Another way to reduce random error is to refine a measurement technique by repeating it until you are more skilled in the technique.

Mistakes

Mistakes, or personal errors, are avoidable errors. Mistakes made during a practical investigation could include:

- misreading the numbers on a scale
- not labelling a sample adequately
- spilling a portion of a sample.

A measurement that involves a mistake must be repeated if possible, or rejected and not included in any calculations. Care in carrying out the experiment will prevent the majority of mistakes occurring.

Repeatability

Repeatability refers to the consistency of the results when the experiment is repeated many times. Maintain your investigation's repeatability by ensuring there is sufficient **replication** of the experiment. In general, you should repeat each experiment at least three times (trials). Three is not necessarily a magic number, and if you can do more trials your results should become more precise, at least because you are becoming more familiar with the method and therefore better at carrying it out.

Reproducibility

Reproducibility is the closeness of the agreement between measurements of the same quantity, carried out under changed conditions of measurements. Reproducibility links closely to validity (described below). Data is reproducible if similar results are obtained by different operators in different laboratories, repeating an experimental procedure and obtaining similar results. A benefit of analysing the reproducibility of data is to help identify possible systematic errors that would affect the accuracy of the experiment.

Resolution

Various types of equipment can be used to measure quantitative data. For example, time could be measured using a stopwatch, phone or using a clock. **Resolution** is the smallest change in the measured quantity that causes a perceptible change in the value shown by the measuring instrument, and has implications for the number of decimal places that should be quoted.

Resolution is determined differently for analogue and digital instruments. For example, if the resolution of a burette is 0.05 mL, then measurement readings of 10.50 mL or 10.55 mL are possible, but a measurement reading of 10.53 mL cannot be claimed. The meniscus of the liquid will either be on the burette line marking, in which case the reading would be 10.50 mL, or it will lie between 10.50 and 10.60, in which case it is measured as 10.55 mL.

Digital measuring equipment has a resolution to the last decimal place. For example, an electronic balance that measures to 0.001 g would have a resolution of 0.001 g.

If we compare analogue and digital devices, the resolution of an analogue clock might be 1 s, the resolution of a stopwatch might be 0.01 s, and the resolution of a phone app might be 0.001 s. Unfortunately, the reaction time of a human is commonly considered to be 0.2 s, so the apparent accuracy of the phone app is limited by the human who is recording the time.

Validity

Validity refers to whether an experiment or investigation is testing the set hypothesis and aims.

To ensure an investigation is valid, it should be designed so that only one variable is being changed at a time. The other variables must remain constant so that meaningful conclusions can be drawn about the effect of each variable.

- *You will now be able to answer key questions 2, 5, 6 and 8.*

TECHNIQUES FOR IMPROVING DATA QUALITY

Designing the method carefully, including selection and use of equipment, will help reduce random and systematic errors. Once you have chosen the appropriate equipment, you will need to calibrate the equipment to increase the accuracy of your measurements. To increase precision with the same instrument, use a larger sample size.

Calibration

Calibration is the process of adjusting an instrument using a reference or standard. Some equipment, such as pH probes, need calibrating before use; for example, to check that a pH probe is giving a true reading of pH 4.0. By using calibrated equipment, you can be more certain that your measured values are accurate. In colorimetry a **calibration curve** is constructed showing the relationship between the concentration of known standards and absorbance.

Equipment

To minimise errors, check the precision of the equipment that you intend to use. Pipettes, burettes and volumetric flasks have greater precision than a measuring cylinder for measuring volumes of liquids. However, you must still use all equipment correctly to reduce error.

Sample size

In general, the larger the sample (for example, the mass of starting material that you are using), the more precise the measured values will be. While wasteful experiments are contrary to the principles of green chemistry, your results will be more precise if you use 5 g of starting material than 0.5 g, given that you will still be using the same electronic balance to determine mass in both cases. Even though any error in the balance will be the same, the percentage error will be smaller for the larger mass.

TECHNIQUES TO SUPPORT YOUR PRACTICAL INVESTIGATION

A variety of techniques can be used for conducting practical investigations. The choice of equipment is important to minimise error, to ensure your measurements are accurate and your results are reproducible and reliable, and to minimise **uncertainty**. Techniques that you might use when conducting practical investigations are outlined in Table 1.3.2 on the following page.

CHEMFILE

Percentage uncertainty and choice of equipment

The calibration of a measuring cylinder varies depending on the volume that can be measured in that measuring cylinder (see figure below). A 10 mL measuring cylinder is marked off in increments of 0.2 mL, so its uncertainty is 0.1 mL, whereas a 50 mL measuring cylinder is marked off in increments of 1 mL, so its uncertainty is 0.5 mL. If you wish to measure 5 mL of a solution, you could do so in the 10 mL or 50 mL measuring cylinder, but to do so in the 50 mL measuring cylinder would introduce a greater uncertainty in the measured volume than in the 10 mL cylinder.

Percentage uncertainty is an expression of the resolution of the equipment compared to the quantity measured, so for the 10 mL cylinder:

$$\text{percentage uncertainty} = \frac{0.1}{5} \times 100 = 2\%$$

whereas for the 50 mL cylinder:

$$\text{percentage uncertainty} = \frac{0.5}{5} \times 100 = 10\%.$$

So, to improve the precision of your experiment, use the smallest measuring cylinder that will hold the volume you want to measure.

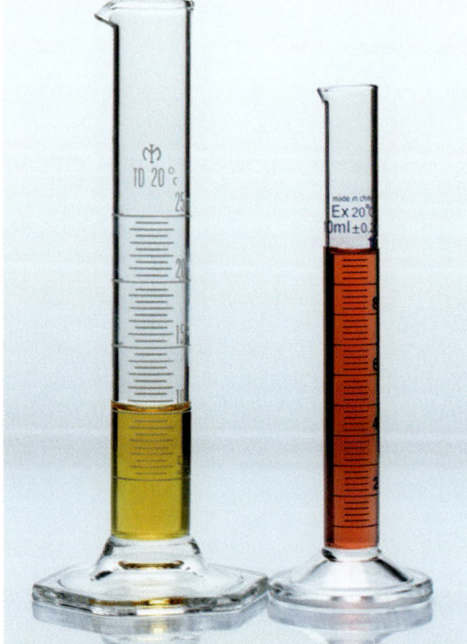

It is more precise to measure 10 mL of liquid in the 10 mL measuring cylinder than in the 25 mL measuring cylinder.

TABLE 1.3.2 Techniques that might be appropriate for a practical investigation

Technique	Purpose	Dependent variable	Example application
constant mass determination	to determine whether the mass of a solid is changing	mass of a solid	determining water of hydration in an ionic salt
titration	to calculate the unknown concentration of a solution	volume of solution (titre)	determining concentration of acids or bases
data logging	to collect data over a defined period of time	change in pH, temperature, conductivity, colour, carbon dioxide concentration, pressure, time	determining temperature or pH to see if a change is occurring
colorimetry	to calculate the unknown concentration of a solution	absorbance using a calibration curve	measuring the concentration of coloured components in solution

Data logging

In recent times, the ability to connect independent sensors to any device has enabled simple and highly effective electronic data acquisition. Data logging in schools is based around sensors, or probes, recording data to a standalone device, computer, tablet or even a phone (Figure 1.3.7). The software will generally graph and allow advanced calculations directly without the need for an additional program. Some probes can enable data collection to occur over an extended period of time. Examples include pH probes, temperature probes, carbon dioxide or oxygen concentration probes, conductivity probes, colorimeters and automatic mass balance equipment.

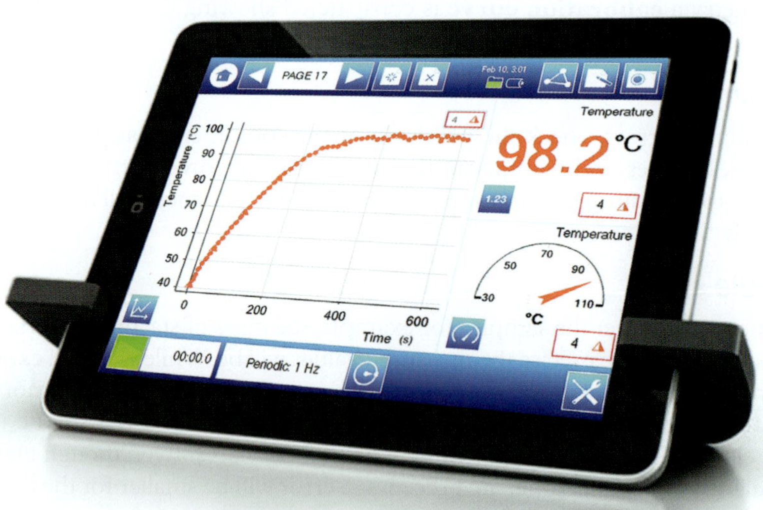

FIGURE 1.3.7 Data acquisition software produces real-time graphs that can be downloaded or printed.

1.3 Review

OA ✓✓

SUMMARY

- Record all information objectively in your logbook, including your data and method of investigation.
- The method is a step-by-step set of instructions that another person can follow to reproduce the experiment.
- Raw data is the data you collect in your logbook.
- Processed data is raw data that has been mathematically analysed.
- Qualitative data is observed and relates to a type or category, such as colour.
- Quantitative data are measured numeric values and have units included.
- Discrete data are values that can be counted or measured, but which can only have certain values.
- Continuous data may be any number value within a given range that can be measured.
- Validity refers to whether your results measure what the investigation set out to measure.
- Repeatability is the consistency of your results when they are repeated many times as trials under the exact same set of experimental conditions.
- Reproducibility is the ability to obtain the same results if an experiment is replicated.
- Accuracy is how close a measurement is to the true value.
- Precision is how closely a set of measurements agree with each other.
- Resolution is the smallest change in the measured quantity that causes a perceptible change in the value shown by the measuring instrument.
- Systematic errors result in errors in the same direction for every measurement—either too high or too low.
- Random errors follow no regular pattern.
- Mistakes or personal errors are avoidable and should not be included in further analysis of results.

KEY QUESTIONS

Knowledge and understanding

1 For each of the following masses, state the number of significant figures.
   - **a** 1.36 g
   - **b** 0.0205 g
   - **c** 1.5000 g
   - **d** 0.004 060 g

2 Fill in the gaps in the following sentences using these terms:
repeatability; reproducibility; resolution; accuracy; precision; true value; validity
   - **a** ________ is how close a measurement is to the ________ , whereas ____________ is how closely a set of measurements agree with each other.
   - **b** ____________ refers to whether your results measure what the investigation set out to measure.
   - **c** ______________ is the consistency of your results when they are repeated many times as trials under the exact same set of experimental conditions, whereas __________ is the ability for another experimenter to obtain the same results if they replicate your experiment.
   - **d** ____________ is the smallest change in the measured quantity that causes a perceptible change in the value shown by the measuring instrument.

3 Explain why all data and observations should be recorded in a logbook.

4 Rewrite the following table, identifying which of the pieces of information about data collected in a constant mass experiment are qualitative and which are quantitative. Place a tick in the appropriate column.

Information	Quantitative	Qualitative
colour of hydrated salt		
colour of dehydrated salt		
initial mass of hydrated salt		
texture of dehydrated salt		
final mass of dehydrated salt		

5 A journal article reported the materials and method used to conduct an experiment. The experiment was repeated three times, and all values were reported in the results section of the article. Repeating an experiment and averaging the quantitative results reduces:
   - **A** precision
   - **B** random errors
   - **C** accuracy
   - **D** systematic errors

continued over page

1.3 Review *continued*

6 Match the following terms with their definitions: repeatability of data; precision; validity of data; accuracy; resolution

a how close a measured value is to the actual value

b the smallest change in the measured quantity that causes a perceptible change in the value

c how many decimal places an experimental value can be reported

d how well an experiment tests a hypothesis; if the data obtained is used to reach a logical conclusion

e how similar the results of an experiment will be when the method is followed by another scientist

Analysis

7 Explain the difference between raw and processed data, using a titration as an example.

8 a State the resolution and measured value for the digital thermometer below.

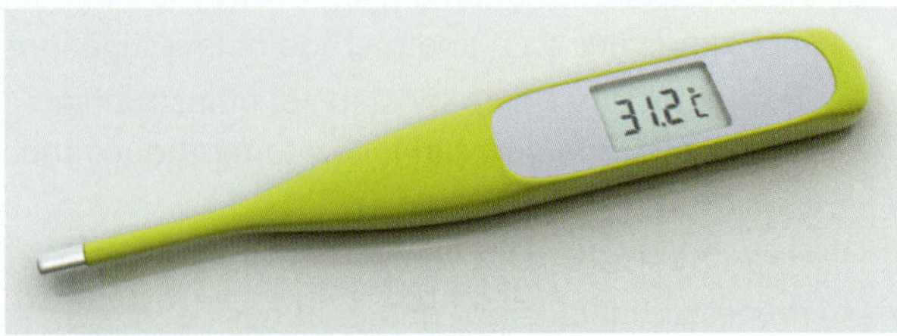

b State the uncertainty and measured value for the measuring cylinder below.

1.4 Data analysis and presentation

In Section 1.3 you learnt about different types of data and the factors that affect data quality. In this section you will learn how to process and analyse your data and present analysed data in graphs.

PROCESSING DATA

Whenever you process data, it is important to show the formula you used and a sample calculation. Even if you don't include all of the information in your report or scientific poster, everything should be recorded in your logbook.

You might have measured the initial and final masses of seashells which were submerged in acidic solutions of different pH values (Table 1.3.2, page 30). You can then process this data further.

For example,

$$\text{change in mass} = \text{final mass} - \text{initial mass}$$

The mean change in mass is found by adding all the changes in mass for a particular pH and dividing them by the number of trials.

$$\text{mean change in mass} = \frac{(\text{change in mass}_1 + \text{change in mass}_2 + \text{change in mass}_3)}{3}$$

When summarising processed data in a table, record the independent variable in the left-hand column and the dependent variable(s) in the column(s) that follow. The processed or calculated values should be recorded in the final column.

In Table 1.4.1, the data in the final two columns is processed data (the change in mass = final mass − initial mass, and the mean of trials).

Calculation of the mean of a number of trials is commonly used in titrations. This is covered in detail in Chapter 14.

TABLE 1.4.1 Summary results table indicating the measured mass loss when varying the pH of a solution in which a seashell is submerged

pH	Trial number	Change in mass (g)	Mean change in mass (g)
6	1	0.221	0.220
	2	0.220	
	3	0.219	
5	1	0.250	0.249
	2	0.245	
	3	0.252	
4	1	0.280	0.278
	2	0.278	
	3	0.277	
3	1	0.310	0.311
	2	0.308	
	3	0.314	

- *You will now be able to answer key question 8.*

Transforming decimal notation to scientific notation

Scientists use **scientific notation (standard form)** to handle very large and very small numbers.

For example, instead of writing 0.000 003 5, scientists would write 3.5×10^{-6}.

A number in scientific notation (also called standard form or power of ten notation) is written as $a \times 10^n$, where:

a is a number equal to or greater than 1 and less than 10: that is, $1 \leq a \leq 10$

n is an integer (a positive or negative whole number)

n is the power that 10 is raised to and is called the **index** (plural, indices).

When large numbers are written in scientific notation, the 10 has a positive index. When very small numbers are written in scientific notation, the 10 has a negative index.

Two examples of how to transform decimal notation into scientific notation are shown in Worked example 1.4.1 on the following page.

Worked example 1.4.1

TRANSFORMING DECIMAL NOTATION TO SCIENTIFIC NOTATION

Transform each of the following numbers into scientific notation:

a 51 400

b 0.004 38

Thinking	Working
a Place a decimal point after the first non-zero digit so that the original number is written as a decimal number greater than 1 but less than 10.	51 400 becomes 5.1400
Determine the index by counting the number of places the decimal point needs to be moved to form the original number again.	5.1400 The decimal has to be moved 4 places.
Multiply the decimal number by the appropriate power of 10, as determined by counting in the step above. If the decimal point is moved to the left, the index will be a positive number. If the decimal point is moved to the right, the index will be a negative number.	The decimal was moved 4 places to the left, so the index number will be 4. 5.1400×10^4
b Place a decimal point after the first non-zero digit so that the original number is written as a decimal number greater than 1 but less than 10.	0.004 38 becomes 4.38
Determine the index number by counting the number of places the decimal point needs to be moved to form the original number again.	0.00438 The decimal has to be moved 3 places.
Multiply the decimal number by the appropriate power of 10, as determined by counting in the step above. If the decimal point is moved to the left, the index will be a positive number. If the decimal point is moved to the right, the index will be a negative number.	The decimal was moved 3 places to the right, so the index number will be –3. 4.38×10^{-3}

Worked example: Try yourself 1.4.1

TRANSFORMING DECIMAL NOTATION TO SCIENTIFIC NOTATION

Transform each of the following numbers into scientific notation:

a 2560

b 0.000 097 1

● *You will now be able to answer key question 1.*

Managing significant figures in calculations

The number of significant figures to which an answer should be quoted depends on what kind of calculation you are doing:

If you are multiplying or dividing, use the smallest number of significant figures provided in the initial values.

If you are adding or subtracting, use the smallest number of decimal places provided in the initial values; (e.g. 12.78 mL + 10.0 mL = 22.78 = 22.8 mL).

When you are processing data using your scientific calculator, be careful to not report every decimal place stated on the calculator. For example, if the least accurate piece of data in your calculation involving multiplication or division is to three significant figures, then the calculated result should likewise be reported to three significant figures. However, if you are undertaking a calculation with several steps, keep the more accurate value with extra decimal places in your calculator.

The use of scientific notation will help you express a final answer to the correct number of significant figures.

The number 2500, in scientific notation, would be written as:

2.5×10^2 (two significant figures),

2.50×10^2 (three significant figures),

2.500×10^2 (four significant figures),

depending on the actual data used to calculate this value.

Note that the significant figures in a molar mass should not limit your significant figures in the answer. The molar mass is not experimental data and, in the VCE Chemistry course, all relative atomic masses are quoted to one decimal place. So, unless you are working with very small elements, such as hydrogen and helium, relative atomic masses will have at least three significant figures.

Only the final calculated result should be reported to the appropriate number of significant figures, as shown in Worked example 1.4.2.

- If you are multiplying or dividing, use the smallest number of significant figures provided in the initial values.
- If you are adding or subtracting, use the smallest number of decimal places provided in the initial values.

Worked example 1.4.2

PRESENTING CALCULATIONS TO THE CORRECT NUMBER OF SIGNIFICANT FIGURES

Calculate the mass of nitrogen, N_2, in a balloon if the volume is 100 L at a pressure of 95 000 Pa and a temperature of 5.0 °C.

Thinking	Working
Record all data, taking note of the significant figures, and convert the pressure to kPa and temperature to kelvin. Remember when adding numbers, the answer should have no more decimal places than the value with the least number of decimal places.	Volume = 100 L (3 sig figs) Pressure = 95 000 Pa = 95.000 kPa (5 sig figs) Temperature = 5.0°C = 273 + 5.0 = 278 K (noting that original temperature is to 2 sig figs) Molar gas constant = 8.31 J mol^{-1} K^{-1}
Calculate the amount of helium, using the universal gas equation, $pV = nRT$. Keep all figures in the calculator for the next step.	$n(N_2) = \frac{pV}{RT} = \frac{95.000 \times 100}{8.31 \times 278} = 4.112$ mol
Calculate the mass of nitrogen, N_2, using $n = \frac{m}{M}$. Note that the significant figures in the molar mass should not limit your significant figures in the answer.	$m(N_2) = n \times M$ $= 4.112 \times (2 \times 14.0)$ $= 115.1425$ g
Round the calculated answer to the lowest number of significant figures, (two significant figures in the temperature), using scientific notation if necessary.	$m(N_2) = 1.2 \times 10^2$ g

Worked example: Try yourself 1.4.2

PRESENTING CALCULATIONS TO THE CORRECT NUMBER OF SIGNIFICANT FIGURES

Calculate the mass of carbon dioxide, CO_2, in a balloon if the volume is 2.50 L at a pressure of 60 000 Pa and a temperature of 15°C.

● *You will now be able to answer key questions 2 and 3.*

DATA ANALYSIS

In the discussion section of your scientific investigation report, the findings of your investigation need to be analysed and interpreted. The following factors should be considered and discussed.

- Whether a pattern, trend or relationship was observed between the independent and dependent variables.
- What kind of pattern it was and under what conditions it was observed.
- Whether there were data points that did not match the trend (**outliers**).
- How did the values you were monitoring change over the course of the experiment?
- How did your results compare to the theoretical results, or those recorded in literature sources?

Graphs

It is easier to observe trends and patterns in data in graphical form than in tabular form. An example of a **line graph** is shown in Figure 1.4.1.

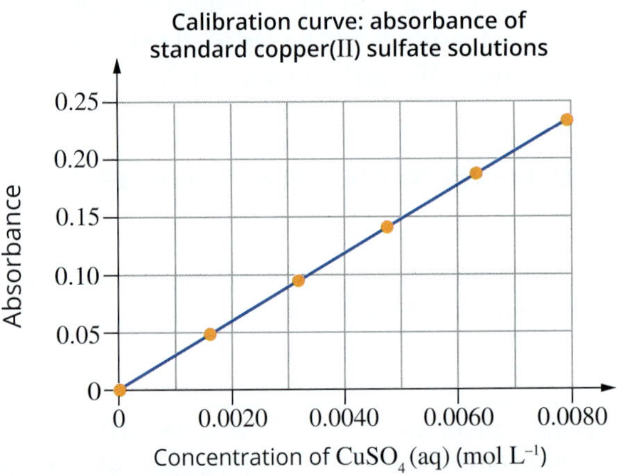

FIGURE 1.4.1 A graph is a good way to observe trends and patterns in data.

General rules to follow when presenting a line graph include the following.

- Use a descriptive title.
- Represent the independent variable on the *x*-axis and the dependent variable on the *y*-axis.
- Clearly label the axes with both the variable and the unit in which they are measured.
- Choose a scale on the axes that will be appropriate for the data and that is consistent.
- If values between plotted points need to be read, then the values between marked points on the scale should be as easy to read as possible; for example, mark the scale off in factors of 2 or 5, rather than 3 or 4.
- If there is a data point at (0,0), then the line of best fit can go through the origin. Otherwise, draw the line of best fit only through the data points on the graph.
- Keep the graph simple and uncluttered.

The independent variable should always be plotted along the *x*-axis and the dependent variable on the *y*-axis of a graph.

Types of graphs

A **scatter graph**, such as Figure 1.4.1, is used when two variables are being considered and one variable is dependent on the other. The independent variable is plotted along the x-axis and the dependent variable is plotted along the y-axis. When an appropriate **line of best fit** is fitted to the data points, the graph should show the relationship between the two variables. When the line of best fit is drawn on the scatter graph, it becomes a line graph.

Bar graphs, also called **column graphs** (Figure 1.4.2), also clearly illustrate relationships between variables, but are most appropriate for discrete rather than continuous data (see Section 1.3, page 24), whereas **pie charts** illustrate percentages effectively (Figure 1.4.3).

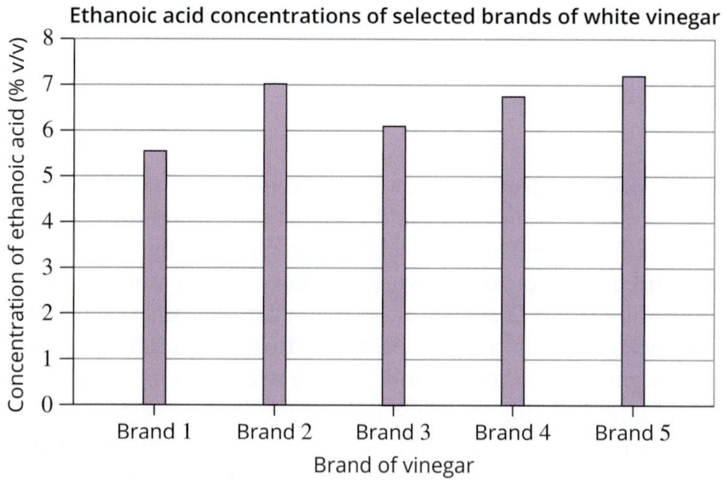

FIGURE 1.4.2 A bar graph can be used to compare discrete data. This graph shows the ethanoic acid concentrations in a number of brands of vinegar.

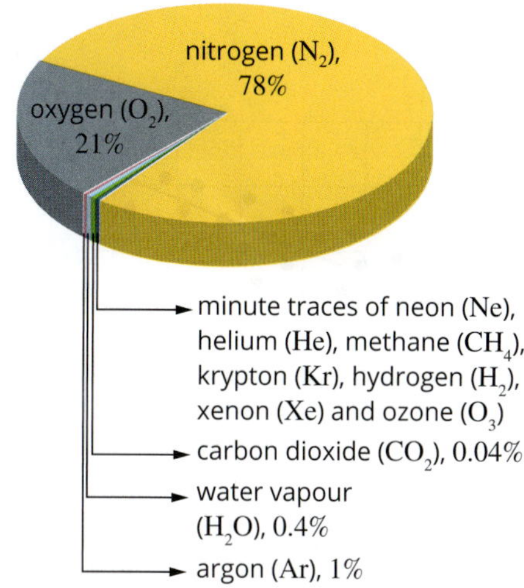

FIGURE 1.4.3 A pie graph is useful for representing proportional data; for example, the chemical composition of air.

● *You will now be able to answer key question 4.*

Trend lines in graphs

Line graphs are drawn to show the relationship, or **trend**, between two variables.

- Variables that change in direct proportion to each other produce a straight trend line, as shown in Figure 1.4.4. This is described as a **linear trend** or relationship.
- Variables that change non-linearly in proportion to each other produce a curved trend line, as shown in Figure 1.4.5. This graph reflects a positive non-linear, or **exponential relationship**; that is, as x increases, y increases non-linearly.
- When there is an **inverse relationship**, one variable increases as the other variable decreases (see Figure 1.4.6).

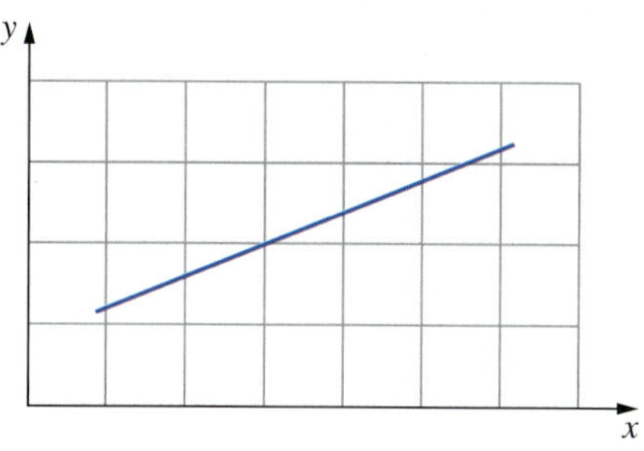

FIGURE 1.4.4 Trend line for variables that change in direct response to each other

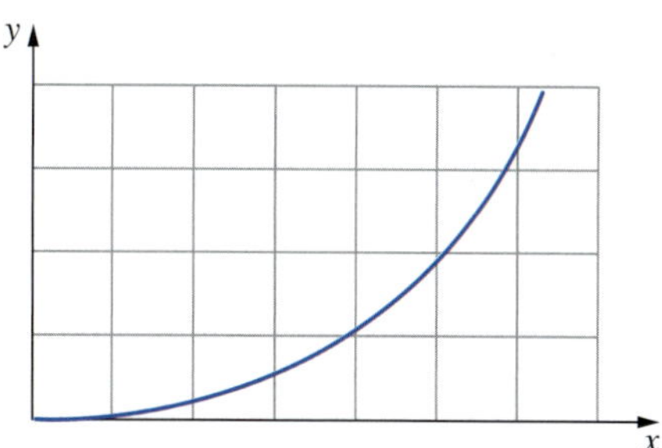

FIGURE 1.4.5 Variables that change in response to each other in a non-linear way

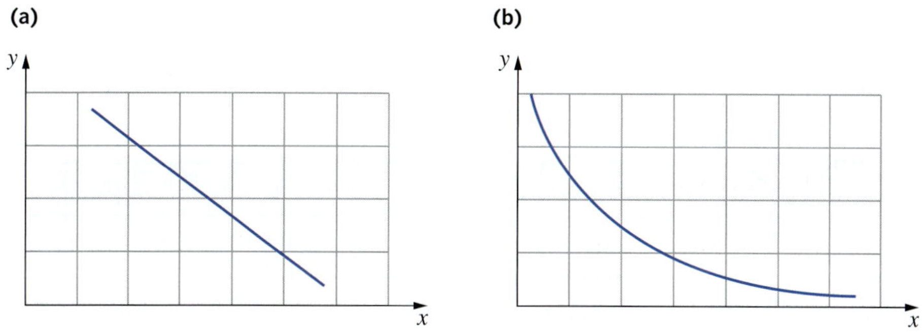

FIGURE 1.4.6 An inverse relationship in which one variable decreases as the other variable increases. The relationship may be (a) direct or (b) non-linear (exponential).

- When there is no relationship between two variables, one variable will not change even if the other changes. A graph in which there is no relationship can be seen in Figure 1.4.7.

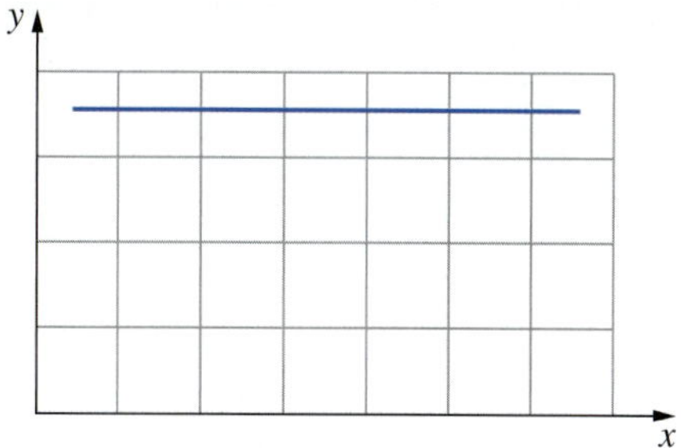

FIGURE 1.4.7 When two variables show no relationship, there is no trend in the graph.

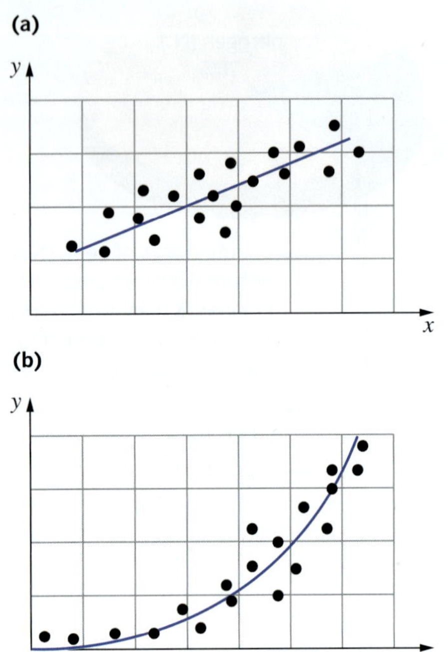

FIGURE 1.4.8 Scatter graphs showing (a) straight and (b) curved trend lines

Scatter graphs and lines of best fit

Scatter graphs are commonly used to display data and can be used to plot raw or processed data. They are used to show the relationship between two variables when one variable is dependent on the other.

The independent variable, which is set by the experimenter, is usually shown on the x-axis. The dependent variable, which is the variable measured in the experiment, is shown on the y-axis. The data is plotted on the graph as a series of points.

In a scatter graph, a trend line is then fitted to the data (Figure 1.4.8). The trend line might be a straight line of best fit or a curved line. It is used to show the overall trend in the data and can be used to predict values between the data points. A line of best fit usually does not pass through every data point. Its position can be estimated by eye (ensuring an equal number of data points are above and below the line when drawn). Alternatively, mathematical graphing programs can be used to accurately determine the line of best fit.

Outliers

Sometimes when you collect data there may be points or observations that differ significantly from other data points and do not fit the trend observed in the data. These are referred to as outliers. An outlier is often caused by a mistake made when data was measured or recorded. To work with your results ethically, any outliers in your data must be further analysed and explained, rather than being automatically dismissed. Repeating trials may be useful in further examining an outlier—for example, to determine whether the outlier is the result of a mistake.

The graph shown in Figure 1.4.9 is a calibration curve for determining the concentration of phosphate in a water sample. The data point that does not fit the trend of the other data points is an outlier. A mistake may have occurred when preparing that particular phosphate standard solution. In this case, you should show the data point, but do not use this point when drawing a line of best fit.

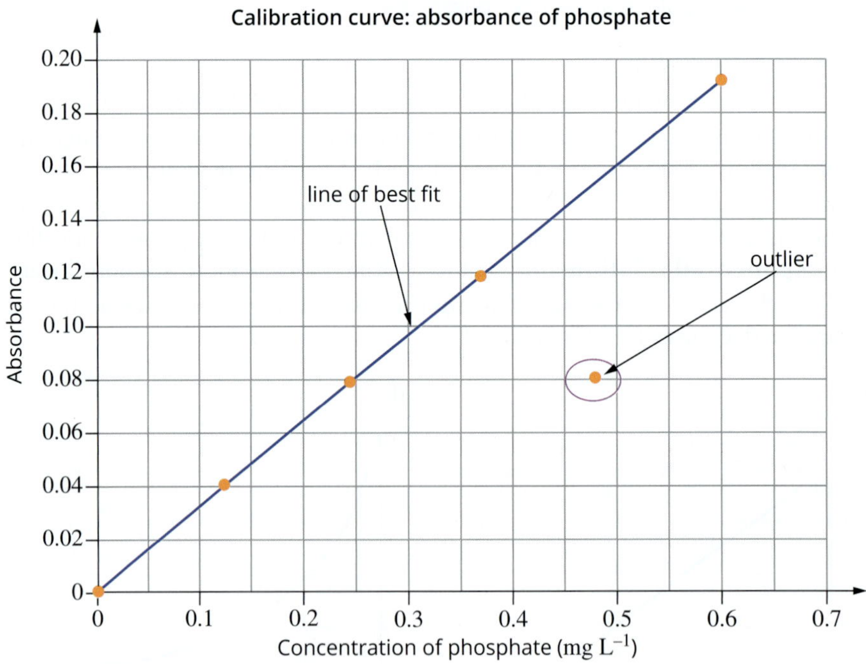

FIGURE 1.4.9 A calibration curve for determining the concentration of phosphate in a water sample

You will now be able to answer key questions 7, 9 and 10.

CASE STUDY ANALYSIS

Rhodamine B in wastewater

Rhodamine B is a fluorescent pink dye which is extensively used in the textiles, paper, plastics, drugs and cosmetics industries. It can be used to make fabric bright pink, add a pink flash to a lipstick, or to add an intense pink colour to a variety of paper types. Rhodamine B may also be used for tracing the rate and direction of flow of water in the environment, such as wastewaters in groundwater systems.

The value of rhodamine B in all of these applications is its high solubility and intense colour, which is achieved with the dissolution of just a tiny mass of the solid. These properties also make **UV–visible spectroscopy** (a technique in which the concentration of a coloured substance is determined by shining UV or visible light through it) a useful way to analyse water samples that have just a tiny concentration of rhodamine B present.

Because rhodamine B is toxic to organisms in the water, its presence in wastewater from the industries that use it in their products is not desirable and so analysis of this wastewater has become of interest to chemists. This dye is supposed to be removed from wastewater by the industries using various separation techniques, before it is released into the environment.

Scientists have recently reported that rhodamine B concentrations as low as 0.49 μg L^{-1} could be detected in tap water using UV–visible spectroscopy at a wavelength of 558 nm. As well as tap water, they also tested samples of nail polish and lipstick.

(a)

(b)

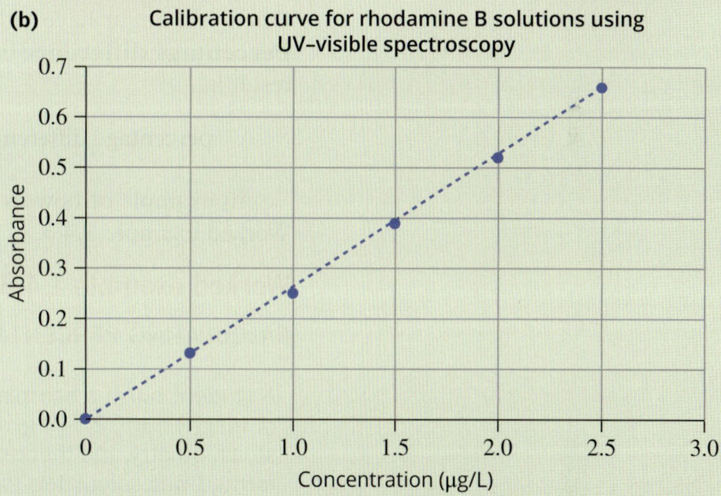

FIGURE 1.4.10 (a) Rhodamine B is a fluorescent pink dye which dissolves readily in water and also in a number of organic solvents. (b) A calibration curve for a solution of rhodamine B, determined using UV–visible spectroscopy.

A bright pink solution of rhodamine B in a school laboratory can be seen in Figure 1.4.10a. In Figure 1.4.10b you can see a calibration curve for concentrations of rhodamine B obtained from UV–visible spectroscopy. The questions that follow give you the opportunity to work out the concentration of a solution similar to the ones that the scientists created and analysed. More details about UV–visible spectroscopy can be found in Section 16.3, page 509.

Analysis

1 A wastewater sample taken from a creek near a textiles company was analysed using UV–visible spectroscopy for the presence of pink rhodamine B dye, as it was suspected that the company was not treating their wastewater correctly. The absorbance of the undiluted sample was 0.44. Use the calibration curve in Figure 1.4.10b to determine the concentration of rhodamine B in the wastewater.

2 In a separate analysis, the chemists diluted a sample of bright pink nail polish with an organic solvent. 1 mL of nail polish was diluted by a factor of 100, so that the sample analysed was 100 times less concentrated than the original. The absorbance of the diluted sample was found to be 0.32.

- **a** Use the calibration curve in Figure 1.4.10b to determine the concentration of rhodamine B in the diluted nail polish sample.
- **b** Calculate the original concentration of rhodamine B in the pink nail polish.

3 Many organic solvents absorb UV light, but the wavelength at which the sample was tested was 558 nm, which corresponds to yellow-green visible light. Use this information to explain why it is acceptable to use the wastewater calibration curve for the absorbance values of the diluted nail polish.

Comparing your data to other values

While you may have taken care in your observations and recording your data, and you have every confidence in your efforts to be precise, you cannot claim your results are accurate unless you compare them to **literature** or **accepted values**. A literature value is one which has been found in published scientific reports, such as journals, or reliable websites, such as those of universities. Accepted values include values that are obtained from published sources and they also include those that you may calculate using a correctly balanced equation for the reaction and stoichiometry. For example, the calculated mass of product when a sample of $BaCl_2.2H_2O$ is heated to constant mass according to the equation

$$BaCl_2.2H_2O(s) \longrightarrow BaCl_2(s) + 2H_2O(l)$$

would be considered an accepted value.

To make a comparison between your results and the accepted or theoretical value, a percentage difference is calculated and can form part of your discussion. If you wish to discuss how your results changed over the duration of the experiment, then a calculation of the percentage change might be appropriate.

Percentage difference

Often you might have a theoretical value to compare to your experimental value. **Percentage difference** is very useful for showing how accurate your experimental result is.

$$\text{percentage difference} = \frac{(\text{experimental value} - \text{theoretical value})}{\text{theoretical value}} \times 100$$

An example of how to calculate and express percentage difference is shown in Worked example 1.4.3.

Worked example 1.4.3

CALCULATING PERCENTAGE DIFFERENCE

A student burnt a sample of magnesium in a crucible, weighing the initial mass of magnesium, 2.548 g, and the final mass of magnesium oxide, 4.022 g. Using stoichiometry, the theoretical mass of magnesium oxide that should have been formed was calculated to be 4.226 g.

Calculate the percentage difference between the mass of magnesium oxide predicted by stoichiometry and the mass obtained in the experiment.

Thinking	Working
Identify the two masses which are to be compared, noting which one is larger. Ignore any information that is not needed for the comparison of results, such as mass of magnesium.	The theoretical mass of magnesium oxide, 4.226 g is larger, and the experimental mass of magnesium oxide, 4.022 g is smaller.
Substitute the masses into the formula: percentage difference = $\frac{(\text{experimental value} - \text{theoretical value})}{\text{theoretical value}} \times 100$	percentage difference $= \frac{(4.022 - 4.226)}{4.226} \times 100$ $= \frac{-0.204}{4.226}$
Ensure that significant figures are correct.	The answer should be expressed to 3 significant figures, due to the subtraction in the first step. $= -4.83\%$
State percentage difference, using the sign (+ or –) to help you compare the experimental and theoretical values clearly.	The experimental value is 4.83% less than the theoretical mass of magnesium oxide.

Worked example: Try yourself 1.4.3

CALCULATING PERCENTAGE DIFFERENCE

A student decomposed a 4.285 g sample of copper(II) carbonate using a crucible and a Bunsen burner, and the final mass of the copper(II) oxide was measured to be 3.000 g. Using stoichiometry, the theoretical mass of copper(II) oxide that should have been formed was calculated to be 2.758 g.

Calculate the percentage difference between the mass of copper(II) oxide predicted by stoichiometry and the mass obtained in the experiment.

$$\text{percentage difference} = \frac{(\text{experimental value} - \text{theoretical value})}{\text{theoretical value}} \times 100$$

Percentage change

Percentage change can be a useful measure of the difference between the initial and final values. The percentage change is calculated using the following mathematical formula:

$$\text{percentage change} = \frac{(\text{final value} - \text{initial value})}{\text{initial value}} \times 100$$

An example of how to calculate and express percentage change is shown in Worked example 1.4.4.

Worked example 1.4.4

CALCULATING PERCENTAGE CHANGE

A student dried a 2.561 g sample of hydrated copper(II) sulfate using a crucible and a Bunsen burner and the final mass of the anhydrous copper(II) sulfate was measured to be 1.639 g. Calculate the percentage change of mass to the correct number of significant figures.

Thinking	Working
Identify the two masses, noting whether there has been an increase or decrease in mass.	The initial mass, 2.561 g is larger, and the final mass, 1.639 g is smaller, so the mass has decreased.
Substitute the masses into the formula: percentage change = $\frac{(\text{final value} - \text{initial value})}{\text{initial value}} \times 100$	percentage change = $\frac{(1.639 - 2.561)}{2.561} \times 100$ $= \frac{-0.922}{2.561} \times 100$
Ensure that significant figures are correct.	The answer should be expressed to 3 significant figures, due to the subtraction in the first step. $= -36.0\%$
State percentage change, using the sign (+ or –) and your initial observation of mass change to guide you as you indicate in words whether there has been an increase or a decrease.	There has been a 36.00% *decrease* in the mass.

Worked example: Try yourself 1.4.4

CALCULATING PERCENTAGE CHANGE

A student decomposed a 2.503 g sample of calcium carbonate using a crucible and a Bunsen burner, and the final mass of the calcium oxide was measured to be 1.403 g. Calculate the percentage change for mass during the experiment to the correct number of significant figures.

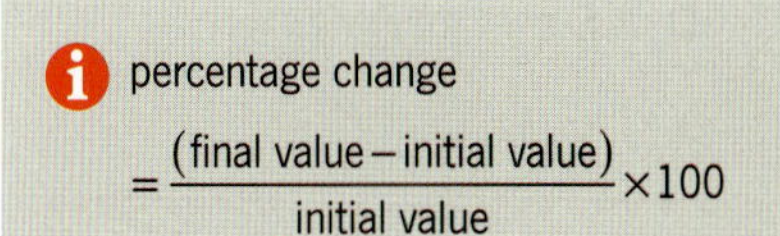

percentage change

$$= \frac{(\text{final value} - \text{initial value})}{\text{initial value}} \times 100$$

● *You will now be able to answer key questions 5 and 6.*

1.4 Review

OA ✓✓

SUMMARY

- The mean is the average of a set of quantitative data.
- When calculating using raw data it is important to correctly state the number of significant figures.
- When adding or subtracting, the final result should be reported to the least number of decimal places.
- When multiplying or dividing, the final result should be reported to the least number of significant figures.
- Tables allow the presentation of summarised calculations, whereas graphs allow trends to be shown more clearly.
- Graphs can be described according to their trend—for example, proportional, linear or inverse.
- Scatter graphs are useful when showing quantitative data where one variable is dependent on another variable. A line of best fit should be included.
- The independent variable is usually plotted along the *x*-axis and the dependent variable on the *y*-axis.
- Line graphs are useful for presenting continuous quantitative data.
- Outliers should be reported and considered, but are not usually included in drawing a line of best fit.
- Bar graphs are useful for comparing discrete data.
- Pie diagrams are useful for showing proportional data.
- Percentage difference can be used as a measure of the difference between an experimental value and the published value.
- Percentage change can be used to describe the change in the value of a measurement during an experiment compared to the initial value.

KEY QUESTIONS

Knowledge and understanding

1 Write each of the following values in scientific notation using the appropriate number of significant figures.

a 0.002

b 2050

c 123.4

d 0.000 032 5

2 A student multiplied 1.22 by 1.364. Which of the options below shows the result of this multiplication with the correct number of significant figures?

A 1.66

B 1.664

C 1.65

D 1.7

3 A student accidentally measured the mass of a watch glass and some copper(II) sulfate on different electronic balances. The mass of the watch glass was 3.57 g. The copper(II) sulfate had a mass of 5.321 g. Which of the following is the combined mass of the watch glass and the copper(II) sulfate to the correct number of significant figures?

A 8.891 g

B 8.89 g

C 8.9 g

D 8.8 g

4 Which graph from the following list would be best to use with each set of data listed here?

pie chart; scatter graph (with line/curve of best fit); bar graph; line graph

a the proportion of the type of pollution found in a sample of wastewater

b a graph showing the mass of a sample of hydrated copper(II) sulfate measured at 30 second intervals while it is being heated to achieve complete dehydration.

c the concentration of citric acid in a range of orange juices

d the relationship between the volume of a gas and the temperature of that gas

5 A pair of students heated 3.13 g of hydrated magnesium sulfate in a crucible until a constant mass of 1.53 g of anhydrous magnesium sulfate was obtained. Calculate the percentage change in mass of the hydrated magnesium sulfate.

6 A student heated 2.138 g of hydrated barium chloride, $BaCl_2.2H_2O$, in a crucible until the white solid has a mass of 1.925 g. When the student carried out the appropriate stoichiometric calculations, the student found that the mass of the anhydrous barium chloride should have been 1.823 g. Calculate the percentage difference between the student's results and the theoretical result.

7 Explain the meaning of the term 'trend' in a scientific investigation and list the types of trends that might exist.

Analysis

8 Consider the following set of results from the titration of a sample of diluted concrete cleaner (active ingredient is HCl) with sodium hydroxide.

	Final burette reading (mL)	Initial burette reading (mL)	Volume of titre (mL)
Trial 1	36.00	17.40	18.60
Trial 2	48.70	29.80	18.90
Trial 3	25.20	0.05	25.15
Trial 4	34.65	15.30	19.35
Trial 5	29.75	10.85	18.90
Trial 6	36.65	17.70	18.95

a List the three results in the above table that would be considered to be concordant.

b Calculate the mean of these **concordant titres**.

c Identify the result which is most likely to be due to a mistake made by the student.

9 The following calibration curve was obtained by colorimetry, but there have been some mistakes made in drawing it. The curve plots the absorbance of iron(III) ions against their concentration. The concentrations of iron(III) ions used to make this calibration curve ranged from 0.00 mg L^{-1} to 0.20 mg L^{-1}.

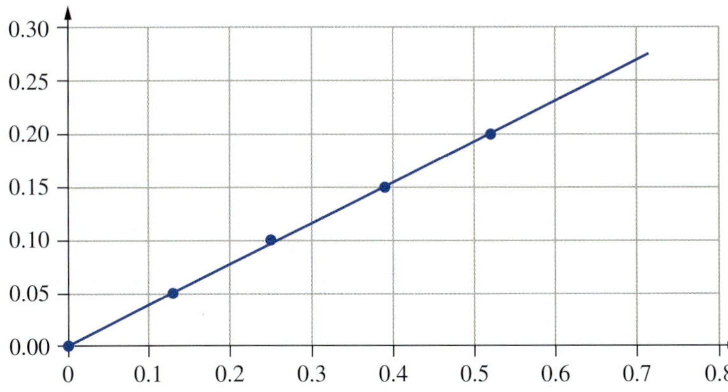

Describe five changes that should be made to improve and correct this calibration curve.

10 Standard nickel(II) nitrate solutions were prepared and analysed using colorimetry. A solution of nickel(II) nitrate of unknown concentration was then inserted into the colorimeter under the same conditions. The following data was obtained.

Nickel (Ni^{2+}), ion concentration (mg L^{-1})	Absorbance
0.00	0.000
0.50	0.034
1.00	0.068
1.50	0.103
2.00	0.140
2.50	0.175
3.00	0.242
3.50	0.247
4.00	0.280
4.50	0.315
5.00	0.352
unknown	0.155

a Plot the data on a scatter plot, using graph paper or a spreadsheet program.

b Identify any outliers in this set of data.

c Draw a trend line.

d Use your graph to determine the concentration of nickel(II) ions in the unknown solution.

1.5 Evaluation and conclusion

In this section, you will learn how to evaluate the method that you used in your investigation and draw evidence-based conclusions in relation to your hypothesis and research question. When you consider the accuracy of your results, ideally by comparing them to the literature or accepted value, you should find a focus for your evaluation of the method.

EVALUATING THE METHOD

It is important to evaluate the method used in the controlled experiment to determine whether it was completely appropriate for your experiment. You may have designed your method with all good intentions, but now your results are not as you expected.

Consider the following:

- What were the sources of systematic errors?
- What were the sources of random errors?
- Was the method valid?
- Was the experiment repeated?
- Was the data precise?
- Would a larger sample or further variations in the independent variable lead to a stronger conclusion?
- What could be done next time to improve the method?

The method should be evaluated to identify its strengths and weaknesses and to consider validity, precision, accuracy and reproducibility.

For example, when measuring the volume of gas produced when calcium carbonate (marble chips) reacted with hydrochloric acid, the following gas collection system was used, which involved the displacement of water (Figure 1.5.1a).

The method was valid—it aimed to measure the volume of carbon dioxide gas produced. However, the method was not accurate. Some of the carbon dioxide gas would dissolve in the water, so the volume of carbon dioxide measured would be less than the true value. This would be a systematic error. This method could be improved by using a gas syringe to collect the gas (Figure 1.5.1b).

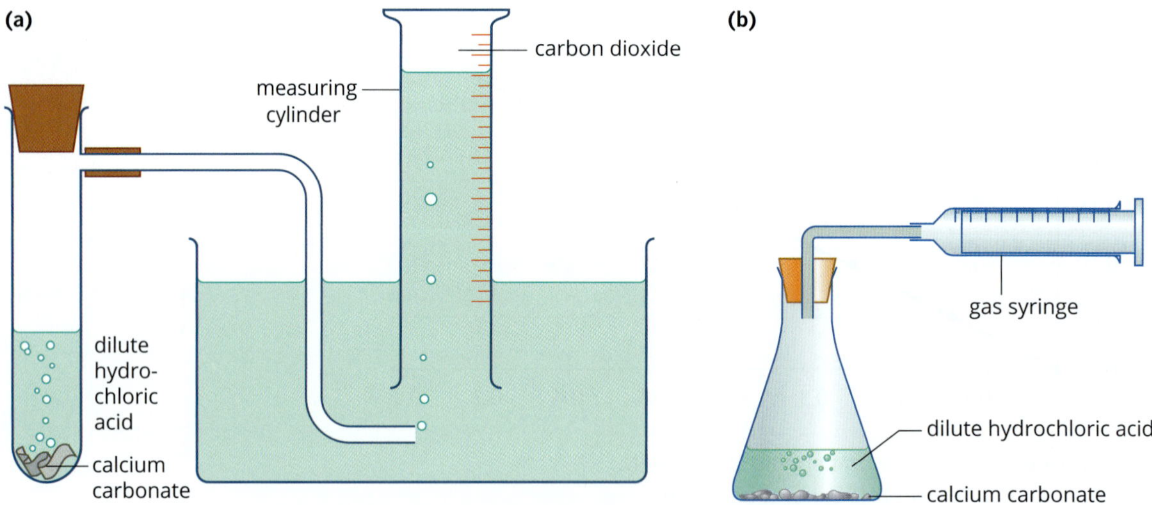

FIGURE 1.5.1 Different methods of carbon dioxide gas collection: (a) using displacement of water and (b) using a gas syringe. Using a gas syringe is usually considered to be a more accurate way of measuring the gas produced.

IDENTIFYING ERRORS

Most practical investigations have errors associated with them. Errors can be classified into two main categories, with mistakes being considered quite separately. An important part of evaluating your method is to consider errors which may have occurred because of the method you have chosen and have impacted on your results. Systematic and random errors were discussed in Section 1.3 (page 27) as were mistakes and outliers. Systematic errors will cause your results to be consistently greater than or less than the accepted value, so these should be identified within your method when your results consistently do not match the accepted value. Random errors may not be as easy to identify, but they will contribute to a lower precision in your results and may be evident when not enough trials have been conducted, or you are not experienced in using the equipment.

What to do about errors

As you consider your results and how they differ from what you predicted in your hypothesis, one of the biggest steps is to suggest what could be done to overcome errors. Random errors can be minimised by repeating trials and averaging the results. Systematic errors can often be avoided by taking care with the use of equipment, but this is not always the case. Mistakes must not be used as valid results.

Systematic errors often are caused by a problem with part of the method. Random errors affect the precision of the results and may be due to a lack of experience.

Your qualitative observations often hold the key to a systematic error. For example, when a hydrocarbon is burning, a yellow, sooty flame suggests that incomplete combustion is occurring. The observation that the test tube became hot during a metal displacement reaction suggests that energy is being lost to the surroundings. You can use the information gained from these observations to identify issues with the method.

When you are analysing your data, be sure to acknowledge contradictions in the data. Rather than ignoring results that don't fit your hypothesis, look for the reasons that these results occurred, working carefully through the method to determine what made these results different.

Sometimes your experimental findings will lead you to formulate new research questions and develop new hypotheses. These can be listed as ways in which the investigation can be extended in the future.

Improving the method

For each methodological issue that you identify, you should suggest a way in which the method could be improved to prevent that issue. While these issues will be particular to any investigation, Table 1.5.1 shows a selection of issues and improvements possible for an investigation with the research question 'How does the concentration of sodium chloride in seawater affect the dissolved oxygen concentration?'

When evaluating the method, make sure that for every weakness you identify, you have suggested a way to improve the method.

TABLE 1.5.1 Methodological weaknesses and suggestions for improvement

Methodological weakness	How this may have influenced the results	Suggestion for improvement
seawater may contain ions other than Na^+ and Cl^-	These other ions, such as K^+, Br^-, Mg^{2+} and O^{2-} may be present and also influence the ability of the seawater to dissolve oxygen.	Deliberately add measured amounts of these ions to the water sample and monitor the effect on the dissolved oxygen concentration.
oxygen may have entered the reaction vessel during analysis	Extra oxygen will increase the dissolved oxygen concentration that is measured.	Practise the steps of the experiment to improve expertise, so that the steps involving an open reaction vessel may be performed rapidly.
volumes of titres were very small (less than 4 mL)	The uncertainty in these measurements represents a large percentage of the volume, so this reduces the precision of the results.	Reduce the concentration of the solution used in the titration by diluting it by a factor of 5, so that the titres are closer to 20 mL and the uncertainty makes less of an impact on the precision.

- *You will now be able to answer key questions 3, 5 and 6.*

A conclusion is a summary of your investigation. It should reflect your aim, results, the validity of the results and whether your hypothesis was supported.

DRAWING EVIDENCE-BASED CONCLUSIONS

A conclusion is a summary of your investigation. It should be possible for a reader to read your conclusion and understand what you did in the investigation, what results you found and how valid the results were. A good conclusion links the collected evidence to your aim and your hypothesis. It uses your results as evidence to provide a justified and relevant response to your research question.

An example of how a conclusion would compare to the aim of the experiment is shown in Table 1.5.2.

TABLE 1.5.2 Writing a conclusion for the determination of the empirical formula of magnesium oxide experiment.

Part of the report		Example of what might be written
Aim:		To determine the empirical formula of magnesium oxide by burning a sample of magnesium in a crucible and describe the physical properties of magnesium oxide.
Conclusion:	**Restatement of the aim:** Instead of starting with 'to', the aim is now rearranged to be in the past tense. It serves as an introduction to the conclusion.	A sample of magnesium was burnt in a crucible and magnesium oxide was produced.
	Statement of your results: These ideally should be the mean of multiple trials values and should include results of all parts of the experiment. Sometimes this will be quite lengthy and you will have to try hard to keep it concise.	The formula, determined by calculations involving the masses of magnesium and oxygen, was found to be Mg_4O_3. The physical properties observed were that the magnesium oxide was a white powder which seemed grey or even black in parts. It dissolved in water with difficulty.
	A comparison of your results to the accepted value or your hypothesis: Some quantitative evaluation of how well your results compared to the accepted value is useful, and this is where a percentage difference may be useful.	The empirical formula agreed quite closely with, although was not exactly the same as, the formula determined by ionic charges, MgO and the physical properties are typical of ionic solids.
	A brief suggestion of what may have caused the results to be different to what was expected.	It is possible that some magnesium oxide was lost to the atmosphere during the heating process, causing the ratio of magnesium to oxygen to be greater than expected.

Language required for conclusions

You will find a full discussion of the language used in a scientific report in Section 1.6, but students often find that the conclusion poses a challenge for them in terms of the style of language required.

Traditionally, the conclusion is written in the past tense and in the third person. This means that you should write the conclusion as though someone else did the work. For example, the sentence 'I found that the mass of the magnesium oxide was less that it should be', should be written as 'The mass of the magnesium oxide was less than was expected'. Similarly, the conclusion can be easily introduced by changing the aim into the past tense and writing it in the third person, as shown in Table 1.5.2.

● *You will now be able to answer key questions 1, 2 and 4.*

1.5 Review

OA ✓✓

SUMMARY

- The method should be evaluated to identify its strengths and weaknesses, and to consider validity, precision, accuracy and reproducibility.
- An evaluation of the method should:
  - consider the effects of random and systematic errors
  - identify issues that may have affected validity, accuracy and precision, repeatability and reproducibility
  - make recommendations for improving the investigation method
- A conclusion should:
  - be introduced by referring to the research question or aim
  - quote the results of the investigation
  - indicate whether the hypothesis was supported or refuted
  - suggest what may have caused the investigation to differ from the expected value.

KEY QUESTIONS

Knowledge and understanding

1 Which of the following options is a suitable conclusion for the research question: 'How does the temperature of the water affect the solubility of carbon dioxide?'

  A I did the experiment and the results proved the hypothesis to be correct.

  B Carbon dioxide dissolves in water at various temperatures. The hypothesis was right.

  C As the temperature of the water increased from 10 °C to 50 °C, the mass of carbon dioxide that dissolved in the water decreased. This supported the hypothesis, which was based on the decreasing solubility of gases in water, with increasing temperature.

  D Temperature increases the solubility of gases such as carbon dioxide.

2 For each of the following aims, construct an introductory sentence for a conclusion which is in the third person and the past tense.

  a To grow crystals of a range of metals and to observe their shapes

  b To observe a polymer with unusual properties

  c To compare the physical properties of a covalent network lattice, a covalent layer lattice and a covalent molecular substance.

Analysis

3 A scientist designed and completed an experiment to test the following hypothesis: 'Increasing the temperature of river water will result in an increase in the measured electrical conductivity of the water'. When the scientist was evaluating their method, they discussed the following points.
For each point describe the impact of the actions on the occurrence of random or systematic errors.

  a The temperature of some of the water samples was measured using a glass thermometer and for some samples a temperature probe was used.

  b The electrical conductivity data-logging equipment was not calibrated, due to a lack of time.

  c The water samples were collected from the river on different days over several weeks.

  d Three water samples from the same place in the river were examined at each temperature.

  e Each water sample was analysed using the same techniques and the measurements were recorded.

4 The scientist testing the hypothesis from Question **3** concluded that there was no relationship between pH and electrical conductivity. Explain why this conclusion is not valid.

5 Complete the following table, which relates to a titration experiment.

Limitation in experiment	Effect on the calculated result	Suggested improvement
A measuring cylinder was used to dispense 25 mL of a solution.		
Only one measurement was obtained.		
Universal indicator was used to measure the end point.		

6 Classify the following as systematic or random errors and propose how to minimise each.

  a consistently overshooting the end point in a titration

  b use of a stopwatch—human reaction times are slower than a stopwatch that can measure to nearest 0.01 s

  c parallax error

  d loss of heat to environment

  e faulty equipment, e.g. a burette that has an air bubble in the tap

1.6 Reporting investigations

Now that you have thoroughly researched your topic, formulated a research question and hypothesis, conducted experiments and collected data, it is time to bring it all together. The final part of an investigation involves summarising the findings in an objective, clear and concise manner for your audience.

Scientists report their findings in a number of ways, including in written peer-reviewed journal articles, on web pages, and with short oral presentations or scientific posters (Figures 1.6.1 and 1.6.2).

FIGURE 1.6.1 Posters at a scientific conference

How does the pH of a 20 mL solution of HCl affect the mass of magnesium that will react?

Lanna Derry

Introduction

pH is a measure of the [H^+] in a solution. Because HCl is a strong acid, the concentration of the acid is considered to be equal to that of the $H^+(aq)$ ions. The reaction between magnesium and hydrochloric acid is: $Mg(s) + 2HCl(aq) \rightarrow MgCl_2(aq) + H_2(g)$ (MacEoin, M. et al., 2022, Ch 11).

In this investigation, the mass of magnesium that reacts with a small volume of HCl in 30 seconds was investigated.

Aim: To determine how the pH of a 20 mL solution of HCl affects the mass of magnesium that reacts at room temperature over a period of 30 seconds.

Hypothesis: As the pH of the solution decreased, it was expected that the mass of magnesium that reacted would increase because a decreasing pH indicates an increasing [H^+].

Methodology and methods

Methodology: A controlled experiment was used.

Method:

1. The pH of the 1.0 M HCl was measured using a pH probe. The value was recorded.
2. A piece of Mg which was 10 cm in length was cut and weighed and the mass was recorded.
3. 20 mL of 1.0 M HCl was measured into a test tube.
4. A piece of Mg was placed in the HCl and left there for 30 seconds.
5. The Mg was removed with tweezers, rinsed in a beaker of water and dried carefully in the air for 10 minutes.
6. The Mg was then weighed again and the mass was recorded.
7. Steps 2 to 6 were repeated two more times
8. Steps 1 to 7 were repeated with 0.5 M HCl, 0.2 M HCl, 0.1 M HCl and 0.05 M HCl.

Risks:

- *HCl(aq) is corrosive, so safety glasses, a lab coat and gloves were worn throughout the investigation.*
- *Any remaining Mg was returned to the Laboratory Prep room for safe disposal or future use.*

Communication statement

As the pH of a solution decreases, the [H^+] increases, so the mass of magnesium that can react with the solution in a given time increases exponentially.

Results

Table 1 Summary table of results

Concentration of HCl (M)	pH	Average mass loss of Mg (g)
0.05	1.3	0.045
0.10	1.0	0.090
0.20	0.70	0.180
0.50	0.30	0.450
1.0	0	0.900

Discussion

Figure 1 Graph of results

Change in mass of magnesium as pH changes

The graph has an exponential trend line because pH is a logarithmic scale.

While these results were particularly favourable, it is recommended that other acids, such as nitric acid and sulfuric acid, are investigated to see if the trend is repeated.

Conclusion

The decreasing pH caused the concentration of $H^+(aq)$ ions to increase, so there were more $H^+(aq)$ ions available in 20 mL of HCl to react with the magnesium. The hypothesis was supported by the results which showed a decreasing mass of magnesium remaining in the test tube as the pH of the acid decreased. This agreed with the equation for the reaction

References:

MacEoin, M. et al. (2022). *Heinemann Chemistry 1* (6th ed.) Pearson Australia

Acknowledgements:

I would like to thank the laboratory technicians, Karen and Tanneale, for their help in preparing the solutions for this investigation.

FIGURE 1.6.2 An example of a student's scientific poster

WRITING A SCIENTIFIC REPORT

Whether an investigation is presented as a poster, written report or oral presentation, an article for a scientific publication, or a multimedia presentation, the same key elements are included in the same sequence: Title, Introduction, Methodology and methods, Results, Discussion, Conclusion, and References and acknowledgements, as summarised in Figure 1.6.3. All of these key elements are described in detail in Sections 1.2, 1.3, 1.4 and 1.5.

A scientific poster format will include the same elements as a scientific report, as outlined in this section. However, a poster is meant to be more succinct and direct in its approach to presenting your findings, and should be appropriate for both technical and non-technical audiences. You should pick the information that you want to present carefully so that the impact of your investigation is best communicated. For example, in analysing your data, you may have produced both a table and a graph. In the poster, it may be best to simply show the graph, which will more clearly represent any trends in your data.

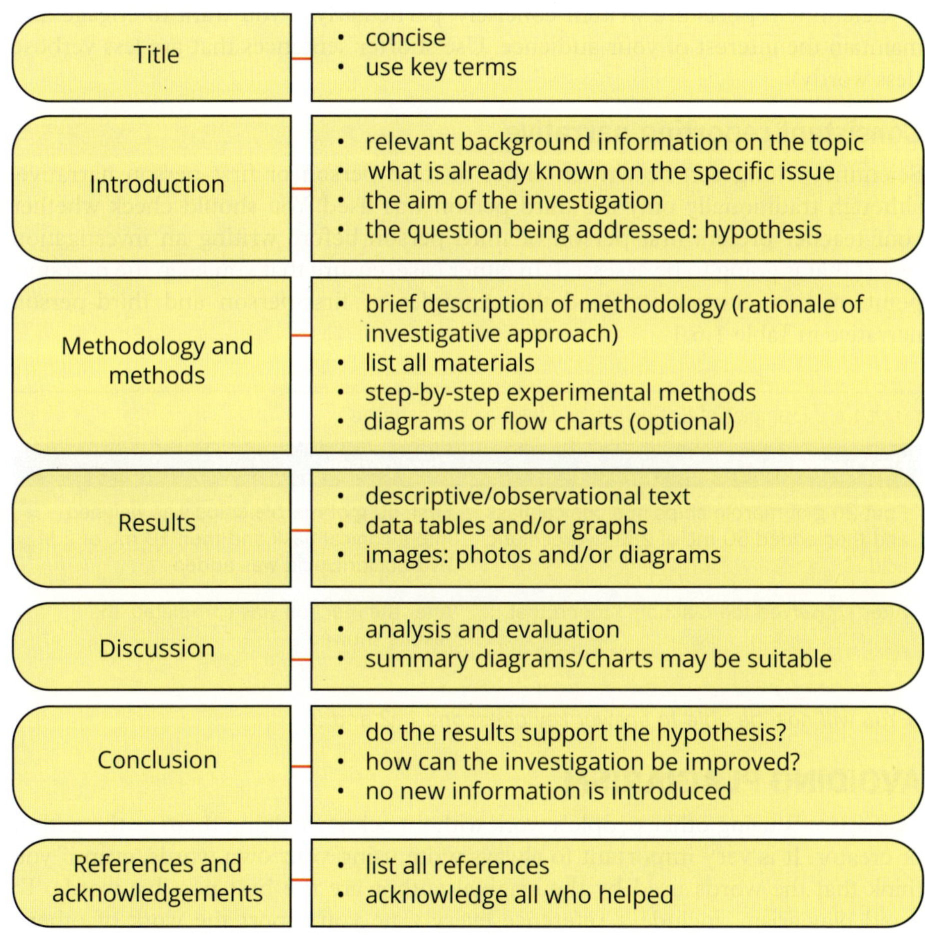

FIGURE 1.6.3 Elements of a scientific report or presentation

● *You will now be able to answer key question 5.*

Visual support

You may be able to identify concepts that can be explained using visual models and information that can be presented in graphs or diagrams. This will not only significantly reduce the word count of your work, but will also make it more accessible for your audience. You can reduce the number of words in your report by:

- including an annotated diagram of the equipment set-up
- using a flow chart to summarise the key steps of the method (a detailed method can appear in the logbook)

- using tables, graphs and schematic diagrams (diagrams that show something complex such as a process in a simple way, often using standardised symbols). Ensure you include:
  - a descriptive title
  - labels, captions or descriptions
  - numbering; for example, Figure 1, Figure 2 or Table 1, Table 2
  - a citation (source) if the work is not your own or is adapted from work that is not your own.

Visual support such as flow charts and graphs can help to convey scientific concepts and processes efficiently.

WRITING FOR SCIENCE

Scientific reports are usually written in an objective or unbiased style. This is in contrast to English writing, which commonly uses the subjective techniques of rhetoric or persuasion. In addition, be careful of words that are absolute, such as always, never, shall, will and proven. Sometimes it may be more accurate and appropriate to use qualifying words, such as may, might, possible, probably, likely and suggests.

Scientific reports are written concisely, particularly if you want to engage and maintain the interest of your audience. Use shorter sentences that are less verbose (less wordy).

Consistent reporting narrative

Scientific writing can be written either in third-person or first-person narrative, although traditionally only the third person was used. You should check whether your teacher prefers first person or third person before writing an investigation report that is going to be assessed. In either case, ensure that you keep the narrative point of view consistent. Read the examples of first-person and third-person narrative in Table 1.6.1.

Scientific writing uses unbiased, objective, accurate, formal language. It should be concise and qualified.

TABLE 1.6.1 Examples of first-person and third-person narrative

First person	Third person
I put 20 g of marble chips in a conical flask and then added 50 mL of 2 M hydrochloric acid.	First, 50 g of marble chips was weighed into a conical flask and then 10 mL of 2 M hydrochloric acid was added.
After I observed the reaction, I found that ...	After the reaction was completed, the results showed ...

● *You will now be able to answer key questions 1, 2 and 3.*

AVOIDING PLAGIARISM

Plagiarism is using other people's work without acknowledging them as the author or creator. It is very important to always write using your own words, even if you think that the words used by the original author are the best possible words. To avoid plagiarism, include a reference every time you report the work of others; for example, at the end of a sentence or following a diagram. If you use a direct quotation from a source, enclose it in quotation marks. This will ensure you give credit to the original author and it will enable the reader to find the original source.

References and bibliography

A bibliography is a list of resources you have consulted during your research, whereas a reference list is a detailed list of references that you have cited in your work. Both lists include details about each resource. To avoid plagiarising the work of others, it is very important that you acknowledge where you have found information. All sources must be listed at the end of the report in alphabetical order (by last name of author or organisation name). The referencing should include enough information for the reader to easily locate the source.

As you gather resources, you should also begin compiling your references in your logbook. This will prevent you from wasting time later trying to find your sources and will be the basis of your bibliography.

APA (American Psychological Association) style is a commonly used referencing style. In the bibliography at the end of your work, you should represent the reference in the following order:

Author (year of publication). Title (edition). Publisher.

This is illustrated in Figure 1.6.4.

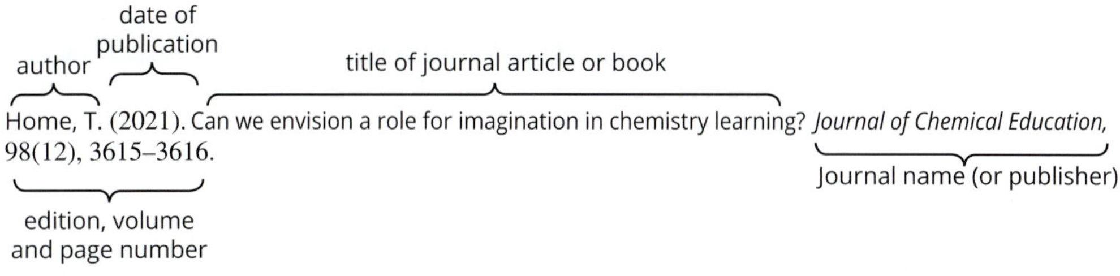

FIGURE 1.6.4 An annotated example of APA referencing of a journal article for a bibliography

In APA referencing, no distinction is made between books, journal articles or internet documents, except where electronic documents do not provide page numbers. Table 1.6.2 gives examples of how to write citations and references of different types of sources using APA academic referencing style.

TABLE 1.6.2 APA academic referencing style

Resource type	Information required	Example
print book	• author's surname and initials • date of publication • title • edition number • page number • publisher's name	Chan, D., Commons, C., Commons, P., Derry, L., Freer, E., Huddart, E., Lennard, L., MacEoin, M., Moylan., M., O'Shea, P., Ross, B. & Vanderkruk, K. 2023. *Heinemann Chemistry 1* (6th ed, pp. 42–44). Pearson Australia.
journal article	• author's surname and initials • date of publication • title of article • journal/magazine title • volume/issue • page numbers	Kizil, N., Soylak, M. & Tüzen, M. (2017). Spectrophotometric detection of rhodamine B in tap water, lipstick, rouge, and nail polish samples after supramolecular solvent microextraction. *Turkish Journal of Chemistry, 41*, 987–994.
internet	• author's surname and initials, or name of organisation or title • year website was written or last revised (if this cannot be found, write 'n.d.' for 'no date') • website title, italicised, or description • website address	University of Canterbury. (n.d.). *Determination of iron by thiocyanate colorimetry.* https://www.canterbury.ac.nz/media/documents/science-outreach/iron_colorimeter.pdf

In-text citations

Each time you write about the findings of other people or organisations, you need to provide an in-text citation and provide full details of the source in a reference list. In APA style, in-text citations include the first author's surname, the year of publication and a page, chapter or section number if you need to be specific, in brackets (author, year, pages). List the full details in your bibliography as shown in Table 1.6.3.

The following examples show the use of in-text citation.

An earlier study (Kizil et al., 2017) reported a method for determining the concentration of rhodamine B by UV–visible spectroscopy.

The procedure for determining the concentration of iron in spinach leaves was adapted from the University of Canterbury method (New Zealand, University of Canterbury, n.d.).

The intention is that in-text citations can be linked to the information in your bibliography, but are more concise, for inclusion in your work at the point where they are most relevant. In Table 1.6.3, three bibliography entries and their corresponding in-text citations are compared.

TABLE 1.6.3 Bibliography entries compared to in-text citations for the same resources

Resource type	Bibliography entry	In-text citation
print book	Chan, D., Commons, C., Commons, P., Derry, L., Freer, E., Huddart, E., Lennard, L., MacEoin, M., Moylan., M., O'Shea, P., Ross, B. & Vanderkruk, K. (2023). *Heinemann Chemistry 1* (6th ed., pp. 42–44). Pearson Australia.	(MacEoin et al. 2023, pp. 42–44)
journal article	Kizil, N., Soylak, M. & Tüzen, M. (2017). Spectrophotometric detection of rhodamine B in tap water, lipstick, rouge, and nail polish samples after supramolecular solvent microextraction. *Turkish Journal of Chemistry, 41*, 987–994.	(Kizil et al., 2017)
internet	University of Canterbury. (n.d.). *Determination of iron by thiocyanate colorimetry*. https://www.canterbury.ac.nz/media/documents/science-outreach/iron_colorimeter.pdf	(University of Canterbury, n.d.)

References and acknowledgements should be present in an appropriate format.

Acknowledgements

It is important to acknowledge the contribution of others in your investigation. This might include acknowledging support from your teacher and preparation of chemicals and materials by the laboratory technician.

For example, statements similar to the following could be used:

- This research was supported by the staff of the Science Faculty at Western High School, Melbourne.
- Special thanks to Mrs Smith for preparing stock solutions.

● *You will now be able to answer key questions 4 and 6.*

1.6 Review

OA ✓✓

SUMMARY

- A practical report should include the following sections:
  - title
  - introduction
  - material and methods
  - results
  - discussion
  - conclusion
  - references
  - acknowledgements.
- The title should give a clear idea of what the report is about, without being too long.
- The introduction sets the context of your report. It should outline relevant chemical ideas, concepts, theories and models, and how they relate to your specific question and hypothesis.
- The methodology and method section should:
  - outline the methodology used and the rationale for using this approach
  - clearly state the materials required and the methods used to collect data during your investigation
  - be presented in a clear, logical order that accurately reflects how you conducted your study.
- The results section should state your results and present them using graphs, figures and tables, but it should not interpret the results.
- The discussion should:
  - interpret data
  - evaluate the investigative method and make recommendations for improving the method
  - explain the link between investigative findings and relevant chemical concepts.
- The conclusion should succinctly link the evidence collected to the hypothesis and research question, indicating whether the hypothesis was supported or refuted.
- References and acknowledgements should be presented in an appropriate format.
- Scientific writing uses unbiased, objective, accurate, formal language. It should be concise and qualified.
- Visual support, such as flow charts and graphs, can help to convey scientific concepts and processes efficiently.

KEY QUESTIONS

Knowledge and understanding

1 Which of the following statements is written in scientific style?
  - **A** The colour was disgusting...
  - **B** The results were awesome...
  - **C** The data in Figure 1 shows...
  - **D** The researchers believed...

2 Which of the following statements is written in third-person narrative?
  - **A** Three trials were conducted...
  - **B** We found that...
  - **C** George's results demonstrated that...
  - **D** I showed that...

3 Rewrite each of the following statements in third-person narrative style.
  - **a** We were watching the beaker for 10 minutes, during which time I noticed that the blue colour of solution faded and a brown solid appeared.
  - **b** I found that the mass of the white solid decreased every minute for 10 minutes after we heated it.
  - **c** I put 2.0 g of magnesium in a test tube and then added 10 mL of 2 M hydrochloric acid.

4 Explain why it is important to reference and acknowledge documents, ideas and quotations in your investigation.

Analysis

5 All of the following information from an investigation should be recorded in your logbook.
  - **a** From this list, select which pieces of information could reasonably be included in a scientific poster, by placing a tick in the second column.
  - **b** For each piece of information that you have selected in part **a**, describe the most concise way to present the information on a poster.

continued over page

1.6 Review *continued*

Section	Could be included in a poster	Concise way to present the information in a poster
communication statement reporting the key finding of the investigation as a one-sentence summary		
research question (title of poster)		
aim		
hypothesis		
variables		
background information		
risk assessment		
detailed materials		
detailed method		
methodology		
summary flow chart of method		
diagram of experimental setup		
summary results table		
detailed results		
sample processed data calculation		
processed data calculations		
graph showing the trend in results		
brief analysis of results and link to chemical theory		
table of limitations and suggested improvements		
conclusion		
acknowledgements		
references		
in-text citations		

6 In the table below all the details of three references are described. Rewrite this information as it should appear in a bibliography using the APA referencing format.

Resource type	Information about the reference	Correct format in a bibliography, using APA referencing format
print book	Title of book: Heinemann Chemistry 1 Edition: 6th edition Authors: Chan, D., Commons, C., Commons, P., Derry, L., Freer, E., Huddart, E., Lennard, L., MacEoin, M., Moylan., M., O'Shea, P., Ross, B. & Vanderkruk, K. Date published: 2023 Section referred to: Pages 387–413 Publisher: Pearson Australia	
journal article	Title of article: Effects of the COVID-19 Pandemic on Student Engagement in a General Chemistry Course Authors: Wu, F. and Teets, T. Journal name: Journal of Chemical Education Volume: Vol 98 Pages: 3633–3642 Date published: November 2021	
internet	Website owner: Royal Society of Chemistry Website title/description: Reactivity of metals video Date posted: no date Website address: https://edu.rsc.org/practical/reactivity-of-metals-practical-videos-14-16-years/4012974.article	

Chapter review

KEY TERMS

accepted value
accuracy
accurate
affiliation
aim
analogue
average
bar graph
bias
biodegradable
calibration
calibration curve
column graph
concordant titres
continuous data
controlled experiment
controlled variable
critical thinking
dependent variable
discrete data
ethics
exponential relationship
green chemistry
hypothesis
independent variable
index
inverse relationship
line graph
line of best fit
linear trend
literature value
logbook
mean
measurement error
meniscus
method
methodology
mistake
objective
observation
outlier
parallax error
peer-reviewed
percentage change
percentage difference
pie chart
precise
precision
primary data
primary source
principle
processed data
qualitative
qualitative data
quantitative
quantitative data
random error
raw data
repeatability
replication
reproducibility
research question
resolution
risk assessment
safety data sheet (SDS)
scatter graph
scientific method
scientific notation
secondary source
significant figures
standard form
subjective
sustainability
systematic error
theory
titration
trend
true value
uncertainty
UV–visible spectroscopy
valid
validity
variable

UNIT 1 How can the diversity of materials be explained?

To achieve the outcomes in Unit 1, you will draw on key knowledge outlined in each area of study and the related key science skills on pages 11 and 12 of the Study Design. The key science skills are discussed in Chapter 1 of this book.

AREA OF STUDY 1

How do the chemical structures of materials explain their properties and reactions?

Outcome 1: On completion of this unit the student should be able to explain how elements form carbon compounds, metallic lattices and ionic compounds, experimentally investigate and model the properties of different materials, and use chromatography to separate the components of mixtures.

AREA OF STUDY 2

How are materials quantified and classified?

Outcome 2: On completion of this unit the student should be able to calculate mole quantities, use systematic nomenclature to name organic compounds, explain how polymers can be designed for a purpose, and evaluate the consequences for human health and the environment of the production of organic materials and polymers.

AREA OF STUDY 3

How can chemical principles be applied to create a more sustainable future?

Outcome 3: On completion of this unit the student should be able to investigate and explain how chemical knowledge is used to create a more sustainable future in relation to the production or use of a selected material.

CHAPTER 02

Elements and the periodic table

In this chapter, you will begin by reviewing your understanding of atoms and elements. You will then extend this knowledge to learn about how electrons behave in atoms: the modern Schrödinger quantum mechanical model. Next, you will look more closely at the elements and see how they are arranged in the periodic table. You will also investigate why some elements are now being described as critical and endangered. Finally, you will use the periodic table to explain the trends that are observed in the properties of the elements within the groups and periods of the table.

Key knowledge

- the definitions of elements, isotopes and ions, including appropriate notation: atomic number; mass number; and number of protons, neutrons and electrons **2.1**
- the periodic table as an organisational tool to identify patterns and trends in, and relationships between, the structures (including shell and subshell electronic configurations and atomic radii) and properties (including electronegativity, first ionisation energy, metallic and non-metallic character and reactivity) of elements **2.2, 2.3, 2.4, 2.5**
- critical elements (for example, helium, phosphorus, rare-earth elements and post-transition metals and metalloids) and the importance of recycling processes for element recovery. **2.4**

2.1 The atomic world

Over time, scientists have gained a deep understanding of the structure of **atoms**, which are the basic building blocks of **matter**. As atoms are too small to be seen with even the most powerful optical microscope, much of what scientists know about atoms has come from theoretical models and indirect observations.

A scientific **model** is a tool used by scientists to understand something they cannot see directly. Using their observations, they are able to construct a theoretical picture of what they are trying to describe. As new data becomes available, the model can develop and become more accurate.

A scientific model is a tool that may be used by scientists to explain something they cannot see directly.

ATOMIC THEORY

In 1802, an English scientist called John Dalton (Figure 2.1.1) presented the first **atomic theory of matter**. Dalton proposed that all matter is made up of tiny spherical particles, which are indivisible and indestructible.

Dalton also accurately described **elements** as materials containing just one type of atom and **compounds** as materials containing different types of atoms in fixed ratios. Subsequent experiments showed that Dalton's atomic theory of matter was mostly correct. However, scientists now know that atoms are not indivisible or indestructible. Atoms are made up of even smaller **subatomic particles**.

Elements are materials that contain just one type of atom. Compounds are materials that contain different types of atoms, in fixed ratios.

FIGURE 2.1.1 John Dalton (1766–1844) proposed that matter was composed of atoms.

CHEMFILE

Viewing atoms

Dalton's atomic theory of matter assumed that atoms are spherical. However, atoms cannot be seen with conventional microscopes. Therefore, there was no way to confirm the shape of atoms until 1981 when a microscope capable of viewing atoms was developed by IBM researchers Gerd Binnig and Heinrich Rohrer. This type of microscope is known as a scanning tunnelling microscope (STM). Using STMs, scientists confirmed that atoms are indeed spherical.

STMs use an extremely sharp metal tip to detect atoms. The tip is scanned, line-by-line, across the surface of a crystal. As it moves, the tip measures minute height differences in the crystal's surface due to the individual atoms. This is similar to the way a person with a vision impairment uses their finger to sense braille on a page. The data from the tip is then sent to a computer, which constructs an image of the atoms. STM images of a lattice of copper atoms and silicon atoms on the surface of a silicon chip are shown in the figures below.

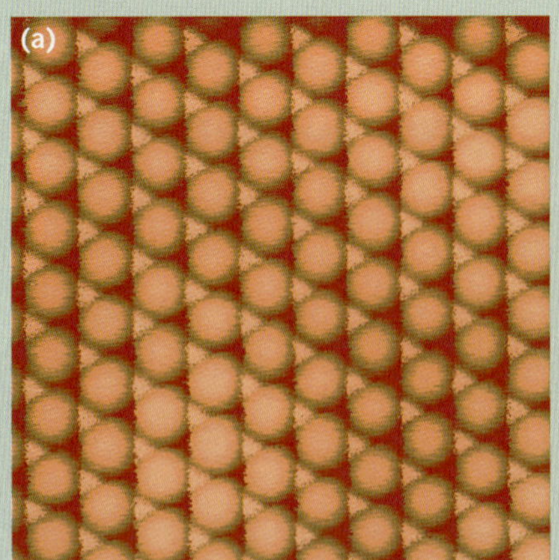

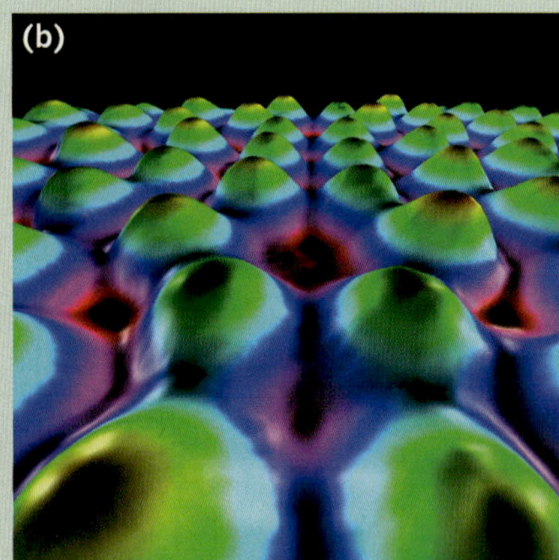

(a) Coloured image of copper atoms produced by a scanning tunnelling microscope (STM).
(b) Silicon atoms on a silicon chip imaged with a scanning tunnelling microscope

STRUCTURE OF ATOMS

Atoms are made up of a small, positively charged nucleus surrounded by a much larger cloud of negatively charged electrons, as shown in Figure 2.1.2. The nucleus is in turn made up of two types of subatomic particle—protons and neutrons. The **protons** are positively charged and the **neutrons** have no charge.

Atoms are made up of protons (positive charge), neutrons (neutral) and electrons (negative charge).

Electrons

Electrons are negatively charged particles. You can imagine them forming a cloud of negative charge around the nucleus. This cloud gives the atom its size and volume.

An electron is approximately 1800 times smaller than a proton or neutron. Therefore, electrons contribute very little to the total mass of an atom. However, the space occupied by the cloud of electrons is 10 000–100 000 times larger than the nucleus.

Negative particles attract positive particles. This is called **electrostatic attraction**. Electrons are bound to the nucleus by the electrostatic attraction to the protons within the nucleus. The charge on an electron is equal but opposite to the charge on a proton. Electrons are said to have a charge of −1, whereas protons have a charge of +1.

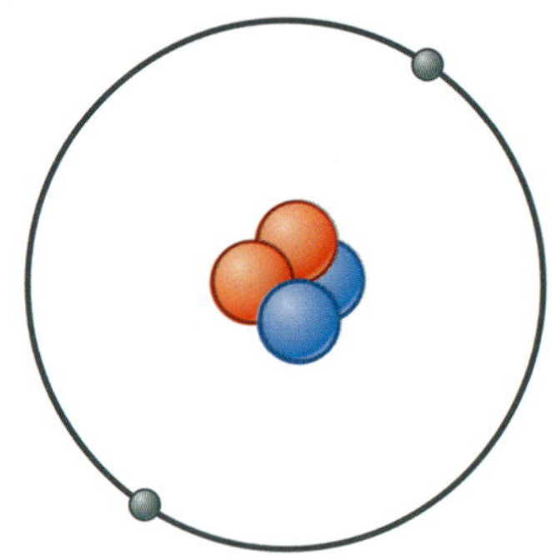

FIGURE 2.1.2 A simplified model of the atom. This shows a helium atom with two protons and two neutrons (in the central nucleus), and two electrons. The ring symbolises that the electrons are held in an orbit; however, evidence suggests they move throughout an area that is more like a cloud.

In some circumstances, electrons can be easily removed from an atom. For example, when you rub a rubber balloon on a woollen jumper or dry hair, electrons are transferred to the balloon and the balloon develops a negative charge. The negative charge is observed as an electrostatic force that can attract hair or even stick the balloon to a wall (Figure 2.1.3). You will look at a different way of removing electrons from atoms, via a chemical reaction, when looking at redox reactions in Chapter 12.

An electron is approximately 1800 times smaller than a proton or a neutron. However, electrons occupy most of the space in an atom.

The electricity that powers lights and appliances is the result of electrons moving in a current through wires. Sparks and lightning are also caused by electrons moving through air.

The nucleus

The **nucleus** of an atom is approximately 10 000–100 000 times smaller than the size of the atom. To put this in perspective, if an atom were the size of the Melbourne Cricket Ground (Figure 2.1.4), then the nucleus would be about the size of a pea in the centre. Nonetheless, the nucleus contributes around 99.97% of the atom's mass. This means that atomic nuclei are extremely dense.

The subatomic particles in the nucleus, the protons and neutrons, are referred to collectively as **nucleons**. Protons are positively charged particles with a mass of approximately 1.673×10^{-27} kg. Neutrons are almost identical in mass to protons.

FIGURE 2.1.3 It is the build-up of electrons, as static charge on the balloon, that attracts this girl's hair to the surface of the balloon.

FIGURE 2.1.4 If an atom were the size of the Melbourne Cricket Ground, then the nucleus would be the size of a pea.

Most of the mass of an atom is concentrated in the extremely dense nucleus.

The charge on an electron is equal but opposite to the charge on a proton.

Protons and neutrons are almost identical in mass.

CHEMFILE

Atoms are mostly empty space

Between 1899 and 1911, New Zealand-born physicist Ernest Rutherford conducted experiments in which he fired a beam of alpha particles (helium nuclei) at a piece of extremely thin gold foil. You can imagine this experiment as a little like throwing pebbles at an object in the dark to deduce its shape. By listening to whether the pebbles hit something or pass straight through, you can build up a picture of the object in front of you. Rutherford noted that most of the alpha particles passed straight through the gold foil (see figure below).

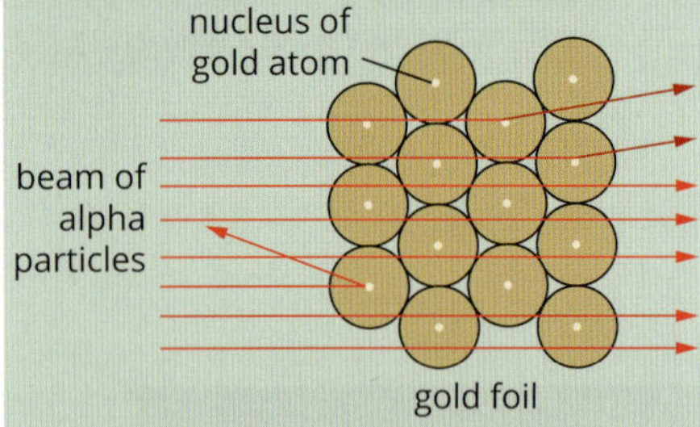

Rutherford's gold foil experiment: only those alpha particles that closely approach the nuclei in the gold foil are deflected significantly. Most particles pass directly through the foil.

Even more surprisingly, Rutherford found that a small number of particles bounced back, some almost directly back at the source. From this observation he deduced that the gold atoms were made up almost entirely of empty space, with a small, extremely dense nucleus.

The atomic number is equal to the number of protons in the nucleus, and is represented by the symbol Z.

The mass number is the sum of all the particles in the nucleus, i.e. protons plus neutrons. This represents the total mass of the atom.

Table 2.1.1 summarises the properties of protons, neutrons and electrons.

TABLE 2.1.1 Properties of the subatomic particles

Particle	Symbol	Charge	Mass relative to a proton	Mass (kg)
proton	p	+1	1	1.673×10^{-27}
neutron	n	0	1	1.675×10^{-27}
electron	e	−1	$\frac{1}{1800}$	9.109×10^{-31}

ELEMENT SYMBOLS

Atoms can be identified by how many protons they have. An element is made up of atoms that all contain the same number of protons in their nucleus. Scientists have discovered 118 different elements, and about 98 of these occur in nature. The other elements have only been observed in the laboratory.

Each element has a name and a unique **chemical symbol**. Table 2.1.2 lists the chemical symbols of some well-known elements.

TABLE 2.1.2 Chemical symbols and names of some well-known elements

Element	Symbol	Element	Symbol
aluminium	Al	mercury	Hg
argon	Ar	nitrogen	N
carbon	C	oxygen	O
chlorine	Cl	potassium	K
copper	Cu	silver	Ag
hydrogen	H	sodium	Na
iron	Fe	uranium	U

The chemical symbol is made up of one or two letters. The first letter is always capitalised and subsequent letters are always lower case.

In many cases, the chemical symbol corresponds to the name of the element. For example, nitrogen has the chemical symbol N, chlorine has the chemical symbol Cl and uranium has the chemical symbol U.

However, some chemical symbols do not seem to correspond to the name of the element. For example, sodium has the chemical symbol Na, potassium has the chemical symbol K and iron has the chemical symbol Fe. This is because the chemical symbols have been derived from the Latin or Greek names of the elements. In Latin, sodium is known as *natrium*, potassium is known as *kalium* and iron is known as *ferrum*.

Representing atoms

The number of protons in an atom's nucleus is known as the **atomic number** and is represented by the symbol Z.

All atoms that belong to the same element must have the same number of protons and therefore have the same atomic number, Z. For example, all hydrogen atoms have Z = 1, all carbon atoms have Z = 6 and all gold atoms have Z = 79.

The total number of protons and neutrons in the nucleus is known as the **mass number** and is represented by the symbol A. The mass number represents the total mass of the nucleus.

As all atoms are electrically neutral, the number of electrons in an atom is equal to the number of protons in an atom. The atomic number therefore tells you both the number of protons and the number of electrons. For example, carbon atoms, with Z = 6, have six protons and six electrons.

The number of protons, neutrons and electrons defines the basic structure of an atom. A standard way of representing an atom is to show its atomic and mass numbers as shown in Figure 2.1.5. This is known as **nuclide notation**.

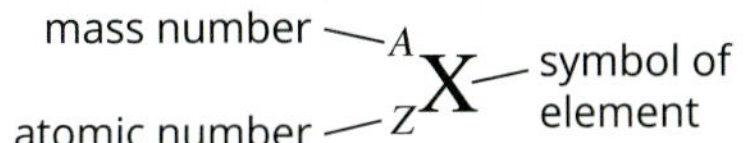

FIGURE 2.1.5 The standard way of representing an atom, showing its atomic number and mass number

For an aluminium atom, this would be written like this: $^{27}_{13}Al$

From this representation, you can determine that:

- the number of protons is 13 because the number of protons is equal to the atomic number (Z)
- the number of neutrons is 14 because the number of neutrons plus the number of protons is equal to the mass number; therefore, you can subtract the atomic number from the mass number to determine the number of neutrons (A – Z)
- the number of electrons is 13 because atoms have no overall charge, therefore the number of electrons must equal the number of protons.

Worked example 2.1.1

CALCULATING THE NUMBER OF SUBATOMIC PARTICLES

Calculate the number of protons, neutrons and electrons for the atom with this nuclide symbol:

$$^{40}_{18}Ar$$

Thinking	Working
The atomic number is equal to the number of protons.	The number of protons = Z = 18
Find the number of neutrons. Number of neutrons = mass number – atomic number	The number of neutrons = A – Z = 40 – 18 = 22
Find the number of electrons. The number of electrons is equal to the atomic number because the total negative charge is equal to the total positive charge.	Number of electrons = Z = 18

Worked example: Try yourself 2.1.1

CALCULATING THE NUMBER OF SUBATOMIC PARTICLES

Calculate the number of protons, neutrons and electrons for the atom with this nuclide symbol:

$$^{235}_{92}U$$

CHEMFILE

Using isotopes to study climate change

One of the ways scientists study climate change is to study frozen bubbles of air, deep within ice that has remained frozen for thousands of years. The first ice-core studies were done using samples from Greenland and Antarctica, but more recently scientists have been searching for suitable ice deposits in the world's warmer regions like Africa and South America. One such expedition took scientists to the summit of Mount Kilimanjaro in Tanzania.

The oldest ice-cores collected in the expedition were 11 700 years old (see figure below). By determining the ratio of water in the ice containing the oxygen-18 isotope versus water containing the oxygen-16 isotope, it is possible to determine the temperature of the air when the water originally fell as rain. If there is a larger amount of oxygen-18, the temperature was higher; if there is a larger amount of oxygen-16, the temperature was lower.

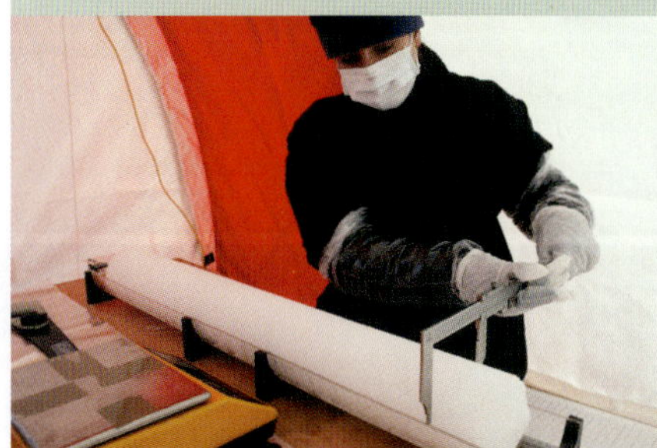

A scientist begins their analysis of an ice-core sample.

Isotopes

Isotopes are atoms with the same atomic number, but with different mass numbers. They contain the same number of protons, but the number of neutrons in the nucleus is different.

All atoms that belong to the same element have the same number of protons in the nucleus and therefore the same atomic number, Z. However, not all atoms that belong to the same element have the same mass number, A. For example, hydrogen atoms can have a mass number of 1, 2 or 3. In other words, hydrogen atoms may contain just a single proton, a proton and a neutron, or a proton and two neutrons as shown in Figure 2.1.6. Atoms that have the same number of protons (atomic number) but different numbers of neutrons (and therefore different mass numbers) are known as **isotopes**.

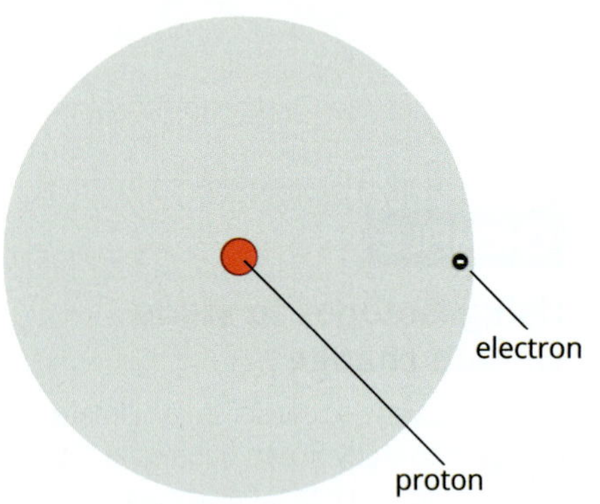

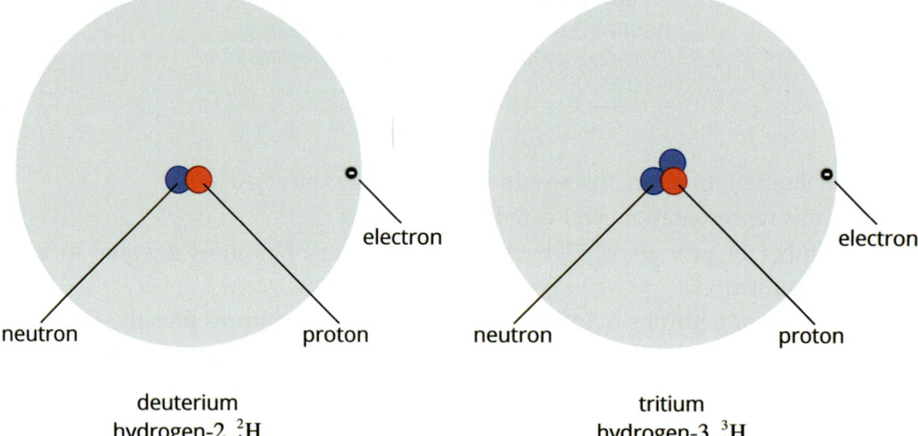

FIGURE 2.1.6 The three isotopes of hydrogen are given special names. A hydrogen atom with just one proton in its nucleus is known as hydrogen or protium. A hydrogen atom with one proton and one neutron is known as deuterium. A hydrogen atom with one proton and two neutrons is known as tritium.

Carbon also has three naturally occurring isotopes. These three isotopes are known as carbon-12, carbon-13 and carbon-14. Carbon-12 atoms have a mass number of 12, carbon-13 atoms have a mass number of 13 and carbon-14 atoms have a mass number of 14. In the 1950s and 1960s, nuclear weapons testing caused a spike in carbon-14 in the atmosphere. This has been declining in the last 50 years. These three carbon isotopes can be represented as shown in Figure 2.1.7.

$^{12}_{6}C$	$^{13}_{6}C$	$^{14}_{6}C$
carbon-12	carbon-13	carbon-14

FIGURE 2.1.7 Ways of representing the three isotopes of carbon

Isotopes have identical chemical properties but different physical properties such as mass and density. Some isotopes are **radioactive**. This means their nucleus is not stable and will break down spontaneously into a more stable form by emitting particles as radiation.

Ions

Nuclide symbols can also be used to represent ions. **Ions** are atoms that have lost or gained one or more electrons. An atom that loses electrons becomes positively charged overall (as the positive charges in the nucleus now outnumber the negatively charged electrons.) Similarly, an atom that gains electrons becomes negatively charged. You will learn more about ions in Chapter 5.

When an atom gains or loses one or more electrons, it becomes an ion. An ion is a charged particle. Atoms that lose electrons become positive ions. Atoms that gain electrons become negative ions.

Worked example 2.1.2

CALCULATING THE NUMBER OF SUBATOMIC PARTICLES IN AN ION

Calculate the number of protons, neutrons and electrons for the ion with this nuclide symbol:

$$^{25}_{12}Mg^{2+}$$

Thinking	Working
The atomic number is equal to the number of protons.	The number of protons = Z = 12
Find the number of neutrons. Number of neutrons = mass number − atomic number	The number of neutrons = A − Z = 25 − 12 = 13
Find the number of electrons in an uncharged atom. The number of electrons is equal to the atomic number.	Number of electrons in an uncharged atom = Z = 12
Find the number of electrons in the ion. The number of electrons is equal to the atomic number minus two, because the total negative charge is two less than the total positive charge.	Number of electrons in ion = Z − (charge) = Z − 2 = 12 − 2 = 10

Worked example: Try yourself 2.1.2

CALCULATING THE NUMBER OF SUBATOMIC PARTICLES IN AN ION

Calculate the number of protons, neutrons and electrons for the ion with this nuclide symbol:

$$^{14}_{7}N^{3-}$$

2.1 Review

OA ✓✓

SUMMARY

- All matter is made of atoms, which are composed of a small, positively charged nucleus surrounded by a negatively charged cloud of electrons.
- The mass of an atom is mostly determined by the mass of the nucleus, while the diameter of an atom is determined by the cloud of electrons.
- The nucleus is made up of two subatomic particles—protons and neutrons. These particles are referred to as nucleons.
- Protons have a positive charge, electrons have a negative charge and neutrons have no charge.
- Protons and neutrons are similar in mass while electrons are approximately 1800 times smaller.
- The charges on protons and electrons are equal but opposite.
- An element contains atoms of the same type and has a chemical symbol that is made up of one or two letters. The first letter is capitalised and the second letter is lower case.
- You can determine the number of subatomic particles in an atom from an element's atomic number and mass number:

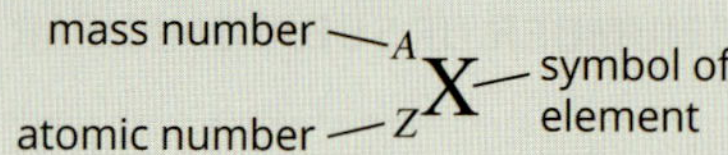

  - Z = number of protons = number of electrons
  - A − Z = number of neutrons
- Isotopes are atoms with the same atomic number but different mass numbers, i.e. they have the same number of protons but different numbers of neutrons.
- Isotopes have the same chemical properties but different physical properties such as mass, density and radioactivity.
- Ions are atoms that have lost or gained electrons to become charged particles.

KEY QUESTIONS

Knowledge and understanding

1 How many times larger is the atom compared to its nucleus?

2 What subatomic particles make up most of the mass of an atom and where are they found?

3 How are electrons held within the cloud surrounding the nucleus?

4 What term is given to the number of protons and neutrons in the nucleus of an atom?

Analysis

5 How many electrons would an atom of $^{65}_{30}Zn^{2+}$ contain?

6 Yttrium-90 (atomic number 39) is used for treatment of cancer, particularly non-Hodgkin lymphoma and liver cancer, and it is being used more widely, including for arthritis treatment.
   a Write the nuclide symbol for yttrium-90.
   b How many neutrons does each atom of this isotope contain?

7 Which of the following nuclide symbols represent isotopes of the same element?
   a $^{14}_{6}X$
   b $^{14}_{7}Y$
   c $^{12}_{6}Z$

8 Suggest why it might be easier to separate $^{46}_{20}Ca$ and $^{46}_{22}Ti$ than $^{46}_{20}Ca$ and $^{40}_{20}Ca$.

2.2 Emission spectra and the Bohr model

When fireworks explode, they create a spectacular show of coloured light (Figure 2.2.1). The light is produced by metal atoms that have been heated by the explosion. This coloured light posed a significant problem for early scientists. The models the scientists were using could not explain the source of the light. However, the light was a clue that ultimately led to a better understanding of the arrangement of electrons in atoms.

FIGURE 2.2.1 The spectacular colours in this New Year's Eve fireworks display are emitted by metal atoms that have been heated to very high temperatures.

EMISSION SPECTRA

When atoms are heated, they can produce coloured light. You may have observed this phenomenon in a flame test. **Flame tests** make use of the fact that some metallic elements can be identified by the characteristic colour produced when a sample is passed through a flame. Figure 2.2.2 shows the characteristic colours produced by some metals.

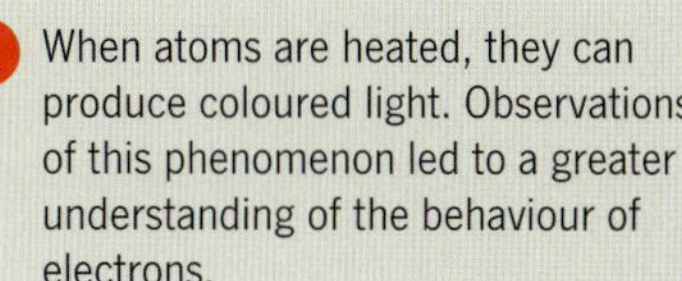

When atoms are heated, they can produce coloured light. Observations of this phenomenon led to a greater understanding of the behaviour of electrons.

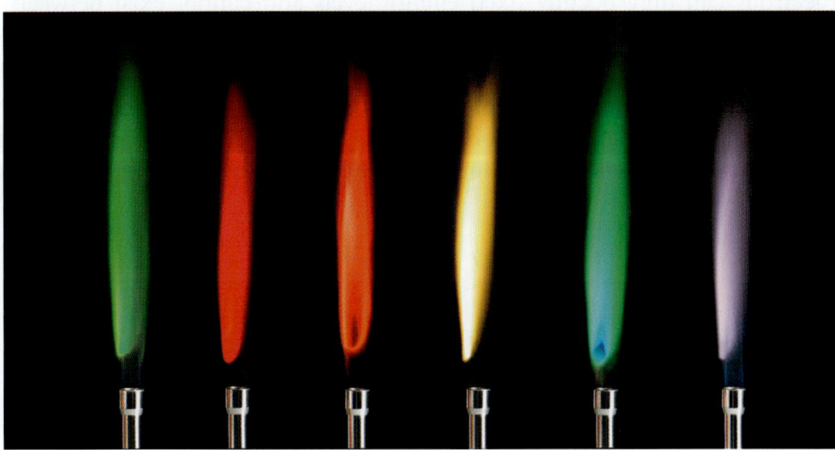

FIGURE 2.2.2 The colour of the flame is determined by the different metal compounds present and can be used to identify these metals. The flame colours shown here are for (from left to right): barium (yellow–green), lithium (crimson), strontium (scarlet), sodium (yellow), copper (green) and potassium (lilac).

If the light from the flame test is passed through a prism, it produces a spectrum with a black background and a number of coloured lines. Figure 2.2.3 shows the apparatus used to produce these spectra.

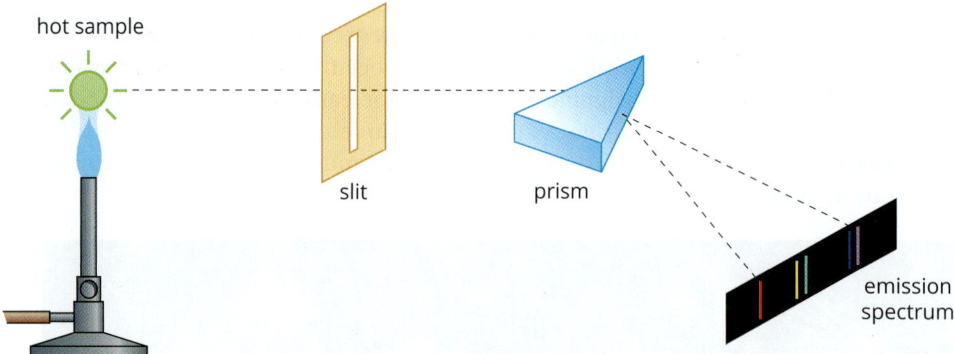

FIGURE 2.2.3 The apparatus used to analyse the coloured light given out when an element is heated. The coloured lines are called an emission spectrum.

These spectra are known as line spectra or **emission spectra**. Each emission spectrum is unique for a particular element and can be used to identify different elements.

The line spectrum produced by helium is shown in Figure 2.2.4.

FIGURE 2.2.4 The emission spectrum of helium is made up of lines ranging from violet to red in colour.

Each line in the spectrum corresponds to light of a different energy. Violet lines correspond to light with high energies. As the colour of the light changes to blue, green, yellow and orange, the energy of the light decreases. Red light is the lowest energy light visible to the human eye. Just as some metals have a characteristic flame colour, elements have a characteristic emission spectrum.

THE BOHR MODEL

In 1913, Niels Bohr developed a new model of the hydrogen atom that explained its emission spectrum. The **Bohr model** proposed the following:

- Electrons revolve around the nucleus in fixed, circular orbits.
- These orbits correspond to specific energy levels in the atom.
- Electrons can only occupy fixed energy levels and cannot exist between two energy levels.
- Orbits of larger radii correspond to energy levels of higher energy.

In the Bohr model, it is possible for electrons to move between the energy levels by absorbing or emitting energy. Bohr's model (Figure 2.2.5) gave close agreement between the calculated energies for lines in the hydrogen spectrum and the observed values in the spectrum.

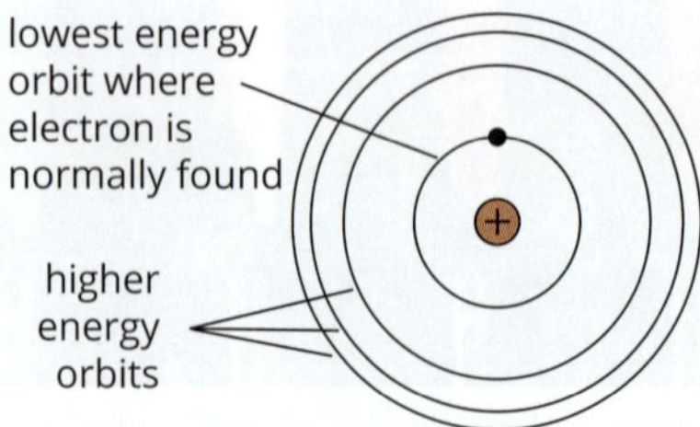

FIGURE 2.2.5 The Bohr model of a hydrogen atom. Bohr suggested that electrons moved in orbits of particular energies.

ELECTRON SHELLS

Scientists quickly extended Bohr's model of the hydrogen atom to other atoms. Scientists proposed that electrons were grouped in different energy levels, called **electron shells**. The electron shells are labelled with the number $n = 1, 2, 3 \ldots$, as shown in Figure 2.2.6.

The orbit in which an electron moved depended on the energy of the electron; electrons with lower energy are in orbits close to the nucleus while high-energy electrons are in outer orbits.

The Bohr model of the atom proposes that electrons are grouped into different energy levels, called shells.

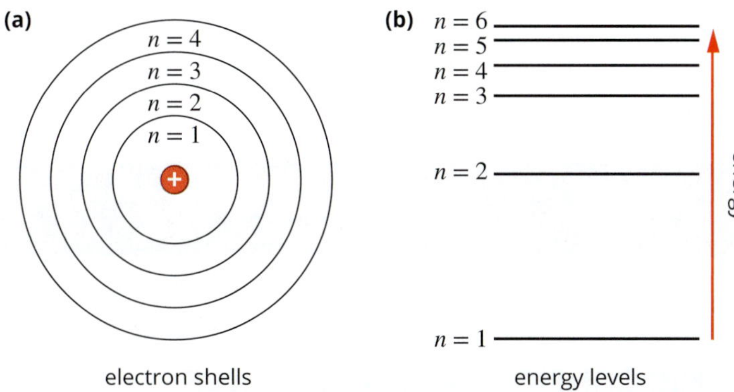

FIGURE 2.2.6 (a) The electron shells of an atom (n) are labelled using integers. The first shell is the shell closest to the nucleus and the radius of each shell increases as the shell number increases. (b) Each shell corresponds to an energy level that electrons can occupy. The first shell has the lowest energy and the energy increases as the shell number increases.

The lowest energy shell is the shell closest to the nucleus and is labelled $n = 1$. Shells with higher values of n correspond to higher energy levels. As the values of n increase, the energy levels get closer together.

LINKING EMISSION SPECTRA TO THE SHELL MODEL

Heating an element can cause an electron to absorb energy and jump to a higher **energy state**. Shortly afterwards, the electron returns to the lower energy state, releasing a fixed amount of energy as light. The electron can return in a number of different ways, some of which are shown in Figure 2.2.7. Each one of the different pathways produces light of a particular colour in the emission spectrum.

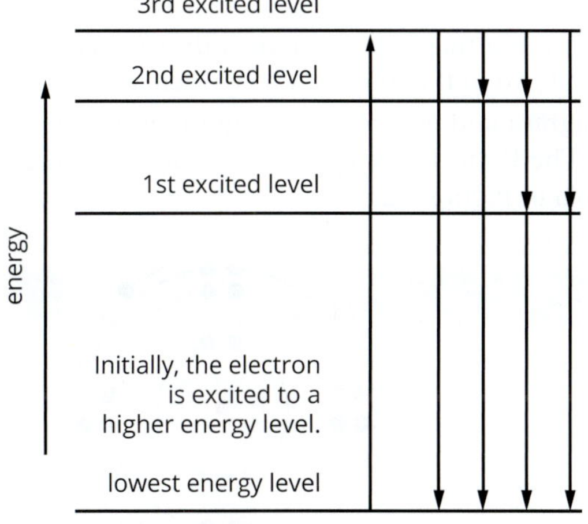

FIGURE 2.2.7 Emission of energy as electrons move from a higher to a lower energy state

Consider a hydrogen atom with one proton and one electron. Usually, the electron exists in the $n = 1$ shell. This is the lowest energy state of the atom and is called the **ground state**.

When the hydrogen atom is heated, the electron absorbs energy and jumps to a higher energy level. This is known as an **excited state**.

Shortly afterwards, the electron returns to the ground state. The electron may return directly to the ground state or may move to other energy levels before returning to the ground state. For example, an electron in the $n = 4$ shell may move to the $n = 2$ shell before returning to the $n = 1$ ground state.

> As an electron falls to a lower energy shell, it emits energy, some of which is in the form of coloured light.

As the electron falls to a lower energy shell, it emits energy, some of which is in the form of coloured light. This energy is exactly equal to the energy difference between the two energy levels. Each coloured line on the spectrum corresponds to a specific amount of energy (Figure 2.2.8).

FIGURE 2.2.8 The emission spectrum of the hydrogen atom has four lines in the visible range—violet, blue, green and red.

ELECTRONIC CONFIGURATION IN SHELLS

> The electronic configuration of an atom is a representation of the arrangement of electrons in energy levels.

The different energy levels or shells can hold different numbers of electrons. The arrangement of these electrons around the nucleus is called the **electronic configuration**. Electron shells can hold a maximum number of electrons, as shown in Table 2.2.1. You can calculate the maximum number of electrons a shell can hold using the formula $2n^2$, where n = shell number. Even though from the third shell on, shells can hold more than eight electrons, a **valence shell** (outer shell) can only hold eight electrons. Once a valence shell reaches eight electrons, the next shell starts filling.

TABLE 2.2.1 The maximum number of electrons held by each shell of an atom. Shell 1 is the shell closest to the nucleus.

Electron shell (n)	Maximum number of electrons
1	2
2	8
3	18
4	32
n	$2n^2$

Electron shells fill in a particular order. The first two electrons go into the first shell. The next eight electrons go into the second shell. The third shell can hold 18 electrons, but once it contains eight electrons, the next two electrons go into the fourth shell. Only then does the third shell fill up. It took the development of Schrödinger's quantum mechanical model to explain these complexities of electronic configurations, as you will read in the next section.

A Bohr diagram is a simple diagram that shows the arrangement of electrons around the nucleus. In such diagrams, only the shells that are occupied are drawn. The Bohr diagram and electronic configuration for lithium are shown in Figure 2.2.9. The electronic configuration can be shown as a series of numbers. For example, the electronic configuration of a magnesium atom with two electrons in the first shell, eight in the second and two in the third can be written as 2,8,2.

Figure 2.2.10 shows a Bohr diagram and electronic configuration for an atom with three electron shells, sodium. The Bohr diagram and electronic configuration for an atom with four shells is shown in Figure 2.2.11.

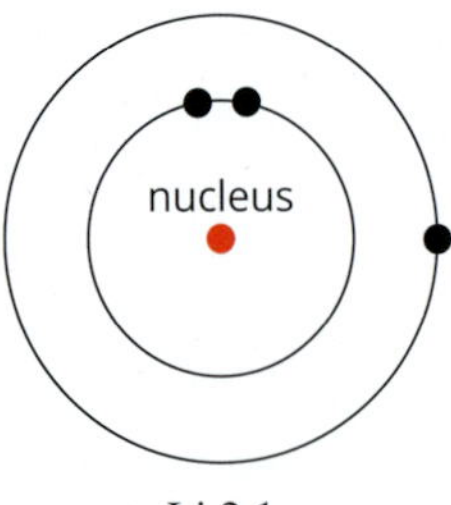

FIGURE 2.2.9 The Bohr diagram and electronic configuration for the lithium atom with three electrons

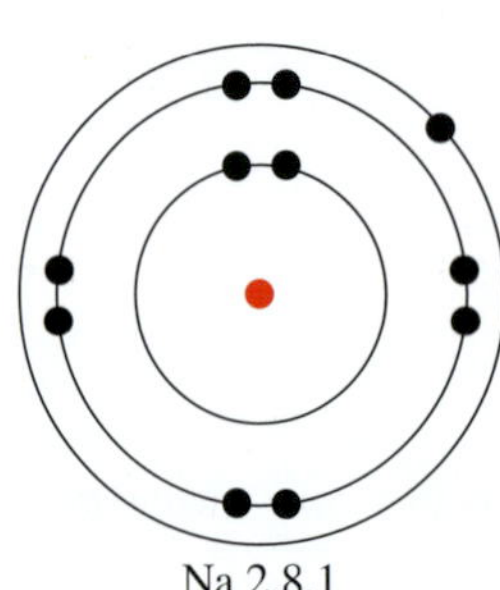

FIGURE 2.2.10 The electronic configuration of a sodium atom

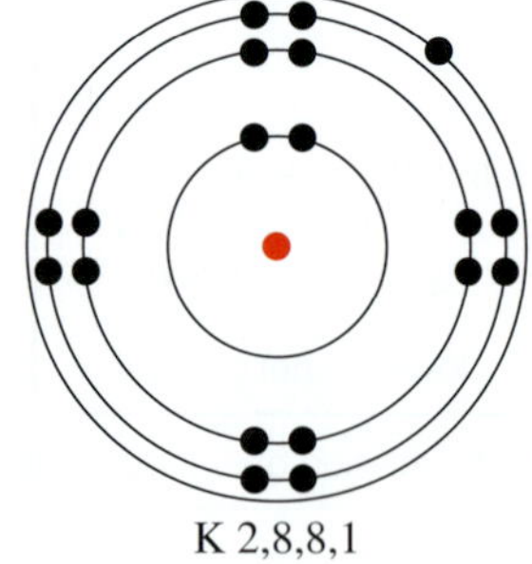

FIGURE 2.2.11 The electronic configuration of potassium

Worked example 2.2.1

ELECTRONIC CONFIGURATION FOR UP TO 36 ELECTRONS

Apply the order of filling of the shell model to determine the electronic configuration of an atom with 28 electrons.

Thinking	Working
Recall the maximum number of electrons that each shell can hold.	Shell (n) Maximum number of electrons 1 2 2 8 3 18 4 32
Place the first 18 electrons in the shells from the lowest energy to the highest energy. Do not exceed the maximum number of electrons allowed.	Shell (n) Electrons in atom 1 2 2 8 3 8 4
Place the next two electrons in the fourth shell.	Shell (n) Electrons in atom 1 2 2 8 3 8 4 2
Continue filling the third shell until it holds up to 18 electrons. Put any remaining electrons in the fourth shell.	Shell (n) Electrons in atom 1 2 2 8 3 16 4 2 The 8 remaining electrons from the previous step have gone into the third shell.
Write the electronic configuration by listing the number of electrons in each shell separated by commas.	The electronic configuration is: 2,8,16,2

Worked example: Try yourself 2.2.1

ELECTRONIC CONFIGURATIONS FOR UP TO 36 ELECTRONS

Apply the order of filling of the shell model to determine the electronic configuration of an atom with 34 electrons.

2.2 Review

OA
✓✓

SUMMARY

- When atoms are heated, they can emit electromagnetic radiation in the form of coloured light. When the light is passed through a prism, it produces a spectrum made up of lines of different colours. These spectra are known as emission or line spectra.
- The Bohr model of the atom was the first atomic model to explain the origin of emission spectra and assumes that electrons can only exist in fixed, circular orbits of specific energies. These orbits later came to be known as energy levels or shells.
- When an electron absorbs energy it can jump to a higher energy shell and when an electron falls back to its original state it can emit energy in the form of coloured light. Each line in the emission spectrum corresponds to a specific electron transition between shells.
- Each shell can hold a maximum number of electrons. This can be calculated using the rule $2n^2$, where n = shell number.
- The valence shell cannot contain more than eight electrons.

KEY QUESTIONS

Knowledge and understanding

1 Explain how emission spectra provide evidence for electron shells in the Bohr model.

2 What form of energy is emitted when an electron moves from a higher energy shell to a lower energy shell?

3 Draw a Bohr diagram and write the electronic configuration for an atom of argon. (Refer to the periodic table at the back of the book to find the atomic number of argon.)

Analysis

4 Determine the maximum number of electrons that the fifth shell can hold.

5 Compare an atom's electronic shell configuration to its position in the periodic table. Describe any patterns you observe.

6 An atom has an electronic configuration 2,6,8. Identify the atom and state why this electronic arrangement is unexpected. Propose a reason for this arrangement.

2.3 The Schrödinger model of the atom

The shell model of the atom that developed from the work of Niels Bohr mathematically explained the lines in the emission spectrum of hydrogen atoms. However, there were some things that the model could not explain. The shell model:

- cannot accurately predict the emission spectra of atoms with more than one electron
- is unable to explain why electron shells can only hold $2n^2$ electrons
- does not explain why the fourth shell accepts two electrons before the third shell is completely filled.

These limitations of the shell model indicate that the model is incomplete. Obtaining a better model of the atom required scientists to think about electrons in an entirely different way.

A QUANTUM MECHANICAL VIEW OF ATOMS

The Bohr model of the atom was revolutionary when it was proposed. Before Bohr, scientists believed that electrons could orbit the nucleus at any distance. This picture of the electrons was based on how scientists observed the world around them. For example, planets can revolve at any distance around the Sun. However, Bohr's theory stated that electrons only occupy specific, circular orbits. This was the first suggestion that the physics inside atoms might be very different to the physics we experience in our daily lives.

Quantum mechanics

The word 'quantum' simply means a specific amount. In the Bohr model, the electrons can only have specific amounts of energy depending on which shell they are in. The energy of the electrons is said to be **quantised**.

In 1926, Erwin Schrödinger proposed that electrons behave as waves around the nucleus. Using a mathematical approach and this wave theory, Schrödinger developed a model of the atom called **quantum mechanics** (Figure 2.3.1). The Schrödinger model of the atom is the model that scientists use today.

Quantum mechanics describes the behaviour of extremely small particles like electrons. You rarely experience quantum mechanics in your everyday life. As a result, the predictions of quantum mechanics are often difficult to visualise. Nonetheless, quantum mechanics accurately predicts the behaviour of electrons in atoms.

> The model of the atom that is used today is called the quantum mechanical or Schrödinger model.

FIGURE 2.3.1 Erwin Schrödinger (1887–1961), the Austrian-Irish physicist whose research into subatomic particles is the basis of quantum mechanics.

THE SCHRÖDINGER MODEL

The fundamental difference between the Bohr model and the Schrödinger model of the atom is the way they view the electrons. Bohr viewed electrons as tiny, solid particles that revolve around the nucleus in circular orbits. Schrödinger viewed electrons as having wave-like properties, just like light. In this model, the electrons occupy a three-dimensional space around the nucleus known as an **orbital**. Figure 2.3.2, on the following page, compares the Bohr model and the Schrödinger model.

By assuming that electrons have wave-like properties, Schrödinger found the following.

- There are major energy levels in an atom that, for historical reasons, are called shells.
- These shells contain separate energy levels of similar energy, called **subshells**, which he labelled *s*, *p*, *d* and *f*. Each subshell can only hold a certain number of electrons.
- The first shell ($n = 1$) contains only an *s*-subshell. The second shell contains *s*- and *p*-subshells. The third shell contains *s*-, *p*- and *d*-subshells and so on. The subshells for the first four shells are summarised in Table 2.3.1 on the following page.

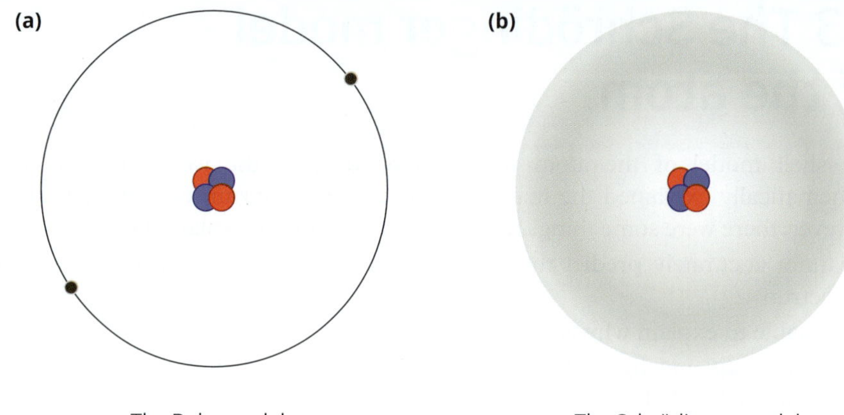

FIGURE 2.3.2 (a) Bohr regarded electrons as particles that travel along a defined path in circular orbits. (b) In Schrödinger's quantum mechanical approach, the electrons behave as waves and occupy a three-dimensional space around the nucleus. The region occupied by the electrons is known as an orbital.

The quantum mechanical model predicts subshells and orbitals within each of the shell levels.

- Each subshell is made up of smaller components known as orbitals. Orbitals can be described as regions of space surrounding the nucleus of an atom in which electrons may be found. Any orbital can hold a maximum of two electrons. An *s*-subshell has just one orbital. A *p*-subshell has three orbitals. A *d*-subshell has five orbitals and an *f*-subshell has seven. The number of orbitals in each subshell for the first four shells is summarised in Table 2.3.1.
- The total number of orbitals in a shell is given by n^2. Each orbital can contain a maximum of two electrons. Therefore, the total number of electrons per shell is given by $2n^2$. For example, the second shell contains *s*- and *p*-subshells and so contains a total of 1 + 3 = 4 orbitals. Each orbital can contain two electrons, so the second shell contains up to 2 × 4 = 8 electrons. The number of electrons in each subshell and shell for the first four shells is summarised in Table 2.3.1.

TABLE 2.3.1 Energy levels within an atom

Shell number (*n*)	Subshells	Number of orbitals in subshell	Maximum number of electrons per subshell	Maximum number of electrons per shell
1	1*s*	1	2	2
2	2*s*	1	2	
	2*p*	3	6	8
3	3*s*	1	2	
	3*p*	3	6	
	3*d*	5	10	18
4	4*s*	1	2	
	4*p*	3	6	
	4*d*	5	10	
	4*f*	7	14	32

Note that the Schrödinger model accurately predicts the maximum number of electrons that each shell can hold. This is something that the Bohr model could not explain.

CHEMFILE

A model built on mathematical probability

In the Schrödinger model, electrons can be regarded as behaving like clouds of charge, where the cloud density at a particular location in the atom indicates the probability of finding an electron at that position. So, this model is based on probability, rather than certainty.

These mathematical regions of probability are called orbitals and are determined using mathematical wave theory. The shape of some of these orbitals may be surprising. The following model represents three *p*-orbitals around a nucleus.

A computer-generated representation of *p*-orbitals around a nucleus. An orbital is a mathematical function.

ELECTRONIC CONFIGURATIONS AND THE SCHRÖDINGER MODEL

The electronic configurations for the Schrödinger model contain more detail than the electronic configurations of the shell model. This is because the Schrödinger model specifies subshells that electrons occupy. The subshell electronic configuration of a sodium atom is shown in Figure 2.3.3.

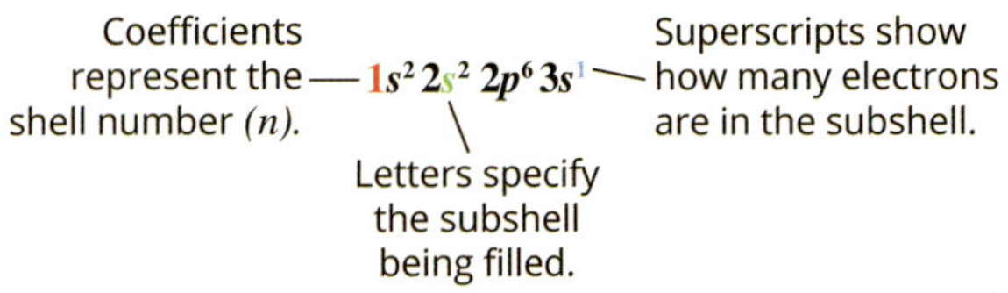

Coefficients represent the shell number (n). — $1s^2 2s^2 2p^6 3s^1$ — Superscripts show how many electrons are in the subshell.

Letters specify the subshell being filled.

FIGURE 2.3.3 The subshell electronic configuration of a sodium atom. It shows that there are 2 electrons in the 1s-subshell, 2 electrons in the 2s-subshell, 6 electrons in the 2p-subshell and 1 electron in the 3s-subshell.

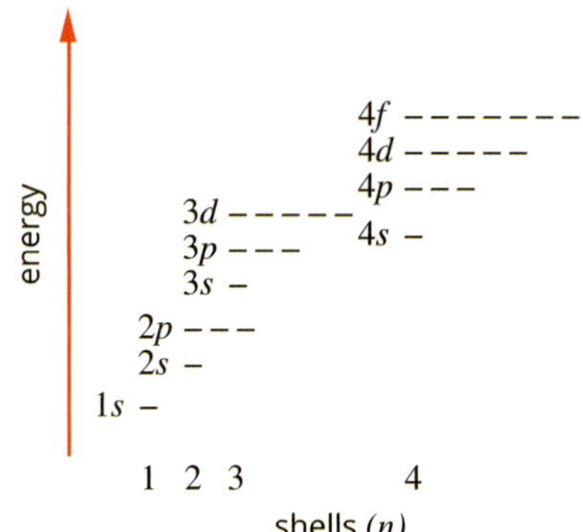

FIGURE 2.3.4 The relative energies of subshells. Each dashed line represents an orbital in a subshell. Each orbital can contain two electrons.

Sodium has 11 electrons. The subshell electronic configuration in Figure 2.3.3 indicates that sodium has:

- two electrons in the s-subshell in the first shell
- two electrons in the s-subshell of the second shell
- six electrons in the p-subshell of the second shell
- one electron in the s-subshell of the third shell.

Although the subshell electronic configurations may look complicated, the rules for constructing them are simple.

- The lowest energy subshells are always filled first.
- Each orbital contains a maximum of two electrons.

The order of energy levels of the subshells is:

$$1s < 2s < 2p < 3s < 3p < 4s < 3d \ldots$$

This is shown in Figure 2.3.4. In this diagram, each dash represents an orbital that can hold two electrons.

The geometric pattern shown in Figure 2.3.5 is a commonly used and convenient way of remembering the order in which the subshells are filled. Note that in this diagram the fourth shell starts filling before the third shell is completely filled. This is because the 4s-orbital is slightly lower in energy than the 3d-orbitals. As a result, the 4s-orbital accepts two electrons after the 3s- and 3p-orbitals are filled but before the 3d-orbital begins filling.

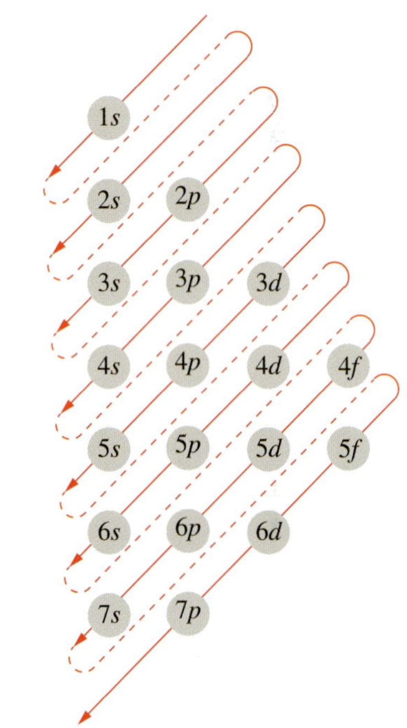

FIGURE 2.3.5 This geometric pattern shows the order in which the subshells are filled.

Figure 2.3.6 shows how the energy levels are filled in a neon atom, which has 10 electrons. The first two electrons fill the 1s-subshell, the second two electrons go into the next highest energy level, the 2s-subshell. The last six electrons then fill the next highest energy level, the 2p-subshell. The electronic configuration is written as $1s^2 2s^2 2p^6$.

Krypton has 36 electrons. According to the order of subshell filling, its electronic configuration is:

$$1s^2 2s^2 2p^6 3s^2 3p^6 4s^2 3d^{10} 4p^6$$

Although the 4s-subshell is filled before the 3d-subshell, the subshells of the third shell are usually grouped together when writing its electronic configuration. Therefore, the electronic configuration for a krypton atom is written as:

$$1s^2 2s^2 2p^6 3s^2 3p^6 3d^{10} 4s^2 4p^6$$

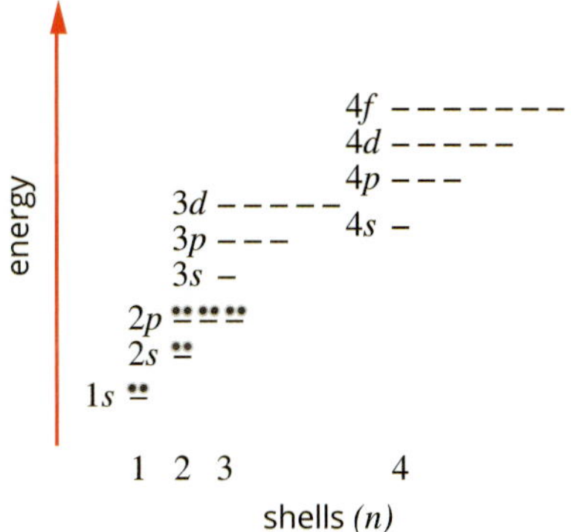

FIGURE 2.3.6 This electron subshell diagram shows how the orbitals in a neon atom are filled to give an electronic configuration.

Condensed electronic configuration

The electronic configuration of phosphorus is $1s^22s^22p^63s^23p^3$. This configuration can be written more simply as [Ne]$3s^23p^3$, where the [Ne] symbol signifies $1s^22s^22p^6$ because this is the electronic configuration of neon. A configuration like [Ne]$3s^23p^3$ is called a **condensed electronic configuration**, or referred to as **noble gas notation**. When writing the condensed electronic configuration of an element, the symbol of the noble gas (with square brackets) in the period before the element is used. This means that the details for the inner shells are not written—they are implied by the use of the noble gas symbol. This is especially useful for elements with multiple shells containing electrons. Some other examples of condensed electronic configuration are shown in Table 2.3.2.

TABLE 2.3.2 Some examples of the use of condensed electronic configuration

Element	Atomic number	Electronic configuration (shell level)	Condensed electronic configuration (subshell level)
Si	14	2,8,4	[Ne]$3s^23p^2$
Zn	30	2,8,18,2	[Ar]$3d^{10}4s^2$
Br	35	2,8,18,7	[Ar]$3d^{10}4s^24p^5$

An advantage of the condensed notation is that it emphasises the valence shell electrons, which are fundamental for understanding how atoms interact. It also enables the detail of subshell notation to be displayed where there is limited space. Periodic tables, like the one on the back page of this book, often use these condensed electronic configurations.

Chromium and copper: exceptions

The electronic configurations for most elements follow the rules and pattern described above. There are two notable exceptions: element 24, chromium, and element 29, copper. Table 2.3.3 shows these exceptions.

TABLE 2.3.3 The electronic configurations for chromium and copper

Element	Electronic configuration predicted according to the pattern above	Actual electronic configuration
chromium, Cr	$1s^22s^22p^63s^23p^63d^44s^2$	$1s^22s^22p^63s^23p^63d^54s^1$
copper, Cu	$1s^22s^22p^63s^23p^63d^94s^2$	$1s^22s^22p^63s^23p^63d^{10}4s^1$

Each orbital can hold two electrons. In the *d*-subshell there are five orbitals and therefore 10 electrons that can be held within them. Scientists have determined that as a subshell fills, a single electron is placed in each orbital first. Then a second electron is entered into the orbitals until the filling process is complete.

Chemists calculate that there is very little difference in energy between 3*d*- and 4*s*-orbitals, and the $3d^54s^1$ configuration for chromium is slightly more stable than the $3d^44s^2$ configuration.

Similarly, for copper, the $3d^{10}4s^1$ arrangement with five completely filled *d*-orbitals is more stable than the $3d^94s^2$.

Worked example 2.3.1

WRITING ELECTRONIC CONFIGURATIONS USING THE SUBSHELL MODEL

Write the subshell electronic configuration for a manganese atom with 25 electrons.

<table>
<tr><th>Thinking</th><th>Working</th></tr>
<tr><td>Recall the order in which the subshells fill by listing them from lowest energy to highest energy and the number of orbitals in each.</td><td>1s, 1 orbital
2s, 1 orbital
2p, 3 orbitals
3s, 1 orbital
3p, 3 orbitals
4s, 1 orbital
3d, 5 orbitals</td></tr>
<tr><td>Fill the subshells by assigning two electrons per orbital, starting from the lowest energy subshells until you have reached the total number of electrons in your atom.</td><td>
<table>
<tr><th>Subshell</th><th>Electrons in subshell</th><th>Progressive total of electrons</th></tr>
<tr><td>1s</td><td>2</td><td>2</td></tr>
<tr><td>2s</td><td>2</td><td>4</td></tr>
<tr><td>2p</td><td>6</td><td>10</td></tr>
<tr><td>3s</td><td>2</td><td>12</td></tr>
<tr><td>3p</td><td>6</td><td>18</td></tr>
<tr><td>4s</td><td>2</td><td>20</td></tr>
<tr><td>3d</td><td>5</td><td>25</td></tr>
</table>
</td></tr>
<tr><td>Write the electronic configuration by writing each subshell with the number of electrons as a superscript. Remember to group subshells from the same shell.</td><td>$1s^22s^22p^63s^23p^63d^54s^2$</td></tr>
</table>

Worked example: Try yourself 2.3.1

WRITING ELECTRONIC CONFIGURATIONS USING THE SUBSHELL MODEL

Write the subshell electronic configuration for a vanadium atom with 23 electrons.

2.3 Review

OA ✓✓

SUMMARY

- The Bohr shell model of the atom was unable to fully explain the properties of atoms and a new model was needed to accurately describe the observed electron behaviour in atoms.
- The Schrödinger (quantum mechanical) model proposes that electrons behave as waves and occupy a three-dimensional space around the nucleus.
- The Schrödinger model describes electronic configurations in terms of shells, subshells and orbitals.
- An orbital can be regarded as a region of space surrounding the nucleus in which an electron may be found. An orbital can hold a maximum of two electrons.
  - Orbitals of similar energy are grouped in subshells that are labelled *s*, *p*, *d* and *f*.
- Each subshell has a different energy in an atom.
- Electrons fill the subshells from the lowest energy subshell to the highest energy subshell.
- The 4*s*-subshell is lower in energy than the 3*d*-subshell, so the fourth shell accepts two electrons before the third shell is completely filled.

Shell	Subshells	Orbitals in subshell	Maximum number of electrons in subshell
1	1*s*	1	2
2	2*s* 2*p*	1 3	2 6
3	3*s* 3*p* 3*d*	1 3 5	2 6 10
4	4*s* 4*p* 4*d* 4*f*	1 3 5 7	2 6 10 14

- Electronic configurations of atoms in the Schrödinger model are more detailed than the shell level configuration, as they specify the number of electrons in each subshell.
- Chromium and copper are exceptions to the predicted order of filling.

KEY QUESTIONS

Knowledge and understanding

1 Copy and complete the following table to write the electronic configuration of each of the atoms listed.

Element (atomic number)	Electronic configuration using the shell model	Electronic configuration using the subshell model
boron (5)	2,3	$1s^2 2s^2 2p^1$
lithium (3)		
chlorine (17)		
sodium (11)		
neon (10)		
potassium (19)		
scandium (21)		
copper (29)		
bromine (35)		

2 Identify the elements shown by the following electronic configurations.

a $1s^2 2s^2 2p^6 3s^2 3p^6 3d^{10} 4s^2 4p^1$

b $1s^2 2s^2 2p^6 3s^2 3p^6 3d^{10} 4s^2 4p^5$

c $1s^2 2s^2 2p^6 3s^2 3p^6 3d^{10} 4s^1$

d $1s^2 2s^2 2p^6 3s^2 3p^6 3d^5 4s^2$

e $1s^2 2s^2 2p^6 3s^2 3p^6 3d^{10} 4s^2 4p^6 4d^{10} 5s^2 5p^2$

f $[Kr]4d^{10} 5s^2 5p^3$

g $[Kr]4d^{10} 5s^2 5p^6$

h $[Ar]4s^1$

Analysis

3 Describe a limitation of the Bohr model of the atom that has been overcome by the Schrödinger model.

4 Describe the essential difference between the Bohr model and the Schrödinger model of the atom in terms of energy levels.

5 Write an electronic configuration of a sulfur atom that has been heated so that its electrons are no longer in the ground state.

2.4 The periodic table

As scientists' understanding of the atom improved and more elements were discovered, a way of organising this knowledge was needed.

The **periodic table** (Figure 2.4.1) is one of the most recognisable symbols of modern science, and more than 150 years after it was first developed by Dimitri Mendeleev, it is still an incredibly useful tool for chemists. It minimises the need to memorise isolated facts about different elements, and provides a framework on which to organise our understanding. By knowing the properties of particular elements and trends within the table, chemists are able to organise what would otherwise be an overwhelming collection of disorganised information.

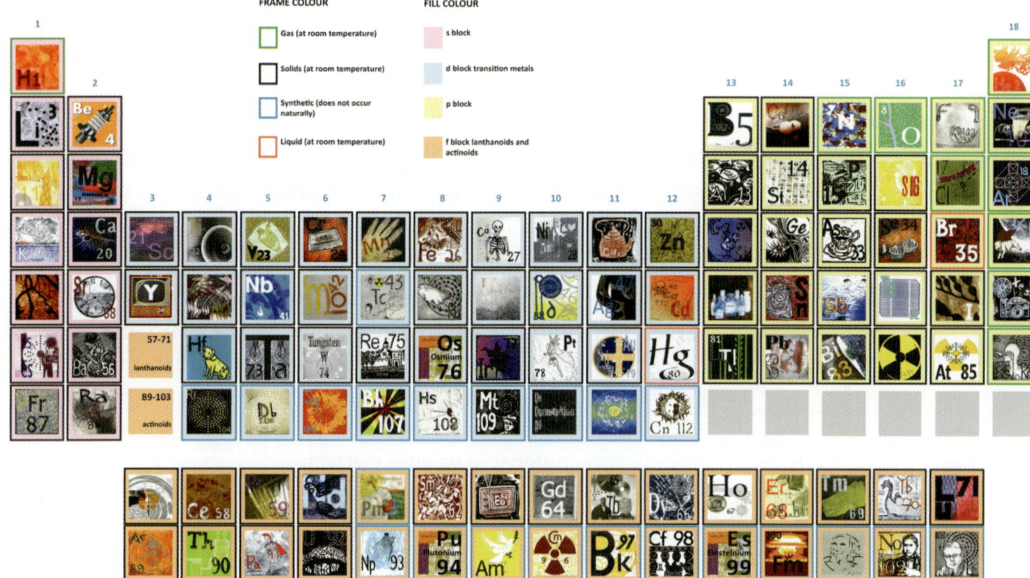

FIGURE 2.4.1 This beautiful periodic table was commissioned by the Royal Australian Chemical Institute (RACI) to commemorate the 150-year anniversary of Mendeleev's original table. Each panel was a collaboration between an artist and a scientist, where the scientist provided information about the properties of the element, and the artist created a visual response to this information.

CHEMFILE

The power of a prediction

The first periodic table was developed by Russian chemist, Dimitri Mendeleev, in 1869 (see figure). When he first proposed his table, based on the observation that atomic properties seemed to vary in repeating patterns, he left spaces in the table for elements that had not yet been discovered. He even went so far as to predict the properties of these yet-undiscovered elements. He was ridiculed and largely ignored by the scientific establishment, and it was not until five years later, when gallium was discovered and matched his predicted properties almost exactly (see table), that the rest of the scientific community took his periodic table seriously.

A comparison between the properties predicted by Mendeleev, and the actual properties for element 31, gallium

Property	Mendeleev's predictions for 'eka-aluminium'	gallium
Atomic mass	68	69.723
Density (g/cm^3)	6.0	5.91
Melting point (°C)	low	29.76

Russian chemist Dimitri Mendeleev (1834–1907) developed the first periodic table in 1869.

One of the remarkable things about the periodic table is that as our understanding of atomic theory has developed this periodic table has remained relevant, even though it is a tool that was developed before protons, neutrons or electrons were even proposed. Each new development in atomic theory has revealed an underlying pattern in the table. In the next section you will see how the shells and valence electrons of Bohr's model correspond to the periods and groups, and you will see how Schrödinger's quantum mechanical model is reflected in the blocks.

THE MODERN PERIODIC TABLE

We now know that the number of protons (the atomic number) is what makes one element fundamentally different from another element. The elements in the modern periodic table are therefore arranged in rows in order of increasing atomic number.

Chemists use the number of electrons in the outer shell (the **valence electrons**), to organise the elements into columns. Shading is often used to highlight different aspects of the table.

The modern periodic table has several key features, as can be seen in Figure 2.4.2.

- The periodic table is arranged in order of increasing atomic number.
- The horizontal rows are known as periods and are labelled 1–7.
- The vertical columns are known as groups and are labelled 1–18.
- **Main group elements** are elements in groups 1, 2 and 13–18.
- The elements in groups 3–12 are known as **transition metals**.

Some periodic tables also indicate other properties of the elements such as boiling point or whether the element is a solid, liquid or gas at room temperature.

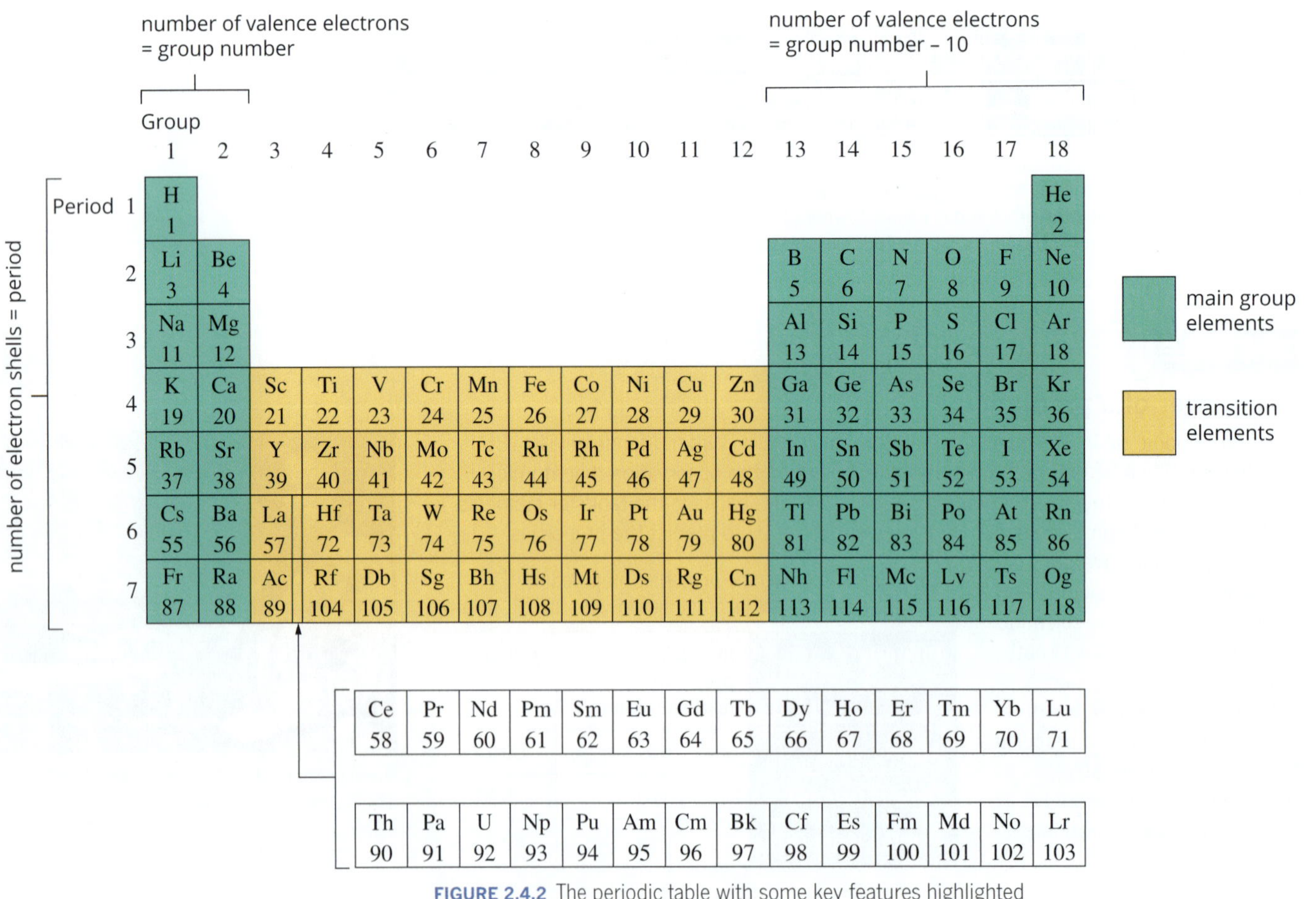

FIGURE 2.4.2 The periodic table with some key features highlighted

Groups

Elements in the periodic table are arranged into vertical columns called **groups**. For main group elements, the group number can be used to determine the number of valence electrons (outer-shell electrons) in an atom of the element.

In groups 1 and 2, the number of valence electrons is equal to the group number. For example, magnesium is in group 2 and therefore has two valence electrons.

In groups 13–18, the number of valence electrons is equal to the group number minus 10. For example, oxygen is in group 16 so oxygen has six outer-shell electrons. Similarly, neon is in group 18 so neon has eight valence electrons. Helium is an important exception. It is located in group 18 but only has two valence electrons. Helium is placed in group 18 because it is unreactive, like other group 18 elements. This information is summarised in Table 2.4.1.

The columns in the periodic table are known as groups. The group number can be used to determine the number of valence electrons in an atom of an element.

TABLE 2.4.1 The number of valence electrons in elements belonging to each group

Group	Number of valence electrons
1	1
2	2
13	3
14	4
15	5
16	6
17	7
18	8*

*Helium has two valence electrons.

The electrons in the outer shell of an atom (the valence electrons) are the electrons that are involved in chemical reactions. As a consequence, the number of valence electrons determines the chemical properties that an element exhibits.

Because elements in the same group have the same number of valence electrons, elements in the same group have similar properties. For example, the **alkali metals** are elements in group 1 (with the exception of hydrogen). They are all relatively soft metals and are highly reactive with water and oxygen. Consider the electronic configurations of the atoms of the first three metals of this group:

Li $1s^2 2s^1$

Na $1s^2 2s^2 2p^6 3s^1$

K $1s^2 2s^2 2p^6 3s^2 3p^6 4s^1$

The valence shell of each atom of each element in group 1 contains one electron in an *s*-subshell. This similarity in the valence shell structure gives these elements similar chemical properties.

Fluorine, chlorine, bromine and iodine are **halogens** (group 17). They are all coloured and highly reactive (Figure 2.4.3 on the following page). Their electronic configurations are:

F $1s^2 2s^2 2p^5$

Cl $1s^2 2s^2 2p^6 3s^2 3p^5$

Br $1s^2 2s^2 2p^6 3s^2 3p^6 3d^{10} 4s^2 4p^5$

I $1s^2 2s^2 2p^6 3s^2 3p^6 3d^{10} 4s^2 4p^6 4d^{10} 5s^2 5p^5$

Notice how all these elements have a highest-energy subshell electronic configuration of s^2p^5.

CHEMFILE

Helium supplies at risk

On the Earth, most helium is found under the Earth's crust with other natural gases. If a deposit of natural gas contains a commercially viable amount of helium, then it is extracted from the mixture. The largest commercial deposits of helium are currently found in the US, Qatar and Algeria.

We all know about the use of helium in balloons, but it has many other uses. For example, it is used in medical research and diagnostic equipment, to cool nuclear reactors and rockets and to provide an unreactive atmosphere for arc welding.

However, once helium is released into the atmosphere it is virtually impossible to recover. This means that we are using up the limited sources of helium, with some scientists estimating that we could run out within 25–30 years.

Helium is one of the most at-risk elements, with some scientists estimating that supplies could run out within 25–30 years.

FIGURE 2.4.3 Three examples of halogens. These conical flasks contain, from left to right, chlorine (Cl, pale green), bromine (Br, red-brown) and iodine (I, purple).

The **noble gases** (group 18) are a particularly interesting group. The noble gases have a very stable electron arrangement: helium and neon have full outer shells and the other members of this group have a stable octet of valence electrons (eight electrons). Chemical reactions involve the rearrangement of valence electrons to achieve a stable outer shell. Noble gases already have a stable electronic configuration, so they do not tend to lose or gain electrons. This means that the noble gases have very low reactivity.

The arrangement of electrons in atoms is responsible for the **periodicity** (periodic pattern) of element properties.

Figure 2.4.4 shows some of the special groups in the periodic table.

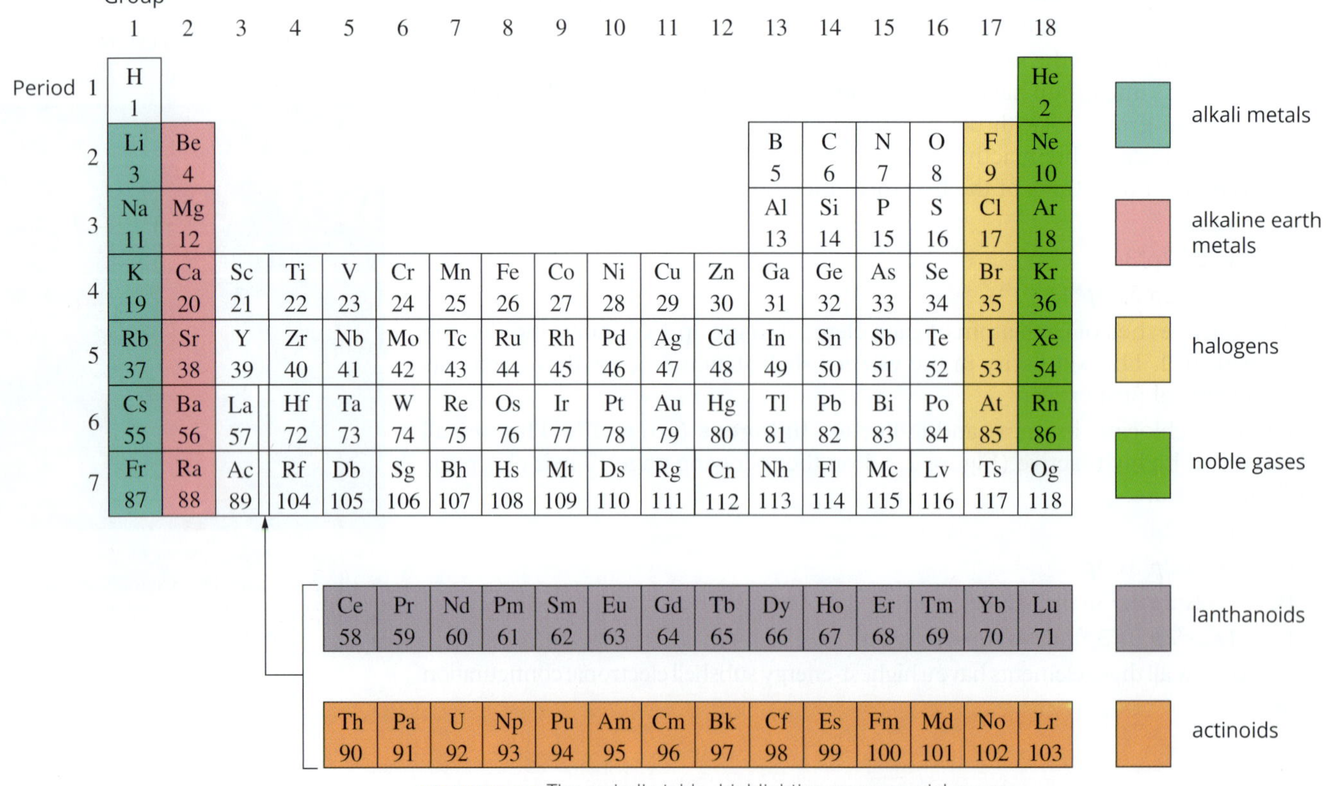

FIGURE 2.4.4 The periodic table, highlighting some special groups

Periods

The horizontal rows in the periodic table are called **periods** and are numbered 1–7. The number of a period gives information about the electronic configuration of an element. The period an element is located within is equal to the number of occupied electron shells in the element's atoms. For example, the outer shell of magnesium and chlorine is the third shell and both of these elements are in period 3:

Mg $1s^22s^22p^63s^2$

Cl $1s^22s^22p^63s^23p^5$

Similarly, the elements in period 5 all have outer shell electrons in the fifth shell.

> The rows of the periodic table are called periods. The period number tells you the number of occupied electron shells in the element's atoms.

Blocks

As understanding of atomic theory developed into Schrödinger's quantum mechanical model, a new pattern emerged in the periodic table. We can now see that it can be divided into four main **blocks**. The elements in each block all have the same type of subshell as the highest energy subshell (*s*, *p*, *d* or *f*).

Figure 2.4.5 shows which groups of elements fall into the four different blocks.

> The blocks in the periodic table contain elements that have the same type of subshell as the highest energy subshell (*s*, *p*, *d* or *f*).

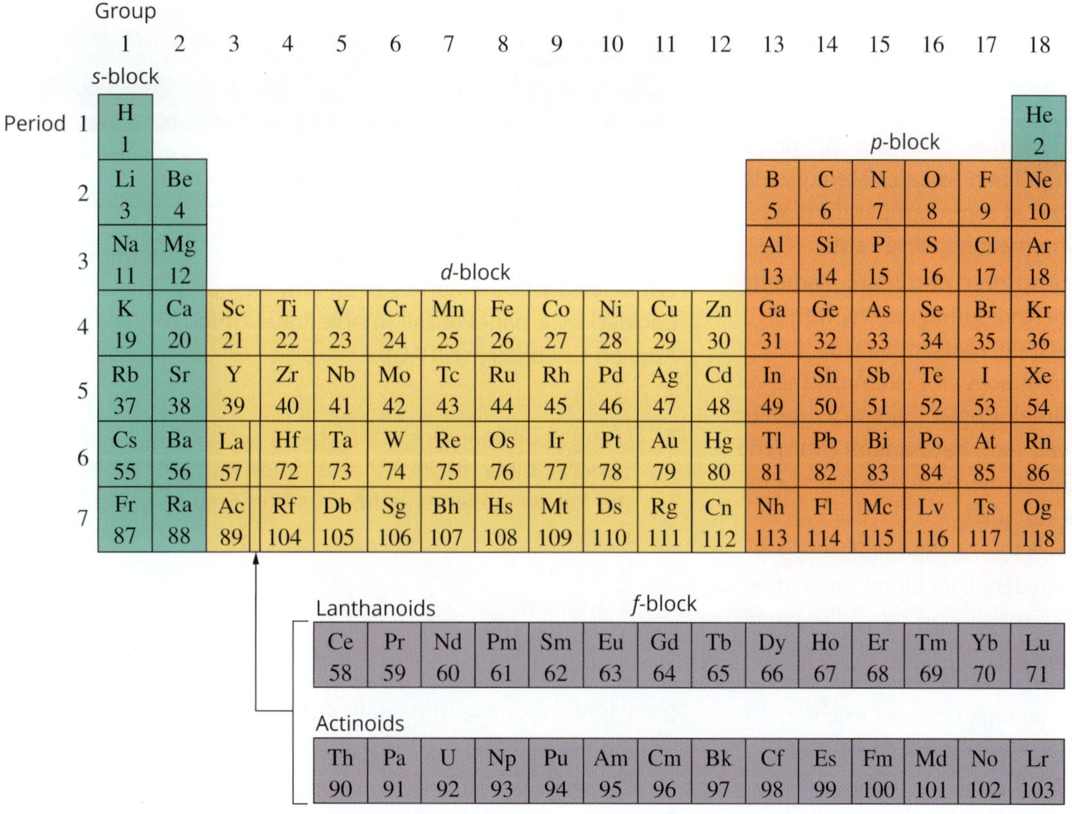

FIGURE 2.4.5 Colour is used to distinguish between the *s*, *p*, *d* and *f*-blocks of elements.

CASE STUDY **ANALYSIS**

Naming elements on the periodic table

Ancient elements

Twelve elements were known in ancient times. These included gold, silver, mercury and sulfur. The original names for these elements were derived from Latin. People at this time did not have an understanding of the modern definition of an element, so materials like soda (sodium oxide) and calomel (mercury chloride) were at first thought to be elemental.

Searching for new elements

Later in the eighteenth century and into the nineteenth century, scientists became able to extract gaseous elements from the air and isolate solid elements from the ground. Elements often took names from the Latin words for where they were found. Such elements included silicon, Si (found in sand, known in Latin as *silex*) and calcium, Ca (found in limestone, known in Latin as *calx*). Mercury, Hg, a silvery metal that is a liquid at room temperature, was given the name *hydrargyrum*, meaning silver water.

As analytical techniques improved and more elements were found, naming became more imaginative, with names being derived from Greek roots, detected characteristics, places or mythology: for example, the inert gas argon (from the Greek *άργόν*, inactive), rubidium (from the Latin *rubidius*, deep red), germanium (from the Latin *germania*, meaning Germany), and vanadium (after the Scandinavian goddess Vanadis).

By 1829, 55 elements were known. The first modern periodic table, constructed in 1869, included 64 elements, and by 1914 a total of 72 elements had been discovered.

Synthetic elements and atomic science

The first synthetic element, technetium, was made in 1936. New elements were discovered after 1940; of these, many were discovered during thermonuclear bomb tests (see Figure 2.4.6), and more recently in particle accelerators. This has brought the total number of known elements in the periodic table to 118. The synthetic elements have large, unstable nuclei and not much is known about their chemistry. Many, such as einsteinium or americium, were named after a scientist or place. The Cold War made naming of these elements contentious, but since 1999 the names of all new elements have been decided by a panel of the International Union of Pure and Applied Chemistry (IUPAC).

FIGURE 2.4.6 Thermonuclear bomb tests like this were the source of many of the first synthetic elements discovered.

Analysis

1 Using your knowledge of element symbols and the periodic table, deduce the common name of each of the following elements from its non-English name.
   - **a** *ferrum*
   - **b** *kalium*
   - **c** *wolfram*
   - **d** *plumbum*
   - **e** *hydrargyrum*

2 IUPAC has a set of guidelines for naming new elements. What are these guidelines? (You will need to research your answer.)

3 The periodic table was most recently updated by IUPAC in both 2012 and 2016. In these two updates, elements 113, 115, 117 ,118 (2016), 114 and 116 (2012) were added. Research the names of these elements and outline why they were given their particular names.

CRITICAL ELEMENTS

Of the 118 elements in the periodic table, many exist in only very small quantities on Earth. In recent decades, some of these rare elements have been used in bulk for the first time to produce new electronics, medicines and catalysts. Unfortunately, our use of these materials is not sustainable, as there is currently very little recycling or recovery of these elements. As a consequence, over 40 elements have been identified as being '**endangered**', with supplies of some elements predicted to run out in less than 100 years. Figure 2.4.7 highlights elements that have been identified as endangered. (Note this classification is constantly changing as our demand for elements changes, and new deposits are found.)

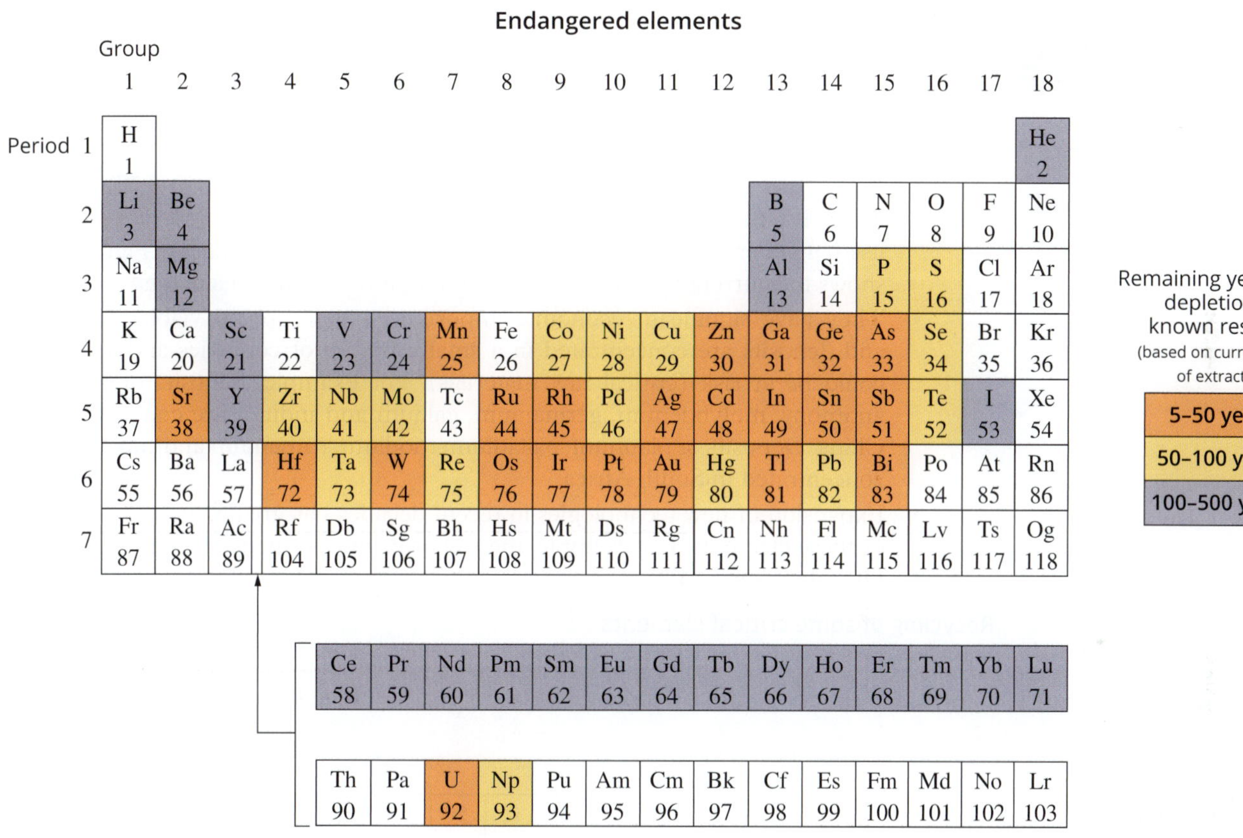

FIGURE 2.4.7 Some of the elements identified as endangered, with estimates for how long they will be available.

It is not just the reserves of an element that are important in ensuring sustainability. In recent years the concept of a critical element has developed. **Critical elements** are elements that are heavily relied on for industry and society in areas such as renewable energy, electronics, food supply and medicine. Some scientists and economists believe that unhindered access to these elements is required for the essential running of society. Critical elements face some form of supply uncertainty. Elements may be considered critical for a number of reasons:

- Only small deposits are available on Earth, and these are fast disappearing (endangered elements). For example, iridium, platinum, osmium and palladium (see Figure 2.4.7).
- Supply of these elements is centred in places of war and conflict (conflict elements). **Conflict elements** include tin, gold, tungsten and tantalum, which are essential in mobile phone production (Figure 2.4.8a on the following page). These elements are mined in areas of war and conflict, and routinely use child labour in their mining, which makes their use non-sustainable (Figure 2.4.8b).

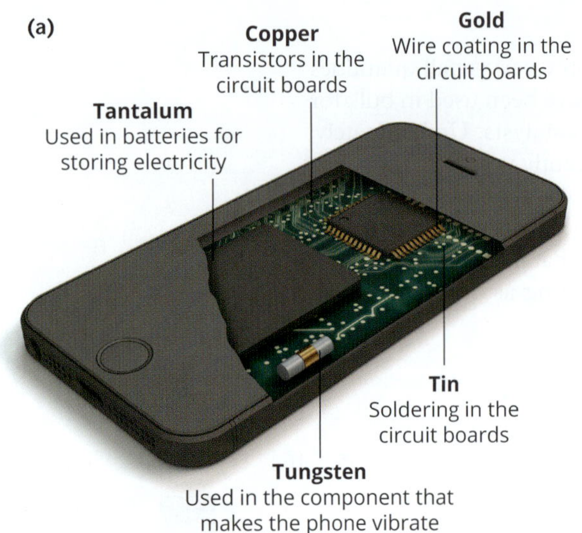

FIGURE 2.4.8 (a) Tin, gold, tungsten and tantalum are all used in mobile phone production. Their use is not sustainable as they are often mined in areas of war and conflict. (b) Child labour is routinely used in the mining of these elements. This young teenage worker is pictured outside a gold mine in Africa.

- There is little to no recycling and recovery of the element, so reserves are being used up. For example, the rare earth elements (Lanthanides). Figure 2.4.9 shows the current rates of recycling for a number of these elements.
- They have significant economic importance, there are no viable substitutes and deposits are concentrated in a small number of countries and so supply could be at risk (critical raw materials). For example, deposits of tungsten, antimony, molybdenum, germanium, gallium and indium are concentrated in China. Deposits of platinum, palladium, rhodium, ruthenium and vanadium are concentrated in South Africa.
- Expansion of new technologies in recent decades has greatly increased our use of these elements. For example, the use of helium in medical technologies and phosphorus in fertilisers.

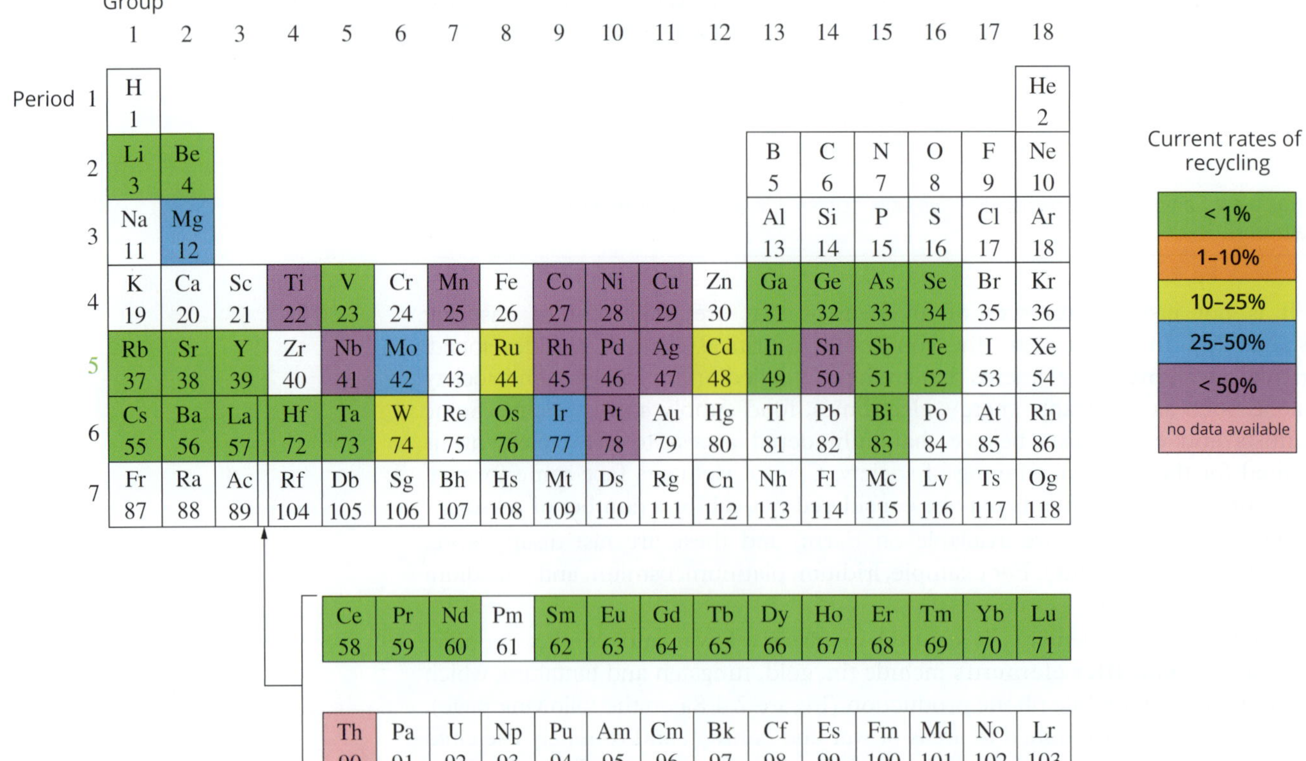

FIGURE 2.4.9 There is little to no recycling of many of the elements currently being used as raw materials.

The concept of critical elements draws attention to the non-sustainability of our consumption. Recycling campaigns have been developed globally to help make products more sustainable. One of the most well known was in the lead-up to the 2020 Tokyo Olympic Games, when the Japanese people collected approximately 79 000 tonnes of old mobile phones and other electronic waste. This was recycled to recover the precious metals inside to make 100% of the gold, silver and bronze Olympic medals presented to the athletes (Figure 2.4.10). The recycling of metals is discussed in more detail in Chapter 4.

Our use of many elements is not currently sustainable. New ways of recovering and recycling these elements are needed.

FIGURE 2.4.10 Emma McKeon of Australia, Sarah Sjöström of Sweden and Pernille Blume of Denmark show off their 2020 Tokyo Olympic medals, made from metal recovered from 79 000 tonnes of electronic waste collected by the Japanese people in the lead-up to the games.

A new approach to recycling

Phosphorus plays a crucial part in the production of the world's food. It is used to create fertilisers that enable greater crop yields (Figure 2.4.11). However, it is estimated that supply of phosphorus will become depleted sometime this century. Experts predict this could have very serious consequences for world food supplies. In order to maintain overall supply of phosphorus, it may even be necessary for it to be recovered from animal and human waste.

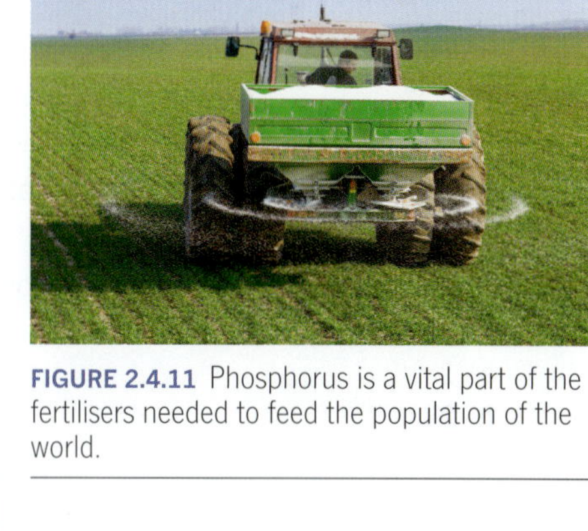

FIGURE 2.4.11 Phosphorus is a vital part of the fertilisers needed to feed the population of the world.

One suggested method of recovery is from human urine, using a toilet with a urine diversion and dehydration unit like that shown in Figure 2.4.12. The phosphorus can eventually be precipitated out of the urine as calcium phosphate.

Interestingly, there is a historical precedent to this method of phosphorus recovery. The element was first discovered in human urine by alchemist Hennig Brand. In 1669, Brand accidentally discovered phosphorus while searching for the mythical 'philosopher's stone'. His method reportedly began with almost 6000 litres of urine (Figure 2.4.13).

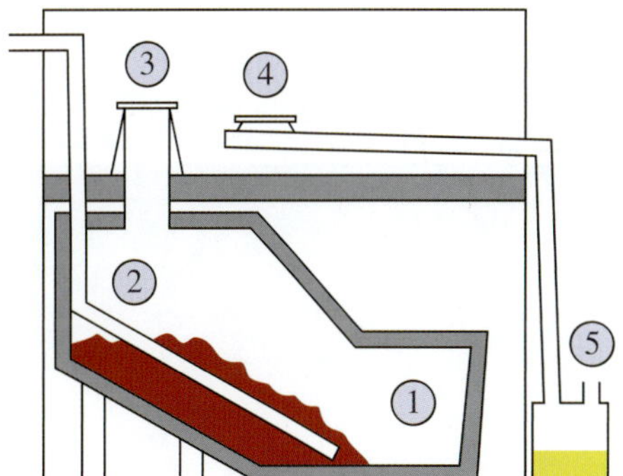

FIGURE 2.4.12 A diagram of a composting toilet with urine diversion and dehydration: 1. solid waste compartment; 2. ventilation pipe; 3. toilet; 4. urinal; 5. urine collection and dehydration

FIGURE 2.4.13 This painting, *The Alchemist in Search of the Philosopher's Stone* depicts the discovery of phosphorus, by alchemist Hennig Brand in 1669, in an experiment that reportedly began with almost 6000 litres of urine.

2.4 Review

OA ✓✓

SUMMARY

- The periodic table is a tool for organising elements according to their chemical and physical properties.
- The elements of the modern periodic table are arranged in order of increasing atomic number.
- Columns in the periodic table are known as groups and are numbered 1–18.
- The main group elements are in groups 1, 2 and 13–18 in the periodic table.
- The number of valence electrons a for main group element can be determined by the group in which it is located.
- Elements in between the main group elements (groups 3–12) are known as transition metals.
- Rows in the periodic table are known as periods and are numbered 1–7.
- The number of occupied electron shells of an atom of an element is equal to the number of the element's period.
- The periodic table has four main blocks of elements; the elements in each block have the same type of subshell (*s*, *p*, *d* or *f*) as their highest energy subshell.
- Our use of many elements is currently non-sustainable, which has led to them being described as critical elements.
- To ensure critical elements are available for future generations, methods of recycling and recovery must be developed.

KEY QUESTIONS

Knowledge and understanding

1 In the periodic table, group is to column as period is to what?

2 Define the terms 'period', 'group' and 'block'.

3 What is the name given to elements in groups 1, 2 and 13–18?

4 State the IUPAC name for the following groups, with reference to the periodic table.

a 1
b 2
c 17
d 18

5 How many valence electrons are in atoms of elements found in the following groups?

a 2
b 13
c 15
d 18

6 What is the subshell electronic configuration of an atom of an element in period 4 and group 1?

Analysis

Use the periodic table in Figure 2.4.2 on page 80 to answer question **7**.

7 **a** In which group of the periodic table will you find the following?

i B
ii Cl
iii Na
iv Ar
v Si
vi Pb

b Determine the name and symbol of the following elements. In addition, write the electronic configuration of each element.

i second element in group 14
ii second element in period 2
iii element that is in group 18 and period 3

c Identify the period of the periodic table in which each of the following elements belongs.

i K
ii F
iii He
iv H
v U
vi P

8 Explain the term 'critical element'. Give three examples of elements that are considered critical elements.

2.5 Trends in the periodic table

The periodic table does not just provide information about an element's electronic configuration. It can also be used as a tool for summarising the relative properties of elements and explaining the trends observed in those properties.

You have already seen how the group number of an element indicates how many valence electrons an atom of that element has. The period indicates how many electron shells are occupied in an atom of an element. Properties such as atomic radius, **electronegativity** (the electron-attracting power of an atom) and **first ionisation energy** (the energy required to remove one electron from an atom) also show common trends in the periodic table.

Periodic trends were observed by Dimitri Mendeleev and formed the basis of the table of the elements that he first published in 1869. Mendeleev described the way the properties of the elements vary as the **periodic law**.

TRENDS IN ELECTRONIC CONFIGURATION

To understand the reason for the periodicity (repeating pattern) of element properties, consider the electronic configurations of the alkali metals (group 1). The elements in this group (lithium, sodium, potassium, rubidium and caesium) are all relatively soft metals and are highly reactive with water and oxygen.

Li $1s^2 2s^1$

Na $1s^2 2s^2 2p^6 3s^1$

K $1s^2 2s^2 2p^6 3s^2 3p^6 4s^1$

Rb $1s^2 2s^2 2p^6 3s^2 3p^6 3d^{10} 4s^2 4p^6 5s^1$

Cs $1s^2 2s^2 2p^6 3s^2 3p^6 3d^{10} 4s^2 4p^6 4d^{10} 5s^2 5p^6 6s^1$

These elements have similar valence shell electronic configurations—all have one electron in an *s*-subshell. This similarity in a group's arrangement of electrons gives elements similar properties and is responsible for the periodicity of element properties.

You can also see that the number of electron shells increases moving down the group. The increase in electron shells means that the valence electrons are in higher energy subshells and have a weaker attraction to the nucleus. The decrease in the attractive force between the valence electrons and the nucleus as you move down a group causes trends in properties to be observed within a group.

EFFECTIVE NUCLEAR CHARGE

The **effective nuclear charge** (sometimes referred to as core charge) of an atom is a measure of the attractive force felt by the valence shell electrons towards the nucleus. Effective nuclear charge can be used to predict the properties of elements and explain trends observed across a period in the periodic table.

Consider an atom of lithium, which has an atomic number of three. It has three protons in its nucleus, two electrons in the first shell and one electron in the second shell (Figure 2.5.1).

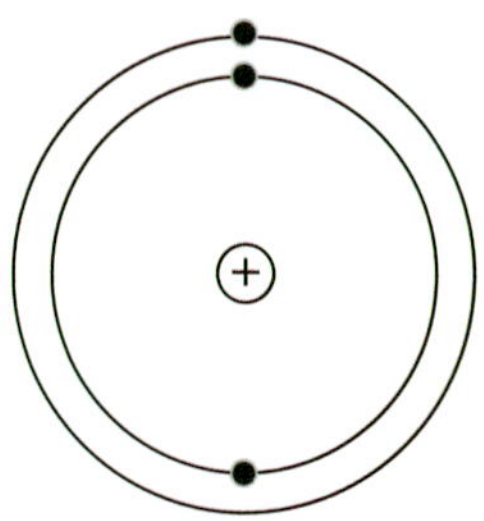

FIGURE 2.5.1 A lithium atom with one valence electron and two electrons in the inner shell. The atom has an effective nuclear charge of +1.

The valence shell electron is attracted to the three positive charges in the nucleus. This electron is also repelled by the two electrons in the inner shell. The electrons in the inner shell shield the valence shell electron from the attraction of the nucleus. The valence shell electron is effectively attracted to the nucleus as if there were a +1 nuclear charge. This atom is therefore said to have an effective nuclear charge of +1.

Similarly, an atom of chlorine has 17 protons and seven valence shell electrons; the number of electrons in the inner shells is 10. The effective nuclear charge of a chlorine atom is $17 - 10 = +7$.

Effective nuclear charge = number of protons in the nucleus − number of total inner-shell electrons

Worked example 2.5.1

EFFECTIVE NUCLEAR CHARGE

Determine the effective nuclear charge of an atom of aluminium.

Thinking	Working
Determine the number of electrons in an atom of the element, using the periodic table as a reference.	The atomic number of aluminium is 13. Therefore, an atom of aluminium has 13 protons and 13 electrons.
Use the number of electrons to determine the electronic configuration.	With 13 electrons the electronic configuration is $1s^22s^22p^63s^23p^1$.
Determine the effective nuclear charge. Effective nuclear charge = number of protons – number of inner-shell electrons	The third shell is the valence shell in this atom. There are 10 inner-shell electrons, which in this atom are electrons in the first and second shell. Effective nuclear charge = 13 – 10 = +3

Worked example: Try yourself 2.5.1

EFFECTIVE NUCLEAR CHARGE

Determine the effective nuclear charge of an atom of fluorine.

TABLE 2.5.1 Effective nuclear charges of main group elements

Group	Effective nuclear charge
1	+1
2	+2
13	+3
14	+4
15	+5
16	+6
17	+7
18*	+8

*Helium has an effective nuclear charge of +2.

The effective nuclear charge experienced by the valence shell electrons in atoms of elements increases from left to right across a period, as you have seen for sodium and chlorine. The effective nuclear charge of an atom of a main group element is equivalent to the number of valence electrons in the atom, as summarised in the Table 2.5.1.

Table 2.5.2 summarises how the attraction between the nucleus and valence electrons changes in the periodic table.

TABLE 2.5.2 The changes in attraction between the nucleus and valence electrons within groups and periods of the periodic table

	Trend in effective nuclear charge	Trend in attraction between the nucleus and valence electrons
Down a group	remains constant	Effective nuclear charge stays constant down a group, but the valence electrons are held less strongly as they are further from the nucleus because there are more shells in the atom.
Left to right across a period	increases	The valence electrons are more attracted to the nucleus as the effective nuclear charge increases.

All of the observed trends in the periodic table can be related to these changes in attraction between the valence electrons and the nucleus.

ELECTRONEGATIVITY

Electronegativity is the ability of an atom to attract electrons towards itself when forming a chemical bond. As the positive pull in an atom comes from the nucleus, when the effective nuclear charge is greater, the electronegativity increases. Figure 2.5.2 on the following page shows the electronegativity of many of the main group elements. Electronegativity values for the noble gases (group 18) are not listed because the elements have a stable outer shell and do not readily form bonds with other atoms.

Electronegativity is the ability of an atom to attract electrons towards itself when forming a chemical bond. It is a measure of how strongly an atom pulls on the electrons of nearby atoms.

Electronegativity increases across a period. →

Electronegativity decreases down a group. ↓

1	2	13	14	15	16	17
Li 1.0	Be 1.6	B 2.0	C 2.6	N 3.0	O 3.4	F 4.0
Na 0.9	Mg 1.3	Al 1.6	Si 1.9	P 2.2	S 2.6	Cl 3.2
K 0.8	Ca 1.0	Ga 1.8	Ge 2.0	As 2.2	Se 2.6	Br 3.0
Rb 0.8	Sr 1.0	In 1.8	Sn 2.0	Sb 2.1	Te 2.1	I 2.7
Cs 0.8	Ba 0.9	Tl 2.0	Pb 2.3	Bi 2.0	Po 2.0	At 2.2
Fr 0.7	Ra 0.9					

FIGURE 2.5.2 The electronegativity of elements generally decreases down a group and increases across a period, from left to right.

The trends observed in the electronegativity of the elements are summarised in Table 2.5.3.

TABLE 2.5.3 Trends in electronegativity in groups and periods of the periodic table

	Trend in electronegativity	Explanation
Down a group	decreases	The effective nuclear charge stays constant and the number of shells increases down a group. Therefore, valence electrons are less strongly attracted to the nucleus as they are further from the nucleus. As a result, electronegativity decreases.
Left to right across a period	increases	The number of occupied shells in the atoms remains constant but the effective nuclear charge increases across a period. Therefore, the valence electrons become more strongly attracted to the nucleus. As a result, electronegativity increases.

> Atomic radius decreases across a period. This is because the effective nuclear charge increases across a period, so the valence shell electrons are pulled in to the nucleus.

ATOMIC RADIUS

Atomic radius is a measurement used for the size of atoms. It can be regarded as the distance from the nucleus to the valence shell electrons. It is usually measured by halving the distance between the nuclei of two atoms of the same element that are bonded together. Figure 2.5.3 depicts the atomic radii of many of the main group elements. As you can see, the atomic radius decreases as you move across a period. This is because as the effective nuclear charge increases, the valence shell electrons are pulled in more tightly towards the nucleus.

Table 2.5.4 explains the trends in atomic radius in the periodic table.

TABLE 2.5.4 Trends in atomic radii in the periodic table

	Trend in atomic radius	Explanation
Down a group	increases	Effective nuclear charge stays constant and the number of shells increases as you move down a group. As a result, atomic radius increases.
Left to right across a period	decreases	As you move across a period, the number of occupied shells in the atoms remains constant but the effective nuclear charge increases. The valence electrons become more strongly attracted to the nucleus, so atomic radius decreases across a period.

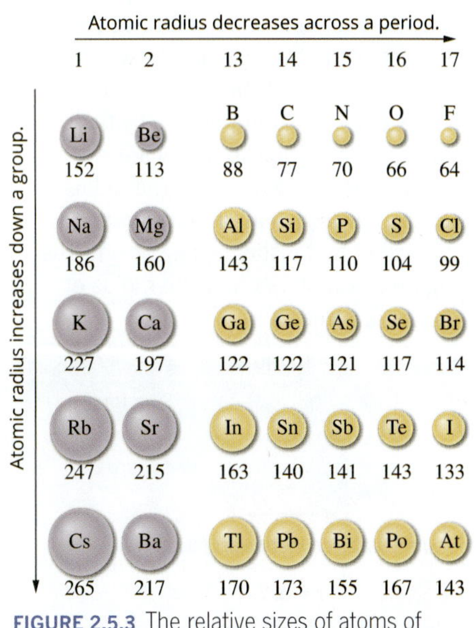

FIGURE 2.5.3 The relative sizes of atoms of selected main group elements. The atomic radii are given in picometres (pm). A picometre is 10^{-12} m.

FIRST IONISATION ENERGY

When an element is heated, its electrons can move to higher energy shells. If an atom is given sufficient energy, an electron can be completely removed from the atom. If this occurs, the atom will now have one less electron than the number of protons in the nucleus, and becomes a positively charged ion.

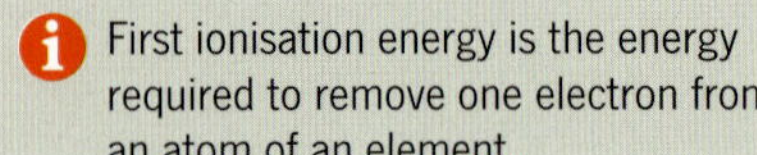

First ionisation energy is the energy required to remove one electron from an atom of an element.

First ionisation energy is the energy required to remove one electron from an atom of an element in the gas phase. For example, the first ionisation energy of sodium is 494 kJ per mole of sodium atoms. (A mole is a way of counting atoms. You will learn how to use the mole in Chapter 8.)

Figure 2.5.4 shows the first ionisation energies of most main group elements.

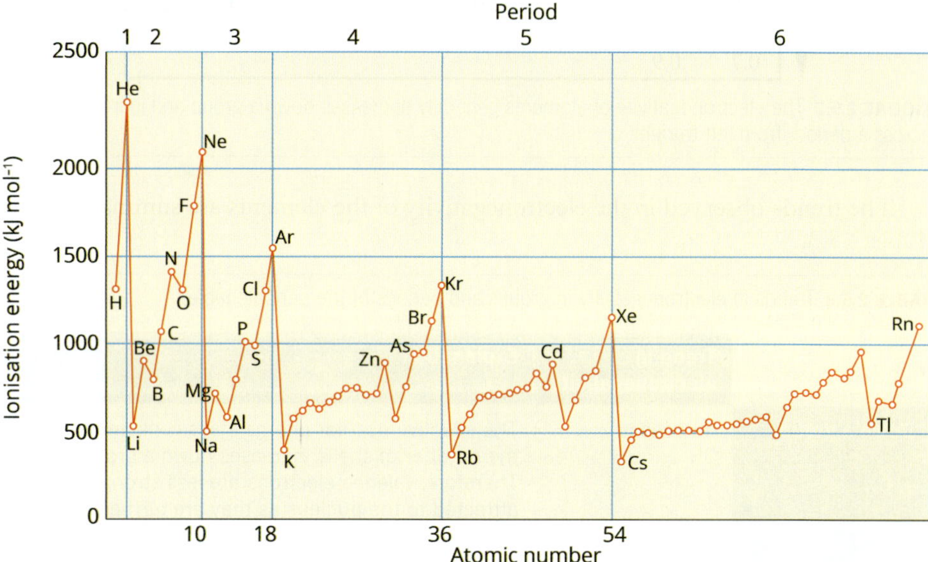

FIGURE 2.5.4 The first ionisation energy generally increases from left to right across a period.

The magnitude of the first ionisation energy reflects how strongly the valence electrons are attracted to the nucleus of the atoms. The more strongly the valence electrons are attracted to the nucleus, the more energy is required to remove one of them from the atom and the higher the first ionisation energy. When this data is compared to position on the periodic table, the trends described in Table 2.5.5 emerge.

TABLE 2.5.5 The trend in ionisation energies in groups and periods of the periodic table

	Trend in first ionisation energy	Explanation
Down a group	decreases	Effective nuclear charge stays constant and the number of shells increases down a group. Therefore, the valence electrons are less attracted to the nucleus as they are further from the nucleus. As a result, the energy required to overcome the attraction between the nucleus and the valence electron is less, and the first ionisation energy decreases down a group.
Left to right across a period	increases	Effective nuclear charge increases and the number of occupied shells remains constant across a period. As a result, the valence electrons become more strongly attracted to the nucleus, and more energy is required to remove an electron. Therefore, first ionisation energy increases across a period.

The *s*- and *p*-block elements of the periodic table follow these patterns. As the effective nuclear charge increases across a period, so too does the ionisation energy. As atomic radius increases down a group (due to the additional electron shell), the electrons are further from the nucleus. Therefore, the valence electrons in elements lower in a group can be removed more easily, meaning the ionisation energy decreases.

CHEMFILE

Measuring atoms

Atoms do not have sharply defined boundaries and so it is not possible to measure their radii directly. One method of obtaining atomic radius, and therefore an indication of atomic size, is to measure the distance between nuclei of atoms of the same element in molecules. For example, in a hydrogen molecule (H_2) the two nuclei are 74 picometres (pm) apart. The radius of each hydrogen atom is assumed to be half of that distance, i.e. 37 pm. The figure below shows how the radius of atoms is determined.

r = Radius

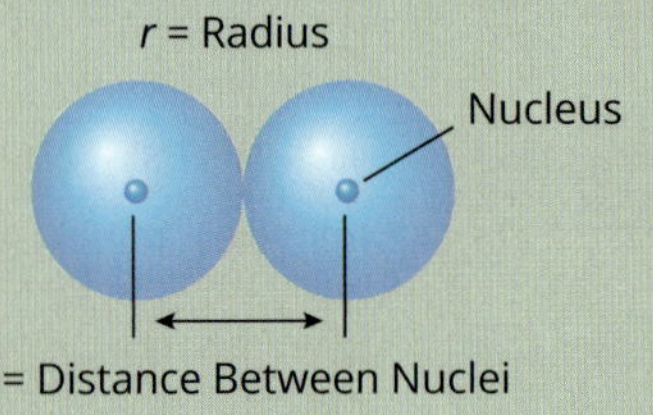

d = Distance Between Nuclei

$r = d/2$

The radius of an atom can be determined by measuring the distance between two nuclei of the same element in a molecule. Half this distance will be the atomic radius.

TRENDS IN METALS AND NON-METALS

One of the most useful distinctions that can be made between elements in the periodic table is the difference between metals, metalloids and non-metals, as can be seen in Figure 2.5.5. **Metalloids** share properties of both metals and non-metals.

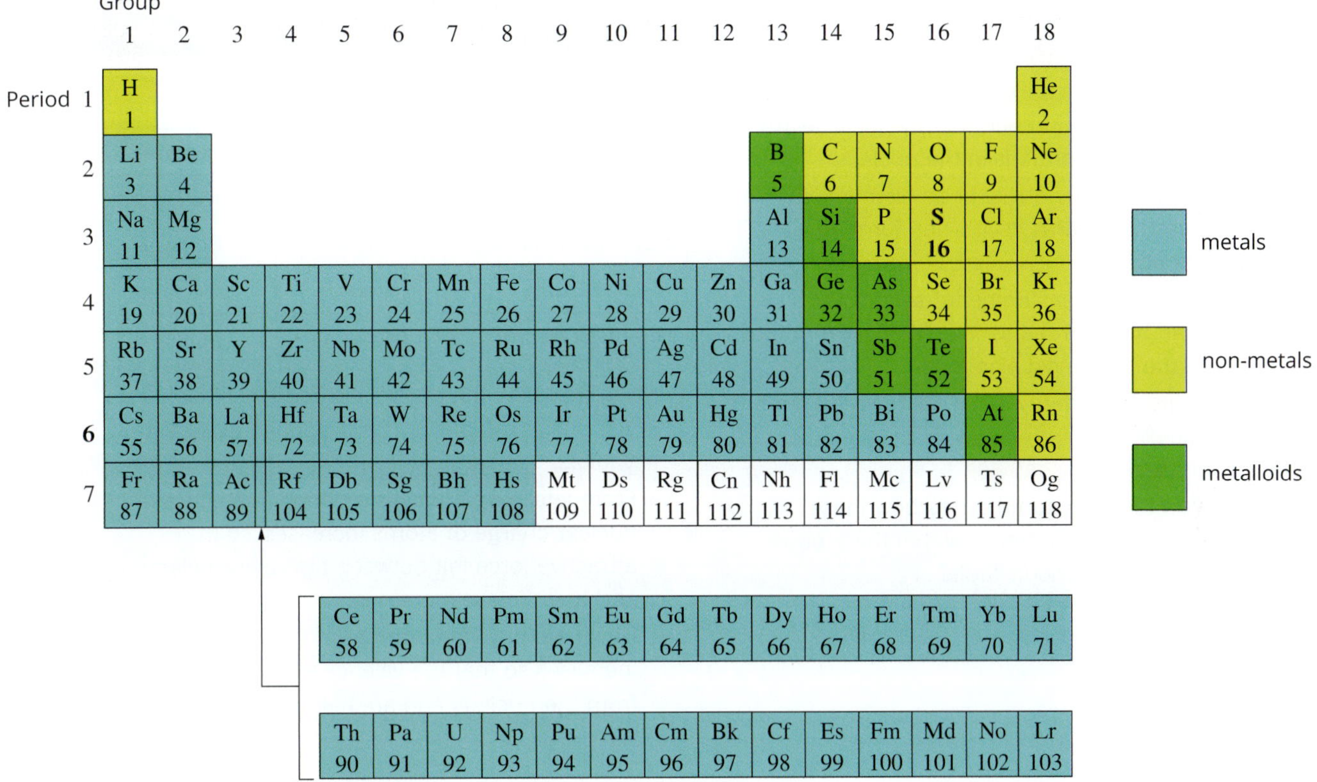

FIGURE 2.5.5 Periodic table highlighting the metals, metalloids and non-metals

As you will discover in the coming chapters, an understanding of whether an element is a metal or non-metal helps you to understand the way it will bond with other elements. The reactivity of both metals and non-metals is also something that follows a trend in the periodic table.

In general, the properties of metals arise from the fact that they readily lose electrons. As you go down a group, the valence shell electrons are further away from the pull of the nucleus and are therefore lost more easily. This means that as you go down a group, metals become more reactive.

Across a period, the opposite occurs, as the effective nuclear charge increases. This means the most reactive metals are in group 1.

Consequently, the most reactive metal in the periodic table is in the bottom left corner: francium.

In contrast, the properties of non-metals arise from the fact that they readily gain electrons. In general, non-metals become less reactive as you go down a group, as the valence electrons become further away from the nucleus. Non-metals become more reactive across a period, as the effective nuclear charge increases. The exception to this are the noble gases, which are inert.

Consequently, the most reactive non-metal is in the top right of the periodic table (excluding the noble gases): fluorine.

Knowing whether an element is a metal, non-metal or metalloid can help you to predict the way it will react and bond with other elements.

2.5 Review

OA ✓✓

SUMMARY

- The effective nuclear charge of an atom is a measure of the attractive force felt by the valence electrons towards the nucleus.
- The effective nuclear charge is calculated by subtracting the total number of inner-shell electrons from the number of protons in the nucleus.
- Electronegativity is the ability of an element to attract electrons towards itself when forming a chemical bond.
- Atomic radius is a measurement used for the size of atoms. It can be regarded as the distance from the nucleus to the outermost electrons.
- The first ionisation energy is the energy required to remove one electron from an atom of an element in the gas phase.
- The table on the right summarises how these properties have specific trends within the groups and periods of the periodic table.

Property	Down a group	Across a period (left to right)
effective nuclear charge	no change	increases
electronegativity	decreases	increases
atomic radius	increases	decreases
first ionisation energy	decreases	increases
metal reactivity	increases	decreases
non-metal reactivity	decreases	increases

- Many trends in the physical properties of elements in the periodic table can be explained using two key ideas.
  - From left to right across a period, the effective nuclear charge of atoms increases, so the attractive force felt between the valence electrons and the nucleus increases.
  - Down a group, the number of shells in an atom increases so that the valence electrons are further from the nucleus and are held less strongly.

KEY QUESTIONS

Knowledge and understanding

1 Define the term 'effective nuclear charge' of an atom and determine the effective nuclear charge of an atom of carbon.

2 State the relationship between electronegativity and effective nuclear charge.

3 Identify whether the following elements are metals, metalloids or non-metals using the periodic table.
- **a** K
- **b** N
- **c** Ge
- **d** Cu
- **e** Sn
- **f** I

Analysis

4 Figure 2.5.2, on page 91, gives electronegativity values for the elements in groups 1, 2 and 13–17 of the periodic table.
- **a** Give the name and symbol of the element that has the:
  - **i** highest electronegativity
  - **ii** lowest electronegativity.
- **b** In which group do you see the following?
  - **i** greatest change in electronegativity as you go down the group
  - **ii** smallest change in electronegativity as you go down the group.

5 Based on your understanding of the trends in the periodic table, sort the following in order of increasing atomic radius:
N, B, Ga, Al, Cl

6 Explain why the number of subatomic particles in an atom increases across a period but the size of the atom decreases.

7 By referring to the periodic table, organise the following elements in order of increasing first ionisation energies:
Na, He, Al, K, S, Ca and P

8 Suggest a reason why caesium is more reactive with water than lithium.

Chapter review

KEY TERMS

alkali metals
atom
atomic number
atomic radius
atomic theory of matter
block (periodic table)
Bohr model
chemical symbol
compound
condensed electronic configuration
conflict element
critical element
effective nuclear charge
electron
electronic configuration
electron shell
electronegativity
electrostatic attraction
element
emission spectrum
endangered element
energy level
excited state
first ionisation energy
flame test
ground state
group (periodic table)
halogen
ion
isotope
main group element
mass number
matter
metalloid
model
neutron
noble gas notation
noble gas
nucleon
nucleus
nuclide notation
orbital
period
periodic law
periodic table
periodicity
proton
quantised
quantum mechanics
radioactive
Schrödinger model
subatomic particle
subshell
transition metal
valence electrons
valence shell

REVIEW QUESTIONS

Knowledge and understanding

1 The maximum number of electrons in the first three shells of an atom is:

A 8,8,18
B 2,8,12
C 2,8,8
D 2,8,18

2 Isotopes of an element contain different numbers of (more than one answer possible):

A protons
B neutrons
C nucleons
D electrons

3 Of the electronic configurations below, which one represents an atom in an excited state?

A $1s^2 2s^2 2p^6 3s^1$
B $1s^2 2s^2 2p^6 3s^2 3p^6$
C $1s^2 2s^2 2p^6 3s^2 3p^5 4s^2$
D $1s^2 2s^2 2p^3$

4 Select the response that best describes how the elements in the periodic table are arranged.

A by the number of neutrons in their nucleus
B by the number of protons in their nucleus
C by increasing mass number
D by increasing atomic mass

5 Identify the most reactive non-metal from the following elements.

A caesium
B chlorine
C fluorine
D oxygen

6 An atom of chromium can be represented by the symbol $^{52}_{24}Cr$.

a State its atomic number and mass number.
b Determine the number of electrons, protons and neutrons in the chromium atom.

7 What is the name of the element that has an electronic shell configuration of 2,8,2?

8 Determine the Schrödinger model of electronic configuration that corresponds to the shell model electronic configuration 2,8,6.

9 Use the periodic table to determine the period and block of the following elements.

a Hydrogen
b Carbon
c Phosphorus
d Copper
e Uranium

10 Select the appropriate word from the following options to complete the sentences below: increase, increases, decrease, decreases

a The force of attraction between the nucleus and valence electrons __________ in a period from left to right.

b Atomic radii of elements __________ in a period from left to right.

c Atomic radii of elements __________ in a group from top to bottom.

d Metallic character of elements __________ from top to bottom in a group.

Application and analysis

11 Two atoms both have 20 neutrons in their nucleus. The first has 19 protons and the other has 20 protons. Are they isotopes? Why or why not?

12 Analyse the following list of atoms.

$^{27}_{13}A$ $^{40}_{20}B$ $^{32}_{16}C$ $^{14}_{7}D$ $^{8}_{3}E$ $^{15}_{7}F$ $^{19}_{9}G$ $^{20}_{9}H$ $^{4}_{2}I$

a Identify which pairs of atoms are isotopes of the same element.

b Identify which atoms have equal numbers of protons and neutrons in the nucleus.

c Identify which is an isotope of sulfur.

d Determine which has one more electron than a magnesium atom.

e Identify how many different elements are shown.

13 Determine which shell the 30th electron of an atom would go into according to the rules for determining the electronic shell configuration of an atom.

14 Determine the subshell electronic configuration of the following ions.

a Na^{+}

b S^{2-}

c Zn^{2+}

15 Explain how the Schrödinger model of the atom explains why the fourth electron shell begins filling before the third shell is completely filled.

16 a Write the subshell electronic configuration of nitrogen.

b What period and group does nitrogen belong to in the periodic table?

c How many valence electrons does nitrogen have?

d What is nitrogen's effective nuclear charge?

17 Determine the period and group of the elements with the following electronic configurations.

a $1s^22s^2$

b $1s^22s^22p^63s^23p^2$

c $1s^22s^22p^63s^23p^63d^{10}4s^24p^1$

18 Consider the elements in period 2 of the periodic table: lithium, beryllium, boron, carbon, nitrogen, oxygen, fluorine. Describe the changes that occur across the period. Consider:

a atom radius

b metallic character

c electronegativity.

19 Explain each of the following briefly.

a The atomic radius of chlorine is smaller than that of sodium.

b The first ionisation energy of fluorine is higher than that of lithium.

c The reactivity of beryllium is less than that of barium.

d There are two groups in the *s*-block of the periodic table.

20 Identify the correct chemical symbol for each of the following.

a the element that is in group 2 and period 4

b a noble gas with exactly three occupied electron shells

c an element from group 14 that is a non-metal

d an element that has exactly three occupied electron shells and is in the *s*-block

e the element in period 2 that has the largest atomic radius

f the element in group 15 that has the highest first ionisation energy

g the element in period 2 with the highest electronegativity

21 Identify the following 23 elements from the clues given, then sort them into periodic table groups.

- third element in group 15; second element in period 5; sixth element in group 17
- elements with the following symbols: C, Ca, Ga, Sb, Te, Rn
- elements with the following atomic numbers: 3,12,13,19,36,53,56
- the lightest noble gas; the third heaviest halogen; the heaviest alkali metal
- elements with the following electronic configurations:
  - $1s^22s^22p^63s^23p^4$
  - $1s^22s^22p^63s^23p^63d^{10}4s^24p^2$
  - $1s^22s^22p^63s^23p^63d^{10}4s^24p^64d^{10}5s^25p^66s^1$
  - $1s^22s^22p^1$

22 Compare the following pairs of elements to predict which requires more energy to remove a valence electron.

a phosphorus or magnesium

b fluorine or iodine

23 Predict which of the following pairs of elements would react together most vigorously. Explain your answer.
Li and Cl_2; Li and Br_2; K and Cl_2; K and Br_2

24 Identify the elements that meet the criteria below using the following representation of the periodic table. (Use the letters given.)

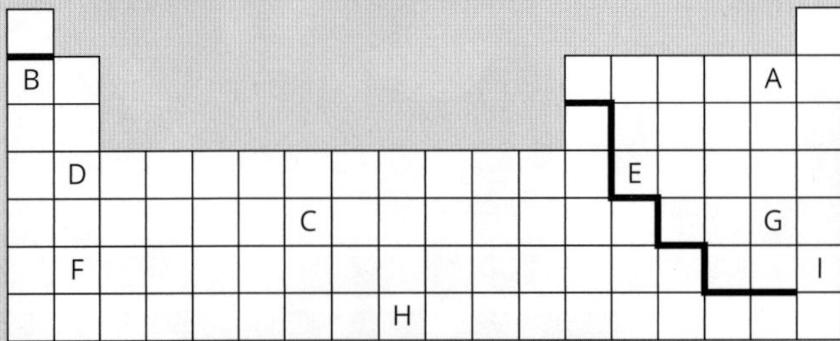

a have the same number of valence electrons

b are in the same period

c has smallest atomic radius

d has smallest ionic radius

e has lowest electronegativity

f is a noble gas

g is classified as a non-metal

h is classified as a metalloid

OA ✓✓

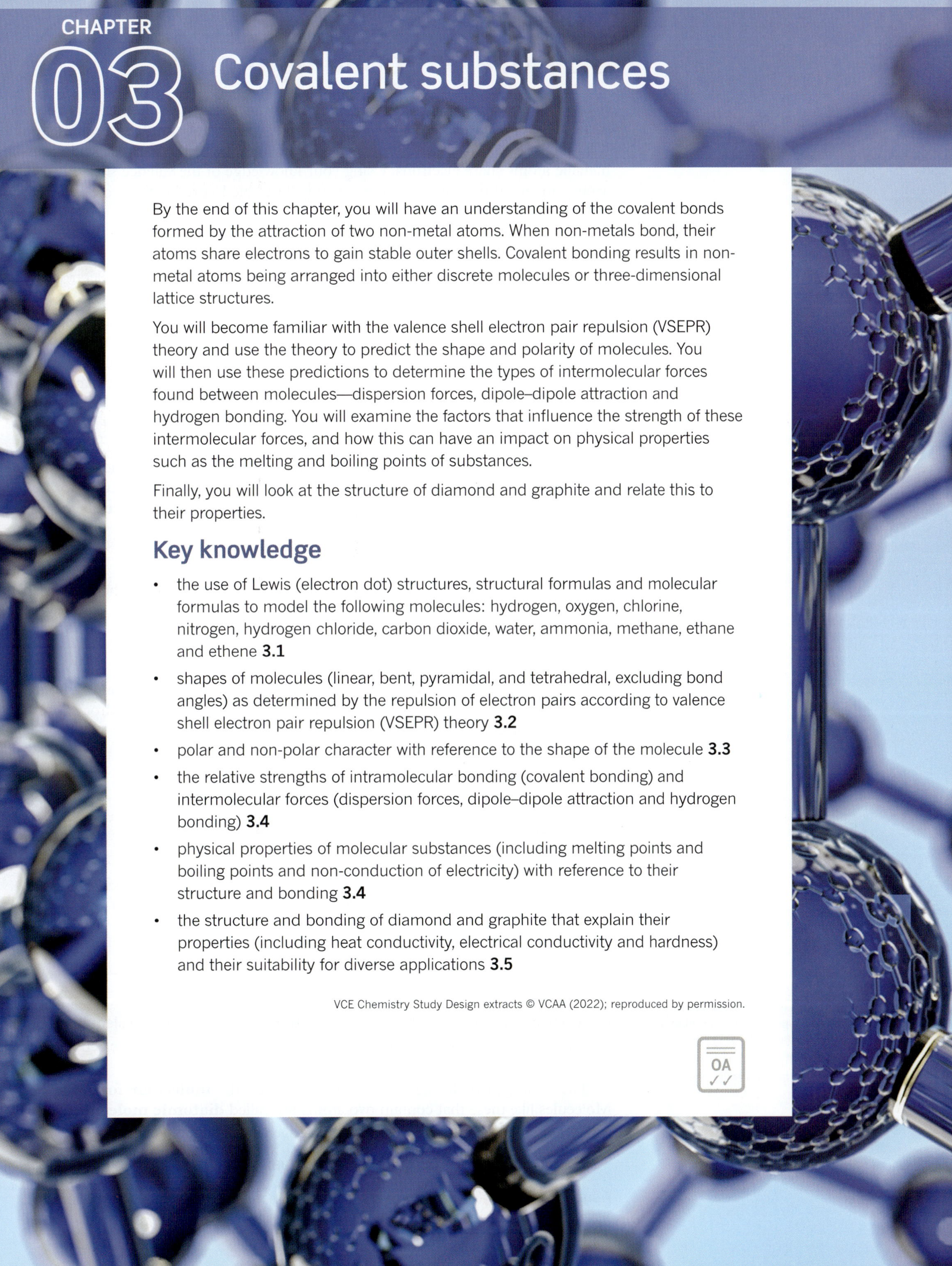

CHAPTER

03 Covalent substances

By the end of this chapter, you will have an understanding of the covalent bonds formed by the attraction of two non-metal atoms. When non-metals bond, their atoms share electrons to gain stable outer shells. Covalent bonding results in non-metal atoms being arranged into either discrete molecules or three-dimensional lattice structures.

You will become familiar with the valence shell electron pair repulsion (VSEPR) theory and use the theory to predict the shape and polarity of molecules. You will then use these predictions to determine the types of intermolecular forces found between molecules—dispersion forces, dipole–dipole attraction and hydrogen bonding. You will examine the factors that influence the strength of these intermolecular forces, and how this can have an impact on physical properties such as the melting and boiling points of substances.

Finally, you will look at the structure of diamond and graphite and relate this to their properties.

Key knowledge

- the use of Lewis (electron dot) structures, structural formulas and molecular formulas to model the following molecules: hydrogen, oxygen, chlorine, nitrogen, hydrogen chloride, carbon dioxide, water, ammonia, methane, ethane and ethene **3.1**
- shapes of molecules (linear, bent, pyramidal, and tetrahedral, excluding bond angles) as determined by the repulsion of electron pairs according to valence shell electron pair repulsion (VSEPR) theory **3.2**
- polar and non-polar character with reference to the shape of the molecule **3.3**
- the relative strengths of intramolecular bonding (covalent bonding) and intermolecular forces (dispersion forces, dipole–dipole attraction and hydrogen bonding) **3.4**
- physical properties of molecular substances (including melting points and boiling points and non-conduction of electricity) with reference to their structure and bonding **3.4**
- the structure and bonding of diamond and graphite that explain their properties (including heat conductivity, electrical conductivity and hardness) and their suitability for diverse applications **3.5**

OA
✓✓

3.1 Covalent bonding model

In this section, you will look at the chemical bonding that occurs when atoms of non-metals combine with each other. By examining a series of simple molecules, you will be introduced to the concept of a covalent bond, which is formed when non-metallic atoms share electrons. Using your knowledge of the valence shell electron arrangements of non-metallic atoms, you will be able to predict the molecules that different elements can form.

COVALENT BONDS

Many atoms become more stable if they obtain an outer shell of eight electrons by combining with other atoms (the **octet rule**).

Commonly, when atoms of non-metals combine, electrons are shared so that each atom has eight electrons in its outer shell. **Molecules** formed in this way are more stable than the separate atoms. A molecule is a discrete (individually separate) group of atoms of known formula, bonded together. Molecular substances are neutral overall. As they contain no free moving charged particles, they are unable to conduct electricity.

Non-metallic atoms have a relatively high number of electrons in their outer shells and they tend to share rather than to transfer electrons. Covalent bonding occurs when electrons are shared between atoms. The **covalent bonds** formed between atoms within a molecule are generally called **intramolecular bonds** (bonds within a molecule).

Single covalent bonds

When atoms share two electrons, one from each atom, the covalent bond formed is called a **single covalent bond**. Two examples of substances that contain single bonds are hydrogen and chlorine.

Example 1: Hydrogen

Hydrogen atoms have one electron. The outer shell for a hydrogen atom can hold a maximum of two electrons. A hydrogen atom can bond to another hydrogen atom to form a molecule of H_2, as shown in Figure 3.1.1.

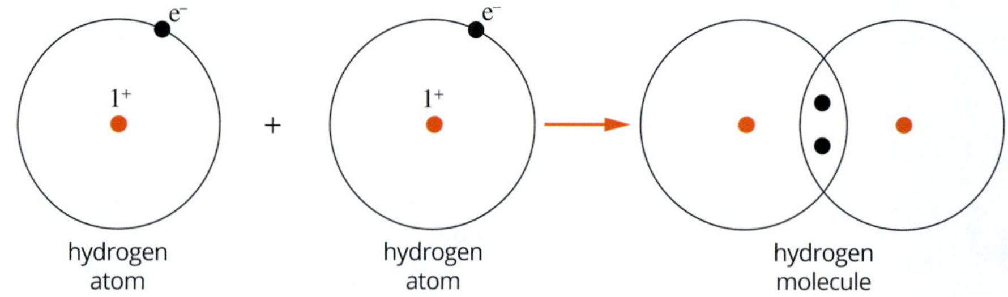

FIGURE 3.1.1 A covalent bond is formed when two hydrogen atoms share two electrons, one from each atom.

The molecular formula indicates the number and type of atoms found in a molecule. For example, H_2O indicates there are two hydrogen and one oxygen atoms in a molecule of water.

In the molecule that is formed:

- two hydrogen atoms each share one electron to form a single covalent bond
- the atoms of hydrogen are now strongly bonded together by two electrons (an electron pair) in their outer shells.

The hydrogen molecule can be represented by the **molecular formula**, H_2. Molecules like these that contain two atoms are called **diatomic molecules**.

Two alternative ways of representing a hydrogen molecule are shown in Figure 3.1.2.

(a) H—H **(b)** H ×• H

FIGURE 3.1.2 The single covalent bond in a hydrogen molecule can be represented by (a) a straight line or (b) a dot and cross.

In a hydrogen molecule, the electron of each atom is attracted to the proton within the neighbouring atom, as well as to its own proton. This means the two electrons will spend most of their time between the two nuclei instead of orbiting their own nuclei. Even though the protons in the two nuclei still repel each other (remember, like charges repel), the electrostatic attraction to the electrons (which are closer) holds the molecule together.

Hydrogen is an example of a covalent molecular substance.

Example 2: Chlorine

A chlorine atom has an electronic configuration of 2,8,7. It requires one more electron to achieve eight electrons in its outer shell.

One chlorine atom can share an electron with another chlorine atom to form a molecule of chlorine with a single covalent bond. As a result, both atoms gain outer shells of eight electrons, as shown in Figure 3.1.3. This is an example of the application of the octet rule.

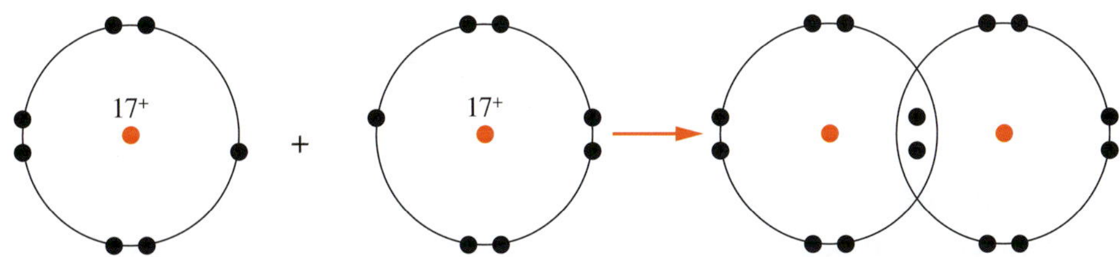

FIGURE 3.1.3 Two chlorine atoms share one electron each to form a chlorine molecule. Note that only outer-shell electrons are shown in these diagrams.

Lewis structures

Chemists often use **Lewis structures** (also known as electron dot structures) to represent molecules.

Lewis structures show the valence shell electrons of an atom, as only these electrons are involved in bonding.

Lewis structures also allow you to distinguish between bonding electrons and non-bonding electrons. A chlorine molecule has one pair of bonding electrons. The outer-shell electrons that are not involved in bonding are called the **non-bonding electrons**. Each chlorine atom has six non-bonding electrons, grouped into three pairs. Pairs of non-bonding electrons are also known as lone pairs.

Figure 3.1.4 shows examples of the Lewis structure for a molecule of chlorine. Electrons can be represented by dots, crosses, lines or a combination of all three.

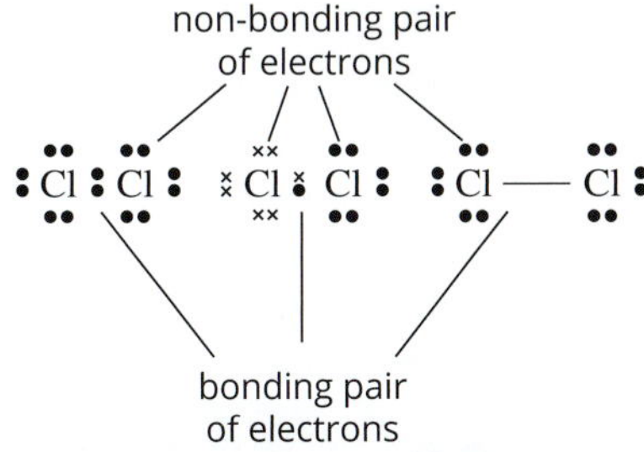

FIGURE 3.1.4 Three examples of Lewis structures for a chlorine molecule showing the bonding and non-bonding pairs of electrons

Double covalent bonds

In a **double covalent bond**, two pairs of electrons are shared between the atoms.

The oxygen molecule contains a double covalent bond. The electronic configuration of an oxygen atom is 2,6. Each oxygen atom requires two electrons to gain a stable outer shell containing eight electrons. Therefore, when one oxygen atom bonds to another, each atom shares two of its electrons.

CHEMFILE

Hydrogen airships

Hydrogen has a low density. This was once thought to make it suitable for use in airships. Zeppelins were a type of rigid airship that were used as a mode of transport during the early 1900s. However, their popularity as a way of travel ended after the hydrogen gas in the Zeppelin *Hindenburg* (see figure below) caught fire in 1937, killing many on board.

The German passenger Zeppelin *Hindenburg* exploded during its attempt to dock at the Lakehurst Naval Air Engineering Station in the United States.

In Lewis structures, electrons can be represented by dots, crosses, lines or a combination of all three.

As you can see in Figure 3.1.5, each oxygen atom in the molecule now has eight valence electrons. Four of these are bonding electrons and four are non-bonding electrons.

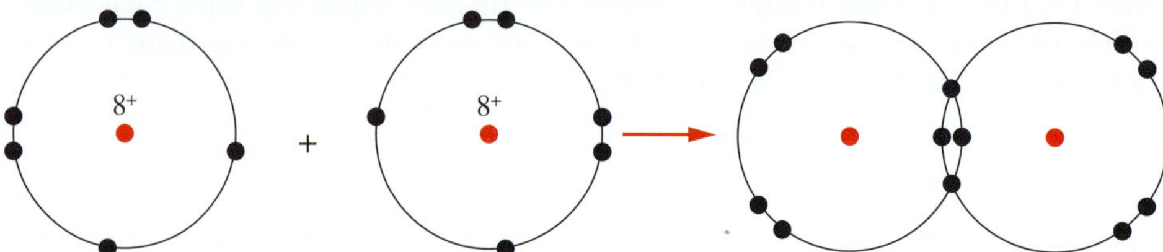

FIGURE 3.1.5 In oxygen molecules, each oxygen atom contributes two electrons to the bond between the atoms.

In Figure 3.1.6, you can see the Lewis structure of an oxygen molecule.

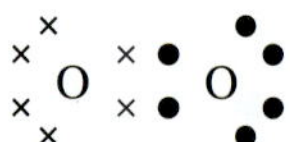

FIGURE 3.1.6 The Lewis structure shows that O_2 has a double covalent bond. Four electrons are shared and each oxygen has two non-bonding electron pairs.

Triple covalent bonds

A **triple covalent bond** occurs when three electron pairs are shared between two atoms. The nitrogen molecule contains a triple covalent bond. The electronic configuration of nitrogen is 2,5. A nitrogen atom requires three electrons to achieve eight electrons in its outer shell. When it bonds to another nitrogen atom, each atom contributes three electrons to the bond that forms, as shown in Figure 3.1.7. The Lewis structure of the nitrogen molecule is shown in Figure 3.1.8.

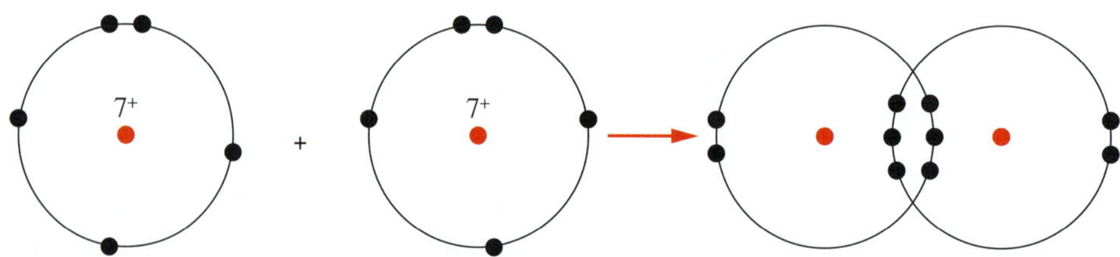

FIGURE 3.1.7 Each nitrogen atoms contributes three electrons to form a triple covalent bond in a molecule of N_2.

MOLECULAR COMPOUNDS

A diatomic molecule contains two atoms. The molecules discussed so far have been diatomic molecules that contain atoms of the same element.

Covalent bonds can also form between atoms of different elements. Hydrogen chloride (HCl) is a simple example (Figure 3.1.9). A hydrogen atom requires one electron to gain a stable outer shell, as does a chlorine atom. They can share an electron each and form a single covalent bond.

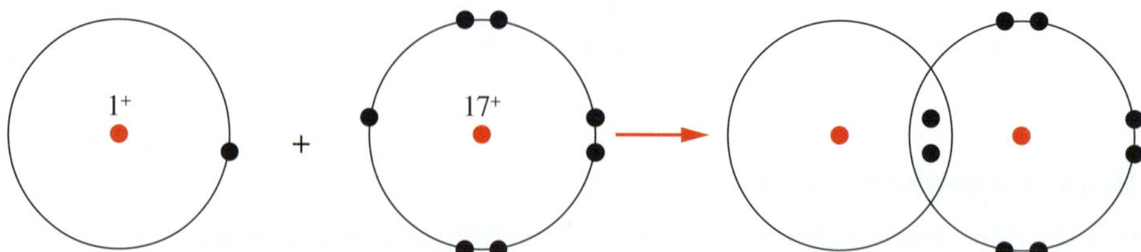

FIGURE 3.1.9 A pair of electrons is shared in the formation of a molecule of HCl.

CHEMFILE

The strong triple covalent bond in N_2

The triple covalent bond in nitrogen gas (N_2) is relatively strong and not easily broken. This means that nitrogen gas is relatively unreactive. Nitrogen is an essential element in living organisms because it is a major component of proteins and other biological molecules. Even though 78% of air is nitrogen gas, very few organisms can make use of the nitrogen because it is so unreactive. Nitrogen-fixing microorganisms are able to convert nitrogen gas into soluble nitrogen-containing compounds. These compounds are then absorbed by plants, allowing nitrogen to be passed up the food chain.

In the early twentieth century, German chemist Fritz Haber invented a process for converting nitrogen gas and hydrogen gas into ammonia, which is used to make synthetic fertilisers (see figure). This allowed humans to grow more food to feed a growing world population.

Nitrogen-rich urea is used in synthetic fertilisers, allowing humans to grow more food to feed a growing world population.

N N

FIGURE 3.1.8 A Lewis structure for N_2

POLYATOMIC MOLECULES

Molecules made up of more than two atoms are called **polyatomic molecules**. Three examples of polyatomic molecules are water, methane and ethene.

Example 1: Water

When a compound forms between hydrogen and oxygen, an oxygen atom bonds with two hydrogen atoms. The oxygen atom shares one electron with each hydrogen atom. Each hydrogen atom shares one electron with the oxygen atom.

As you can see in the Lewis structure (Figure 3.1.10), a water molecule contains:

- two single covalent bonds, each containing a shared electron pair
- two non-bonding pairs of electrons on the oxygen atom.

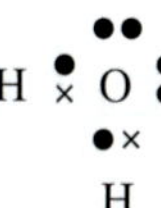

FIGURE 3.1.10 A water molecule has two single covalent bonds and two non-bonding pairs of electrons.

Example 2: Methane

When a compound forms between carbon and hydrogen, four hydrogen atoms are needed to provide the four electrons required to have eight electrons in the outer shell of a carbon atom (Figure 3.1.11). The molecule is called methane and has a molecular formula of CH_4.

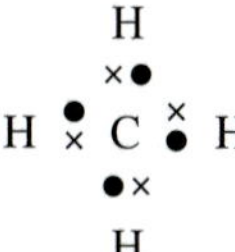

FIGURE 3.1.11 In a methane molecule, a carbon atom shares one electron with each of four hydrogen atoms to gain eight electrons in its outer shell.

Example 3: Ethene

Ethene (C_2H_4) is another example of a compound containing carbon and hydrogen. Each carbon atom shares two electrons with the other carbon atom, forming a double covalent bond. Two hydrogen atoms each share one electron with each carbon atom, forming two single covalent bonds. Figure 3.1.12 shows the Lewis structure of ethene.

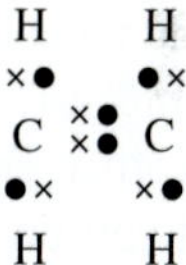

FIGURE 3.1.12 The Lewis structure of ethene contains a double covalent bond.

Worked example 3.1.1

LEWIS STRUCTURES

Draw the Lewis structure of hydrogen sulfide (H_2S).

Thinking	Working
Write the shell electronic configuration of the atoms in the molecule.	S electronic configuration: 2,8,6 H electronic configuration: 1
Determine how many electrons each atom requires for a stable outer shell.	S requires 2 electrons H requires 1 electron
Draw a Lewis structure of the likely molecule, ensuring that each atom has a stable outer shell. Electrons not involved in bonding will be in non-bonding pairs.	Draw a Lewis structure of the molecule. H S H

Worked example: Try yourself 3.1.1

LEWIS STRUCTURES

Draw the Lewis structure of ammonia (NH_3).

CASE STUDY

Historical development of bonding in molecules

Before the twentieth century, chemists had a limited understanding of the bonds formed between atoms in molecules. A very early theory was that some atoms had 'hooks' and others had 'eyes' and a bond formed when one atom hooked onto another (Figure 3.1.13).

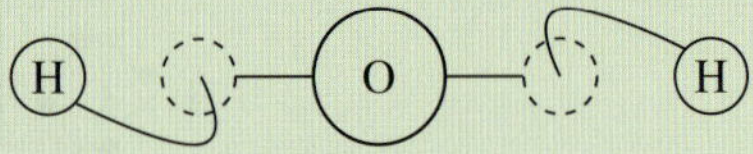

FIGURE 3.1.13 Hydrogen atoms with 'hooks' looped in the 'eyes' of the oxygen atom

By the nineteenth century, chemists understood that atoms of elements combined together in set proportions. For example, two hydrogen atoms combined with one oxygen atom to form a water molecule. In the mid-nineteenth century, German chemist Friedrich August Kekulé and Scottish chemist Archibald Scott Couper proposed that carbon formed four bonds. Kekulé is known for first proposing the structure for benzene (C_6H_6) (Figure 3.1.14). However, these chemists were far from understanding how atoms bonded together.

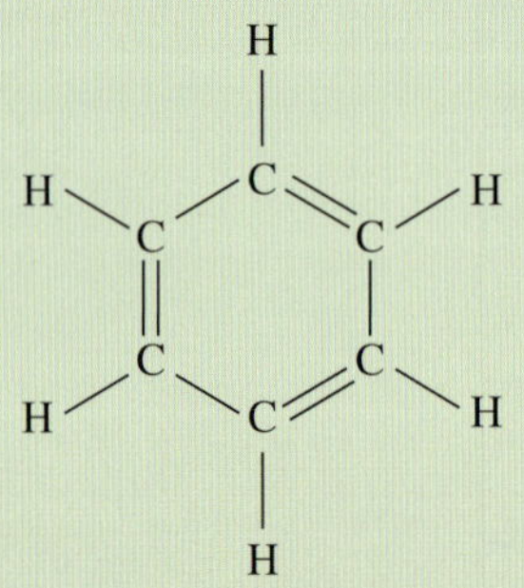

FIGURE 3.1.14 Kekulé structure for benzene

In the early twentieth century, the Bohr model of the atom was developed (see Chapter 2). It proposed that electrons were found in 'shells' around the nucleus of an atom. Gilbert Newton Lewis was an American chemist who, in his lectures to his university students, used dots to represent the electrons in atoms. He is the one who proposed the octet rule. Atoms were pictured as having electrons located in the corners of a cube. Cubes could combine in a way to satisfy the octet rule. For example, a single bond was formed when two atoms shared two electrons along an edge (Figure 3.1.15a) or a double bond was formed when atoms shared four electrons on a face (Figure 3.1.15b). These ideas where shared in a 1916 article called 'The atom and the molecule', which also contained the first examples of Lewis structures.

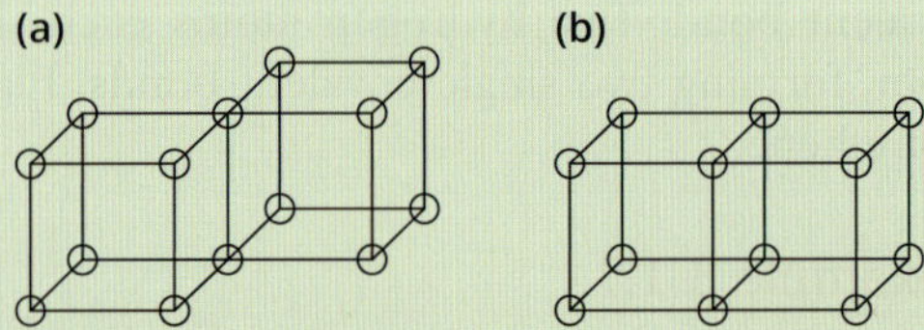

FIGURE 3.1.15 Reproductions of Lewis's cubic atoms from the 1916 article 'The atom and the molecule'. (a) A single bond formed from two atoms sharing an edge. (b) A double bond formed from two atoms sharing a face.

3.1 Review

OA ✓✓

SUMMARY

- Covalent bonds form between non-metallic atoms, often enabling the atoms to obtain outer shells containing eight electrons (except hydrogen, which obtains an outer shell containing two electrons).
- A covalent bond involves the sharing of electrons.
- A single covalent bond forms when two atoms share a pair of electrons.
- A double covalent bond forms when two atoms share two pairs of electrons.
- A triple covalent bond forms when two atoms share three pairs of electrons.
- Valence electrons that are not involved in bonding are called non-bonding pairs.
- Lewis structures show the valence electron arrangements of atoms in a molecule.

KEY QUESTIONS

Knowledge and understanding

1 Define the term 'molecule'.

2 How many covalent bonds are formed between atoms in these diatomic molecules?
   - **a** H_2
   - **b** N_2
   - **c** O_2
   - **d** F_2

3 What is the maximum number of covalent bonds formed by an atom of each of the following elements?
   - **a** H
   - **b** S
   - **c** P
   - **d** Si
   - **e** Br
   - **f** Ar

Analysis

4 Draw Lewis structures for each of the following molecules.
   - **a** fluorine (F_2)
   - **b** hydrogen fluoride (HF)
   - **c** water (H_2O)
   - **d** tetrachloromethane (CCl_4)
   - **e** phosphine (PH_3)
   - **f** carbon dioxide (CO_2).

5 When oxygen forms covalent molecular compounds with other non-metals, the Lewis structures that represent the molecules of these compounds all show each oxygen atom with two non-bonding pairs of electrons. Why are there always two non-bonding pairs?

6 Suggest the most likely molecular formula of the compound formed between the following pairs of elements.
   - **a** C, F
   - **b** P, Cl
   - **c** C, S
   - **d** Si, H
   - **e** N, Br

3.2 Shapes of molecules

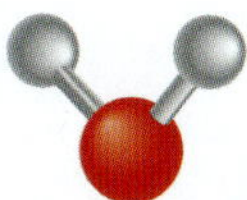

FIGURE 3.2.1 Water (H_2O) molecules have a distinctive shape that is responsible for many of its properties. The valence shell electron pair repulsion theory accurately predicts the shape of water molecules.

The shapes of molecules are critical in determining many physical properties of covalent molecular substances. In particular, molecular shape affects melting point, boiling point, hardness and solubility. This is because the shape of a molecule determines how it interacts with other molecules.

The shape of small molecules can be predicted using a relatively simple model known as the **valence shell electron pair repulsion (VSEPR) theory**. In this section, you will see how VSEPR theory can be used to predict the shape of molecules, such as the water molecule shown in Figure 3.2.1.

VALENCE SHELL ELECTRON PAIR REPULSION THEORY

An electron group can either be covalent bonds or a non-bonding electron pair.

Lewis structures represent the arrangement of valence electrons in the atoms of a molecule. As the name suggests, the VSEPR theory uses the arrangement of the valence electrons as shown in Lewis structures to predict the shape of the molecule. These valence electrons can be arranged into **electron groups**—either as different types of covalent bonds or as non-bonding pairs. The VSEPR theory is based on the principle that negatively charged electron groups around an atom repel each other. As a consequence, these electron groups are arranged as far away from each other as possible.

ELECTRON GROUP REPULSION

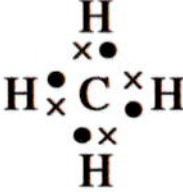

FIGURE 3.2.2 This Lewis structure of a methane (CH_4) molecule shows the four electron pairs surrounding the central carbon.

From the previous section, you learnt that atoms in covalent molecules are most stable when they have eight electrons in their valence shell. This is known as the octet rule. These eight electrons are arranged into four pairs of electrons.

Using the Lewis structure of methane (CH_4) as an example (Figure 3.2.2), you can see the carbon atom shares a pair of electrons with each hydrogen atom.

Each single covalent bond is an example of an electron group. The VSEPR theory states that the electron groups in methane repel each other so that they are as far apart as possible. This repulsion between the electron groups results in a **tetrahedral** shape, as shown in Figure 3.2.3.

FIGURE 3.2.3 In a methane (CH_4) molecule, the four electron groups repel each other. The repulsion forces the single covalent bonds as far apart as possible, leading to a tetrahedral molecular shape.

Non-bonding pairs of electrons

In the VSEPR theory, non-bonding pairs of electrons are considered another type of electron group.

In the ammonia molecule shown in Figure 3.2.4, the nitrogen atom has a stable octet made up of one non-bonding pair of electrons and three single bonds. The four electron groups repel each other to form a tetrahedral arrangement. However, the shape of the molecule is only determined by the position of the single bonds (and hence the position of the atoms). The non-bonding pair of electrons influences the shape but is not a part of it. The three hydrogen atoms are described as forming a **pyramidal** shape with the nitrogen atom.

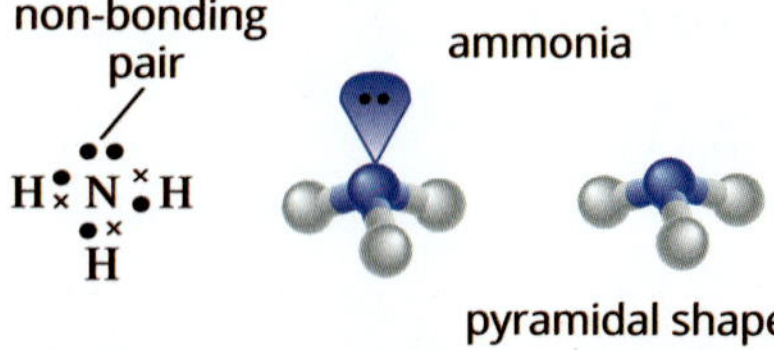

FIGURE 3.2.4 The Lewis structure of an ammonia (NH_3) molecule shows that the nitrogen atom has one non-bonding pair of electrons and three single bonds. These four electron groups repel each other to form a tetrahedral arrangement around the nitrogen atom. The result is a pyramidal shaped molecule.

In water molecules, the oxygen atom has a stable octet made up of two non-bonding pairs and two single bonds. The four electron groups repel each other to form a tetrahedral arrangement. This causes the two hydrogen atoms to form a bent arrangement with the oxygen atom, as shown in Figure 3.2.5.

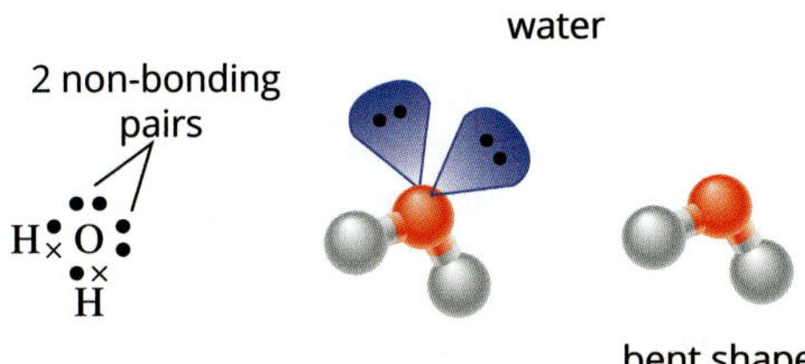

FIGURE 3.2.5 The Lewis structure of a water (H_2O) molecule shows that the oxygen atom has two non-bonding pairs of electrons and two single bonds. These four electron groups repel each other to form a tetrahedral arrangement around the oxygen atom. The result is a bent-shaped molecule.

In a hydrogen fluoride molecule, the fluorine atom has a stable octet made up of three non-bonding pairs and one single bond. The four electron groups repel each other to form a tetrahedral arrangement. The hydrogen and fluorine atoms form a linear molecule, as you can see in Figure 3.2.6.

Non-bonding pairs of electrons influence a molecule's shape but are not considered a part of the shape.

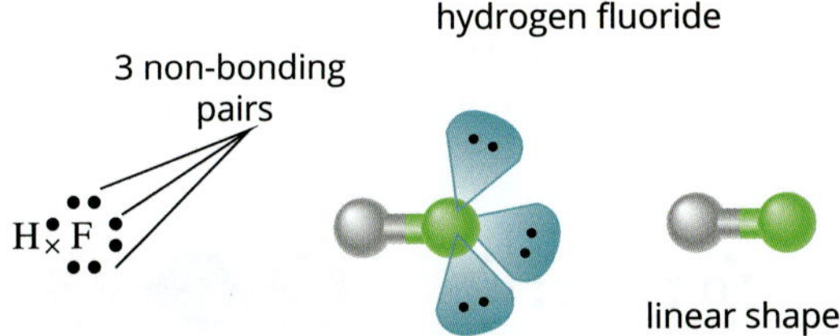

FIGURE 3.2.6 The Lewis structure of a hydrogen fluoride (HF) molecule shows that the fluorine atom has three non-bonding pairs of electrons and one single bond. These four electron groups repel each other to form a tetrahedral arrangement around the fluorine atom. The result is a linear molecule.

Worked example 3.2.1

PREDICTING THE SHAPE OF MOLECULES

Predict the shape of a molecule of phosphine (PH_3).

Thinking	Working
Draw the Lewis structure for the molecule.	H : P : H H
Count the number of electron groups around the central atom.	There are four electron groups (three single bonds and one non-bonding pair).
Determine how the electron groups will be arranged to get maximum separation.	Because there are four electron groups, they will be arranged in a tetrahedral arrangement.
Deduce the shape of the molecule by considering the arrangement of just the atoms.	The phosphorus and hydrogen atoms are arranged in a pyramidal shape.

Worked example: Try yourself 3.2.1

PREDICTING THE SHAPE OF MOLECULES

Predict the shape of a molecule of hydrogen sulfide (H_2S).

Molecules with fewer than four electron groups

Molecules where the central atom forms double or triple covalent bonds tend to have fewer than four electron groups. VSEPR theory treats double and triple bonds in the same way that it treats single bonds and non-bonding pairs of electrons as single electron groups.

When using VSEPR theory to determine the shape of molecules, double and triple bonds each count as one electron group.

For example, if a central atom has two single bonds and one double bond, then the three electron groups will repel each other to get maximum separation. This results in a molecular shape known as **trigonal planar** because the atoms form a triangle in one plane. An example of this structure is the methanal (CH_2O) molecule shown in Figure 3.2.7.

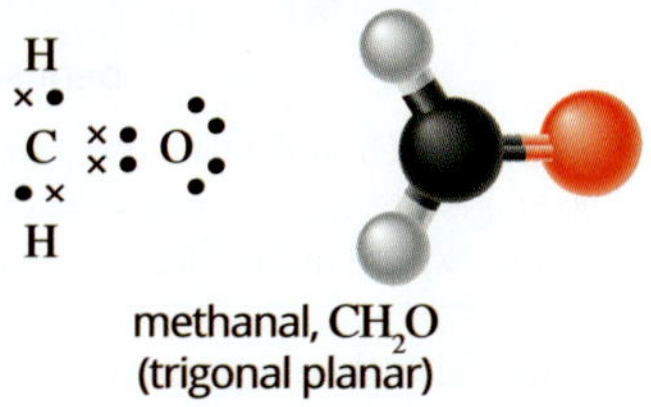

methanal, CH_2O
(trigonal planar)

FIGURE 3.2.7 Methanal has a central carbon atom that forms a double bond with an oxygen atom and single bonds with two hydrogen atoms. The three electron groups repel each other to form a trigonal planar arrangement.

If the central atom has two double bonds, then the two electron groups repel each other. This results in a linear molecule like carbon dioxide (CO_2), shown in Figure 3.2.8.

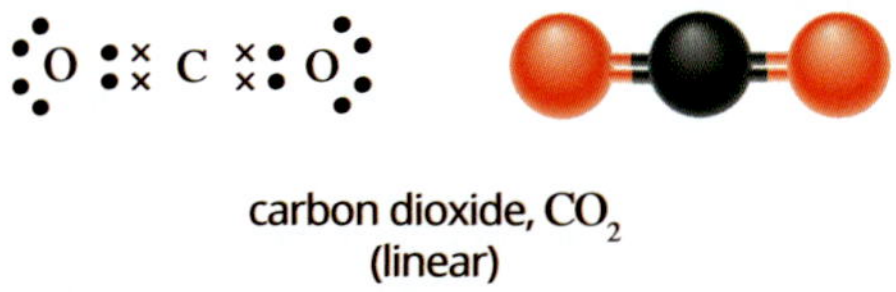

carbon dioxide, CO_2
(linear)

FIGURE 3.2.8 In a carbon dioxide molecule, the carbon atom forms double bonds with two oxygen atoms. The two electron groups repel each other. This results in a linear molecule.

Finally, if the central atom has a single bond and a triple bond, as in hydrogen cyanide (HCN), then the molecule also has two electron groups, and is linear (Figure 3.2.9).

hydrogen cyanide, HCN
(Linear)

FIGURE 3.2.9 The hydrogen cyanide molecule is linear. In this case, the central carbon atom forms a triple bond with the nitrogen atom and a single bond with the hydrogen atom.

STRUCTURAL FORMULAS

In the previous section, you were introduced to Lewis structures as a way of representing how valence shell electrons of atoms in a molecule are arranged. A limitation of a Lewis structure is that it is not always drawn in a way that shows the shape of the molecule. After using VSEPR to determine a molecule's shape, it can be represented using a **structural formula**. In a structural formula, each bonding pair of electrons is shown as a line. Non-bonding electrons are not shown. For example, Figure 3.2.10a shows the Lewis structure of nitrogen trifluoride (NF_3) and Figure 3.2.10b is its structural formula, showing its pyramidal shape. You will see more examples of how structural formulas are used to represent carbon-based molecular compounds in Chapter 8.

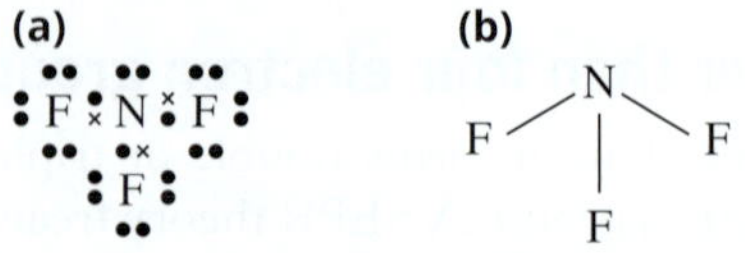

(a) (b)

FIGURE 3.2.10 (a) Lewis structure and (b) structural formula of nitrogen trifluoride (NH_3)

CHEMFILE

The molecular shape of sulfur hexafluoride

Sulfur hexafluoride (SF_6) is made up of a central atom of sulfur with six valence electrons. The sulfur atom forms single bonds to six fluorine atoms. These six electron groups repel each other to form an octahedral shape (see figure below). Sulfur hexafluoride is a gas at room temperature and is denser than air. It is mainly used as an insulating gas in high voltage equipment. As a greenhouse gas, it is 23 900 times worse than carbon dioxide, so its use is highly regulated.

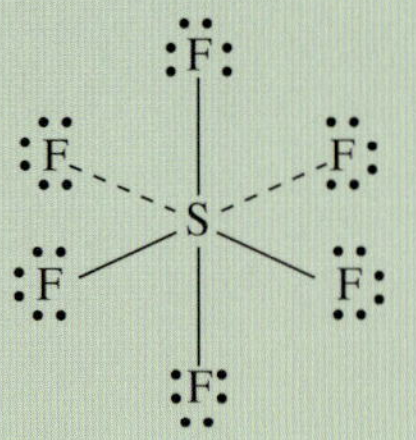

Bond angles in SF_6

3.2 Review

SUMMARY

- The shapes of simple molecules can be predicted by the valence shell electron pair repulsion (VSEPR) theory.
- The VSEPR theory is based on the principle that electron groups around an atom repel each other. As a consequence, these electron groups are arranged as far away from each other as possible.
- Electron groups can be covalent bonds or non-bonding pairs of electrons.
- Non-bonding pairs influence a molecule's shape, but are not considered a part of the shape.
- Structural formulas are used to represent the shape and bonds within molecules.

TABLE 3.2.1 Summary of shapes of molecules

Number of electron groups	No non-bonding pairs	1 non-bonding pair	2 non-bonding pairs	3 non-bonding pairs
4	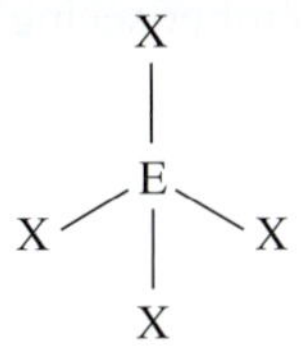 tetrahedral (EX_4)	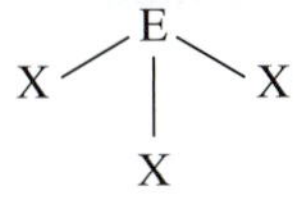 pyramidal (EX_3)	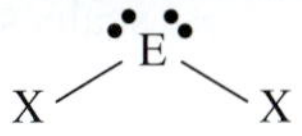 bent (EX_2)	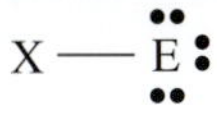 linear ($X—E$)
3	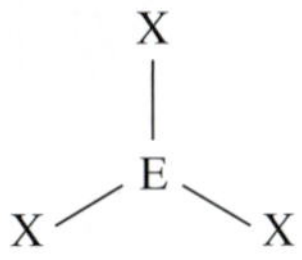 trigonal planar (EX_3)	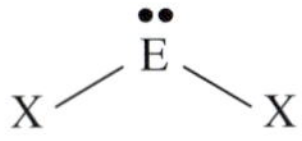 angular or bent (EX_2)	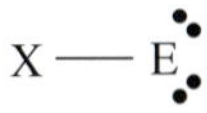 linear ($X—E$)	
2	 linear ($X—E—E$)	 linear ($X—E$)		

KEY QUESTIONS

Knowledge and understanding

1 Explain VSEPR theory and how it is used to determine the shape of molecules.

2 How many electron groups are there around the fluorine atom in a hydrogen fluoride molecule?

Analysis

3 Draw the Lewis structure for each of the following molecules.

a H_2S
b HI
c CCl_4
d PH_3
e CS_2
f SiH_4

4 Identify the shape of each of the molecules in Question **3**.

5 Draw the structural formula of each of the molecules in Question **3**.

3.3 Polarity in molecules

FIGURE 3.3.1 A stream of water bending near a statically charged balloon

The covalent molecules examined so far in this chapter are neutral. They do not have an overall charge because the number of protons equals the number of electrons in the molecules. However, a substance such as water can behave as though it is charged. When a statically charged balloon is held near a stream of water, it bends towards the balloon (Figure 3.3.1). This behaviour is due to the uneven distribution of electrons within the water molecules, causing them to be partially charged. The following section examines how the shape of a molecule and the electronegativity of its atoms can cause these uneven electron distributions.

ELECTRONEGATIVITY AND POLAR BONDS

Electronegativity is the key factor that determines the electron distribution in diatomic molecules. Electronegativity is the ability of an atom to attract electrons towards itself when forming a chemical bond. Electronegativity increases from left to right across the periods of the periodic table and decreases down the groups of the table, as shown in Figure 3.3.2. (Note that the electronegativity scale is a relative scale, so there are no units.) You will remember seeing these patterns in Chapter 2.

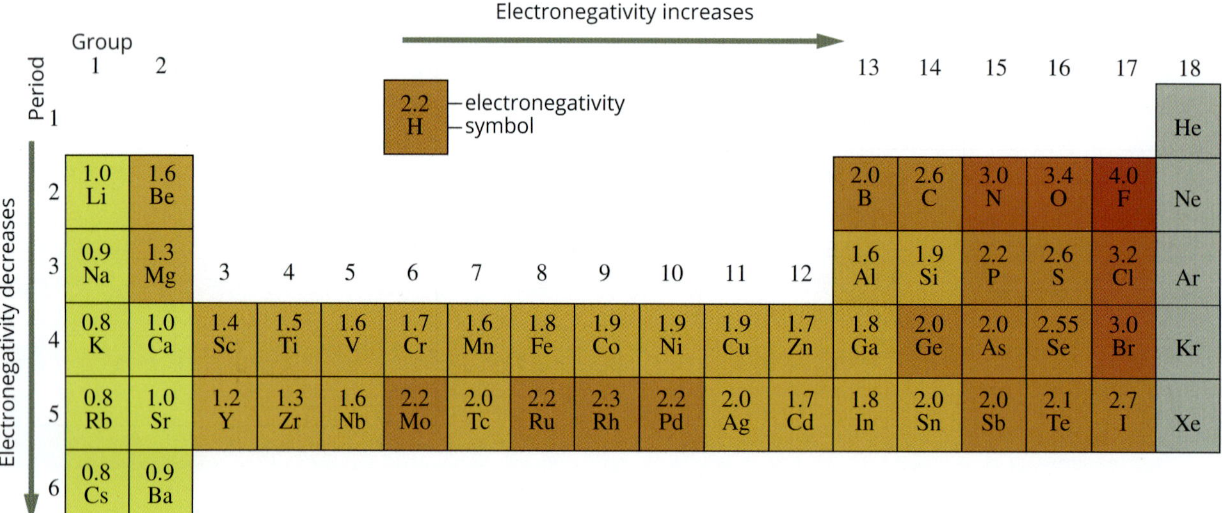

Period	Group 1	2	3	4	5	6	7	8	9	10	11	12	13	14	15	16	17	18
1																		He
2	1.0 Li	1.6 Be											2.0 B	2.6 C	3.0 N	3.4 O	4.0 F	Ne
3	0.9 Na	1.3 Mg											1.6 Al	1.9 Si	2.2 P	2.6 S	3.2 Cl	Ar
4	0.8 K	1.0 Ca	1.4 Sc	1.5 Ti	1.6 V	1.7 Cr	1.6 Mn	1.8 Fe	1.9 Co	1.9 Ni	1.9 Cu	1.7 Zn	1.8 Ga	2.0 Ge	2.0 As	2.55 Se	3.0 Br	Kr
5	0.8 Rb	1.0 Sr	1.2 Y	1.3 Zr	1.6 Nb	2.2 Mo	2.0 Tc	2.2 Ru	2.3 Rh	2.2 Pd	2.0 Ag	1.7 Cd	1.8 In	2.0 Sn	2.0 Sb	2.1 Te	2.7 I	Xe
6	0.8 Cs	0.9 Ba																

FIGURE 3.3.2 Table of electronegativity values. This periodic table shows the electronegativities of the atoms of each element. The electronegativities generally increase from left to right across the periods and decrease down the groups.

Non-polar bonds

When two atoms form a covalent bond, you can regard the atoms as competing for the electrons being shared between them. If the two atoms in a covalent bond are the same (i.e. have identical electronegativities), then the electrons are shared equally between the two atoms. This is the case for diatomic molecules, such as chlorine (Cl_2), oxygen (O_2), hydrogen (H_2) and nitrogen (N_2).

Bonds with an equal distribution of bonding electrons are said to be **non-polar** because there is no charge on either end of the bond.

Electron density is the measure of the probability of an electron being present at a particular location within an atom. In molecules, areas of electron density are commonly found around the atom and its bonds.

Figure 3.3.3 shows the electron distribution in the non-polar fluorine (F_2) molecule. The molecule has a high electron density between the two fluorine atoms, forming the covalent bond. The bonding electrons are distributed evenly between the two atoms, making the bond and the molecule non-polar.

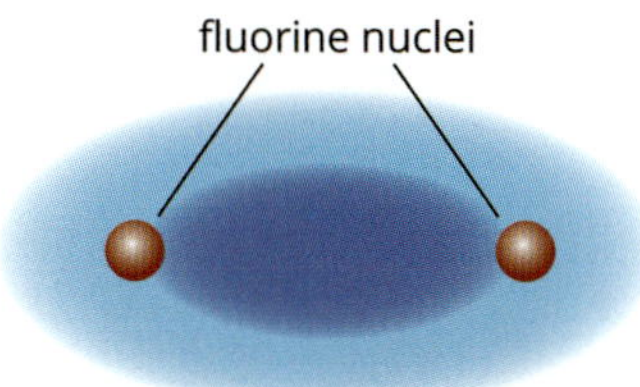

F_2 contains a non-polar bond.

FIGURE 3.3.3 Fluorine molecules have a symmetric distribution of electrons and are therefore non-polar.

Polar bonds

If the covalent bond is between atoms of two different elements, then the electrons will stay closer to the most electronegative atom, as it has a stronger pull on the electrons in the bond. An example is the hydrogen fluoride (HF) molecule, shown in Figure 3.3.4. Bonds with an imbalanced electron distribution are said to be **polar**.

A fluorine atom is more electronegative than a hydrogen atom. Therefore, in a hydrogen fluoride molecule the electrons tend to stay closer to the fluorine atom. The fluorine atom is described as having a partial negative charge, which is represented with the Greek letter delta (lowercase) as δ–. The hydrogen atom is described as having a partial positive charge, δ+. The separation into positive and negative charges is known as a **dipole**, as they have two oppositely charged poles at each end of the molecule.

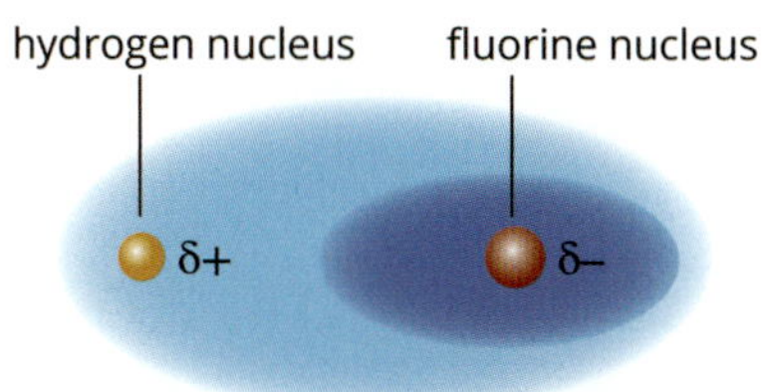

HF contains a polar bond; fluorine is more electronegative than hydrogen.

FIGURE 3.3.4 The electron distribution in hydrogen fluoride is asymmetric because of the different electronegativities of the hydrogen and fluorine atoms. Hydrogen fluoride is an example of a polar molecule.

Hydrogen fluoride has a permanent dipole due to the different electronegativities of the two atoms. All diatomic (two-atom) molecules that are made up of different elements are polar to some extent. The level of **polarity** will depend on the difference between the electronegativities of the two atoms. The greater the difference between the electronegativities, the greater the polarity of the molecule.

It is not just the covalent bonds in diatomic molecules that can be polar. The polarity of any covalent bond can be compared by examining the difference in the electronegativities of the atoms involved in the bond.

Worked example 3.3.1

COMPARING THE POLARITY OF COVALENT BONDS

Compare the polarity of the bond in hydrogen fluoride (HF) and carbon monoxide (CO).

Thinking	Working
Use the table of electronegativity values in Figure 3.3.2 to find the electronegativities of the atoms in each bond.	HF: hydrogen 2.2; fluorine 4.0 CO: carbon 2.6; oxygen 3.4
For each bond, subtract the lowest electronegativity value from the highest value.	HF: 4.0 – 2.2 = 1.8 CO: 3.4 – 2.6 = 0.8
Determine which bond has the biggest difference in electronegativity to determine the more polar bond.	The bond in HF is more polar than in CO.

Worked example: Try yourself 3.3.1

COMPARING THE POLARITY OF COVALENT BONDS

Compare the polarity of the bond in nitrogen monoxide (NO) and hydrogen chloride (HCl).

The range of bond types

The type and polarity of a chemical bond can be predicted by considering the electronegativity difference between the atoms involved in bond formation. There is a range (continuum) of bond types, from non-polar covalent to polar covalent to ionic (a bond formed when an electron is transferred from one atom to another). The type of bond depends upon the extent to which electrons are shared. This, in turn, depends upon the electronegativity difference between the atoms involved in the bond.

When the electronegativity difference between two atoms is zero, the bonding electrons are shared equally, and the bond formed between the two atoms is covalent and non-polar. For example, in a fluorine (F_2) molecule both atoms have the same electronegativity and, as a result, the covalent bond between the two fluorine atoms is non-polar. If two different elements have the same electronegativity, the covalent bond between them will also be non-polar. For example, the electronegativity of both carbon and sulfur is 2.6. The covalent bonds between atoms of carbon and sulfur in carbon disulfide (CS_2) are non-polar.

As you have learnt in this section, any difference in electronegativity between two bonded atoms will result in an unequal sharing of electrons and the bond will be polar. For example, in a hydrogen fluoride (HF) molecule, fluorine is more electronegative than hydrogen. As a result, HF has a dipole, and the bond in HF is described as polar covalent.

The polar nature of a bond between two atoms increases as the electronegativity difference between these atoms increases. The bonding in HF is more polar than in HCl because the fluorine atom is more electronegative than the chlorine atom. Electrons will be transferred between two atoms if the electronegativity difference between them is great enough. As you will discover in Chapter 7, electron transfer between metallic and non-metallic atoms results in the formation of ionic bonds. A fluorine atom is much more electronegative than a sodium atom. When fluorine and sodium react, the sodium atom's valence electron is transferred to the outer shell of the fluorine atom. This results in the formation of Na^+ and F^- ions. The bonding between the ions is described as ionic. The range of bond types is illustrated in Figure 3.3.5.

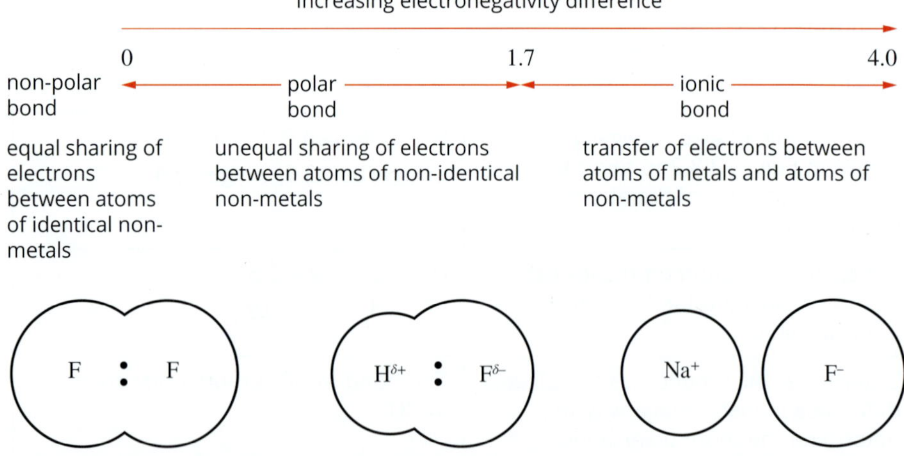

FIGURE 3.3.5 The range of bond types. There is a transition of bond type from non-polar through to polar, then ionic, as the difference in electronegativity increases.

An electronegativity difference of 1.7 is often used as the approximate cut-off point between polar covalent bonding and ionic bonding (Table 3.3.1). In fact, there is no sharp distinction between polar covalent and ionic bonding at this level of electronegativity difference. Compounds such as aluminium chloride ($AlCl_3$), which have an electronegativity difference between their atoms of around 1.7, should be considered as having both polar covalent and ionic bonding characteristics.

TABLE 3.3.1 Electronegativity difference and bond type

Electronegativity difference	Distribution of bonding electrons	Type of bond	Example
zero	electrons shared equally	non-polar covalent	H_2, Cl_2, CS_2
less than 1.7	electrons attracted to the more electronegative atom	polar covalent	NH_3, H_2O, HCl
greater than 1.7	electrons transferred to the more electronegative atom	ionic	NaCl, CaF_2

POLARITY OF POLYATOMIC MOLECULES

Determining the polarity of molecules with more than two atoms is a little more complicated. This is because the polarity of polyatomic molecules depends on the shape of the molecule, as well as the polarity of its covalent bonds. A molecule can possess polar bonds yet still be non-polar.

- Molecules that contain only non-polar bonds are non-polar molecules.
- **Symmetrical molecules** (molecules that contain polar bonds that are evenly distributed) are non-polar, as the bond dipoles cancel each other out.
- **Asymmetrical molecules** that contain polar bonds are polar molecules, as a net dipole is created in the molecule.

Non-polar molecules

Even molecules with polar covalent bonds can be non-polar if the arrangement of these bonds within the molecule is symmetrical.

In methane, the carbon atom is slightly more electronegative than the hydrogen atoms. Therefore, the carbon atom has a partial negative charge, leaving hydrogen with a partial positive charge (Figure 3.3.6a). However, the methane molecule has a tetrahedral shape and is therefore symmetrical. The symmetry of the molecule means that the individual dipoles of the covalent bonds (represented by arrows) cancel each other out perfectly (Figure 3.3.6b). The result is a molecule with no overall dipole. It is non-polar.

The methane molecule shown in Figure 3.3.6 is an example of a non-polar molecule with polar covalent bonds.

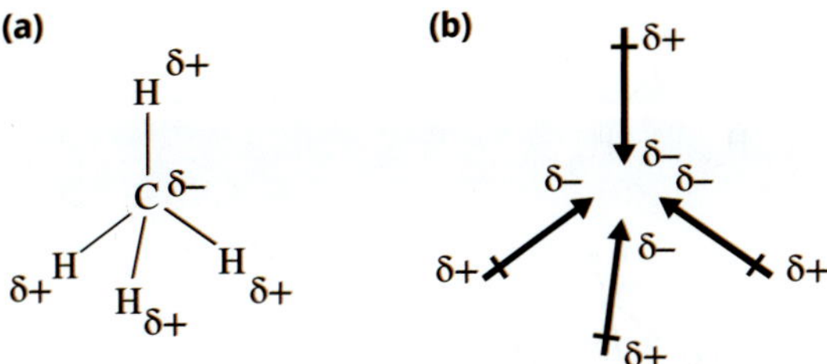

FIGURE 3.3.6 (a) Structure of a methane molecule showing the partial charges on the atoms. (b) The individual bond dipoles are distributed symmetrically around the molecule.

Polar molecules

In asymmetrical molecules, the individual dipoles of the covalent bonds do not cancel each other out. This results in a net dipole, making the overall molecule polar.

The chloromethane molecule shown in Figure 3.3.7 is an example of an asymmetric molecule. The chlorine atom is more electronegative than the carbon atom. Therefore, the chlorine atom attracts electrons from the carbon atom while the carbon atom attracts electrons from the hydrogen atoms (Figure 3.3.7a). The individual dipoles of the covalent bonds are shown in Figure 3.3.7b. These add to give the molecule a net dipole (Figure 3.3.7c).

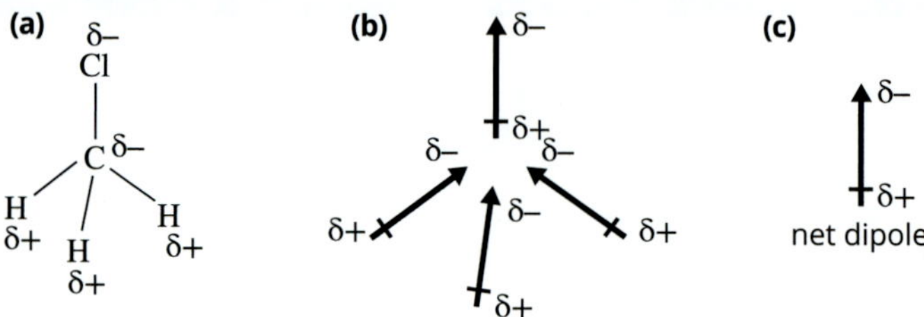

FIGURE 3.3.7 (a) Structure of a chloromethane molecule showing the partial charges on the atoms. (b) The individual dipoles are added together. (c) The result is a net dipole.

Table 3.3.2 shows some more examples of how symmetry determines the polarity of covalent molecules.

TABLE 3.3.2 Examples of polar and non-polar covalent molecules

Molecule	Structure	Symmetrical/ asymmetrical	Polar/non-polar
methanal	$\overset{\delta+}{H}$, $\overset{\delta+}{H}$ bonded to C; C=$\overset{\delta-}{O}$	asymmetrical	polar
carbon dioxide	$\overset{\delta-}{O}=\overset{\delta+}{C}=\overset{\delta-}{O}$	symmetrical	non-polar
tetrafluoromethane	$\overset{\delta+}{C}$ bonded to four $\overset{\delta-}{F}$	symmetrical	non-polar
water	$\overset{\delta-}{O}$ bonded to $H_{\delta+}$ and $H_{\delta+}$	asymmetrical	polar
ammonia	$\overset{\delta-}{N}$ bonded to three $\overset{\delta+}{H}$	asymmetrical	polar

CHEMFILE

How a microwave oven works

Microwave ovens use the polarity of molecules (especially water molecules) to heat food. The microwave oven irradiates the food with microwaves. The microwaves produce an electric field that interacts with polar molecules. The electric field causes the molecules to rotate up and down billions of times per second. This gives the molecules extra kinetic energy. As the kinetic energy of the molecules increases, the temperature increases. Therefore, the molecules heat and cook the food.

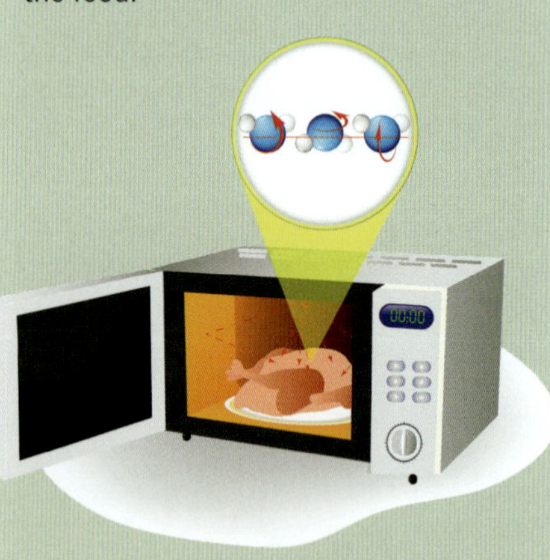

Food heats up in a microwave because polar molecules, such as water molecules, rotate billions of times per second in response to microwave radiation.

3.3 Review

OA ✓✓

SUMMARY

- As the difference in electronegativities of two atoms increases, a covalent bond increases in polarity.
- Diatomic molecules containing the same type of atom are non-polar.
- In general, symmetrical molecules are non-polar and asymmetrical molecules with polar bonds are polar.
- The polarity of polyatomic molecules depends on the electronegativity of the atoms in the molecule and the asymmetry of the molecule.

KEY QUESTIONS

Knowledge and understanding

1 Define each of the following.

- **a** non-polar bond
- **b** polar bond

2 Covalent bonds can form between the following pairs of elements in a variety of compounds. Use the electronegativity values given in Figure 3.3.2 to identify the atom in each pair that would have the largest share of bonding electrons.

- **a** S and O
- **b** C and H
- **c** C and N
- **d** N and H
- **e** F and O
- **f** P and F

3 The greater the differences in electronegativity between two atoms, the more polar the bond formed between them.

- **a** Which of the examples in Question **2** would be the most polar bond?
- **b** Which of the examples in Question **2** would be the least polar bond?

Analysis

4 Use the electronegativity values in Figure 3.3.2 on page 110 to order the following diatomic molecules from least to most polar: HCl, N_2, HBr, NO.

5 Label the following structural formula a phosphorus trifluoride molecule to show the partial charges that would occur on each atom.

P
F F
F

6 Determine whether each of the following molecules is polar or non-polar.

- **a** CF_4
- **b** CHF_3
- **c** CH_2F_2
- **d** CH_3F
- **e** CH_4

7 For each of the following pairs of molecules:

- **i** draw the structure of each molecule
- **ii** determine which is a polar molecule
- **iii** on the polar molecule, indicate the polarities of the bonds and the direction of the net dipole.

- **a** $CHCl_3$ or CCl_4
- **b** NH_3 or CH_4

3.4 Intermolecular forces

Covalent bonds between atoms within a molecule (intramolecular bonds) are much stronger than the intermolecular forces between molecules.

When liquid water is heated (Figure 3.4.1), the water molecules gain kinetic energy. Some molecules gain enough energy to break free from the others and escape from the surface of the liquid. The water vapour formed still contains molecules of water, so the covalent bonds between atoms (the intramolecular bonds) have not been broken. Instead, it is the **intermolecular forces** between water molecules that have been disrupted. This indicates that the covalent bonds between atoms are much stronger than the intermolecular forces between molecules.

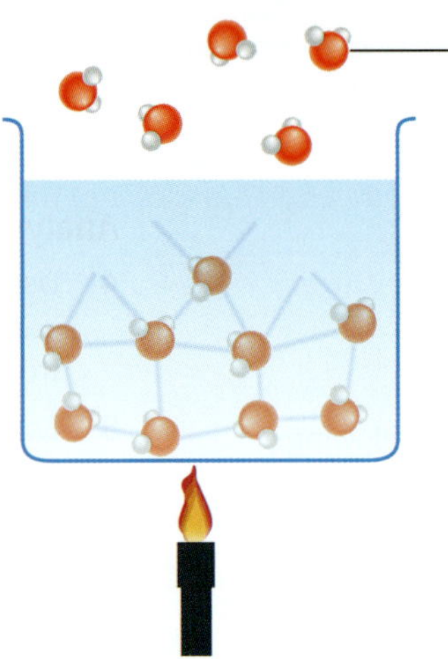

FIGURE 3.4.1 Changes occur to the forces between water molecules when it starts to boil.

Many factors determine the strength of intermolecular forces, including the size, shape and polarity of molecules. These factors not only determine the strength of the intermolecular forces in a substance, they also determine the types of intermolecular forces.

There are three main types of intermolecular forces:

- dispersion forces
- dipole–dipole attraction
- hydrogen bonding.

In this section, you will examine the nature of these three types of intermolecular forces and their role in determining the physical properties of covalent molecular substances.

DISPERSION FORCES

Dispersion forces are forces that exist between all molecules, whether they are polar or non-polar. Dispersion forces are caused by **temporary dipoles** in the molecules that are the result of random movement of the electrons surrounding the molecule. These temporary dipoles are also known as **instantaneous dipoles**.

Dispersion forces are always present between molecules, as electrons are constantly in motion within atoms (Figure 3.4.2). As they are caused by temporary dipoles, dispersion forces are the weakest of the three types of intermolecular forces.

Strength of dispersion forces

The strength of dispersion forces increases as the relative molecular mass of the molecule increases. Larger molecules have more electrons. It is easier to produce temporary dipoles in molecules with large numbers of electrons. Since larger molecules have stronger dispersion forces, they have higher melting and boiling points.

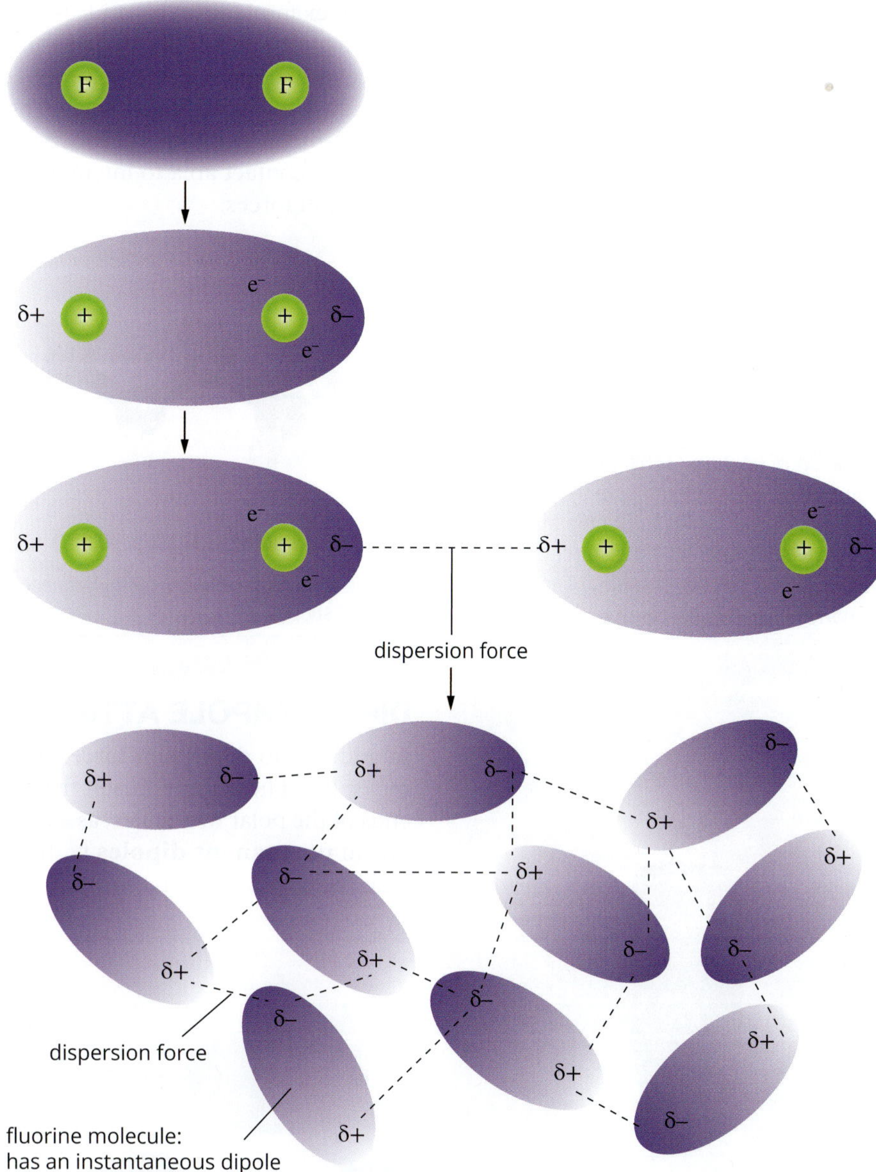

FIGURE 3.4.2 How dispersion forces form within non-polar substances. These diagrams show the forces forming between fluorine molecules.

Table 3.4.1 shows the boiling points of the halogens (group 17), which all form non-polar, diatomic molecules. The only forces between their molecules are dispersion forces. You can see that as molecular mass increases and the dispersion forces become stronger, the boiling points of the substances increase.

TABLE 3.4.1 The effect of dispersion forces on the boiling points of the halogens

Molecule	Molecular mass	Number of electrons	Boiling point (°C)
fluorine (F_2)	38.0	18	–188
chlorine (Cl_2)	71.0	34	–35
bromine (Br_2)	159.8	70	59
iodine (I_2)	253.8	106	184

The shape of a molecule also influences the strength of the dispersion forces. Molecules that form long chains will tend to have stronger dispersion forces than more compact molecules with similar molecular masses.

For example, butane and 2-methylpropane (Figure 3.4.3) both contain four carbon atoms and 10 hydrogen atoms. The boiling point of butane is –0.5°C, while the boiling point of 2-methylpropane is –11°C. The higher boiling point of butane is because of the different shapes of the two molecules; butane is a long molecule while 2-methylpropane is branched. Being longer and un-branched means butane has more contact area to interact with its neighbouring molecules to form stronger dispersion forces.

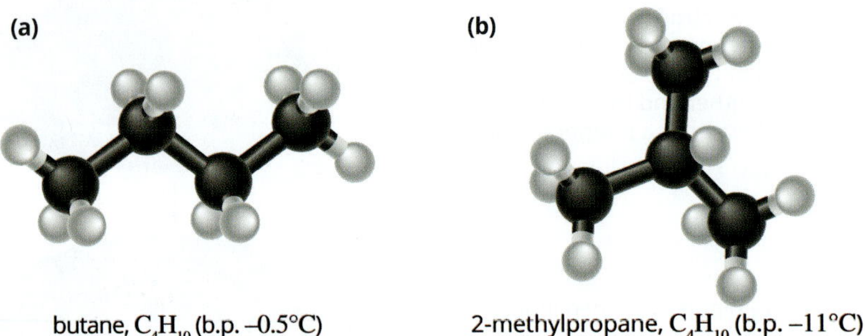

FIGURE 3.4.3 Butane and 2-methylpropane have different boiling points because their molecules are different shapes.

CHEMFILE

Use of beeswax by Indigenous Australians

Indigenous Australians have traditionally used the properties of beeswax to make useful objects, such as the *bathi marrin* (sacred dilly bag) below, which belongs to the clans of the Dhurili nation of the Yolnu people in north-east Arnhem Land. The *bathi marrin* is connected to Gandjalala, the *mokuy* spirit man of the Dhurili clans, who uses it to collect honey. The *bathi marrin* is also used ceremonially.

Beeswax is used to seal this *bathi marrin*.

Beeswax is a soft substance, and it melts at about 65°C. It is made up of a mixture of long-chain molecules containing many carbon and hydrogen atoms. The typical formula of a molecule of beeswax is $C_{46}H_{92}O_2$. Beeswax will soften when warmed in your hand, making it easy to mould into a desired shape. Dispersion forces are the main type of bonding between wax molecules. As the wax is warmed, the dispersion forces are weakened, enabling the long-chain molecules to slide past one another. As the wax cools, new dispersion forces are formed between adjacent molecules, locking the piece of wax into a new shape.

DIPOLE-DIPOLE ATTRACTION

In addition to dispersion forces, **dipole–dipole attraction** occurs between polar molecules. These forces result from the attraction between the positive and negative ends of the polar molecules, as shown in Figure 3.4.4. As dipole–dipole attraction is due to **permanent dipoles** within molecules, dipole–dipole attraction is stronger than the dispersion forces found between molecules.

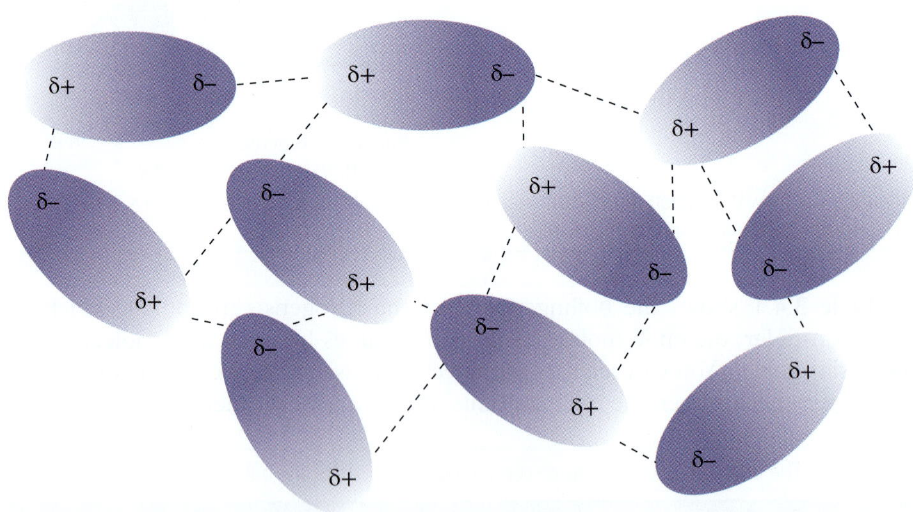

FIGURE 3.4.4 The positive and negative ends of polar molecules attract each other.

The more polar a molecule is, the stronger the dipole–dipole attraction. The molecular dipole will be stronger when there is a large difference in the electronegativities of the atoms and a large asymmetry in the shape of the molecule. Molecules that are more polar have stronger dipole–dipole attraction between them.

The melting and boiling points of a substance are directly related to the strength of the dipole–dipole attraction between molecules. The stronger the dipole–dipole attraction, the higher the melting and boiling points. This is because it takes more energy (i.e. higher temperatures) to break the stronger dipole–dipole attraction when the substance changes from a solid to a liquid or a liquid to a gas.

For example, compare methanal (CH_2O) and ethane (CH_3CH_3), shown in Figure 3.4.5. Methanal molecules are asymmetrical and polar. This results in dipole–dipole attraction between molecules and methanal has a relatively high boiling point of –19°C. On the other hand, ethane molecules are symmetrical and non-polar. There are only weaker dispersion forces between molecules, so ethane has a much lower boiling point of –88.5°C.

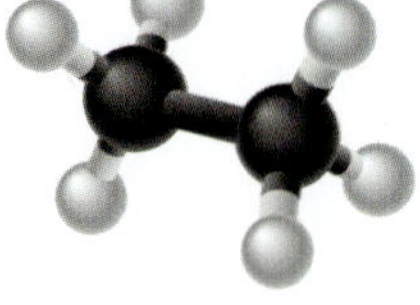

methanal, b.p. –19°C (polar) | ethane, b.p. –88.5°C (non-polar)

FIGURE 3.4.5 Dipole–dipole attraction exists between molecules of methanal, giving it a much higher boiling point than ethane.

HYDROGEN BONDING

Hydrogen bonding is a particularly strong form of dipole–dipole attraction. Hydrogen bonding only occurs between highly polar molecules in which one of the molecules has a hydrogen atom covalently bonded to a nitrogen, oxygen or fluorine atom.

Nitrogen, oxygen and fluorine atoms are small and highly electronegative. When bonded with a hydrogen atom, they strongly attract the electron pair in the covalent bond and a large dipole forms. Remember that hydrogen atoms only have one electron and it is pulled towards the highly electronegative atom in the bond. The hydrogen nucleus (a proton) is therefore left exposed and is attracted to a non-bonding pair of electrons on the nitrogen, oxygen or fluorine atom of a neighbouring molecule. The small size of the hydrogen atom allows the neighbouring molecule to closely approach and the resulting attractive force is relatively strong.

This intermolecular force is known as a **hydrogen bond**. It is approximately ten times stronger than a dipole–dipole attraction, but about one-tenth the strength of an ionic or a covalent bond. Figure 3.4.6 shows examples of hydrogen bonding among molecules where at least one of the molecules contains a hydrogen atom covalently bonded to either an N, O or F atom.

hydrogen bonding between ammonia molecules

hydrogen bonding between water molecules

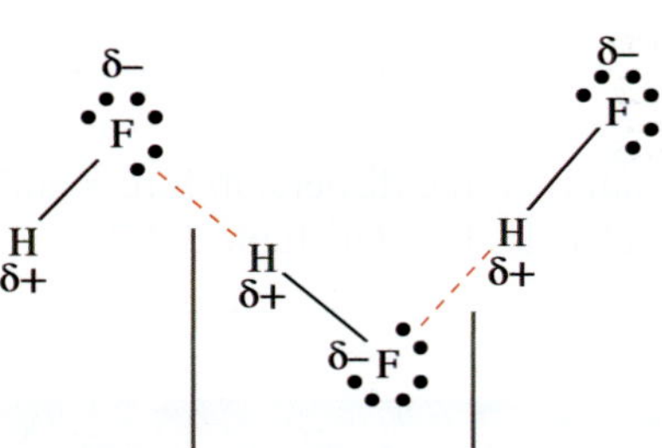

hydrogen bonding between hydrogen fluoride molecules

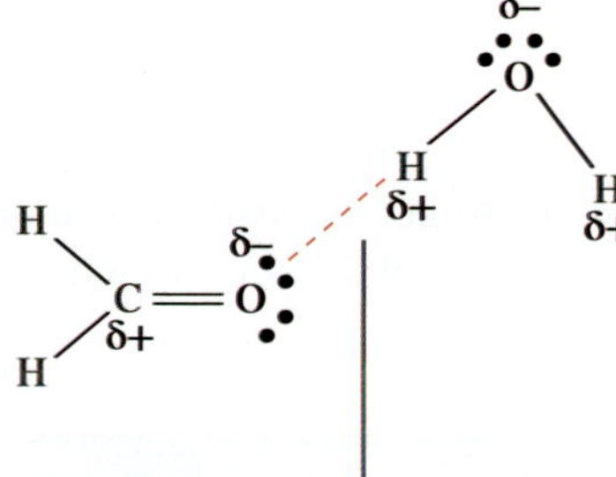

hydrogen bonding between water and the oxygen atom of methanal

FIGURE 3.4.6 Examples of polar hydrogen bonding between molecules

The presence of hydrogen bonds results in higher melting and boiling points. Figure 3.4.7 demonstrates the effect of hydrogen bonding on boiling point by comparing methanol to ethane and methanal. Recall that ethane only has weak dispersion forces between non-polar molecules. Methanal contains polar molecules that attract each other through dipole–dipole attraction but not hydrogen bonds. Methanol, however, contains molecules where a hydrogen atom is attached to an oxygen atom. This part of the methanol molecule is highly polar and can form a hydrogen bond with neighbouring methanol molecules. As a result, the boiling point of methanol (64.7°C) is significantly higher than the boiling points of methanal (–19°C) and ethane (–88.5°C).

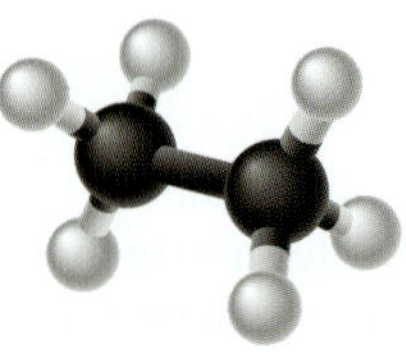

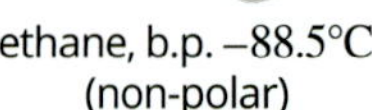

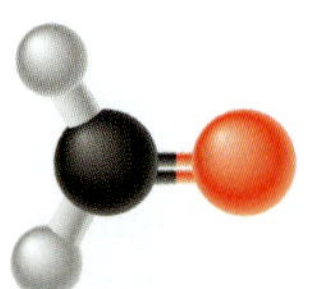

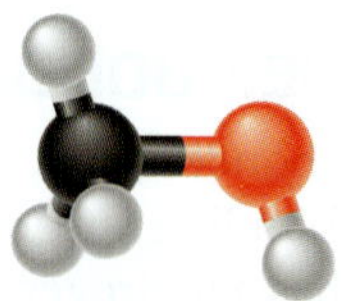

FIGURE 3.4.7 Hydrogen bonding exists between molecules of methanol, giving it a much higher boiling point than methanal and ethane.

For hydrogen bonding to occur, the molecules must have:

1 a hydrogen atom covalently bonded to a nitrogen, oxygen or fluorine atom

2 a non-bonding pair of electrons on the nitrogen, oxygen or fluorine atoms on neighbouring molecules.

There are two key requirements for hydrogen bonding:

1 a hydrogen atom covalently bonded to a nitrogen, oxygen or fluorine atom

2 a non-bonding pair of electrons on the nitrogen, oxygen or fluorine atoms of neighbouring molecules.

You may wonder why other highly electronegative atoms do not form hydrogen bonds. For example, chlorine atoms have a high electronegativity, but they are not involved in hydrogen bonding. This is because they are larger atoms and the electron density is more spread out and less concentrated. This results in weaker dipole–dipole attraction with the hydrogen atoms on neighbouring molecules.

MOLECULAR SIZE AND INTERMOLECULAR FORCES

It is important to remember that dispersion forces occur between all molecules; those that are polar as well as those that are non-polar. In substances with large molecular masses, the dispersion forces between molecules can even be stronger than dipole–dipole attraction and hydrogen bonding.

Table 3.4.2 shows the boiling points of four hydrogen halides: hydrogen fluoride (HF), hydrogen chloride (HCl), hydrogen bromide (HBr) and hydrogen iodide (HI). Of the last three hydrogen halides shown, hydrogen chloride is the most polar of these molecules and therefore has the strongest dipole–dipole attraction. However, hydrogen iodide has the highest boiling point. This is because it has the largest molecular mass and can form stronger dispersion forces, which outweigh the effects of the dipole–dipole attraction seen in hydrogen chloride.

The boiling point of hydrogen fluoride is 19.5°C, which is much higher than any of these other compounds. This is because the hydrogen bonding between hydrogen fluoride molecules is much stronger than both the dispersion forces and the dipole–dipole attraction between the other molecules listed in Table 3.4.2.

TABLE 3.4.2 Comparison of the boiling points of the first four hydrogen halides

Hydrogen halide	Molecular mass	Number of electrons	Boiling point (°C)
hydrogen fluoride (HF)	20.0	10	19.5
hydrogen chloride (HCl)	36.5	18	–85.1
hydrogen bromide (HBr)	80.9	36	–66.8
hydrogen iodide (HI)	127.9	54	–35.4

3.4 Review

OA ✓✓

SUMMARY

- Covalent bonds (intramolecular bonds) are much stronger than intermolecular forces.
- The melting and boiling points of covalent molecular substances increase as the strength of the intermolecular forces increase.
- There are three main types of intermolecular forces: dispersion forces, dipole–dipole attraction and hydrogen bonding.
- Dispersion forces exist between all molecules and are the result of attraction between temporary dipoles that form in molecules.
- Dispersion forces are stronger between larger molecules because it is easier to create temporary dipoles in molecules with a larger number of electrons.
- Dispersion forces are stronger between linear molecules than between highly branched molecules.
- Dipole–dipole attraction is only present between polar molecules and are the result of the attraction between the partial positive and negative ends of the molecules.
- The greater the polarity of a molecule, the stronger the dipole–dipole attraction.
- Hydrogen bonds are the strongest of the three main types of intermolecular forces.
- Hydrogen bonding occurs between highly polar molecules in which one of the molecules has a hydrogen atom covalently bonded to a nitrogen, oxygen or fluorine atom. The other molecule has a non-bonding pair of electrons on a nitrogen, oxygen or fluorine atom.

KEY QUESTIONS

Knowledge and understanding

1 Identify the types of intermolecular forces that exist between:
  a polar molecules
  b non-polar molecules

2 Identify which of the following substances would contain dipole–dipole forces between their molecules: fluorine (F_2), hydrogen iodide (HI), methane (CH_4), tetrafluoromethane (CF_4), fluoromethane (CH_3F).

Analysis

3 In ice, each water molecule is surrounded, at equal distances, by four other water molecules. In each case, there is an attraction between the partial positive hydrogen atom on one water molecule and a non-bonding pair associated with the oxygen atom of another water molecule.
  a Draw a diagram to show the arrangement of four water molecules around another water molecule.
  b Identify the strongest type of intermolecular force present in ice.

4 'Cloudy ammonia' is often used as a cleaning solution in bathrooms. This solution contains ammonia (NH_3) dissolved in water. Draw a diagram to represent hydrogen bonding between a water molecule and an ammonia molecule.

5 Consider the following substances. Identify the strongest type of intermolecular force found between their molecules.
  a PH_3
  b $CHCl_3$
  c CH_3Cl
  d F_2O
  e CO_2
  f HBr
  g H_2S
  h HF
  i CH_3OH
  j H_2

6 For each of the following pairs of substances, predict which has the higher boiling point and explain in terms of the intermolecular forces found in each substance.
  a CHF_3 or CF_4
  b O_2 or CO_2
  c NH_3 or CH_4

7 When sugar is gently heated, it turns into a clear liquid. If the liquid is heated strongly, it turns black and a gas is produced. Explain what is happening to the bonds in sugar when it is heated. Use the terms 'intramolecular bonds' and 'intermolecular forces' in your answer.

3.5 Covalent lattices

In the previous sections, you were introduced to covalent bonding in the context of substances made up of molecules. Alternatively, covalent bonding also occurs between non-metal atoms in a way that results in a continuous three-dimensional **covalent lattice** structure. This section will focus on two key examples of substances with a covalent lattice structure: diamond and graphite.

ALLOTROPES

Allotropes are different forms of the same element.

Some elements can exist with their atoms in several different structural arrangements called **allotropes**. In different allotropes, the atoms are bonded to each other in different, specific ways. This gives them significantly different properties from other allotropes of the same element.

Oxygen forms allotropes. Oxygen gas consists of diatomic molecules with the formula O_2. Each oxygen atom in this arrangement is bound to one other oxygen atom. Ozone is another molecule containing only oxygen. Ozone molecules have the formula O_3 and consist of a central oxygen atom bound to two other oxygen atoms. Figure 3.5.1 shows the structure of these two molecules. As both contain only oxygen atoms, they are allotropes of oxygen. Diamond and graphite, the focus of this section, are both different allotropes of carbon.

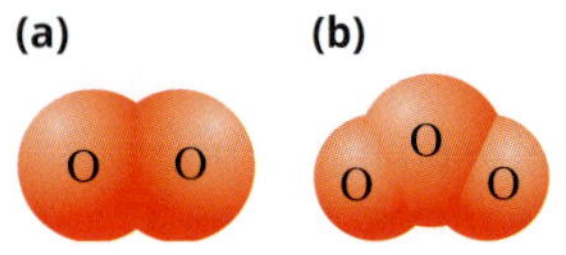

(a) (b)

O O O O O

FIGURE 3.5.1 (a) O_2 oxygen molecule (b) O_3 ozone molecule

ALLOTROPES OF CARBON

Diamonds (Figure 3.5.2) might be a 'girl's best friend', but it is unlikely that graphite (Figure 3.5.3) will ever be held in the same esteem. Both of these minerals are made of the same single element—carbon.

FIGURE 3.5.2 Diamond is the hardest naturally occurring substance.

FIGURE 3.5.3 Natural graphite is soft and black.

Table 3.5.1 summarises some information about the structure, properties and uses of the three most common allotropes of carbon: diamond, graphite and amorphous carbon.

TABLE 3.5.1 Comparison of properties of some of the allotropes of carbon

Allotrope	Structure	Properties	Uses
diamond	Covalent network lattice, each carbon surrounded by four other carbon atoms in a tetrahedral arrangement	• very hard • sublimes • non-conductive • brittle	• jewellery • cutting tools • drills
graphite	Covalent layer lattice, each carbon bonded to three other carbons, one delocalised electron per carbon atom	• conductive • slippery • soft • greasy material	• lubricant • pencils • electrodes • reinforcing fibres
amorphous carbon	Irregular structure of carbon atoms. Many varieties exist with many different, non-continuous packing arrangements	• conductive • non-crystalline • cheap	• printing ink • carbon black filler • activated charcoal • photocopying

Diamond

Diamond is the hardest naturally occurring substance known.

Diamond does not contain small, discrete (individual) molecules. Instead, the carbon atoms bond to each other to form a continuous three-dimensional structure called a **covalent network lattice**. There are no weak intermolecular forces present, only strong covalent bonds. This is what gives diamond its strength.

In general, substances that have a network lattice structure have very high melting points or decomposition temperatures. They are also very hard because the atoms are held firmly in fixed positions in the lattice.

As you saw in previously in Section 3.2, when an atom has four electron pairs in its outer shell, the electron pairs position themselves as far away from each other as possible in a tetrahedral shape. In the covalent network lattice for diamond, shown in Figure 3.5.4, you can also see that individual atoms within diamond form single covalent bonds to four other carbon atoms in a tetrahedral arrangement.

The properties of diamond are directly related to its structure.

- Single covalent bonds between carbon atoms are strong bonds. The entire structure of a diamond consists of a continuous network of these bonds, making diamond very hard and rigid.

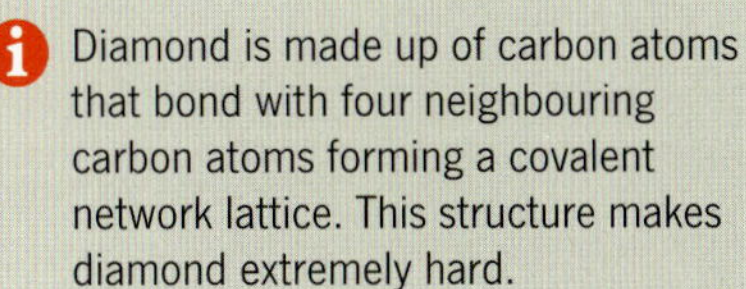
Diamond is made up of carbon atoms that bond with four neighbouring carbon atoms forming a covalent network lattice. This structure makes diamond extremely hard.

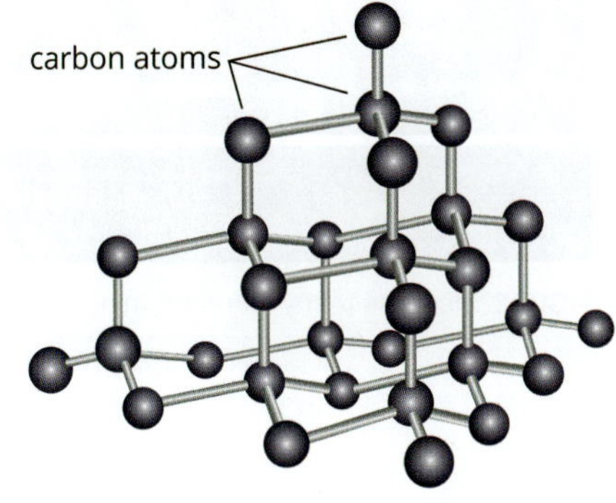

FIGURE 3.5.4 The structure of diamond showing each carbon atom with four single covalent bonds to neighbouring atoms

FIGURE 3.5.5 Diamond-tipped drills used to drill through rock in the oil mining industry

- Since there are only strong covalent bonds between carbon atoms, this makes diamond's **sublimation point** (the point at which a substance goes directly from the solid phase to the gaseous phase, without passing through a liquid phase) very high (around 3500°C).
- The rigidity means that diamonds are brittle and break rather than bend.
- Diamond does not conduct electricity because it does not contain any charged particles that are free to move.
- Because the atoms in diamonds are held together very strongly, the heat conductivity is extremely high. It is five times greater than that of copper, leading to some specialty electronic uses where diamond is used to transfer heat away from important electrical components.

The crystalline appearance of diamonds and their high refractive index make them sparkle and has made them extremely popular as jewellery, but the hardness of diamond also lends itself to industrial uses. Many industrial cutting and drilling tools for working with tough materials are diamond tipped. The drill tips in Figure 3.5.5 are used to drill through rock in the oil mining industry. They contain small pieces of diamond that improve the hardness and durability of the tool.

CHEMFILE

Impact diamonds

'We are speaking about trillions of carats', trumpeted the 2012 headline from the British *Daily Mail*. It was in reference to a 100 km wide meteorite crater, the Popigai Crater, in Russia that could supply world markets with diamonds for 3000 years. The now closed Mirny mine, shown in Figure 3.5.6, is also in Russia. This open-cut mine is over 500 metres deep and has yielded diamonds worth more than $20 billion since 1951.

It is thought that the impact of a large meteorite created enough heat and pressure in the Popigai Crater to make diamonds. Russian scientists are reported to have known of this deposit since 1971, but kept details hidden until supplies from other sources began to run out. Diamonds formed from a meteorite strike, like those in Figure 3.5.7, are referred to as 'impact diamonds'.

FIGURE 3.5.6 The Mirny diamond mine is over 500 metres deep.

FIGURE 3.5.7 High-quality 'impact diamonds' can be almost the size of a 20-cent coin.

CASE STUDY ANALYSIS

Mined versus synthetic diamonds

You might have seen or heard advertisements trying to convince you of the idea that mined diamonds, which might be referred to as 'real' or 'natural' diamonds, are much better than synthetic diamonds that are grown under specific laboratory conditions.

Mined diamonds are produced naturally in a process that takes billions of years. Around 120–150 km deep underground, under high temperature (around 1100°C) and pressure conditions, carbon is crystallised to form diamonds. Then, through volcanic activity, the diamonds are brought closer to the Earth's surface. The diamonds are extracted in large open-cut mines such as the Argyle mine in Western Australia (which closed in November 2020) (Figure 3.5.8).

FIGURE 3.5.8 An aerial photo of the Argyle diamond mine in Western Australia

Synthetic diamonds are grown in laboratories under controlled conditions that match the high temperature and pressure conditions found deep underground. This allows the diamonds to be grown in weeks instead of billions of years. Synthetic diamonds are chemically identical to mined diamonds—they are both made of carbon and have the same covalent network lattice structure. As the conditions for growing synthetic diamonds is highly controlled, they are very pure, which gives them excellent clarity (one of the key indicators of a diamond's value). There are also very few flaws (as these are due to the inclusion of atoms of other elements in the lattice structure). One key difference between the two types of diamond is price. Synthetic diamonds are about a quarter the price of mined diamonds.

Due to the almost identical chemical composition and crystal structure of mined and synthetic diamonds, experts cannot tell the two apart just by looking at them (Figure 3.5.9). More complex analytical techniques are needed to differentiate the two types of diamonds.

FIGURE 3.5.9 Mined and synthetic diamonds look identical.

Analysis

1 Outline the similarities and differences in the chemical structure and composition of mined and synthetic diamonds.

2 Using online resources, outline some of the environmental and social issues linked to the production of mined diamonds.

3 Evaluate the statement 'natural diamonds are real and synthetic diamonds are fake'.

Graphite

In graphite, each carbon atom is covalently bonded to three other carbon atoms. The layered network structure contains delocalised electrons. Bonds within the layers are strong, but bonds between layers are weak dispersion forces.

Graphite is a very different form of carbon. As you can see in Figure 3.5.10, the carbon atoms in graphite are in layers. There are strong covalent bonds between the carbon atoms in each layer. However, there are weak dispersion forces between the layers. As a consequence, it is hard in one direction, but quite slippery and soft in another direction. The structure of graphite is referred to as a **covalent layer lattice**.

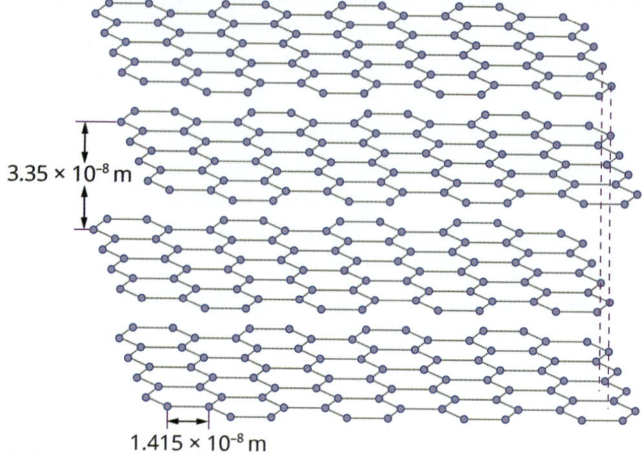

FIGURE 3.5.10 Graphite has a covalent layer lattice structure. The carbon atoms within each layer are covalently bonded to each other. Weak dispersion forces exist between the layers.

The covalent layer lattice structure of graphite also explains some of its other properties.

- The strong covalent bonds between the atoms in each layer explain graphite's resistance to heat. Graphite sublimes at a temperature of about 3600°C.
- Each carbon atom is bonded to three other carbon atoms. The fourth valence electron from each atom is able to move within the layer. The electrical conductivity of graphite is due to these delocalised electrons.

The conductivity of graphite makes it suitable for applications such as battery electrodes where conductivity is required but a metal is not suitable.

Graphite can also be used as a lubricant. The weak dispersion forces between layers allow these layers to slide over each other and to reduce the friction between moving parts, such as in locks or machinery.

Graphite is also used as an additive to improve the properties of rubber products and it can be woven into a fibre. This helps to reinforce plastics. Figure 3.5.11 shows spun graphite fibre, which can be used to make strong composite materials, such as those used in tennis racquets, fishing rods and racing car shells.

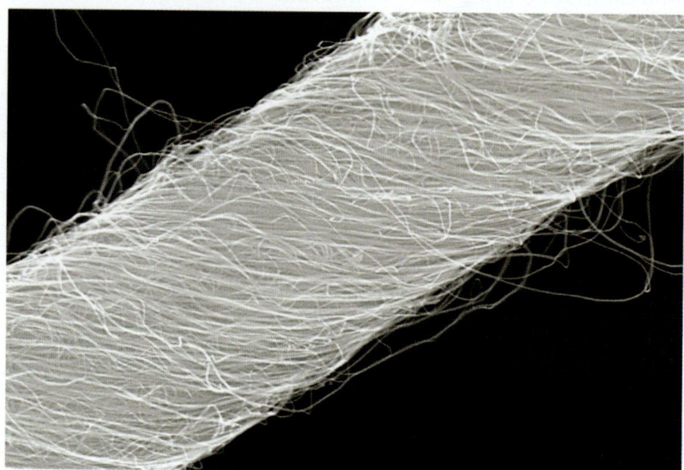

FIGURE 3.5.11 Graphite fibre can be used to reinforce plastics.

CHEMFILE

Black-lead pencils

In 1564, a very pure deposit of graphite was discovered in England. The graphite was so stable that it could be cut into thin, square sticks that could be used for writing. String was wrapped around the graphite to make the first pencils. Later the string was replaced with wood.

The pencils were so effective that during the Napoleonic Wars, the English were considered to have a technological advantage, because their pencil-written communications were far more effective than the French equivalents. Napoleon commissioned a French inventor, Nicholas-Jacques Conte, to develop an alternative to pure graphite. The mixtures of clay and powdered graphite that he designed are the basis for the 'lead' in modern pencils.

An early pencil consisting of a strip of graphite placed between pieces of wood

PA 3

Amorphous forms of carbon

Charcoal (Figure 3.5.12) and carbon black (Figure 3.5.13) are examples of **amorphous** carbon that has no consistent structure. It contains irregularly packed, tiny crystals of graphite and other non-uniform arrangements. Lumps of charcoal are produced for use as a fuel, while carbon black is used to make printer toner ink.

Amorphous carbon can be formed from the **combustion** of wood and other plant matter when there is a limited supply of air. There are several other types of amorphous carbon, including soot, which can be seen in Figure 3.5.14 being emitted from an industrial chimney. The distinctions between the different forms of amorphous carbon are blurred.

FIGURE 3.5.12 Lumps of charcoal are produced for use as a fuel.

FIGURE 3.5.14 Soot is emitted from an industrial chimney.

FIGURE 3.5.13 Carbon black is used in printer toner ink.

Each form of amorphous carbon has its uses and some have been used by society for centuries. Since the Middle Ages it has been common to produce charcoal in ovens. Figure 3.5.15 shows a number of beehive-shaped ovens that were used to produce charcoal from timber. These ovens were built between 1876 and 1879.

FIGURE 3.5.15 These ovens in Nevada, USA, were built between 1876 to 1879 to make charcoal.

Uses of carbon black

Carbon black is a refined type of amorphous carbon in which the particle size is more uniform. Most carbon black is used to reinforce rubber products such as tyres and hoses, causing their black appearance. The surface interaction between the fine carbon particles and the rubber molecules increases the strength and toughness of the product.

Many printer and photocopier toners contain carbon black particles mixed with a binder polymer and other additives. More than 9 million tonnes of carbon black is used annually worldwide.

3.5 Review

OA ✓✓

SUMMARY

- Allotropes are different forms of the same element.
- Carbon can be found in the Earth's crust in the form of diamond, graphite or charcoal. The structures and properties of these allotropes are very different.
- In diamond, each carbon atom is covalently bonded to another four carbon atoms in a tetrahedral shape, forming a covalent network lattice structure. Diamond sublimes at a high temperature, is extremely hard, is a good conductor of heat and has a sparkling, crystalline appearance. Due to its properties, diamonds have diverse applications, such as jewellery and industrial cutting and drilling tools.
- In graphite, each carbon atom is covalently bonded to three other carbon atoms. The layered network structure contains delocalised electrons. Bonds within the layers are strong, but bonds between layers are weak dispersion forces. Graphite is slippery, conducts electricity and sublimes at a high temperature. Due to its properties, graphite has diverse applications, such as an electrical conductor, a lubricant and as a component of strong composite materials.
- Amorphous carbon products, such as carbon black, soot and charcoal, are formed from the combustion of plant and animal matter in a limited supply of air. Amorphous carbon has no consistent structure.

KEY QUESTIONS

Knowledge and understanding

1 How many covalent bonds are formed by each carbon atom in diamond and graphite?

2 **a** What is meant by the word sublime?
b Explain why diamond and graphite only sublime at temperatures over 3500°C.

3 List three uses for amorphous carbon.

Analysis

4 Explain the following properties of diamond in terms of its bonding and structure.
a Hardness or softness
b Ability or inability to conduct electricity
c Ability or inability to conduct heat

5 Explain the following properties of graphite in terms of its bonding and structure.
a Hardness or softness
b Ability or inability to conduct electricity

Chapter review

KEY TERMS

allotrope
amorphous
asymmetrical molecule
combustion
covalent bond
covalent lattice
covalent layer lattice
covalent network lattice
diamond
diatomic molecule
dipole
dipole–dipole attraction
dispersion forces
double covalent bond
electron density
electron group
electronegativity
graphite
hydrogen bond
instantaneous dipole
intermolecular force
intramolecular bond
Lewis structure
molecular formula
molecule
non-bonding electron
non-polar
octet rule
permanent dipole
polar
polarity
polyatomic molecule
pyramidal
sublimation point
single covalent bond
structural formula
symmetrical molecule
temporary dipole
tetrahedral
trigonal planar
triple covalent bond
valence shell electron pair repulsion (VSEPR) theory

REVIEW QUESTIONS

Knowledge and understanding

1 Select the statement that best describes the way hydrogen atoms bond to each other.

A One hydrogen atom donates an electron to another hydrogen atom to form a molecule.
B Hydrogen atoms form a lattice with delocalised electrons.
C Hydrogen atoms share electrons to obtain a complete outer shell of eight electrons.
D Two hydrogen atoms share an electron each to form a hydrogen molecule.

2 The formula of a molecule is XY_4. Select the alternative that could match this formula.

A OH_4
B CH_4
C HBr_4
D CO_4

3 Oxygen forms a compound with fluorine with the molecular formula OF_2. Identify the correct shape of the molecule.

A bent
B linear
C pyramidal
D tetrahedral

4 The following substances all contain carbon atoms. Which one will have the highest boiling point (or sublimation point)?

A carbon dioxide
B methane
C graphite
D methanol

5 Solid ammonia (NH_3) has a melting point of –73°C. Explain what happens to the bonds in ammonia when it melts. Use the terms 'intramolecular bonds' and 'intermolecular forces' in your answer.

6 Explain why neon atoms do not form covalent bonds.

7 All of the following molecules have four electron groups around the central atom. Classify the molecular shapes as bent, pyramidal or tetrahedral.

a

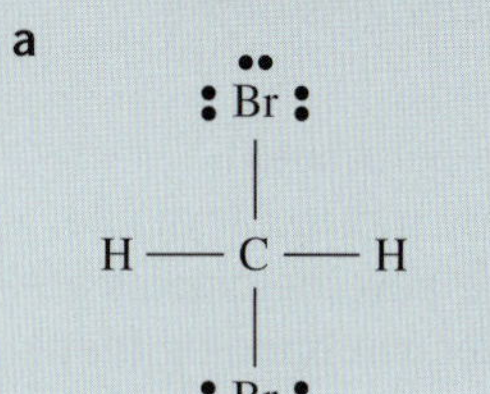

b

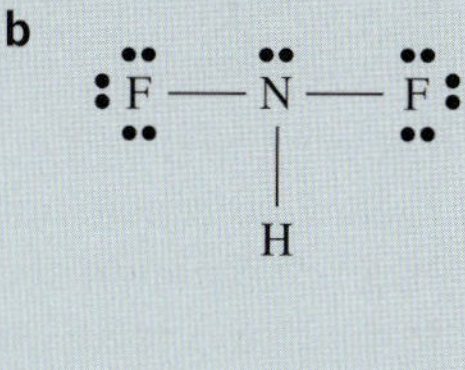

c

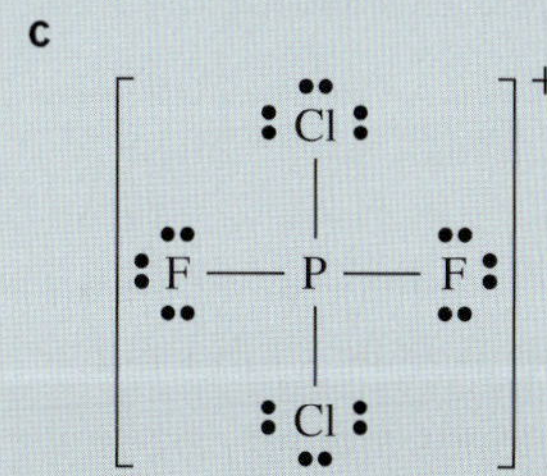

d

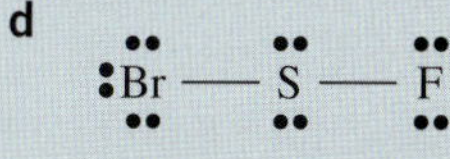

e

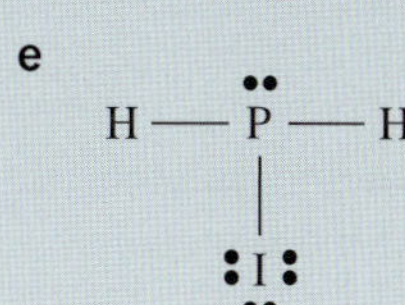

8 Match the molecular formula to the correct molecular shape.

Molecular formula	Molecular shape
nitrogen tribromide (NBr_3)	tetrahedral
water (H_2O)	linear
difluoromethane (CH_2F_2)	bent
hydrogen cyanide (HCN)	pyramidal

9 Use the electronegativities from Figure 3.3.2 to order the following covalent bonds from least to most polar. Si–O, N–O, F–F, H–Br, O–Cl

10 Use the electronegativities from Figure 3.3.2 to determine which of the following molecules contains the most polar bond.

a CO_2
b H_2O
c H_2
d H_2S
e NH_3

11 Water is a polar molecule. Explain how this fact shows that water is not a linear molecule.

12 Hydrogen chloride (HCl) exists as a gas at room temperature. What can you conclude about the strength of the intermolecular forces in pure hydrogen chloride?

13 At room temperature, CCl_4 is a liquid, whereas CH_4 is a gas.

a Which substance has the stronger intermolecular forces?
b Explain the difference in the strengths of the intermolecular forces.

14 Explain the difference between a permanent molecular dipole and a temporary molecular dipole. Your explanation should describe how the dipoles are formed and the type of intermolecular bonding that results.

15 'Carbon forms several allotropes.' Explain the meaning of this statement.

16 Why does diamond have such a high sublimation point?

17 Explain why graphite sublimes at a high temperature, conducts electricity and can be used as a lubricant.

18 Describe the geometry of the bonds around carbon atoms in diamond and graphite.

Application and analysis

19 Examine the following Lewis structure and use the VSEPR theory to predict the shape of the molecule.

H — Ö — Br:

20 Are the following molecules polar or non-polar? Draw structural formulas to help you decide.

a CS_2
b Cl_2O
c SiH_4
d CH_3Cl
e CH_3CH_3
f CCl_4

21 For each of the following structures, state whether:

i the molecule is polar or non-polar
ii the strongest intermolecular forces of attraction between molecules of each type would be dispersion forces, dipole–dipole attraction or hydrogen bonding.

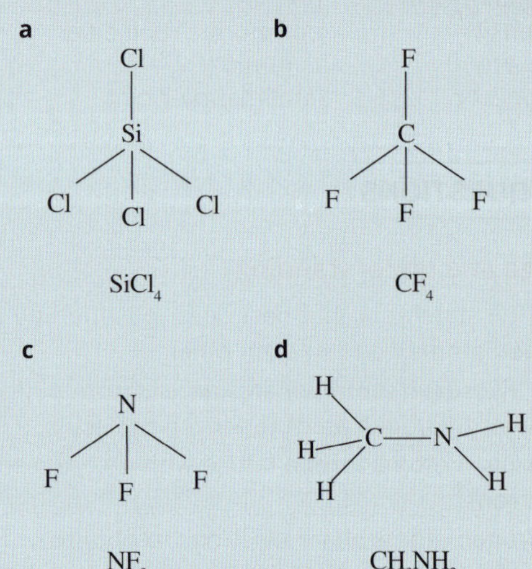

22 Consider solid samples of the following compounds. In which cases are there only be dispersion forces between molecules? (Hint: First draw the structural formula for each molecule, then determine whether each one is polar or non-polar.)

a tetrachloromethane (CCl_4)
b methanal (CH_2O)
c carbon dioxide (CO_2)
d hydrogen sulfide (H_2S)

23 Draw Lewis structures for each of the following molecules and identify the number of bonding and non-bonding electrons in each molecule.

a HBr
b CCl_2F_2
c C_2H_6
d PF_3

24 Identify whether each of the following statements about carbon dioxide (CO_2) is true or false.

a Carbon dioxide is a molecular compound.

b A molecule of carbon dioxide contains three atoms.

c The bonds between the carbon and oxygen atoms in carbon dioxide are intramolecular bonds.

d There are two single covalent bonds in a molecule of carbon dioxide.

e There are four non-bonding pairs of electrons in a molecule of carbon dioxide.

25 Differentiate between the non-metallic elements argon (Ar) and chlorine (Cl) in terms of their electronic configurations and the types of bonds their atoms form with other atoms.

26 The melting points of four halogens are given in the table below. Describe and explain the trend in melting points of these elements.

Halogen	Melting point (°C)
fluorine (F_2)	–220
chlorine (Cl_2)	–101
bromine (Br_2)	–7
iodine (I_2)	114

27 Consider the two compounds OF_2 and CF_4. OF_2 has a boiling point of –145°C and CF_4 has a boiling point of –128°C. Between molecules of which compound would the intermolecular forces of attraction be greater? Explain your answer in terms of the polarity of each of the molecules and the relative strength of the intermolecular forces in each substance.

28 The mass of a hydrogen fluoride molecule is similar to the mass of a neon atom. However, the boiling points of these substances are very different. The boiling point of hydrogen fluoride is 19.5°C whereas that of neon is –246°C. Explain the difference in this property of the two substances.

29 The structures of methane and diamond are shown in the image below. Each carbon atom in methane (CH_4) has a tetrahedral arrangement of atoms around it. A carbon atom in diamond also has a tetrahedral arrangement. However, the two substances have very different properties.

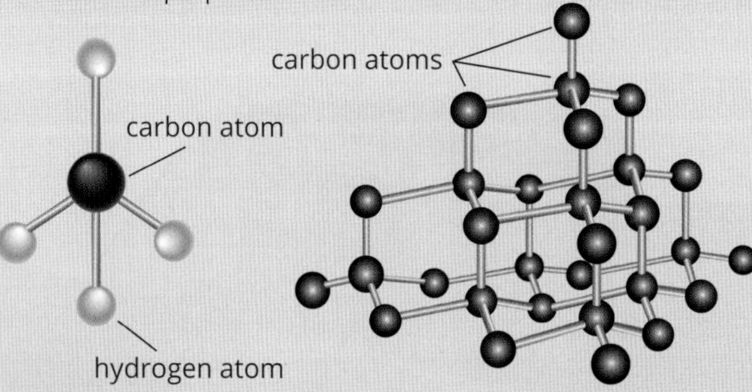

a Describe all of the types of bonding that would be present in each substance.

b Use the types of bonding present in each substance to explain the different properties you would expect each to have.

30 The table below lists some key properties of diamond and graphite.

Material	Hardness (Moh scale)	Electrical conductivity, S m^{-1}	Thermal conductivity, W m^{-1} K^{-1}
diamond	10	10^{-13}	2000
graphite	<1	3×10^5	200

a Explain the difference in hardness and electrical conductivity of diamond and graphite, referring to the structure and bonding of each substance.

b When discussing the conductivity of diamond and graphite, why it is important to be specific as to what type of conductivity you are referring to?

c One method used by jewellers to test the quality of diamonds is to heat them to 1000°C and time the rate of cooling. How will a high-quality diamond compare to a low-quality one in this test?

OA ✓✓

CHAPTER

04 Metals

At the end of this chapter, you will be able to describe the properties and uses of metals. You will see that the properties of metallic elements differ from those of non-metals. You will also understand how these properties are used to make products for many different purposes.

You will learn how chemists have been able to relate these properties to the structure of metals and be able to explain their structure in terms of a metallic bonding model.

The reactivities of metals can be determined on the basis of their reactions with water, acids and oxygen, and these reactions can be used to compare their reactivities.

Finally, you will examine how and why metals are recycled, and also investigate some of the developments taking place in the recycling industry to achieve more sustainable economic and industrial development.

Key knowledge

- the common properties of metals (lustre, malleability, ductility, melting point, heat conductivity and electrical conductivity) with reference to the nature of metallic bonding and the existence of metallic crystals **4.1**
- experimental determination of a reactivity series for metals based on their relative ability to undergo oxidation with water, acids and oxygen **4.2**
- metal recycling as an example of a circular economy where metal is mined, refined, made into a product, used, disposed of via recycling and then reprocessed as the same original product or repurposed as a new product. **4.3**

OA
✓✓

4.1 Metallic properties and bonding

More than 80% of the elements in the periodic table are metals (Figure 4.1.1).

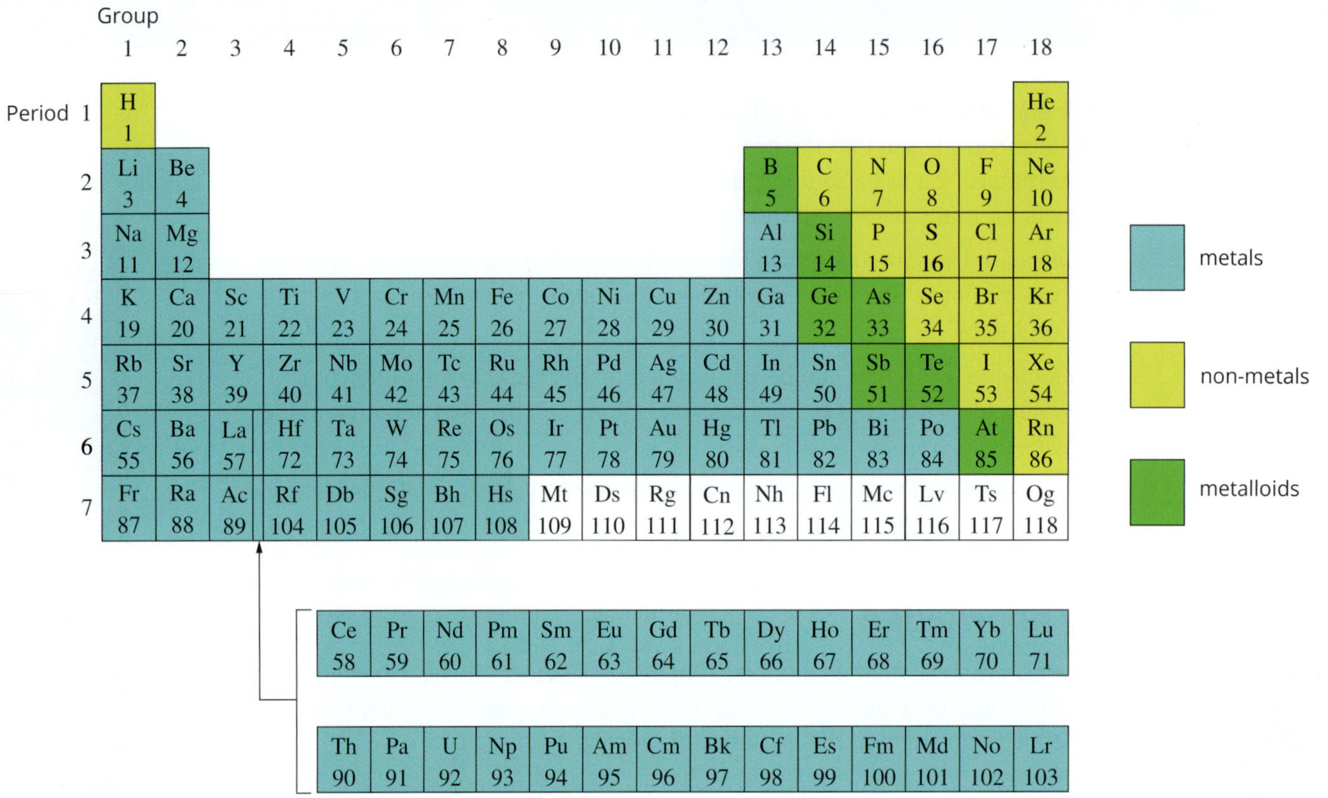

Period \ Group	1	2	3	4	5	6	7	8	9	10	11	12	13	14	15	16	17	18
1	H 1																	He 2
2	Li 3	Be 4											B 5	C 6	N 7	O 8	F 9	Ne 10
3	Na 11	Mg 12											Al 13	Si 14	P 15	S 16	Cl 17	Ar 18
4	K 19	Ca 20	Sc 21	Ti 22	V 23	Cr 24	Mn 25	Fe 26	Co 27	Ni 28	Cu 29	Zn 30	Ga 31	Ge 32	As 33	Se 34	Br 35	Kr 36
5	Rb 37	Sr 38	Y 39	Zr 40	Nb 41	Mo 42	Tc 43	Ru 44	Rh 45	Pd 46	Ag 47	Cd 48	In 49	Sn 50	Sb 51	Te 52	I 53	Xe 54
6	Cs 55	Ba 56	La 57	Hf 72	Ta 73	W 74	Re 75	Os 76	Ir 77	Pt 78	Au 79	Hg 80	Tl 81	Pb 82	Bi 83	Po 84	At 85	Rn 86
7	Fr 87	Ra 88	Ac 89	Rf 104	Db 105	Sg 106	Bh 107	Hs 108	Mt 109	Ds 110	Rg 111	Cn 112	Nh 113	Fl 114	Mc 115	Lv 116	Ts 117	Og 118

Ce 58	Pr 59	Nd 60	Pm 61	Sm 62	Eu 63	Gd 64	Tb 65	Dy 66	Ho 67	Er 68	Tm 69	Yb 70	Lu 71
Th 90	Pa 91	U 92	Np 93	Pu 94	Am 95	Cm 96	Bk 97	Cf 98	Es 99	Fm 100	Md 101	No 102	Lr 103

FIGURE 4.1.1 The distribution of metals in the periodic table

The development of many civilisations has been measured by the way we have used metals. For example, in many parts of the world, the Copper Age (5000–3000 BCE) was followed by the Bronze Age (3000–1000 BCE) and the Iron Age (from 1000 BCE–800 CE).

Gold, silver and copper can be found on Earth in an almost pure form. These metals were used by prehistoric humans to make ornaments, tools and weapons. As humans' knowledge of metallurgy (the science of modifying metals) has developed, metals have played a central role in fields as diverse as construction, agriculture, art, medicine and transport.

The diverse properties of different metals make them suitable for many purposes. Table 4.1.1 shows the uses of some metals. For example, titanium (Figure 4.1.2) is a very strong, relatively unreactive metal with a low density close to that of bone. Consequently, it is used in surgical implants that can last up to 20 years with little effect on the body. Titanium is also used in the aerospace industry, in art and architecture, and in sporting products such as golf clubs.

TABLE 4.1.1 Properties and uses of some metals

Metal	Properties	Uses
iron	soft, malleable, magnetic, good thermal and electrical conductor, fairly reactive, readily forms alloys	can corrode and is usually converted to more stable steel, which is used in buildings and bridges, automobiles, machinery and appliances
aluminium	low density, relatively soft when pure, excellent thermal and electrical conductor, malleable and ductile, good reflector of heat and light, readily forms alloys	saucepans, frying pans, drink cans, cooking foil, food packaging, roofing, window frames, appliance trim, decorative furniture, electrical cables, aircraft and boat construction
titanium	very strong, high melting point, low density, low reactivity, readily forms alloys	medical devices within the body, wheelchairs, computer cases; lightweight alloys are used in high-temperature environments such as spacecraft and aircraft
gold	shiny gold appearance, excellent thermal and electrical conductor, unreactive, readily forms alloys	electrical connections, jewellery, monetary standard, dentistry

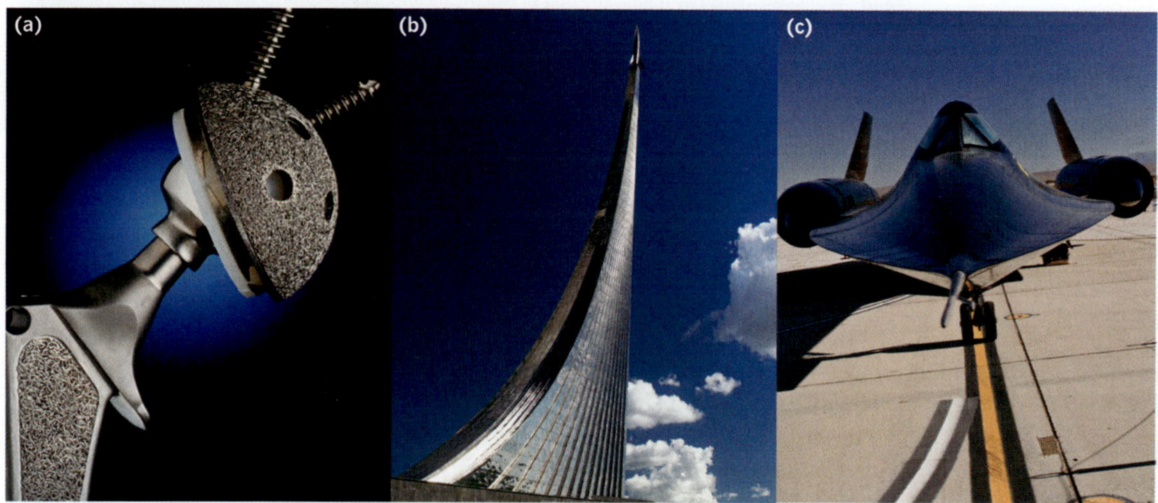

FIGURE 4.1.2 Titanium has many uses: (a) a replacement hip socket, (b) the spectacular curved space museum building in Moscow and (c) the SR-71 Blackbird reconnaissance aircraft.

In this section, you will examine the properties of metals and learn about the bonding model that chemists have developed to explain these properties. This model has helped chemists and materials engineers to understand why metals behave the way they do.

GENERAL PROPERTIES OF METALS

Table 4.1.2 gives the properties of some metals and non-metals. Despite the varying properties of metals, most metals:

- have relatively high melting points and boiling points
- are good **conductors** of electricity
- are good conductors of heat
- generally have high **densities** (mass per unit volume).

Not all metals have all of these properties. Mercury is a liquid at room temperature—it has an unusually low melting point. The group 1 elements (the **alkali metals**) have some properties that make them different from most other metals. They are all soft enough to be cut with a knife and they react vigorously with water to produce hydrogen gas. Both mercury and the group 1 elements exhibit most of the other properties listed above and are classified as metals.

TABLE 4.1.2 Properties of some metallic and non-metallic elements

Element	Melting point (°C)	Boiling point (°C)	Electrical conductivity (MS m^{-1})*	Thermal conductivity (J s^{-1} m^{-1} K^{-1})†	Density (g mL^{-1})
Metals					
gold	1063	2970	45	310	19.3
iron	1540	3000	9.6	78	7.86
mercury	–39	357	1	8.4	13.5
potassium	64	760	14	100	0.86
silver	961	2210	60	418	10.5
sodium	98	892	21	135	0.97
Non-metals					
carbon (diamond)	3550	‡	10^{-17}	–	3.51
oxygen	–219	–183	–	0.026	1.15 (liquid)

*MS m^{-1} = megasiemens per metre; the unit of electrical conductivity. Conductivity is inversely related to resistance, so silver conducts electricity well and offers little resistance to the flow of electricity.
†Thermal conductivity measures the conductance of heat.
‡Diamond sublimes (changes straight from a solid to a gas) when heated.

FIGURE 4.1.3 This power transmission tower relies on the strength of iron in steel for its structural integrity. The electricity cables are made from aluminium, utilising its ductility and electrical conductivity.

Metals also generally have the following characteristics in common. They:

- are **malleable** (can be shaped by beating or rolling)
- are **ductile** (can be drawn into a wire)
- are **lustrous** (reflective when freshly cut or polished)
- are often hard, with high **tensile strength** (a measure of the force needed to break something or cause it to fail, such as a cable or a piece of metal in the wing of a plane)
- have low **ionisation energies** and electronegativities (which means they readily lose electrons).

These properties can allow different metals to be used together to solve many engineering problems. The power transmission tower in Figure 4.1.3 is made of a variety of metals to take advantage of their different properties.

Between group 2 and group 13 in the periodic table is a block of particularly useful elements, known as the **transition metals**. (See the periodic table at the end of the book.) They include metals such as iron and nickel, which are used to build cities, bridges, cars and railway lines, and precious metals such as silver and gold, which have ornamental and economic uses. Most transition metals are silver-coloured and are similar in appearance, as can be seen in Figure 4.1.4.

Metals are often shaped for use in different applications by hammering, exploiting their malleability. Some metals, such as gold, copper and aluminium, are very malleable at room temperature. Other metals, such as iron, must be heated before they can be shaped. Compared to the main group metals, transition metals tend to be harder, have higher densities and melting points, and some of them have strong magnetic properties.

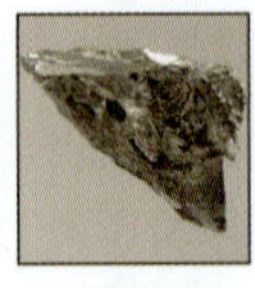

Sc Ti V Cr Mn Fe Co Ni Cu Zn

FIGURE 4.1.4 The elements in the first row of the transition metals

Often, metals are mixed with small amounts of another substance, usually another metal or carbon. The substances are melted together, mixed and then allowed to cool. The resultant solid is called an **alloy**. By varying the composition of alloys, you can obtain materials with specific properties. Generally, an alloy is harder and melts at a lower temperature than a pure metal. For example, transition metals and carbon are combined with iron to produce an alloy called **steel**, which is stronger and has other more desirable characteristics than pure iron.

Copper is one of the few metals that is mainly used in its pure form. It has a high electrical conductivity and so is used for most of the millions of kilometres of electrical wires that enable the transmission of electric energy for heating, lighting and to power the many devices used in our homes.

Most metals are similar in appearance, being lustrous (reflective) and silvery-grey. Gold and copper are notable exceptions. Gold is a yellow-coloured metal; copper is reddish.

CHEMFILE

Beryllium

Beryllium is the fourth element in the periodic table. It is one of the lightest metals, its density being two-thirds the density of aluminium. Beryllium is non-magnetic and has six times the stiffness of steel. Both the Space Shuttle and the Spitzer Space Telescope use beryllium due to its strength and low weight. NASA's next-generation James Webb Space Telescope (see figure below), which was launched in 2021, depends on a 6.5-metre mirror constructed using beryllium to see objects 200 times fainter than those previously visible.

Some of the 18 mirror segments of the James Webb Space Telescope. The mirrors are supported by beryllium ribs that maintain the mirror's shape under extreme conditions.

FORMING METAL IONS

Metallic elements are found on the left-hand side of the periodic table. The atoms of metals are generally larger than the atoms of non-metallic elements within a period and the **effective nuclear charge** of their atoms is lower. It takes less energy to remove electrons from an outer shell when an atom is large, so the ionisation energy of metals is usually lower than for non-metals in the same period. As a consequence, metal atoms tend to lose their outer-shell electrons to form positive ions, called **cations**.

Atoms of metals typically have one, two or three electrons in their outer shell. The cations that are formed when these metal atoms lose these valence electrons have a stable noble gas electronic configuration, with eight electrons in their outer shell.

Worked example 4.1.1 shows how you can find the charge on a metal cation based on the electronic configuration of its uncharged atom. Note that the charge on many metal cations can be quickly found from their group number in the periodic table:

- group 1 metals (e.g. Na) form 1+ charge cations (e.g. Na^+)
- group 2 metals (e.g. Mg) form 2+ charged cations (e.g. Mg^{2+})
- group 13 metals (e.g. Al) form 3+ charged cations (e.g. Al^{3+})

Worked example 4.1.1

DETERMINING CHARGES

Write the electronic configuration of a calcium atom and hence determine the charge of a calcium cation.

Thinking	Working
Unreacted calcium atoms have the same number of protons and electrons.	Atomic number (Z) of calcium is 20: number of protons is 20, number of electrons is 20
The electrons in an atom are in shells.	Shell configuration of calcium: 2,8,8,2
Only the outer-shell electrons will be lost.	Outer shell contains two electrons, 20 – 2 = 18 electrons remaining
Cation charge = number of protons – number of electrons	Cation charge = 20 – 18 = 2+ *(Since Ca is in group 2 of the periodic table, its cation charge will be 2+.)*

Worked example: Try yourself 4.1.1

DETERMINING CHARGES

Write the electronic configuration of an aluminium atom and hence determine the charge of an aluminium cation.

CASE STUDY **ANALYSIS**

Colourful transition metal compounds

Transition metal compounds display a wide range of different colours. They are extensively used as pigments in paints, and to colour glass, ceramics and enamel. In Figure 4.1.5, the colours used by the artist are a result of the different transition metals present. *Five Bells* was painted by John Olson in 1963, and the colours are still as vivid today as when they were painted.

FIGURE 4.1.5 *Five Bells* (1963) by Australian artist John Olson

FIGURE 4.1.6 *Untitled* (2010) by Indigenous Australian artist Mavis Ngallametta

Some of the pigments that are widely used by artists are:

- cobalt blue, $CoAl_2O_4$
- cadmium yellow, CdS
- Prussian blue, $Fe_7(CN)_{18}$
- chrome green, CrO_3
- chrome yellow, $CdPbO_4$
- cerulean blue, Co_2SnO_4

Ochre is a type of hard clay that contains iron oxides and hydroxides, such as Fe_2O_3 and FeO(OH). Ochre occurs naturally in many colours, including red, pink, white and yellow. When ground into a powder and mixed with liquids, ochre forms a paste that has been used for millennia by Aboriginal and Torres Strait Islander peoples for body decoration, cave painting, bark painting and other artwork (Figure 4.1.6).

The colours of many gemstones are also due to the presence of transition metals. For example, sapphires (Figure 4.1.7) contain traces of titanium and iron in a crystal lattice of aluminium oxide, Al_2O_3.

The colours arise when electrons within the metal ions in the compounds absorb light of particular wavelengths and move to higher energy levels. By contrast, compounds of other metals do not usually absorb wavelengths of visible light and are therefore colourless.

FIGURE 4.1.7 Blue sapphires get their colour from trace amounts of titanium and iron.

Analysis

1. Use Figure 4.1.1, on page 134, and the periodic table at the back of the book to answer the following questions.
   a. Which elements in the paint pigments listed above are metals?
   b. Which metals in the paint pigments are transition metals and therefore responsible for the colours of the pigments?
2. Both rubies and sapphires have a crystal lattice of aluminium oxide, Al_2O_3, but rubies are red whereas sapphires are blue. Research the transition metal component that gives rubies their red colour.
3. Diamonds are composed of the element carbon and are usually colourless. Suggest why diamonds are not coloured like most gemstones.

CONNECTING PROPERTIES AND STRUCTURE

Some properties of metals are listed in Table 4.1.3. Each of these properties gives some information about the structure and bonding of particles in metals.

TABLE 4.1.3 The physical properties of metals and resulting conclusions about metal structure and bonding

Property	What this tells us about structure
Metals are usually hard and tend to have high boiling points.	The forces between the particles must be strong.
Metals conduct electricity in the solid state and in the molten liquid state.	Metals have charged particles that are free to move.
Metals are malleable and ductile.	The attractive forces between the particles must be stronger than the repulsive forces between the particles when the layers of particles are moved.
Metals generally have high densities.	The particles are closely packed in a metal.
Metals are good conductors of heat.	There must be a way of quickly transferring energy throughout a metal object.
Metals are lustrous or reflective.	Free electrons are present, so metals can reflect light and appear shiny.
Metals tend to react by losing electrons.	Electrons must be relatively easily removed from metal atoms.

Chemists have developed a model for the structure of metals to explain all the properties that have been mentioned so far. You can deduce from the information in Table 4.1.3 that the **metallic bonding model** must include:

- charged particles that are free to move and conduct electricity
- strong forces of attraction between particles throughout the metal structure
- some electrons that are relatively easily removed.

The model that has been developed describes the arrangement of particles within a single metal **crystal**. A crystal is a region in a solid where the particles are arranged in a regular way. A sample of a solid metal consists of an extremely large number of small crystals which are not usually visible to the naked eye.

METALLIC BONDING MODEL

Electrons are the particles that enable metals to conduct electricity. Negatively charged electrons can be lost from the outer shell of metal atoms, forming positive ions (cations). These valence electrons are able to move within the **lattice** (tightly packed regular arrangement) of cations in the crystal. As shown in Figure 4.1.8, the freed electrons **delocalise** (spread through a large area) to form a 'sea' of electrons throughout the entire metal structure.

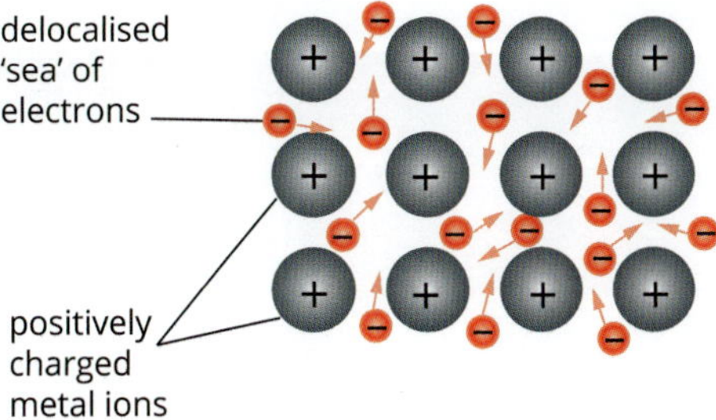

FIGURE 4.1.8 The metallic bonding model. Positive metal cations in a lattice are surrounded by a mobile sea of delocalised electrons. This diagram shows just one layer of ions in a metal crystal.

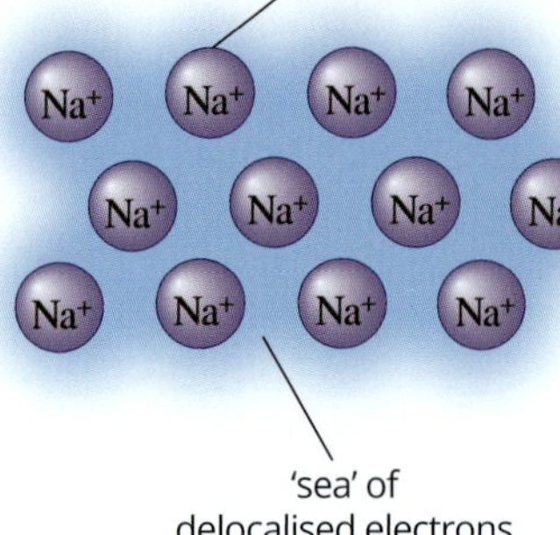

FIGURE 4.1.9 A representation of a sodium metal lattice. Each sodium atom loses its one valence electron. This electron is shared with all atoms in the lattice to form a sea of delocalised electrons.

> In the metallic bonding model, positive metal cations are surrounded by a sea of delocalised electrons.

The metallic bonding model suggests that, in a solid sample of a metal:

- positive ions are arranged in a closely packed structure within crystals. This structure is described as a regular, three-dimensional lattice of positive ions. The cations occupy fixed positions in the lattice
- negatively charged electrons move freely throughout the crystal lattice. These electrons are called **delocalised electrons** because they belong to the lattice as a whole, rather than staying in the shell of a particular atom
- the delocalised electrons come from the outer shells of the atoms. Inner-shell electrons are not free to move throughout the lattice and remain firmly bonded to individual cations
- the positive cations are held in the lattice by the strong electrostatic force of attraction between these cations and the delocalised electrons. This attraction extends throughout the lattice and is called **metallic bonding**.

Together, these ideas make up the metallic bonding model. An example of how a metal, such as sodium, could be represented using this model is shown in Figure 4.1.9.

Explaining the properties of metals

Table 4.1.4 shows how the metallic bonding model is consistent with the relatively high boiling point, electrical conductivity, malleability and ductility of metals.

TABLE 4.1.4 Physical properties of metals and explanations from metallic bonding model

Property	Explanation	
Metals are hard and have relatively high boiling points.	Strong electrostatic forces of attraction between positive metal ions and the sea of delocalised electrons holds the metallic lattice together.	
Metals are good conductors of electricity.	Free-moving delocalised electrons will move towards a positive electrode and away from a negative electrode in an electric circuit.	
Metals are malleable and ductile.	When a force causes metal ions to move past each other, layers of ions are still held together by the delocalised electrons between them.	

Other properties of metals

Metals generally have a high density. The cations in a metal lattice are closely packed. The density of a metal depends on the mass of the metal ions, their radius and the way in which they are packed in the lattice.

Metals are also good conductors of heat. When the delocalised electrons bump into each other and into the metal ions, they transfer energy to their neighbour. Heating a metal gives the ions and electrons more energy and they vibrate more rapidly. The electrons, being free to move, transmit this energy rapidly throughout the lattice.

Because of the presence of free electrons in the lattice, metals reflect light of all wavelengths and appear shiny, or lustrous.

Metals also tend to react by losing electrons. The delocalised electrons in metals may participate in reactions anywhere on the metal's surface. The reactivity of a metal depends on how easily electrons can be removed from its atoms. This is covered in more detail in Section 4.2.

Limitations of the metallic bonding model

Although this model of metallic bonding explains many properties of metals, some cannot be explained so simply. These include the:

- range of melting points, hardness and densities of different metals
- differences in electrical conductivities of metals
- magnetic nature of metals such as cobalt, iron and nickel.

Scientists have developed other models that explain these properties of metals, but these models are beyond the scope of this book.

Worked example 4.1.2

ELECTRONIC CONFIGURATION OF ALUMINIUM

With reference to the electronic configuration of aluminium, explain why solid aluminium can conduct electricity.

Thinking	Working
Using the atomic number of the element, determine the electronic configuration of its atoms. (You may need to refer to the periodic table.)	Al has an atomic number of 13. This means a neutral atom of aluminium has 13 electrons, with an electronic configuration of 2,8,3.
From the electronic configuration, find how many outer-shell electrons are lost to form cations that have a stable, noble gas electronic configuration. These electrons become delocalised.	Al has 3 electrons in its outer shell. Al atoms will tend to lose these 3 valence electrons to form a cation with a charge of 3+. The outer-shell electrons become delocalised and form the sea of delocalised electrons within the metal lattice.
An electric current is possible when there are free-moving charged particles.	If the Al is part of an electric circuit, the delocalised electrons are able to move through the lattice towards a positively charged electrode.

Worked example: Try yourself 4.1.2

ELECTRONIC CONFIGURATION OF MAGNESIUM

With reference to the electronic configuration of magnesium, explain why solid magnesium can conduct electricity.

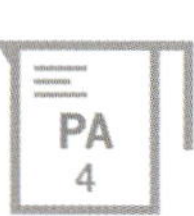

4.1 Review

OA ✓✓

SUMMARY

- Metals have the following characteristic properties:
  - high boiling points
  - good conductors of electricity in solid and liquid states
  - malleable and ductile
  - high densities
  - good conductors of heat
  - lustrous
  - low electronegativities
  - low ionisation energies
  - react by losing electrons.
- Metallic bonding is the electrostatic force of attraction between a lattice of positive ions and delocalised valence electrons. The lattice of cations is surrounded by a sea of delocalised electrons.
- The metallic bonding model can be used to explain the properties of metals, including their malleability, thermal conductivity, generally high melting point and electrical conductivity.

KEY QUESTIONS

Knowledge and understanding

1 List three properties common to most metals.

2 **a** Use the information in Table 4.1.2 on page 136 to decide which metals you would select if you wanted a good electrical conductor.
  b What other factors might influence your choice?

3 Explain the meaning of the term 'ductile' when referring to metals. Identify how this property is different from malleability.

4 The properties of calcium mean that it is classed as a metal.
  a Draw a diagram to represent a calcium metal lattice.
  b Describe the forces that hold this lattice together.

5 Barium is an element in group 2 of the periodic table. It has a melting point of 850°C and conducts electricity in the solid state. Describe how the properties of barium can be explained in terms of its bonding and structure.

Analysis

6 Use the information in Table 4.1.2 on page 136 to answer the following questions.
  a Potassium is classed as a metal. Which of its properties are similar to those of the metal gold? In what ways is it different?
  b Identify another element in the table that has similar properties to potassium.
  c Identify another metal in the table that has similar properties to gold.
  d Where are these four metals (from parts **b** and **c**) in the periodic table?

7 Determine the charge of the cations formed from the following metals if they lost all of their outer-shell electrons.
  a Li
  b Mg
  c Ga
  d Ba

8 Explain how the metallic properties of metals and their first ionisation energy are related. The trends in ionisation energies are described in Table 2.5.5 in Chapter 2 (page 92).

4.2 Reactivity of metals

In the previous section, you learned that metals have many common properties. Metallic elements can also have very different properties. These include their **reactivity** with water, oxygen and acids. 'Reactivity' is a term used to describe the ease with which a chemical can undergo reaction. Some metals are extremely reactive and others are much less so. Gold and platinum are essentially unreactive and are described as **inert**.

This section will look at how the relative reactivity of different metals can be determined experimentally.

DETERMINING THE REACTIVITY OF METALS

To determine the relative reactivity of each metal you can look at what happens when they are exposed to water, oxygen or acids.

Reactivity with water

Figure 4.2.1 shows the reaction of potassium, a group 1 metal, with water. Potassium hydroxide and hydrogen gas are produced. Enough heat is generated to instantly melt the potassium and ignite the hydrogen. The vigour of the reaction is an indication of the reactivity of the metal. Potassium has a high reactivity with water, which is characteristic of the group 1 metals.

$$\text{potassium} + \text{water} \longrightarrow \text{potassium hydroxide} + \text{hydrogen gas}$$

$$2K(s) + 2H_2O(l) \longrightarrow 2KOH(aq) + H_2(g)$$

FIGURE 4.2.1 When water is dropped onto metallic potassium, potassium hydroxide and hydrogen gas are produced.

Trends in the reactivity of metals with water

Table 4.2.1 describes the reaction of some metals with water.

As the table indicates, some metals are so reactive that they must be handled with great care. Sodium and potassium, for example, need to be stored under oil to prevent the metal coming into contact with moisture in the atmosphere. On the other hand, transition metals are generally less reactive with water and oxygen. Most react only slowly or not at all. For example, iron reacts with water and oxygen over an extended period of time to form hydrated iron oxide, which we call rust.

TABLE 4.2.1 Reaction of selected metals with water

Period	Group	Element	Reaction with water
2	1	lithium	Lithium floats on water, producing a steady 'fizz' of hydrogen gas and becoming smaller, until it eventually disappears.
3	1	sodium	Sodium reacts vigorously, producing enough energy to melt the sodium, which fizzes and skates on the water surface.
4	1	potassium	Potassium reacts violently, making crackling sounds as the heat evolved ignites the hydrogen produced by the reaction.
5	1	rubidium	Rubidium explodes violently on contact with water, producing hydrogen gas.
3	2	magnesium	Magnesium does not react with water at room temperature, but will react with steam.
4	2	calcium	Calcium reacts slowly with water at room temperature.
5	2	strontium	Strontium reacts more vigorously than calcium with water. The strontium metal sinks and after a short while bubbles of hydrogen are evident.
3	13	aluminium	Aluminium metal rapidly develops a thin layer of aluminium oxide that prevents the metal from reacting with water. It reacts with steam to form hydrogen gas.
4	8 transition metal	iron	Over an extended period, iron reacts with water and oxygen to form hydrated iron oxide, which is called rust.
4	11 transition metal	copper	Copper does not react appreciably with water or steam.
6	11 transition metal	gold	Gold does not react with water, even as steam.

Reactivity with oxygen

Many metals also react with oxygen to form metal oxides. The group 1 metals all react rapidly with oxygen. Figure 4.2.2 shows sodium metal burning in a container of pure oxygen. The sodium atoms and oxygen molecules rearrange to form a new compound, sodium oxide. The word and formula equations for this reaction are:

$$\text{sodium metal} + \text{oxygen gas} \longrightarrow \text{solid sodium oxide}$$

$$4Na(s) + O_2(g) \longrightarrow 2Na_2O(s)$$

The group 2 metals also react with oxygen to form oxides, although not as rapidly as group 1 metals. Heat is usually required to start the reaction.

While the transition metals are less reactive with oxygen than the metals in groups 1 and 2, their reactions are also important. Iron forms rust (hydrated iron oxide) when exposed to oxygen and water over a period of time. Many transition metals needed by society cannot be found in nature as a pure element, but often exist as oxides.

FIGURE 4.2.2 Sodium burning in pure oxygen

Iron, copper, titanium and aluminium are all mined as oxides and must be processed to obtain the finished metal. Figure 4.2.3 shows copper metal embedded in a cluster of crystals of cuprite, an oxide **mineral** from which copper is extracted. (Minerals are naturally occurring solid substances with a definite chemical composition, structure and properties.)

FIGURE 4.2.3 Crystals of orange-coloured copper embedded in cuprite, a form of copper oxide

Gold and platinum, which are much less reactive than most other metals, are found in the Earth's crust in their pure form. Gold is often found in rock formations called seams alongside quartz, as shown in Figure 4.2.4.

FIGURE 4.2.4 Gold, an unreactive metal, exists in the Earth's crust in its metallic elemental form.

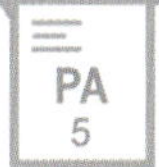

FIGURE 4.2.5 Different metals in equal amounts of dilute acid. Magnesium ribbon (left) reacts more rapidly than iron filings (centre). Copper turnings (right) do not react.

Reactivity with acids

The reactivity of different metals with acids follows the same general patterns as the reactivity of metals with water and oxygen. Metals are normally more reactive with acids than with either water or oxygen, and the reactions tend to be more energetic.

The reactions of magnesium, iron and copper with an acid are shown in Figure 4.2.5. The large amount of bubbling and the mist produced by magnesium show that it is the most reactive metal of the three; iron filings react less vigorously and copper does not react at all. Hydrogen gas is produced in the reactions involving magnesium and iron.

REACTIVITY SERIES OF METALS

Chemists have used experimental data from the reactions of metals to produce a **reactivity series of metals**, as shown in Figure 4.2.6. Group 1 metals are at the top of the series, while transition metals appear at the bottom.

In general, the reactivity of main group metals increases going down a group in the periodic table and decreases from left to right across a period. This trend in reactivity can be explained in terms of the relative attractions of valence electrons to the nucleus of atoms.

When metals react, their atoms tend to form positive ions by donating one or more of their valence electrons to other atoms. The metal atoms that require less energy to remove electrons tend to be most reactive. Therefore, the most reactive metals tend to be those with the lowest ionisation energies, which are found in the bottom left-hand corner of the periodic table. Transition metals are usually less reactive than other metals.

The reactivity of metals therefore follows periodic trends, increasing down a group and decreasing across a period.

Metal
potassium, K
sodium, Na
calcium, Ca
magnesium, Mg
aluminium, Al
zinc, Zn
iron, Fe
nickel, Ni
lead, Pb
copper, Cu
silver, Ag
gold, Au
platinum, Pt

increasing reactivity (arrow pointing up)

FIGURE 4.2.6 The reactivity series of some common metals

CHEMFILE

Aluminium: A reactive element

Although aluminium is widely used for products that are exposed to air and water, such as cooking utensils, window frames and aeroplane bodies, aluminium is in fact quite a reactive metal. The reason it does not react with water or oxygen is that it has already reacted; it forms a thin layer of aluminium oxide which protects the metal from further attack, as shown on the left in the figure below. However, if aluminium is treated with mercury it forms a surface amalgam (alloy) that no longer protects the metal and 'whiskers' of aluminium oxide are quickly formed, as shown on the right. The amalgamated aluminium can spontaneously ignite in moist air.

Untreated aluminium foil (left) and aluminium treated with mercury (right)

4.2 Review

OA ✓✓

SUMMARY

- The reactivity of different metals with water, acid and oxygen can be determined experimentally.
- Group 1 metals are very reactive with water.
- Metals tend to be more reactive with acids than with water.
- The experimental results of metals reacting with water, acid and oxygen are used to place metals in an order of reactivity called the reactivity series.
- The most reactive metals are found at the bottom left-hand side of the periodic table.
- Although the reactivities of transition metals vary, they are usually less reactive than the main group metals.

KEY QUESTIONS

Knowledge and understanding

1 Name the products formed when:

a calcium is burnt in oxygen

b sodium reacts with water.

2 Which one of the following metals is most likely to react with cold water to form hydrogen gas?

A lead

B sodium

C silver

D copper

3 Using the reactivity series of some common metals (see Figure 4.2.6 on page 146), determine whether calcium, platinum or aluminium would react most vigorously with oxygen in the air.

Analysis

4 Pieces of iron, zinc and gold metal were each placed into test tubes containing hydrochloric acid. The table on the right summarises the observations from each test. On the basis of the reactions, list the metals in order of reactivity.

Summary of observations from tests in which iron, zinc and gold were added to hydrochloric acid

Metal	Evidence of reaction
iron	bubbles of gas are slowly produced
	iron metal disappears after a very long time
zinc	bubbles of gas are rapidly produced
	zinc metal disappears
gold	no signs of a chemical reaction
	gold does not change appearance

5 Use the data in Table 4.2.1 and the periodic table at the back of the book to answer the following questions.

a **i** What is the trend in the reactivity of metals going down a group in the periodic table?

ii What is the trend in the reactivity of metals going across a period in the periodic table?

b Compare the reactivity of transition metals with the elements in groups 1 and 2.

c How does the trend in metal reactivity compare with the periodic trends in first ionisation energy (see Table 2.5.5 on page 92).

6 Explain why the reactivity of potassium is greater than the reactivity of both sodium and calcium.

4.3 Producing and recycling metals

The only metals in the Earth's crust that are found entirely in their elemental state are the inert elements platinum and gold. Most metals in the Earth's crust are in the form of mineral compounds, such as oxides, sulfides or silicates. Minerals are naturally occurring solid substances with a definite chemical composition, structure and properties. When the concentrations of minerals in rock are high enough to be economically extracted for use, the rock is referred to as an **ore**.

METAL PRODUCTION FROM ORES

The process of obtaining pure metal from an ore has three main steps: mining, ore processing and metal extraction.

1 Mining—Surface mining involves the removal of surface vegetation, soil and bedrock to reach the ore deposits below the Earth's surface. Open-pit mining is a common surface mining technique. Over time, the mine progressively becomes a wider and deeper hole in the ground. Large vehicles are usually used to transfer the ore from the pit to the processing facility.

   By contrast, underground mining involves digging a vertical shaft into the ground, from which tunnels are dug to reach the ore. The ore is initially transferred to vehicles and then onto conveyors to carry the ore to the shaft, before being lifted to the surface for further processing.

   Some metals, including rare earth elements and uranium, are mined by **in-situ leaching**. This involves injecting solutions into natural or manmade holes in the ore deposit to dissolve the minerals, and pumping the solution to the surface where the minerals can be recovered. This technique avoids the costs associated with conventional mining processes.

2 Ore processing—Once the ore is mined, it is processed to separate the valuable minerals from the waste rock. The processing method used to separate the metal-containing mineral from the rest of the ore depends on the nature of the ore. The ore is usually crushed into pebble-sized pieces or ground into a fine powder. Some ores, like iron ore, do not need further processing. Other ores may require treatment using one or more techniques based on the particular properties of the mineral, such as its density and magnetism.

3 Metal extraction—The method employed to extract a metal from a mineral depends upon the relative reactivity of the metal. All metals can be extracted from their compounds by passing an electric current through the **molten** (liquefied) compound. This technique is called **electrolysis**. However, large amounts of electrical energy are needed to do this, so electrolysis is expensive and is only used to extract very reactive elements, such as aluminium, sodium and potassium.

   Iron can be extracted from iron oxide ores by heating it at high temperatures with carbon, in a process called **smelting**. Copper, lead and zinc are found mainly as sulfide ores. These ores are first **roasted**—a process involving heating in air—to convert them to the metal oxides. The metals can then be produced from the oxides by smelting with carbon. In the case of copper and zinc, the impure metal obtained by this process is further purified by electrolysis.

Environmental issues

The mining, processing and extraction of metals contribute to environmental problems on a global scale as metals are manufactured in large quantities and their production is very energy intensive, using about 8% of the total global energy supply. The problems can be divided into their impact on land, water, air and biodiversity.

- Land—The use of land for mining is associated with clearing of vegetation and erosion. Considerable quantities of waste can be produced during mining and processing, and the management of this waste has been described as one of the world's largest environmental issues. Coarser waste is often stockpiled in long-term storage dumps, whereas storage ponds are used for tailings (finely ground rock).
- Water—Large quantities of water are consumed; there is the potential for toxic substances to be released into groundwater and surface water.
- Air—The use of fossil fuel energy sources and widespread use of carbon as a reactant in the smelting process results in substantial greenhouse gas emissions, principally of carbon dioxide and sulfur dioxide, which contribute to global warming. Dust and particulates are also of local concern due to their damaging effects on ecosystems and human health.
- Biodiversity—Degradation of landscape and ecosystems is linked to the decline in the number and variety of species in a given area.

METAL RECYCLING AND THE CIRCULAR ECONOMY

Innovation and change are taking place in metal production industries to achieve more sustainable economic and industrial growth. Metal recycling has a key role in these developments.

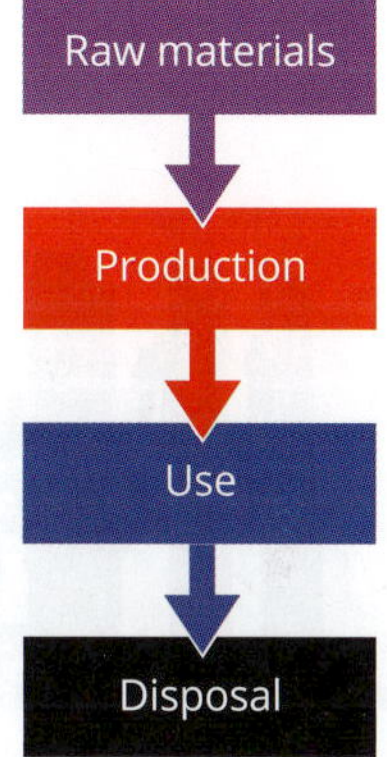

FIGURE 4.3.6 Representation of a linear economy. The products in a linear economy are ultimately destined to become waste.

Linear and circular economies

In the past our economy was based on a 'linear' approach to the use of resources for manufacturing products. In a **linear economy**, raw materials in the form of natural resources are used to make a product and after its use the article is ultimately thrown away. This type of economy is often described as operating with a 'take-make-dispose' or a 'take-make-waste' model (Figure 4.3.6).

However, there is growing momentum for society and industry to adopt the **circular economy** model that is represented in Figure 4.3.7. In a circular economy, manufacturers design products to be reusable and able to be recycled. The reasons for this change of approach include:

- resources are becoming scarcer and, in some cases, their quality is declining
- the prices of raw materials are increasing
- there is an increasing demand for manufactured goods by a growing, more prosperous global population
- a desire to minimise environmental impacts (air and water pollution, climate change, land degradation, biodiversity loss).

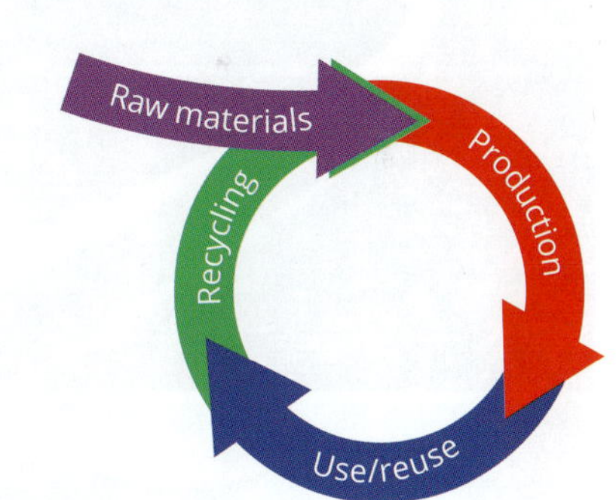

FIGURE 4.3.7 Representation of a circular economy. A circular economy aims for optimal use and reuse of resources from the extraction of raw materials, to production to consumption.

A circular economy aims to create a sustainable society by eliminating waste and pollution, keeping products and materials in use, and regenerating natural systems. To achieve this, processes and technologies need to be developed with reusing and recycling in mind. For example, electrical devices would be designed in such a way that they would last longer and be easier to repair. In the case of plastic products, once used the products would be recycled to make new plastic items.

The efficient and extensive recycling of metals is seen as a key component of initiatives to transition to a circular economy, both within Australia and globally.

 A circular economy is an economic system that seeks to minimise climate change, waste, pollution and loss of biodiversity.

CASE STUDY

Iron production

Modern society is very dependent on iron. Iron is blended with other transition metals and carbon to produce the steel used for construction and many other purposes (Figure 4.3.1).

FIGURE 4.3.1 Some uses of steel are (a) building frames in construction, (b) train tracks and (c) surgical instruments.

Australia is the world's largest exporter of iron ore. The massive deposits of iron ore in the Pilbara region of Western Australia contain iron in the form of haematite (Fe_2O_3) (Figure 4.3.2).

FIGURE 4.3.2 The mineral haematite (Fe_2O_3) is mined as the main source of iron. This sample has been partially polished, removing most of the red oxide coating that forms upon exposure to air.

Mining

Iron ore is mined by open-pit mining methods (Figure 4.3.3).

FIGURE 4.3.3 Iron ore is sourced from open-cut mines in Australia.

Coal and **limestone** are also mined for the extraction of iron. The coal is converted to **coke**, a solid which contains 80–90% carbon, by strongly heating the coal in air-tight ovens for about 15 hours. Limestone is a sedimentary rock which is mainly composed of calcium carbonate ($CaCO_3$).

Iron extraction

Extraction of iron from ore is performed at very high temperatures in a tall, bottle-shaped tower called a **blast furnace** (Figure 4.3.4).

Figure 4.3.5 shows a diagram of a blast furnace. Hot air is blasted into the bottom of the furnace, while solid 'charges' (scoops) of iron ore, coke and limestone are added to the top.

As the air rises through the furnace and meets the descending charge, oxygen reacts with the coke to ultimately produce carbon monoxide. Extraction of iron from iron oxide occurs in a series of steps summarised by the single chemical equation:

carbon monoxide + iron oxide ⟶ iron + carbon dioxide

$$3CO(g) + Fe_2O_3(s) \longrightarrow 2Fe(l) + 3CO_2(g)$$

FIGURE 4.3.4 A blast furnace used to extract iron from iron ore. Iron ore is added continuously to the top of the furnace by the conveyor belt on the left-hand side.

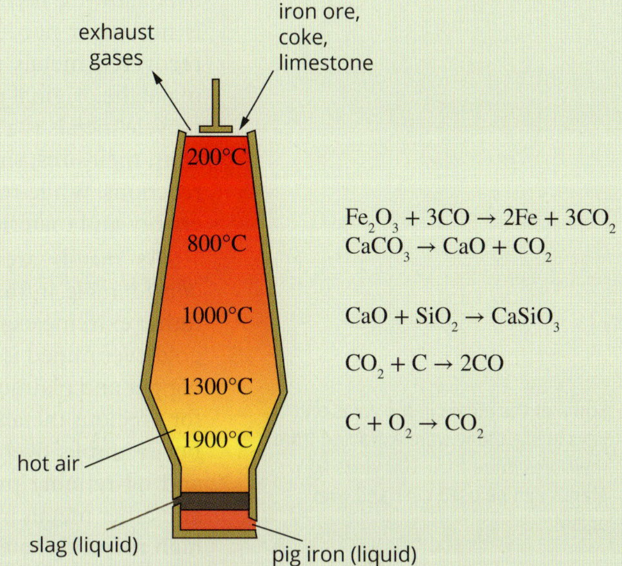

FIGURE 4.3.5 A diagram of a blast furnace. Hot air jets blast air in at the bottom. Iron ore, coke and limestone are added at the top.

Impurities such as silica (silicon dioxide) in the iron oxide are removed by adding limestone. It breaks down in the furnace to form calcium oxide (CaO) and carbon dioxide. Calcium oxide can react with the impurities to form materials such as calcium silicate ($CaSiO_3$) which are called **slag**.

Holes at the base of the furnace are opened and the molten iron and slag are drained out and separated. The iron contains dissolved carbon and, to make the steel, the molten iron is usually transferred to another furnace where oxygen is blown through the metal to reduce its carbon content. Other elements are also added to form the steel alloy.

The iron and steel industry produces nearly 10% of all global greenhouse gas emissions, in the form of carbon dioxide gas, from blast furnaces and steel-making furnaces. New methods are being explored that use hydrogen gas instead of coke. Hydrogen reacts with iron oxide to form iron metal and the main byproduct is water vapour. If hydrogen used in this process is derived from renewable or low-carbon sources, the steel-making process could become almost completely emission-free in the future.

Metal recycling

Metals are present in numerous products that have both short-term and long-term use. Metals can be re-melted and reshaped and do not lose their properties after use, allowing them to be recycled repeatedly. Once a product containing metals is no longer needed, in most cases it can be reprocessed relatively easily and the recovered metals returned to the production process. The metal can be used to make the original product or repurposed as a new product. The energy required to recycle metals is a relatively small fraction of the energy required to produce metals from their ores, since energy is required largely only for melting and not chemical reactions. While recycling consumes energy, this represents a more efficient use of energy and materials.

As metals are so valuable, there is a long history of metal recycling and considerable infrastructure is already in place for their recovery and reuse. As a rather extreme example, it is estimated that more than 90% of the 180 000 tonnes of gold ever mined is still in use. Metals that are used in their pure form, such as gold, copper and platinum, tend to have high rates of recovery. The average recycling rate for precious metals is above 50%, but there are large differences across applications. For example, over 90% of the precious metals from the catalysts used in chemical and oil-refining processes are recovered, whereas the recycling rate for platinum and other metals in automotive catalysts is about 65% (Figure 4.3.8). There are also high recycling rates for aluminium and steel.

Some of the atoms in the coins and keys in use today were extracted from their ores more than 100 years ago. Metals can be used over and over again!

FIGURE 4.3.8 A cross-section of a new car catalyst. The honeycomb structure is coated with a catalyst made from an alloy of platinum and iridium.

While the energy required to extract a metal from an ore depends on several factors, broadly speaking it increases as you go up the reactivity series. As a consequence, the energy savings for metal recycling compared to mining and extraction also tend to increase as you go up the series. Typical energy savings reported for metal recycling are aluminium 95%, copper 84%, zinc 75%, lead 65% and steel 60%.

The highly reactive group 1 and 2 metals, such as sodium, potassium and calcium, are found in abundance in the Earth's crust. Since there are more uses for their compounds than the metals themselves, these metals are not usually extracted in recycling processes.

Despite the relatively high recycling rates for some metals, it is widely acknowledged that there is plenty of room for improvement. Insufficient collection of consumer goods and inefficient handling within the recycling chain are two of the biggest challenges. Much more metal could be recovered from industrial waste and from consumer goods that have reached the end of their useful life (such as vehicles, electronic devices and rechargeable batteries), which at present go to landfill. Metal recovery from electronic waste is less than 15% (see the case study on page 155). Efforts are being made to move towards a more efficient circular economy by increasing recovery rates, thereby reducing the amount of raw resources that are consumed and minimising the waste that is dumped.

The contributions of metal recycling to the circular economy and the environment can be summarised as follows. Metal recycling:

- saves up to 20 times the energy needed to extract metals directly from their ores. This has economic benefits because the recycled materials are less costly for manufacturers to use
- avoids using metals as landfill. Dumping metal-containing waste is not only a loss of valuable raw materials, but can also affect the environment if metal compounds leach into water courses
- reduces CO_2 emissions (since energy consumption is less) and the impact on land and water associated with mining and ore extraction. Using recycled metal is estimated to reduce air pollution by approximately 80%, water pollution by 75%, and water use by 40%.

Metal recycling helps to close the loop within production processes, reducing both the amount of raw materials consumed and the amount of waste that goes into landfill.

It should also be mentioned that mining and metal extraction industries are implementing other more 'circular' (sustainable) approaches in their operations, as well as recycling. For example, there is increased use of energy sources such as solar and wind power, and industries are introducing water treatment and collection programs that aim to avoid the discharge of waste water. Similarly, the smelting and refining phases of metal production are being further optimised to minimise wastes and increase efficiency. Large 'industrial parks' involving several industries have been created in Kwinana, Western Australia, and Gladstone, Queensland, to allow one industry to reuse the wastes of another. Furthermore, as improvements in energy efficiency have become increasingly difficult to achieve for established technologies, new production methods are being investigated. Fortescue Metals has announced plans to build a facility in Gladstone that produces hydrogen from renewable sources. The hydrogen would then be used instead of coal to produce steel without carbon emissions.

The recycling process

The recycling of metals is often referred to as secondary production. Today, more aluminium and lead is produced by secondary production than directly from their ores, and very large quantities of steel and copper are also manufactured in this manner.

The metals recovered for recycling are referred to as '**scrap metal**', or simply as 'scrap', and come from three sources:

- waste from the initial manufacture and processing of the metal
- waste from the fabrication of the metals into products
- discarded metal-based product itself.

There are four steps involved in the recycling process.

1 Collection

Scrap metal yards are widely used as collecting centres for metals.

2 Preparation for recovery

Scrap metal is classified into two main groups: **ferrous** (containing iron, such as steel) and **non-ferrous** (other metals). Ferrous metals are magnetic and are easily pulled out of mixed waste. Sensors based on X-ray and infra-red scanning methods are used in large sorting facilities to identify metals and improve metal recovery rates. New methods for efficiently separating metals are under development.

CHEMFILE

Lithium-ion batteries: a challenge to the environment and economy

Lithium-ion batteries are being enthusiastically adopted worldwide as a core source of portable energy. In particular, the numbers of electric cars powered by these batteries are expected to grow spectacularly. This is cause for environmental concern because less than 9% of the spent batteries are recycled—in Australia the figure is less than 3%. The batteries contain valuable metals, like cobalt, nickel, copper and manganese, and recovering these metals has been the main focus of recycling to date. Lithium metal is rarely recycled.

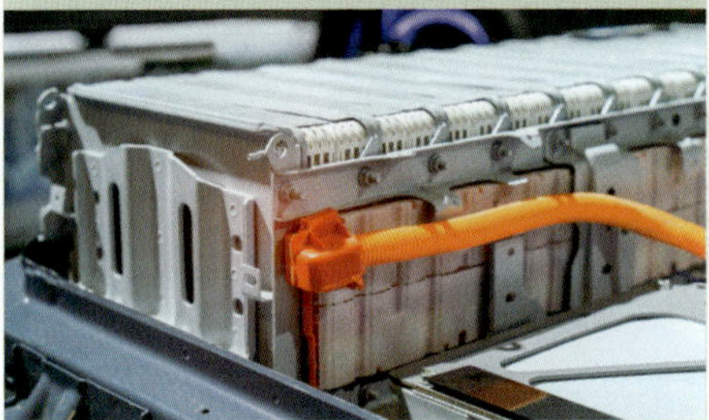

A lithium-ion battery pack and power connections in an electric car

The demand for lithium is rapidly increasing and consumption has more than doubled over the last decade. The element is not evenly distributed around the world and is in limited supply, which is encouraging the growth of technologies to recycle lithium from the batteries. Lithium is a reactive group 1 element and existing processes are complex and consume a large amount of energy. Improved methods for its recovery are currently being developed.

The recycled metal is then compacted and shredded because small pieces of metal can be melted using less energy than larger pieces of metal.

3 Smelting

The scrap metal may be fed into one of two types of smelter to extract the metal by the application of heat:

- a **primary smelter**—one that is also used for extracting the metal from its ores, or
- a **secondary smelter**—one designed especially for recycled metal.

4 Purification

After the melting process is complete, the impure metal is further refined to ensure that the final product is free of impurities. Metals are purified using different methods. Other chemicals may then be added to the molten metal to form alloys with desirable properties.

The sustainable nature of the process is further enhanced if other metals are also recovered from this process. For example, copper producers may recover not only copper, but also a wide range of other metals, such as gold and silver.

Recycling aluminium

More aluminium is used in manufacturing than any other metal except iron. It is employed in a huge variety of products, including cans, foils, window frames and aeroplane parts. Other elements, such as copper, manganese and magnesium, are often mixed with aluminium to form alloys because aluminium itself is not particularly strong. Globally, it is estimated that about 50% of new aluminium products are made from recycled aluminium. Seventy-five per cent of all aluminium produced since commercial production started in 1886 is thought to still be in use today.

Aluminium is recycled at low cost using secondary smelters. The scrap metal is placed in a furnace and the furnace is heated with gas or oil burners. The molten aluminium is run off and solidified.

Some metals such as aluminium, which are usually manufactured as alloys, require careful handling and processing during recycling. In the case of aluminium, the different forms of waste are collected and sorted before being placed in the smelter. This allows new aluminium cans to be made from old cans composed of the particular alloys of aluminium that are most suitable for this purpose. Further increasing the efficiency of this process, gases produced when labels on the cans burn can be used as fuel for melting the scrap metal. Recycling a single aluminium can may save about 95% of the energy it would take to make a brand new one.

CASE STUDY **ANALYSIS**

E-waste

E-waste is a term used for electronic waste and describes discarded electrical or electronic devices. E-waste is considered the fastest-growing source of hazardous waste in the world, with tens of millions of tonnes generated annually. Less than 20% of e-waste is recycled; most of it ends up in landfill, potentially damaging the environment with the leakage of hazardous substances such as mercury, lead and cadmium.

A typical mobile phone contains over 40 different metals. The printed circuit boards in e-waste contain precious metals such as gold, silver and platinum and other metals such as copper, iron and aluminium. Computer circuit boards contain around 200 grams per tonne ($g\ t^{-1}$) of gold and around 80 $g\ t^{-1}$ palladium; mobile phone handsets contain up to 350 $g\ t^{-1}$ gold and 130 $g\ t^{-1}$ palladium. This is significantly higher than the metal content in their ores (usually less than 10 $g\ t^{-1}$). One approach to their processing is to melt the circuit boards, burn the cable sheathing to recover copper wire and use acid to dissolve and separate metals of value.

Recycling is considered to be an essential component of managing e-wastes. It offers the potential of reducing environmental damage and reducing the use of natural resources. However, since e-wastes are complex mixtures of materials and contain relatively small concentrations of metals, their processing can be challenging. It is probable that recycling will need to be encouraged by local authorities and regulation. A number of countries have enacted legislation that requires electronic manufacturers to pay for recycling and disposal of their electronic products.

The e-waste recycling industry is growing. One Japanese smelter extracts hundreds of thousands of tonnes of gold, silver, copper, palladium and other valuable metals each year from the circuit boards of discarded home appliances, personal computers and mobile phones.

FIGURE 4.3.9 E-waste from thousands of electronic devices, such as televisions, computers and mobile phones, was used to make the medals for the Tokyo Olympics held in Japan in 2021.

In the two years before the Olympic Games held in Tokyo in 2021, small unwanted electronic devices were collected from the Japanese people so that all the medals awarded to the athletes were made from recycled metals (Figure 4.3.9). The ability to efficiently recycle e-waste can be seen as 'closing a loop' of a section of the economy.

Analysis

1 Use the internet to research why the mercury, lead and cadmium found in e-waste are described as hazardous substances.

2 **a** Explain the term 'closing the loop' with reference to the objectives of a circular economy.

   b Does the recycling process described for mobile phones meet all the aims of a circular economy?

3 Consider society's current approach to the purchase and use of mobile phones. Suggest ways that the aims of a circular economy could be better met by changing current practices.

4.3 Review

OA ✓✓

SUMMARY

- In a linear economy, raw materials in the form of natural resources are used to make a product and the article is thrown away after its use.
- In a circular economy, manufacturers design products to be reusable and able to be recycled.
- Economies are tending to change from linear to circular as resources become scarcer, the prices of raw materials increase, the demand for manufactured goods increases, and in order to minimise environmental damage.
- Metal recycling is a component of a circular economy where the metal is mined, refined, made into a product, used, disposed of via recycling and then either reprocessed to make the original product or repurposed as a new product.
- Metals can be recycled because they can be re-melted and reshaped countless times, without altering their properties.
- There is a long history of metal recycling and considerable infrastructure is already in place for the reuse, remanufacture and recycling of metals.
- Precious metals tend to have high rates of recovery. Aluminium and steel are also highly recycled.
- There is room for improvement in the recovery rates of many metals.
- Metal recycling contributes to the circular economy and reduces negative impacts on the environment by:
  - using less energy than is needed to extract metals directly from their ores.
  - avoiding the use of metals as landfill
  - reducing CO_2 emissions compared to mining
  - minimising the impact of mining on land and water.
- The recycling process involves smelting to extract the metal from waste by the application of heat.

KEY QUESTIONS

Knowledge and understanding

1 Explain the differences between a linear economy and a circular economy.

2 In what ways does metal recycling meet the objectives of a circular economy?

3 Describe some of the factors that currently cause metal recycling to be less than 100%.

4 Explain why a higher proportion of gold and platinum is recycled than copper and aluminium.

Analysis

5 The lead in the lead–acid batteries used in cars can be recovered by dismantling spent batteries and feeding the metal electrodes into a smelter. High temperature, open-air furnaces are used as smelters in some developing countries. Research what problems could arise with this lead recycling process.

Chapter review

KEY TERMS

alkali metal
alloy
blast furnace
cation
circular economy
coke
conductor
crystal
delocalise
delocalised electron
density
ductile
e-waste
effective nuclear charge
electrolysis
ferrous
in-situ leaching
inert
ionisation energy
lattice
limestone
linear economy
lustrous
malleable
metallic bonding
metallic bonding model
mineral
molten
non-ferrous
ore
primary smelter
reactivity
reactivity series of metals
roasting
scrap metal
secondary smelter
slag
smelting
steel
tensile strength
transition metal

REVIEW QUESTIONS

Knowledge and understanding

1 Some metals are found as elements in nature; others are found as compounds combined with other elements such as oxygen and sulfur in ores. Australia has natural reserves of many metals and ores. The nation's mining and metal extraction industries produce large quantities of metals such as aluminium, copper, gold, iron and silver.

- **a** Give the chemical symbol for each of the metals listed above.
- **b** In which group and period of the periodic table are each of these metals found?
- **c** Which two of these metals are found in nature as elements rather than only as compounds?
- **d** Which of these metals are transition elements?
- **e** Which one of these elements is the rarest?

2 What do you think is the most important property of each of the following metals that has led to its widespread use?

- **a** Aluminium
- **b** Copper
- **c** Iron

3 Which property most clearly distinguishes the metals from the non-metals listed in Table 4.1.2 on page 136?

4 Using your knowledge of trends in the periodic table, which one of the following lists contains metals which would all have similar properties to beryllium?

- **A** Ca, Cs, Cu
- **B** Mg, Zn, Sr
- **C** Mg, Ca, Sr
- **D** K, Pb, Cu

5 Use a diagram to describe what is meant by the term 'metallic lattice'.

6 Consider the metallic bonding model used to describe the structure and bonding of metals.

- **a** What is meant by the following terms?
  - **i** Delocalised electrons
  - **ii** A lattice of cations
  - **iii** Metallic bonding
- **b** Which electrons are delocalised in a metal?

7 Use the metallic bonding model to explain each of the following observations.

- **a** Copper wire conducts electricity.
- **b** A metal spoon used to stir a boiling mixture becomes too hot to hold.
- **c** Iron has a high melting point, 1540°C.
- **d** Copper can be drawn out to form a wire.

8 When a reactive metal is added to water, fizzing or bubbles can be observed. Explain the reason why this occurs.

9 Decide if the following statements about the reactivity of metals are true or false.

- **a** All metals react with acid to produce hydrogen gas.
- **b** Hydrogen is a flammable gas.
- **c** The most reactive metals are located at the top of a group in the periodic table.
- **d** Metals are normally more reactive with acids than with either water or oxygen.

Application and analysis

10 The atomic number of calcium is 20. How many electrons are in an atom of calcium and in a Ca^{2+} cation?

A 20; 20
B 20; 18
C 18; 20
D 20; 2

11 Which one of the following is the electronic configuration of an aluminium atom and the electronic configuration of its most stable cation?

A 2,8,2; 2,8
B 2,8,3; 2,8,2
C 2,8,2; 2,8,1
D 2,8,3; 2,8

12 Draw a concept map using the following terms:

metal; cation; delocalised electron; lattice; electrostatic attraction

13 The boiling points of three metals—sodium, potassium and calcium—are given in the table below.

Metal	Boiling point (°C)
Na	892
K	760
Ca	1490

a In which group and period of the periodic table are these metals found?

b Write the electronic subshell configuration for each of the three metals.

c Use the metallic bonding model to suggest why:

i sodium has a higher boiling point than potassium

ii calcium has a much higher boiling point than potassium

14 Which one of the following metals would you expect to be the least reactive with water?

A Aluminium
B Sodium
C Rubidium
D Chromium

15 Look at the periodic table at the end of the book. Which one of the following metals would have a similar reactivity to rubidium? Explain your answer.

A Zn
B K
C Ba
D Na

16 The following image shows similar-sized pieces of iron and silver in test tubes of sulfuric acid of the same concentration. Observe the reactions taking place in the two test tubes. What conclusions can you make about the reactivity of the two metals involved? Based on these conclusions, determine the identity of the metal on the left and on the right.

17 Observations of the reaction of metals A, B and C with dilute hydrochloric acid are summarised below. Identify which metal is aluminium, copper and sodium and explain your answer.

- Metal A does not react with dilute hydrochloric acid.
- Metal B is stored in oil because of its high reactivity, and it undergoes a violent reaction with dilute hydrochloric acid.
- Metal C reacts very slowly with dilute hydrochloric acid.

18 The recycling rates for many metals are considerably higher than that for plastic. The metals are recycled by feeding waste into a high temperature smelter. Explain why metals can be more easily recycled in this way than plastic.

19 a Describe three benefits of a transition from a linear economy to a circular economy.

b Describe three ways that the recycling of metals contributes to a circular economy.

c Identify three challenges to increasing the extent of metal recycling.

20 Explain why the metals at the top of the reactivity series are not usually recycled whereas the metals at the bottom of the series are almost completely recovered after use and reprocessed.

CHAPTER

Ionic compounds

Imagine what the Earth's solid land surface would look like with all living things stripped away. This is what it would have been like before the first life forms appeared, more than 3.5 billion years ago. There would have been just rocks—and soil and sand formed from the weathering of rocks. Rocks and soil form most of the outer crust of our Earth today. Rocks, molten lava and even dinner plates belong to a group of substances called ionic compounds. At the end of this chapter you will be able to explain the structure and properties of these compounds.

Ionic compounds are made by the chemical combination of metallic and non-metallic elements. You will see that the properties of ionic compounds are a direct result of the bonding between the particles within these compounds.

The writing of chemical formulas and naming ionic compounds are other important skills that you will learn in this chapter.

Key knowledge

- the common properties of ionic compounds (brittleness, hardness, melting point, difference in electrical conductivity in solid and molten liquid states), with reference to the nature of ionic bonding and crystal structure **5.1**
- deduction of the formula and name of an ionic compound from its component ions, including polyatomic ions (NH_4^+, OH^-, NO_3^-, HCO_3^-, CO_3^{2-}, SO_4^{2-} and PO_4^{3-}) **5.2**
- the formation of ionic compounds through the transfer of electrons from metals to non-metals, and the writing of ionic compound formulas, including those containing polyatomic ions and transition metal ions **5.2**
- the use of solubility tables to predict and identify precipitation reactions between ions in solution, represented by balanced full and ionic equations including the state symbols: (s), (l), (aq) and (g). **5.3**

5.1 Properties of ionic compounds

In this chapter, you will study the structure and properties of a group of substances called **ionic compounds.** Ionic compounds are made by the chemical combination of metallic and non-metallic elements.

These materials are very common in the natural world because the Earth's crust is largely made up of complex ionic compounds. Most rocks, minerals and **gemstones** (Figure 5.1.1) are ionic compounds. Soil contains weathered rocks mixed with decomposed organic material, so soil contains large quantities of ionic compounds. **Ceramics,** kitchen crockery and bricks are made from clays. Clays are formed by the weathering of rocks, so these materials also contain ionic compounds. Kitchen crockery and bricks contain mixtures of different ionic compounds. Table salt (sodium chloride) is a pure ionic compound.

If you think about the characteristics of rocks, kitchen crockery and table salt, you will recognise that these materials, and therefore ionic compounds, have some properties in common.

FIGURE 5.1.1 Many gemstones are made from ionic compounds.

Table 5.1.1 lists some properties of typical ionic compounds. These compounds can be found in materials you might encounter in everyday life. Note that the compounds listed are simple ionic compounds, whereas rocks, ceramics and bricks contain more complex ionic compounds.

TABLE 5.1.1 Properties of typical ionic compounds

Ionic compound	Melting point (°C)	Conductive as solid	Conductive as liquid	Conductive in aqueous solution (0.1 mol L^{-1})	Solubility in water at 25°C (g/100 g water)	Example of commercially available product containing the compound
copper(II) sulfate	decomposes 110	no	yes	yes	22	bluestone spray (used to kill pathogens on fruit)
sodium chloride	801	no	yes	yes	36	food salt
calcium carbonate	1339*	no	yes	–	0.0013	main component in marble
zinc oxide	1975	no	yes	–	insoluble	zinc cream
sodium hydroxide	318	no	yes	yes	114	oven cleaner

*Melting point determined under pressure to prevent decomposition of compound.

DEDUCING THE STRUCTURE OF A COMPOUND FROM ITS PROPERTIES

The main properties of ionic compounds are generalised and summarised in the first column of Table 5.1.2. Attempts to explain why ionic compounds have these properties led to some deductions about what the particles in these compounds were like and how they were arranged. Taking sodium chloride as an example, these are summarised in the second column of Table 5.1.2.

TABLE 5.1.2 Properties of sodium chloride and the information this provides about its structure

Property of sodium chloride	What the properties tell us about structure of sodium chloride and other ionic compounds
high melting point	forces between particles are strong
hard but brittle	forces between particles are strong, but particles break apart when a force is applied
conducts electricity in molten state or in solution	free-moving charged particles are present
does not conduct electricity in solid state	charged particles may be present, but they are not free-moving in the solid state

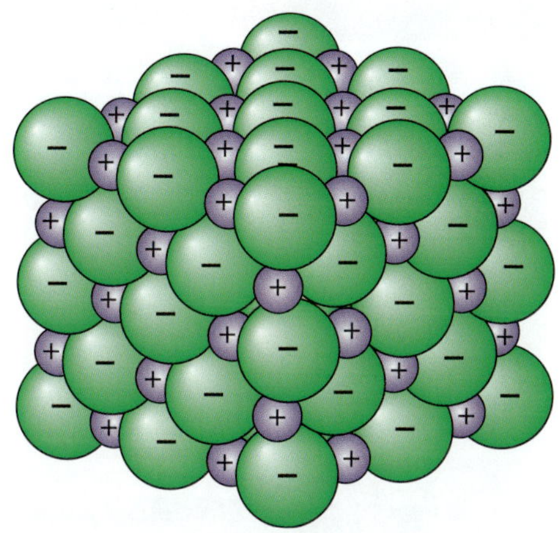

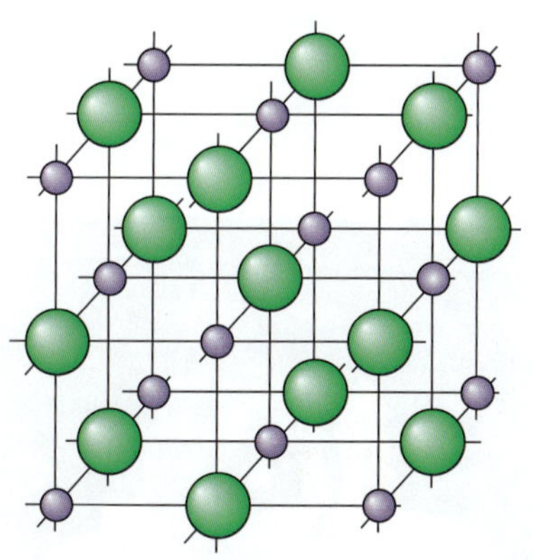

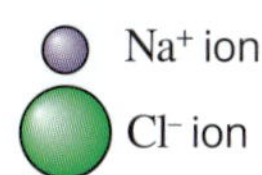

FIGURE 5.1.2 Two representations of part of the crystal lattice of the ionic compound sodium chloride. Forces of attraction between oppositely charged ions result in strong bonding.

THE IONIC BONDING MODEL

Now that the key structural features of ionic compounds have been identified, the next step is to devise a model that accounts for these features.

The model that has been developed to describe this arrangement is known as the **ionic bonding model** (Figure 5.1.2). The features of this model are described below.

When metallic and non-metallic atoms react to form ionic compounds, the following steps occur:

- Metal atoms lose electrons to non-metal atoms and so become positively charged metal ions (called **cations**).
- Non-metal atoms gain electrons from metal atoms and so become negatively charged non-metal ions (called **anions**).

Cations and anions then arrange themselves in the following way:

- Large numbers of cations and anions combine to form a three-dimensional ionic lattice. An ionic lattice is a regularly repeating three-dimensional arrangement of positively and negatively charged ions.
- The three-dimensional lattice is held together strongly by electrostatic forces of attraction between the oppositely charged ions. The electrostatic force of attraction holding the ions together is called **ionic bonding**.
- In sodium chloride, each sodium ion is surrounded by six chloride ions and each chloride ion is surrounded by six sodium ions, thus maximising the forces of attraction (Figure 5.1.2).
- Even though each chloride ion is close to another chloride ion, the attractive force they have towards the sodium ions outweighs the repulsive force from the chloride ions, so the lattice is held together quite strongly.

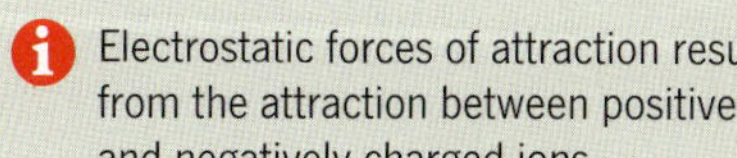

Electrostatic forces of attraction result from the attraction between positively and negatively charged ions.

The formula of sodium chloride

The **chemical formula** of sodium chloride is written as NaCl. However, it is important to note that in a solid sample of an ionic compound, such as sodium chloride, individual pairs of sodium and chloride ions do not exist. The solid is also not built up of discrete NaCl molecules.

Instead, the solid is made up of a continuous lattice of alternating Na^+ and Cl^- ions. All sodium ions are an equal distance from six chloride ions and all chloride ions are an equal distance from six sodium ions. The overall ratio of sodium ions to chloride ions in the lattice is 1 : 1, so that is why the formula is written as NaCl. A formula that shows the simplest, whole number ratio of particles, like this, is known as an **empirical formula**. You will learn more about these in the next chapter.

FIGURE 5.1.3 Bricks made from the ionic compound magnesium oxide are used to line furnaces and kilns.

USING THE IONIC BONDING MODEL TO EXPLAIN PROPERTIES

In Chapter 4, you saw that the metallic bonding model represents the structure of metals as a lattice of positively charged metal ions held together by delocalised electrons. This model explains many of the properties of metals, such as why metals generally have high melting points and conduct electricity in the solid state.

In a similar way, you will see how the ionic bonding model explains the properties of ionic compounds.

High melting points

To melt an ionic solid such as sodium chloride, you must provide energy to allow the ions to break free and move. Sodium chloride has a high melting point (801°C). This indicates that a large amount of energy is needed to overcome the electrostatic attraction between oppositely charged ions and allow them to move freely. Therefore, the ionic bonds between the positive sodium ions and negative chloride ions must be strong, which explains why a high temperature is required to melt solid sodium chloride.

The high melting point of ionic compounds is put to use in the bricks that line furnaces and kilns (Figure 5.1.3) and in the ceramic materials used to make brake discs for high-performance cars (Figure 5.1.4). A ceramic brake disc contains ionic compounds that have very high melting temperatures and withstand the heat produced by braking better than metals. In the wheel shown in the photograph, brake pads are housed inside the yellow caliper assembly. When the brakes are applied, the pads are pressed against the rotating disc, slowing the vehicle.

FIGURE 5.1.4 Ceramic brake discs work more effectively than steel ones at high temperatures.

Hardness and brittleness

There are strong electrostatic forces of attraction between ions in an ionic compound, so a strong force is needed to disrupt the **crystal lattice**. Therefore, one of the properties of ionic compounds is that they are hard. This means that a sodium chloride crystal cannot be scratched easily.

The strength of house bricks, concrete bridges and cobbled streets can be attributed to the ionic bonding within their structures.

CASE STUDY **ANALYSIS**

How fluoride ions make tooth enamel harder

Tooth enamel is the hardest material in the human body (Figure 5.1.5). It is harder than iron.

The main material responsible for the hardness of tooth enamel is the ionic compound hydroxyapatite, which has the formula $Ca_{10}(PO_4)_6(OH)_2$.

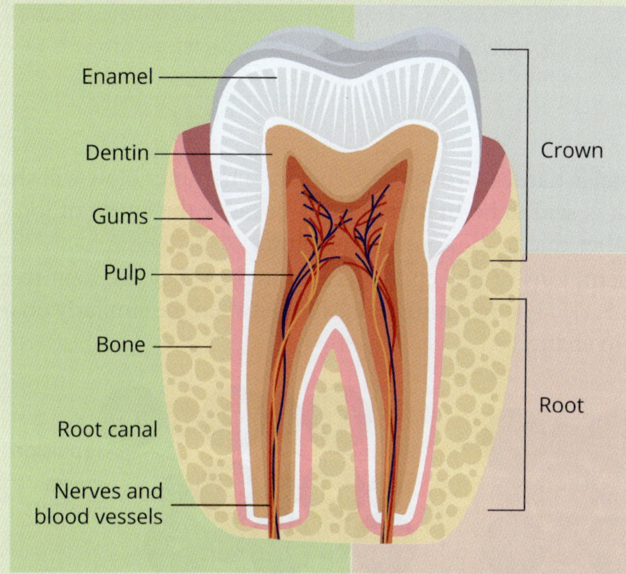

FIGURE 5.1.5 Cross-section view of a tooth showing the enamel

Although this compound is extremely hard, it is also vulnerable to chemical attack by acids. Acids can react with the hydroxyl groups in hydroxyapatite, and this can lead to breakdown of the ionic lattice and the beginning of tooth decay.

A major source of the acids causing tooth decay are sugary foods and drinks. Bacteria in the mouth are able to convert the sugar and other carbohydrate molecules to carboxylic acids and it is these compounds which are able to react with hydroxyapatite.

Regular brushing of teeth can reduce or prevent the occurrence of tooth decay by preventing the build-up of plaque and acids around teeth and gums. Toothpastes contain a mixture of ingredients that contribute to the prevention of tooth decay (Figure 5.1.6). These include:

- antibacterial agents to kill bacteria
- detergents which help to wash away acids and food particles.
- flavourants to mask the taste of detergents and other ingredients
- abrasives
- fluorides.

FIGURE 5.1.6 List of ingredients on a carton containing a tube of toothpaste

Abrasives include ionic compounds, such as sodium bicarbonate, calcium carbonate, aluminium oxide ($Al_2O_3 3H_2O$) and sodium pyrophosphate ($Na_4P_2O_7$). These compounds are used to gently remove plaque and clean the surface of the tooth enamel during the brushing process.

The use of fluoride compounds in toothpaste is an important factor in preventing tooth decay. Fluoride ions from these compounds can react with the hydroxyapatite in tooth enamel by replacing hydroxyl ions. The resulting ionic compound, fluoroapatite ($Ca_{10}(PO_4)_6F_2$), is considerably more resistant to acid attack and so helps reduce or prevent tooth decay. Fluoride compounds added to toothpastes include sodium fluoride, sodium monofluorophosphate (Na_2PO_3F) and stannous fluoride (SnF_2).

Adding fluoride compounds to the water supply of a city has also proven to be a very effective way of reducing tooth decay in the population. The city of Melbourne has had fluoridated water since 1977. Compounds used to fluoridate water supplies include sodium fluoride (NaF), and sodium fluorosilicate (Na_2SiF_6).

Analysis

1 State the property of ionic compounds that is relevant to each of the following.
 a the structure of tooth enamel
 b the use of abrasives in toothpaste

2 Fluoroapatite ($Ca_{10}(PO_4)_6F_2$) contains three separate charged ions (Ca^{2+}, PO_4^{3-} and F^-). Explain why, overall, it has zero charge.

3 Sodium fluorosilicate (Na_2SiF_6) contains sodium ions and fluorosilicate ions. Write a formula for the fluorosilicate ion.

CHEMFILE

Early tools

Some of the earliest tools used by humans were axes, spearheads and coarse needles used for weaving. Each application required a material that was hard and could be shaped. Certain types of rocks that are composed of ionic compounds served this purpose well.

The figure on the right shows a primitive axe used during the Stone Age. The Stone Age ended at different times in different parts of the world, as humans learned to smelt (fuse or melt) metals such as copper from their ores to create more refined and lighter tools. However, their metallic nature meant that they were not as hard as the original tools made of rock composed of ionic compounds.

The hardness of ionic compounds is the reason why axes were once made from rocks.

CHEMFILE

Porcelain

Porcelain is a type of chinaware made from a clay called kaolin. The figure below shows a dinner setting made from porcelain. Clay is weathered rock and consists of a mixture of complex ionic compounds. The chemical formula for kaolin is $Al_2Si_2O_5(OH)_4$. The ions in kaolin are aluminium (Al^{3+}), silicate ($Si_2O_5^{2-}$) and hydroxide (OH^-).

Porcelain is made by moulding the object from kaolin and then heating it to about 1300°C in a kiln. Because it is made from ionic compounds, the resulting porcelain is hard but brittle. Be careful not to drop the family heirloom chinaware as it will most likely shatter!

Porcelain cups and plates are made from clay, which contains ionic compounds. Therefore, porcelain is hard and brittle.

Although a salt crystal is hard, a strong force such as a hammer blow will shatter the crystal. Therefore, it is said to be **brittle**. This is because the layers of ions will move relative to each other due to the force of the blow.

During this movement, ions of like charge are shifted so they are next to each other, as seen in Figure 5.1.7. The resulting repulsion between the similarly charged ions causes the crystal to shatter.

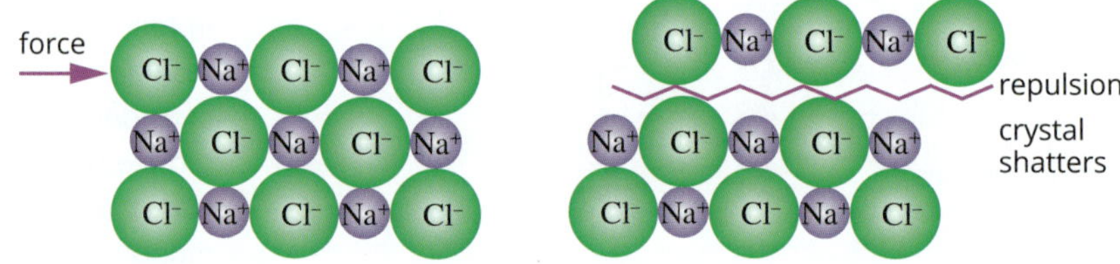

FIGURE 5.1.7 A lattice of an ionic compound shattering. Note that just before shattering, the Cl^- ions are adjacent to other Cl^- ions and the Na^+ ions are also next to each other.

Materials made from clay, such as kitchen crockery (Figure 5.1.8), ceramic tiles and bricks are hard, but they are also brittle.

FIGURE 5.1.8 A ceramic coffee cup is hard but brittle.

Electrical conductivity

In the solid form, ions in sodium chloride are held in the crystal lattice and are not free to move, so solid sodium chloride does not conduct electricity. Remember that for a substance to conduct electricity, it must contain charged particles that are free to move. Figure 5.1.9 shows how the particles are arranged in an ionic compound in solid form.

The force of attraction between oppositely charged ions is strong, so ionic compounds are hard and have high melting points.

In the solid state, oppositely charged ions are held strongly within the lattice and cannot move. Solid ionic compounds do not conduct electricity.

FIGURE 5.1.9 The arrangement of ions within a solid ionic compound form a crystal lattice.

For an object to conduct electricity, it must contain charged particles that are free to move. Solid ionic compounds are made up of a crystal lattice, so the particles are not free to move and they cannot conduct electricity. When molten (melted) or dissolved in water, the charged particles are free to move and conduct.

The non-conducting property of ionic compounds is used in the ceramic insulators that keep high-voltage power lines insulated from electricity poles and electric fence wires (Figure 5.1.10).

When solid ionic compounds melt, the ions become free to move, enabling the cations and anions in the molten compound to conduct electricity.

Similarly, when ionic compounds **dissolve** in water, ionic bonds in the lattice are broken and the ions are separated and move freely in solution.

When an electric current is applied to either a molten ionic compound or a solution of the compound in water, positive ions move towards the negatively charged electrode and negative ions move towards the positively charged electrode, resulting in an electric current, as shown in Figure 5.1.11.

A solution or molten substance that conducts electricity by means of the movement of ions is called an **electrolyte**.

FIGURE 5.1.10 A ceramic insulator on the post of an electric fence

Solubility

Some ionic compounds are very soluble in water, meaning that they dissolve very readily. Other ionic compounds are very insoluble. The term **solubility** refers to how soluble a compound is in a solvent.

When a soluble ionic compound is added to water, the ions break away from the **ionic lattice** and mix with the water molecules. If an insoluble compound is added to water, the ions remain bonded together in the ionic lattice and do not form a solution.

Whether an ionic compound is soluble or insoluble depends on the relative strength of the forces of attraction between:

- the positive and negative ions in the lattice
- the water molecules and the ions.

You will look at the solubility of ionic compounds in water in more detail in Section 5.3.

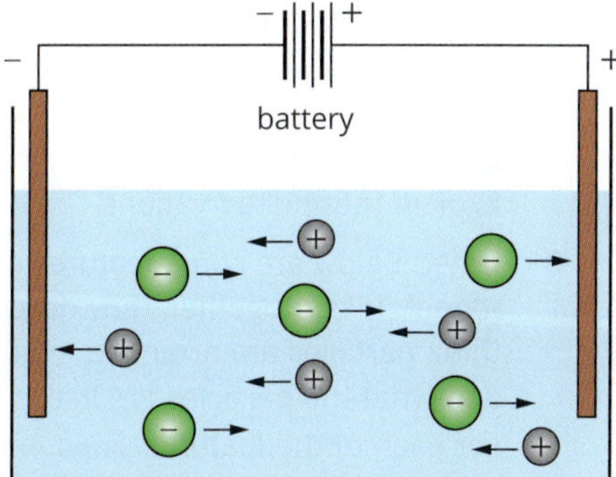

FIGURE 5.1.11 A molten ionic compound or an ionic compound dissolved in water will conduct an electric current.

5.1 Review

OA ✓✓

SUMMARY

- Metal and non-metal elements react together to form ionic compounds.
- Ionic compounds form a crystal lattice made up of positively charged cations and negatively charged anions.
- A three-dimensional ionic lattice is held together strongly by electrostatic forces of attraction between the cations and anions. The electrostatic forces of attraction are called ionic bonding.
- Ionic compounds are hard and have high melting and boiling points. This is because of the strong forces of attraction between the positively and negatively charged ions in the ionic lattice.
- Ionic compounds are brittle. This is because, when an ionic compound is hit, the ions move within the lattice so that like-charged ions line up opposite each other. The ions then repel, causing the lattice to be disrupted.
- Ionic compounds do not conduct electricity in the solid state because, although there are charged particles present, they are not free to move.
- When ionic compounds are dissolved in water or are in molten form, the charged particles (anions and cations) are free to move, which means they can conduct electricity.
- In water, ionic compounds vary from very soluble to insoluble.

KEY QUESTIONS

Knowledge and understanding

1 Copy and complete the sentences below by choosing words from the following list and placing them in the appropriate spaces. Note that not all the words will be used.

magnetic; cations; negatively; lose;
electrostatic; anions; gain; positively

Ionic compounds are formed when non-metal atoms react with metal atoms. In this process non-metal atoms __________ electrons to form __________ charged ions called __________ and metal atoms __________ electrons to form __________ charged ions called __________. The ions formed pack together in a three-dimensional lattice held strongly together by __________ forces of attraction.

2 Sodium chloride has a melting point of 801°C and will not conduct electricity at room temperature. If sodium chloride is dissolved in water it will conduct electricity. It will also conduct electricity if heated to 900°C. Explain these observations.

3 Listed below are some features of the ionic bonding model. Taken together, they describe how chemists think particles are arranged in an ionic compound, and what the particles are like.
For each of the features listed, name a physical property of ionic compounds that provides evidence for that feature.
   a The forces between the particles in ionic compounds are strong.
   b Ionic compounds contain charged particles.
   c Positively and negatively charged particles alternate in the three-dimensional lattice of ionic compounds.

4 What happens to an ionic compound when it is hit with a hammer? Use diagrams to help you with your explanation.

Analysis

5 In this question you will compare the properties of metals and ionic compounds. Listed below are some properties. Decide whether each of these properties apply to 'metals only', 'ionic compounds only' or 'metals and ionic compounds'.
   a brittle
   b conduct electricity in the molten state
   c malleable
   d high boiling point
   e conduct electricity in the solid state

6 The table below contains some data about three different substances. Use the information to deduce the nature of the bonding (metallic or ionic) in each of the three substances. Give reasons for your answers.

Substance	Melting point (°C)	Electrical conductivity		Solubility in water
		At 100°C	At 1000°C	
A	797	does not conduct	conducts	insoluble
B	660	conducts	conducts	insoluble
C	770	does not conduct	conducts	soluble

5.2 Formation of ionic compounds

Some of the reactions that occur between metals and non-metals to form ionic compounds are extremely vigorous. The reaction between sodium and chlorine to form sodium chloride produces a lot of heat. You will remember from Chapter 4 that sodium is very reactive. At high temperatures, the production of sodium chloride from sodium metal and chlorine gas is highly explosive, producing a flame and large amounts of energy.

In the previous section, you saw that the positive and negative ions formed in this reaction are arranged to form a three-dimensional lattice. In this section you will learn how these positive and negative ions are formed from the atoms. You will also learn how the ratio of each type of atom in the compound is determined.

FORMING IONS

As you learned in Section 5.1, when metal atoms react with non-metal atoms to form an ionic compound two things occur.

- Metal atoms lose electrons to form positively charged ions (cations).
- Non-metal atoms gain electrons to form negatively charged ions (anions).

From Chapter 2 you will remember that most metals have low electronegativities, whereas non-metal atoms are usually more electronegative. In other words, non-metals have a stronger attraction for electrons than metals. In reactions that form ionic compounds, non-metal atoms take one or more electrons from the outermost shell of metal atoms.

The octet rule and ionic compounds

The tendency for elements to react in such a way that their atoms end up with eight electrons in their outer shell (the most stable valence shell configuration) is known as the octet rule. This feature of chemical reactions has been discussed in Chapter 3.

Noble gases (group 18) are elements that already have the most stable valence shell configuration. Therefore, another way of thinking about the octet rule is that atoms tend to gain or lose electrons to obtain a stable electronic configuration identical to that of the noble gas nearest to them on the periodic table. The formation of stable ions is a powerful driving force in reactions between metals and non-metals when they produce ionic compounds.

For example, when sodium reacts with chlorine, each sodium atom loses one electron and each chlorine atom gains one electron.

After the reaction, the:

- sodium ion has the stable electron shell configuration of 2,8 (the same as a neon atom)
- chloride ion has the stable electron shell configuration of 2,8,8 (the same as an argon atom).

When an atom loses electrons it becomes more positively charged as the number of protons no longer equals the number of electrons. The ion formed is written with a superscript + sign indicating the charge. If there are two electrons lost, the charge is written as 2+. When an atom gains electrons, in the case of anions, they become negatively charged. This is written with a superscript –. If three electrons are gained, the charge is written as 3– in superscript.

ELECTRON TRANSFER DIAGRAMS

Figure 5.2.1 illustrates how, when sodium reacts with chlorine, an electron is lost by a sodium atom and gained by a chlorine atom. A diagram of this type is called an **electron transfer diagram**.

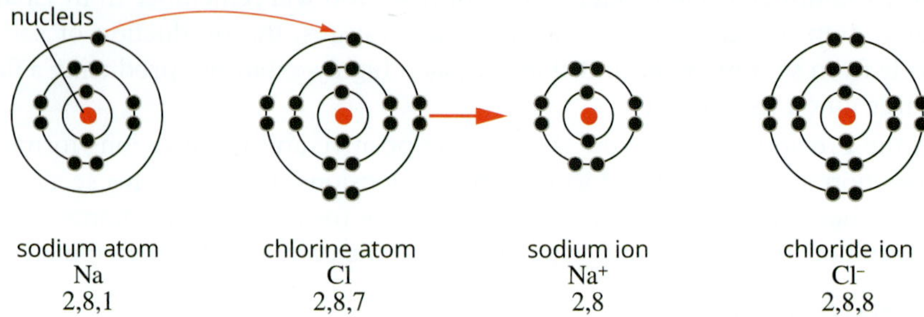

FIGURE 5.2.1 An electron transfer diagram showing the formation of sodium and chloride ions

Group 17 non-metals, such as chlorine, have seven electrons in their valence shell. They readily gain one electron to fill the valence shell according to the octet rule. This means they form anions with a charge of 1−. In the reaction shown in Figure 5.2.1, a chlorine atom has gained an electron from a sodium atom to form a chloride ion, Cl^-. The sodium atom then becomes a sodium ion, Na^+. A sodium ion has the same electron configuration as the noble gas neon.

The reaction between lithium and oxygen atoms is illustrated in Figure 5.2.2. Since oxygen is in group 16, it has six electrons in its valence shell. In this reaction, an oxygen atom needs to gain two electrons to have eight electrons in its outer shell and form a stable ion with a charge of 2−. To allow this to happen, one oxygen atom will react with two lithium atoms, taking one electron from each atom.

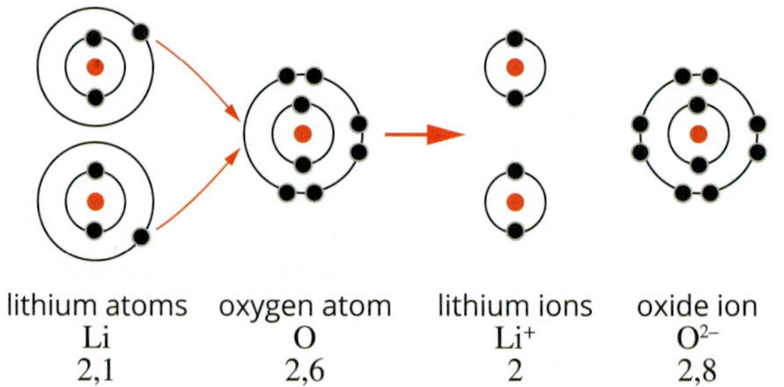

FIGURE 5.2.2 An electron transfer diagram showing the formation of lithium and oxide ions

Although it is generally true that atoms are at their most stable when they have a valence shell containing eight electrons, you can see that lithium is an exception. As lithium has only one electron in the second shell, it loses this and the first shell becomes the valence shell. The first shell can only hold two electrons, so a lithium ion is stable with an electronic configuration of 2. This configuration is the same as that of the noble gas closest to it, helium.

Metals in group 2 of the periodic table, such as magnesium, have electronic configurations with two electrons in their valence shells. Therefore, they readily form ions with a charge of 2+ as they lose these electrons. Figure 5.2.3 shows an electron transfer diagram for the reaction of magnesium with oxygen. (In this diagram, the charge on the nucleus is also shown.)

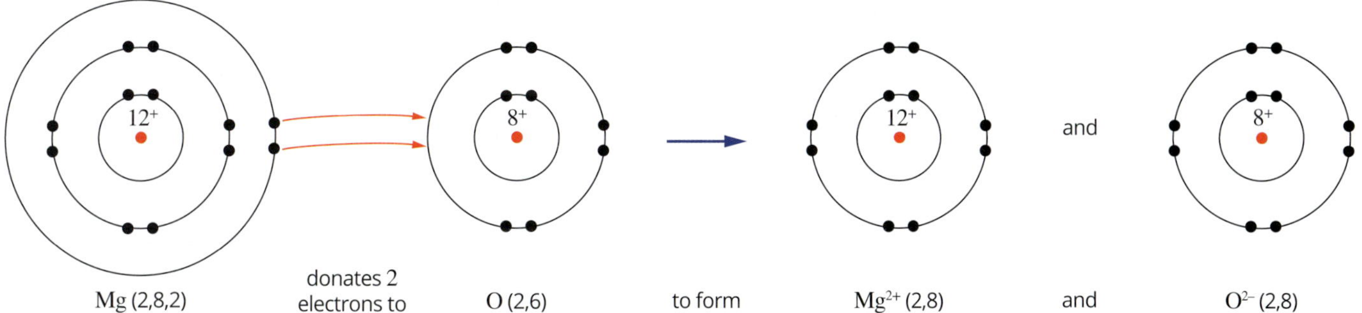

FIGURE 5.2.3 An electron transfer diagram showing the transfer of electrons when magnesium atoms react with oxygen atoms

Figure 5.2.4 shows the diagram for the reaction of magnesium with chlorine to form magnesium chloride, containing Mg^{2+} and Cl^- ions.

Note that although Figure 5.2.4 shows a single Mg^{2+} ion placed between two Cl^- ions, this is not a true indication of how the ions are arranged in a solid sample of magnesium chloride. Like all ionic solids, magnesium chloride consists of a lattice of alternating positively and negatively charged ions. In the $MgCl_2$ lattice the ratio of Mg^{2+} ions to Cl^- ions is 1 : 2.

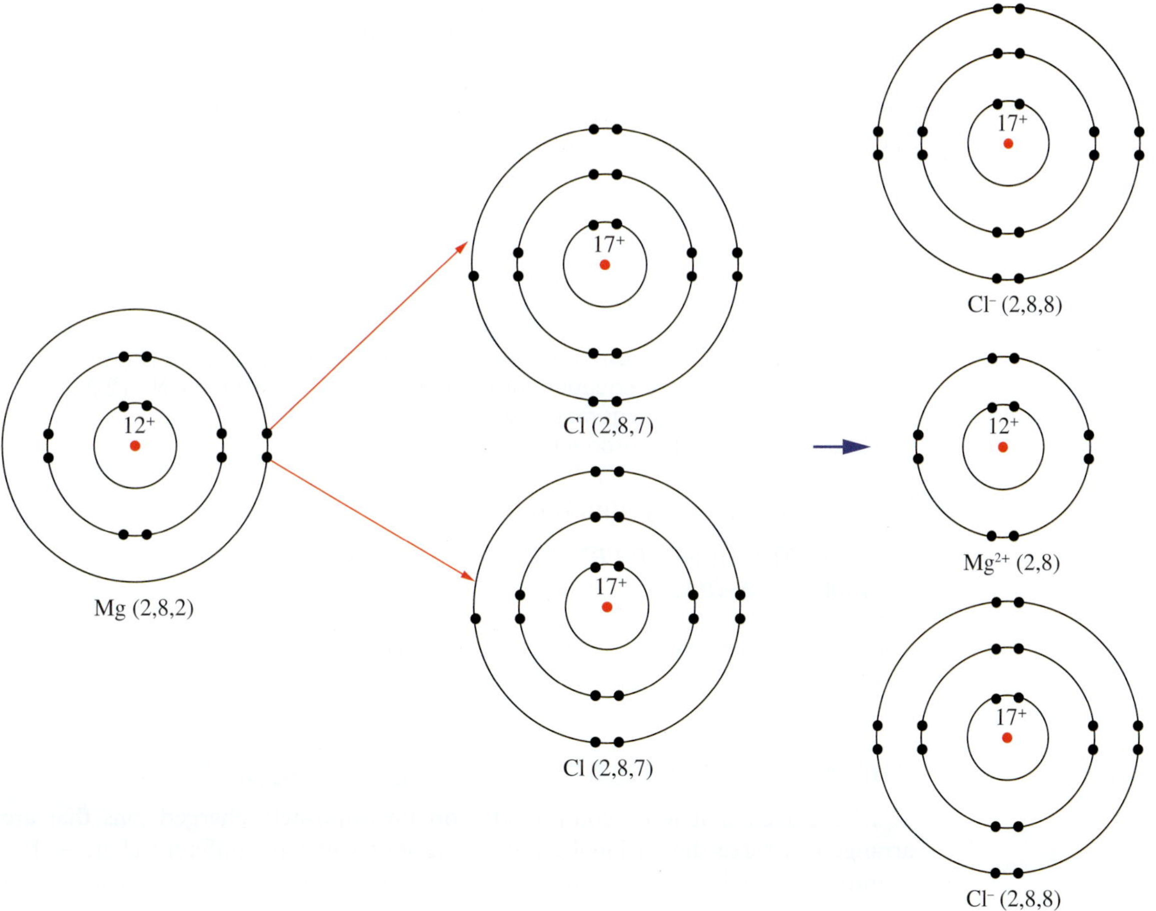

FIGURE 5.2.4 An electron transfer diagram showing the formation of the ionic compound magnesium chloride

To make sure you understand how ions are formed when metals and non-metals react, it can be useful to write equations for reactions that include the electronic configurations of the reactants and products, as shown in Worked example 5.2.1.

Worked example 5.2.1

WRITING EQUATIONS FOR REACTIONS BETWEEN METALS AND NON-METALS ATOMS

Write an equation for the reaction between lithium and nitrogen atoms. Show the electronic configurations for each element before and after the reaction.

Thinking	Working
Write the symbol and electronic configuration for the metal atom.	Li (2,1)
How many electrons will the metal atom lose from its outer shell when it reacts?	1
Write the symbol and electronic configuration of the metal ion that will be formed.	Li^{+} (2)
Write the symbol and electronic configuration for the non-metal atom.	N (2,5)
How many electrons will the non-metal atom gain in its outer shell when it reacts?	3
Write the symbol and electronic configuration of the non-metal ion that will be formed.	N^{3-} (2,8)
The total number of electrons lost by metal atoms must equal the total number of electrons gained by non-metal atoms. What is the lowest number ratio of metal atoms to non-metal atoms that will allow this to happen?	metal atom: non-metal atom = 3 : 1
Using the ratio of metal ion : non-metal ion calculated above, write a balanced equation for the reaction. Show the electronic configurations for both the reactant atoms and the product ions.	3Li (2,1) + N (2,5) $\rightarrow$ $3Li^{+}$ (2) + N^{3-} (2,8)

Worked example: Try yourself 5.2.1

WRITING EQUATIONS FOR REACTIONS BETWEEN METALS AND NON-METALS ATOMS

Write an equation for the reaction between calcium and phosphorus atoms. Show the electronic configurations for each element before and after the reaction.

CHEMICAL FORMULAS OF IONIC COMPOUNDS

You have seen that ionic compounds contain oppositely charged ions that are arranged in three-dimensional lattices. The ions can have different charges. For example, aluminium forms ions with a 3+ charge, whereas oxygen forms an ion with a 2– charge.

In this section, you will learn how to use your knowledge of the charges on ions to write an overall formula for an ionic compound.

Writing formulas of simple ionic compounds

Because ionic compounds are electrically neutral, the total number of positive charges on the metal ions must equal the total number of negative charges on the non-metal ions. This is the most important guiding principle when you are trying to work out the formula of an ionic compound.

This can be seen with the formula of the ionic compound sodium chloride.

- A sodium ion (Na^+) has a 1+ charge.
- A chloride ion (Cl^-) has a 1– charge.
- Therefore, in a crystal of sodium chloride, the ratio of sodium ions to chloride ions is 1 : 1 and the formula of sodium chloride is NaCl.

Using the same steps, you can work out the formula of magnesium chloride.

- A magnesium ion (Mg^{2+}) has a 2+ charge.
- A chloride ion (Cl^-) has a 1– charge.
- Therefore, in a crystal of magnesium chloride, two chloride ions are needed to provide two negative charges so that they balance the 2+ charge on every magnesium ion. Therefore, the ratio of magnesium ions to chloride ions in the crystal is 1 : 2 and the formula of magnesium chloride is $MgCl_2$.

Figure 5.2.5 illustrates how the formulas for some other ionic compounds can be determined.

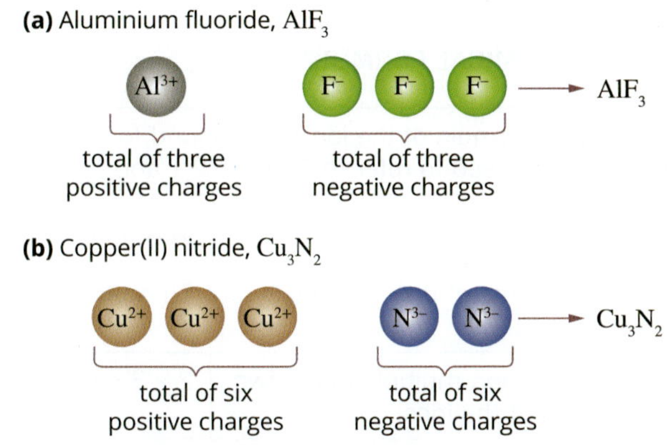

FIGURE 5.2.5 How to deduce chemical formulas from the charges on ions

Tables 5.2.1 and 5.2.2 list some of the more common positively and negatively charged ions. You may use these when you are writing formulas for ionic compounds.

TABLE 5.2.1 Names and formulas of some common cations

Charge			
1+	**2+**	**3+**	**4+**
caesium, Cs^+ copper(I), Cu^+ gold(I), Au^+ lithium, Li^+ potassium, K^+ rubidium, Rb^+ silver, Ag^+ sodium, Na^+	barium, Ba^{2+} cadmium(II), Cd^{2+} calcium, Ca^{2+} cobalt(II), Co^{2+} copper(II), Cu^{2+} iron(II), Fe^{2+} lead(II), Pb^{2+} magnesium, Mg^{2+} manganese(II), Mn^{2+} mercury(II), Hg^{2+} nickel, Ni^{2+} strontium, Sr^{2+} tin(II), Sn^{2+} zinc, Zn^{2+}	aluminium, Al^{3+} chromium(III), Cr^{3+} gold(III), Au^{3+} iron(III), Fe^{3+}	lead(IV), Pb^{4+} tin(IV), Sn^{4+}

TABLE 5.2.2 Names and formulas of some common anions

Charge		
1–	**2–**	**3–**
bromide, Br^- chloride, Cl^- fluoride, F^- iodide, I^-	oxide, O^{2-} sulfide, S^{2-}	nitride, N^{3-}

Rules for writing chemical formulas

Here are some simple rules to follow when you are writing chemical formulas.

- Write the symbol for the positively charged ion first.
- Use subscripts to indicate the number of each ion in the formula. Write the subscripts after the ion they refer to.
- If there is just one ion present in the formula, omit the subscript '1'.
- Do not include the charges on the ions in the balanced formula.

These rules are illustrated in Figure 5.2.6.

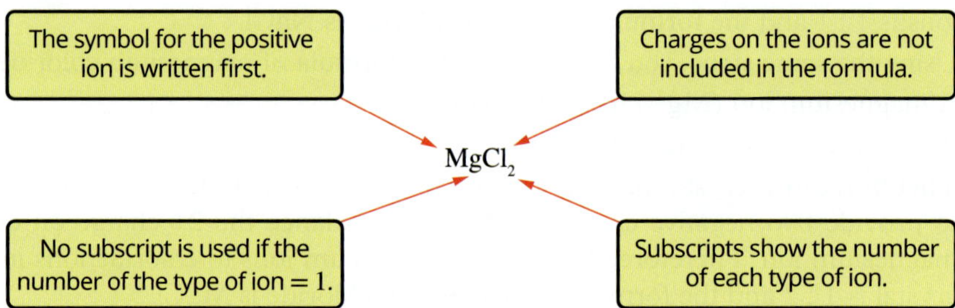

FIGURE 5.2.6 This diagram summarises the information provided by a chemical formula.

Worked example 5.2.2

STEPS IN WRITING A CHEMICAL FORMULA

Determine the chemical formula of the ionic compound formed between zinc and nitride ions. You may need to refer to Tables 5.2.1 and 5.2.2 on page 171.

Thinking	Working
Write the symbol and charge of the two ions forming the ionic compound.	Zn^{2+} and N^{3-}
Calculate the lowest common multiple of the two numbers in the charges of the ions.	$2 \times 3 = 6$
Calculate how many positive ions are needed to equal the lowest common multiple.	Three Zn^{2+} ions
Calculate how many negative ions are needed to equal the lowest common multiple.	Two N^{3-} ions
Use the answers from the previous two steps to write the formula for the ionic compound. Write the symbol of the positive ion first. (Note that 1 is not written as a subscript.)	Zn_3N_2

Worked example: Try yourself 5.2.2

STEPS IN WRITING A CHEMICAL FORMULA

Determine the chemical formula of the ionic compound formed between barium and fluoride ions. You may need to refer to Tables 5.2.1 and 5.2.2 on page 171.

WRITING FORMULAS OF MORE COMPLEX IONIC COMPOUNDS

The chemical formulas you have written for ionic compounds so far contain simple ions—ions that contain only one atom of an element. However, other ions contain two or more atoms, which may be of different elements. These ions are called **polyatomic ions**.

In polyatomic ions:

- if different elements are present, they are combined in a fixed ratio
- the group of atoms behaves as a single unit with a specific charge
- subscripts are used to indicate the number of each kind of atom in the ion.

For example, a carbonate ion (CO_3^{2-}) contains one carbon atom and three oxygen atoms combined together to form an ion. The carbonate ion has a charge of 2–. Other polyatomic ions are nitrate (NO_3^-), hydroxide (OH^-), hydrogen carbonate, HCO_3^- and phosphate (PO_4^{3-}). The formulas of a number of polyatomic ions can be seen in Table 5.2.3.

> Polyatomic ions are ions that are made up of two or more different atoms that have an overall charge. For example OH^-, NO_3^-, SO_4^{2-}, NH_4^+.

> The alternative, common name for the ethanoate ion is acetate. The alternative name for the hydrogen carbonate ion is bicarbonate.

TABLE 5.2.3 Common polyatomic cations and anions

Charge			
1+	1–	2–	3–
ammonium, NH_4^+	cyanide, CN^- dihydrogen phosphate, $H_2PO_4^-$ ethanoate, CH_3COO^- hydrogen carbonate, HCO_3^- hydrogen sulfide, HS^- hydrogen sulfite, HSO_3^- hydrogen sulfate, HSO_4^- hydroxide, OH^- nitrite, NO_2^- nitrate, NO_3^- permanganate, MnO_4^-	carbonate, CO_3^{2-} chromate, CrO_4^{2-} dichromate, $Cr_2O_7^{2-}$ hydrogen phosphate, HPO_4^{2-} oxalate, $C_2O_4^{2-}$ sulfite, SO_3^{2-} sulfate, SO_4^{2-}	phosphate, PO_4^{3-}

If more than one polyatomic ion is required in a formula to balance the charge, it is placed in brackets with the required number written as a subscript after the brackets. Some examples are:

- magnesium nitrate, $Mg(NO_3)_2$
- aluminium hydroxide, $Al(OH)_3$
- ammonium sulfate, $(NH_4)_2SO_4$.

Note that brackets are not required for the formula of sodium nitrate ($NaNO_3$), where there is only one NO_3^- ion present for each sodium ion.

A formula that is expressed in terms of the simplest whole-number ratio of particles (in this case the particles are ions) is called an empirical formula. You will learn more about empirical formulas in Chapter 7.

The formulas of two other ionic compounds containing polyatomic ions are shown in Figure 5.2.7.

(a) Magnesium nitrate, $Mg(NO_3)_2$

Mg^{2+} — total of **two** positive charges

NO_3^- NO_3^- — total of **two** negative charges

→ $Mg(NO_3)_2$

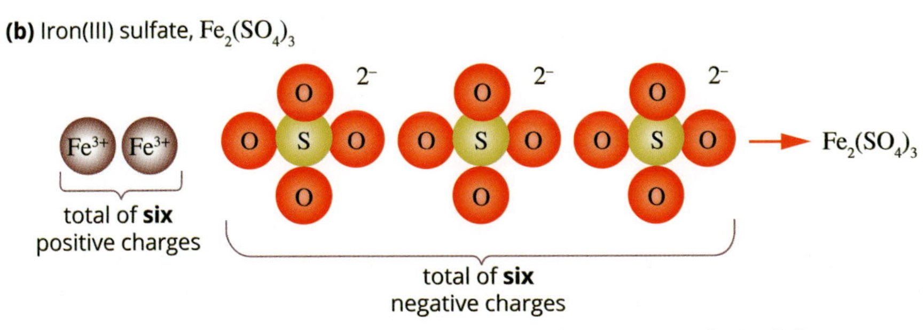

FIGURE 5.2.7 The chemical formulas of two ionic compounds containing polyatomic ions

Formulas involving elements with multiple electrovalencies

The **electrovalency** of an ion is the charge on the ion. Some transition metals, including copper, lead, iron and tin, can form ions with several electrovalencies (see Table 5.2.1). For example, copper can form Cu^{+} ions with a charge of 1+ and Cu^{2+} ions with a charge of 2+.

Some other metals with ions of variable electrovalency are:

- lead (Pb^{2+} and Pb^{4+})
- iron (Fe^{2+} and Fe^{3+})
- tin (Sn^{2+} and Sn^{4+}).

> Some transition metals can form ions with variable charges. To indicate the charge on the metal, write a Roman numeral in brackets immediately after the metal in the name of the compound.

For compounds of these metals, you need to specify the charge on the ion when naming the compound. This is done by placing a Roman numeral (in brackets) immediately after the metal in the name of the compound. For example:

- iron(II) chloride contains the Fe^{2+} ion, and so the formula is $FeCl_2$
- iron(III) chloride contains the Fe^{3+} ion, and so the formula is $FeCl_3$
- copper(I) sulfide contains the Cu^{+} ion, and so the formula is Cu_2S.

NAMING IONIC COMPOUNDS

There are some basic conventions that are followed when naming ionic compounds:

- The name of a positively charged metal ion (cation) is the same as the name of the metal. For example, the cation of a sodium atom is called a sodium ion; the cation of an aluminium atom is an aluminium ion.
- For simple non-metal ions (anions), the name of the ion is similar to that of the atom, but ends in '-ide'. For example, the anion of the chlorine atom is chloride; the anion of the oxygen atom is oxide.
- For polyatomic anions containing oxygen, the name of the ion will usually end in '-ite' or '-ate'. For example, the NO_2^- ion is called a nitrite ion; the NO_3^- ion is called a nitrate ion. (For two different ions of the same element with oxygen, the name of the ion with the smaller number of oxygen atoms usually ends in '-ite' and the one with the larger number of oxygen atoms ends in '-ate'.)

See Tables 5.2.1, 5.2.2 and 5.2.3 for more examples of how to name ions correctly.

5.2 Review

OA ✓✓

SUMMARY

- During the formation of ionic compounds:
  - metal atoms lose electrons to form positively charged ions (cations)
  - non-metal atoms gain electrons to form negatively charged ions (anions).
- The ions present in an ionic compound have a stable electronic configuration identical to that of the noble gas nearest to them on the periodic table.
- Electron transfer diagrams can be used to represent the formation of an ionic compound from its elements.
- When an ionic compound is formed from cations and anions, the ions combine in proportions that produce an ionic compound with an overall charge of zero.
- When considering the formula of an ionic compound, the total number of positive charges on the metal ions must equal the total number of negative charges on the non-metal ions.
- When writing formulas of ionic compounds:
  - the symbol for the positively charged ion is written first
  - subscripts are used to indicate the number of each ion in the formula
  - the charges on the ions are not included in the balanced formula.
- Ions that contain two or more atoms of different elements are called polyatomic ions (e.g. NO_3^-). When a chemical formula contains more than one polyatomic ion, the formula of the ion is placed in brackets with the number of ions written as a subscript after the brackets (e.g. $Mg(NO_3)_2$)
- The charge of an ion is called its electrovalency.
- When naming ionic compounds, the following rules apply:
  - The name of the metal ion is the same as the name of the metal.
  - For metals that form ions with different charges, the charge on the ion is shown by placing a Roman numeral after the name of the metal (e.g. copper(II) sulfate).
  - Simple non-metal ions take the name of the atom, but end in '-ide'.
  - Polyatomic anions containing oxygen usually end in '-ite' or '-ate'.

KEY QUESTIONS

Knowledge and understanding

1 Write equations to show the formation of ions when the following ionic compounds dissolve in water.

- **a** potassium bromide, KBr
- **b** calcium nitrate, $Ca(NO_3)_2$
- **c** sodium sulfide, Na_2S
- **d** iron(III) chloride, $FeCl_3$
- **e** aluminium sulfate, $Al_2(SO_4)_3$

2 Which of the following compounds are insoluble in water? Use Tables 5.3.1 and 5.3.2 on page 178 to help you answer this question.

- **A** magnesium hydroxide
- **B** sodium carbonate
- **C** tin(II) ethanoate
- **D** zinc sulfide
- **E** lead(II) sulfate
- **F** iron(III) chloride

3 Copy and complete the following sentences. Choose from the words or symbols listed below to fill in the blanks. Not all of the words and symbols in the list will be needed.

metallic; three; krypton; negatively; noble gas; gain two; argon; positively; lose one; non-metallic one; S^-; S^{2-}; S^+; S^{2+}

Potassium and sulfur will react together to form a compound, potassium sulfide. During this process each sulfur atom will ______ ______ electron(s) to form a __________ charged sulfide ion with the symbol __________. Each sulfide ion will have the same stable electron configuration as an atom of __________, which is the ______ ______ element nearest to it on the periodic table. Also during the reaction, each potassium atom will ______ _____ electron(s) to achieve the same stable electron configuration as an atom of _______,

continued over page

5.2 Review *continued*

4 Use the information in Tables 5.2.1 and 5.2.2 on page 171 to write formulas for the following ionic compounds.
- **a** Zinc chloride
- **b** Potassium oxide
- **c** Strontium nitride
- **d** Sodium carbonate
- **e** Aluminium sulfate
- **f** Zinc phosphate
- **g** Copper(I) chloride
- **h** Iron(III) oxide
- **i** Chromium(III) sulfate

5 Use the information in Tables 5.2.1 and 5.2.2 on page 171 and Table 5.2.3 on page 173 to name the following ionic compounds.
- **a** MgS
- **b** K_2O
- **c** $FeSO_4$
- **d** $Ba(NO_3)_2$
- **e** Cu_2SO_4
- **f** $Fe(CN)_3$
- **g** $Au_2(Cr_2O_7)_3$
- **h** $Pb_3(PO_4)_4$

Analysis

6 Indicate whether the following atoms will form cations or anions and explain why. For each part of the question, write the formula of the ion which is formed.
- **a** Calcium
- **b** Nitrogen
- **c** Fluorine
- **d** Aluminium
- **e** Phosphorus

7 Use electron transfer diagrams similar to Figure 5.2.2 on page 168 to show the formation of ions in the reactions between:
- **a** potassium and fluorine
- **b** magnesium and sulfur
- **c** aluminium and fluorine
- **d** sodium and oxygen
- **e** aluminium and oxygen.

8 Using the technique demonstrated in Worked example 5.2.1, write an equation for the reaction between the following metal and non-metal atoms. Show the electronic configurations for each element before and after the reaction.
- **a** Sodium and chlorine atoms
- **b** Magnesium and oxygen atoms
- **c** Aluminium and sulfur atoms

5.3 Precipitation reactions

A **precipitation reaction** occurs if ions in a solution combine to form a new compound that is insoluble in water. The insoluble compound formed in such a reaction is called a **precipitate**.

Precipitation reactions occur naturally in undersea hydrothermal vents. The vents release superheated solutions containing sulfide ions, which then combine with metal ions to form precipitates of mineral sulfides, creating the chimney-like structures seen in Figure 5.3.1. The areas around these chimneys are biologically rich, often hosting complex communities of aquatic organisms fuelled by the chemicals dissolved in the vent fluids.

FIGURE 5.3.1 Undersea hydrothermal vents release superheated water containing sulfide ions, which form precipitates with metal ions.

Precipitation reactions are used to remove minerals from drinking water, to remove heavy metals from wastewater and in the purification plants of reservoirs.

In this section you will look at what takes place during precipitation reactions.

IONS IN SOLUTION

In Section 5.1 you learned that some ionic compounds are soluble in water and some are not. Sodium chloride, NaCl, and copper(II) sulfate, $CuSO_4$, are examples of compounds that will dissolve readily in water to form **aqueous** solutions of the ions. When they dissolve, the ions in each of the compounds separate from each other and move freely in the solution. We can write equations to represent the dissolving process for these compounds:

$$NaCl(s) \xrightarrow{H_2O} Na^+(aq) + Cl^-(aq)$$

$$CuSO_4(s) \xrightarrow{H_2O} Cu^{2+}(aq) + SO_4^{2-}(aq)$$

The ways in which compounds dissolve will be discussed in more detail in Chapter 6.

If solutions of sodium chloride and copper(II) sulfate were added to a beaker, the mixture would consist of separate Na^+, Cl^-, Cu^{2+} and SO_4^{2-} ions moving about independently, but not combining with each other.

REACTIONS BETWEEN IONIC COMPOUNDS IN SOLUTION

FIGURE 5.3.2 Mixing aqueous solutions of sodium chloride and silver nitrate produces a solid, called a precipitate.

Solubility tables can be used to predict whether the mixing of two solutions of different ionic compounds will result in the formation of a precipitate.

In some cases, a reaction occurs when two solutions of ionic compounds are mixed. For example, when a colourless solution of silver nitrate is mixed with a colourless solution of sodium chloride, the solution turns cloudy because a very fine-powdered white solid has formed. This is shown in Figure 5.3.2.

To understand what happens in the reaction between silver nitrate and sodium chloride, you need to identify the ions present in the reactant solutions and how they interact with each other.

- In the silver nitrate solution, there are dissolved silver ions (Ag^+) and nitrate ions (NO_3^-).
- In the sodium chloride solution, there are dissolved sodium ions (Na^+) and chloride ions (Cl^-).
- When one solution is added to the other, the mixture formed will contain all of the ions.

In the mixture of the two solutions all the ions are moving around independently, as in the mixture of sodium chloride and copper(II) sulfate solutions, discussed at the start of this section. As the ions move in the solution, they will collide with one another. If positive and negative ions collide, they may join together to form a new, insoluble precipitate. Some of the ions have done this in the mixture of sodium chloride and silver nitrate solutions.

Two new combinations of positive and negative ions are possible:

- sodium and nitrate ions (to form sodium nitrate)
- silver and chloride ions (to form silver chloride).

You can decide which one of these two compounds is the precipitate by referring to **solubility tables**. Table 5.3.1 lists ionic compounds that are soluble in water. (Note that there are exceptions for some ions.) Insoluble compounds are listed in Table 5.3.2.

TABLE 5.3.1 Relative solubilities of soluble ionic compounds

Soluble in water (>0.1 mol dissolves per L at 25°C)	Exceptions: insoluble (<0.01 mol dissolves per L at 25°C)	Exceptions: slightly soluble (0.01–0.1 mol dissolves per L at 25°C)
most chlorides (Cl^-), bromides (Br^-) and iodides (I^-)	AgCl, AgBr, AgI, PbI_2	$PbCl_2$, $PbBr_2$
all nitrates (NO_3^-)	no exceptions	no exceptions
all ammonium (NH_4^+) salts	no exceptions	no exceptions
all sodium (Na^+) and potassium (K^+) salts	no exceptions	no exceptions
all ethanoates (CH_3COO^-)	no exceptions	no exceptions
most sulfates (SO_4^{2-})	$SrSO_4$, $BaSO_4$, $PbSO_4$	$CaSO_4$, Ag_2SO_4

TABLE 5.3.2 Relative solubilities of insoluble ionic compounds

Insoluble in water	Exceptions: soluble	Exceptions: slightly soluble
most hydroxides (OH^-)	NaOH, KOH, $Ba(OH)_2$, NH_4OH*, AgOH†	$Ca(OH)_2$, $Sr(OH)_2$
most carbonates (CO_3^{2-})	Na_2CO_3, K_2CO_3, $(NH_4)_2CO_3$	no exceptions
most phosphates (PO_4^{3-})	Na_3PO_4, K_3PO_4, $(NH_4)_3PO_4$	no exceptions
most sulfides (S^{2-})	Na_2S, K_2S, $(NH_4)_2S$	no exceptions

*NH_4OH does not exist in significant amounts in an ammonia solution. Ammonium and hydroxide ions readily combine to form ammonia and water.

†AgOH readily decomposes to form a precipitate of silver oxide and water.

From Table 5.3.1 we can see that, in general:

- all compounds containing a nitrate ion are soluble in water
- all compounds containing a sodium ion are soluble in water
- most compounds containing a chloride ion are soluble in water, but silver chloride is an exception.

Therefore, the precipitate shown in Figure 5.3.2 must be silver chloride.

The process for the precipitation reaction between sodium chloride and silver nitrate is shown in Figure 5.3.3. When the hydrated Ag^+ and Cl^- ions come into contact an ionic lattice of AgCl is formed.

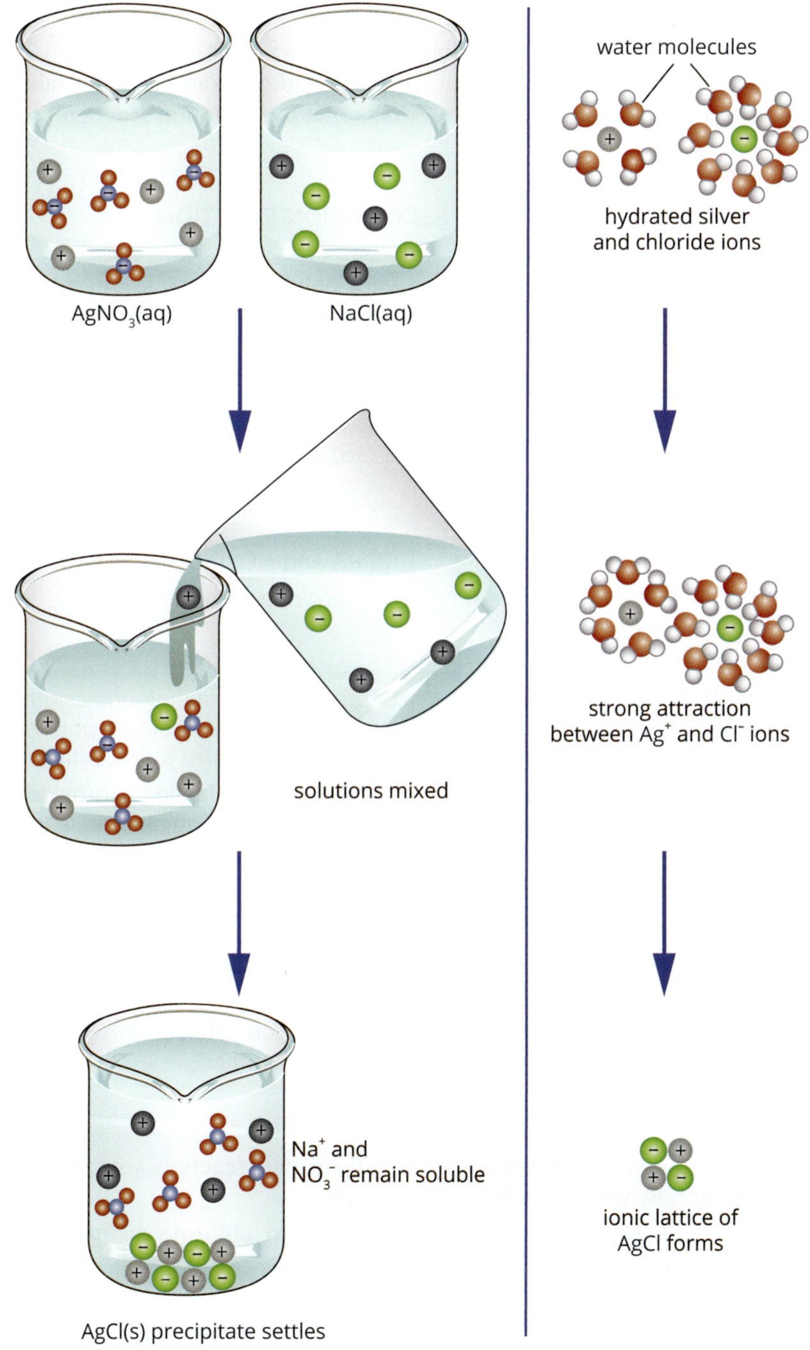

FIGURE 5.3.3 Pictorial representation of mixing aqueous solutions of sodium chloride and silver nitrate to produce a precipitate of silver chloride

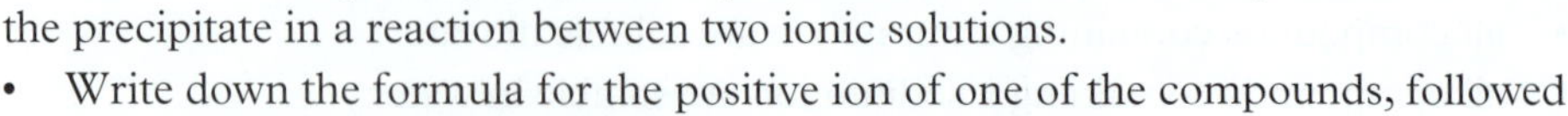

There is a simple way that allows you to work out which compound will form the precipitate in a reaction between two ionic solutions.

- Write down the formula for the positive ion of one of the compounds, followed by its negative ion. Repeat the process for the second compound. For example, Ag^+, NO_3^-, Na^+, Cl^-.
- Then draw two lines. The first line joins the positive ion of the first solution to the negative ion of the second. The second line joins the negative ion of the first solution to the positive ion of the second (Figure 5.3.4).
- Finally, use solubility tables to work out which of the two new combinations of ions will result in an insoluble compound. This will be the precipitate. The other ions will remain in solution.

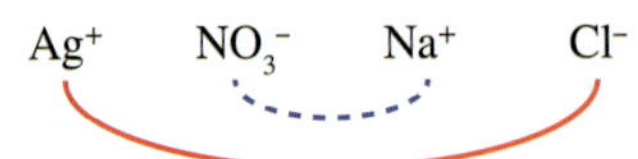

FIGURE 5.3.4 Working out which compound is the precipitate in a reaction

It is important for chemists to be able to predict whether a precipitate will form in a reaction and what this precipitate will be. Worked example 5.3.1 takes you through the process of predicting the products of a precipitation reaction.

Worked example 5.3.1

PREDICTING THE PRODUCTS OF A PRECIPITATION REACTION

What precipitate, if any, will be produced when solutions of potassium hydroxide and lead(II) nitrate are added together? You will need to refer to the solubility tables (Tables 5.3.1 and 5.3.2 on page 178) to complete this question.

Thinking	Working
Identify which ions are produced by each of the ionic compounds in the mixture.	$K^+(aq)$, $OH^-(aq)$, $Pb^{2+}(aq)$ and $NO_3^-(aq)$
Identify which two new combinations of positive and negative ions are possible in the mixture of the solutions.	$K^+(aq)$ and $NO_3^-(aq)$ $Pb^{2+}(aq)$ and $OH^-(aq)$
Use the solubility tables to check which, if any, of these combinations will produce an insoluble compound.	Compounds containing potassium ions are usually soluble, so potassium nitrate will not form a precipitate. Compounds containing hydroxide ions are usually insoluble, so lead(II) hydroxide will form a precipitate.

Worked example: Try yourself 5.3.1

PREDICTING THE PRODUCTS OF A PRECIPITATION REACTION

What precipitate, if any, will be produced when solutions of sodium sulfide (Na_2S) and copper(II) nitrate ($Cu(NO_3)_2$) are added together?

CHEMFILE

Limescale accumulation

Have you ever wondered where that flaky white build-up on the heating element of your kettle comes from?

When you boil water in the kettle, ions present in the water can precipitate out, leaving a white coating called limescale on the element (see figure below).

The accumulation of limescale on domestic kettles is the result of precipitation of calcium carbonate as water is repeatedly boiled.

The amount of build-up depends on the type of water treatment in your area. Areas that have hard water (that is, with high levels of dissolved ions) have a bigger problem with limescale.

Limescale mostly consists of calcium carbonate ($CaCO_3$) that precipitates out as a crystalline solid when the water is boiled.

Limescale can be a big problem in the home. A coating as thin as 1.5 mm over a heating element can reduce its efficiency by as much as 12%.

Many modern houses use ion filters or water conditioners to remove dissolved ions from the water and reduce the accumulation of limescale.

Writing equations for precipitation reactions

Now that you can identify the precipitate that forms in a reaction, you can show the complete reaction by writing a chemical equation.

The reaction between silver nitrate and sodium chloride solutions can be summarised in words as:

silver nitrate solution + sodium chloride solution ⟶
silver chloride solid + sodium nitrate solution

An alternative representation is an equation that uses formulas. This type of equation is called a **full equation**. The complete formulas of all the reagents and products are shown in the reaction. The state symbols for each of the species in the chemical reaction must be shown:

$$AgNO_3(aq) + NaCl(aq) \longrightarrow AgCl(s) + NaNO_3(aq)$$

Although 'sodium nitrate' or '$NaNO_3(aq)$' is written as a product in these equations, the sodium ions and nitrate ions are not combined with each other. They move freely through the solution. They are present at the start of the reaction and they are still there, as separate ions, at the end of the reaction.

Because the sodium and nitrate ions have not been involved in forming a precipitate, they are said to be **spectator ions**.

Spectator ions do not undergo a chemical change in the reaction. In a precipitation reaction, they will always start as aqueous (aq) ions and will remain as aqueous ions after the reaction is complete.

Worked example 5.3.2 looks at the process of writing full equations for precipitation reactions and identifying spectator ions

Worked example 5.3.2

WRITING EQUATIONS FOR PRECIPITATION REACTIONS

Write a balanced equation for the reaction between iron(III) nitrate and sodium sulfide, in which the precipitate is iron(III) sulfide. Identify the spectator ions in this reaction.

Thinking	Working
Write an incomplete, unbalanced equation showing the reactants and the precipitate product. Include symbols of state.	$Fe(NO_3)_3(aq) + Na_2S(aq) \longrightarrow Fe_2S_3(s)$
Add to the equation above the formula of the other compound formed in the reaction.	$Fe(NO_3)_3(aq) + Na_2S(aq) \longrightarrow Fe_2S_3(s) + NaNO_3(aq)$
Balance the equation.	$2Fe(NO_3)_3(aq) + 3Na_2S(aq) \longrightarrow Fe_2S_3(s) + 6NaNO_3(aq)$
Write the formulas of the ions that do not form a precipitate in the reaction. These are the spectator ions.	$Na^+(aq)$ and $NO_3^-(aq)$ are spectator ions.

Worked example: Try yourself 5.3.2

WRITING EQUATIONS FOR PRECIPITATION REACTIONS

Write a balanced equation for the reaction between copper(II) sulfate and sodium hydroxide, in which the precipitate is copper(II) hydroxide. Identify the spectator ions in this reaction.

Writing ionic equations for precipitation reactions

The essential feature of the reaction between silver nitrate and sodium chloride is the combination of silver ions and chloride ions to form a precipitate. This reaction can be summarised in an **ionic equation**. Note that spectator ions are not included in an ionic equation. Only the species that combine to form the precipitate are included.

An ionic equation can be thought of as a full equation with the spectator ions removed. So full and ionic equations for the reaction between silver nitrate and sodium chloride solutions are:

Full equation: $AgNO_3(aq) + NaCl(aq) \longrightarrow AgCl(s) + NaNO_3(aq)$

Ionic equation: $Ag^+(aq) + Cl^-(aq) \longrightarrow AgCl(s)$

A simple way to write an ionic equation for a precipitation reaction is shown in the Worked example 5.3.3 on the following page.

Worked example 5.3.3

WRITING IONIC EQUATIONS FOR PRECIPITATION REACTIONS

Write an ionic equation for the precipitation reaction between solutions of aluminium nitrate and sodium sulfide, in which the precipitate is aluminium sulfide.

Thinking	Working
Write the formula of the precipitate on the right-hand side of the page. Place an arrow to the left of the formula.	$\rightarrow Al_2S_3$
To the left of this formula, add the formulas of the two types of ions that form the precipitate, using the ratio of ions shown in the precipitate formula.	$2Al^{3+} + 3S^{2-} \rightarrow Al_2S_3$
Add symbols of state to the equation and check that it is balanced.	$2Al^{3+}(aq) + 3S^{2-}(aq) \rightarrow Al_2S_3(s)$

Worked example: Try yourself 5.3.3

WRITING IONIC EQUATIONS FOR PRECIPITATION REACTIONS

Write an ionic equation for the precipitation reaction between solutions of sodium hydroxide and barium nitrate, in which the precipitate is barium hydroxide.

CASE STUDY ANALYSIS

The chemistry of colour

If you have ever walked through an art supplies store, you will have noticed displays selling paints with names that sound as though they belong in a chemistry laboratory. Names such as lead yellow, titanium white and cobalt blue are just a few of the colour names still in use today (Figure 5.3.5).

The names are not just for show. The colours that they represent have their basis in chemistry. All paints consist of a pigment (the colour) and a binder that holds the pigments in a suspension. Historically, raw materials for paints were collected from minerals and other sources and then sold as powdered pigments, like those in Figure 5.3.6.

Artists would mix these pigments with their own binders, such as linseed oil, to produce the required paints. Ancient civilisations, like the Egyptians, would grind minerals such as lapis lazuli and ochre from the earth to make pigments. Early alchemists were able to manufacture synthetic pigments using precipitation reactions, collecting the coloured precipitates through filtration.

FIGURE 5.3.5 Paint colours are often named after the pigments that were once used to make them. Many of these pigments were the product of precipitation reactions.

continued on next page

CASE STUDY ANALYSIS

FIGURE 5.3.6 An array of different-coloured pigments

However, in the nineteenth century a better knowledge of chemistry was used to manufacture pigments more reliably and in a larger variety of colours.

Prussian blue was a very popular colour throughout the nineteenth century and is made from a precipitate of iron(III) hexacyanoferrate(II). It has been used extensively in ceramics and painting, including the famous Japanese block print from the 1830s, *The Great Wave off Kanagawa* (Figure 5.3.7).

FIGURE 5.3.7 *The Great Wave off Kanagawa* (c. 1830–1833). This woodblock print is by the Japanese artist Hokusai. The deep blue tones were painted with a colour known as Prussian blue, made from iron(III) hexacyanoferrate(II).

Pigments based on ionic precipitates provided a much wider array of colours for artists. This paved the way for the more flamboyant use of colour seen in the works of the French Impressionists. Cadmium sulfide was used to make red pigments and cadmium red is still used widely today. Viridian, a deep green pigment, is manufactured from chromium(III) oxide dihydrate.

So the next time you visit an art gallery, take a moment to think about the complex chemistry that went into bringing those colours to life.

Analysis

The table below shows the chemical constituents of several artists' pigments.

Pigment	Chemical formula
cadmium yellow	CdS
chrome yellow	$PbCrO_4$
mosaic gold	SnS_2
titanium white	TiO_2
vermilion	HgS
zinc yellow	$ZnCrO_4$

1 What is the charge on the tin ion in mosaic gold?

2 The formula of a chromate ion is CrO_4^{2-}. Write a full equation for the precipitation reaction in which chrome yellow (lead(II)chromate) is formed from a mixture of sodium chromate and lead(II) nitrate solutions.

3 Cadmium yellow can be made in a precipitation reaction using solutions of sodium sulfide and a cadmium compound. Suggest a suitable cadmium compound for this purpose. Tables 5.3.1 and 5.3.2 on page 178 may help with answering this question.

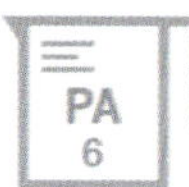

5.3 Review

OA ✓✓

SUMMARY

- Soluble ionic compounds dissolve in water to form ions.
- A precipitation reaction occurs when two solutions of compounds are mixed and a solid product is formed. The solid product is called a precipitate.
- Solubility tables can be used to predict which compound, if any, will precipitate when two solutions of ionic compounds are added together.
- Ions that are not directly involved in the formation of the precipitate during a precipitation reaction are called spectator ions.
- Full and ionic equations can be written for precipitation reactions. Chemical equations include state symbols, e.g. (s), (aq).
- Ionic equations do not include spectator ions.

KEY QUESTIONS

Knowledge and understanding

1 Write equations to show the formation of ions when the following ionic compounds dissolve in water.

a Potassium bromide, KBr
b Calcium nitrate, $Ca(NO_3)_2$
c Sodium sulfide, Na_2S
d Iron(III) chloride, $FeCl_3$
e Aluminium sulfate, $Al_2(SO_4)_3$

2 **a** Write the formula of the precipitate formed, if any, when the following solutions are mixed.

i $Ca(NO_3)_2$ and K_2CO_3
ii $MgSO_4$ and $Cu(NO_3)_2$
iii Na_2S and $MgSO_4$
iv $FeCl_2$ and NH_4OH
v Na_3PO_4 and $AgNO_3$

b Write the formulas of any spectator ions in the questions in part **a**.

3 **a** Name the precipitate formed when aqueous solutions of the following compounds are mixed.

i K_2S and $MgCl_2$
ii $CuCl_2$ and $AgNO_3$
iii KOH and $AlCl_3$
iv $MgSO_4$ and NaOH

b Write a balanced full chemical equation for each reaction.

4 For each of the following combinations of solutions, represent the reactions that occur by:

i a full equation
ii an ionic equation

a $NH_4Cl + AgNO_3$
b $Cu(NO_3)_2 + K_2CO_3$
c $K_3PO_4 + MgSO_4$
d $Ca(OH)_2 + FeCl_2$
e $Ba(NO_3)_2 + (NH_4)_2SO_4$
f $Pb(CH_3COO)_2 + Na_2SO_4$

5 For each of the parts in Question **4**, identify the spectator ions.

Analysis

6 Aqueous solutions of potassium sulfate, barium ethanoate and magnesium nitrate were added to a beaker and a precipitate formed.

a Name the precipitate.
b Write a full equation for the reaction.
c Write an ionic equation for the reaction.

Chapter review

KEY TERMS

anion
aqueous
brittle
cation
ceramic
chemical formula
crystal lattice
dissolve
electrolyte
electron transfer diagram
electrovalency
empirical formula
full equation
gemstone
ionic bonding
ionic bonding model
ionic compound
ionic equation
ionic lattice
polyatomic ion
precipitate
precipitation reaction
solubility
solubility table
spectator ion

REVIEW QUESTIONS

Knowledge and understanding

1 Some properties of four different substances are described below. Which substance is most likely to be an ionic compound?

A Substance A has a melting point of 420°C and will be flattened but not shattered when hit with a hammer.

B Substance B has a melting point of 135°C. It does not conduct electricity at 120°C or at 150°C.

C Substance C has a melting point of 181°C and will conduct electricity at 25°C.

D Substance D is a white solid that melts at 770°C and readily dissolves in water at 80°C.

2 Which one of the following is the formula of a polyatomic ion?

A OH^-

B H_2O

C Br^-

D Mn^{2+}

3 The spectator ions present when a solution of barium nitrate is added to a solution of sodium sulfate are:

A Ba^{2+} and Na^+ ions

B NO_3^- and Ba^{2+} ions

C SO_4^{2-} and Na^+ ions

D Na^+ and NO_3^- ions

4 Use the ionic bonding model to explain the following properties of ionic compounds.

a They generally have high melting points.

b They are hard and brittle.

c They do not conduct electricity in the solid state but will conduct when molten or dissolved in water.

5 Here are four statements about the structure of an ionic compound.

Statement 1: In the molten state, the ions in the compound are able to move freely.

Statement 2: In the solid state, the ions are held in a lattice and are not free to move.

Statement 3: The electrostatic force of attraction between positive and negative ions in the ionic lattice is very strong.

Statement 4: In the solid state, if the ionic compound is struck with a hammer, some layers in the ionic lattice may move so that ions with like charge become adjacent to each other.

For which property of ionic compounds does each of the above statements provide an explanation? (Some statements may provide an explanation for more than one property.)

Statement 1: ______________________

Statement 2: ______________________

Statement 3: ______________________

Statement 4: ______________________

6 Give the electronic configurations of the following ions.

a Na^+

b O^{2-}

c Mg^{2+}

d N^{3-}

7 Write the chemical formula for the ionic compound formed in the reaction between:

a potassium and bromine

b magnesium and iodine

c calcium and oxygen

d aluminium and fluorine

e calcium and nitrogen.

8 Consider the following ions: SO_4^{2-}, PO_4^{3-}, Br^-, S^{2-}. Which one or more of these ions would combine with $Fe^{2+}(aq)$ to give a precipitate?

9 Write the chemical formulas for the following ionic compounds.
- a Copper(I) nitrate
- b Chromium(II) fluoride
- c Potassium carbonate
- d Magnesium hydrogen carbonate
- e Nickel(II) phosphate

10 Name the ionic compounds with the following chemical formulas.
- a $(NH_4)_2CO_3$
- b $Cu(NO_3)_2$
- c $Cu(NO_2)_2$
- d $CrBr_3$
- e $Sn(H_2PO_4)_2$
- f $Pb(HSO_3)_4$

11 Create diagrams like those in Figure 5.2.2 on page 168 to show the electron transfers that occur when the following elements react with each other.
- a lithium reacts with chlorine
- b magnesium reacts with fluorine
- c potassium reacts with sulfur
- d magnesium reacts with nitrogen

12 Using Worked example 5.3.1 on page 180 as a guide, write equations for the reactions between the following elements, showing the shell electronic configuration for each element before and after the reaction.
- a Magnesium and chlorine atoms
- b Aluminium and oxygen atoms
- c Sodium and phosphorus atoms

13 What do subscripts in the formula of an ionic compound tell you about the metal and non-metal ions?

14 Explain why elements in group 17 of the periodic table are likely to forms ions with a 1– charge.

15 What precipitate will be formed (if any) when the following solutions are mixed?
- a Barium nitrate and sodium sulfate
- b Sodium chloride and copper(II) sulfate
- c Magnesium sulfate and lead(II) nitrate
- d Potassium chloride and barium nitrate

Application and analysis

16 An atom of a non-metal, X, has 6 electrons in its outermost shell. An atom of a metal, Y, has 3 electrons in its outermost shell.
When atoms of X and Y react to form a compound, the correct formula of the compound will be:
- A X_2Y_3
- B Y_2X_3
- C X_3Y_2
- D Y_3X_2

17 A_2B is an ionic compound. Both of the ions in A_2B have the same electronic configuration as an argon atom. What is the identity of A_2B?
- A Calcium chloride
- B Potassium sulfide
- C Calcium sulfide
- D Sodium oxide

18 Potassium fluoride will conduct electricity at 860°C but not at 840°C. Calcium oxide will conduct electricity at 2580°C but not at 2520°C.
- a Write chemical formulas for the ions present in samples of both potassium fluoride and calcium oxide.
- b Comment on the relative strength of the forces between particles in both potassium fluoride and calcium oxide to help you explain the difference between the melting points of the compounds.

19 A student compares the structure and bonding in metals and ionic compounds and makes the following statements.
- a In both metals and ionic solids there are metal atoms which have lost the electrons from their outermost shells.
- b In both metals and ionic solids there are forces of repulsion between particles of opposite charge.
- c In both metals and ionic solids negatively charged particles are held in fixed positions around positively charged particles.
- d Both solid metals and molten ionic compounds can conduct an electric current. In each of these cases it is only moving negatively charged particles that are responsible for the conduction of the electric current.
- e Metals are ductile because, unlike ionic compounds, they contain negatively charged particles that are free to move.

Comment on each of these statements, explaining clearly why you either agree or disagree.

20 Construct a concept map to show the connection between the terms: metals, non-metals, atoms, valence electrons, anions, cations, electrostatic attraction, ionic compounds.

21 In the table below is a list of elements whose atoms can form ions by reacting with atoms of other elements.
In the second column of the table, indicate whether each of the elements will form ions by losing or gaining electrons.
In the third column of the table, write the name of the noble gas that will have the same electron configuration as the ion formed by the element.

Element	Electrons lost or gained when forming an ion?	Noble gas with same electron configuration as ion formed
phosphorus		
lithium		
oxygen		
aluminium		
potassium		
bromine		
sulfur		

22 The electronic configurations of some metallic and non-metallic elements are given. (The symbols shown for the elements are not their real ones.) Write formulas for the compounds they are most likely to form if they react together.

a C: 2,8,3 D: 2,7
b E: 2,8,8,2 F: 2,8,6
c G: 2,8,8,1 H: 2,5
d K: 2,8,2 L: 2,6

23 Write full balanced chemical equations and ionic equations for each of the following precipitation reactions.

a $NH_4Cl(aq) + AgNO_3(aq) \longrightarrow$
b $FeCl_2(aq) + Na_2S(aq) \longrightarrow$
c $Fe(NO_3)_3(aq) + KOH(aq) \longrightarrow$
d $CuSO_4(aq) + NaOH(aq) \longrightarrow$
e $Ba(NO_3)_2(aq) + Na_2SO_4(aq) \longrightarrow$

24 Copy and complete the following table. Identify which reaction mixtures will produce precipitates and write their formulas.

	NaOH	$(NH_4)_3PO_4$	NaI	$MgSO_4$	$BaCl_2$
$Pb(NO_3)_2$					
KI					
$CaCl_2$					
Na_2CO_3					
Na_2S					

25 The formula of sodium perchlorate is $NaClO_4$ and that of potassium ferrocyanide is $K_4Fe(CN)_6$. Using this information, write formulas for the following compounds.

a calcium perchlorate
b aluminium ferrocyanide
c iron(III) perchlorate
d ammonium ferrocyanide

26 The elements X, Y and Z can form ionic compounds. The formulas of three examples of these compounds are: $X(NO_3)_2$; Fe_2Y_3 and $Z(H_2PO_4)_3$.

a What is the electrovalency (charge) of the ion formed by:
  i element X?
  ii element Y?
  iii element Z?
b Use these charges on the ions of elements X, Y and Z to write correct chemical formulas for the:
  i nitride salt of X
  ii lead(IV) salt of Y
  iii dichromate salt of Z
  iv ionic compound formed between Y and Z

27 Refer to the periodic table at the end of the book and, for each general formula given, identify two elements that will react to form an ionic compound with that formula. (Remember the metal ion, as represented by X, is written first in each formula.)

a XY_2
b XY
c X_2Y
d X_3Y
e XY_3

28 Describe an experiment you could carry out to demonstrate each of the following properties of the compounds given. In each case:
  i sketch the equipment you would use
  ii describe what you would expect to observe.

a Solid magnesium chloride does not conduct electricity.
b A solution of sodium chloride in water does conduct electricity.
c Solid sodium chloride is hard and brittle.

OA ✓✓

CHAPTER 06 Separation and identification of components of mixtures

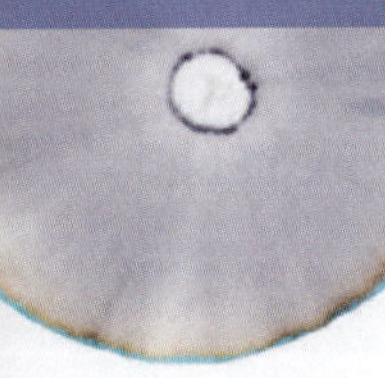

Water is often described as the universal solvent because it can dissolve a wide range of substances. The Earth's oceans, rivers and lakes are aqueous solutions, containing many different dissolved minerals and gases. In this chapter, you will learn how different types of substances dissolve in water and other liquids, and how to predict whether a compound is likely to be soluble in a particular solvent. The polarity of the solvent plays a critical role in determining the compounds that can be dissolved in it.

Differences in the solubilities of compounds form the basis of a laboratory technique called chromatography, which is commonly used to separate and analyse chemical mixtures. There are several different types of chromatography and this chapter describes the principles that underpin two of the simpler methods, paper chromatography and thin-layer chromatography.

Key knowledge

- polar and non-polar character with reference to the solubility of polar solutes dissolving in polar solvents, and non-polar solutes dissolving in non-polar solvents **6.1**
- experimental application of chromatography as a technique to determine the composition and purity of different types of substances, including calculation of R_f values. **6.2**

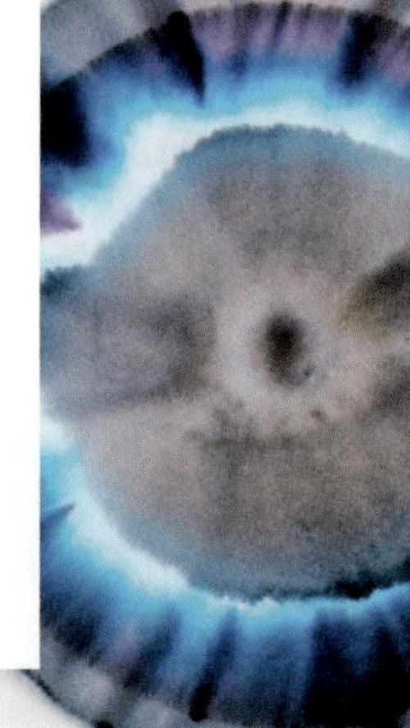

6.1 How substances dissolve

Water is an excellent solvent. This is one of its most important properties. Almost all biological processes and many industrial processes occur in water. These systems are known as **aqueous** environments.

When substances are dissolved in water, the particles are free to move throughout the **solution**. In Figure 6.1.1, you can see how the deep purple colour of potassium permanganate spreads through the water as the solid dissolves. Eventually the liquid will appear completely purple as the particles continue to mix and move.

This section looks at the process of dissolving and the types of forces involved when molecular and ionic substances dissolve in water and other liquids.

FIGURE 6.1.1 When solid potassium permanganate is added to water, it dissolves. The particles disperse into the solution and move around freely.

CHARACTERISTICS OF A SOLUTION

A solution is formed when a solid, liquid or gas is dissolved in a liquid. The substance being dissolved is called the **solute** and the liquid in which the substance is dissolved is the **solvent**.

If the solvent is water, the solution is called an **aqueous solution**. The solute of an aqueous solution can change, but the solvent will always be water. For example, you can have an aqueous solution of salt (saline) or an aqueous solution of sugar. Table 6.1.1 lists some common aqueous solutions.

TABLE 6.1.1 Some everyday aqueous solutions

Solution	Solute/s	Solvent
saline solution (for use with contact lenses)	sodium chloride	water
soft drink	carbon dioxide, sugar, flavour, colour	water
coffee	coffee, sugar, milk	water

In a homogeneous solution, the solute and solvent particles cannot be distinguished from each other visually. Every part of the solution is the same as any other part.

All solutions have the following characteristics:

- The solute and the solvent cannot be distinguished from each other visually. This means the solution is **homogeneous**.
- The dissolved particles are too small to see.
- The amount of dissolved solute can vary from one solution to another.

THE PROCESS OF DISSOLVING

The process of a substance dissolving in another substance is called **dissolution**.

During dissolution the following processes occur:

- Solute particles are attracted to some of the solvent particles.
- The particles of the solute are separated from one another.
- Some of the solvent particles are separated from one another to allow the solute particles to disperse throughout the liquid.
- Solvent particles not attached to solute particles will still be attracted to other solvent particles.

For a solution to form, the solute particles must interact with the solvent molecules. The solute particles are surrounded by solvent molecules and carried throughout the solution.

Forces involved in dissolving

For a substance to dissolve, there must be a change in the way particles in the solute and solvent interact. This means that you need to look at the forces of attraction that occur between the particles.

It is useful to think of three different forces of attraction when considering if and how a substance will dissolve in water. (Figure 6.1.2). These are:

- the forces holding the particles of the substance (solute) together before it is added to the solvent
- the forces holding the solvent molecules together; in water, these forces are **hydrogen bonds** and **dispersion forces**
- the forces that can form between the solute particles and the solvent molecules.

For a substance to dissolve, the attractive forces that form between the solute and solvent particles must be similar to, or greater than, the forces between the particles in the solute and the forces between the solvent molecules.

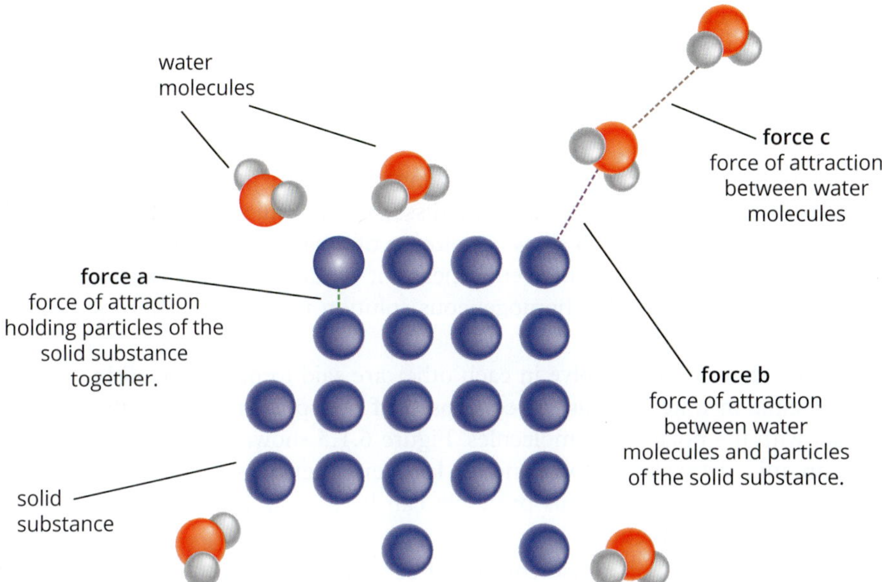

FIGURE 6.1.2 Forces present when a substance dissolves in water. For the solute to dissolve in water the strength of force b must be greater than, or about the same as, the strength of forces a and c.

Figure 6.1.3 shows that as a solute dissolves, the solute particles separate and become evenly distributed in the solvent. If the attraction between the solute and solvent particles is not strong enough, the substance will not readily dissolve.

For a substance to be soluble, the attraction between the solvent and the solute particles must be similar to, or stronger than, the attraction between solute particles and the attraction between solvent particles.

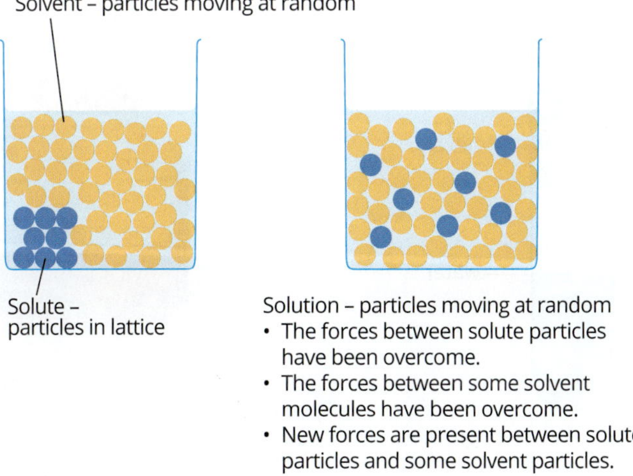

FIGURE 6.1.3 The rearrangement of particles when a solute dissolves in a solvent. The solute particles separate and become evenly distributed in the solvent.

CHEMFILE

Water cannot cool a mouthful of chilli

The chemical that gives a chilli its heat is capsaicin. Capsaicin (see figure below) is a non-polar molecule that cannot dissolve in water. When you drink water after eating chilli, it will not rinse the capsaicin away, so the burning sensation remains.

Some common advice given to cool your mouth after eating chillies is to drink milk. As milk is largely aqueous, the capsaicin does not dissolve in the milk, but a protein in milk breaks the bonds between capsaicin and pain receptors in the mouth.

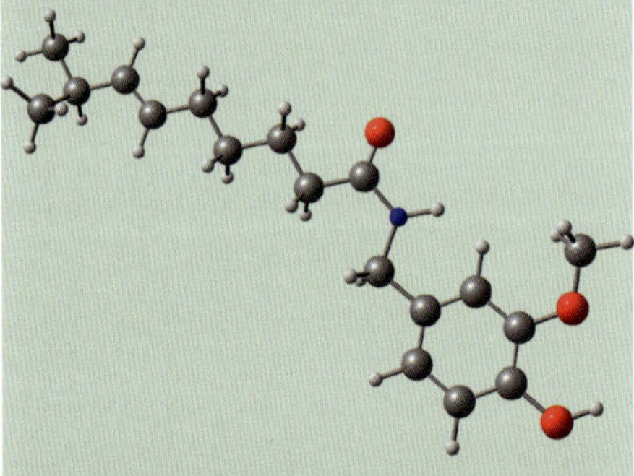

The structure of capsaicin contains mostly non-polar groups. The molecule will not interact with polar water molecules and so it is not soluble.

Like dissolves like

You can use a 'like dissolves like' rule to predict whether a substance is likely to dissolve in another substance. The general statement that a solvent will only dissolve 'like' solutes tells you that:

- **polar** solvents will generally dissolve substances consisting of polar molecules or ions, but will not dissolve solutes made up of **non-polar** molecules
- non-polar solvents can dissolve substances consisting of non-polar molecules, but will not dissolve ones with polar molecules or ions.

Wax and other non-polar molecular substances do not dissolve well in water because the only intermolecular forces in these substances are dispersion forces, whereas the strong intermolecular forces of hydrogen bonding exist between the water molecules. These hydrogen bonds between water molecules are much stronger than the dispersion forces that could occur between molecules of wax and water. As a result, the attraction between the water molecules cannot be overcome and the water molecules do not separate to form a solution with the wax molecules.

When the two substances in the dissolving process are liquids, you can say they are **miscible**. You can observe the 'like dissolves like' rule with the miscibility of different liquids. Figure 6.1.4 shows a solution of ethanol (alcohol) and water. The polar nature of the ethanol molecule means it readily dissolves in water, which is also polar. When mixed, a homogeneous solution is formed with no separation between solute and solvent.

Liquids that do not dissolve in each other are said to be **immiscible**. Hexane is immiscible in water. Hexane is composed of non-polar molecules that will not interact with the polar water molecules. Figure 6.1.5 shows that when hexane and water are mixed, two layers form and the less dense hexane sits on top of the water. An everyday illustration of this principle can be seen in a bottle of home-made salad dressing, as in Figure 6.1.6.

Like hexane, olive oil is also composed of non-polar molecules. When olive oil and hexane are mixed with each other, the 'like dissolves like' rule applies and the two non-polar liquids readily mix with each other to form a homogeneous solution (Figure 6.1.7 on the following page).

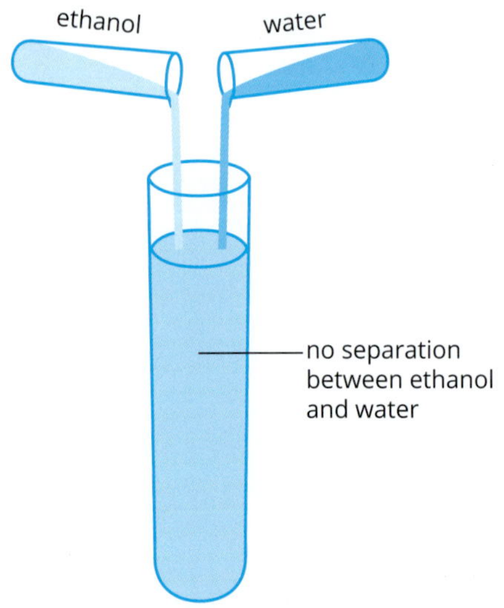

FIGURE 6.1.4 The polar compound ethanol is completely miscible in water. A homogeneous solution is formed on mixing.

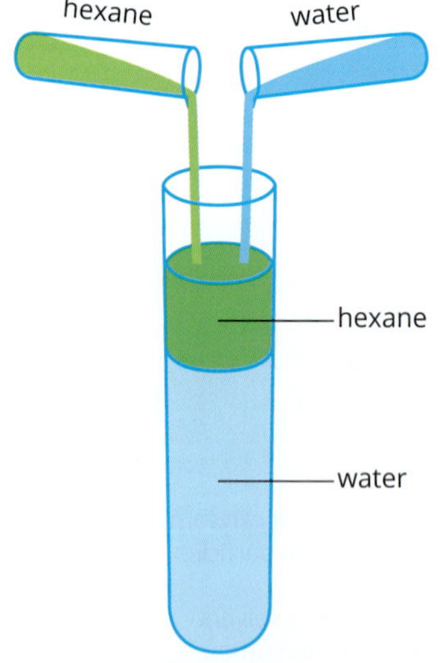

FIGURE 6.1.5 Hexane is a non-polar molecular compound that is immiscible in water. Two layers are formed with the non-polar liquid sitting on top of the water layer.

FIGURE 6.1.6 This home-made salad dressing contains vinegar and olive oil. Vinegar is an aqueous solution and so this polar liquid is immiscible with the non-polar olive oil. The oil is less dense than water and forms a layer on top of the vinegar.

DIFFERENT WAYS COMPOUNDS DISSOLVE IN WATER

The way a compound dissolves depends on the bonding present in the substance. Most molecular substances are insoluble (or only very slightly soluble) in water. However, some smaller molecules, such as ammonia (NH_3), hydrogen chloride (HCl) and sugar ($C_{12}H_{22}O_{11}$), dissolve well in water. Solutions in which a molecular substance is at least one of the solutes include brick cleaner, various liquid fertilisers and drinks such as wine and cordial (Figure 6.1.8).

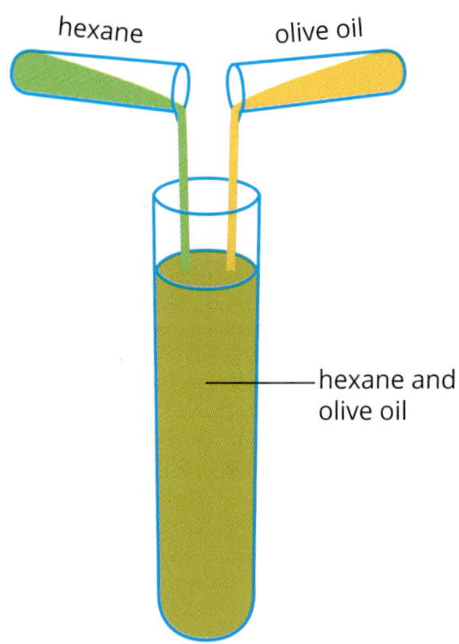

FIGURE 6.1.7 Hexane and olive oil are both non-polar molecular compounds and so are completely miscible.

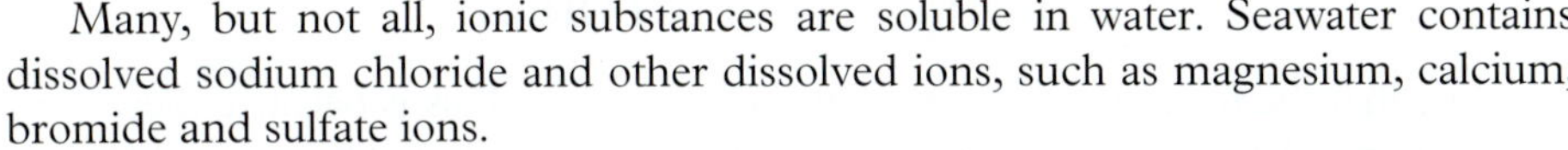

FIGURE 6.1.8 Lemon cordial is a solution that contains several dissolved molecular substances.

The 'like dissolves like' principle can be used to predict if something will dissolve in a solvent.

Many, but not all, ionic substances are soluble in water. Seawater contains dissolved sodium chloride and other dissolved ions, such as magnesium, calcium, bromide and sulfate ions.

Chemical compounds dissolve in water in different ways, depending on the nature of the compound. There are three main ways that compounds dissolve in water:

1 Some molecular compounds dissolve by forming hydrogen bonds with water.
2 Other molecular compounds dissolve in a process that involves the formation of ions.
3 Ionic compounds dissolve in a process called dissociation.

In this section we will consider, in turn, each of these ways of dissolving.

Molecular compounds that form hydrogen bonds with water

One way a molecular compound might dissolve in water is if its molecules form hydrogen bonds with water molecules. An example of such a molecule is ethanol.

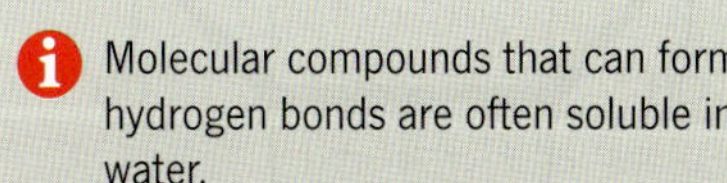

Molecular compounds that can form hydrogen bonds are often soluble in water.

Ethanol (C_2H_5OH) is a liquid at room temperature. Its molecules contain the polar –OH group, with lone pairs of electrons on the oxygen atom. The presence of the hydrogen atom bonded to the electronegative oxygen atom allows an ethanol molecule to form hydrogen bonds.

Figure 6.1.9 shows how hydrogen bonds form between molecules in pure ethanol.

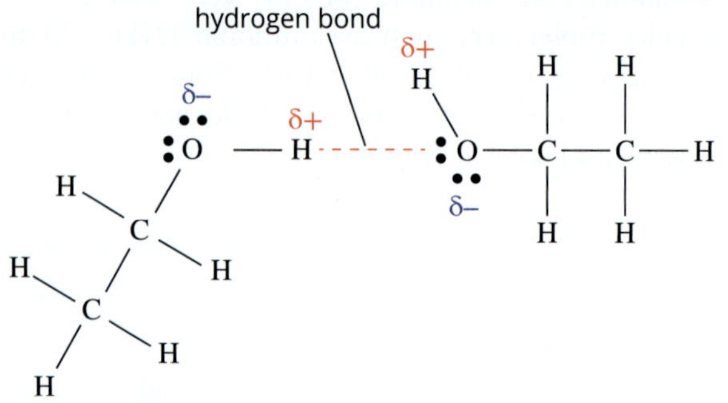

FIGURE 6.1.9 Hydrogen bonding in pure ethanol. The intermolecular hydrogen bond is formed between the lone pair electrons on the oxygen atom of one ethanol molecule and the electron-deficient hydrogen atom of an adjacent ethanol molecule.

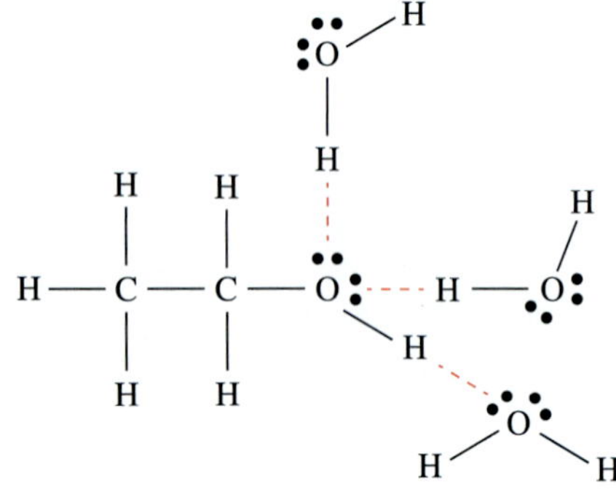

FIGURE 6.1.10 Hydrogen bonding between ethanol and water

When ethanol is added to water, it dissolves. The two liquids are miscible. Figure 6.1.10 is a representation of how hydrogen bonds form between the ethanol and surrounding water molecules.

Because the strength of the solute–solute intermolecular forces is similar to the strength of the solute–solvent intermolecular forces, the two substances can readily interact with each other. Therefore, water and ethanol molecules mix freely with each other, held together in solution by hydrogen bonds.

In summary, when ethanol dissolves in water:

- hydrogen bonds between water molecules break
- hydrogen bonds between ethanol molecules break
- hydrogen bonds form between ethanol molecules and water molecules.

An equation for the dissolution of ethanol can be written to represent this process:

$$C_2H_5OH(l) \xrightarrow{H_2O(l)} C_2H_5OH(aq)$$

When a polar molecular substance dissolves in water by forming hydrogen bonds, the equation for dissolution places water above the arrow.

Note that the formula of water sits above the arrow. This is because there is no direct chemical reaction between the water and the ethanol. The two substances simply mix together. No chemical change occurs; in the equation only the state symbol for ethanol is altered from (l) to (aq), indicating that it is now dissolved in water.

Like ethanol, sugars such as glucose and sucrose also contain the polar –OH group and can therefore dissolve in water by forming hydrogen bonds. The structure of glucose is shown in Figure 6.1.11. Each of the –OH groups in the molecule can form hydrogen bonds with water molecules. Because each molecule has many –OH groups, glucose is very soluble in water.

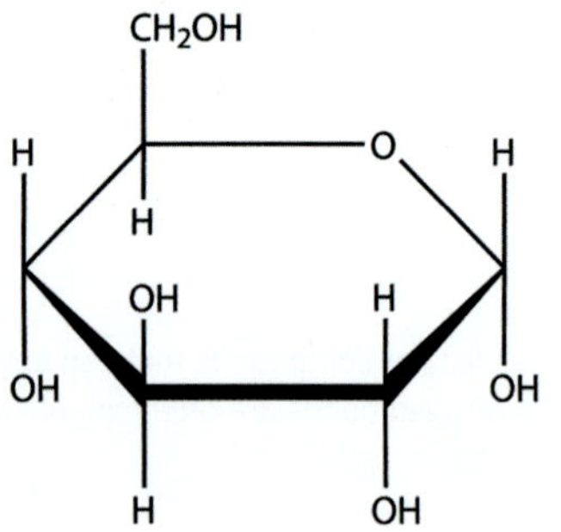

FIGURE 6.1.11 The molecular structure of glucose, the sugar found in our blood. Sugars can dissolve in water because of the presence of polar –OH groups.

The more polar the molecules of a molecular compound are, the more likely the compound is to dissolve in water. Some molecules have a polar section and a non-polar section. In general, the larger the non-polar section of the molecule, the less soluble it is in water.

Non-polar molecular substances do not have charged ends, so there is no significant attraction to water molecules. The only forces of attraction that exist between non-polar substances and water are weak dispersion forces, which are not strong enough to overcome the relatively strong hydrogen bonding between water molecules.

CASE STUDY **ANALYSIS**

Vitamin C and vitamin A: Similar but different

Vitamin C is an essential nutrient required for the growth, development and repair of body tissues and is important for the function of the immune system. Vitamin A is important for growth and development, the function of the immune system, and also for good vision. Both can be taken as supplements, but most people obtain the vitamins needed by the body by eating a variety of healthy, unprocessed foods.

Table 6.1.2 compares the structures and properties of vitamin C and vitamin A. Both molecules contain the polar –OH group, but only vitamin C is soluble in water. The higher proportion of polar –OH groups on the vitamin C molecule allows it to form sufficient hydrogen bonds with water to overcome the strong attraction between water molecules, therefore dissolving it to form a solution.

TABLE 6.1.2 Comparison of the solubility and structures of vitamin C and vitamin A

	Vitamin C	Vitamin A
Structure	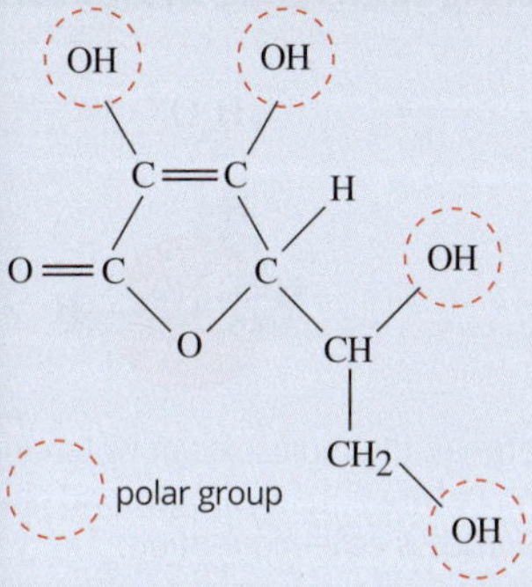 polar group Vitamin C: • contains four polar –OH groups • molecule quite polar.	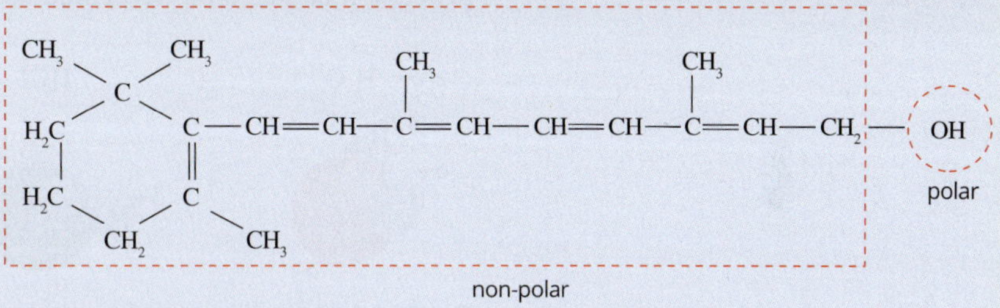 non-polar; polar Vitamin A: • contains one polar –OH group • molecule largely non-polar.
Solubility	highly soluble in water, insoluble in fats	soluble in fats, insoluble in water
Biological significance	The high solubility of vitamin C means it is excreted in urine, so it must form a regular part of the diet.	Vitamin A is not excreted in urine, but is stored in body fat. The body can tolerate a diet low in vitamin A for a limited period.

Analysis

1 Vitamin D is classed as a fat-soluble vitamin, whereas vitamin B is water soluble. How would you expect the polarity of vitamin D to compare with that of vitamin B?

2 If a person consumes too much vitamin supplement, over time they can develop a potentially dangerous condition called hypervitaminosis (literally, too much vitamin in the body). Suggest why this condition is more likely to occur for vitamin A than vitamin C.

3 Vitamin E is a compound found in the leaves of green vegetables and some plant oils. It has the molecular formula $C_{29}H_{50}O_2$ and each molecule contains a long hydrocarbon chain and one –OH group. Is the vitamin likely to be water soluble or fat soluble? Explain your answer.

Molecular compounds that ionise in water

Some compounds have molecules with one or more covalent bonds that are so polar they break when the compound is placed in water. Hydrogen chloride is such a compound. Hydrogen chloride (HCl) is a gas at room temperature. Chlorine is much more electronegative than hydrogen, so the H–Cl covalent bond is highly polar; the molecule forms a dipole.

Some highly polar molecular compounds will form ions when dissolved in water.

When hydrogen chloride is added to water, the hydrogen atom in HCl forms such a strong attraction to the oxygen atom in a water molecule that the H–Cl covalent bond breaks. The two electrons that made up the H–Cl covalent bond remain with the more electronegative Cl atom, and the newly formed hydrogen ion (H^+) joins the water molecule.

Figure 6.1.12 shows that when the hydrogen ion (H^+) bonds to the water molecule, it forms a new ion known as the **hydronium ion** (H_3O^+).

Since the Cl atom has gained an electron, it has a negative charge, forming a chloride ion (Cl^-).

The HCl is said to have become **ionised**; that is, it has produced ions. The process is called **ionisation**. Since the HCl molecule has broken apart in the process, it can also be described as having undergone **dissociation**.

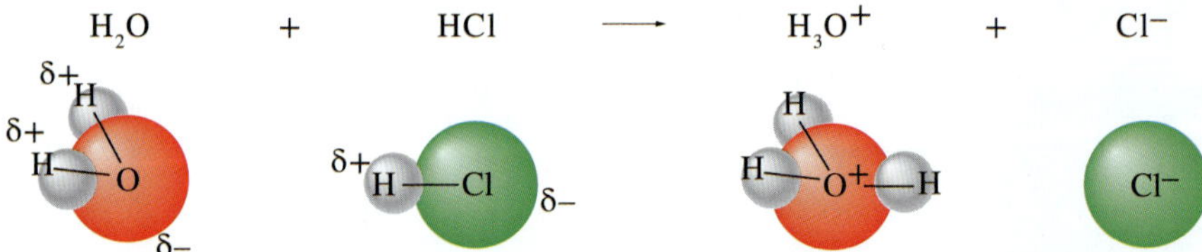

FIGURE 6.1.12 The dipole–dipole attraction between the molecules of water and hydrogen chloride leads to the breaking of the polar covalent bond between the hydrogen and chlorine atoms. New ions, hydronium and chloride, are formed in a process called ionisation.

Figure 6.1.13 shows how the two ions produced in the reaction of HCl with water (Cl^- and H_3O^+) become hydrated. The charged ions are surrounded by other polar water molecules. The polar water molecules form **ion–dipole attractions** to the Cl^- ions and hydrogen bonds to the H_3O^+ ions. The ions are described as being **hydrated**.

Around the chloride ions, the δ+ charges on the hydrogen atoms of surrounding water molecules are attracted to the negative charge of the chloride ion.

Around the hydronium ions the δ– charges on the oxygen atoms of water molecules are attracted to the positive charge of the hydronium ion.

In the diagram, the ion–dipole attractions and hydrogen bonds are represented by red dashed lines. (Ion–dipole attractions will be discussed again later in this section.)

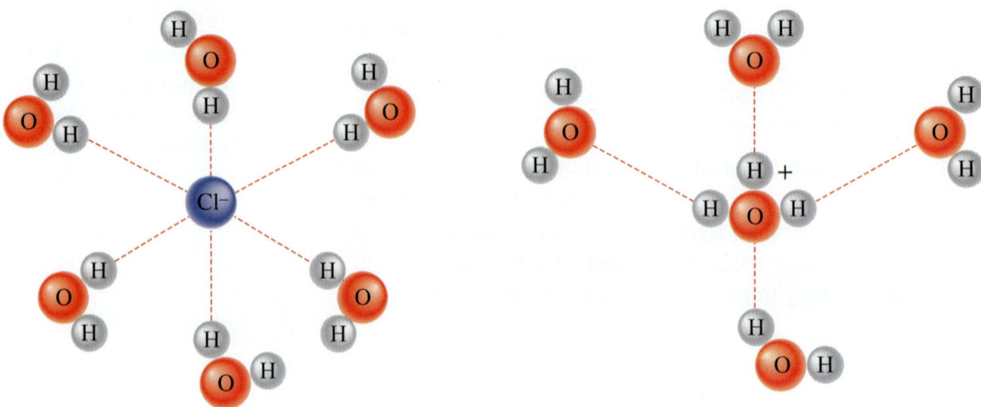

FIGURE 6.1.13 When HCl molecules ionise in water, the newly formed ions are surrounded by adjacent water molecules. The polar water molecules form ion–dipole attractions to the Cl^- ions and hydrogen bonds to the H_3O^+ ions. These bonds are represented in the diagram by dashed lines.

In summary, when hydrogen chloride dissolves in water:

- polar covalent bonds within hydrogen chloride molecules break, producing hydrogen ions (H^+) and chloride ions (Cl^-)
- a covalent bond forms between each H^+ and an H_2O molecule, forming H_3O^+ ions
- ion–dipole attractions form between polar water molecules and Cl^-
- hydrogen bonds form between polar water molecules and H_3O^+ ions.

An equation can be written to represent this process:

$$HCl(g) + H_2O(l) \rightarrow H_3O^+(aq) + Cl^-(aq)$$

You should note two important points about this equation:

- Water is included as a reactant (it is not above the arrow) because there has been a rearrangement of atoms to form new substances.
- The aqueous state of the H_3O^+ and Cl^- ions tells you that they are hydrated in solution.

Other compounds that dissolve in water by ionising include hydrobromic acid (HBr) and the common acids nitric acid (HNO_3) and sulfuric acid (H_2SO_4). These compounds ionise in a similar way to that of hydrochloric acid.

Dissociation of soluble ionic compounds in water

Sports drinks are advertised to athletes as a way to replace the electrolytes lost in sweat during exercise (Figure 6.1.14). The electrolytes are dissolved ionic substances such as sodium chloride and potassium phosphate.

FIGURE 6.1.14 Sports drinks are used by athletes to replace water and dissolved ionic solutes.

Many ionic compounds dissolve readily in water. Sodium chloride is a typical ionic compound that exists as a solid at room temperature. In Figure 6.1.15, you can see the arrangement of sodium cations (Na^+) and chloride anions (Cl^-) in a three-dimensional ionic lattice. The ions are held together by strong electrostatic forces between the positive and negative charges of the ions.

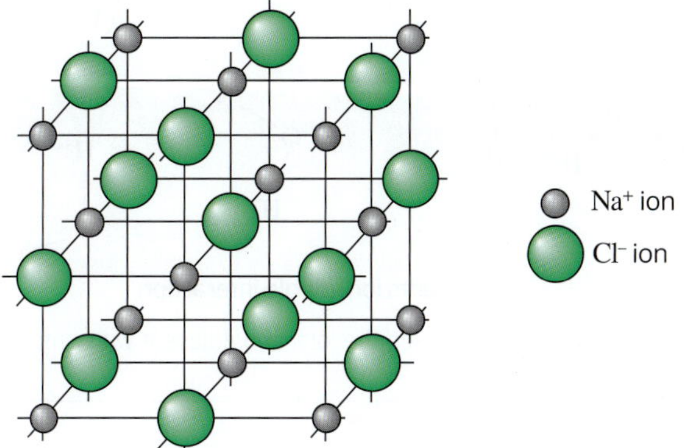

FIGURE 6.1.15 A representation of the crystal lattice of sodium chloride. The ions are held together by strong ionic bonds.

When an ionic compound such as sodium chloride is added to water, the positive ends of the water molecules are attracted to the negatively charged chloride ions and the negative ends of the water molecules are attracted to the positively charged sodium ions (Figure 6.1.16). The attraction between an ion and a polar molecule such as water is described as an ion–dipole attraction.

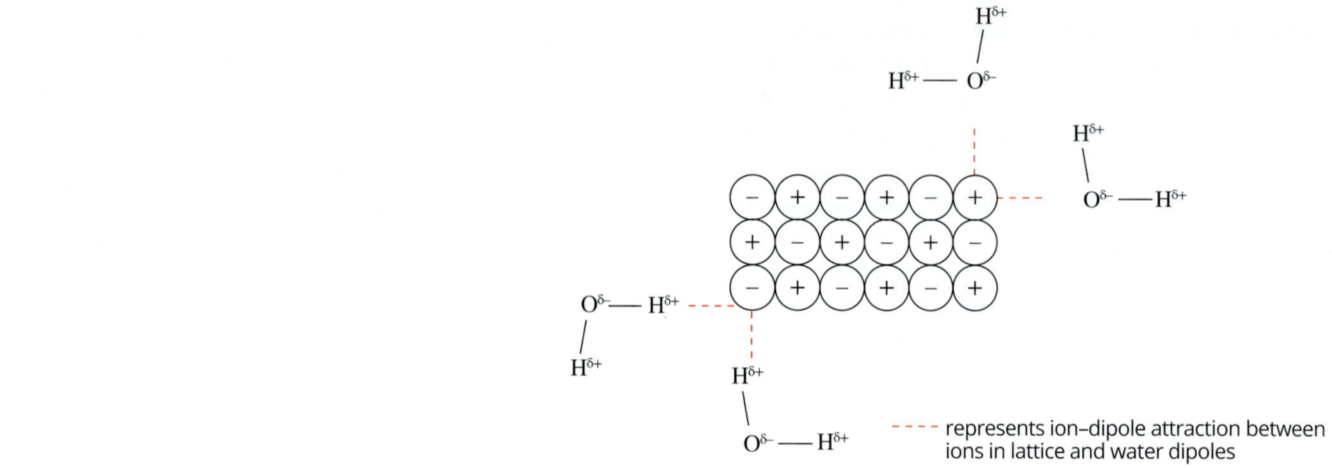

---- represents ion–dipole attraction between ions in lattice and water dipoles

FIGURE 6.1.16 A representation of sodium chloride dissolving in water. Electrostatic attraction occurs between the negative chloride ions in an NaCl lattice and the hydrogen atoms in polar water molecules. Electrostatic attraction also occurs between the positive sodium ions in the lattice and the oxygen atoms in water molecules.

Water molecules are in a continuous state of random motion. If the ion–dipole attractions between the ions and the water molecules are strong enough, the water molecules can pull the sodium and chloride ions on the outer part of the crystal out of the lattice and into the surrounding solution.

Sodium ions and chloride ions pulled out of the lattice become surrounded by water molecules. As you saw earlier, these ions can be described as being hydrated (Figure 6.1.17). Note the different arrangements of the water molecules around the positive and negative ions. Hydrogen atoms in the water molecule are more positive so some of them are orientated towards the negative chloride ion. The positive sodium ion is surrounded by the more negative oxygen atom of the water molecules.

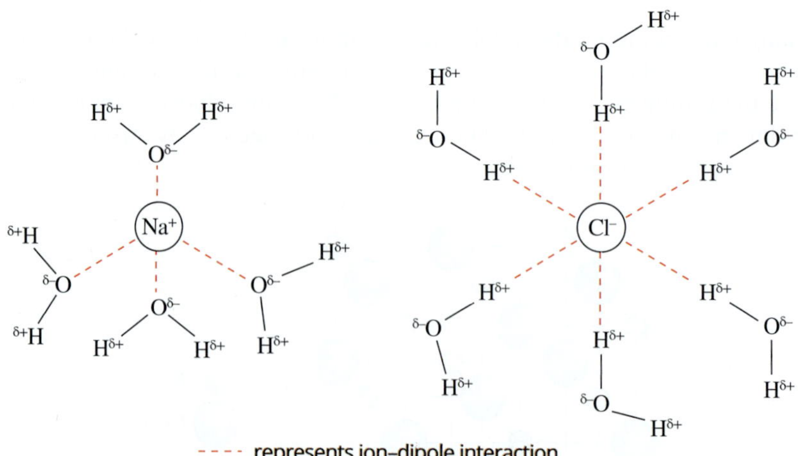

FIGURE 6.1.17 Ion–dipole attractions between the ions and adjacent water molecules form hydrated sodium and chloride ions.

The process of separating positive and negative ions from a solid ionic compound to form hydrated ions when an ionic compound dissolves in water is called dissociation.

Although the ionic bonds within the lattice are strong, the ions can be pulled away from the lattice by the interactions of many water molecules.

In summary, when sodium chloride dissolves in water:

- ionic bonds within the sodium chloride lattice are broken
- hydrogen bonds between water molecules are broken
- ion–dipole attractions between ions and polar water molecules are formed.

An ionic equation can be written to represent the dissociation process:

$$NaCl(s) \xrightarrow{H_2O(l)} Na^+(aq) + Cl^-(aq)$$

In this instance, the formula of water sits above the arrow because there is no direct reaction between the water and the sodium chloride. No chemical change occurs; the ions are separated and have the state symbol (aq), indicating they are now dissolved in water.

It is important to note that dissociation of ionic compounds is simply freeing ions from the lattice so that they can move freely throughout the solution. This is different from the ionisation and dissociation of molecular compounds where new ions are formed by the reaction of the molecule with water.

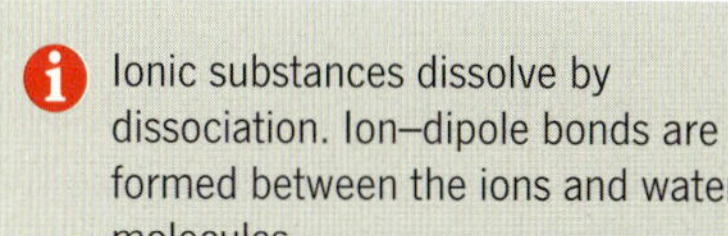

Ionic substances dissolve by dissociation. Ion–dipole bonds are formed between the ions and water molecules.

Insoluble ionic compounds

As you learnt in Chapter 5, not all ionic compounds are soluble in water. For example, limestone ($CaCO_3$) is relatively insoluble in water. Limestone caves, such as the one in Figure 6.1.18, are formed over a long period of time as $CaCO_3$ is dissolved and re-deposited. The ionic compound $Ca_{10}(PO_4)_6(OH)_2$, which gives strength to bones and teeth, is also (fortunately) insoluble in water.

FIGURE 6.1.18 Limestone caves are formed by limestone (calcium carbonate) dissolving over very long periods of time. This photograph is of a limestone cave at Loch Ard Gorge, Victoria.

Insoluble ionic compounds do not dissolve in water because the forces of attraction between the ions in the lattice (ionic bonds) are much stronger than the forces of attraction between water molecules and the ions in the lattice (ion–dipole attractions).

Substances are rarely ever completely 'soluble' or 'insoluble'. The solubility of substances ranges from highly soluble to almost insoluble.

Although substances are often described as 'soluble' or 'insoluble', this is a generalisation. Substances that are described as 'insoluble' tend to dissolve very slightly. Those that are described as 'soluble' dissolve to varying extents.

6.1 Review

OA ✓✓

SUMMARY

- A solution is a homogeneous mixture in which molecules or ions are evenly dispersed throughout a solvent.
- Solutions in which water is the solvent are called aqueous solutions.
- A solution can form when the bonds between the solute and solvent particles are similar to or greater than the attractive forces between the solute particles and between the solvent particles.
- 'Like dissolves like' is a rule that predicts polar solutes will dissolve in polar solvents and non-polar solutes will dissolve in non-polar solvents.
- Water is a good solvent for some polar molecular compounds.
  - Some polar molecular compounds dissolve by forming hydrogen bonds with water.
  - Some molecular compounds are so polar that they dissolve in water by ionising to form hydrated ions.
- Soluble ionic compounds dissociate in water to form hydrated ions.
- In a hydrated ion, water molecules are attracted to the central ion by ion–dipole attractions.
- Table 6.1.3 summarises the three main ways that compounds dissolve in water.

TABLE 6.1.3 Summary of three different ways that compounds can dissolve in water

Type of solute	Example of solute	Bond broken in the solute	Bonds formed with water	Equation
polar covalent molecule that can hydrogen bond	ethanol, C_2H_5OH	hydrogen bonds between ethanol molecules	hydrogen bonds between ethanol and water molecules	$C_2H_5OH(l) \xrightarrow{H_2O(l)} C_2H_5OH(aq)$
polar molecules that ionise	hydrogen chloride, HCl	covalent bond between hydrogen and chlorine atoms in the HCl molecule	Covalent bond formed between H^+ from HCl and oxygen atom in water molecule, forming H_3O^+ ions. H_3O^+ and Cl^- ions form ion–dipole bonds with water molecules.	$HCl(g) + H_2O(l) \rightarrow H_3O^+(aq) + Cl^-(aq)$
soluble ionic compounds	sodium chloride (NaCl)	ionic bonds between Na^+ and Cl^- ions	ion–dipole attractions between dissociated ions and polar water molecules	$NaCl(s) \xrightarrow{H_2O(l)} Na^+(aq) + Cl^-(aq)$

KEY QUESTIONS

Knowledge and understanding

1 Identify each of the following components of a bottle of soft drink as solute, solvent or solution.
  - **a** flavour
  - **b** soft drink
  - **c** water
  - **d** sweetener

2 Classify the following substances as likely to be soluble or insoluble in water.
  - **a** nitric acid (HNO_3)
  - **b** nitrogen gas (N_2)
  - **c** glucose ($C_6H_{12}O_6$)
  - **d** canola oil
  - **e** ethane (C_2H_6)
  - **f** tetrachlormethane (CCl_4)
  - **g** ethanoic acid (CH_3COOH)

3 Which one or more of the following substances are likely to dissolve in water by forming hydrogen bonds with water molecules?
  - **A** potassium chloride (KCl)
  - **B** lead (Pb)
  - **C** nitrogen gas (N_2)
  - **D** glycerol ($C_3H_5(OH)_3$)
  - **E** ethane (C_2H_4)
  - **F** propanol (C_3H_7OH)

4 Methanol (CH_3OH), like ethanol, will dissolve in water without ionising. Nitric acid (HNO_3), like HCl, will ionise when it dissolves in water. Write equations to represent the dissolving process for each of these compounds.

5 This question contains a list of statements that describe what happens when a substance is dissolved. Use the following terms to fill in the gaps in the following sentences.

hydrated; electrostatic; ionic bond;
ions; ion–dipole attraction; lattice

Sodium chloride is an ionic compound consisting of sodium and chloride ______. In solid sodium chloride the two different ions form a/an ______ which is held together by a/an _______force of attraction called a/an ______. When water is added to solid sodium chloride water molecules attach themselves to ions in the solid by forces of ______. When the lattice breaks up and the solid dissolves, the sodium and chloride ions are surrounded by water molecules and are said to be _______.

6 Write equations to show what happens when the following ionic compounds dissolve in water.

a magnesium sulfate

b copper(II) nitrate

c ammonium sulfide

d aluminium sulfate

e sodium phosphate

Analysis

7 Copy the table below and fill in the empty spaces to make a summary of the three main ways that chemical compounds dissolve in water.

	Ionising molecular compounds	**Ionic compounds**	**Non-ionising molecular compounds**
Examples	sulfuric acid, H_2SO_4	$Ca(OH)_2$	propanol, C_3H_7OH
Type of particles present before dissolving occurs		ions	
Type of particles present after dissolving occurs			
Equation for dissolving process		$Ca(OH)_2(s) \xrightarrow{H_2O} Ca^{2+}(aq) + 2OH^-(aq)$	

8 Methanol (CH_3OH) and hydrogen chloride (HCl) are both soluble in water. Are the following statements true for methanol only, hydrogen chloride only, both methanol and hydrogen chloride or neither methanol nor hydrogen chloride?

a Ion-dipole forces of attraction exist in the solution of the dissolved compound.

b The only forces of attraction between the particles of the compound and water in solution are dispersion forces.

c Hydrogen bonds are formed when the compound dissolves in water.

d Atoms within the solute molecules are separated from each other during the dissolving process.

6.2 Principles of chromatography

Chromatography is an analytical technique that is based on the differences in the solubility of compounds described in Section 6.1. The technique can be used to separate the different substances present in a mixture. It can also be used to identify numerous inorganic and organic substances, such as contaminants in water, toxic gases in air, impurities in food, and drugs present in blood.

In this section, you will learn about the underlying principles of chromatography and details about the operation of two simple forms of the technique.

HOW CHROMATOGRAPHY WORKS

All methods of chromatography have:

- a **stationary phase**
- a **mobile** (moving) **phase**.

You can perform a simple chromatography experiment by dipping the end of a stick of chalk into water-soluble black ink and then standing the chalk in a beaker containing a small amount of water, as seen in Figure 6.2.1.

FIGURE 6.2.1 The pattern of bands produced when this stick of chalk is dipped in black ink and then placed in water is called a chromatogram.

As the water carries the ink up the chalk, you will see that the ink separates into bands of different colours. Each band contains one of the substances present in the ink mixture. The pattern of bands or spots is called a **chromatogram**.

In this simple chalk-and-ink exercise, the stationary phase is the chalk and the mobile phase is the water.

The different substances present in the ink are called its **components**. As the components in the ink are swept upwards over the stationary phase by the solvent, they undergo a continual process of **adsorption** (adhering onto the solid stationary phase), followed by **desorption** (dissolving into the mobile phase). The ability of the components to stick to the stationary phase will depend upon the polarity of the stationary phase and the component molecules. Similarly, the attraction of the components to the solvent molecules is determined by their polarity.

CHEMFILE

Colour writing

While studying the coloured materials (pigments) in plants, Russian botanist Mikhail Tsvet developed the separation technique known as chromatography. The word chromatography means 'colour writing'. It is a way by which a chemist can separate the components in mixtures.

Passionate about botany, Tsvet discovered that different pigments appeared as different coloured bands when analysed using chromatography. In 1903, chlorophyll and xanthophyll were the only known plant pigments. Tsvet produced chromatograms similar to the one shown in the image below and discovered two forms of chlorophyll and eight other pigments.

Chromatography is such an important technique that several Nobel prizes have been awarded for research based mainly on this analytical method. Erika Cremer was one of a number of pioneer scientists who developed a more advanced form of chromatography called gas chromatography, which is highly sensitive and capable of detecting miniscule amounts of a compound. Gas chromatography is widely used, including for quality control in the manufacture of pharmaceuticals and for forensic investigations.

(a) Mikhail Tsvet was the inventor of chromatography.
(b) Chromatogram of a plant sample similar to the chromatograms obtained by Mikhail Tsvet.

The rate of movement of each component depends mainly upon:

- how strongly the component adsorbs onto the stationary phase
- how readily the component dissolves (its solubility) in the mobile phase.

The components separate because they undergo the processes of adsorption and desorption to different degrees. In the example involving the chalk and ink shown in Figure 6.2.1 on page 202, the blue dye in the ink is more soluble in the mobile phase than the red dye, and bonds less strongly than the red dye with the stationary phase. The blue dye in the ink has moved faster up the piece of chalk than the red dye, resulting in their separation.

Water, the mobile phase, is a polar solvent. The blue dye moves more quickly with the water up the stationary phase than the red dye, indicating that the blue dye is more polar than the red dye.

PAPER AND THIN-LAYER CHROMATOGRAPHY

In the laboratory, **paper chromatography** is performed on high-quality absorbent paper, similar to filter paper, as the stationary phase. **Thin-layer chromatography** (TLC) is very similar to paper chromatography. In this case the stationary phase is a thin layer of a fine powder, such as alumina (aluminium oxide) or silica (silicon dioxide), spread on a glass or plastic plate. Both techniques are useful for qualitative analysis.

Paper and thin-layer chromatography in practice

In both paper and thin-layer chromatography, a small spot of the solution of the sample to be analysed is placed on one end of the chromatography paper or plate.

The position of this spot is called the **origin**. The paper or plate is then placed in a container with solvent. The origin must be a little above the level of the solvent so that the components can be transported up the paper or plate and not dissolve into the liquid in the container. As the solvent rises up the paper or plate, the components of each sample separate, as shown in Figure 6.2.2.

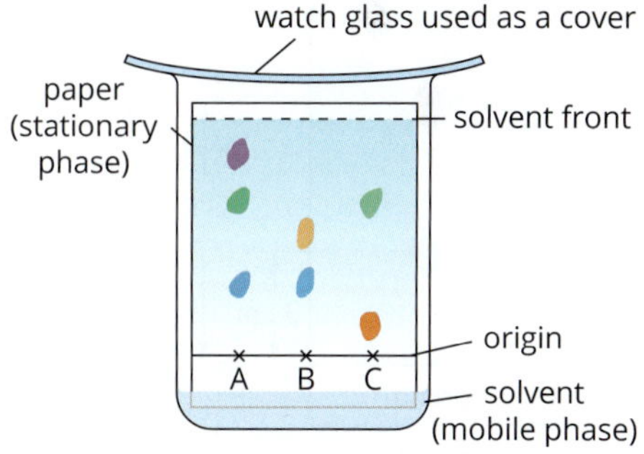

FIGURE 6.2.2 Paper chromatography of three different food colours (A, B and C)

Paper or thin-layer chromatography can be used to quickly and cheaply determine the **purity** of a sample. Direct comparison is done between the chromatogram of the sample and one of the pure material; impurities are observed as extra spots.

Identifying the composition of a mixture

The components in a mixture can be identified by chromatography in one of two ways:

1 by including **standards** of known chemicals on the same chromatogram as the unknown sample (a standard is a substance with an identity and concentration that are precisely known)

2 by calculating the **retardation factor** ($\boldsymbol{R_f}$) of the sample.

The determination of the chemical composition of a sample is called **qualitative analysis**. Other forms of analysis that are described in later chapters allow a chemist to measure the *amount* of different chemical components present in a given sample. This type of analysis is called **quantitative analysis**.

Method 1: Using standards

In this method, you need to know what chemicals might be present in the sample. For example, if you wish to find out whether a vitamin tablet contains vitamins A and D, a sample of the vitamin tablet can be placed alongside pure samples of each of vitamins A and D on the same chromatogram. If spots from the tablet sample move the same distance from the origin as the spots from the pure samples, then the tablet is likely to contain the vitamins.

The sample and standards are 'run' on the same chromatogram because the distances moved from the origin will depend on the distance moved by the solvent front. The further the solvent front is allowed to travel, the further the spots travel and the greater the separation between them.

Method 2: Calculating R_f values

Another way of identifying the components of a mixture is by comparing the distance they travel up the stationary phase to the distance travelled by the solvent front. This is expressed as a retardation factor, R_f, for a component:

$$R_f = \frac{\text{distance the component travelled from the origin}}{\text{distance the solvent front travelled from the origin}}$$

You can see from the chromatogram in the case study (Figure 6.2.5) that:

- R_f values will always be less than one
- the component most strongly adsorbed onto the stationary phase moves the shortest distance and has the lowest R_f value.

Each component has a characteristic R_f value for the conditions under which the chromatogram was obtained. By comparing the R_f values of components of a particular mixture with the R_f values of known substances determined under identical conditions, you can identify the components present in a mixture.

In this method, the distance moved by the solvent front is no longer critical as the proportion of the distance moved from the origin (the R_f value) stays the same, provided the conditions under which the chromatogram is obtained are the same.

This means the R_f values of unknown spots can be compared against a table of R_f values of common materials. However, changes in the temperature, the type of stationary phase, the amount of water vapour around the paper or plate and the type of solvent will all change the R_f value for a particular chemical.

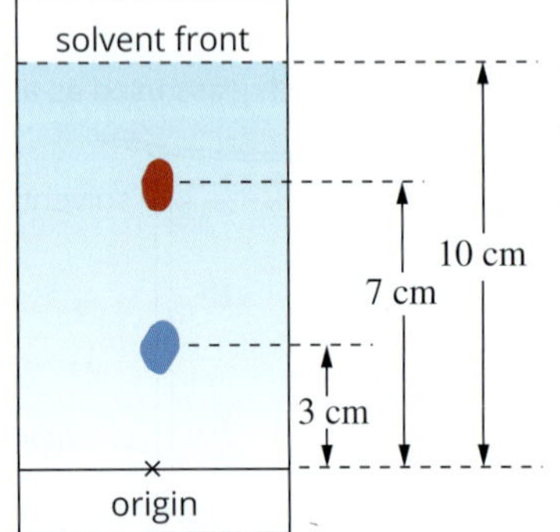

FIGURE 6.2.3 A chromatogram of a sample that consists of two components. The distances from the origin enable R_f calculations.

Worked example 6.2.1

CALCULATING R_f VALUES

Calculate the R_f value of the red component in Figure 6.2.3.

Thinking	Working
Record the distance the component has moved from the origin and the distance the solvent front has moved from the origin.	Distance from origin of red component = 7 cm Distance from origin of solvent front = 10 cm
$R_f = \frac{\text{distance of component from origin}}{\text{distance of solvent front from origin}}$	R_f(red component) $= \frac{7}{10} = 0.7$

Worked example: Try yourself 6.2.1

CALCULATING R_f VALUES

Calculate the R_f value of the blue component in Figure 6.2.3.

In chromatograms of plant pigments and food dyes, the components can be seen easily; however, most compounds are colourless and must be made visible. Many organic compounds fluoresce and appear blue when viewed under ultraviolet light. For other compounds, the chromatogram can be sprayed with a chemical that reacts to form coloured or fluorescent compounds. For example, amino acids can be sprayed with ninhydrin to give blue- and brown-coloured compounds.

The choice between paper and thin-layer chromatography depends upon the sample being analysed. Table 6.2.1 lists the advantages of each method.

For a particular combination of stationary phase and mobile phase, many different chemicals may have similar R_f values. Paper and thin-layer chromatography are only guides to the identity of a chemical. Further testing using other forms of chromatography, such as high-performance liquid chromatography (HPLC) or gas chromatography (GC) may be required to confirm the identity of a chemical. You will learn more about HPLC in Unit 4 of this chemistry course.

TABLE 6.2.1 A comparison of paper and thin-layer chromatography

Paper chromatography	Thin-layer chromatography
cheap	detects smaller amounts
little preparation	better separation of less polar compounds
more efficient for polar and water-soluble compounds	corrosive materials can be used
easy to handle and store	a wide range of stationary phases is available

CASE STUDY ANALYSIS

Investigating the ingredients of whipped cream

Whipped cream from aerosol cans is a popular topping on deserts such as apple pie, ice cream sundaes or hot chocolate (Figure 6.2.4). It is usually made from only a few ingredients—cream, of course, and the food additive E471.

FIGURE 6.2.4 Whipped cream from an aerosol can is a popular addition to many desserts.

E471 is the code for a naturally-occurring class of food additive. In fact, E471 is actually composed of two molecules called glyceryl monostearate and glyceryl distearate. It is manufactured from glycerine, which can be obtained from either animal fats or soy bean oil.

Cream is made from milk and is a complex mixture of oil and fat, water and other proteins. As you know, oil and water do not usually mix because oil molecules are non-polar and water molecules are polar. However, the proteins in milk act as emulsifiers. Emulsifiers are chemicals composed of large molecules with polar and non-polar ends. When an emulsifier is added to a mixture of water and oil, the emulsifier molecules arrange themselves on the interface of the two substances, anchoring their polar ends into water and their non-polar ends into oil. As a consequence, the oil and fat in milk remain suspended in the water rather than separating, giving milk its familiar uniformly white appearance. Milk is an example of an emulsion—a mixture of two or more immiscible liquids with one liquid distributed throughout the other as droplets of microscopic size.

Why is E471 added to whipped cream? In this case, there are insufficient natural emulsifiers in the cream to allow it to be squirted from an aerosol can and form appealing shapes for dessert toppings. This is where the addition of E471, which is also an emulsifier, is needed.

Scientists have recently reported that they have developed a new method to reliably analyse the concentration of E471 in different brands of whipped cream (Figure 6.2.5). The method is based on a form of thin layer chromatography in which the TLC plates were coated with silica and the chromatograms run using a liquid mixture of pentane, hexane and diethyl ether. The positions of the components on the plates were detected using ultraviolet light and their concentrations determined by comparing the fluoresence of the components with that of standards.

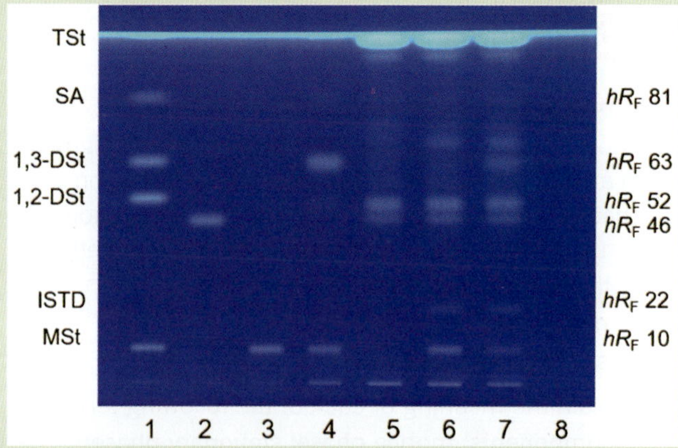

FIGURE 6.2.5 A photo of a chromatogram taken under UV light, showing mixtures of standards (1–4) and those of aerosol whipped cream samples (5–8).

Analysis

1 **a** Identify the mobile and stationary phases used in the TLC analysis of whipped cream.

b Suggest why it is necessary to use ultraviolet light to detect the spots on the TLC plates.

c Suggest why the scientists might use a mixture of liquids as the solvent, rather than a single liquid.

2 An emulsifier molecule can be represented by the following shape:

polar end ●〰〰 non-polar end

Draw a diagram to show how emulsifier molecules are arranged at the interface of an oil droplet suspended in water in cream.

6.2 Review

OA ✓✓

SUMMARY

- Chromatography is a technique commonly used to determine the composition and purity of different types of substances; it can be used to separate and identify the components in a mixture.
- Paper and thin-layer chromatography are simple forms of this technique.
- All chromatographic techniques involve a mobile phase and a stationary phase.
- Components separate during chromatography as a consequence of how strongly they adsorb to the stationary phase and desorb back into the mobile phase.
- Components in a mixture have differing affinities for the mobile and stationary phases.
- In paper and thin-layer chromatography, the components in a mixture can be identified by comparison with known standards or determination of R_f values.

KEY QUESTIONS

Knowledge and understanding

1 Paper chromatography is used to separate the pigments in a plant leaf. A spot of each pigment is placed on a sheet of chromatography paper and the chromatogram is run inside a closed jar that is partly filled with ethanol. In this experiment, what is the:

a mobile phase?

b stationary phase?

2 Explain the meaning of each of the following terms:

components; mobile phase; stationary phase; adsorption; desorption

Analysis

3 An extract from a plant was analysed by thin-layer chromatography with a non-polar solvent. The chromatogram obtained is shown in the following diagram. The following table gives the R_f values of some chemicals commonly found in plants measured under the same conditions.

solvent front

origin

Chemical	R_f
xanthophyll	0.67
β-carotene	0.82
chlorophyll a	0.48
chlorophyll b	0.35
leutin	0.39
neoxanthin	0.27

a Measure and record the distance from the origin to the centre of each band, and the distance of the solvent front from the origin.

b Calculate the R_f value of each band.

c Compare R_f values for the bands with the R_f values in the table and name the chemicals present in the extract.

d If water had been used as the solvent, would the chromatogram be likely to have a similar appearance? Explain.

4 Phenacetin was once an ingredient in analgesic drugs, but it is not used now because it causes liver damage. It is soluble in chloroform. A chemist wishes to analyse a brand of analgesic by thin-layer chromatography to determine whether it contains phenacetin. Outline the steps in the analysis. (Assume that a sample of pure phenacetin is available to the chemist.)

Chapter review

KEY TERMS

adsorption
aqueous
aqueous solution
chromatogram
chromatography
component
desorption
dispersion forces
dissociation
dissolution
homogeneous
hydrated
hydrogen bonds
hydronium ion
immiscible
ion–dipole attractions
ionisation
ionised
miscible
mobile phase
non-polar
origin
paper chromatography
polar
purity
qualitative analysis
quantitative analysis
retardation factor (R_f)
solute
solution
solvent
standards
stationary phase
thin-layer chromatography

REVIEW QUESTIONS

Knowledge and understanding

1 Explain why water is such a good solvent for polar and ionic substances.

2 Propanol (C_3H_7OH) is soluble in water, but propane (C_3H_8) is not. Refer to the structure and bonding in each of these molecules to explain why this is so.

3 **a** What is the name given to the process that ionic solids undergo when dissolving in water?

b What ions will be produced when the following compounds are added to water?

- **i** $Cu(NO_3)_2$
- **ii** $ZnSO_4$
- **iii** $(NH_4)_3PO_4$

4 Methanol (CH_3OH) and sucrose ($C_{12}H_{22}O_{11}$) dissolve in water without ionising. Write chemical equations to represent the dissolving process for each of these compounds.

5 Write equations to show the dissociation of the following compounds when they are added to water.

- **a** Magnesium sulfate
- **b** Sodium sulfide
- **c** Potassium hydroxide
- **d** Copper(II) ethanoate
- **e** Lithium sulfate

6 What term is used to describe the attraction of water molecules to a potassium ion in a solution of potassium bromide? Describe the arrangement of water molecules around the potassium ion in the solution.

7 Which bonds or forces of attraction would not be present between atoms or particles in an aqueous solution of methanol?

- **A** Hydrogen bonds
- **B** Dispersion forces
- **C** Ion-dipole attractions
- **D** Covalent bonds

8 Match the type of compound with the way it is likely to behave in water.

- **i** Ionic compound
- **ii** Compound composed of polar molecules with –OH groups
- **iii** Compound composed of small polar molecules in which a hydrogen atom is covalently bonded to an atom of a group 17 element
- **iv** Non-polar molecular compound
- **v** Compound composed of covalent molecules with a large non-polar end and one –OH group.

- **a** Does not dissolve in water because a large proportion of the molecule is non-polar
- **b** Dissolves in water by ionising, then forming ion–dipole bonds with water
- **c** Does not dissolve in water
- **d** Dissolves in water by forming hydrogen bonds with water molecules
- **e** Dissolves in water by dissociating, then forming ion–dipole bonds with water

9 Describe what happens to the forces between solute and solvent when propan-1-ol ($CH_3CH_2CH_2OH$) dissolves in water.

10 Use the following terms to complete the sentences about paper chromatography

R_f values; components; spots; composition; stationary phase; purity; mobile phase

Thin layer chromatography is a technique that allows you to determine the ___________ and ______________ of different types of substances. In this technique, a thin layer of a solid ______________ is applied to a plate. The ______________ of the sample are carried over the surface of the stationary phase by the solvent, or ______________ . The components separate, depending on the relative attractions of compounds towards the two phases. The individual components are seen as ______________ on the plate. which can be identified by calculating their ______________ .

11 **a** Use the terms 'adsorbed' and 'absorbed' correctly in each of the sentences below.

- **i** Water was ______________ by the towel as the wet swimmer dried himself.
- **ii** A thin layer of grease _________________ onto the cup when it was washed in the dirty water.

b Explain the difference between the terms 'adsorbed' and 'absorbed'.

Application and analysis

12 Indicate whether the following substances will dissolve in water or not.

$C_6H_{12}O_6$, HI, I_2, C_2H_4, C_3H_7OH, HNO_3, CH_4

Group them as:

a insoluble, or

b dissolve via hydrogen bonding with water, or

c dissolve by ionising

The structures of $C_6H_{12}O_6$ and C_3H_7OH are shown to assist you.

13 Which one of the following substances is most likely to dissolve best in the non-polar solvent benzene (C_6H_6)? Explain your answer.

potassium chloride (KCl); glycerol ($C_3H_5(OH)_3$); ethanol (C_2H_5OH); hexane (C_6H_{14})

14 A student carried out some solubility trials on three different liquids: methanol (CH_3OH), pentane (C_5H_{12}) and butanol (C_4H_9OH). Two solvents were used in the trials, water and hexane (a non-polar liquid).

a Give the order of solubility of the three liquids in water, from lowest to highest.

b Give the order of solubility of the three liquids in hexane, from lowest to highest.

c Explain your answers to **a** and **b**.

15 Give concise explanations for the following observations.

a Ammonia (NH_3) and methane (CH_4) are both covalent molecular substances. Ammonia is highly soluble in water, but methane is not.

b Glucose ($C_6H_{12}O_6$) and common salt (NaCl) are very different compounds. Glucose is a covalent molecular substance, whereas common salt is ionic, yet both of these substances are highly soluble in water.

16 The chromatogram of a dye is shown below. Calculate the R_f values for each of the blue, purple and yellow components. Give answers to one decimal place.

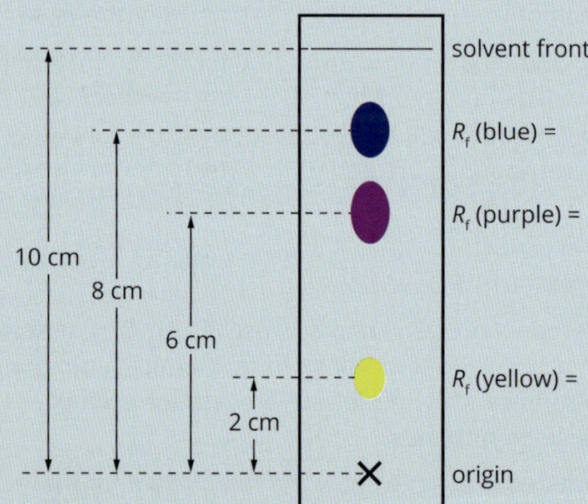

17 A sample of brown dye from a lolly is placed at the origin on a chromatography plate. The solvent front moves 9.0 cm from the origin. A blue component of the dye moves 7.5 cm and a red component 5.2 cm in the same time. Calculate the R_f values of the two components.

18 Consider the paper chromatogram of three food colours in Figure 6.2.2 on page 203.

a Why must the level of the solvent be lower than the origin where spots of the mixture are originally placed?

b Why are R_f values always less than one?

c How many different components have been used to make colour B?

d Which components present in colours B and C are also in colour A? Explain.

e Which component of colour A is least strongly adsorbed on the stationary phase?

f Calculate the R_f values of each component of colour C.

19 One component of a mixture is near the top of a paper chromatogram, whereas another component is near the bottom. Compare the two components in terms of their solubility in the mobile phase and adsorption to the stationary phase.

20 A black dye used to colour jeans was shown by thin-layer chromatography to contain yellow, orange, purple and green components. Using water as solvent, the R_f values of these substances are 0.21, 0.80, 0.61 and 0.30, respectively.

a How far apart would the purple and yellow components be after the solvent front had moved 8.0 cm from the origin?

b When the green component had travelled 5.0 cm from the origin, how far would the orange component have travelled?

c Sketch the chromatogram of the ink to scale after the solvent front had moved 15 cm from the origin.

21 The diagram below shows a thin-layer chromatogram of amino acids in a medicine.

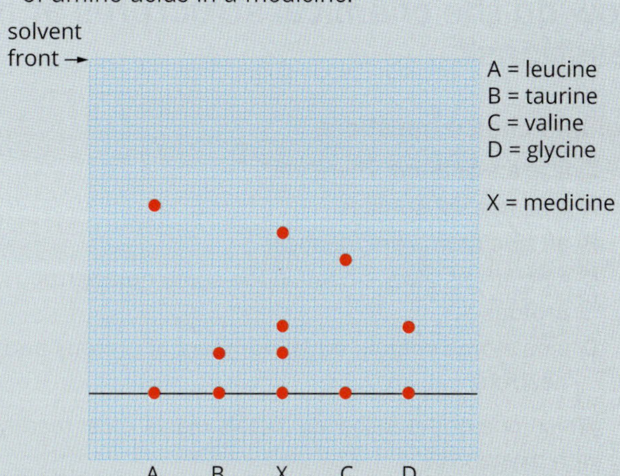

a Which amino acids are in the medicine?

b Amino acids are colourless. What technique could be used to make the amino acid spots visible?

c Calculate the R_f value of taurine.

d Which amino acid is bonded least strongly to the stationary phase?

OA ✓✓

REVIEW QUESTIONS

How do the chemical structures of materials explain their properties and reactions?

Multiple-choice questions

1 Zinc is an element. Therefore:

A zinc has no isotopes
B all zinc atoms are identical
C zinc atoms always contain the same number of protons
D zinc atoms contain equal numbers of protons and neutrons

2 Which one of the following metals reacts most readily with oxygen?

A Na
B K
C Mg
D Ca

3 Which one of the following statements about ionic bonding is not correct?

A When molten, ionic compounds are conductors of electricity.
B An ionic lattice contains both cations and anions in fixed positions.
C Ionic bonding involves the sharing of electrons between two different atoms.
D Compounds held together by ionic bonding generally have high melting temperatures.

4 An ion that contains 11 protons, 12 neutrons and 10 electrons will have a mass number and charge corresponding to:

	Mass number	Charge
A	11	1–
B	11	1+
C	23	1–
D	23	1+

5 The ground state electronic configuration for an ion of sulfur, S^{2-}, is:

A $1s^22s^22p^63s^23p^4$
B $1s^22s^22p^63s^23p^5$
C $1s^22s^22p^63s^23p^6$
D $1s^22s^22p^63s^23p^44s^2$

6 The 3*d* subshell has:

A 3 orbitals and can hold up to 3 electrons
B 3 orbitals and can hold up to 6 electrons
C 5 orbitals and can hold up to 10 electrons
D 5 orbitals and can hold up to 15 electrons

The following information relates to questions 7 and 8.

The atomic number, mass number and electronic configuration of four particles, W, X, Y and Z, are given below:

	Atomic number	Mass number	Electronic configuration
W	17	37	$1s^22s^22p^63s^23p^6$
X	19	39	$1s^22s^22p^63s^23p^6$
Y	20	40	$1s^22s^22p^63s^23p^64s^2$
Z	19	40	$1s^22s^22p^63s^23p^64s^1$

7 Which one of the following alternatives lists particles that are isotopes of the same element?

A W and X
B X and Z
C Y and Z
D W, X and Y

8 Which one of the following statements about particles W, X, Y and Z is correct?

A W is a noble gas.
B X is a positively charged ion.
C Y is in group 4 of the periodic table.
D Z is a negatively charged ion.

9 Element X is in group 14 of the periodic table. Element Y is in group 16. The most likely formula for a compound between X and Y is:

A X_4Y_{16}
B $X_{16}Y_4$
C XY_4
D XY_2

10 Predict which one of the following ionic compounds will have the highest melting point.

A MgO
B $MgCl_2$
C Na_2O
D NaCl

11 Which one of the equations below best represents table sugar ($C_{12}H_{22}O_{11}$) dissolving in water?

A $C_{12}H_{22}O_{11}(s) + H_2O(l) \rightarrow C_{12}H_{23}O_{11}^+(aq) + OH^-$
B $C_{12}H_{22}O_{11}(s) + H_2O(l) \rightarrow C_{12}H_{21}O_{11}^-(aq) + H_3O^+(aq)$
C $C_{12}H_{22}O_{11}(s) \xrightarrow{H_2O(l)} C_{12}H_{22}O_{11}(aq)$
D $C_{12}H_{22}O_{11}(s) \xrightarrow{H_2O(l)} C_{12}H_{22}O_{11}(l)$

12 Which of the following gives the correct shape for each of the molecules listed?

	Linear	Bent	Tetrahedral
A	CO_2	H_2S	CH_4
B	H_2	CO_2	NH_3
C	HF	H_2O	NH_3
D	H_2O	NH_3	CH_4

13 Which one of the following alternatives correctly describes the intermolecular forces in pure samples of F_2, HF and CH_3F?

	F_2	HF	CH_3F
A	dispersion forces only	dispersion forces and hydrogen bonds	dispersion forces and dipole–dipole attraction
B	dispersion forces and hydrogen bonds	dispersion forces and hydrogen bonds	dispersion forces and hydrogen bonds
C	dispersion forces only	dispersion forces and hydrogen bonds	dispersion forces and hydrogen bonds
D	dispersion forces	dispersion forces and dipole–dipole attraction	dispersion forces and dipole–dipole attraction

14 Which one of the following describes the types of bonds broken in the solute and formed with water when hydrogen chloride dissolves in water?

	Bonds broken	Bonds formed
A	covalent	hydrogen and dipole–dipole
B	dipole–dipole	covalent and ion–dipole
C	dipole–dipole	hydrogen and dipole–dipole
D	covalent	covalent, hydrogen and ion–dipole

15 Addition of which one of the following substances to a solution of copper(II) sulfate will not result in the formation of a precipitate?

A $BaCl_2(aq)$
B $NH_4Cl(aq)$
C $Na_2CO_3(aq)$
D $Pb(NO_3)_2(aq)$

Short-answer questions

16 Magnesium is a commonly used structural metal.

a Write the electronic configuration of a magnesium atom, using subshell notation.

b i The radius of a magnesium atom is 160 pm ($1 \text{ pm} = 10^{-12} \text{ m}$). What is the radius of the magnesium atom in nanometres? ($1 \text{ nm} = 10^{-9} \text{ m}$)

ii Would you predict the radius of a sodium atom to be smaller or larger than 160 pm? Explain your answer.

c i Describe the model commonly used to describe the structure of metals such as magnesium and the nature of the bonding between its particles. You may include a labelled diagram in your answer.

ii Use the metallic bonding model to explain why magnesium is a good conductor of electricity.

d i Magnesium reacts readily with dilute hydrochloric acid. What would you observe if a small piece of magnesium was added to some hydrochloric acid? Write a chemical equation for this reaction.

ii Give the symbol for one metal that would be expected to react more vigorously with hydrochloric acid than magnesium.

17 **a** With the aid of a periodic table, identify the correct chemical symbol for each of the following.

i the element that is in group 2 and period 4
ii a noble gas with exactly three occupied electron shells
iii an element from group 14 that is a non-metal
iv an element that has exactly three occupied electron shells and is in the s-block
v the element in period 2 that has the largest atomic radius
vi the element in group 15 that has the highest first ionisation energy
vii the element in period 2 with the highest electronegativity

b Describe the trend in chemical reactivity of elements going down group 1 of the periodic table and give a brief explanation for this trend.

18 Give a concise explanation for each of the following.

a The atomic radius of chlorine is smaller than that of sodium.
b The first ionisation energy of fluorine is higher than that of lithium.
c The reactivity of Be is less than that of Ba.
d There are two groups in the *s*-block of the periodic table.

19 Some excess potassium chloride solution is added to a solution containing lead(II) nitrate. A white precipitate forms.

a Write a full chemical equation for the formation of the precipitate.
b Write an ionic equation for the formation of the precipitate.
c For this reaction, give the name of the:
i precipitate
ii spectator ions.

20 Provide concise explanations for each of the following observations.

a The melting temperature of ice (solid H_2O) is 0°C, but a temperature of over 1000°C is needed to decompose water molecules to hydrogen and oxygen gases.

b A molecule of ethyne (C_2H_2) is linear, but a molecule of hydrogen peroxide (H_2O_2) is not. (You will need to draw the structure of these two molecules first).

21 Ammonia (NH_3) is a constituent of many cleaning products for bathrooms.

a Draw a Lewis structure of an ammonia molecule, including non-bonding electron pairs.

b Draw a Lewis structure for two ammonia molecules. Clearly show, and give the name of, the shape of these molecules. On your diagram, label the type of forces/bonds that exist between the:

- atoms within each ammonia molecule
- two ammonia molecules.

c Draw a Lewis structure for a molecule of:

i nitrogen gas

ii carbon dioxide gas.

d i Explain why the forces between nitrogen molecules and those between molecules of carbon dioxide are of the same type, even though the bonds inside these molecules differ in strength and polarity.

ii Explain why the forces between ammonia molecules are different from those between nitrogen molecules or carbon dioxide molecules.

22 Consider the following molecules: oxygen (O_2) and hydrogen peroxide (H_2O_2).

a Draw a Lewis structure of each molecule.

b Which one has the stronger intramolecular (within the molecule) bonds? Explain your choice.

c i What sort of interactions exist between molecules of oxygen?

ii What sort of interactions exist between molecules of hydrogen peroxide? Are these stronger or weaker than those between oxygen molecules?

d Unlike nitrogen atoms, oxygen atoms do not form triple bonds in molecules. Explain why, in terms of the electronic configurations of these two atoms.

23 Diamond and graphite are allotropes of carbon. Although they are similar in some respects, they are very different in their structure and uses.

a Give a brief definition of the term 'allotrope'.

b Describe the similarity in the bonding of these two materials.

c Compare the structures of diamond and graphite.

d Graphite is an excellent conductor of electricity while diamond is unable to conduct electricity. Explain this difference in terms of the structure of the lattices.

e Graphite is used as a lead in pencils. With reference to the bonding in the lattice explain why graphite can be used this way.

24 Small amounts of solid magnesium chloride and liquid ethanol are dissolved in separate beakers of water.

a i Write an equation for the dissolving process of magnesium chloride in water.

ii What type of bonds need to break in the solid magnesium chloride in order for it to dissolve?

iii What bonds are formed when magnesium chloride dissolves in water?

b i Write an equation for the dissolving process of ethanol (C_2H_5OH) in water.

ii What type of bonds need to break in the liquid ethanol in order for it to dissolve?

iii What bonds are formed when ethanol dissolves in water?

c Which solution would you expect to be the better conductor of electricity? Justify your answer.

25 You are given five solutions. They are not labelled, but are known to be sodium carbonate, potassium nitrate, magnesium nitrate and copper(II) nitrate. You have been set the task to identify the solutions. From previous experiments, you know that copper(II) ions usually form a blue/green precipitate, and that precipitates containing magnesium ions are usually white.

a Predict what you would observe if the samples were added together, as listed below. (If no observable reaction is predicted, write 'no change'.)

i sodium carbonate and potassium nitrate

ii sodium carbonate and magnesium nitrate

iii sodium carbonate and copper(II) nitrate

iv potassium nitrate and magnesium nitrate

v potassium nitrate and copper(II) nitrate

vi magnesium nitrate and copper(II) nitrate

b Write a balanced ionic equation for each of the reactions predicted in part **a**.

c Construct a results table and use this to describe how the solutions can be identified by mixing each solution with each of the other three solutions.

26 A student sets up a paper chromatogram and places a spot of green food dye on the origin. In one experiment the solvent moved 12 cm and a blue spot moved 9 cm from the origin. When the experiment was repeated the blue spot moved 15 cm. Determine how far (in cm) from the origin the solvent is likely to be in the second experiment.

27 Phosphine is a hydride of phosphorus with the formula PH_3.

a Draw a Lewis dot diagram of a PH_3 molecule.

b Explain whether PH_3 is a polar or non-polar molecule.

c Explain why a PH_3 molecule does not have a trigonal planar shape.

d Determine the relative molecular mass of phosphine (PH_3) and ammonia (NH_3).

e Explain why ammonia has a much higher melting point than phosphine.

f Phosphorus has several isotopes. State the symbol, using nuclide notation, for the radioactive phosphorus-32 isotope.

g Write the electronic configuration for phosphorus.

28 The amino acids present in a sample of fruit juice can be detected by thin-layer chromatography. R_f values of some amino acids in two separate solvents are given in the table below.

Amino acid	Solvent 1 R_f	Solvent 2 R_f
lysine	0.12	0.55
leucine	0.58	0.82
proline	0.39	0.88
valine	0.40	0.74
2-aminobutyric acid	0.28	0.58
threonine	0.21	0.49
hydroxyproline	0.21	0.67
β-phenylalanine	0.50	0.86
isoleucine	0.57	0.81
alanine	0.24	0.55
serine	0.19	0.34
glutamic acid	0.25	0.33
glycine	0.20	0.40
arginine	0.13	0.60
taurine	0.12	0.33
tyrosine	0.38	0.62

To achieve better separation of the complex mixture of substances present in the juice, a 'two-way' chromatogram was prepared. The first step in this procedure was to run a chromatogram using solvent 1.

The results of this chromatogram are shown in the following figure.

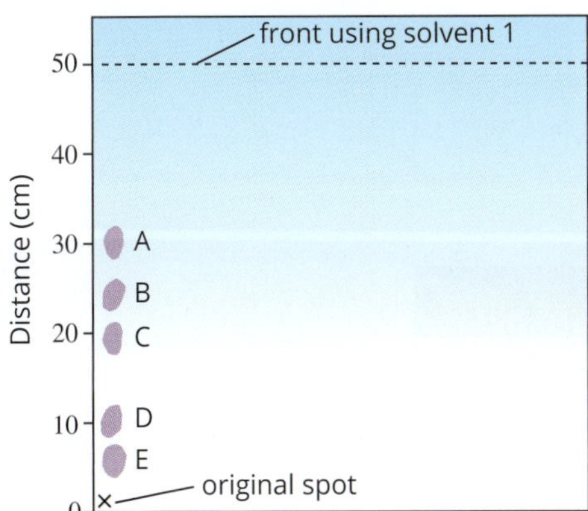

a Calculate the R_f value of each spot on the chromatogram.

b Use the information in the table to determine the amino acids responsible for each spot.

The TLC plate was then turned around so that it lay at a right angle to the original and a second chromatogram was produced using solvent 2. The figure below shows the appearance of the TLC plate after some time.

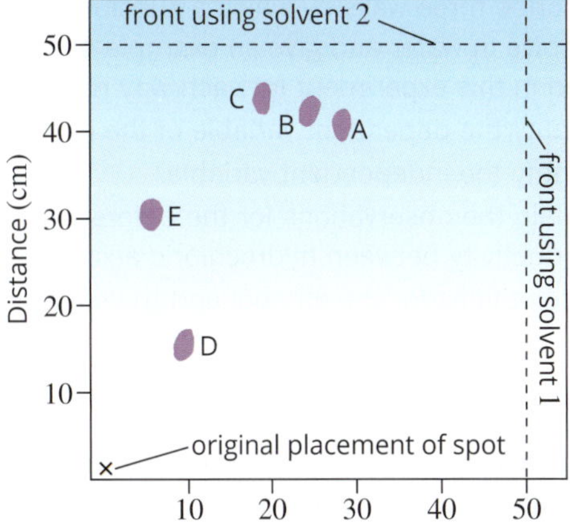

c Using the information in the table, determine each component in the mixture.

d Evaluate the advantage of a two-way chromatogram.

29 A student designed an experiment to investigate how different substances dissolve in water. The student's report is provided below, but it contains many faults.

Title: How do substances dissolve in water?

Introduction: Different compounds have different ways of dissolving in water depending on their bonding and ability to form interactions with water.

Aim: To investigate the different ways substances dissolve into water.

Hypothesis: That if a substance is more soluble, it will create a solution with a greater conductivity.

Method:

Add the following substances to 200 mL of tap water:

2.0 g sugar ($C_{12}H_{22}O_{11}$)

1.0 g salt (NaCl)

10.0 mL ethanol (CH_3CH_2OH)

2 mL of 1.0 M hydrochloric acid (HCl)

10 mL of 0.1 M vinegar (CH_3COOH)

2.0 g magnesium hydroxide ($Mg(OH)_2$)

Use a conductivity metre to measure the conductivity of the solution in μS cm^{-1}.

Results:

Substance	Conductivity (μS cm^{-1})
sugar ($C_{12}H_{22}O_{11}$)	49
salt (NaCl)	864
ethanol (CH_3CH_2OH)	51
hydrochloric acid (HCl)	495
vinegar (CH_3COOH)	59
magnesium hydroxide ($Mg(OH)_2$)	53

> **Conclusion:**
> Sugar and ethanol were the least soluble, as they had the lowest conductivity. The most soluble substance was salt, as it had the highest conductivity.

a Describe three ways by which a substance may dissolve in water and give an example of a substance used in this experiment for each way named.
b Identify the dependent variable of the experiment.
c Identify the independent variable.
d Explain the observations for the difference in conductivity between hydrochloric acid and vinegar.
e Observations for the ethanol and the sugar indicated that these substances had both dissolved. Explain why conductivity is not a good indicator of solubility for these substances.
f Considering the purpose of the experiment, discuss the validity of the conclusion
g Suggest two improvements to the experiment.

30 Helium is used in a number of ways: as a buoyancy device for party balloons and airships (Zeppelins), to provide a safe atmosphere for chemical reactions or welding, as a carrier gas in instruments for chemical analysis, and as liquid helium to provide extremely low temperatures for medical and scientific research.

- Despite the fact that helium is the second most abundant element in the universe (after hydrogen), there are fears that the supply of helium on Earth is running critically low.
- On Earth, helium is produced naturally by the radioactive decay of other elements. It is then captured from sources of natural gas, where over time enough helium has collected to make its recovery viable.
- Once we use helium, it is almost impossible to recover and recycle, as it is difficult to capture by forming compounds with other elements and it easily escapes our atmosphere. This makes it difficult to turn the extraction and use of helium from a linear process to a circular one.
- One suggested replacement for helium in some applications is hydrogen.

Some properties of helium and hydrogen are provided in the table below:

	Helium	Hydrogen
Melting point °C	−272.20	−259.16
Boiling point °C	−268.93	−252.88
Density (at standard conditions) g L^{-1}*	0.1786	0.089 88
Stable isotopes (and % abundance)	helium-3 (0.0002%) helium-4 (99.9998%)	hydrogen-1 (99.98%), hydrogen-2 (0.02%)

*(Note: density of air is 1.225 g L^{-1})

a Determine the subshell electronic configuration of helium.
b Give the location of helium in the periodic table.
c Using your knowledge of trends in the periodic table, predict the chemical reactivity of helium.
d In terms of atomic structure, explain the difference between helium-3 and helium-4
e Using your knowledge of intermolecular forces, explain why helium has such a low melting and boiling point.
f Explain how the properties of helium are linked to its uses. (Give two examples.)
g **i** Explain the difference between the linear economy model and the circular economy model.
ii What properties of helium make it difficult to process, use, reuse and recycle in a way that would fit the description of a circular economy?
h Hydrogen is often suggested as an alternative to helium.
i What properties of hydrogen would make it a suitable replacement for helium in some situations?
ii What are some problems you could predict when using hydrogen in some situations?

CHAPTER

07 Quantifying atoms and compounds

By the end of this chapter, you will have a greater understanding of the way chemists measure quantities of chemicals; in particular, the way the number of particles in samples of elements and compounds can be accurately determined by weighing them. You will also learn about empirical and molecular formulas and how they are used to describe the components of a molecule or compound, as well as how these types of formulas can be determined. This is essential knowledge for designing and producing a range of materials, including cosmetics, fuels, fertilisers, pharmaceuticals and building materials.

Key knowledge

- the relative isotopic masses of isotopes of elements and their values on the scale in which the relative isotopic mass of the carbon-12 isotope is assigned a value of 12 exactly **7.1**
- determination of the relative atomic mass of an element using mass spectrometry (details of instrument not required) **7.1**
- Avogadro's constant as the number 6.02×10^{23} indicating the number of atoms or molecules in a mole of any substance; determination of the amount, in moles, of atoms (or molecules) in a pure sample of known mass **7.2, 7.3**
- determination of the molar mass of compounds, the percentage composition by mass of covalent compounds, and the empirical and molecular formula of a compound from its percentage composition by mass. **7.4**

WS 13

OA ✓✓

7.1 Relative mass

Chemists in fields as diverse as environmental monitoring, pharmaceuticals and fuel production routinely carry out chemical reactions in their work. It is important for chemists to be able to measure specific quantities of reactants quickly and easily, in part because the amount of desired products formed depends on the amount of reactants used in the reaction. Atoms and molecules are so small that they cannot be counted individually or even in groups of thousands or millions. It is knowledge of the masses of atoms and molecules that chemists use to measure out the specific quantities they need.

The mass of a single atom is incredibly small. For example, one atom of carbon has a mass of approximately 2×10^{-23} g. Figure 7.1.1a shows a ball and stick model of a single glucose **molecule**. The teaspoon of glucose shown in Figure 7.1.1b contains approximately 1.4×10^{22} glucose molecules. One glucose molecule contains 6 carbon atoms, 12 hydrogen atoms and 6 oxygen atoms bonded together, and has an actual mass of 3×10^{-22} g. Such small masses are not easily measured and can be inconvenient to use in calculations. In this section you will learn about more convenient measures used to determine quantities in a sample.

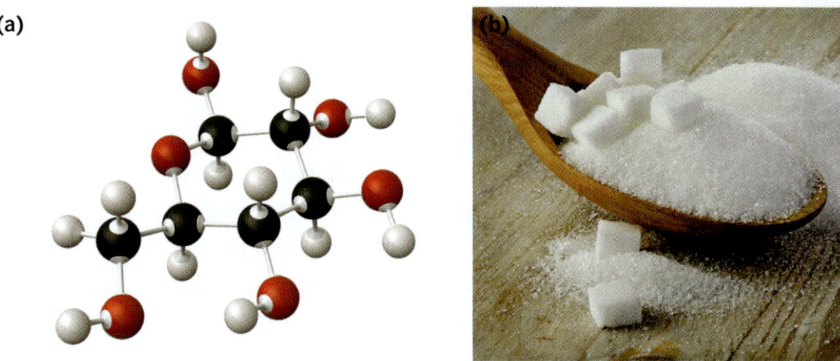

FIGURE 7.1.1 (a) The single glucose molecule pictured has a mass of 3×10^{-22} g. (b) A teaspoon of glucose crystals contains an incredibly large number of extremely small glucose molecules.

RELATIVE MASSES

The masses used most frequently in chemistry are **relative masses**, rather than actual masses. A relative mass is based on comparison to a standard mass. The standard to which all relative masses used in chemistry are compared is the common **isotope** of carbon, **carbon-12** or ^{12}C, as represented in Figure 7.1.2. This isotope is assigned a mass value of exactly 12.

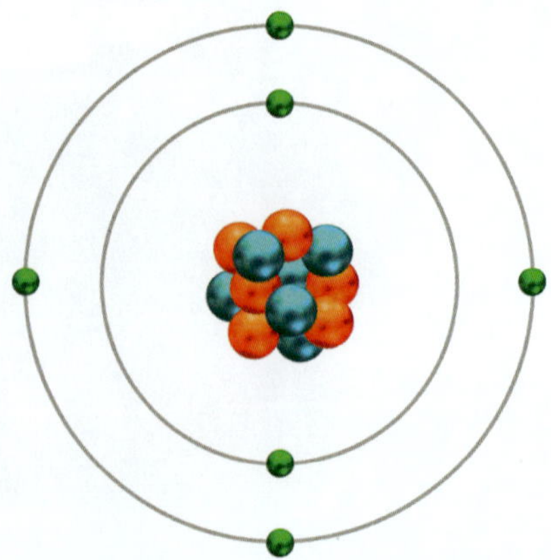

FIGURE 7.1.2 An atom of the carbon-12 isotope with six protons, six neutrons and six electrons

CHEMFILE

Selection of carbon-12 as the reference standard

Carbon-12 was selected as the standard for determining relative masses in 1961. Before then, oxygen was used as the standard. Physicists and chemists could not agree on a way of assigning a standard mass to oxygen; chemists assigned a mass of exactly 16 to the average mass of oxygen atoms whereas physicists assigned a mass of exactly 16 to the oxygen-16 isotope. This resulted in two different tables of slightly different atomic masses. Choosing carbon-12 as the standard was a compromise between physicists and chemists of the day. Assigning carbon-12 a mass of exactly 12 created a new scale, which was then adopted universally.

RELATIVE ISOTOPIC MASSES

The individual isotopes of each element have a **relative isotopic mass**. Table 7.1.1 shows the approximate mass of various isotopes relative to the carbon-12 isotope, being given the value of 12 units exactly.

You will remember from Chapter 2 that isotopes are atoms of the same element that have different numbers of neutrons in their nucleus. So, isotopes are atoms with the same atomic number but different mass number.

TABLE 7.1.1 Approximate mass of various isotopes relative to the ^{12}C isotope, which is given the value of exactly 12 units

Isotope	Representation of nucleus	Number of protons in nucleus	Number of neutrons in nucleus	Total number of protons and neutrons	Approximate mass of atom relative to carbon-12 isotope
hydrogen $^{1}_{1}H$		1	0	1	$\frac{1}{12}$ of 12 = 1
helium $^{4}_{2}He$		2	2	4	$\frac{4}{12}$ of 12 = 4
lithium $^{7}_{3}Li$		3	4	7	$\frac{7}{12}$ of 12 = 7
carbon $^{12}_{6}C$		6	6	12	12

As shown in Figure 7.1.3, there are two isotopes of the element chlorine:

- $^{35}_{17}Cl$, which contains 17 protons and 18 neutrons
- $^{37}_{17}Cl$, which contains 17 protons and 20 neutrons.

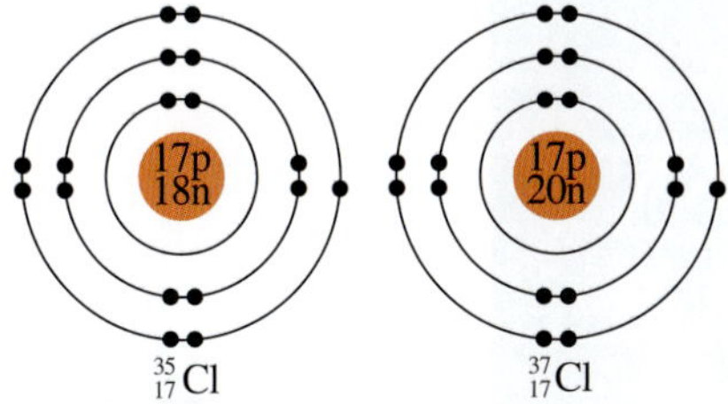

FIGURE 7.1.3 The two isotopes of the element chlorine, one with 18 neutrons and the other with 20

The isotopes have different masses because of the different numbers of neutrons. Remember that the mass of an atom is mainly determined by the number of protons and neutrons in their nucleus, since the mass of electrons is relatively small.

The relative isotopic masses of the two chlorine isotopes are experimentally determined to be 34.969 (^{35}Cl) and 36.966 (^{37}Cl). Since the masses of a proton and a neutron are similar and close to 1 on the $^{12}C = 12$ scale, the relative isotopic mass of an isotope is almost, but not exactly, equal to the number of protons plus neutrons in the nucleus.

The relative isotopic mass of an isotope is the mass of an atom of that isotope relative to the mass of an atom of carbon-12 (^{12}C) taken as 12 units exactly.

Relative isotopic abundance

Naturally occurring chlorine is made up of the two isotopes shown in Figure 7.1.3. Of all the chlorine atoms that exist, 75.80% are of the lighter isotope and 24.20% are of the heavier isotope. This composition is virtually the same no matter the source of the chlorine. The percentage abundance of an isotope in the natural environment is called its **relative isotopic abundance**.

Most elements, like chlorine, are a mixture of two or more isotopes. Details of naturally occurring isotopes of some common elements are shown in Table 7.1.2.

TABLE 7.1.2 Isotopic composition of some common elements

Element	Isotopes	Relative isotopic mass	Relative isotopic abundance (%)
hydrogen	^{1}H	1.008	99.986
	^{2}H	2.014	0.014
	^{3}H	3.016	0.0001
oxygen	^{16}O	15.995	99.76
	^{17}O	16.999	0.04
	^{18}O	17.999	0.20
silver	^{107}Ag	106.9	51.8
	^{109}Ag	108.9	48.2

Relative isotopic masses of elements and their isotopic abundances are determined using an instrument called a **mass spectrometer**, which was invented by Francis Aston in 1919. A mass spectrometer, like the one in Figure 7.1.4, is a laboratory instrument used to detect isotopes of an element and determine both the relative mass and abundance of each isotope.

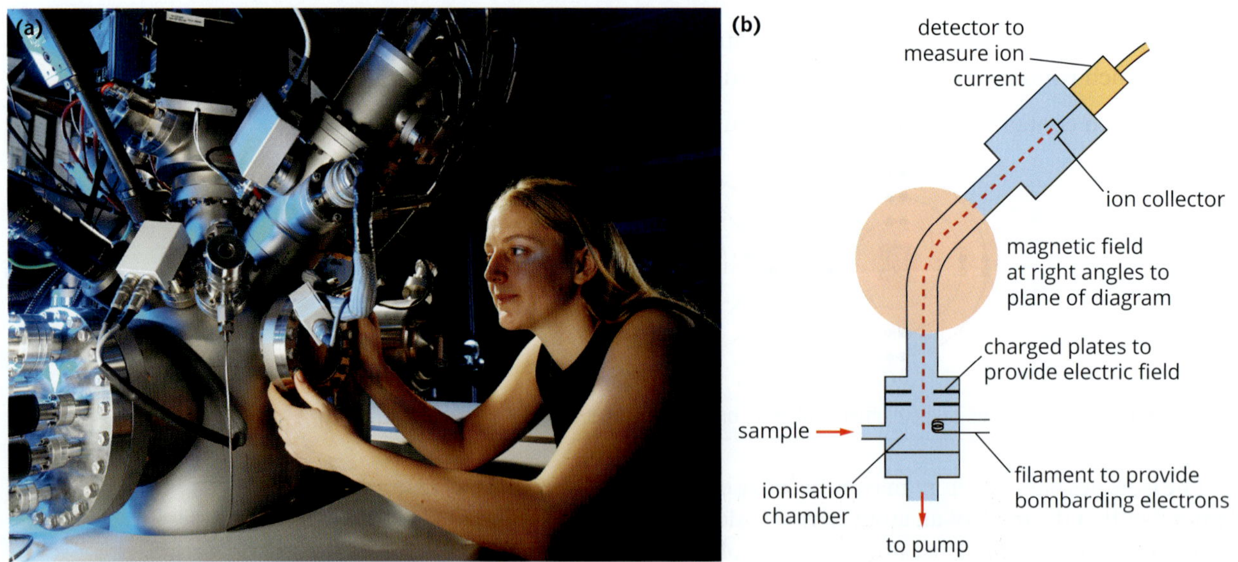

FIGURE 7.1.4 (a) A mass spectrometer is used to determine the relative masses of different isotopes. (b) The construction of a mass spectrometer. The sample is ionised and the ions are deflected as they pass through a magnetic field before reaching the detector.

CASE STUDY

How is a mass spectrometer used to determine abundances of isotopes?

Inside a mass spectrometer, a sample of an element is ionised and then the particles are deflected according to their mass-to-charge ratio. As this differs for the different isotopes, their presence and abundance can be detected.

Scientists can use mass spectrometry to determine the relative abundance of one isotope to another in tissues taken from a living, or once living, organism. This can help scientists determine important features about that organism's life.

For example, mass spectrometry can be used to gain insight into the migratory history of animals in marine environments, such as the whale shown in Figure 7.1.5. Scientists analyse the oxygen isotopes present in barnacle shells attached to the whale. As the barnacles grow, the type and abundance of the oxygen isotopes in their shells reflects the salinity (salt concentration) of the water, as well as the water temperature in which they are swimming. As the whale moves through parts of the ocean that differ in temperature and salinity, that information will be carried in their barnacles.

In another example, archaeologists can analyse the fossil jawbone from an extinct short-necked giraffe (Figure 7.1.6) to study the giraffe's diet or to determine the habitat in which the animal lived. For example, the oxygen isotopic compositions of the carbonate and phosphate in its bones and teeth are determined primarily by the surface water it drank or the water in the food it ate. The abundances of different isotopes in teeth and bones represent the ratios of the region where the giraffe was born and raised.

FIGURE 7.1.5 A North Atlantic right whale with barnacles around its mouth and above its eye. Isotopic analysis of the oxygen in the barnacle shells provides scientist with migratory information about the whale.

FIGURE 7.1.6 Ancient fossil jawbone from an extinct short-necked giraffe. Isotopic analysis using a mass spectrometer provides archaeologists with information about the organism's life.

Figure 7.1.7 shows the **mass spectrum** of chlorine. Key points that can be gained from the spectrum are as follows:

- the number of peaks indicates the number of isotopes present; in this case there are two isotopes
- the horizontal axis indicates the relative mass of each isotope present according to the isotope's mass-to-charge ratio, which is given the symbol *m/z* or *m/e*. (The charge on most ions reaching the detector is 1+, so the mass of an isotope can be read directly from the horizontal axis.) The two isotopes of chlorine have relative masses of approximately 35 and 37.
- the vertical axis indicates the abundance of each isotope in the sample: 75% ^{35}Cl, and 25% ^{37}Cl.

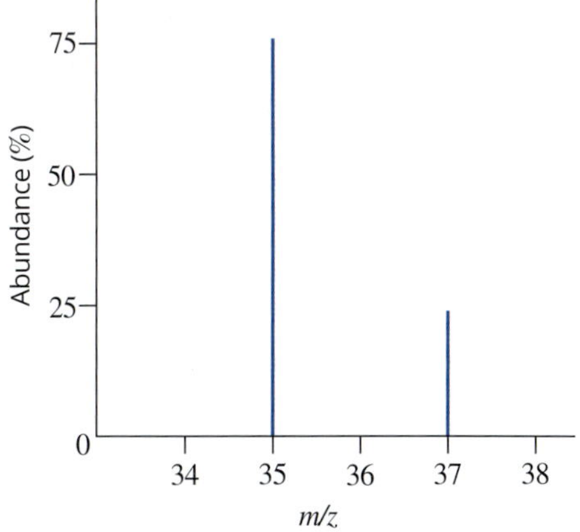

FIGURE 7.1.7 The mass spectrum of chlorine. The two peaks indicate there are two isotopes. The most abundant isotope has a relative isotopic mass of approximately 35 and an abundance of 75%. The other isotope has a relative isotopic mass of approximately 37 and an abundance of 25%.

Relative atomic mass

Most elements consist of a mixture of isotopes, each with a different relative mass. For the purpose of calculations, it is convenient to know the average relative mass of an atom in the mixture of isotopes. The average relative mass of an element is called the **relative atomic mass**, and given the symbol A_r. The relative atomic mass of an element is the weighted average of the relative masses of the isotopes of the element on the ^{12}C scale.

To calculate the average of the relative masses of the isotopes that exist in a naturally occurring mixture of an element, you must consider the relative abundances of each isotope.

Data obtained from the mass spectrum of chlorine shown in Figure 7.1.7 is summarised in Table 7.1.3.

TABLE 7.1.3 The isotopic composition of chlorine

Isotope	Relative isotopic mass	Relative abundance (%)
^{35}Cl	34.969	75.80
^{37}Cl	36.966	24.20

The data about the two isotopes can be used to calculate the relative atomic mass of chlorine. A weighted average mass is calculated by using the relative isotopic masses and abundances to find the total mass of 100 atoms. This mass is then divided by 100 to find the average mass of one atom.

The formula for this calculation is:

$$A_r = \frac{(\text{relative isotopic mass} \times \%\ \text{abundance}) + (\text{relative isotopic mass} \times \%\ \text{abundance})}{100}$$

Most periodic tables include the relative atomic mass of each element, calculated by taking into account the relative abundances of all the natural isotopes of each element.

Worked example 7.1.1

CALCULATING RELATIVE ATOMIC MASS FROM ISOTOPIC MASSES AND PERCENTAGE ABUNDANCES

Determine the relative atomic mass of chlorine from the data obtained from its mass spectrum. (Refer to Figure 7.1.7 and Table 7.1.3.)

Thinking	Working
Determine the relative isotopic masses and abundances of each isotope.	Two peaks on the spectrum indicate two isotopes: First isotope: relative isotopic mass 34.969; abundance 75.80% Second isotope: relative isotopic mass 36.966; abundance 24.20%
Substitute the relative isotopic masses and abundances into the formula for calculating relative atomic mass: $A_r = \frac{(\text{relative isotopic mass} \times \%\ \text{abundance}) + (\text{relative isotopic mass} \times \%\ \text{abundance})}{100}$	$A_r = \frac{(34.969 \times 75.80) + (36.966 \times 24.20)}{100}$
Calculate the relative atomic mass.	$A_r = \frac{2650.65 + 894.58}{100}$ $= 35.452$
Express the answer to the correct number of significant figures.	$A_r(Cl) = 35.45$

Worked example: Try yourself 7.1.1

CALCULATING RELATIVE ATOMIC MASS FROM ISOTOPIC MASSES AND PERCENTAGE ABUNDANCES

Boron has two isotopes. Their relative isotopic masses and percentage abundances are provided. Calculate the relative atomic mass of boron.

Isotope	Relative isotopic mass	Relative abundance (%)
^{10}B	10.013	19.91
^{11}B	11.009	80.09

Worked example 7.1.2

CALCULATING PERCENTAGE ABUNDANCES FROM RELATIVE ATOMIC MASS AND THE RELATIVE ISOTOPIC MASSES

The relative atomic mass of rubidium is 85.47. The relative isotopic masses of its two isotopes are 84.95 and 86.94. Calculate the relative abundances of the isotopes in naturally occurring rubidium.

Thinking	Working
State the relative abundances of the isotopes in terms of x, where x is the abundance of the lighter isotope. Abundance of lighter isotope = x. The abundance of heavier isotope must equal 100 – x.	Abundance of 84.95 isotope = x Abundance of 86.94 isotope = 100 – x
Substitute the relative isotopic masses, relative abundances and relative atomic mass into the formula: $A_r = \frac{(\text{relative isotopic mass} \times \% \text{ abundance}) + (\text{relative isotopic mass} \times \% \text{ abundance})}{100}$	$85.47 = \frac{84.95x + (86.94 \times (100 - x))}{100}$
Expand the top line of the equation.	$85.47 = \frac{84.95x + 8694 - 86.94x}{100}$
Solve the equation to find x, the relative abundance of the lightest isotope.	$8547 = 84.95x + 8694 - 86.94x$ $8547 - 8694 = 84.95x - 86.94x$ $-147 = -1.99x$ $x = 73.87\%$
Determine the abundance of the heavier isotope.	Abundance of 86.94 isotope $= 100 - x$ $= 100 - 73.87$ $= 26.13\%$

Worked example: Try yourself 7.1.2

CALCULATING PERCENTAGE ABUNDANCES FROM RELATIVE ATOMIC MASS AND THE RELATIVE ISOTOPIC MASSES

The relative atomic mass of copper is 63.54. The relative isotopic masses of its two isotopes are 62.95 and 64.95. Calculate the relative abundances of the isotopes in naturally occurring copper.

Using significant figures in relative mass calculations

You will remember from Chapter 1 that when doing calculations in chemistry, the accuracy of an answer is limited by the accuracy of the information given. An answer must have the same number of significant figures as there are in the least accurate piece of information used in a calculation. Thus, if relative isotopic masses have five significant figures, but relative abundances have only four significant figures, the calculated answer to a question using this data will include four significant figures.

7.1 Review

OA ✓✓

SUMMARY

- Most elements consist of a mixture of isotopes.
- The most common isotope of carbon, carbon-12, is used as the reference standard to determine relative masses of isotopes.
- The relative isotopic mass of an isotope is the mass of a single atom of the isotope relative to the mass of a single atom of carbon-12 being given a value of exactly 12.
- The relative isotopic mass and relative abundance of an isotope can be determined from a mass spectrometer.
- The relative atomic mass, A_r, of an element is a weighted average of its isotopic masses. Relative atomic mass is calculated using the percentage abundances of relative isotopic masses.

KEY QUESTIONS

Knowledge and understanding

1 Select the reference standard used to determine the relative masses of isotopes.

A ^{13}C assigned a mass of 13.

B ^{12}C assigned a mass of 12.

C ^{16}O assigned a mass of 16.

D The average mass of O, which is assigned a mass of 16.

2 Three isotopes of potassium exist: ^{39}K, ^{40}K and ^{41}K. Given the relative atomic mass of potassium is 39.1, which isotope would be most abundant? Give a reason for your answer.

3 There are two common isotopes of chlorine. They have relative isotopic masses of 34.969 and 36.966. The relative atomic mass of chlorine is 35.45. Select the correct statement about the relative atomic mass of chlorine.

A The precise relative atomic mass of chlorine was determined by adding the two relative isotopic masses together and dividing by two.

B The relative atomic mass of chlorine is based on the scale where carbon-12 has a mass of exactly 12.

C The lighter isotope is less abundant.

D The reason the isotopes have different masses is because they have different numbers of protons.

4 Use the data in Table 7.1.2 on page 218 to calculate the relative atomic mass of hydrogen.

Analysis

5 The element lithium has two isotopes:

- ^{6}Li with a relative isotopic mass of 6.02
- ^{7}Li with a relative isotopic mass of 7.02.

The relative atomic mass of lithium is 6.94. Calculate the percentage abundance of the lighter isotope.

6 The mass spectrum of zirconium is shown in the following graph.

a Use the relative peak heights to calculate the percentage abundance of each zirconium isotope.

b Use the percentage abundances calculated in part **a** to determine the relative atomic mass of zirconium.

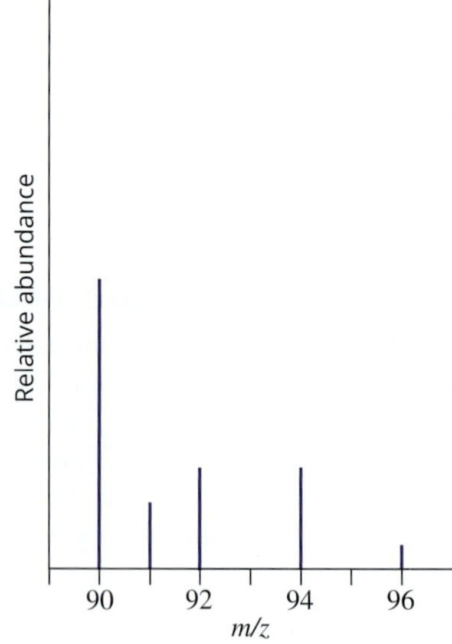

7.2 Avogadro's constant

The particles in elements and compounds are so small that it would be difficult to count atoms, ions or molecules individually or even by the thousands of millions. Furthermore, if it were possible to count individual particles, the numbers in even very small samples would be huge and very inconvenient to work with. For example: the ice cubes shown in Figure 7.2.1 each contain more than 10^{23} water molecules (H_2O). As each water molecule is composed of two hydrogen atoms and one oxygen atom, the number of individual atoms in each ice cube is greater than 10^{23}.

Chemists require a counting unit that will allow them to accurately measure amounts of extremely small particles.

FIGURE 7.2.1 A single ice cube contains more than 10^{23} water (H_2O) molecules.

THE CHEMIST'S COUNTING UNIT

Quantity units that are used to conveniently count specific numbers are often used in everyday life. Figure 7.2.2 shows the quantity units of pair and dozen, which you may be familiar with. If one dozen is equal to 12, then two dozen equals 24, twenty dozen equals 240 and half a dozen equals 6.

FIGURE 7.2.2 Convenient quantity units that you may be familiar with. A pair equals two and a dozen equals twelve.

A dozen is a useful unit for counting eggs or roses; however, chemists require a unit for counting atoms, ions and molecules that represents a much larger number.

The accepted convenient quantity used by chemists is called the **mole**. The mole is often referred to as the 'amount of substance' and is given the symbol n, and the unit mol.

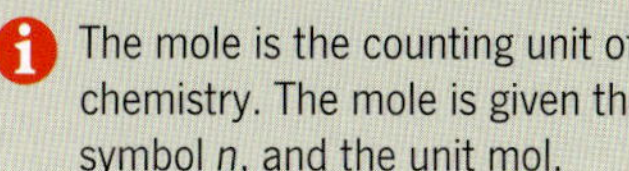

The mole is the counting unit of chemistry. The mole is given the symbol n, and the unit mol.

So n(glucose) = 2 mol means 'the amount of glucose is 2 moles'.

Chemists use the mole as a counting unit. Chemists know that one mole is equivalent to a certain number and that 2 moles, 20 moles and half a mole are all multiples of that number.

THE NUMBER IN A MOLE

One mole of any substance is defined as containing exactly $6.022\,140\,76 \times 10^{23}$ particles. That's 602 214 076 000 000 000 000 000 particles!

This number is commonly rounded to 6.02×10^{23} and is referred to as **Avogadro's constant**. It is given the symbol N_A.

Avogadro's constant is written using scientific notation, which is a way of writing very large or very small numbers.

Avogadro's constant is an enormous number, but the extremely small size of atoms, ions and molecules means that one mole of most elements and compounds does not take up a great deal of mass or volume.

For example, in Figure 7.2.3 you can see a balloon that contains one mole of a gas, a beaker that contains one mole of nickel(II) chloride and a flask that contains one mole of copper(II) sulfate dissolved in one litre of water. Each contains 6.02×10^{23} of the substance.

If it is known that 1 mol of a substance contains 6.02×10^{23} particles, then you can calculate the number of particles in different amounts of moles of the substance:

- 2 mol of a substance contains $2 \times (6.02 \times 10^{23}) = 1.204 \times 10^{24}$ particles
- 0.3 mol of a substance contains $0.3 \times (6.02 \times 10^{23}) = 1.81 \times 10^{23}$ particles
- 4.70×10^{23} particles $= \dfrac{4.70 \times 10^{23}}{6.02 \times 10^{23}} = 0.781$ mol
- 7.35×10^{24} particles $= \dfrac{7.35 \times 10^{24}}{6.02 \times 10^{23}} = 12.2$ mol

As you can see from Figure 7.2.4, a mathematical relationship exists between the number of particles, N, and the amount of substance in moles, n.

$N_A = 6.02 \times 10^{23}$ mol^{-1}
Therefore, 1 mol of any substance contains 6.02×10^{23} particles.

FIGURE 7.2.3 A balloon, a flask and a beaker all contain one mole of a substance.

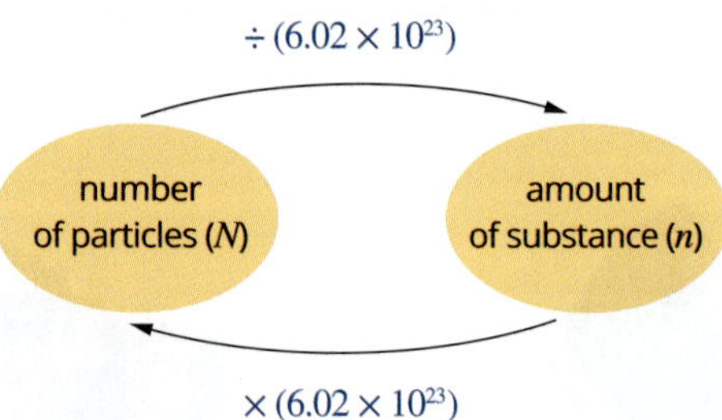

FIGURE 7.2.4 Relationship between number of particles and amount of substance in moles

CHEMFILE

Changing mole definitions

The definition of the mole as containing an exact number of particles was adopted in November 2018. Prior to this date, one mole was defined as containing the same amount of substance as there were carbon atoms in exactly 12.000 g of ^{12}C. The new definition emphasised that moles are to do with counting particles, rather than measuring the mass of a sample of another substance. Practically, the new definition does not change the way the mole is used in calculations.

The amount of substance vs the amount of atoms

When referring to a mole of a substance, it is important to indicate which particle is being specified. For example, an oxygen molecule contains two oxygen atoms joined by a double covalent bond. Figure 7.2.5 shows four different ways that chemists use to represent the oxygen molecule. The oxygen molecule is most commonly described by the **molecular formula** O_2.

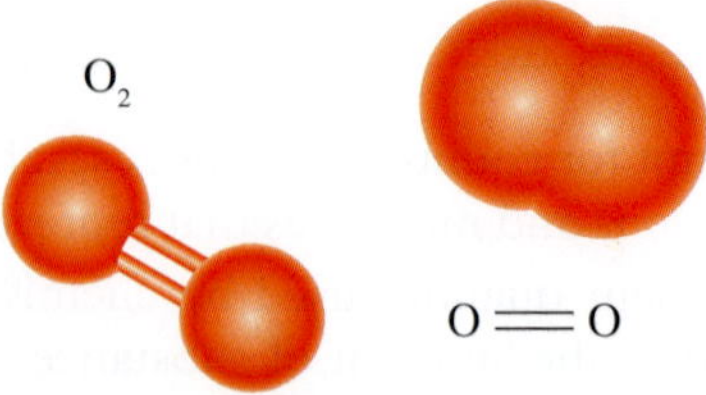

FIGURE 7.2.5 Four different ways chemists represent the oxygen molecule: a molecular formula (O_2), a structural formula (O=O) and two coloured molecular models. Each formula or model shows that one oxygen molecule contains two oxygen atoms.

The expression, 'one mole of oxygen' is ambiguous because it could describe one mole of oxygen atoms, O, or one mole of oxygen molecules, O_2. As there are two atoms in each oxygen molecule, one mole of oxygen molecules will contain two moles of oxygen atoms.

The **empirical formula** of an ionic compound indicates the number of each ion in one formula unit of the compound. For example, in one mole of aluminium chloride, $AlCl_3$, there is one mole of aluminium ions and three moles of chloride ions.

Some other examples of using the mole as a counting unit for elements, molecules and ionic compounds, and the particles found in each of them, are provided in Table 7.2.1.

TABLE 7.2.1 Examples of the use of the mole as a counting unit

Number of moles of substance	Information that can be obtained about numbers of particles
1 mole of hydrogen atoms (H)	1 mole of hydrogen atoms (H)
1 mole of hydrogen molecules (H_2)	1 mole of hydrogen molecules (H_2) 2 moles of hydrogen atoms (H)
2 moles of aluminium atoms (Al)	2 moles of aluminium atoms (Al)
2 moles of calcium chloride ($CaCl_2$)	2 moles of Ca^{2+} ions 4 moles of Cl^- ions
10 moles of glucose ($C_6H_{12}O_6$) molecules	10 moles of glucose ($C_6H_{12}O_6$) molecules 60 moles of carbon atoms 120 moles of hydrogen atoms 60 moles of oxygen atoms

Calculations using the mole and Avogadro's constant

Three quantities have been introduced so far:

- the mole, which is given the symbol n and the unit mol
- Avogadro's constant, which is given the symbol N_A and has the value 6.02×10^{23}
- the actual number of particles (atoms, ions or molecules), which is given the symbol N.

The mathematical relationship that links the three quantities is $n = \frac{N}{N_A}$, where:

n is the number of moles, unit mol
N is the actual number of particles
N_A is Avogadro's constant and is equal to 6.02×10^{23}

CHEMFILE

Using Sydney Harbour as a counting unit

Astronomers and engineers also deal with very large numbers and have developed units to handle the huge distances and volumes they deal with. An unusual Australian unit of measurement is a 'Sydharb', which is a volume defined in the Macquarie Dictionary as equal to the amount of water in Sydney Harbour at high tide (see figure below).

News reports often discuss the size of a flood or the capacity of a reservoir in multiples of Sydney Harbour, and one Sydharb is approximately 5×10^{11} L. In the same way that it is easier to talk about a mole than 6.02×10^{23} particles, this unit is an easy way to visualise and discuss large numbers.

A Sydharb is an unusual unit of volume equal to the amount of water in Sydney Harbour at high tide. Floods and reservoir capacity are often measured in Sydharbs.

The following formula is used to calculate the amount or number of specified particles in a sample:

$$n = \frac{N}{N_A}$$

where
n is the number of moles, unit mol
N is the actual number of particles
N_A is Avogadro's constant and is equal to 6.02×10^{23}

Worked example 7.2.1

CALCULATING THE NUMBER OF MOLECULES GIVEN THE AMOUNT OF A SUBSTANCE

Calculate the number of molecules in 3.5 moles of water (H_2O).

Thinking	Working
List the data given in the question next to the appropriate symbol. Include units.	The number of water molecules is the unknown, so: $N(H_2O) = ?$ $n(H_2O) = 3.5$ mol $N_A = 6.02 \times 10^{23}$
Rearrange the formula to make the unknown the subject.	$n = \frac{N}{N_A}$ so $N(H_2O) = n \times N_A$
Substitute in data and solve for the answer.	$N(H_2O) = n \times N_A$ $= 3.5 \times 6.02 \times 10^{23}$ $= 2.1 \times 10^{24}$ molecules

Worked example: Try yourself 7.2.1

CALCULATING THE NUMBER OF MOLECULES GIVEN THE AMOUNT OF A SUBSTANCE

Calculate the number of molecules in 1.6 moles of carbon dioxide (CO_2).

Worked example 7.2.2

CALCULATING THE NUMBER OF MOLES OF ATOMS GIVEN THE NUMBER OF MOLES OF MOLECULES

Calculate the amount, in mol, of hydrogen atoms in 3.6 mol of sulfuric acid (H_2SO_4).

Thinking	Working
List the data given in the question next to the appropriate symbol. Include units.	The number of mol of hydrogen atoms is the unknown, so: $n(H) = ?$ $n(H_2SO_4) = 3.6$ mol
Calculate the amount, in mol, of hydrogen atoms from the amount of sulfuric acid molecules and the molecular formula.	$n(H) = n(H_2SO_4) \times 2$ $= 3.6 \times 2$ $= 7.2$ mol

Worked example: Try yourself 7.2.2

CALCULATING THE NUMBER OF MOLE OF ATOMS GIVEN THE NUMBER OF MOLE OF MOLECULES

Calculate the amount, in mol, of hydrogen atoms in 0.75 mol of water (H_2O).

Worked example 7.2.3

CALCULATING THE NUMBER OF ATOMS GIVEN THE AMOUNT OF A SUBSTANCE

Calculate the number of oxygen atoms in 2.5 mol of oxygen gas (O_2).

Thinking	Working
List the data given in the question next to the appropriate symbol. Include units.	The number of oxygen atoms is the unknown, so: $N(O) = ?$ $n(O_2) = 2.5$ mol $N_A = 6.02 \times 10^{23}$
Calculate the amount, in mol, of oxygen atoms from the amount of oxygen molecules and the molecular formula.	$n(O) = n(O_2) \times 2$ $= 2.5 \times 2$ $= 5.0$ mol
Rearrange the formula to make the unknown the subject.	$n = \frac{N}{N_A}$ so $N(O) = n \times N_A$
Substitute in data and solve for the answer.	$N(O) = n \times N_A$ $= 5.0 \times 6.02 \times 10^{23}$ $= 3.0 \times 10^{24}$ atoms

Worked example: Try yourself 7.2.3

CALCULATING THE NUMBER OF ATOMS GIVEN THE AMOUNT OF A SUBSTANCE

Calculate the number of hydrogen atoms in 0.35 mol of methane (CH_4).

Worked example 7.2.4

CALCULATING THE NUMBER OF ATOMS GIVEN THE AMOUNT OF AN IONIC COMPOUND

Calculate the number of oxygen atoms in 1.5 mol of iron (III) sulfate, $Fe_2(SO_4)_3$.

Thinking	Working
List the data given in the question next to the appropriate symbol. Include units.	The number of oxygen atoms is the unknown, so: $N(O) = ?$ $n(Fe_2(SO_4)_3) = 1.5$ mol $N_A = 6.02 \times 10^{23}$
Calculate the amount, in mol, of oxygen atoms from the amount of oxygen molecules and the molecular formula.	$n(O) = n(Fe_2(SO_4)_3) \times 12$ $= 1.5 \times 12$ $= 18$ mol
Rearrange the formula to make the unknown the subject.	$n = \frac{N}{N_A}$ so $N(O) = n \times N_A$
Substitute in data and solve for the answer.	$N(O) = n \times N_A$ $= 18 \times 6.02 \times 10^{23}$ $= 1.1 \times 10^{25}$ atoms

Worked example: Try yourself 7.2.4

CALCULATING THE NUMBER OF ATOMS GIVEN THE AMOUNT OF AN IONIC COMPOUND

Calculate the number of hydrogen atoms in 0.85 mol of aluminium hydroxide ($Al(OH)_3$).

Worked example 7.2.5

CALCULATING THE NUMBER OF MOLES OF PARTICLES GIVEN THE NUMBER OF PARTICLES

Calculate the amount, in mol, of ammonia molecules (NH_3) represented by 2.5×10^{22} ammonia molecules.

Thinking	Working
List the data given in the question next to the appropriate symbol. Include units.	The number of mol of ammonia molecules is the unknown, so: $n(NH_3) = ?$ $N(NH_3) = 2.5 \times 10^{22}$ molecules $N_A = 6.02 \times 10^{23}$
Rearrange the formula to make the unknown the subject.	$n = \frac{N}{N_A}$ n is the unknown so rearrangement not required
Substitute in data and solve for the answer.	$n(NH_3) = \frac{N}{N_A}$ $= \frac{2.5 \times 10^{22}}{6.02 \times 10^{23}}$ $= 0.042$ mol

Worked example: Try yourself 7.2.5

CALCULATING THE NUMBER OF MOLES OF PARTICLES GIVEN THE NUMBER OF PARTICLES

Calculate the amount, in mol, of magnesium atoms represented by 8.1×10^{20} magnesium atoms.

7.2 Review

OA ✓✓

SUMMARY

- A mole is a convenient quantity for counting particles. The mole is given the symbol n and the unit mol.
- One mole is defined as the amount of substance that contains exactly $6.022\,140\,76 \times 10^{23}$ constituent particles.
- The number of particles in 1 mol is given the symbol N_A. This is known as Avogadro's constant and is often rounded to a value of 6.02×10^{23}.
- The following formula can be used or rearranged to calculate the amount or number of specified particles in a sample:

$$n = \frac{N}{N_A}$$

KEY QUESTIONS

Knowledge and understanding

1 What is the exact number of particles in Avogadro's constant?

2 What are the quantities represented by the following symbols?

- **a** n
- **b** N
- **c** N_A

3 Explain why the expression 'one mole of hydrogen' is ambiguous.

4 Calculate the number of:

- **a** atoms in 3.0 mol of sodium atoms (Na)
- **b** atoms in 1.5×10^{-2} mol of iron atoms (Fe)
- **c** molecules in 2.85×10^{-5} mol of CO_2 molecules.

Analysis

5 Calculate the amount of substance, in mol, represented by:

- **a** 6.0×10^{23} molecules of water (H_2O)
- **b** 2.5×10^{23} atoms of neon (Ne)
- **c** 3.2×10^{25} molecules of ethanol (C_2H_5OH).

6 Calculate the amount, in mol, of:

- **a** chlorine atoms in 0.40 mol of chlorine molecules (Cl_2)
- **b** hydrogen atoms in 1.2 mol of methane molecules (CH_4)
- **c** oxygen atoms in 1.5 mol of sodium sulfate (Na_2SO_4).

7 Calculate the amount, in mol, of:

- **a** oxygen atoms represented by 4.0×10^{23} oxygen molecules
- **b** hydrogen atoms represented by 3.5×10^{22} methane molecules
- **c** chlorine atoms represented by 1.0×10^{20} chlorine molecules.
- **d** chloride ions present in 2.0×10^{23} of magnesium chloride ($MgCl_2$)

7.3 Molar mass

Although a mole is a convenient counting unit for chemists, a mole of a substance cannot be simply counted out in a laboratory. Fortunately, chemical laboratories always contain a simple balance, like the one in Figure 7.3.1, which is used for weighing quantities of a substance. If a chemist knows that a specific mass of a substance always contains a specific number of moles, it is possible to easily weigh a sample of the substance and calculate the exact number of moles present in the sample.

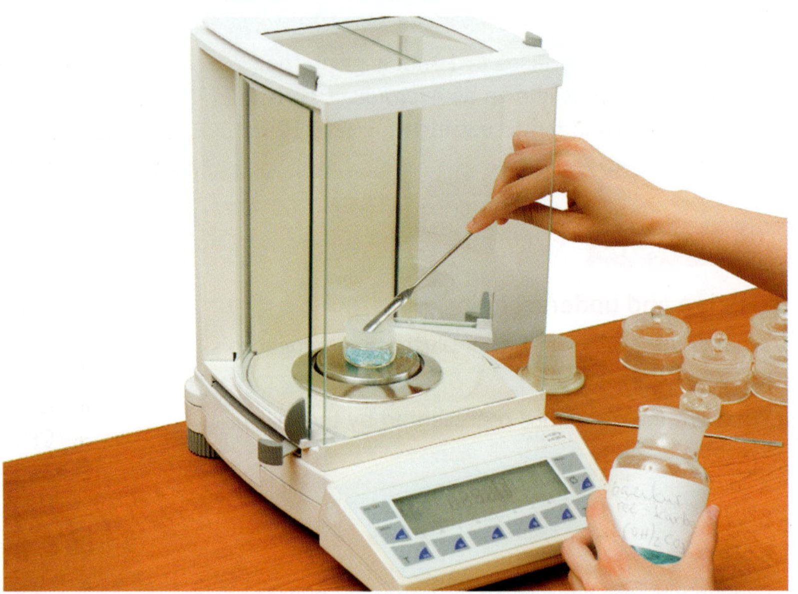

FIGURE 7.3.1 A digital balance is a simple piece of laboratory equipment used for weighing.

The particles of different elements and compounds have different masses. Therefore, the masses of one mole of different elements or compounds will also be different. This is like saying that the mass of one dozen oranges will be greater than the mass of one dozen mandarins because one orange is heavier than one mandarin.

In this section, you will learn how the amount of a substance, in moles, is related to the mass of the substance.

MOLAR MASSES OF ELEMENTS

The molar mass of an element, M = relative atomic mass of the element expressed in g mol^{-1}.

The mass, in grams, of one mole of a particular element is known as its **molar mass**. It is given the symbol M and the unit g mol^{-1}. The molar mass of an element is obtained simply by adding the unit g mol^{-1} to the relative atomic mass of the element. This very convenient relationship works because the seemingly random figure of Avogadro's constant was chosen as it was determined experimentally that 12 g of ^{12}C contains exactly 6.02×10^{23} atoms. Table 7.3.1 shows the relationship between the relative atomic mass and molar mass of several elements.

TABLE 7.3.1 Relationship between relative atomic masses and molar mass of an element

Element	Relative atomic mass, A_r	Molar mass, M
O	16.0	16.0 g mol^{-1}
Fe	55.8	55.8 g mol^{-1}
Ag	107.9	107.9 g mol^{-1}

MOLAR MASSES OF MOLECULES AND IONIC COMPOUNDS

Molecules contain a fixed number of two or more atoms. The atoms may be the same; for example, O_2, or different, for example, H_2O. The composition of an ionic compound is also given with a fixed formula, for example, $NaNO_3$. The mass of one mole of an atom, molecule or ionic compound is equal to the sum of the relative atomic masses of each atom in the formula of the substance. Table 7.3.2 shows you how to calculate the molar masses of some common substances.

TABLE 7.3.2 The molar mass of a molecule or ionic compound is determined by adding the relative atomic masses of each atom in its formula.

Substance	Relative atomic masses	Molar mass of substance
O_2	O: 16.0	$= 2 \times 16.0$ $= 32.0$ g mol^{-1}
H_2O	H: 1.0 O: 16.0	$= (2 \times 1.0) + 16.0$ $= 18.0$ g mol^{-1}
CO_2	C: 12.0 O: 16.0	$= 12.0 + (2 \times 16.0)$ $= 44.0$ g mol^{-1}
$NaNO_3$	Na: 23.0 N: 14.0 O: 16.0	$= 23.0 + 14.0 + (3 \times 16.0)$ $= 85.0$ g mol^{-1}

FIGURE 7.3.2 A collection of beakers, each containing one mole of a substance. Each has a different weight and volume. Clockwise from top left the substances are: sucrose, potassium permanganate, copper(II) sulfate, copper turnings, iron filings and nickel(II) chloride.

From the calculations in Table 7.3.2 and the photograph of one mole of some common substances in Figure 7.3.2, you can see that one mole of each substance has a different mass.

Worked example 7.3.1

CALCULATING THE MOLAR MASS OF MOLECULES

Calculate the molar mass of carbon dioxide (CO_2).

Thinking	Working
Use the periodic table to find the relative atomic mass for the elements represented in the formula.	$A_r(C) = 12.0$ $A_r(O) = 16.0$
Determine the number of atoms of each element present, taking into consideration any brackets in the formula.	1 × C atom 2 × O atoms
Determine the molar mass by adding the appropriate relative atomic masses.	$M_r = (1 \times A_r(C)) + (2 \times A_r(O))$ $= (1 \times 12.0) + (2 \times 16.0)$ $= 44.0$ g mol^{-1}

Worked example: Try yourself 7.3.1

CALCULATING THE MOLAR MASS OF MOLECULES

Calculate the molar mass of nitric acid (HNO_3).

CHEMFILE

Measuring the molar mass of a compound without a mass spectrometer

When chemists were first determining the molar masses of molecules in the nineteenth century, a common method was to dissolve the molecule in a solvent and measure the decrease in the solvent's freezing point. This is called a freezing point depression experiment. It was a successful way to measure a molar mass because the freezing point of a solution depends on the number of particles dissolved in the solution, but not on their identity.

When water is the solvent, this relationship is written as:

$$T_f\text{(solution)} - m\text{(water)} = -1.86n$$

where:

T_f(solution) is the freezing point of the solution, in °C

m(water) is the mass of water, in kg

n is the number of moles of the dissolved substance.

Therefore, if you know the mass of the substance dissolved in the solution, you can use the formula $M = n \times m$ to work out its molar mass.

> The following formula is used to calculate the amount, in mol, of a substance from its mass.
>
> $n = m/M$
>
> where
>
> n is the number of moles, unit mol
>
> m is the mass, unit g
>
> M is the molar mass, unit g mol^{-1}

Calculations using moles and molar mass

A useful formula, shown in Figure 7.3.3, links the amount of a substance (n), in mol, its molar mass (M), in g mol^{-1}, and the given mass of the substance (m), in grams. This allows the number of moles of an element or compound to be determined from its mass, if the molar mass is known.

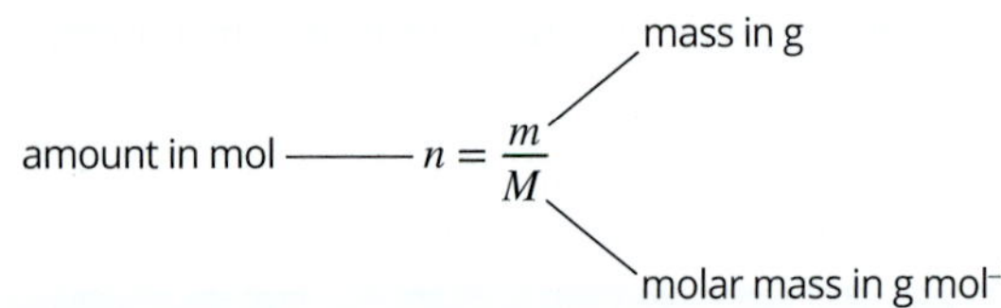

FIGURE 7.3.3 The relationship between the amount of substance, n, its molar mass, M, and the given mass of a substance, m

Alternatively, the mass of a particular number of moles of substance can be determined by rearranging the equation to:

$$m = n \times M$$

Worked example 7.3.2

CALCULATING THE AMOUNT OF SUBSTANCE FROM MASS

Calculate the amount, in mol, of 250 g of sodium chloride (NaCl).

Thinking	Working
Find the molar mass of sodium chloride using relative atomic masses sourced from the periodic table.	$A_r(Na) = 23.0$ $A_r(Cl) = 35.5$ $M(NaCl) = 23.0 + 35.5$ $= 58.5$ g mol^{-1}
Calculate the amount, in mol, of sodium chloride using: $n = \frac{m}{M}$	$n(NaCl) = \frac{m}{M}$ $= \frac{250}{58.5}$ $= 4.27$ mol

Worked example: Try yourself 7.3.2

CALCULATING THE AMOUNT OF SUBSTANCE FROM MASS

Calculate the amount, in mol, of 100 g of methane gas (CH_4).

Worked example 7.3.3

CALCULATING THE MASS OF A SUBSTANCE FROM NUMBER OF MOLES

Calculate the mass of 0.35 mol of magnesium nitrate ($Mg(NO_3)_2$).

Thinking	Working
List the data given to you in the question. Remember that whenever you are given a formula, you can calculate the molar mass by referring to the periodic table for relative atomic masses.	$m(Mg(NO_3)_2) = ?$ g $n(Mg(NO_3)_2) = 0.35$ mol $M(Mg(NO_3)_2) = 24.3 + (2 \times 14.0) + (6 \times 16.0)$ $= 148.3$ g mol^{-1}
Calculate the mass of magnesium nitrate using: $m = n \times M$	$m = n \times M$ $m(Mg(NO_3)_2) = 0.35 \times 148.3$ $= 52$ g

Worked example: Try yourself 7.3.3

CALCULATING THE MASS OF A SUBSTANCE FROM NUMBER OF MOLES

Calculate the mass of 4.68 mol of sodium carbonate (Na_2CO_3).

CASE STUDY ANALYSIS

The sting of a bee

The formula for pentyl ethanoate ($C_7H_{14}O_2$) is represented by the structure shown in Figure 7.3.4.

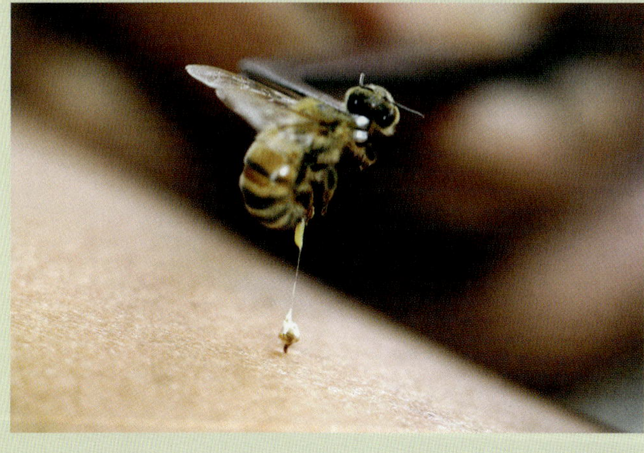

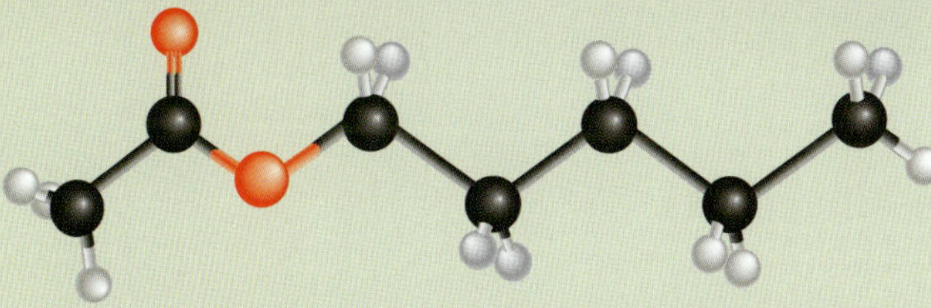

FIGURE 7.3.4 Pentyl ethanoate is a compound released when bees sting.

Pentyl ethanoate is the compound that gives bananas their characteristic odour. It is also a pheromone released by bees when they sting. A pheromone is a chemical produced by an animal or insect that changes the behaviour of another animal or insect of the same species. Pheromones are detected in a similar way to smell. Bees release pentyl ethanoate as an alarm pheromone to alert other bees to danger. The compound is released near the sting shaft and attracts other bees to the area, where the group behaves defensively, charging and stinging. Smoke masks the alarm pheromone, interrupting the defensive response and calming the bees, allowing beekeepers to work with beehives.

Each time a bee stings, one-thousandth of a milligram (1.0×10^{-6} g) of pentyl ethanoate is released. Knowing the mass of pentyl ethanoate released in a bee sting enables chemists to calculate the number of pentyl ethanoate molecules in each sting using the relationship between mass and mole.

Analysis

1. Calculate the molar mass of pentyl ethanoate.
2. Calculate the amount, in mol, of pentyl ethanoate released by one bee sting.
3. Calculate the number of pentyl ethanoate molecules released by one bee sting.

Calculations using moles, mass and Avogadro's constant

The mole is the quantity used by chemists to count the number of particles in a substance. Mass is used to measure out samples in a laboratory using a balance. The number of particles in a specified mass of a substance can be determined by using the two formulas $n = \frac{m}{M}$ and $n = \frac{N}{N_A}$.

Worked example 7.3.4

CALCULATING THE NUMBER OF MOLECULES

Calculate the number of CO_2 molecules present in 22 g of carbon dioxide.

Thinking	Working
List the data given to you in the question. Convert mass to grams, if required. Remember that whenever you are given a formula you can calculate the molar mass.	$N(CO_2) = ?$ $M(CO_2) = 12.0 + (2 \times 16.0) = 44.0$ g mol^{-1} $m(CO_2) = 22$ g
Calculate the amount, in mol, of CO_2 using: $n = \frac{m}{M}$	$n(CO_2) = \frac{m}{M}$ $= \frac{22}{44.0}$ $= 0.50$ mol
Calculate the number of CO_2 molecules using: $n = \frac{N}{N_A}$	$n = \frac{N}{N_A}$ so $N = n \times N_A$ $N(CO_2) = 0.50 \times 6.02 \times 10^{23}$ $= 3.0 \times 10^{23}$ molecules

Worked example: Try yourself 7.3.4

CALCULATING THE NUMBER OF MOLECULES

Calculate the number of sucrose molecules in a teaspoon (4.2 g) of sucrose ($C_{12}H_{22}O_{11}$).

7.3 Review

OA ✓✓

SUMMARY

- The molar mass of an element or compound is the mass, in grams, of one mole of that element or compound. Molar mass is given the symbol M and the unit g mol^{-1}.
- The molar mass of an element is equal to the relative atomic mass of the element expressed in g mol^{-1}.
- The molar mass of a molecule or ionic compound is equal to the sum of the relative atomic masses of the atoms in the formula of the molecule or compound.
- The formula $n = \frac{m}{M}$ can be used or rearranged to calculate the mass, amount or molar mass of an element or compound.

KEY QUESTIONS

Knowledge and understanding

1 Describe how you can determine the molar mass of a compound.

2 State the formula used to calculate the mass of a substance from a given number of moles.

Analysis

3 Calculate the molar mass of:

a chlorine (Cl_2)

b vitamin C (ascorbic acid, $C_6H_8O_6$)

c hydrated copper(II) sulfate ($CuSO_4{\cdot}5H_2O$).

4 Calculate the mass of:

a 3.0 mol of oxygen molecules (O_2)

b 1.5 mol of methane molecules (CH_4)

c 2.5 mol of aluminium oxide (Al_2O_3).

5 Calculate the amount, in mol, of:

a H_2 molecules in 5 g of hydrogen (H_2)

b Aluminium chloride in 50 g of aluminium chloride ($AlCl_3$)

c CH_4 molecules in 4.5 g of methane (CH_4)

6 Calculate the number of atoms in:

a 23 g of sodium (Na)

b 4.0 g of argon (Ar)

c 0.243 g of magnesium (Mg)

7 Calculate the amount, in mol, of:

a H atoms in 5 g of hydrogen (H_2)

b Nitrate ions in 100 g of magnesium nitrate ($Mg(NO_3)_2$)

c P atoms in 1.2×10^{-3} g of phosphorus molecules (P_4).

8 Calculate the:

a number of molecules in:

i 48 g of oxygen (O_2)

ii 50 g of nitrogen (N_2)

b number of oxygen atoms in 3.2 g of sulfur dioxide (SO_2)

c total number of atoms in 170 g of ammonia (NH_3).

7.4 Percentage composition, empirical and molecular formulas

CHEMFILE

Elemental analysis of magnetite

The mineral magnetite is one of the three most commonly occurring oxides of iron. The most magnetic of all naturally occurring minerals on Earth, magnetite was used by the Chinese to make the first magnetic compass.

Found in many different types of igneous and metamorphic rocks, large deposits are located in the Pilbara region of Western Australia. Elemental analysis allowed chemists to determine the percentage composition of magnetite as 72.4% iron and 27.6% oxygen (see figure below). The percentage composition can then be used to determine the chemical formula, which is Fe_3O_4.

Elemental analysis of the mineral magnetite has allowed chemists to determine that the percentage composition is 72.4% iron and 27.6% oxygen.

Compounds are substances that contain two or more different elements. The relative proportions of each element in a compound can be expressed as:

- a percentage in terms of the mass contributed by each element
- a formula showing either the ratio of atoms contributed by each element in a compound or the actual numbers of atoms in a molecule.

The formulas of compounds are an important part of the language of chemistry. Formulas are used to represent chemicals in equations and on numerous other occasions.

PERCENTAGE COMPOSITION BY MASS

Percentage composition by mass describes the proportion of each element in a compound in terms of mass. The proportion of each element is expressed as a percentage of the total mass of the compound. For example, Figure 7.4.1 shows that 82% of the mass of the covalent molecular compound ammonia is due to nitrogen and 18% is contributed by hydrogen.

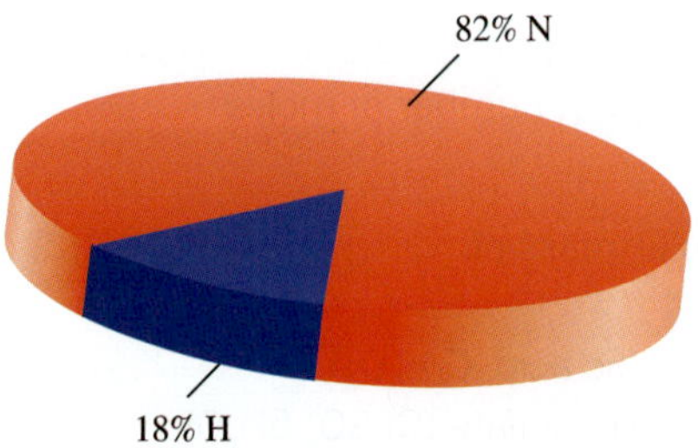

FIGURE 7.4.1 Percentage composition, by mass, of ammonia (NH_3)

The percentage composition by mass of an element in a compound can be determined from the chemical formula of the compound using the molar mass of the element and the molar mass of the compound.

$$\text{\% by mass of element} = \frac{\text{mass of the element in 1 mol of the compound}}{\text{molar mass of the compound}} \times 100$$

Worked example 7.4.1

CALCULATING PERCENTAGE COMPOSITION BY MASS

Calculate the percentage by mass of carbon in carbon dioxide (CO_2).

Thinking	Working
Find the molar mass of the compound.	$M(CO_2) = (12.0) + (2 \times 16.0)$ $= 44.0\ g\ mol^{-1}$
Find the total mass of the element in one mole of the compound.	mass of C in 1 mol = 12.0 g
Find the percentage by mass of the element in the compound.	% by mass of C in CO_2 $= \frac{\text{mass of C in 1 mol of } CO_2}{\text{molar mass of } CO_2} \times 100$ $= \frac{12.0}{44.0} \times 100$ = 27.3%

Worked example: Try yourself 7.4.1

CALCULATING PERCENTAGE COMPOSITION BY MASS

Calculate the percentage by mass of nitrogen in ammonium nitrate (NH_4NO_3).

EMPIRICAL FORMULA

While percentage mass provides a ratio of elements in a compound by mass, the number of atoms or ions present in compounds are given in a chemical formula. Atoms or ions are always present in compounds in fixed whole-number ratios. The empirical formula of a compound gives the simplest whole-number ratio of each type of element in the compound.

- The formula of an ionic compound is always an empirical formula as it provides the simplest ratio of ions in the ionic lattice.
- For molecular compounds, the empirical formula can be the same as the molecular formula, or a simplification of the molecular formula.

  See Table 7.4.1 for some examples.

TABLE 7.4.1 Empirical formulas of some common compounds

Compound	Empirical formula	Simplest whole-number ratio of elements in the compound
water (H_2O)	H_2O	H:O 2 : 1
ethene (C_2H_4)	CH_2	C:H 1 : 2
ethane (C_2H_6)	CH_3	C:H 1 : 3
calcium carbonate	$CaCO_3$	Ca:C:O 1 : 1 : 3

Determining empirical formula

The empirical formula of a compound is determined from the mass of each element present in a given mass of the compound. If percentage composition is known, it can be used to determine the mass of each element in 100 g of the compound. The masses can also be determined experimentally.

Once the masses of elements in a compound are known, the steps in Figure 7.4.2 are followed to convert masses to a mole ratio, that is, a ratio by number of atoms, and then to an empirical formula.

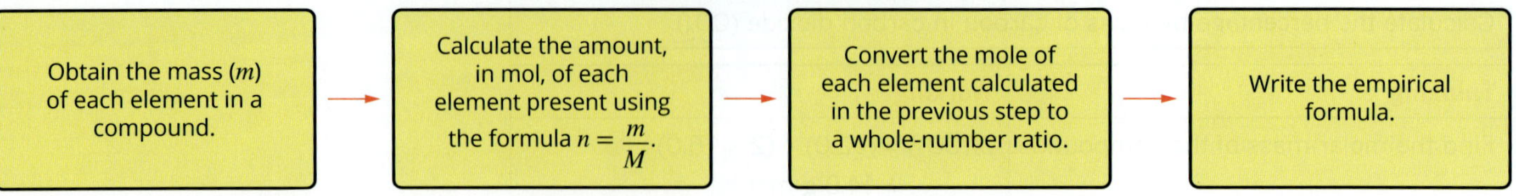

FIGURE 7.4.2 Steps to follow to calculate an empirical formula from masses of a compound

Worked example 7.4.2

DETERMINING THE EMPIRICAL FORMULA FROM PERCENTAGE COMPOSITION

An organic compound is found to be composed of 52.2% carbon, 13.0% hydrogen and the remainder is oxygen. Calculate the empirical formula of the compound.

Thinking	Working
Write down the mass, in g, of all elements present in the compound. If masses are given as percentages, assume that the sample weighs 100 g, then the percentages become masses in grams.	$m(\text{C}) = 52.2$ g $m(\text{H}) = 13.0$ g $m(\text{O}) = 100 - 52.2 - 13.0 = 34.8$ g
Calculate the amount, in mol, of each element in the compound using: $n = \frac{m}{M}$	$n(\text{C}) = \frac{52.2}{12.0}$ $= 4.35$ mol $n(\text{H}) = \frac{13.0}{1.0}$ $= 13.0$ mol $n(\text{O}) = \frac{34.8}{16.0}$ $= 2.18$ mol
Simplify the ratio by dividing each number of moles by the smallest number of moles calculated in the previous step. This gives you a ratio of the number of atoms of each element.	$\text{C} = \frac{4.35}{2.18}$ $= 2.0$ $\text{H} = \frac{13.0}{2.18}$ $= 6.0$ $\text{O} = \frac{2.18}{2.18}$ $= 1.0$
Find the simplest whole-number ratio.	C : H : O 2 6 1
Write the empirical formula.	C_2H_6O

Worked example: Try yourself 7.4.2

DETERMINING THE EMPIRICAL FORMULA FROM PERCENTAGE COMPOSITION

Chemical analysis of an organic compound present in the gaseous emissions from a factory shows that its percentage composition is 40.0% carbon, 6.7% hydrogen and the remainder is oxygen. Find its empirical formula.

CASE STUDY **ANALYSIS**

Analysing a life-saving rat poison

Warfarin is frequently prescribed to treat blood clots and prevent strokes in people. Warfarin is an anticoagulant, meaning it stops blood from clumping together to form clots. It is derived from a chemical called coumarin, which is found in plants such as tonka beans and cinnamon. The first large-scale commercial use of warfarin was as rat poison.

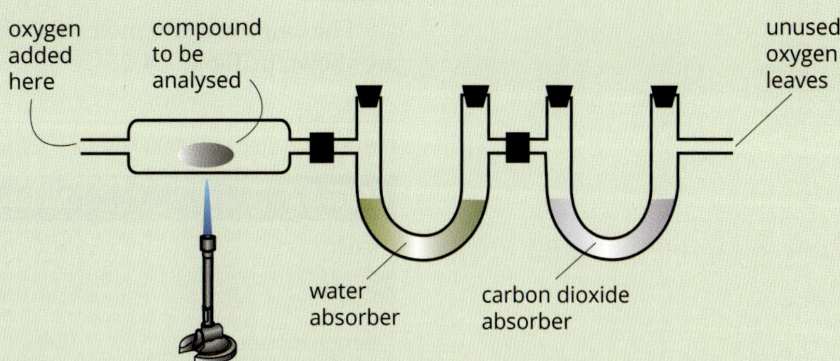

FIGURE 7.4.3 Experimental set-up for the elemental analysis of an organic compound

The empirical formula of warfarin can be determined if the mass of each element in a sample of the compound is known in order to calculate percentage composition. That information can come from an experiment called elemental microanalysis.

Elemental microanalysis of an organic compound can be performed by burning the compound in air. The combustion products, carbon dioxide and water, are absorbed by chemicals that have been previously weighed in order to determine their concentration.

- Carbon dioxide can be absorbed by sodium hydroxide solution.
- Water is absorbed by a drying agent such as magnesium perchlorate, $Mg(ClO_4)_2$.

The experimental apparatus is shown in Figure 7.4.3.

The empirical formula of a hydrocarbon, or a compound containing only carbon, hydrogen and oxygen, can be calculated from the masses of carbon dioxide and water obtained by elemental analysis.

DETERMINING THE MASS OF CARBON IN THE ORGANIC COMPOUND

Suppose an empirical formula calculation is being carried out on an organic compound containing only carbon, hydrogen and oxygen, using data from elemental analysis. From the mass of carbon dioxide produced, the amount, in mol, of carbon dioxide produced can be calculated:

$$n(CO_2) = \frac{m}{M} = \frac{m}{44.0} \text{ mol}$$

Since all the carbon in the organic compound ends up in the carbon dioxide produced, the amount of carbon in the organic compound is equal to the amount of carbon dioxide produced:

$$n(C) = n(CO_2) \text{ mol}$$

The mass of carbon in the organic compound can then be determined:

$$m(C) = n \times M = n \times 12.0 \text{ g}$$

DETERMINING THE MASS OF HYDROGEN IN THE ORGANIC COMPOUND

From the mass of water produced, the amount, in mol, of water produced can be calculated:

$$n(H_2O) = \frac{m}{M} = \frac{m}{18.0} \text{ mol}$$

All the hydrogen in the organic compound ends up in the water produced. The amount of hydrogen in the organic compound is twice the amount of water produced because there are two hydrogen atoms in each molecule of water:

$$n(H) = 2 \times n(H_2O) \text{ mol}$$

The mass of hydrogen in the organic compound can then be determined:

$$m(H) = n \times M = n \times 1.0 \text{ g}$$

Warfarin contains only carbon, hydrogen and oxygen. During elemental microanalysis, 20.900 g of CO_2 and 3.600 g of H_2O were formed upon the complete combustion of 7.700 g of warfarin.

Analysis

1. In the microanalysis above, the masses of carbon and hydrogen have been determined directly from the masses of carbon dioxide and water. If the mass of the original compound is known, and the masses of carbon and hydrogen have been found, explain how the mass of oxygen in the compound could be found.
2. Create a flow chart to show the steps that are followed in order to determine the empirical formula of the organic compound using elemental microanalysis of the compound.
3. Determine the empirical formula of warfarin.

MOLECULAR FORMULA

Molecular compounds have a **molecular formula** in addition to an empirical formula. The molecular formula gives the actual number of atoms of each element present in a molecule, rather than the simplest whole-number ratio.

The molecular formula can be the same as or different from the empirical formula.

The empirical and molecular formulas of some common molecular compounds are shown in Table 7.4.2.

TABLE 7.4.2 Empirical and molecular formulas of some common molecular compounds

Molecule	Molecular formula	Empirical formula
water	H_2O	H_2O
ethane	C_2H_6	CH_3
carbon dioxide	CO_2	CO_2
glucose	$C_6H_{12}O_6$	CH_2O

Ionic compounds do not have molecular formulas because they do not exist as molecules.

Determining molecular formula

A molecular formula can be determined from the empirical formula of a compound if the molar mass of the compound is also known. Remember that the molar mass of a compound is the mass of one mole of the compound. For example, the molar mass of water (H_2O) is 18 g mol^{-1}.

The molecular formula of a molecule is always a whole number multiple of the empirical formula. The number of the multiple is determined by the following formula.

$$\text{Number of empirical formula units in a molecule} = \frac{\text{molar mass of the compound}}{\text{molar mass of one empirical formula unit}}$$

The general steps in the determination of a molecular formula are shown in Figure 7.4.4.

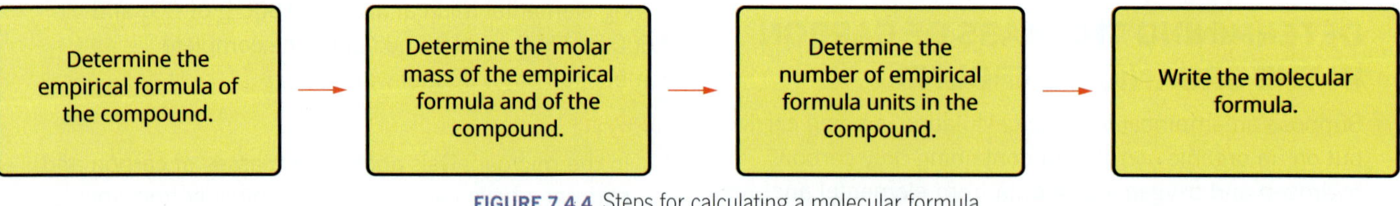

FIGURE 7.4.4 Steps for calculating a molecular formula

Worked example 7.4.3

DETERMINING MOLECULAR FORMULA

A compound has the empirical formula CH. The molar mass of this compound is 78 g mol^{-1}. What is the molecular formula of the compound?

Thinking	Working
Calculate the molar mass of one unit of the empirical formula.	Molar mass of a CH unit = 12.0 + 1.0 = 13.0 g mol^{-1}
Determine the number of empirical formula units in the molecular formula.	Number of CH units = $\frac{78}{13.0}$ = 6
Determine the molecular formula of the compound.	Molecular formula = 6 × CH = C_6H_6

Worked example: Try yourself 7.4.3

DETERMINING MOLECULAR FORMULA

A compound has the empirical formula C_2H_5. The molar mass of this compound was determined to be 58 g mol^{-1}. What is the molecular formula of the compound?

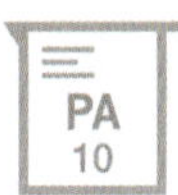

CHEMFILE

Caffeine

Many of the chemicals you use in daily life are large molecules with molecular formulas that are different from their empirical formulas. Caffeine is a molecule found in tea, coffee, cola drinks and chocolate. A naturally occurring stimulant, caffeine increases the activity of your brain and nervous system. Some students rely on it to stay awake whilst studying! Caffeine is found naturally in the leaves and seeds of many plant species, of which coffee beans, cocoa beans and tea leaves are the most well known. The structure of the caffeine molecule is shown below. The molecular formula for caffeine is $C_8H_{10}N_4O_2$ and its empirical formula is $C_4H_5N_2O$.

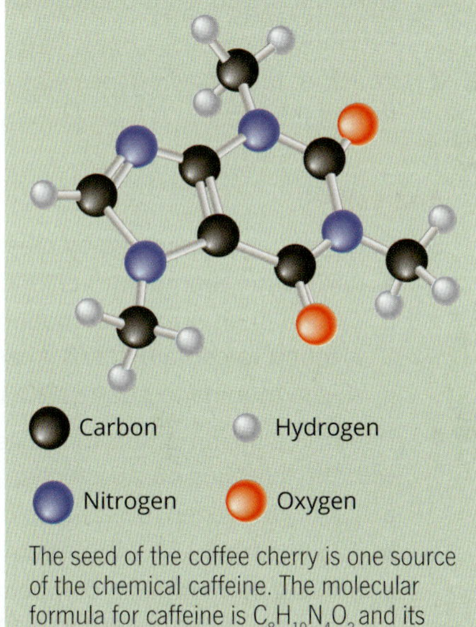

The seed of the coffee cherry is one source of the chemical caffeine. The molecular formula for caffeine is $C_8H_{10}N_4O_2$ and its empirical formula is $C_4H_5N_2O$.

7.4 Review

OA ✓✓

SUMMARY

- The percentage composition by mass of an element in a compound can be calculated from the mass of the element in one mole of the compound and the molar mass of the compound.
- The empirical formula of a compound gives the simplest whole-number ratio of atoms or ions in the compound. The empirical formula of an ionic compound can be determined from its percentage composition by mass.
- Ionic compounds only have an empirical formula.
- Molecular compounds have a molecular formula that gives the actual number of atoms of each element in the molecule. It may be the same as, or different to, the empirical formula.

KEY QUESTIONS

Knowledge and understanding

1 Describe the concept of percentage composition of a compound.

2 Which of the following formulas is not an empirical formula? Explain.

H_2SO_4 H_2O $C_6H_{12}O_6$

Analysis

3 Calculate the percentage by mass of:

a nitrogen in ammonium chloride (NH_4Cl)

b carbon in an organic compound in which 2.450 g is found to contain 1.278 g carbon

c silicon in mineral quartz (SiO_2)

4 Determine the empirical formulas of the compounds with the following compositions:

a 2.74% hydrogen, 97.26% chlorine

b 42.9% carbon, 57.1% oxygen

c 6.966 g of an organic compound that is found to contain 4.104 g of carbon, 0.682 g of hydrogen and the rest is oxygen.

d 3.2 g of a hydrocarbon that contains 2.4 g of carbon

5 A hydrocarbon contains 85.7% carbon with the remainder hydrogen. Determine the empirical formula of the hydrocarbon.

6 Determine the molecular formula of the following compounds.

	Empirical formula	Relative molecular mass
a	CH	78
b	HO	34
c	CH_2O	90
d	NO_2	46
e	CH_2	154

7 A sample of the carbohydrate glucose contains 1.8 g carbon, 0.3 g hydrogen and 2.4 g oxygen.

a Calculate the empirical formula of the compound.

b Deduce its molecular formula given that its relative molecular mass is 180.

Chapter review

KEY TERMS

Avogadro's constant
carbon-12
empirical formula
isotope
mass spectrometer
mass spectrum
molar mass
mole
molecular formula
molecule
percentage composition
relative atomic mass
relative isotopic abundance
relative isotopic mass
relative mass

REVIEW QUESTIONS

Knowledge and understanding

1 Select the correct definition for relative atomic mass from the options below.

A The relative atomic mass is the mass of 1 mol of a compound.

B The relative atomic mass is the mass of one isotope of an atom relative to the mass of a ^{12}C atom taken as 12 units exactly.

C The relative atomic mass is the weighted average of the atomic masses of isotopes of the element on the ^{12}C scale.

D The relative atomic mass is the percentage abundance of an isotope in the natural environment.

2 How does relative atomic mass different to relative isotopic mass? Use carbon as an example.

3 What is the unit used for molar mass (M) of a compound?

4 Using suitable examples, clearly distinguish between:

a relative isotopic mass

b relative atomic mass

c molar mass.

5 Explain the difference between empirical formula and molecular formula.

Application and analysis

6 A sample of palladium is placed in a mass spectrometer, and a mass spectrum obtained. The relative isotopic masses of particles present and their corresponding percentage abundances are given in the table below.

Relative isotopic mass	Abundance (%)
101.9049	0.9600
103.9036	10.97
104.9046	22.23
105.9032	27.33
107.9039	26.71
109.9044	11.80

Calculate the relative atomic mass of palladium.

7 The table below gives isotopic composition data for argon and potassium.

Element	Atomic number	Relative isotopic mass	Relative abundance (%)
argon	18	35.978	0.307
		37.974	0.060
		39.974	99.633
potassium	19	38.975	93.3
		39.976	0.011
		40.974	6.69

a Determine the relative atomic masses of argon and potassium.

b Explain how the relative atomic mass of argon can be greater than that of potassium, even though potassium has a larger atomic number.

8 The mass spectrum of chromium is shown in the graph below.

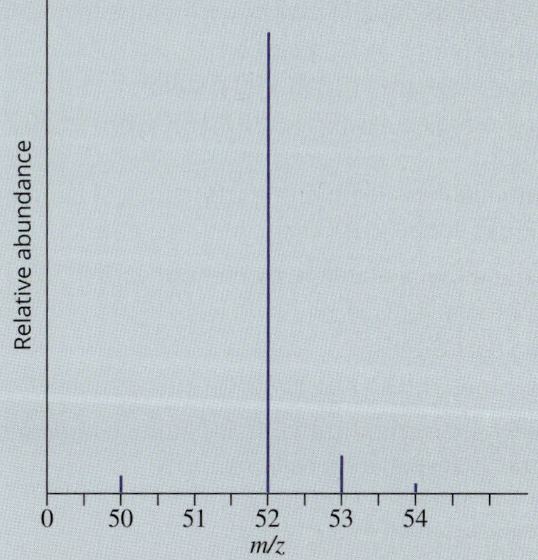

a Measure the peak heights to calculate the percentage abundance of each chromium isotope.

b Use the percentage abundances calculated in part **a** to determine the relative atomic mass for chromium. The mass number is a good approximation to the relative isotopic mass.

9 For each of the following numbers of molecules, calculate the amount of substance, in mol.
a 4.50×10^{23} molecules of water (H_2O)
b 9.00×10^{24} molecules of methane (CH_4)
c 2.3×10^{28} molecules of chlorine (Cl_2)
d 1.0 molecule of sucrose ($C_{12}H_{22}O_{11}$)

10 Which one of the following quantities does not contain one mole of oxygen atoms?
A 0.50 mol of O_2
B 1.0 mol of CO_2
C 1.0 mol of CO
D 1.0 mol of H_2O

11 For each of the following amounts of molecular substances calculate the:
i number of molecules
ii total number of atoms.
a 1.45 mol of ammonia (NH_3)
b 0.576 mol of hydrogen sulfide (H_2S)
c 0.0153 mol of hydrogen nitrate (HNO_3)
d 2.5 mol of sucrose ($C_{12}H_{22}O_{11}$)

12 What is the molar mass (*M*) of each of the following?
a Sodium (Na)
b Nitric acid (HNO_3)
c Magnesium nitrate ($Mg(NO_3)_2$)
d Hydrated iron(III) chloride ($FeCl_3 \cdot 6H_2O$)

13 What is the mass of each of the following?
a 0.080 mol of ethane (C_2H_6)
b 0.45 mol of glucose ($C_6H_{12}O_6$)
c 2.8×10^{-3} mol of urea (NH_2CONH_2)
d 5.35 mol of copper atoms (Cu)

14 What is the amount, in mol, of each of the following?
a Carbon atoms in 12 g carbon
b Sulfur molecules (S_8) in 100 g sulfur
c Aspirin molecules ($C_6H_4(OCOCH_3)COOH$) in 500 mg aspirin
d Aluminium oxide (Al_2O_3) in 2.8 tonnes of aluminium oxide (1 tonne = 1000 kg)

15 Determine the percentage by mass of carbon in:
a naphthalene ($C_{10}H_8$)
b urea (NH_2CONH_2)
c aspirin ($C_6H_4(OCOCH_3)COOH$)

16 Determine the empirical formulas of the compounds with the following compositions.
a 42.9% carbon, 57.1% oxygen
b 27.2% carbon, 72.8% oxygen
c 54.5% carbon, 36.4% oxygen, 9.1% hydrogen
d 9.4 g carbon, 4.7 g chlorine, 0.65 g hydrogen

17 A compound used as a solvent for dyes has the following composition by mass: 32.0% carbon, 6.7% hydrogen, 18.7% nitrogen and 42.6% oxygen. Find the empirical formula of the compound.

18 Caffeine contains 49.48% carbon, 5.15% hydrogen, 28.87% nitrogen and the rest oxygen.
a Determine the empirical formula of caffeine.
b If 0.20 mol of caffeine has a mass of 38.8 g, what is the molar mass of a caffeine molecule?
c Show working to justify that the molecular formula of caffeine is $C_8H_{10}N_4O_2$.
d How many moles of caffeine molecules are in 1.00 g caffeine?

19 The relative atomic mass of europium is 151.96. The relative isotopic masses of its two isotopes are 150.92 and 152.92. Calculate the relative abundances of the isotopes in naturally occurring europium.

20 a If 6.0×10^{23} atoms of calcium have a mass of 40.1 g, what is the mass of one calcium atom?
b If 1 mol of water molecules has a mass of 18 g, what is the mass of one water molecule?
c What is the mass of one molecule of carbon dioxide?

21 What mass of iron (Fe) would contain as many iron atoms as there are molecules in 20.0 g water (H_2O)?

22 For each of the following ionic substances calculate the amount of:
i substance, in moles
ii each ion, in moles.
a 5.85 g of NaCl
b 45.0 g of $CaCl_2$
c 1.68 g of $Fe_2(SO_4)_3$

23 a If 0.50 mol of a substance has a mass of 72 g, what is the mass of 1.0 mol of the substance?
b If 6.0×10^{22} molecules of a substance have a mass of 10 g, what is the molar mass of the substance?

24 Which of the following metal samples has the greatest mass?
A 150 g copper
B 2.0 mol of iron atoms
C 1.2×10^{24} atoms of silver
D 1.5 mol of sodium atoms

25 A new antibiotic has been isolated and only 2.0 mg is available. The molar mass is found to be 12.5 kg mol^{-1}.
a Express the molar mass in g mol^{-1}.
b Calculate the amount of antibiotic, in mol.
c How many molecules of antibiotic have been isolated?

26 Find the relative atomic mass of nickel if 3.370 g nickel was obtained by reduction of 4.286 g of the oxide (NiO).

27 A clear liquid extracted from fermented lemons was found to consist of carbon, hydrogen and oxygen. Analysis showed it to be 52.2% carbon and 34.8% oxygen.

a Find the empirical formula of the substance.

b If 2.17 mol of the compound has a mass of 100 g, find the molecular formula of the compound.

28 The empirical formula of a metal oxide can be found by experimentation as shown in the figure on the right. The mass of the oxygen that reacts with the mass of the metal must be determined. Steps A–F form the experimental method.

A Ignite a burner and heat the metal.

B Allow the crucible to cool, then weigh it.

C Continue the reaction until no further change occurs.

D Clean a piece of metal with emery paper to remove any oxide layer.

E Place the metal in a clean, weighed crucible and cover with a lid.

F Weigh the metal and record its mass.

a Place the steps in the correct order by letter.

b Wan and Eric collected the following data:
Mass of the metal = 0.542 g
Mass of the empty crucible = 20.310 g
Mass of the crucible and metal oxide = 21.068 g

They found from this data that the metal oxide had a 1:1 formula, i.e. *MO*, where *M* = metal. Copy and complete the table below, using the data given.

	Metal	Oxygen
Mass (g)		
Relative atomic mass		16.0
Moles		
Ratio		

c Deduce the metal that was used in the experiment.

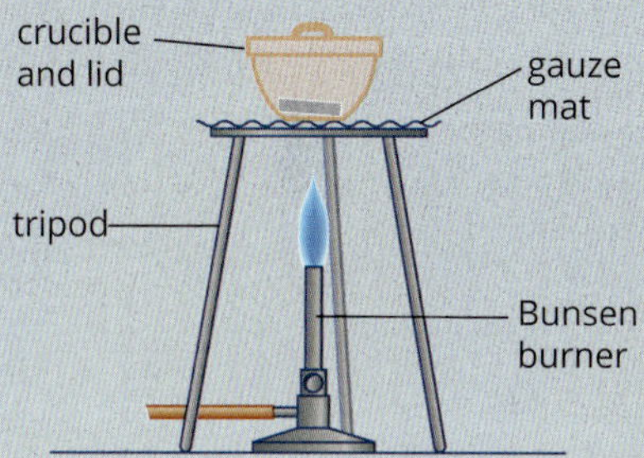

OA ✓✓

CHAPTER

08 Organic compounds

At the end of this chapter, you will appreciate the significance of the compounds that are formed between carbon, hydrogen and other elements. Carbon is only the eleventh most abundant element in the universe, yet it forms more compounds than all other elements except hydrogen. Carbon is present in all living things and many non-living things.

The prevalence of carbon in living things is why the study of the compounds of carbon and hydrogen is known as organic chemistry. You make use of these compounds in so many aspects of your life: from the plastic case of your telephone to the pasta that you may eat for dinner tonight.

In this chapter, you will start to explore some of the many families of carbon compounds.

Key knowledge

- the grouping of hydrocarbon compounds into families (alkanes, haloalkanes, alkenes, alcohols, carboxylic acids) based upon similarities in their physical and chemical properties, including general formulas and general uses based on their properties **8.2, 8.3, 8.4**
- representations of organic compounds (structural formulas, semi-structural formulas) and naming according to the International Union of Pure and Applied Chemistry (IUPAC) systematic nomenclature (limited to non-cyclic compounds up to C8, and structural isomers up to C5) **8.2, 8.3, 8.4**
- plant-based biomass as an alternative renewable source of organic chemicals (for example, solvents, pharmaceuticals, adhesives, dyes and paints) traditionally derived from fossil fuels **8.1**
- materials and products used in everyday life that are made from organic compounds (for example, synthetic fabrics, foods, natural medicines, pesticides, cosmetics, organic solvents, car parts, artificial hearts), the benefits of those products for society, and the health and/or environmental hazards they pose. **8.1**

8.1 Organic materials

Over 90% of known compounds contain carbon. The ability of carbon to form this vast range of compounds has led to an equally vast range of living things composed of carbon-based molecules. Because the major source of carbon compounds is living or once-living material, it was originally thought that only living things were able to produce carbon-based compounds, hence the name 'organic'.

Organic chemistry studies the chemistry of carbon-containing compounds. However, some carbon compounds, such as carbon monoxide (CO), carbon dioxide (CO_2), carbonic acid (H_2CO_3), and carbonate compounds such as calcium carbonate ($CaCO_3$) and sodium carbonate (Na_2CO_3), are considered to be inorganic compounds. This historical distinction goes back to Jöns Joseph Berzelius (1779–1848), who defined inorganic compounds as originating from non-living systems.

Organic compounds are carbon-based compounds, excluding such compounds as carbon monoxide, carbon dioxide and metal carbonate compounds.

Carbon-based molecules, known as **organic compounds**, are all around you and inside you. These compounds typically also have hydrogen and oxygen in them. In addition, they may contain nitrogen, sulfur, phosphorus and halogens (fluorine, chlorine, bromine, iodine). Caffeine, petrol, plastics and biological molecules, such as proteins, fats and carbohydrates, are all carbon-based compounds. Many non-biological organic compounds are produced from crude oil. Crude oil is produced by the effects of heat and pressure on dead animals, plants and microorganisms trapped in the Earth's crust, buried beneath sediment formed over millions of years. This section compares crude oil as a source of organic compounds to the alternative, renewable source of these compounds: plant-based biomass.

CHEMFILE

We are made of stardust

Our Earth is a planet that was formed when a giant star more than 50 times the size of our Sun exploded. The elements, including carbon, that had been formed in that star by a series of nuclear fusion reactions were flung out in a supernova explosion into the Universe. These elements were taken up into new clouds of gas and dust, such as the Crab Nebula in the figure below, and the process continued. Eventually planets, such as Earth, were formed during these supernova explosions. Over the course of billions of years and multiple star lifetimes, the elements, such as carbon, oxygen and nitrogen that make up our bodies, were formed. However, astronomers believe that it's also possible that some of the hydrogen in our bodies originated directly from the Big Bang.

The Crab Nebula, as seen from the Hubble Space telescope, is a cloud of gas and dust—the remains of a supernova which was observed from Earth in 1054.

ORGANIC COMPOUNDS

Organic chemistry is central to understanding the chemistry (and biology) of living systems, as well as so many compounds which are of huge importance to society, such as fuels, **polymers** (plastics), paints, medicines and cosmetics.

The amount of carbon on Earth is essentially fixed, with most of the carbon atoms on Earth having been here for billions of years. It is the location of these carbon atoms that changes over time. Whether they are currently part of a sunflower seed, a molecule of carbon dioxide, a human being or a plastic chair, you can be certain that the carbon atoms have been in a different compound previously.

SOURCES OF ORGANIC COMPOUNDS

When prehistoric marine microorganisms, such as bacteria and plankton, died and were buried by sands millions of years ago, these organisms accumulated as organic sediment and gradually became part of the Earth's crust. Over millions of years, this organic material was affected by high temperatures and pressures, causing the oils and fats to be converted into **hydrocarbons** (compounds of hydrogen and carbon) and other organic compounds. This mixture of hydrocarbons is called **crude oil**.

Crude oil

Crude oil is not used in its raw state. It is transported from oil fields to oil refineries where it undergoes **fractional distillation**. During fractional distillation, the various hydrocarbons in crude oil are separated, according to their **boiling points** (the temperature at which the liquid becomes a gas). The longer the hydrocarbon chain, the greater its intermolecular forces, and so the higher its boiling point will be. The components of crude oil that are obtained by fractional distillation are used for a wide range of purposes, but presently over 90% of them are used for fuels. For example, petrol is obtained from the gasoline (C_5 to C_{12}) fraction, while diesel may be obtained from either the kerosene (C_{10} to C_{18}) or diesel/gas oil (C_{12} to C_{20}) fractions.

Crude oil is a fossil fuel, containing carbon that has been locked away in the Earth's crust for millions of years. It is a **non-renewable** source for organic materials and products, as it is being used up at a far greater rate than it is being made. In comparison, a **renewable** source can be constantly replenished.

While many products can be made from the components of crude oil, the finite availability of this resource means that it will eventually be used up if it is our only source of hydrocarbon raw materials. There are many ways to ensure that this does not happen. These include:

- recycling materials that are made from hydrocarbons, such as polymers and making them into new products
- replacing petrol and diesel-fuelled cars with electric cars
- reusing materials derived from crude oil, rather than disposing of them after one use
- finding renewable replacement materials for those derived from crude oil.

These renewable replacement materials may be sourced from plants.

Materials and products made from crude oil are not sustainable, as they rely on non-renewable source materials.

Plant-sourced biomass

Products that have been sourced from plants and animals have always served as an inspiration for the development of new pharmaceuticals, pesticides and dyes. Acetylsalicylic acid, otherwise known as aspirin, is closely related to the active ingredient in willow bark, obtained from willow trees, which has been used for pain relief for more than 2000 years. The term **plant-sourced biomass** refers to a carbon-based material that has come from plants.

In recent times, it has become apparent that plants can be an alternative renewable source of organic chemicals that can be transformed into a range of products. They are able to fix carbon dioxide from the atmosphere into glucose via a process called **photosynthesis** (see the equation below), making a plant source of carbon a renewable source.

$$6CO_2(g) + 6H_2O(l) \xrightarrow{\text{UV light}} C_6H_{12}O_6(aq) + 6O_2(g)$$

Making polyethene from sugar

Ethanol, C_2H_5OH, is a compound that is used as a starting material for many other organic compounds, and in particular can be used to make the polymer, polyethene. A polymer is commonly known as a plastic, and polyethene is used for making a range of common products such as clingwrap, toys such as LEGO®, and food containers. Polyethene derived from plant sources is called bio-polyethene. The first step in making bio-polyethene is to grow a suitable crop that is rich in sugar. This is often sugar cane, as seen in Figure 8.1.1, but the process can start with other crops.

FIGURE 8.1.1 Harvesting sugar cane in Queensland. Sugar cane can be a source of the raw materials for the production of ethanol.

CHEMFILE

Fossil fuels

Coal, oil and natural gas are known as fossil fuels because they come from the remains of plants and animals which died millions of years ago. Carbon, in the form of carbon dioxide from the atmosphere, was stored by plants, and the animals ate plants and other animals. As a result, fossil fuels are made up of the same elements as those found in plants and animals: carbon, hydrogen, oxygen and sulfur.

Crude oil has a low density, which means it could migrate upwards through the crust, where it often became trapped beneath impervious (unable to be passed through) rock, as in the diagram below. The accumulation of oil and gas under the rock creates an oil field from which we mine these valuable, but limited, resources.

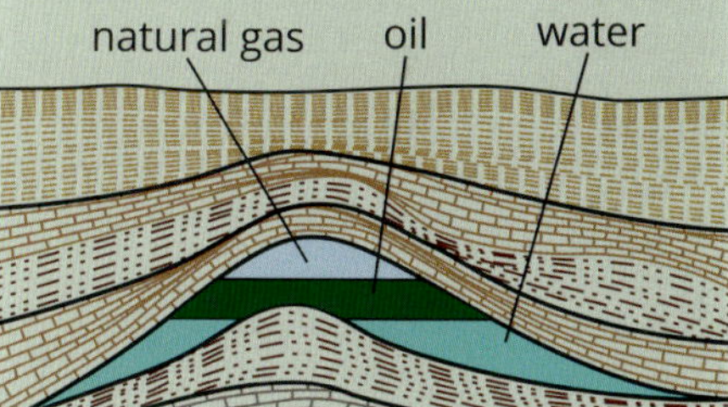

Typical structure of the impervious rock that traps oil and natural gas underground and creates an oil field

Plant-sourced biomass provides a supply of organic compounds that are renewable.

Once the plant is harvested, the sugar is extracted from the cane and then fermented to make a solution of ethanol in water, similar to the way in which alcoholic beverages are made. Ethanol made in this way is called **bioethanol** and this reaction can be written as:

$$C_6H_{12}O_6(aq) \rightarrow 2C_2H_5OH(aq) + 2CO_2(g)$$

These steps in the process indirectly convert CO_2 from the atmosphere into ethanol, and it is very important to note that more carbon dioxide is absorbed through photosynthesis than emitted through fermentation. This is in contrast to the production of ethene from crude oil, which emits very large amounts of CO_2.

The ethanol is removed from the solution by distilling it in an industrial refinery like the one shown in Figure 8.1.2.

Bioethanol is further processed to make bioethene, which is identical to ethene derived from crude oil and can be used straight away in manufacturing processes that use ethene as a reactant. Many companies are using **bio-derived** polyethene rather than polyethene derived from crude oil to make their products more environmentally friendly. For example, about 150 elements (particular shaped pieces) used in LEGO sets are made of bio-polyethene. This includes all leaves and plants in the LEGO® Ideas Treehouse you can see in Figure 8.1.3.

FIGURE 8.1.2 Ethanol refinery in Manildra at Nowra, New South Wales

FIGURE 8.1.3 All 185 leaves and plants in the LEGO Ideas Treehouse were made from polyethene derived from Brazilian sugar cane.

CASE STUDY

Everyday products from organic compounds

While sitting at your desk, contemplating a happy hour or so of chemistry study, it is easy to notice many items that are made from organic chemicals. Most of these are polymers, such as those used to make the keyboard and mouse of your computer, the calculator sitting ready to use or the cover on your mobile phone, which of course is turned off while you study.

The keyboard, mouse and calculator body are likely to have been made from ABS (Acrylonitrile Butadiene Styrene), a polymer that can be softened by heating, and shaped by injection moulding into many useful products. ABS is also used to make the surroundings of light switches and electrical wall sockets, some interior parts of cars, the LEGO pieces that are not yet sustainably produced, and even medical equipment.

The three monomers used for making the ABS polymer, shown in Figure 8.1.4, are all derived from crude oil. These non-renewable materials are strong and durable and resistant to reaction with many chemicals, making them excellent for the uses listed above. In particular, their use in medical applications requires them to be unreactive and therefore safe in the medical context. Unfortunately, as is the case with many polymers, their durability and lack of chemical reactivity makes these materials an environmental nightmare, if these products are disposed of in landfill. Like many modern polymers, ABS is 100% recyclable, in Australia. However, ABS is classified as recycling number 7 for household collection, which means that it is grouped together with many other polymers and requires special sorting to actually be recycled. It is clear that the possibility of effective recycling is on the horizon, but the practicalities are lagging behind.

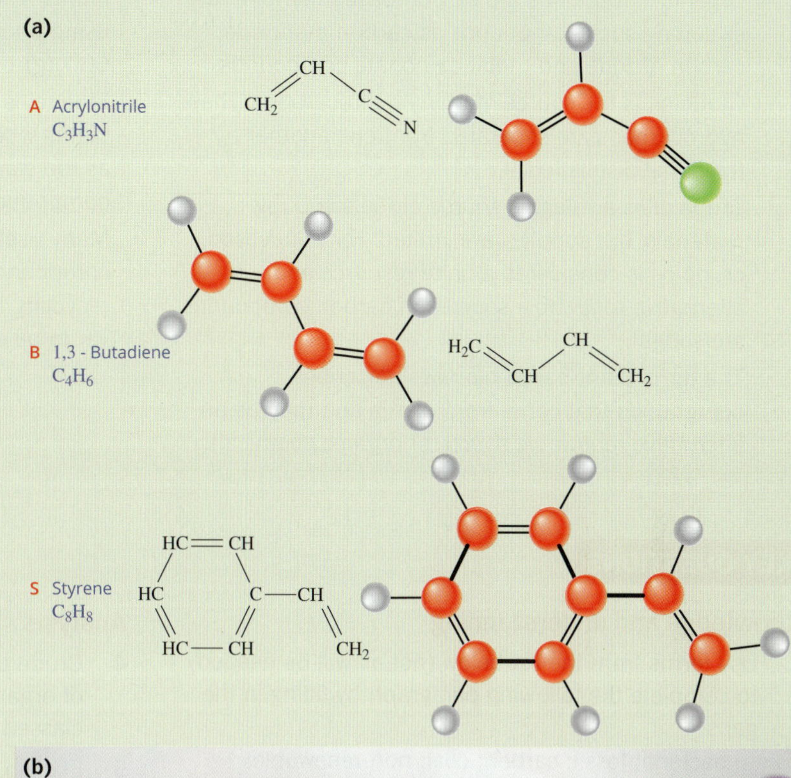

FIGURE 8.1.4 (a) The three monomers used for making ABS polymers, acrylonitrile, 1,3-butadiene and styrene are all derived from the hydrocarbons in crude oil. (b) ABS filaments for use in 3D printing.

8.1 Review

OA ✓✓

SUMMARY

- Organic compounds are carbon-based compounds, excluding such compounds as carbon monoxide, carbon dioxide and metal carbonate compounds.
- Crude oil is the source of many hydrocarbons. These hydrocarbons are separated from one another by fractional distillation.
- Crude oil is a valuable source of fuels and raw materials, but supplies are limited. For this reason, crude oil is considered a finite resource and alternative renewable sources of carbon compounds are sought.
- The term 'plant-based biomass' describes compounds that come from plants and may be an alternative renewable source of organic chemicals.
- Plants fix carbon dioxide in the form of glucose using photosynthesis:
  $6CO_2(g) + 6H_2O(l) \xrightarrow{\text{UV light}} C_6H_{12}O_6(aq) + 6O_2(g)$
- Ethanol is a compound which can be derived from plants when glucose is fermented. It can be used as the starting material for making many organic compounds.
- Many useful everyday items are made from organic compounds, but the difficulties encountered in recycling them once they are no longer needed are yet to be effectively overcome.

KEY QUESTIONS

Knowledge and understanding

1 Use terms from the list below (not all will be needed) to complete the following paragraph by filling in the gaps:

bacteria; fossil; carbon; coal; non-renewable; crude oil; humans; renewable

Compounds such as petrol, polymers and cosmetics are _________-based compounds. Many of these compounds are currently produced from _________ ______, which is a mixture made up of the remains of marine microorganisms, such as ________ and plankton that died millions of years ago. The great age of these deposits explains why petrol is called a ________ fuel. Crude oil is a __________ resource because no more carbon is being added to the environment.

2 Explain what is meant by the term 'organic compounds'.

3 Name and describe the process by which carbon dioxide is changed by plants into glucose.

Analysis

4 Crude oil is used as the starting material for a range of organic chemicals. Explain why these chemicals are classed as non-renewable.

5 Explain why polyethene that has been made from ethanol derived from glucose can be described as renewable.

6 Consider each of the following proposals for change and explain why the renewability of the material is improved by the change.

a Materials made from hydrocarbons, such as polymers, should be recycled or reused rather than throwing them away.

b Governments should legislate for petrol and diesel-fuelled cars to be replaced by electric cars.

c Chemical industries should find replacement materials for organic compounds derived from crude oil.

8.2 Hydrocarbons

Carbon forms more compounds than any other element. This is because:

- carbon has four valence electrons, so carbon can potentially form covalent bonds with four different atoms
- carbon atoms can form strong covalent bonds with other carbon atoms
- the covalent bonds formed can be a combination of single, double or triple bonds.

Because of this, carbon can bond to itself to form molecules of varied length and shape. The different structures have different properties and applications. Compounds that are made up only of carbon and hydrogen are known as hydrocarbons.

NAMING HYDROCARBONS

The naming of hydrocarbons follows set conventions, which are determined by the International Union of Pure and Applied Chemists (IUPAC). Under the IUPAC system, the name of the compound provides details of its structure. The **stem name**, or parent name, indicates the number of carbons in that compound. For example, the stem name of propane (prop-) indicates there are three carbon atoms in the molecule. The stem names used for molecules with up to ten carbon atoms are listed in Table 8.2.1.

TABLE 8.2.1 Stem names used for molecules with up to ten carbon atoms

Stem (parent) name	Number of carbon atoms
meth-	1
eth-	2
prop-	3
but-	4
pent-	5
hex-	6
hept-	7
oct-	8
non-	9
dec-	10

ALKANES

Hydrocarbons can be classified into several groups or series. Methane is the first of a series of compounds known as the **alkanes**. Alkanes are hydrocarbons containing molecules in which all the carbon–carbon bonds are single covalent bonds. Molecules such as this are said to be **saturated**.

Each member of the alkane series differs from the previous member by a $-CH_2-$ unit. A series of molecules in which each member differs by $-CH_2-$ from the previous member is known as a **homologous series**.

Compounds that are members of the same homologous series have:

- a similar structure
- a pattern to their physical properties
- similar chemical properties
- the same general formula.

Alkanes are named by adding -ane after the stem name. For example, an alkane that contains five carbon atoms is called pentane. The **molecular formula** of pentane is C_5H_{12}. Alkanes have the **general formula** C_nH_{2n+2}, where n stands for the number of carbon atoms. If an alkane molecule has 12 carbon atoms, the number of hydrogen atoms is $2n + 2 = 2 \times 12 + 2 = 26$. The molecular formula is $C_{12}H_{26}$.

The name of an alkane ends in -ane and alkanes have the general formula C_nH_{2n+2}.

Writing formulas of alkanes

Sometimes we want to show more information about a molecule than just its overall composition. You can use a variety of ways to write the formulas of carbon compounds. Table 8.2.2 shows different ways of representing ethane and butane.

TABLE 8.2.2 Different ways of representing alkanes

Alkane	Molecular formula	Electron dot diagram (Lewis structure)	Semi-structural formula	Structural formula
ethane	C_2H_6	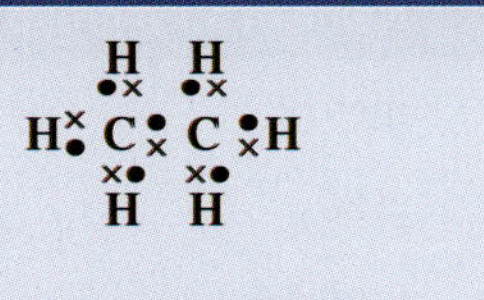 [electron dot diagram]	CH_3CH_3	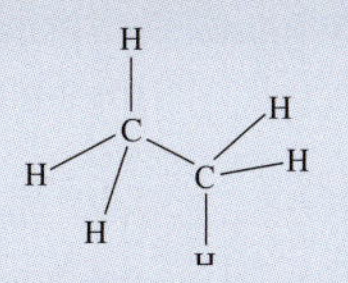 [structural formula]
butane	C_4H_{10}	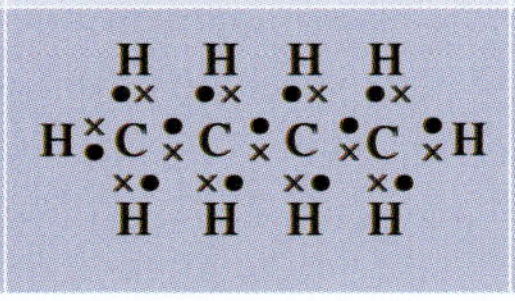 [electron dot diagram]	$CH_3CH_2CH_2CH_3$ or $CH_3(CH_2)_2CH_3$	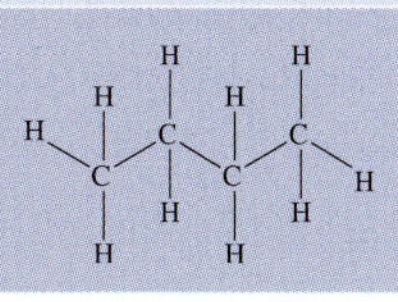 [structural formula]

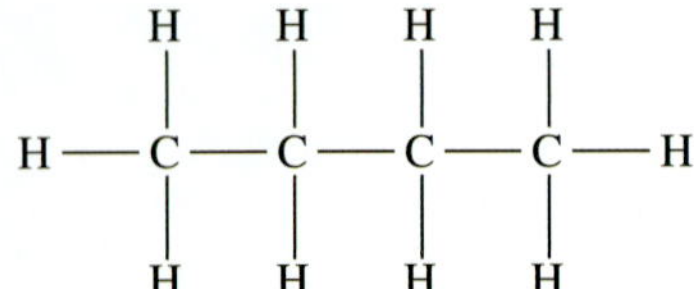

FIGURE 8.2.1 Molecules such as butane are often drawn in two dimensions for simplicity.

Structural formulas show the arrangement of the atoms in a molecule and they show all the bonds in that molecule. **Semi-structural formulas** are also known as **condensed structural formulas** and show the arrangement of the carbon atoms and the other atoms that are bonded to each carbon atom, but do not show the bonds. There can be more than one correct semi-structural formula for a molecule; for example, butane can be written as $CH_3CH_2CH_2CH_3$ or $CH_3(CH_2)_2CH_3$.

Representing three-dimensional molecules on a two-dimensional page or computer screen is a challenge for chemists. For example, each carbon in an alkane has four bonds in a tetrahedral arrangement. When the shape of the molecule is not important, it is often drawn in a simplified two-dimensional format, as shown in Figure 8.2.1.

Naming structural isomers of alkanes

There is only one molecule that can be formed with the molecular formula of methane (CH_4). This is also the case for ethane (C_2H_6) and propane (C_3H_8). However, alkanes that have four or more carbon atoms have more than one possible structure.

Figure 8.2.2 shows two different molecules that have the molecular formula C_4H_{10}. Molecules that have the same molecular formula but have different arrangements of atoms are said to be **structural isomers** of each other. The more atoms in the molecule, the more possible isomers there are.

Figure 8.2.2a shows the straight-chain isomer of butane, so named because the four carbon atoms are bonded in a continuous chain. Figure 8.2.2b is called the branched isomer because you could consider it as a straight-chain alkane with a $-CH_3$ side chain, or branch, attached. A $-CH_3$ side chain is called a methyl group, as its structure is similar to that of methane with one less hydrogen. A $-CH_2CH_3$ side chain is called an ethyl group.

Therefore, Figure 8.2.2b is named 2-methylpropane, as it can be regarded as a propane molecule with a methyl group attached to the second carbon in the chain. Methyl and ethyl groups are examples of **alkyl groups** or **alkyl side chains**. Alkyl groups are named after the alkane from which they are derived, with a -yl ending. They have one less hydrogen atom than the corresponding alkane of the same name, so the general formula of an alkyl group is $-C_nH_{2n+1}$.

Alkyl groups are named after the alkane from which they are derived with a -yl ending and general formula C_nH_{2n+1}.

(a) $CH_3(CH_2)_2CH_3$

(b) $(CH_3)_3CH$ or $CH_3CH(CH_3)_2$

FIGURE 8.2.2 There are two structural isomers that have the molecular formula C_4H_{10}: (a) butane and (b) 2-methylpropane.

Table 8.2.3 lists the names of the alkyl groups with 1–3 carbon atoms.

TABLE 8.2.3 Names of the alkyl groups with 1–3 carbons

Alkyl group	Corresponding alkane	Name of alkyl group
$-CH_3$	methane	methyl
$-CH_2CH_3$	ethane	ethyl
$-CH_2CH_2CH_3$	propane	propyl

When writing the semi-structural formula of a branched alkane, the alkyl groups that make up the branches are written in brackets (Figure 8.2.3).

Structural isomers are different compounds with different physical and chemical properties. Table 8.2.4 shows the isomers of hexane and their respective **melting points** and boiling points.

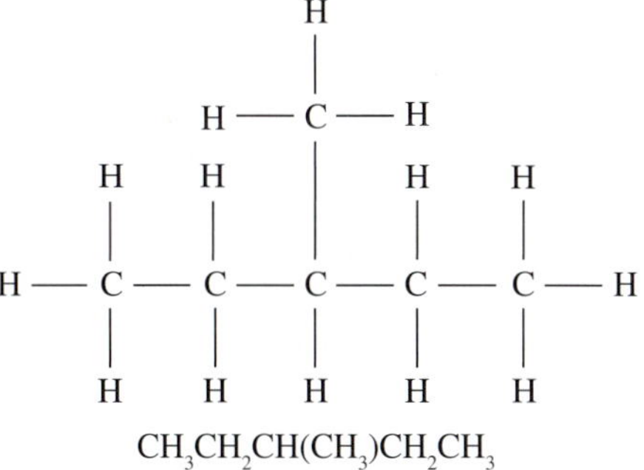

$CH_3CH_2CH(CH_3)CH_2CH_3$

FIGURE 8.2.3 The structural and semi-structural formula of a branched alkane

TABLE 8.2.4 Melting and boiling points of the isomers of hexane

Isomer	Melting point (°C)	Boiling point (°C)
$CH_3-CH_2-CH_2-CH_2-CH_2-CH_3$ hexane	-95.3	68.7
CH_3 | $CH_3-CH_2-CH-CH_2-CH_3$ 3-methylpentane	-118.0	63.3
CH_3 | $CH_3-CH_2-CH_2-CH-CH_3$ 2-methylpentane	-153.7	60.3
CH_3 CH_3 | | $CH_3-CH-CH-CH_3$ 2,3-dimethylbutane	-128.6	58
CH_3 | $CH_3-CH_2-C-CH_3$ | CH_3 2,2-dimethylbutane	-99.8	49.7

Rules for systematic naming of alkanes

Under the IUPAC system, the following rules apply when naming alkanes.

1 Identify the longest unbranched carbon chain.

2 Number the carbon atoms in the chain from the end of the chain that will give the smallest numbers to branching side chains.

3 Name the alkyl side chains according to the alkane from which they are derived.

4 Place the number and position of each of the alkyl side chains at the beginning of the compound's name.

5 If two identical side chains are present, use 'di-' as a prefix; for three use 'tri-'.

6 If there are alkyl side chains of different lengths on the molecule, list them in alphabetical order at the start of the name, with their numbers to indicate their respective positions.

When naming an alkane, carefully check that you have identified the longest unbranched carbon chain. Sometimes the longest carbon chain is not drawn in a straight line.

FIGURE 8.2.4 The IUPAC system will be used to name this isomer of hexane.

$CH_3—CH_2—CH—CH_2—CH_3$ (with CH_3 on the CH)

3-methylpentane

$CH_3—CH_2—C—CH_2—CH_3$ (with CH_3 above and CH_3 below the C)

3,3-dimethylpentane

$CH_3—CH—CH—CH_3$ (with CH_3 on each CH)

2,3-dimethylbutane

FIGURE 8.2.5 IUPAC systematic names for three alkanes. Note that there are no spaces in the names of these compounds.

The following steps show the process of naming the isomer of hexane shown in Figure 8.2.4.

1 Identify the longest unbranched carbon chain.

2 Number the carbons, starting from the end closest to the side chain.

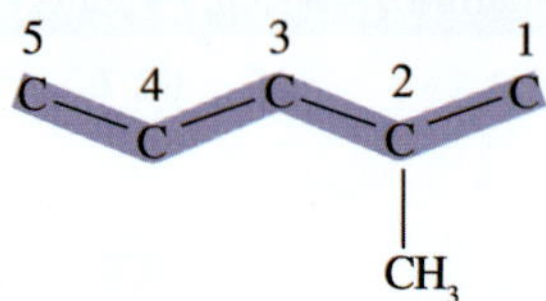

3 Name the side chains and main chain.

5 carbon atoms = pentane

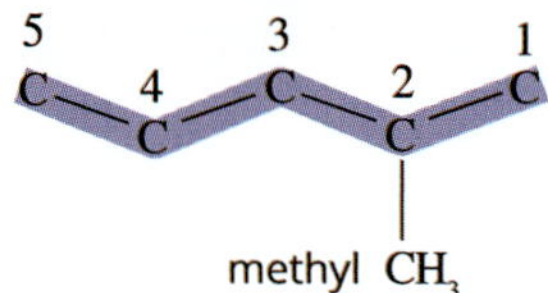

4 Combine all components to write the full name, including the position of the side chain indicated by a number: 2-methylpentane.

Figure 8.2.5 shows examples of applications of these rules. Note the use of the prefix 'di-' to indicate the presence of two methyl side chains and the numbering to indicate their position along the longest continuous carbon chain (for example, 3,3- indicates that both side chains come off the third carbon atom).

Worked example 8.2.1

IUPAC NAMING SYSTEM FOR ALKANES

Write the systematic name for the following molecule.

$CH_3—CH_2—CH_2—CH—CH_2—CH_3$ (with CH_3 on the CH)

Thinking	Working
Identify the longest carbon chain in the molecule. The stem name of the molecule is based on this longest chain.	There are 6 carbons in the longest chain. The stem name is based on hexane.
Number the carbon atoms, starting from the end closest to the side chain.	$CH_3—CH_2—CH_2—CH—CH_2—CH_3$ (with CH_3 on the CH) 6 5 4 3 2 1
Identify the side chain and its location.	The side chain is a methyl group on carbon number 3.
Combine all components.	The name of the molecule is 3-methylhexane.

Worked example: Try yourself 8.2.1

IUPAC NAMING SYSTEM FOR ALKANES

Write the systematic name for the following molecule.

$$CH_3-CH_2-\underset{\Large|}{\overset{CH_3}{}}\!\!CH-CH_2-CH_3$$

PHYSICAL AND CHEMICAL PROPERTIES OF ALKANES

The **physical properties** of an element or compound are properties that can be measured without changing the substance into another substance/s. The physical properties of organic compounds follow a pattern within each homologous series. Melting and boiling points, as well as solubility, are considered physical properties and are determined by intermolecular forces. In comparison, **chemical properties** are more closely related to the strength or polarity of the covalent bonds within the organic compound, although the arrangement of the carbon atoms in the molecule can sometimes influence chemical properties.

Physical properties of alkanes

Alkanes are non-polar molecules. As a result, they are insoluble in water and the only attractive forces between the molecules are dispersion forces. As the number of carbons and the size of the molecules increase within any homologous series, the strength of the dispersion forces increases, so more energy is needed to separate the molecules. As a result, the melting and boiling points of the alkanes increase as the number of carbons in the molecules increases.

Table 8.2.5 shows the first three members of the alkane series. Note that the bonds around each carbon atom adopt a tetrahedral shape and the boiling points increase as the molecules become larger.

Alkanes are non-polar and have relatively low boiling points.

TABLE 8.2.5 Structure, properties and some uses of the first three alkanes

Name and molecular formula	Structural formula	Properties	Uses
methane, CH_4	[tetrahedral structure: C bonded to four H]	non-polar, gas, boiling point (BP) −164°C	cooking, Bunsen burners, gas heating
ethane, C_2H_6	[structure: two C, each bonded to three H]	non-polar, gas, BP −87°C	conversion to ethene
propane, C_3H_8	[structure: three C chain with eight H]	non-polar, gas, BP −42°C	liquid petroleum gas (LPG)

CHEMFILE

Incomplete combustion and Bunsen burner flames

If a fuel is burning in an environment that is lacking in oxygen, complete combustion may not occur. In this case, the reaction is called **incomplete combustion**. The products of incomplete combustion may be carbon monoxide and water, or they may be carbon and water, depending on the amount of oxygen available. Incomplete combustion of an alkane releases less energy than complete combustion and often occurs in situations where the air supply is limited, such as in a Bunsen burner when the air holes are closed. This results in the classic yellow, or safety, flame of the Bunsen burner, as shown in the figure below. Two equations for incomplete combustion reactions of methane can be written as:

$$2CH_4(g) + 3O_2(g) \longrightarrow 2CO(g) + 4H_2O(l)$$

$$CH_4(g) + O_2(g) \longrightarrow C(s) + 2H_2O(l)$$

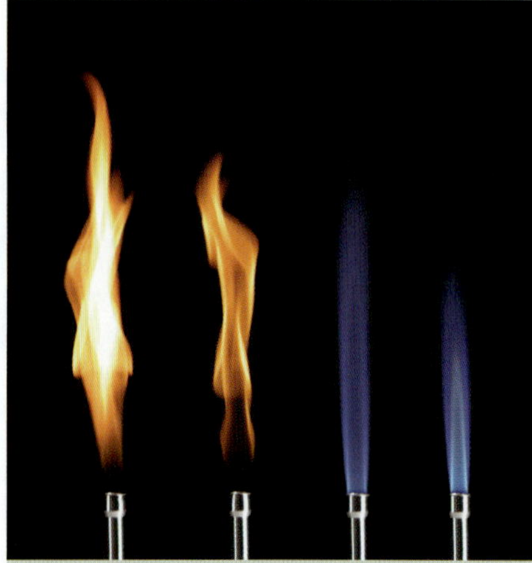

The yellow flame of a Bunsen burner is due to incomplete combustion and produces carbon. The blue flame is the hotter flame that occurs when complete combustion is occurring.

Chemical properties of alkanes

Chemical properties are determined by the way in which an element or compound reacts with another substance in a chemical reaction. Alkanes tend to be relatively unreactive, although like most hydrocarbons, they can be used as a fuel. The reaction between a fuel and oxygen is known as **combustion**. The burning of petrol in a car engine and the use of natural gas for cooking are examples of combustion reactions.

If the supply of oxygen is plentiful, the products of combustion will be carbon dioxide and water. This is known as **complete combustion**. The complete combustion of an alkane releases significant amounts of energy, which is why alkanes are used as fuels. For example, the blue flames of the Bunsen burners pictured in the Chemfile show methane undergoing complete combustion. The open air-holes of the Bunsen burner enable enough oxygen to reach the methane. The equation for the complete combustion of methane is:

$$CH_4(g) + 2O_2(g) \longrightarrow CO_2(g) + 2H_2O(l)$$

ALKENES

Carbon forms many compounds with hydrogen in which there are double bonds between the carbon atoms. You will recall from Chapter 3 that a double bond is formed when two pairs of electrons are shared. These compounds are called **unsaturated** hydrocarbons. Some of the most useful carbon compounds are unsaturated. **Alkenes** are a homologous series of unsaturated hydrocarbons with one carbon–carbon double bond. The general formula for the alkenes with one double bond is C_nH_{2n}. Alkenes are more reactive than alkanes, which do not have any carbon–carbon double bonds.

The simplest alkene is ethene (C_2H_4). The next member in the series is propene (C_3H_6), which has an additional $–CH_2–$ unit.

Writing formulas of alkenes

Like alkanes, there are a variety of ways for writing the formulas of alkenes. Table 8.2.6 shows different ways of representing ethene and propene. The double bond can be shown in the semi-structural formula, although it is also correct to write a semi-structural formula of an alkene without the double bond. For example, ethene can be written as $CH_2{=}CH_2$ or CH_2CH_2. The double bond must always be shown in the structural formula.

TABLE 8.2.6 Different ways of representing alkenes

Alkene	Molecular formula	Semi-structural formula	Structural formula
ethene	C_2H_4	$CH_2{=}CH_2$	H H C=C H H
propene	C_3H_6	$CH_2{=}CHCH_3$	H H H C—H C=C H H

Naming structural isomers of alkenes

Structural isomers exist for alkene molecules that contain more than three carbon atoms. As you can see in Figure 8.2.6, isomers may result from branches in the carbon chain, or if the carbon–carbon double bond is in a different position.

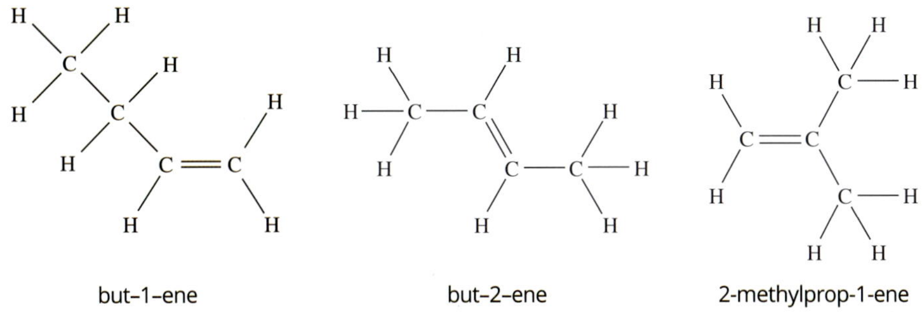

but-1-ene　　but-2-ene　　2-methylprop-1-ene

FIGURE 8.2.6 Isomers of C_4H_8. Isomers of alkenes can have branches in the carbon chain or the double bond at different points in the carbon chain.

When naming alkene isomers, the carbons are numbered from the end of the carbon chain that gives the lowest number to the first carbon in the double bond. The location of the double bond in the molecule is indicated by this number. In Figure 8.2.6, you can see that the double bond in but-1-ene is between carbons 1 and 2. The double bond in but-2-ene is between carbons 2 and 3. Hyphens are used to separate numbers from the rest of the name.

Although it has been common practice to omit numbers when a structure is unambiguous (e.g. propene, not prop-1-ene), the IUPAC rules are more specific than this. A **locant** (number indicating the location of a functional group) may only be omitted in specified situations. For the molecules studied in this course, these can be summarised as:

- A locant can only be omitted if it is a '1' *and* the location in the molecule is unambiguous.
- If one locant is needed, then all locants should be specified for that molecule.

For example, the locant is omitted in the following molecules: butanoic acid, chloromethane, ethanol, propene.

In contrast, locants are required in 2-methylpropane (not methylpropane) and 2-methylprop-1-ene (not methylpropene or 2-methylpropene), despite the fact that these structures are unambiguous without locants.

You may find these molecules commonly referred to as methylpropane and methylpropene, but this is not the correct systematic IUPAC name.

Rules for systematic naming of alkenes

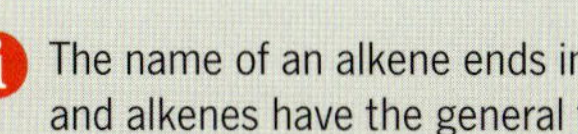

The name of an alkene ends in -ene and alkenes have the general formula C_nH_{2n}.

The following rules apply when naming alkenes.

1 Identify the longest unbranched carbon chain. This chain must include the double bond.

2 Number the carbon atoms in the chain from the end of the chain that will give the smallest numbers to double-bonded carbon atoms.

3 Name any alkyl side chains according to the alkane from which they are derived.

4 Identify the position of the double bond by the number of the first carbon atom involved in the bond. Use the suffix '-ene' to indicate the presence of a double bond, e.g. pent-2-ene.

5 List the number and position of each of the alkyl side chains at the beginning of the compound's name in alphabetical order.

6 If two identical side chains are present, use 'di-' as a prefix; for three use 'tri-'.

The following steps show the process of naming the molecule shown in Figure 8.2.7.

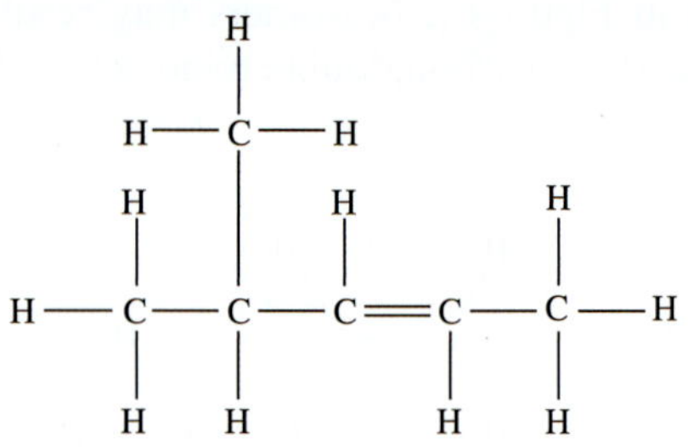

FIGURE 8.2.7 The IUPAC system will be used to name this molecule.

1 Identify the longest carbon chain that contains the double bond. The name of the molecule is based on this chain.

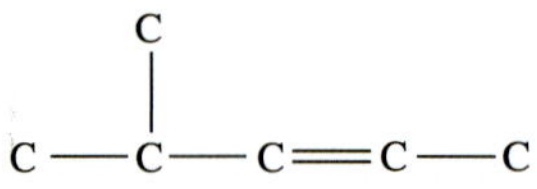

C—C(C)—C=C—C

2 Number the carbons, starting from the end closest to the double bond. Note the position of the double bond.

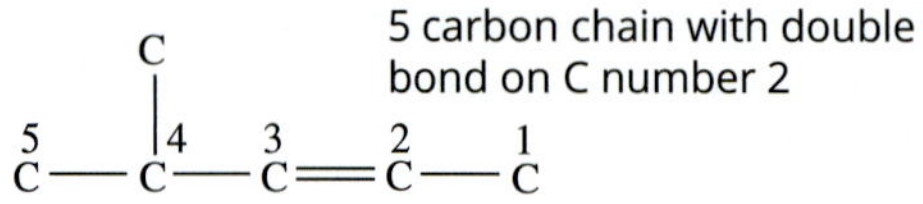

5 carbon chain with double bond on C number 2

It is a 5-carbon chain with double bond starting at C number 2.

3 Name each side chain and the number of the carbon that it is on.

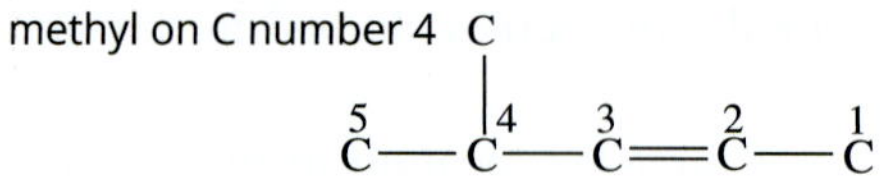

methyl on C number 4

There is a methyl side chain on C number 4.

4 Combine all components to write the full name.

4-methylpent-2-ene

Figure 8.2.8 shows examples of applications of these rules.

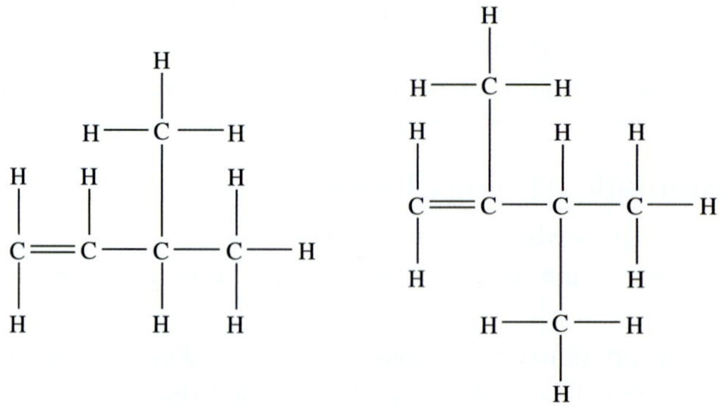

3-methylbut-1-ene

2,3-dimethylbut-1-ene

FIGURE 8.2.8 IUPAC systematic names for two alkenes

Worked example 8.2.2

IUPAC NAMING SYSTEM FOR ALKENES

Write the systematic name for the following molecule.

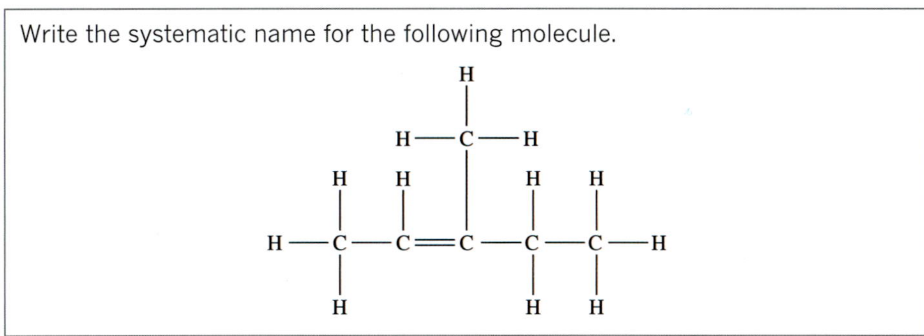

Thinking	Working
Identify the longest carbon chain in the molecule that contains the double bond. The name of the molecule is based on this longest chain.	There are 5 carbons in the longest chain with the double bond. The name is based on pentene.
Number the carbon atoms, starting from the end closest to the double bond. Note the position of the double bond.	There is a double bond on carbon number 2, so the longest chain is pent-2-ene.
Identify each side chain and the number carbon that it is on.	The side chain is a methyl group and it is on carbon number 3.
Combine all components.	The name of the molecule is 3-methylpent-2-ene.

Worked example: Try yourself 8.2.2

IUPAC NAMING SYSTEM FOR ALKENES

Write the systematic name for the following molecule.

CHEMFILE

Blackening bananas

Ripening fruit gives off ethene, which triggers further ripening. Damaged fruit produces extra ethene, causing any nearby fruit to ripen more quickly, hence the phrase 'one rotten apple can spoil the whole barrel'. Banana growers recommend that to keep newly purchased bananas fresh, they should be stored more than one metre away from any ripe bananas, such as those in the figure below, due to the amount of ethene that this fruit is releasing.

Artificially introducing ethene increases the rate of the normal ripening process. For example, exposing 1 kg of tomatoes to as little as 0.1 mg of ethene for 24 hours will ripen them, but they do not taste as nice as those slowly ripened on the vine! The use of ethene in this way allows fruit growers to harvest fruit while it is still hard and green, then transport and cool-store it until it can be ripened shortly before sale.

Ripening bananas produce ethene, which will make other bananas ripen more quickly.

 Like alkanes, alkenes are non-polar and have low boiling points.

Physical and chemical properties of alkenes

Because alkenes differ from the alkanes of the same chain length by only two hydrogen atoms, their physical properties are not very different to alkanes. In contrast, the chemical properties of alkenes are strongly influenced by the presence of the carbon–carbon double bond, making them some of the most useful hydrocarbons that we derive from crude oil.

Physical properties and uses of alkenes

All alkenes are non-polar and hence do not dissolve in water. Like alkanes, as the length of the carbon chain increases, the melting and boiling points increase due to the increasing strength of the dispersion forces between the molecules. The first three members of the alkene homologous series, their physical properties and uses, are shown in Table 8.2.7. The uses of alkenes usually centre around the reactivity of the alkenes and their ability to form other products.

TABLE 8.2.7 Structure, properties and some uses of the first three alkenes

Name and molecular formula	Structural formula	Physical properties	Uses
ethene, C_2H_4	$H_2C{=}CH_2$	non-polar, gas, boiling point (BP) −78.4°C	in the manufacture of a wide range of chemicals such as polyethene
propene, C_3H_6	$H_2C{=}CH{-}CH_3$	non-polar, gas, BP −47.7°C	in the manufacture of propene oxide and polymers
but-1-ene, C_4H_8	$H_2C{=}CH{-}CH_2{-}CH_3$	non-polar, gas, BP −6.3°C	in the manufacture of butanol and polymers

Chemical properties of alkenes

Like alkanes, all alkenes undergo complete combustion in a plentiful supply of oxygen. For example, ethene can undergo complete combustion according to the following equation:

$$C_2H_4(g) + 3O_2(g) \rightarrow 2CO_2(g) + 2H_2O(l)$$

The greater reactivity of a carbon–carbon double bond compared with a single bond enables alkenes to also take part in **addition reactions**. In addition reactions, part of a reactant becomes bonded to one carbon in the double bond and the other part of the reactant becomes bonded to the other carbon atom in the double bond. The double bond is broken and a single C–C bond is formed.

Table 8.2.8 lists some of the more important addition reactions of ethene. The first reaction shown in the table describes the reaction of an alkene with bromine water (Br_2 dissolved in aqueous solution). Bromine water is orange and when it reacts with an alkene in an addition reaction, the bromine solution loses its colour. This test is shown in Figure 8.2.9 and is often used to determine if an organic compound is unsaturated.

FIGURE 8.2.9 Test for unsaturation. Adding a few drops of orange-coloured bromine solution to hexane (right) produces no reaction. In sunflower oil (left), the colour of the bromine solution disappears almost immediately because molecules in the sunflower oil contain carbon–carbon double bonds, which undergo addition reactions with bromine (Br_2).

TABLE 8.2.8 Addition reactions of ethene

Reactants	Products	Name and uses of products
$H_2C{=}CH_2 + Br{-}Br \longrightarrow$	$H{-}CH(Br){-}CH(Br){-}H$	1,2-Dibromoethane: reaction can be used as to test for unsaturation of organic compounds
$H_2C{=}CH_2 + H{-}Cl \longrightarrow$	$H{-}CH_2{-}CH_2{-}Cl$	Chloroethane: once used as a refrigerant and aerosol spray propellant
$H_2C{=}CH_2 + H_2C{=}CH_2 + H_2C{=}CH_2 + \ldots \longrightarrow$	$[{-}CH_2{-}CH_2{-}]_n$	Polyethene: common plastic

8.2 Review

SUMMARY

- Hydrocarbons are compounds containing carbon and hydrogen only.
- Alkanes are an example of a homologous series. Members of a homologous series have similar structures and chemical properties and the same general formula. In a homologous series, each member has one more $-CH_2-$ unit than the previous member.
- Alkanes are saturated hydrocarbon molecules that contain only single bonds and have a general formula C_nH_{2n+2}.
- Alkanes are relatively unreactive, although they do undergo combustion reactions.
- Hydrocarbon molecules can be drawn using structural formulas or semi-structural formulas.
- Structural formulas show the arrangement of the atoms in a molecule and all the bonds, e.g.

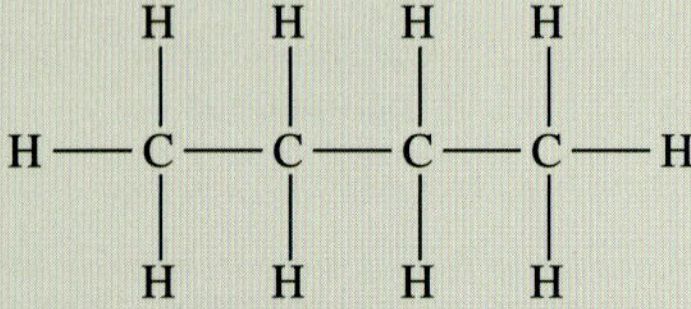

- Semi-structural formulas generally do not show the bonds, but instead show the arrangement of the atoms in the molecule e.g. $CH_3CH_2CH_2CH_3$.
- Molecular formulas group all the atoms of each type together, with no attempt to show the arrangement of the atoms in the molecule, e.g. C_4H_{10}.
- Structural isomers are molecules with the same molecular formula but different arrangements of atoms.
- The IUPAC naming system is used to provide systematic names for hydrocarbon molecules. Names are based on the longest unbranched carbon chain.
- Alkyl side chains are named after the alkane from which they are derived, with a -yl ending and general formula C_nH_{2n+1}.
- Compounds containing carbon and hydrogen only, with a carbon–carbon double bond, are called unsaturated hydrocarbons.
- Alkenes are more reactive than alkanes.
- Alkenes undergo addition reactions, where part of the reactant bonds to one carbon in the double bond and the other part of the reactant bonds to the other carbon in the double bond. The double bond breaks and a single C–C bond is formed.
- Alkenes have the general formula C_nH_{2n}.
- Names of alkenes are based on the longest unbranched carbon chain. The position of the double bond determines which end to start numbering from.

KEY QUESTIONS

Knowledge and understanding

1 A hydrocarbon has the molecular formula, C_3H_8.

- **a** State the name of this hydrocarbon.
- **b** Draw the structural formula of this hydrocarbon.
- **c** Write the semi-structural formula of this hydrocarbon.
- **d** Write a balanced chemical equation for the complete combustion of this hydrocarbon, which is a gas at room temperature.

2 **a** Write the general formulas for alkanes and alkenes.

- **b** Use the general formulas you have written to explain the difference in structure and bonding between these two families of hydrocarbons.

3 **a** Explain how structural isomers of an alkane, such as C_5H_{12}, differ from one another.

- **b** Draw the structural formulas and state the systematic names of all the possible structural isomers of C_5H_{12}.

4 **a** Name the compound that is represented by the following structural formula.

- **b** Write the semi-structural formula for this compound.

5 Draw the structural formulas of these unsaturated hydrocarbons.

- **a** but-2-ene
- **b** 4-methylpent-1-ene

Analysis

6 Write the systematic names of the following alkanes.

a $CH_3CH_2CH_2CH_2CH_2CH_3$

b $CH_3CH_2CH(CH_3)CH_2CH_2CH_3$

c $CH_3CH(CH_3)CH_2CH(CH_3)CH_2CH_3$

d $CH_3C(CH_3)_2CH_2CH_3$

7 Draw the structural formulas of the following alkanes.

a hexane

b 3-methylhexane

c 3,3-dimethylpentane

d 3-ethyl-2-methylpentane

8 **a** Draw the structure of 3-methylbut-1-ene

b State the molecular formula of 3-methylbut-1-ene.

c There are four other structural isomers for this alkene. Draw the structures of these isomers and name them.

d Explain why 3-methylbut-2-ene is not the correct name of this isomer.

e Explain why 2,2-dimethylpropene is not a possible isomer of this alkene.

9 The following names of some alkanes have been written incorrectly.

i Draw the molecule that seems to be described by the name.

ii Identify what is wrong with the name.

iii Write the correct name or names if more than one is possible.

a 1-methylpropane

b dimethylbutane

c 3-methylbutane

10 The following semi-structural formulas have been incorrectly written. For each alkene, identify the mistake and write the correct semi-structural formula.

Alkene	Incorrect semi-structural formula	Mistake	Correct semi-structural formula
but-2-ene	$CH_2CHCH_2CH_3$		
2-methylprop-1-ene	$CH_2CH(CH_3)_2$		
2,3-dimethylpent-2-ene	$CH_3C(CH_3)CCH_2CH_3$		

8.3 Haloalkanes

> A functional group is an atom or a group of atoms that gives a characteristic set of chemical properties to a molecule containing those atoms.

Many organic compounds can be regarded as alkanes which have one or more hydrogen atoms replaced by other atoms or groups of atoms called a **functional group**. When this occurs, a new homologous series is created. A functional group is an atom or a group of atoms that gives a characteristic set of chemical properties to a molecule containing those atoms. The presence of a particular functional group in a molecule gives a substance certain physical and chemical properties.

In this section, you will learn about a homologous series in which at least one hydrogen atom is replaced by a halogen atom, the **haloalkanes**.

WRITING FORMULAS OF HALOALKANES

The **halogen** elements are in group 17 of the periodic table. All halogen atoms have seven valence electrons, which means they can form one single covalent bond with a carbon atom. The halogen elements that commonly form halo functional groups in organic compounds are chlorine, bromine and iodine, with fluorine being less commonly used. Haloalkanes are compounds with the general formula $C_nH_{2n+1}X$, where X is a halogen. Table 8.3.1 shows different ways of representing two haloalkanes.

> The name of a haloalkane starts with the name of the halogen, shortened and ending with 'o', and haloalkanes have the general formula $C_nH_{2n+1}X$, where X is F, Cl, Br or I.

TABLE 8.3.1 Different ways of representing haloalkanes

Alkene	Molecular formula	Semi-structural formula	Structural formula
bromoethane	C_2H_5Br	CH_3CH_2Br	Br H H—C—C—H H H
1-chloropropane	C_3H_7Cl	$CH_3CH_2CH_2Cl$	H H Cl H—C—C—C—H H H H

NAMING STRUCTURAL ISOMERS OF HALOALKANES

Haloalkanes with more than three carbon atoms, or with two or more carbon atoms and more than one halogen atom, have structural isomers. As you can see in Figure 8.3.1, isomers may result from branches in the carbon chain, or if the halogen atoms are in different locations.

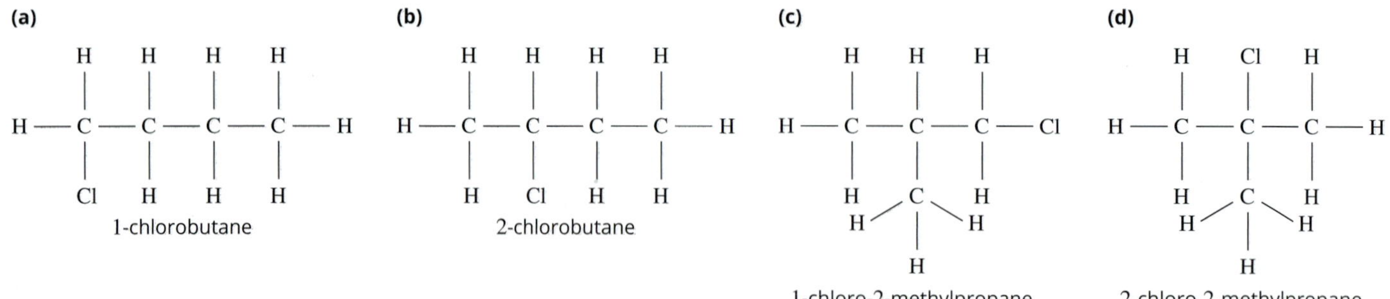

FIGURE 8.3.1 Isomers of C_4H_9Cl. Isomers of haloalkanes can have branches in the carbon chain or the halo group can be at different points in the carbon chain.

When naming haloalkane isomers, the carbons are numbered from the end of the carbon chain that is closest to the first halogen in the molecule. The location of the halogen in the molecule is indicated by this number. In Figure 8.3.1a, you can see that the chlorine atom is bonded to the **terminal** (end) **carbon**, so this carbon atom is numbered 1 and the molecule is called 1-chlorobutane. More details about the systematic naming of these haloalkanes can be found below.

RULES FOR SYSTEMATIC NAMING OF HALOALKANES

The names of haloalkane functional groups are derived from the name of the halogen, as shown in Table 8.3.2.

The rules for naming haloalkanes follow the rules for naming alkanes (refer to Section 8.2). In addition, the following conventions are applied.

1 Place the name of the specific halo functional group at the start of the parent alkane's name.

2 Number the carbons of the parent chain, beginning at the end closest to the first halo group.

3 If there is more than one of the same type of halogen atom, use the prefix 'di-' or 'tri-'

4 If more than one type of halo functional group or alkyl group is present, list them in alphabetical order.

The following steps show the process of naming the isomer of C_4H_8BrCl, shown in Figure 8.3.2.

1 Identify the longest carbon chain that contains the halogens. The name of the molecule is based on this chain.

```
      Cl
      |
1     |2    3     4
C --- C --- C --- C
      |     |
      Br    C
```

It is a 4-carbon chain.

2 Number the carbons, starting from the end closest to the first halogen. Note the position of the halogen atom/s.

```
      Cl
      |
1     |2    3     4
C --- C --- C --- C
      |     |
      Br    C
```

4 carbon chain with Cl and
Br on C number 2

It is a 4-carbon chain with two halogen atoms on carbon number 2.

3 Name each side chain and the number of the carbon that it is on.

```
      Cl
      |
1     |2    3     4
C --- C --- C --- C
      |     |
      Br    C
```

methyl on C number 3

There is a methyl side chain on C number 3.

4 Combine all components to write the full name.

```
     H    Cl    H    H
     |    |     |    |
H -- C -- C --- C -- C -- H
     |    |     |    |
     H    Br    C    H
             H/ | \H
                H
```

2-bromo-2-chloro-3-methylbutane

TABLE 8.3.2 Names of the haloalkane functional groups

Halogen	Functional group name
fluorine	fluoro-
chlorine	chloro-
bromine	bromo-
iodine	iodo-

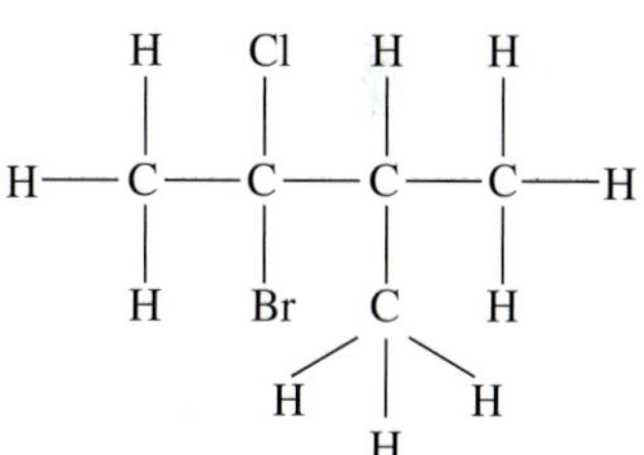

FIGURE 8.3.2 The IUPAC system will be used to name this haloalkane.

When numbering the carbons in a haloalkane, the numbering should start from the end nearest the halogen.

Worked example 8.3.1

IUPAC NAMING SYSTEM FOR HALOALKANES

Write the systematic name for the following molecule.

```
         H     H     Cl    H
         |     |     |     |
    H —  C  —  C  —  C  —  C  — H
         |     |     |     |
         H     H     |     H
               H  —  C  —  H
                     |
                     H
```

Thinking	Working
Identify the longest carbon chain in the molecule that contains the halogen atom/s. The name of the molecule is based on this longest chain.	There are 4 carbons in the longest chain with the halogen atom. The name is based on butane.
Number the carbon atoms, starting from the end closest to the first halogen atom. Note the position of the halogen atom/s.	H H Cl H H — C(4) — C(3) — C(2) — C(1) — H H H H H — C — H H There is a chlorine atom on carbon number 2, so the longest chain is 2-chlorobutane.
Identify each side chain and the number carbon that it is on.	The side chain is a methyl group and it is on carbon number 2.
Combine all components, remembering to list them in alphabetical order.	The name of the molecule is 2-chloro-2-methylbutane

Worked example: Try yourself 8.3.1

IUPAC NAMING SYSTEM FOR HALOALKANES

Write the systematic name for the following molecule.

```
                     H
                     |
               H  —  C  —  H
         H     H     |     H     H
         |     |     |     |     |
    H —  C  —  C  —  C  —  C  —  C  — H
         |     |     |     |     |
         H     Br    H     H     H
```

PHYSICAL AND CHEMICAL PROPERTIES OF HALOALKANES

In contrast to alkanes and alkenes, the presence of the halogen atom in a haloalkane molecule makes the molecule polar. This influences the physical properties of the haloalkanes, as the intermolecular forces will be stronger than in alkanes and alkenes. The presence of a halogen atom also affects the chemical properties of these compounds. In general, these compounds are more reactive than the alkanes and alkenes.

Physical properties and uses of haloalkanes

Most haloalkanes are polar, as is shown in Figure 8.3.3; however, if the haloalkane molecule has more than one halogen of the same type and is symmetrical, as is the case for the molecule in Figure 8.3.4, then it is non-polar. Their polarity allows smaller haloalkanes to dissolve in water, but as the length of the carbon chain increases, the solubility in water decreases. Like alkanes, as the length of the carbon chain increases, the melting and boiling points increase due to the increasing strength of the dispersion forces between the molecules, and the presence of a polar carbon–halogen bond will introduce dipole–dipole attractions and will further increase the melting and boiling points. The size of the halogen atom will also influence the strength of the dispersion forces between haloalkane molecules.

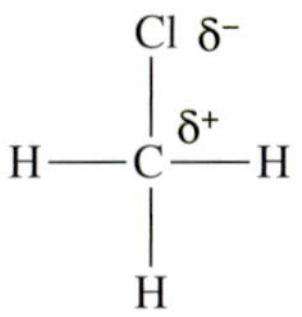

Cl δ−
δ+
H—C—H
H

FIGURE 8.3.3 A molecule of chloromethane is polar because the chlorine atom is more electronegative than the carbon atom, and the molecule is asymmetrical.

Cl
Cl—C—Cl
Cl

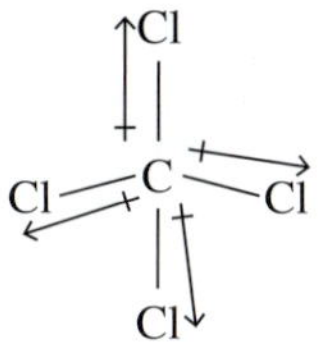

FIGURE 8.3.4 A molecule of tetrachloromethane is non-polar because the molecule is symmetrical and there is no overall dipole.

Three haloalkanes with different halogen atoms, their physical properties and uses are shown in Table 8.3.3. You can see that the boiling point of each haloalkane increases significantly as the size of the halogen increases, from chlorine to bromine to iodine.

Haloalkanes are often, but not always polar. If the molecule is asymmetrical it will be polar and will have dipole–dipole attractions between the molecules.

TABLE 8.3.3 Names, structures, physical properties and uses of three haloalkanes with different halogen atoms.

Name	Structural formula	Semi-structural formula	Boiling point (°C)	Solubility in water	Use
chloromethane	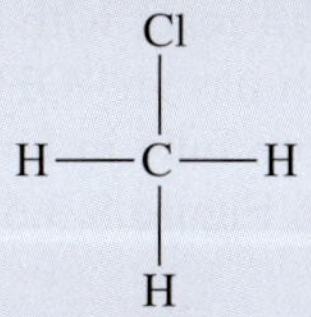 Cl above H—C—H, H below	CH_3Cl	–24	slightly soluble	refrigerant
bromomethane	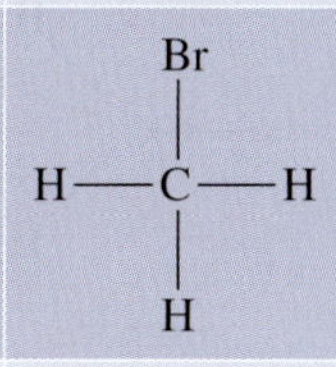 Br above H—C—H, H below	CH_3Br	4	slightly soluble	kills pests, rats, insects and fungi
iodomethane	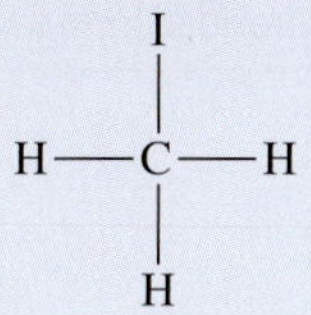 I above H—C—H, H below	CH_3I	33	slightly soluble	pesticide

Table 8.3.4 shows the effect of the increasing carbon chain length on the boiling point of three haloalkanes with the same halogen. While none of the chloroalkanes would be described as very soluble in water, the solubility decreases as the chain length increases, due to the increasing influence of the non-polar part of the molecule.

TABLE 8.3.4 Names, structures and boiling points of three haloalkanes with increasing carbon chain length

Name	Structural formula	Semi-structural formula	Boiling point (°C)	Solubility in water (g/100 mL)
chloroethane	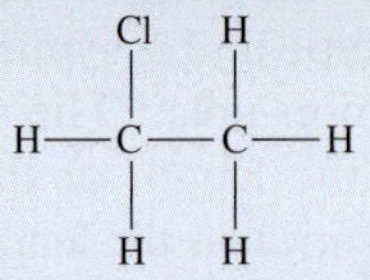 Cl H H—C—C—H H H	CH_3CH_2Cl	12	0.57
1-chloropropane	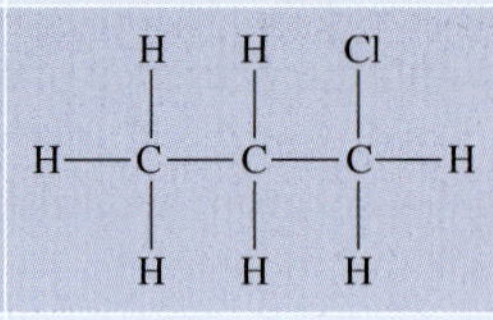 H H Cl H—C—C—C—H H H H	$CH_3CH_2CH_2Cl$	46	0.27
1-chlorobutane	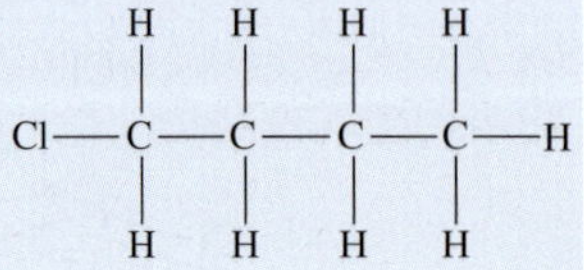 H H H H Cl—C—C—C—C—H H H H H	$CH_3CH_2CH_2CH_2Cl$	79	0.05

Haloalkanes are used widely in industry as flame retardants, refrigerants, propellants, pesticides, solvents and pharmaceuticals. Some haloalkanes (in particular, chlorofluorocarbons or CFCs) are ozone-depleting chemicals and their use has been phased out in many applications where previously they were simply released to the environment when they were no longer needed.

Chemical properties of haloalkanes

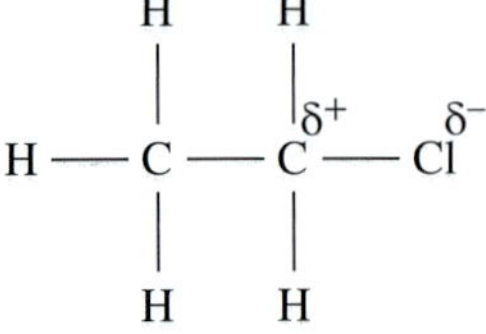

FIGURE 8.3.5 Chloroethane. The presence of the halogen atom in a haloalkane makes the carbon–halogen bond polar.

In addition to increasing the boiling points of haloalkanes, the polarity of the carbon–halogen bond affects the chemical reactivity of haloalkanes. The very electronegative chlorine atom in a molecule like chloroethane makes the carbon atom that is bonded to it positively charged, whereas carbon and hydrogen have almost the same electronegativities and a C–H bond can be regarded as non-polar. In Figure 8.3.5 you can see the difference that the chlorine atom makes to the polarity of chloroethane.

The result of this altered polarity is to make haloalkanes significantly more reactive than alkanes. For example, a haloalkane can react with hydroxide ions in a **substitution reaction** to make an alcohol, which has an –OH group instead of the halogen atom. This reaction will be studied more fully in Unit 4, but for your interest, the equation for the reaction can be seen in Figure 8.3.6.

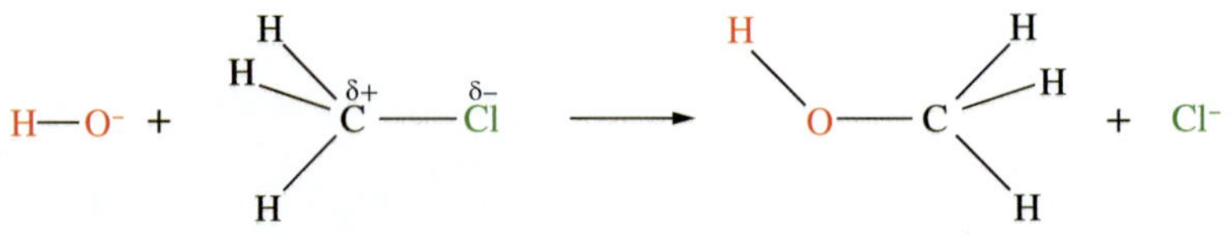

$$H{-}O^- + H_3C^{\delta+}{-}Cl^{\delta-} \longrightarrow H{-}O{-}CH_3 + Cl^-$$

chloromethane methanol

FIGURE 8.3.6 Reaction of a chloromethane molecule with a hydroxide ion to form methanol

CASE STUDY **ANALYSIS**

Haloalkanes and the ozone layer

Ozone, O_3, is an unstable form of oxygen. While it has harmful effects on humans when encountered at ground level, ozone forms a layer in the upper atmosphere, the stratosphere, that protects us from much of the UV radiation from the Sun.

In the ozone layer, reactions are constantly occurring which break down the ozone and reform it. The bonds in ozone and oxygen are both broken by various wavelengths of UV light (UVB and UVC) and these reactions can be represented as follows.

Dissociation of ozone:

$$O_3(g) \xrightarrow{UVB} O\cdot(g) + 2O_2(g)$$
$$O_3(g) + O\cdot(g) \rightarrow 2O_2(g)$$

Formation of ozone:

$$O_2(g) \xrightarrow{UVC} 2O\cdot(g)$$
$$O\cdot(g) + O_2(g) \rightarrow O_3(g)$$

The symbol O· represents a species known as a free radical, with an unpaired electron making it highly reactive.

Chlorofluorocarbons (CFCs) are haloalkanes which were used widely as coolants in refrigeration and air conditioners, as foaming agents in fire extinguishers and as propellants in aerosols during the twentieth century. Their use as propellants and foaming agents resulted in them routinely being released into the atmosphere, while their use in refrigeration units meant that they were often released into the atmosphere when servicing of these units occurred. A CFC molecule can exist in the atmosphere for more than 100 years, so while their use has been limited since the 1980s, they still persist in large concentrations in the stratosphere.

In the stratosphere, the chlorine–carbon bond in CFCs can be broken by high energy UV radiation and chlorine radicals, Cl·, are released. The following reaction gives an example of how this can occur.

$$CCl_2F_2(g) \xrightarrow{\text{UV light}} CClF_2\cdot(g) + Cl\cdot(g)$$

These chlorine radicals catalyse the decomposition of ozone, changing the ozone into oxygen, O_2.

$$Cl\cdot(g) + O_3(g) \rightarrow O_2(g) + ClO\cdot(g)$$
$$ClO\cdot(g) + O\cdot(g) \rightarrow O_2(g) + Cl\cdot(g)$$

As the ozone has been broken down over time, the ozone layer gradually thinned, and eventually reached a point where the 'hole' covered an area of 22 million square kilometres, much of which impacted on Australia (Figure 8.3.7). The reduction in protection has resulted in an increase in the incidence of skin cancer and cataracts, among other detrimental effects on human health and the environment.

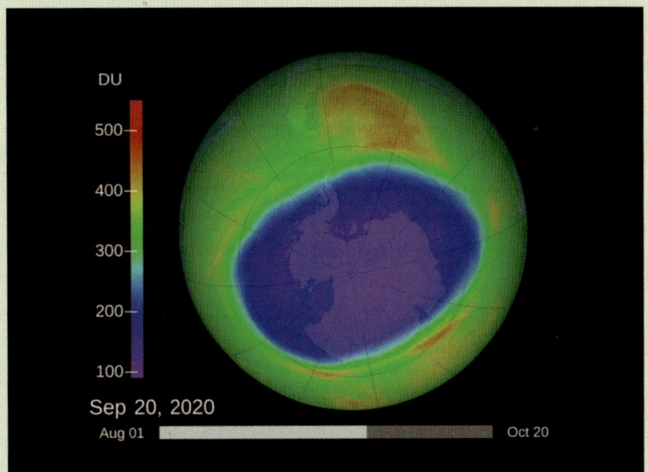

FIGURE 8.3.7 Satellite image of the extent of the ozone hole (purple and blue) over Antarctica on 20 September 2020. The amount of ozone-depleting substances in the Earth's atmosphere has decreased from peak levels in 2000. (DU stands for Dobson Unit, and is the unit of measurment for ozone thickness.)

Analysis

1 Using your understanding of covalent bonding, explain why a free radical is so reactive.

2 The energy required to break the bond between oxygen atoms in O_2 is greater than that required to break the bond between the oxygen atoms in ozone, O_3. Use this information to compare the strength of the bonds in ozone to the covalent double bond in an oxygen molecule, O_2.

3 When UV light breaks a bond in a CFC, the carbon–chlorine bond breaks, rather than the carbon–fluorine bond. Explain what this means about the relative strength of the carbon–chlorine bond and the carbon–fluorine bond.

4 The chlorofluorocarbon, CCl_2F_2 was used as an example in this case study. What is the systematic name of this chlorofluorocarbon?

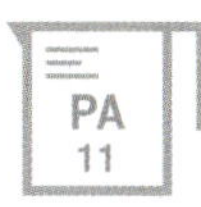

8.3 Review

SUMMARY

- Haloalkanes are compounds with the general formula $C_nH_{2n+1}X$, where X is a group 17 atom (halogen).
- When naming haloalkanes, the name is based on the longest unbranched carbon chain. The name of the halogen is added as a prefix, e.g. chloro-, bromo-, iodo-. The position of the halogen atom must be indicated by a number before the prefix.
- The halogen–carbon bond is polar, making the haloalkane molecule polar in many cases, with dipole–dipole attractions between the molecules.
- The boiling points of haloalkanes are higher than for alkanes of the same carbon chain length.
- Haloalkanes are more reactive than alkanes.
- Haloalkanes undergo substitution reactions, where the halo functional group is replaced by another functional group, such as –OH.

KEY QUESTIONS

Knowledge and understanding

1 Write the semi-structural formula for each of the following haloalkanes.

a chloroethane

b bromomethane

c 2-iodopropane

2 Draw the structural formula for each of the following haloalkanes.

a 1-bromopropane

b 2-chlorobutane

c 1,2-dichloroethane

d 1,1-dibromoethane

3 Name each of the following haloalkanes.

a

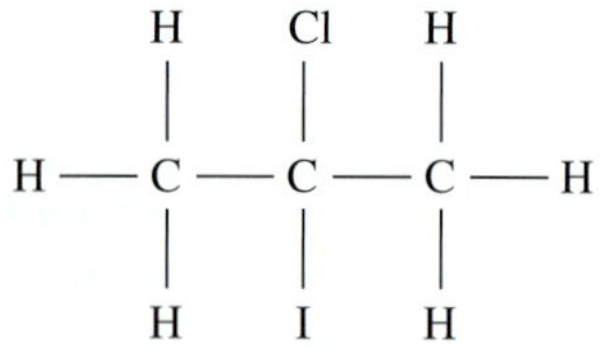

```
    H   Cl   H
    |   |    |
H — C — C —  C — H
    |   |    |
    H   I    H
```

b

```
   H        Br
    \      /
Br — C — C — Br
    /      \
   H        H
```

c

```
      Cl
      |
      C
    / | \
  Cl  |  Cl
      Cl
```

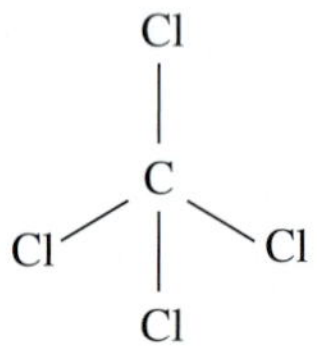

Analysis

4 Name the four isomers of C_4H_9Cl.

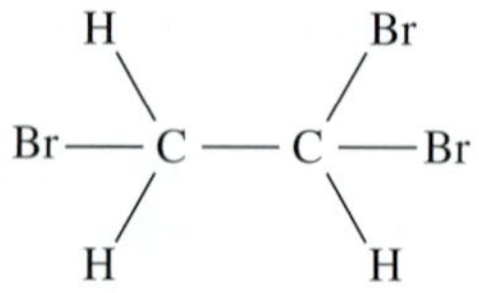

5 Explain why a molecule cannot simply be named bromobutane.

6 Explain why there are more isomers of $C_3H_6Cl_2$ than there are of C_3H_7Cl. As part of your answer, you should write the semi-structural formulas of the isomers, and name the isomers for both molecules.

7 Consider the polarity of the bonds in each of the following molecules and hence determine whether the molecules are polar or non-polar.

a

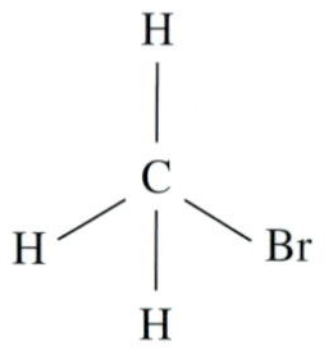

```
      H
      |
      C
    / | \
   H  |  Br
      H
```

b

```
   H        H
    \      /
H — C — C — H
    /      \
  Cl        H
```

c

```
   H        Br
    \      /
H — C — C — Br
    /      \
   H        H
```

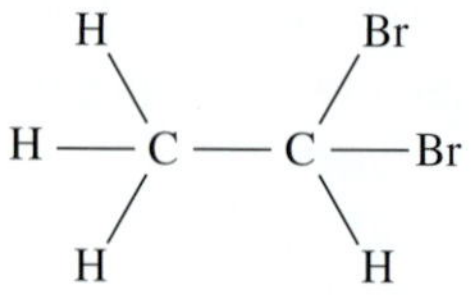

d

```
   I        Cl
    \      /
H — C — C — H
    /      \
   H        H
```

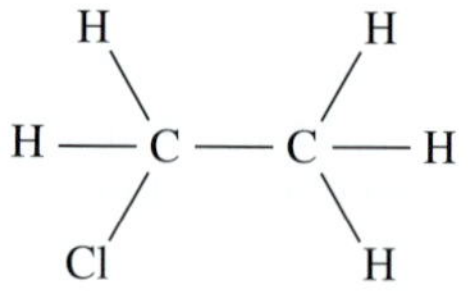

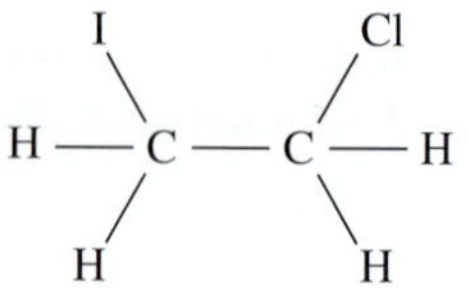

8.4 Alcohols and carboxylic acids

A vast range of organic molecules contain other atoms as well as carbon and hydrogen. The presence of different atoms in organic compounds increases their chemical reactivities compared to that of alkanes. It also explains why different functions can be performed by carbon-based compounds.

For example, the presence of a hydroxyl, –OH, group in a molecule can enable the compound to dissolve in water and changes its boiling point. Such compounds are called **alcohols**. Compounds with a carboxyl, –COOH, functional group are called **carboxylic acids**. In this section you will learn about the properties, uses and structures of the homologous series of alcohols and carboxylic acids.

FUNCTIONAL GROUPS

While carbon atoms are covalently bonded only to other carbon and hydrogen atoms in hydrocarbons, such as alkanes and alkenes, it is possible for carbon to form covalent bonds with other atoms or groups of atoms called functional groups. Homologous series, such as alkenes, haloalkanes, alcohols and carboxylic acids, are characterised by the presence of a particular functional group.

The carbon–carbon double bond that is present in the alkene molecules described in Section 8.2 is considered the functional group in alkenes. In Section 8.3 you learnt that halogen atoms are the functional groups in haloalkanes.

Some examples of compounds that contain two carbon atoms and a particular functional group are shown in Table 8.4.1. Although all have two carbon atoms, they have different physical and chemical properties and are members of different homologous series.

Functional groups include the carbon–carbon double bond, halogen atoms, hydroxyl groups and the carboxyl group.

TABLE 8.4.1 Examples of two-carbon compounds with different functional groups

Homologous series	Name and semi-structural formula	Functional group	Ball and stick model	Additional information
alkenes	ethene, $CH_2{=}CH_2$	C=C double bond		Ethene is a gas which is an important natural plant hormone. It is used in agriculture to force fruit to ripen.
haloalkanes (chloroalkanes)	chloroethane, CH_3CH_2Cl	Cl atom		Chloroethane is a gas which was used as a refrigerant and aerosol spray propellant in the past.
alcohols	ethanol, C_2H_5OH	hydroxyl group –OH		Ethanol is the alcohol found in alcoholic drinks.
carboxylic acids	ethanoic acid, CH_3COOH	carboxyl group –COOH		Ethanoic acid dissolves in water to form vinegar, giving its distinctive sour taste and pungent smell.

Members of the same homologous series contain the same functional group. The presence of the same functional group in these molecules means that they have similar, although not identical, physical and chemical properties.

ALCOHOLS

The functional group in alcohols is made up of an oxygen atom bonded to a hydrogen atom. This –OH group is known as a **hydroxyl group** and replaces one hydrogen in the structure of an alkane. Successive members of this homologous series differ by a $-CH_2-$ unit. The general formula of alcohols is $C_nH_{2n+1}OH$.

Formulas and structures of alcohols

When naming alcohols, a similar systematic naming process to the one used for alkanes and alkenes applies. The number of carbon atoms in the molecule is indicated by the stem name, e.g. meth-, eth-, prop-, but-. However, the suffix (ending) of the name of an alcohol is always '-ol'. Table 8.4.2 shows different ways to represent alcohols.

TABLE 8.4.2 Different ways of representing alcohols

Alcohol	Molecular formula	Semi-structural formula	Structural formula
methanol	CH_4O	CH_3OH	H H—C—O—H H
ethanol	C_2H_6O	CH_3CH_2OH	H H H—C—C—O—H H H
propanol	C_3H_8O	$CH_3CH_2CH_2OH$	H H H H—C—C—C—O—H H H H

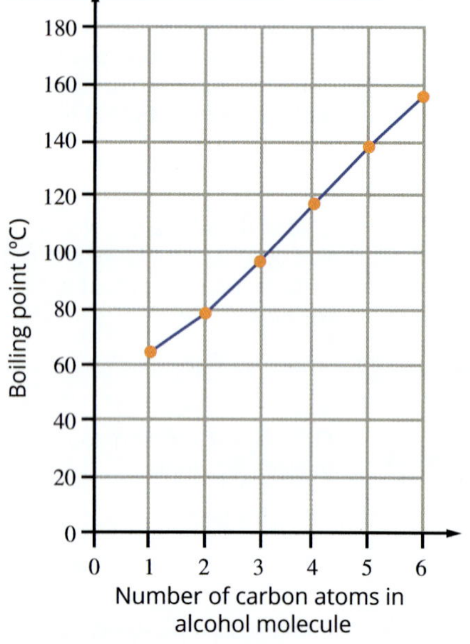

FIGURE 8.4.1 As the number of carbon atoms in an alcohol molecule increases, the boiling point of the alcohol increases.

Physical properties and uses of alcohols

Many of the small alcohols are useful as fuels. Ethanol can be used as a fuel on its own or mixed with petrol. The combustion of ethanol produces carbon dioxide and water:

$$C_2H_5OH(l) + 3O_2(g) \rightarrow 2CO_2(g) + 3H_2O(l)$$

The boiling points of alcohols increase as the size of the alcohol molecule increases. However, the boiling points of the alcohols are all higher than those of the corresponding alkanes. All the alcohols are liquids at room temperature, in contrast to the alkanes and alkenes, many of which are gases. The trend in boiling points of the first six alcohols can be seen in Figure 8.4.1.

The higher boiling point of alcohols is due to the presence of the –OH group, which allows hydrogen bonding to occur between molecules and strengthens the intermolecular bonding.

Hydrogen bonding also influences the solubility of alcohols. The presence of the –OH group allows hydrogen bonds to form between water molecules and alcohol molecules (Figure 8.4.2). For the smaller alcohols, such as methanol and ethanol, this allows the alcohol to dissolve readily in water. The solubility of the other alcohols decreases as the length of the carbon chain attached to the –OH group increases. A longer carbon chain means that more of the molecule is non-polar and the molecule becomes less polar overall.

FIGURE 8.4.2 Hydrogen bonds form between the hydroxyl group of a methanol molecule and water.

Structural isomers of alcohols

The position of the hydroxyl (–OH) functional group influences the chemical and physical properties of alcohols. Alcohols with more than two carbon atoms have more than one position where the hydroxyl functional group may be found. For example, the hydroxyl group in propanol (C_3H_7OH) can be bonded to the first or the second carbon atom. This gives two different isomers: $CH_3CH_2CH_2OH$ and $CH_3CH(OH)CH_3$. The name of the alcohol must reflect the structure of the molecule, so a system of nomenclature is used.

Naming alcohols

To name an alcohol, start with the name of the parent alkane, remove the 'e' from the end and add the suffix '-ol'. The atoms in the longest carbon chain are numbered from the end that is closest to the functional group. The number of the atom that the hydroxyl group is bonded to is shown before the -ol ending and is separated by hyphens.

CHEMFILE

Making champagne

The solubility of ethanol in water is essential in the production of alcoholic drinks. Champagne and wine are made by the fermentation of glucose ($C_6H_{12}O_6$) obtained from grapes, with the aid of yeast. Ethanol and carbon dioxide are formed:

$$C_6H_{12}O_6(aq) \xrightarrow{\text{yeast}} 2C_2H_5OH(aq) + 2CO_2(g)$$

When a carefully selected mixture of grapes is fermented in bottles to make champagne, the ethanol and some carbon dioxide dissolve in the aqueous solution. Because the sparkle of the carbon dioxide is required in the final product, the yeast must be removed from the bottles very carefully while keeping the carbon dioxide in solution.

When it is time to remove the yeast from the fermented champagne solution, the bottle is inverted and the yeast is frozen in the neck of the bottle (see figure). The stopper and yeast are then removed quickly and the stopper is replaced.

An inverted bottle of champagne in the Moet & Chandon champagne cellar, Epernay, France. The yeast can be seen near the temporary cap of the bottle

Worked example 8.4.1

IUPAC NAMING SYSTEM FOR ALCOHOLS

Write the systematic name for the following molecule.

```
      H   H   H
      |   |   |
  H — C — C — C — O — H
      |   |   |
      H   H   H
```

Thinking	Working
Identify the longest carbon chain in the molecule. The name of the molecule is based on this longest chain.	There are 3 carbons in the longest chain. The name is based on propane.
Identify the functional group that is present.	There is a hydroxyl group present.
Number the carbon atoms, starting from the end closest to the functional group.	H H H / H—C(3)—C(2)—C(1)—O—H / H H H
Identify the position(s) and type(s) of side chains.	There are no side chains in this molecule.
Combine all components. Place the number for the position of the side chain in front of the prefix, and the number for the position of the hydroxyl group in front of the -ol ending.	The name of the molecule is propan-1-ol.

Worked example: Try yourself 8.4.1

IUPAC NAMING SYSTEM FOR ALCOHOLS

Write the systematic name for the following molecule.

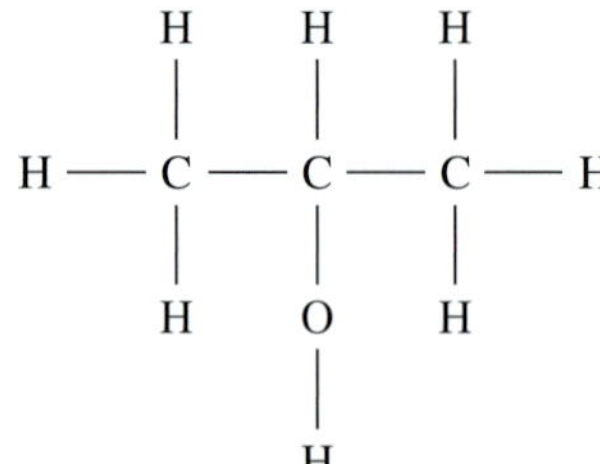

Worked example 8.4.2

IUPAC NAMING SYSTEM FOR ALCOHOLS

Write the systematic name for the following molecule.

```
                   CH3
                    |
CH3 — CH  —  CH — CH3
       |
       OH
```

Thinking	Working
Identify the longest carbon chain in the molecule. The name of the molecule is based on this longest chain.	There are 4 carbons in the longest chain. The name is based on butane.
Identify the functional group that is present.	There is a hydroxyl group present.
Number the carbon atoms, starting from the end closest to the functional group.	1 CH_3 — 2 CH — 3 CH(CH_3) — 4 CH_3, with OH on C2
Identify the position(s) and the type(s) of side chains.	There is a methyl ($-CH_3$) group on C3 so the prefix 'methyl' will be used.
Combine all components. Place the number for the position of the side chain in front of the prefix, and the number for the position of the hydroxyl group in front of the -ol ending.	The name of the molecule is 3-methylbutan-2-ol.

The name of an alcohol ends in -ol and the functional group in an alcohol is the hydroxyl group .

Worked example: Try yourself 8.4.2

IUPAC NAMING SYSTEM FOR ALCOHOLS

Write the systematic name for the following molecule.

```
CH3 — CH — CH — CH — CH3
      |     |     |
      CH3   OH    CH3
```

CARBOXYLIC ACIDS

The carboxylic acids are an important class of organic compounds. These compounds are **weak acids**, which are often present in food, giving it a sour taste. As shown in Figure 8.4.3, carboxylic acids are also found in some insect venoms. At other times they are formed when food deteriorates, such as when wine becomes sour on exposure to air.

FIGURE 8.4.3 These ants are attacking the intruder from another colony by spraying it with a jet of methanoic (formic) acid from their abdomens. Methanoic acid is the smallest carboxylic acid molecule.

Formulas and structures of carboxylic acids

Carboxylic acids are identified by the presence of a **carboxyl group** (–COOH). In this functional group, a carbon atom has a double bond to one oxygen atom and a single bond to a second oxygen atom. This second oxygen atom is also bonded to a hydrogen atom. The structure of the carboxyl functional group is shown in Figure 8.4.4a and the three-dimensional structure is illustrated by a model in Figure 8.4.4b.

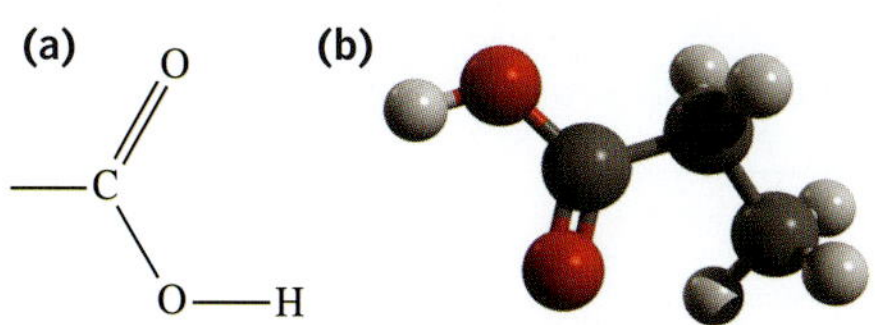

FIGURE 8.4.4 (a) The structure of a carboxyl functional group (b) The three-dimensional shape of a carboxylic acid

Naming carboxylic acids

The prefixes that are used to indicate the number of carbon atoms in alkanes are also used with carboxylic acids. Other features of carboxylic acid nomenclature are listed below.

- Names of carboxylic acids end with the suffix '-oic acid'.
- The carbon in the functional group is counted in the chain used to name the carboxylic acid. For example, C_2H_5COOH has three carbons and is called propanoic acid.
- The carboxyl carbon is always carbon number 1.

The name of a carboxylic acid ends in -oic acid. The functional group is the carboxyl group, with the formula –COOH.

The general formula of a carboxylic acid is often represented as RCOOH, where R is an alkyl group such as $-CH_3$ and $-C_2H_5$. The first three members of the carboxylic acid homologous series are listed in Table 8.4.3. Successive members of the series differ by a $-CH_2^-$ unit.

TABLE 8.4.3 Different ways of representing carboxylic acids

Name	Molecular formula	Condensed structural formula (semistructural formula)	Structural formula
methanoic acid	CH_2O_2	HCOOH	H—C(=O)—O—H
Ethanoic acid	$C_2H_4O_2$	CH_3COOH	H—C(H)(H)—C(=O)—O—H
Propanoic acid	$C_3H_6O_2$	CH_3CH_2COOH	H—C(H)(H)—C(H)(H)—C(=O)—O—H

Carboxylic acids are named according to the total number of carbons in the molecule, so C_3H_7COOH is butanoic acid because it has 4 carbons.

Structural isomers of carboxylic acids

The carboxyl group can only occur on the end of a molecule because the carbon atom in the carboxyl group has a double bond to one oxygen and a single bond to another oxygen. Carboxylic acids can form isomers with different branches, but the carbon atom in the carboxyl group is always carbon number 1, therefore no number is needed to indicate the location of the functional group in the systematic name.

Worked example 8.4.3

IUPAC NAMING SYSTEM FOR CARBOXYLIC ACIDS

Write the systematic name for the following molecule.
CH_3—CH_2—CH(—CH_3)—C(=O)—OH

Thinking	Working
Identify the functional group that is present.	There is a carboxyl group present.
Identify the longest carbon chain that includes the carboxyl carbon. This atom will be C1. The stem name of the molecule is based on this longest chain.	There are 4 carbons in the longest chain, so the stem name is based on butane.
Number the carbon atoms, starting from the end incorporating the functional group.	4 CH_3—3 CH_2—2 CH(—CH_3)—1 C(=O)—OH
Identify the position(s) and the type(s) of side chains.	There is a methyl group on C2.
Combine all components. Place the number for the position of the side chain in front of the prefix and use the ending -oic acid.	The name of the molecule is 2-methylbutanoic acid.

Worked example: Try yourself 8.4.3

IUPAC NAMING SYSTEM FOR CARBOXYLIC ACIDS

Write the systematic name for the following molecule.
CH_3—CH(—CH_3)—CH_2—CH_2—CH_2—C(=O)—OH

WS 21

Properties and uses of carboxylic acids

Carboxylic acids are organic acids. They are commonly found in nature, giving a sour taste to lemon juice and vinegar, or adding a sting to an injury from a stinging nettle or an ant bite. Figure 8.4.6 shows ball-and-stick models of the structures of citric acid and methanoic acid.

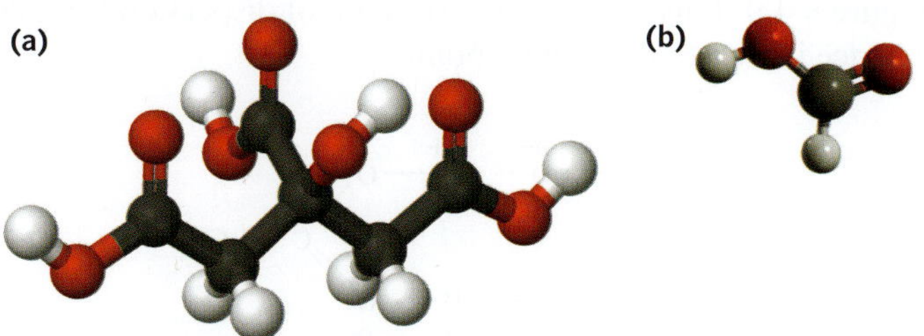

FIGURE 8.4.6 Ball-and-stick models of two carboxylic acids: (a) citric acid, found in lemon juice, and (b) methanoic (formic) acid, found in stinging nettles and ant stings

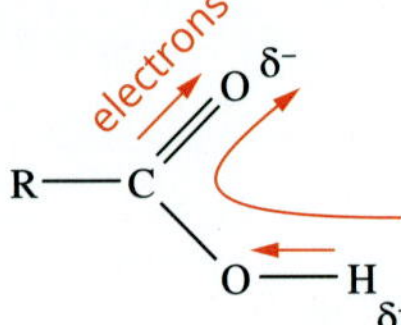

FIGURE 8.4.7 Electrons are drawn away from the hydrogen of the carboxyl functional group, allowing the hydrogen to be donated as a H^+ ion in an acid–base reaction.

As shown in Figure 8.4.7, the carboxyl functional group is made up of a carbonyl group (C=O) and a hydroxyl (–OH) group. Both of these groups are polar, with oxygen being much more electronegative than carbon and hydrogen. The electrons are drawn away from the hydrogen atom in the hydroxyl group, enabling it to react in water to form an $H^+(aq)$ ion, so the functional group can act as an acid (the nature of acids will be discussed in more detail in Chapter 11).

Carboxylic acids are weak acids and are often found in food. When foods go bad, carboxylic acids can be formed. For example, ethanoic acid is produced when wine is left open to oxygen in the atmosphere. We describe the taste of this wine as 'vinegary' because it actually has changed to vinegar. This reaction is used deliberately in the manufacture of the many different varieties of vinegar, such as apple cider vinegar.

The acid–base reaction of carboxylic acids with water produces a carboxylate ion, with the functional group R-COO⁻. The systematic names of **carboxylate ions** have the ending -oate. The equation below shows the formation of an ethanoate ion, CH_3COO^-, when ethanoic acid reacts with water:

$$\underset{\text{ethanoic acid}}{CH_3COOH(aq)} + H_2O(l) \rightleftharpoons \underset{\text{ethanoate ion}}{CH_3COO^-(aq)} + H_3O^+(aq)$$

Carboxylic acids have higher boiling points than you might expect from their molecular masses. (See Figure 8.4.8 for a comparison between the boiling points of carboxylic acids and alcohols.)

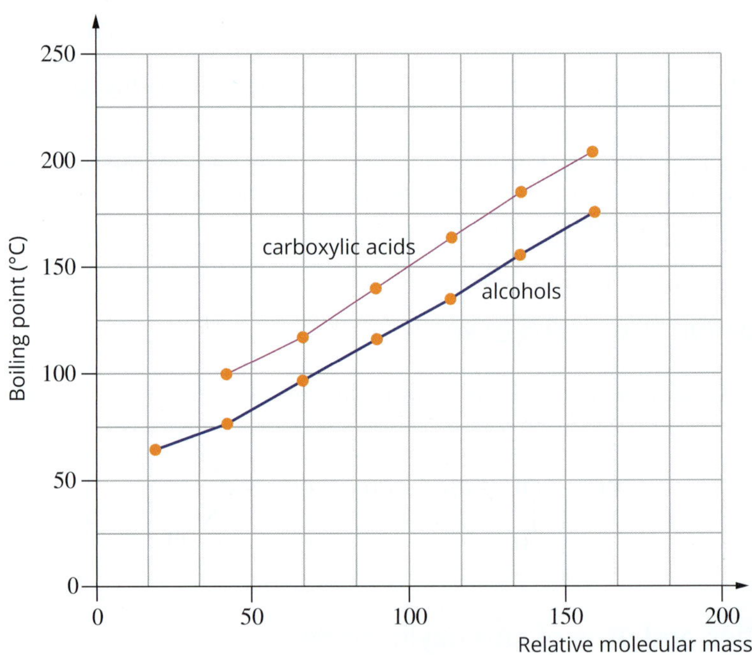

FIGURE 8.4.8 Carboxylic acids have higher boiling points than alcohols with similar relative molecular masses.

Carboxylic acids form hydrogen bonds between pairs of molecules which are then called dimers. This causes the boiling point to be much higher than expected, due to stronger dispersion forces.

Hydrogen bonding between two carboxylic molecules results in the two molecules forming a **dimer** (two identical molecules bonded together). You can see this in Figure 8.4.9. This increases the strength of dispersion forces between the dimers and hence increases the boiling point.

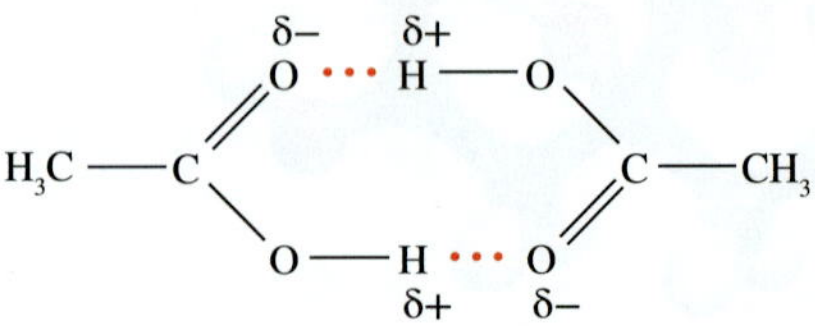

FIGURE 8.4.9 Hydrogen bonding between two ethanoic acid molecules results in the formation of a dimer.

As shown in Figure 8.4.10, when dissolved in water, hydrogen bonding occurs between the carboxyl group of carboxylic acids and water molecules, making carboxylic acids more soluble than alcohols in water. The high solubility of carboxylic acids explains why they are frequently found in solutions such as citric acid in orange and lemon juice.

FIGURE 8.4.10 Hydrogen bonds form between carboxylic acid molecules and water molecules.

The solubility of the carboxylic acids decreases as they increase in size. A longer carbon chain means that more of the molecule is non-polar. This is shown in Table 8.4.4.

TABLE 8.4.4 Solubility of carboxylic acids with 1–8 carbon atoms

Name	Formula	Number of carbon atoms	Solubility (g per 100 g H_2O)
methanoic acid	$HCOOH$	1	unlimited
ethanoic acid	CH_3COOH	2	unlimited
propanoic acid	CH_3CH_2COOH	3	unlimited
butanoic acid	$CH_3(CH_2)_2COOH$	4	unlimited
pentanoic acid	$CH_3(CH_2)_3COOH$	5	5.0
hexanoic acid	$CH_3(CH_2)_4COOH$	6	1.1
heptanoic acid	$CH_3(CH_2)_5COOH$	7	0.25
octanoic acid	$CH_3(CH_2)_6COOH$	8	0.07

CHEMFILE

Goat acids

While we are now familiar with the systematic names of hydrocarbons and, in this section, carboxylic acids, previously these compounds were known by common names which may be called 'trivial' names. Interestingly, there are three carboxylic acids which have trivial names that take their inspiration from *Capra*, the genus to which goats belong. These acids, caproic acid, $C_5H_{11}COOH$, caprylic acid, $C_7H_{15}COOH$, and capric acid, $C_9H_{19}COOH$, are found in high concentrations in goat's milk and they give the milk its characteristic 'goaty' flavour.

A goat being milked by hand. Goat's milk contains three different carboxylic acids that are all named after the genus name for goats, *Capra*.

8.4 Review

OA ✓✓

SUMMARY

- A functional group is an atom or group of atoms that gives a characteristic set of chemical properties to a molecule.
- The uses of alcohols relate to the properties of the alcohols, such as their ability to dissolve in water.
- All alcohols have a hydroxyl (–OH) functional group.
- Alcohols can dissolve in water because the hydroxyl group can form hydrogen bonds with water molecules. However, the solubility of alcohols decreases as the length of the carbon chain attached to the hydroxyl group increases.
- When naming an alcohol:
  - identify the longest carbon chain to give the stem name of the molecule
  - number the carbons from the end nearest to the hydroxyl group and use the same set of numbers for any side chains that occur
  - indicate the presence of the hydroxyl functional group with the suffix '-ol'.
- Carboxylic acid molecules contain the functional group –COOH.
- The –COOH group is called a carboxyl functional group and its presence is shown in the name of carboxylic acids with the suffix '-oic acid'.
- Carboxylic acids are weak acids with relatively high boiling points.
- Hydrogen bonds exist between carboxylic acid molecules, resulting in the formation of dimers.
- Smaller carboxylic acids are soluble in water because they can form hydrogen bonds with water molecules and less of the molecule is non-polar.
- When naming a carboxylic acid:
  - identify the longest carbon chain containing the carboxyl group to determine the stem name of the molecule
  - number from the carboxyl group and assign appropriate numbers to branches.

KEY QUESTIONS

Knowledge and understanding

1 Explain why alcohol molecules such as methanol and ethanol can dissolve in water.

2 Explain why the boiling points of alcohols

a are greater than those of alkanes with the same number of carbon atoms

b increase as the length of the carbon chain increases.

3 **a** Write a general formula which represents the alcohol homologous series.

b Use your answer to part **a** to evaluate whether it possible for the molecule $C_5H_{12}OH$ to be a member of the alcohol homologous series.

4 Small carboxylic acids, such as ethanoic acid, are very soluble in water. Draw a labelled diagram showing the intermolecular forces that occur between an ethanoic acid molecule and surrounding water molecules.

5 Write semi-structural formulas of the carboxylic acids based on their systematic names.

a butanoic acid

b 2-methylpropanoic acid

c 2,3-dimethylpentanoic acid

Analysis

6 Write the systematic names of the compounds shown in the following diagrams.

(a)

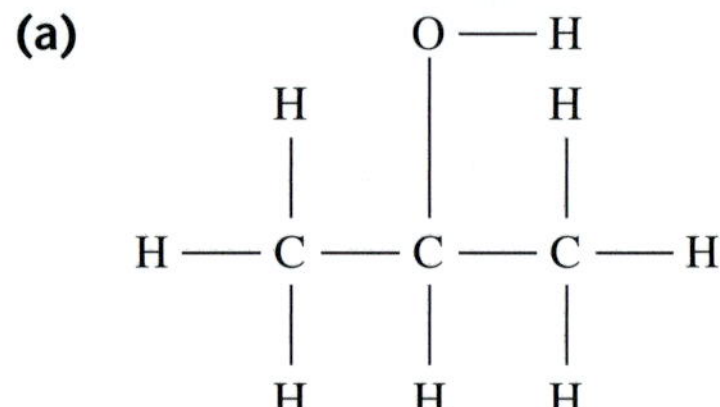

(b)

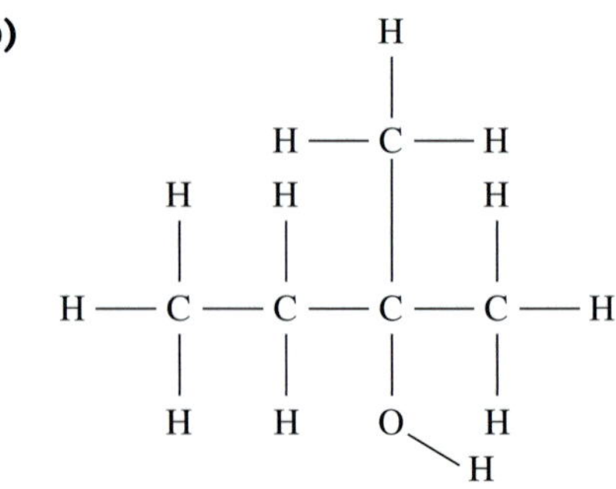

(c)

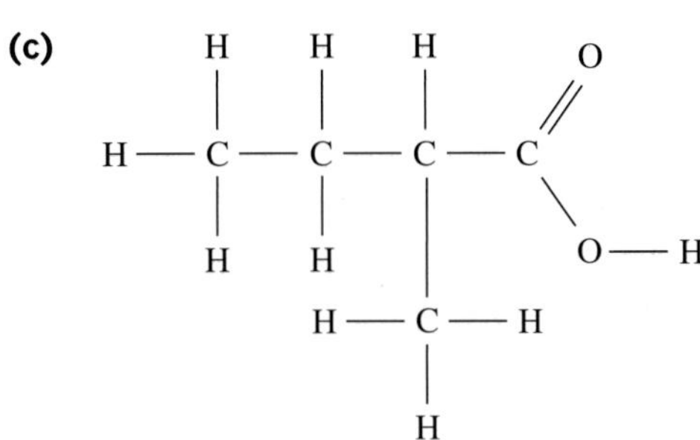

(d)

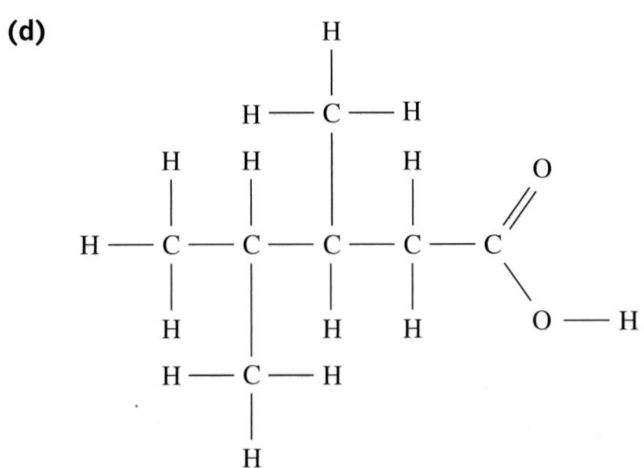

7 Draw the structural formulas of the following compounds.

- **a** pentan-3-ol
- **b** heptanoic acid
- **c** 3-methylpentan-1-ol
- **d** 2,5-dimethylhexan-3-ol
- **e** 3-methylbutanoic acid

8 Using the incorrect systematic names as a guide, draw the structures of the following compounds and state their correct systematic names.

- **a** butan-4-ol
- **b** 1,1-dimethylhexan-2-ol
- **c** 2-methylbutan-4-oic acid
- **d** 2,3-dimethylbutan-3-ol

Chapter review

KEY TERMS

addition reaction
alcohol
alkane
alkene
alkyl group
alkyl side chain (branch)
bio-derived
bioethanol
boiling point
carboxyl group
carboxylate ion
carboxylic acid
chemical property
combustion
complete combustion
condensed structural formula
crude oil
dimer
fractional distillation
functional group
general formula
haloalkane
halogen
homologous series
hydrocarbon
hydroxyl group
incomplete combustion
locant
melting point
molecular formula
non-renewable
organic chemistry
organic compound
photosynthesis
physical property
plant-sourced biomass
polymer
renewable
saturated
semi-structural formula
stem name
structural formula
structural isomer
substitution reaction
terminal carbon
unsaturated
weak acid

REVIEW QUESTIONS

Knowledge and understanding

1 Which of the following products is not derived from crude oil?

A petrol
B diesel
C coal
D asphalt

2 Which of the following compounds could be described as unsaturated?

A chloroethane
B 2-methylbutane
C 2,2-dimethylpropane
D but-2-ene

3 Which of the following statements is *not* correct?

A A functional group affects the chemical properties of a molecule.
B Alkanes have the general formula C_nH_{2n+1}.
C Alkene molecules are non-polar.
D Haloalkane molecules can be polar or non-polar, depending on the structure of the molecule.

4 Which of the following shows the correct order of increasing boiling point (lowest to highest) of alkanes, haloalkanes, alcohols, and carboxylic acids when molecules with the same number of carbon atoms are compared?

A carboxylic acids, alcohols, haloalkanes, alkanes
B alkanes, carboxylic acids, alcohols, haloalkanes
C alcohols, haloalkanes, carboxylic acids, alkanes
D alkanes, haloalkanes, alcohols, carboxylic acids

5 Draw the structural formulas and give the systematic names of:

a CH_3CH_3
b $CH_3CH(CH_3)_2$
c $CH_3CH_2CH(CH_3)CH_2CH_3$
d $CH_3(CH_2)_3CH_3$

6 The formula of a hydrocarbon is $C_{16}H_{34}$.

a To which homologous series does it belong?
b What is the formula of the next hydrocarbon in the homologous series?
c What is the formula of the previous hydrocarbon in the same homologous series?
d If this molecule had one C=C double bond in it, what would its molecular formula become?

7 Classify each of the following hydrocarbons as alkanes or alkenes.

a C_2H_4
b C_4H_{10}
c C_6H_{12}
d C_5H_{12}
e $C_{20}H_{42}$

8 Explain the following.

a The first member of the alkene homologous series is ethene, not methene.
b Carbon compounds usually have four covalent bonds around each carbon atom.

9 Name all the structural isomers of $C_3H_6Br_2$.

10 Identify two uses for alkanes, two uses for alkenes and one use for haloalkanes. In each instance you should identify which particular compound is put to that use.

11 Write a balanced chemical equation for the complete combustion of the following alkanes. Remember to include the states of the reactants and products.

a gaseous methane

b liquid hexane

c gaseous butane

12 Explain why fossils fuels such as petrol and diesel are not described as renewable.

13 Using a labelled diagram to illustrate your answer, explain why carboxylic acids have higher boiling points than the alcohols with the same number of carbon atoms.

Application and analysis

14 Which one of the following semi-structural formulas correctly matches the structural formula below?

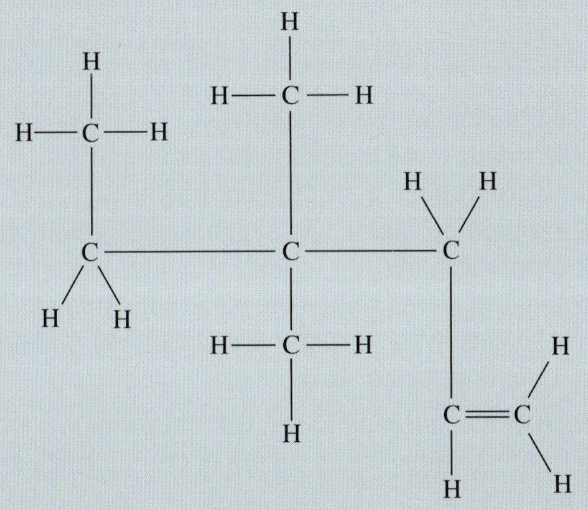

A $CH_3CH_2C(C_2H_6)CH_2CHCH_2$

B $CH_3CHCH_2CH(CH_3)_2CH_2CH_2$

C $CH_3CH_2C(CH_3)_2CH_2CHCH_2$

D $CH_3CH_2C(CH_3)_2CH_2CH_2CH_3$

15 Consider the following statement released by the LEGO company.

In 2018, we started making a range of sustainable LEGO elements from sugarcane to create polyethylene, a soft, durable and flexible plastic ... More than 80 LEGO elements are made from sustainably sourced polyethylene. Although these represent just 2% of the 3 600 elements available for designers, it is the first important step out of many on the journey towards using sustainable materials by 2030.

Explain why the LEGO company would believe that it is important to make this change to the material used for their building blocks.

16 Write the systematic names of the following compounds based on their structural formulas.

a

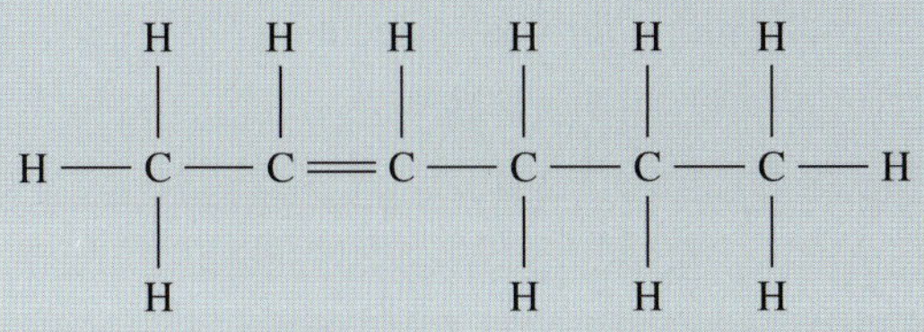

b

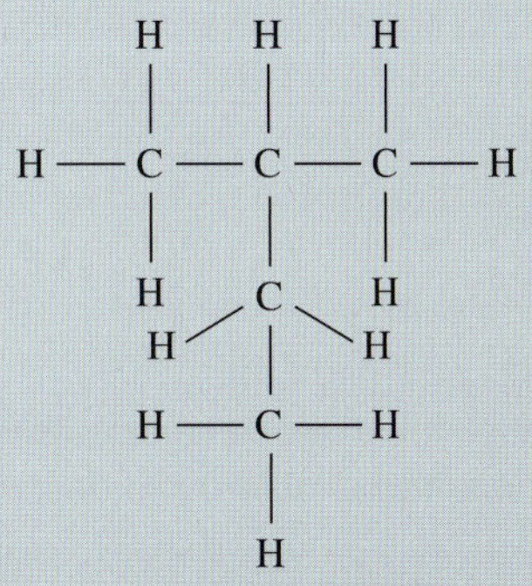

c

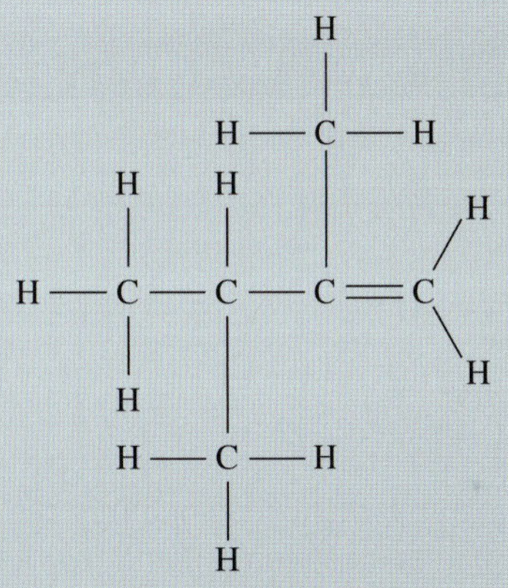

d

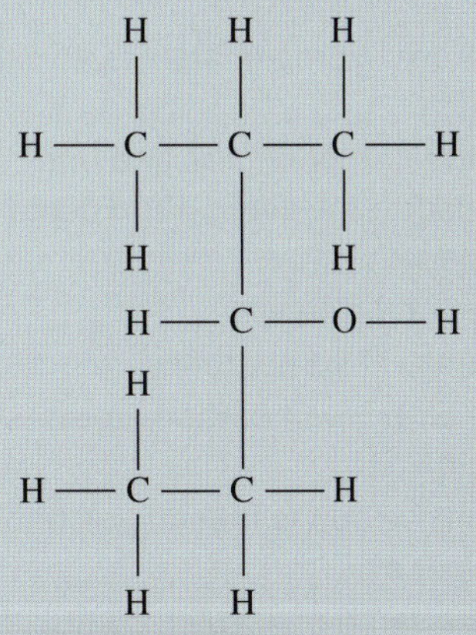

e

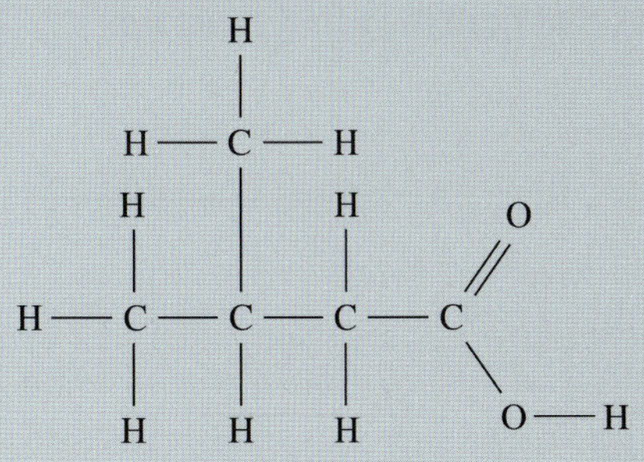

17 Explain why the systematic name for this molecule is not 3-methylpent-3-ene and state the correct systematic name for this molecule.

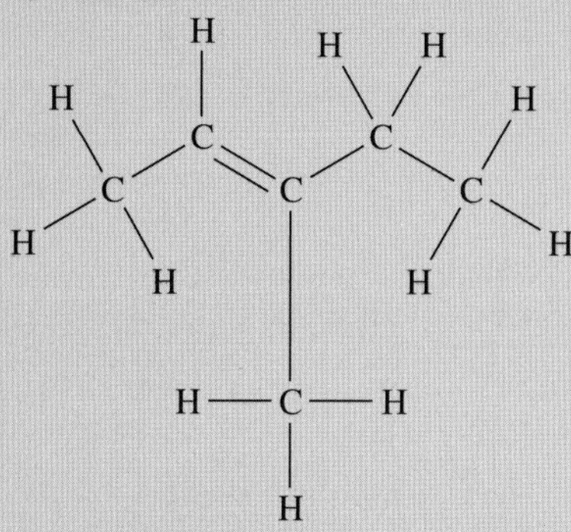

18 Complete the following summary table which compares the physical properties of alkanes, alkenes, haloalkanes and alcohols. You should try to describe trends or important aspects of each property for each class of hydrocarbons.

	Alkanes	Alkenes	Haloalkanes	Alcohols
Solubility in water				
Boiling point				
Bonding between molecules				

19 Each of the following statements has one mistake in it. Identify the mistake in each statement and rewrite the statement correctly.

- **a** Alkenes have two more hydrogen atoms per carbon atom than alkanes.
- **b** A haloalkane with five carbon atoms and one chlorine atom bonded to the end carbon could be called 1-chloropropane.
- **c** Pentane has 4 structural isomers.
- **d** Alkanes are unsaturated hydrocarbons.
- **e** The carboxylic acid with seven carbon atoms is called septanoic acid.
- **f** The alcohol functional group is found in alcohols and has the formula –OH.
- **g** Compounds with the same molecular formula have molecules that have the same structure.

20 The table below lists the boiling points of the first five alkanes and alcohols (with an hydroxyl group on carbon number 1).

Numbers of carbon atoms	Boiling point of alkane (°C)	Boiling point of alcohol (°C)
1	−162	65
2	−89	79
3	−42	97
4	0	117
5	36	138

- **a** Explain why the boiling points of the alkanes increases as the number of carbon atoms increases.
- **b** Identify all the homologous series described in this chapter that would have a higher boiling point than their corresponding alkane (with the same number of carbon atoms) and explain why this is the case.

21 Explain the following statements.

- **a** a '1' needs to be included in the name of 1-chloropropane but not in chloroethane.
- **b** 2,2-dibromopropane has a higher boiling point than 2-bromopropane.
- **c** octan-1-ol is less soluble in water than ethanol.
- **d** a '1' needs to be included in the name of butan-1-ol but not in butanoic acid.

22 In each of the following cases, use the written description to draw the structure of the molecule and name it.

- **a** A straight chain, saturated hydrocarbon with 12 hydrogen atoms.
- **b** A symmetrical unsaturated hydrocarbon with 4 carbon atoms.
- **c** An alkane with a methyl side chain on the second carbon and the third carbon, and a total of 14 hydrogen atoms.
- **d** A haloalkane with two carbon atoms and a molar mass of 108.9.

OA
✓✓

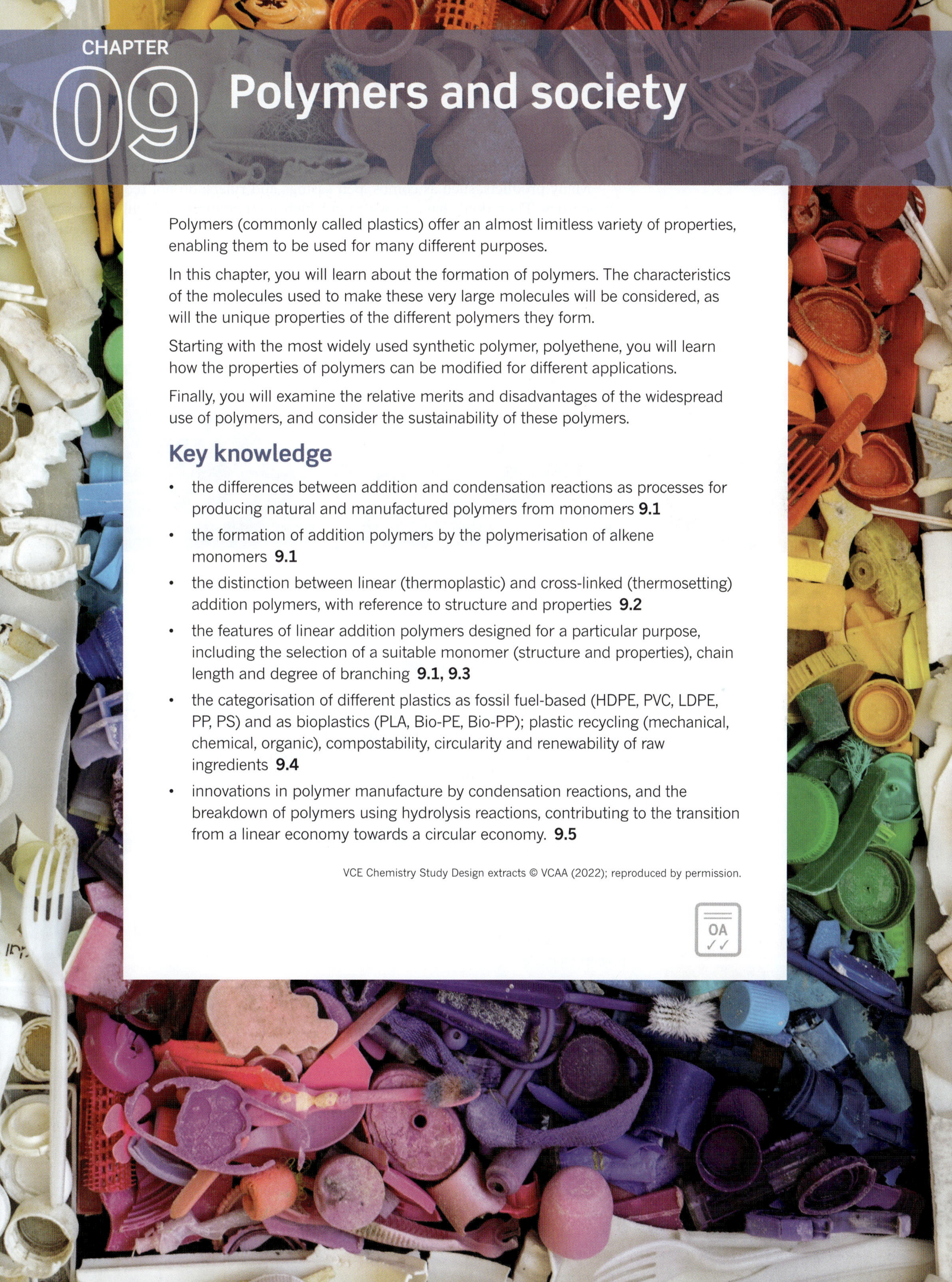

CHAPTER 09

Polymers and society

Polymers (commonly called plastics) offer an almost limitless variety of properties, enabling them to be used for many different purposes.

In this chapter, you will learn about the formation of polymers. The characteristics of the molecules used to make these very large molecules will be considered, as will the unique properties of the different polymers they form.

Starting with the most widely used synthetic polymer, polyethene, you will learn how the properties of polymers can be modified for different applications.

Finally, you will examine the relative merits and disadvantages of the widespread use of polymers, and consider the sustainability of these polymers.

Key knowledge

- the differences between addition and condensation reactions as processes for producing natural and manufactured polymers from monomers **9.1**
- the formation of addition polymers by the polymerisation of alkene monomers **9.1**
- the distinction between linear (thermoplastic) and cross-linked (thermosetting) addition polymers, with reference to structure and properties **9.2**
- the features of linear addition polymers designed for a particular purpose, including the selection of a suitable monomer (structure and properties), chain length and degree of branching **9.1, 9.3**
- the categorisation of different plastics as fossil fuel-based (HDPE, PVC, LDPE, PP, PS) and as bioplastics (PLA, Bio-PE, Bio-PP); plastic recycling (mechanical, chemical, organic), compostability, circularity and renewability of raw ingredients **9.4**
- innovations in polymer manufacture by condensation reactions, and the breakdown of polymers using hydrolysis reactions, contributing to the transition from a linear economy towards a circular economy. **9.5**

OA ✓✓

9.1 Polymer formation

Polymers are often referred to by the general term **plastics**. You can probably identify many items that are made of polymers. Polymers are used in the construction of many different objects because they are cheap, versatile and easy to manufacture.

Many products, such as combs, pen casings and rulers, do not require special properties. They don't have to withstand high temperatures or highly corrosive environments. These products can be made from cheap, lightweight polymer materials.

In Figure 9.1.1, you can see a range of familiar polymers. The polymers that make up these objects are selected for their strength or flexibility or other properties. In this chapter, you will learn that scientists have developed very sophisticated polymers with high performance properties.

FIGURE 9.1.1 (a) The polymer used to make the toy soccer players in this game was selected for its strength and how easy it is to mould. (b) The polymer banknotes used in Australia are strong and flexible. (c) The polymers in the bike helmet, gloves and bottle are similar, but differences in processing have given them very different properties.

POLYMER STRUCTURE

Polymers are covalent molecular substances composed of many small molecules all joined together. The word is made up of two parts, which come from Greek words. *Poly* means 'many', and *mer* means 'part'. They are formed by joining together thousands of smaller molecules, called **monomers** (*mono* means 'one') through a process called **polymerisation**, as shown in Figure 9.1.2.

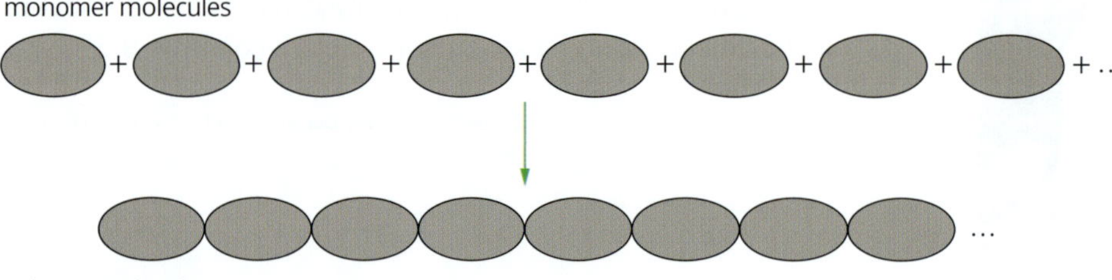

FIGURE 9.1.2 Monomers join to each other to form polymers.

Plastics and polymers

The word 'plastic' is frequently used to describe many items such as plastic wrap and detergent bottles. However, for chemists, the word 'plastic' describes a property of a material, not the material itself. A substance is described as being plastic if it can be moulded into different shapes readily. This is because the material from which it is made is a plastic material. Figure 9.1.3 shows two examples of objects made from polymers. The polymer used to make the basket has plastic properties because upon heating, the polymer would melt, allowing it to be reshaped. However, the saucepan handle is hard and brittle and will not melt when heated.

FIGURE 9.1.3 A plastic basket has plastic properties, whereas a polymer frying pan handle does not.

ADDITION POLYMERISATION

As you saw in Chapter 8, addition reactions involve the reaction of an alkene with another molecule. All of the atoms of both molecules are present in the final molecule. Under some conditions, alkenes undergo an addition reaction with themselves to produce long chains. The reaction of the monomer ethene with itself to form polyethene, shown in Figure 9.1.4, is an example of the **addition polymerisation** process. Several thousand ethene monomers usually react to make one molecule of polyethene.

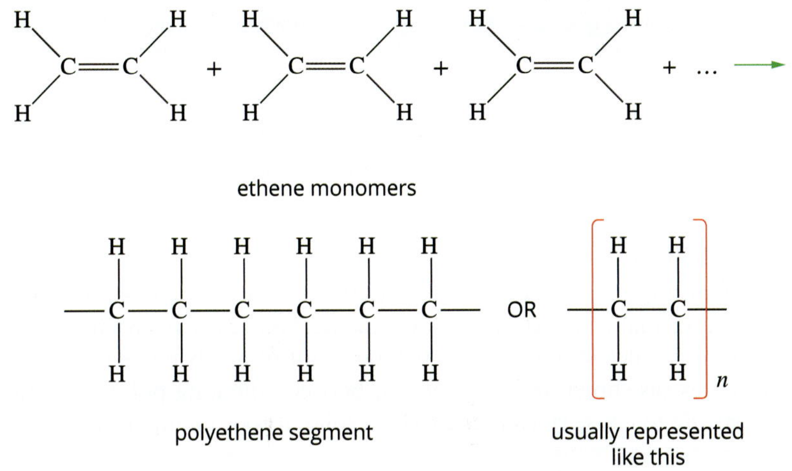

$$H_2C{=}CH_2 + H_2C{=}CH_2 + H_2C{=}CH_2 + \ldots \rightarrow$$

ethene monomers

$$-CH_2-CH_2-CH_2-CH_2-CH_2-CH_2- \quad \text{OR} \quad \left[-CH_2-CH_2-\right]_n$$

polyethene segment

usually represented like this

FIGURE 9.1.4 Thousands of ethene monomers join together to make one chain of polyethene. The standard notation shown simplifies the drawing of such a large chain.

Large square brackets and the subscript n are used to simplify the drawing of long polymer molecules. The value of n may vary within each polymer molecule, but the average molecular chain formed might contain as many as 20 000 carbon atoms. Polymers really are very large molecules!

Since all the atoms of the monomers are present in an **addition polymer**, the empirical formula of the monomer is the same as that of the polymer. Figure 9.1.5 provides an alternative representation of a polyethene chain segment, called a ball-and-stick model.

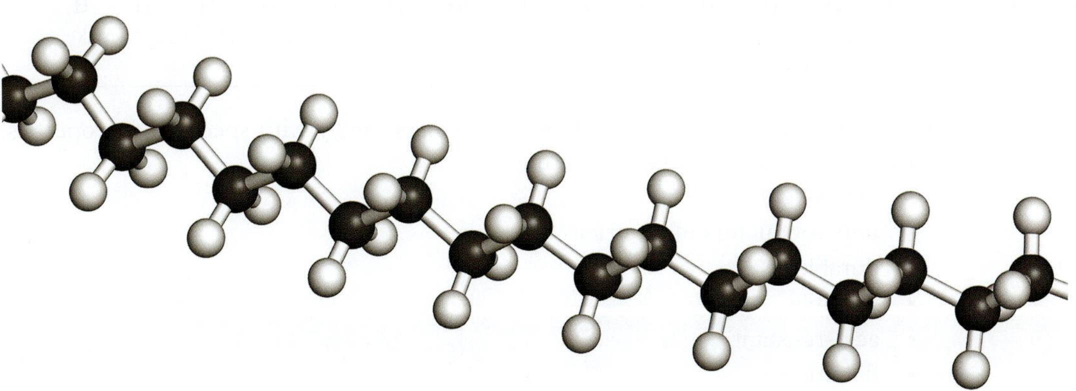

FIGURE 9.1.5 A ball-and-stick representation of a segment of polyethene

Ethene is an **unsaturated** molecule because it contains a carbon–carbon double bond. When ethene polymerises, the double bonds break and new covalent bonds are formed between carbon atoms on nearby monomers. The polyethene formed does not contain any double bonds. The name of a polymer formed through addition polymerisation will often include the monomer that was used to make it. The names of three common addition polymers and their monomers are listed in Table 9.1.1.

TABLE 9.1.1 Monomer and polymer names

Monomer	Polymer
Ethene	Polyethene
Propene	Polypropene
Tetrafluoroethene	Polytetrafluoroethene

Polymer properties

It is the length of polymer molecules that gives them many of their useful properties. Polyethene is essentially an extremely long alkane. You know from Chapter 8 that as the size of molecules increases, the melting point of a substance increases. The dispersion forces (as shown in Figure 9.1.6) between the long polymer chains are sufficiently strong to cause polyethene to be a solid at room temperature.

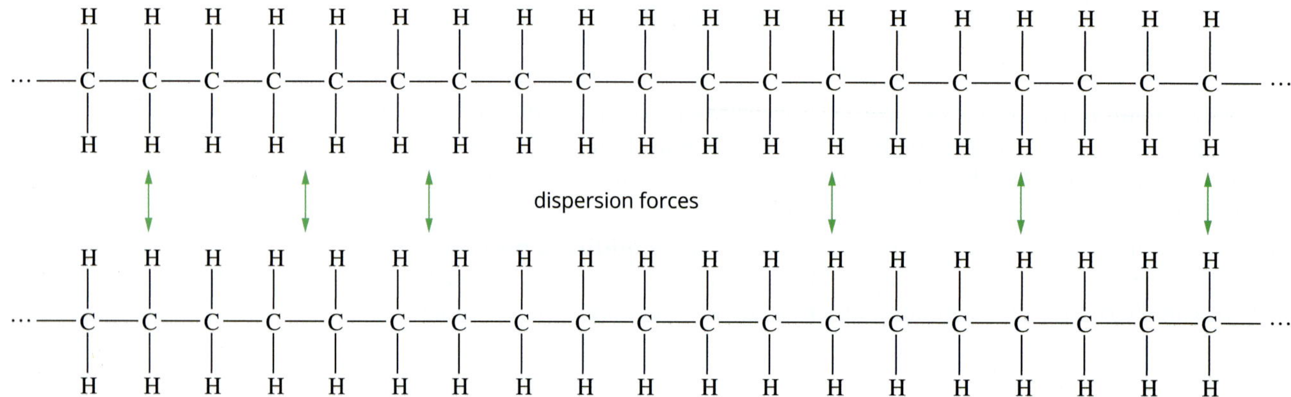

FIGURE 9.1.6 Dispersion forces between polyethene molecules are sufficiently strong to make it a solid at room temperature.

There are thousands of different polymers, many with specialised properties. However, in general, polymers are:

- lightweight
- non-conductors of electricity
- durable
- versatile
- acid-resistant
- flammable.

CASE STUDY

History of polymers

Wool, cellulose and proteins are naturally occurring polymers. However, most commercial polymers are synthetic and it is only in the last 100 years or so that their use has become widespread.

It is said that Native Americans were playing with crude rubber balls when European settlers first arrived in North America. These balls were made from the sap of rubber trees. American Charles Goodyear introduced large-scale production of rubber in 1839 when he invented the process of vulcanisation. Vulcanisation involves heating a polymer, usually rubber, with a small amount of sulfur. Goodyear realised that the properties of rubber were improved by this process. Goodyear is still a famous name associated with tyres.

The first completely synthetic polymer was released in 1909 by Leo Baekeland, a Belgian-born American chemist. He reacted the chemicals phenol and formaldehyde to form a hard material that he called Bakelite. Bakelite continues to be used to make bowling balls and saucepan handles. Table 9.1.2 shows a number of significant milestones in the history of polymers.

The two contrasting photos of cyclists in Figure 9.1.7 highlight the rapid developments made by the polymer industry over the last 100 years. Not only have the materials that the bicycles are made from changed, but so has the attire of the cyclists.

TABLE 9.1.2 Timeline for the development of some early polymers

Year	Polymer	Significance
1869	Celluloid (cellulose nitrate)	Billiard balls, photographic film and table-tennis balls
1907	Bakelite (phenol formaldehyde)	Light switches, saucepan handles
1927	Nylon	Created a shopping frenzy when used to make stockings in 1939
1927	PVC (polyvinyl chloride)	Low flammability and low electrical conductivity
1933	Perspex (polymethyl methacrylate)	Transparency enabled it to take the place of glass during World War II
1937	Polyurethane	Invented in Germany by Professor Otto Bayer. First used to replace rubber
1938	Teflon (polytetrafluoroethene)	Extremely difficult to handle due to its lack of 'stickiness'
1951	Polypropene (PP)	Second-most used polymer in the world
1972	Kevlar	Very strong and lightweight polymer. Flameproof
1980	Polyacetylene	Conductive polymer
1990	Polylactic acid	Biodegradable polymer

(a)

(b)

FIGURE 9.1.7 (a) Cyclists and metal bicycles from around 1920. (b) Australian cyclist, Paige Greco, competing at the 2021 Tokyo Paralympics.

CHEMFILE

The discovery of polyethene

The first practical method for the synthesis of polyethene was discovered by accident in 1933 in the laboratory of ICI in Cheshire, England, when some oxygen was accidently added to a container of ethene. The oxygen initiated the polymerisation reaction between the ethene molecules.

'When it first happened, it was a fluke,' recalled Frank Bebbington, a young laboratory assistant who was involved in the discovery. He assembled a reaction vessel to see if it was possible to reproduce this unexpected outcome, only to watch the pressure slowly fall. 'We thought there was a small leak in the system. I felt embarrassed,' he said.

His colleagues went to lunch and he continued to top up the reaction vessel with more ethene. After they returned, the vessel was opened and they found that they had made a new material. This material was later named polyethene.

Commercial use of the polymer flourished during World War II, when it was used to replace much heavier components in planes and ships.

Low-density polyethene

The earliest method of producing polyethene involved high temperatures (around 300°C) and extremely high pressures. Under these harsh conditions, the polymer is formed too rapidly for the molecules to be arranged in a neat and symmetrical manner. Figure 9.1.8 shows that the product contains many small chains that divide off from the main polymer, called **branches**.

```
                     CH3
                      |
                     CH2
                      |
                     CH2
                      |
                     CH2                           CH3
                      |                             |
···—CH—CH2—CH—CH2—CH2—CH—CH—CH2—···
     |                              |
    CH3                            CH2
                                    |
                                   CH3
```

FIGURE 9.1.8 Polyethene made under high pressure and at high temperatures has short branches off the main chain.

The presence of these branches impacts upon the properties of the polymer as the molecules cannot pack closely together. The dispersion forces between molecules are weaker when the molecules are further apart. The arrangement of the polymer molecules can be described as **amorphous** or non-crystalline. This arrangement of polymer molecules gives the material a relatively low density, so it is known as **low-density polyethene**, or LDPE. Its structure and properties are described in Figure 9.1.9.

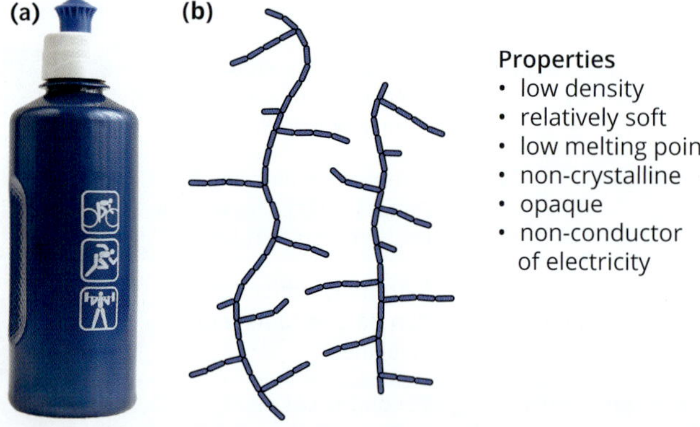

FIGURE 9.1.9 Low-density polyethene. (a) A bottle made from LDPE. (b) LDPE branched structure.

High-density polyethene

A low-pressure method of producing polyethene was developed by Union Carbide, an American chemical corporation, in the late 1960s. Highly specialised transition metal **catalysts**, known as Ziegler–Natta catalysts, are used to avoid the need for high pressures. The polymer molecules are produced under much milder conditions and there are very few branches.

The lack of branches allows the molecules to pack together tightly, increasing the density and the hardness of the polymer formed. The arrangement of the polymer molecules is more ordered, resulting in crystalline sections. This form of polyethene is known as **high-density polyethene**, or HDPE. Its properties and uses are summarised in Figure 9.1.10 on the following page.

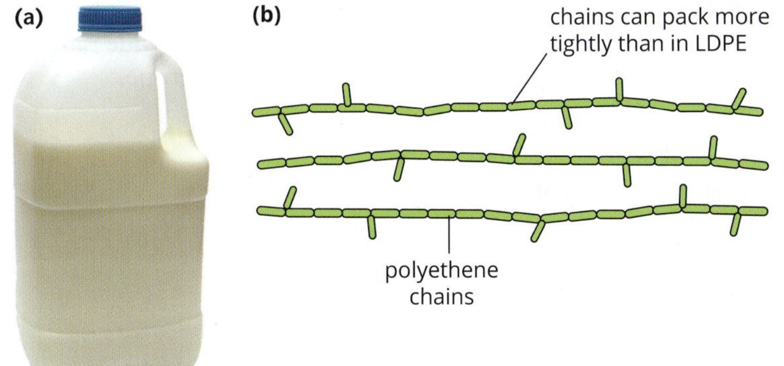

Properties
- high density
- hard
- higher melting point than LDPE
- crystalline sections
- non-conductor of electricity

FIGURE 9.1.10 Properties and structure of HDPE. (a) A bottle made from HDPE. (b) HDPE structure.

In some polymer materials, the entire solid is amorphous. Amorphous polymers are usually less rigid, weaker and often transparent (see-through). Table 9.1.3 compares the structure and properties of the highly crystalline HDPE with the relatively amorphous LDPE. The diagrams in the table contrast a thin, opaque HDPE freezer bag with the softer LDPE film.

TABLE 9.1.3 A comparison of a highly crystalline HDPE with an amorphous LDPE

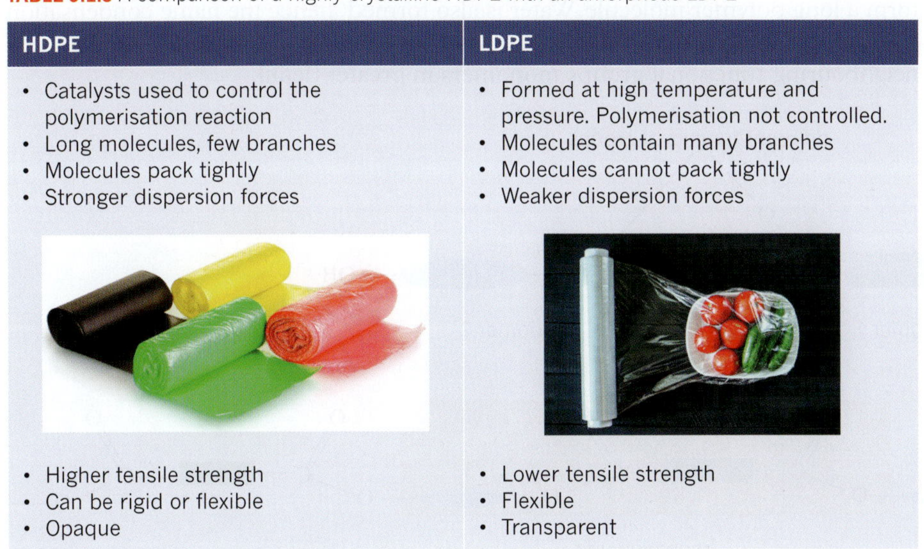

HDPE	LDPE
• Catalysts used to control the polymerisation reaction • Long molecules, few branches • Molecules pack tightly • Stronger dispersion forces	• Formed at high temperature and pressure. Polymerisation not controlled. • Molecules contain many branches • Molecules cannot pack tightly • Weaker dispersion forces
• Higher tensile strength • Can be rigid or flexible • Opaque	• Lower tensile strength • Flexible • Transparent

Other addition polymers

Ethene is only one of many possible monomers that can form useful products. Monomers need to be unsaturated to undergo addition polymerisation. Figure 9.1.11 shows in detail the reaction between bromoethene monomers to form polybromoethene. This is a specialty polymer used when a flame-resistant polymer is required.

$$n\,\mathrm{H_2C{=}CHBr} \rightarrow -\mathrm{CH_2-CHBr-CH_2-CHBr-CH_2-CHBr}- \rightarrow \left[-\mathrm{CH_2-CHBr}-\right]_n$$

bromoethene — polybromoethene

FIGURE 9.1.11 Addition polymerisation of bromoethene

One of the few polymers made in Australia is polypropene (PP), manufactured at LyondellBasell's Geelong plant (Figure 9.1.12). Polypropene has a wide range of uses, including synthetic sporting fields, microwave containers and rope.

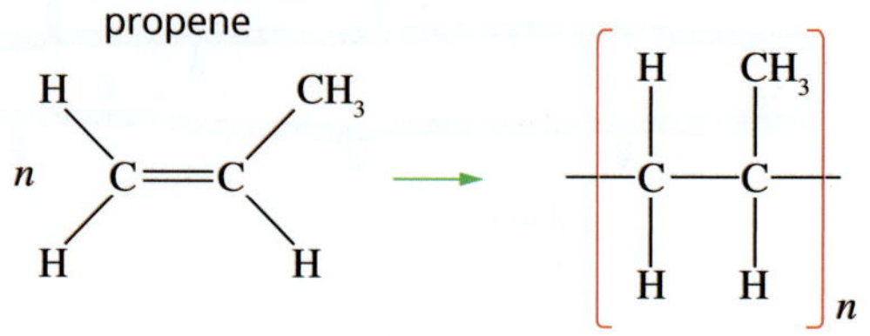

propene

$$n\,\mathrm{CH(H)\!=\!C(H)CH_3} \rightarrow \left[\mathrm{-CH_2-CH(CH_3)-}\right]_n$$

WS 19 PA 13

FIGURE 9.1.12 Propene undergoes addition polymerisation to form polypropene.

CONDENSATION POLYMERISATION

Another method for producing polymers is to use monomers with functional groups on each end of the molecule. The monomers join when the functional groups react with each other. Polymers formed in this way are called **condensation polymers**.

The formation of polyester is shown in a simplified way in Figure 9.1.13a. The carboxyl group on monomer 1 reacts with the hydroxyl group on monomer 2 to form what is known as an **ester** link. Since there are functional groups on both ends of each monomer, the same reaction can continue between further monomers to form a long polymer molecule. Water is also formed, hence the name condensation polymerisation. Figure 9.1.13b shows the condensation reaction between neighbouring functional groups monomers in greater detail.

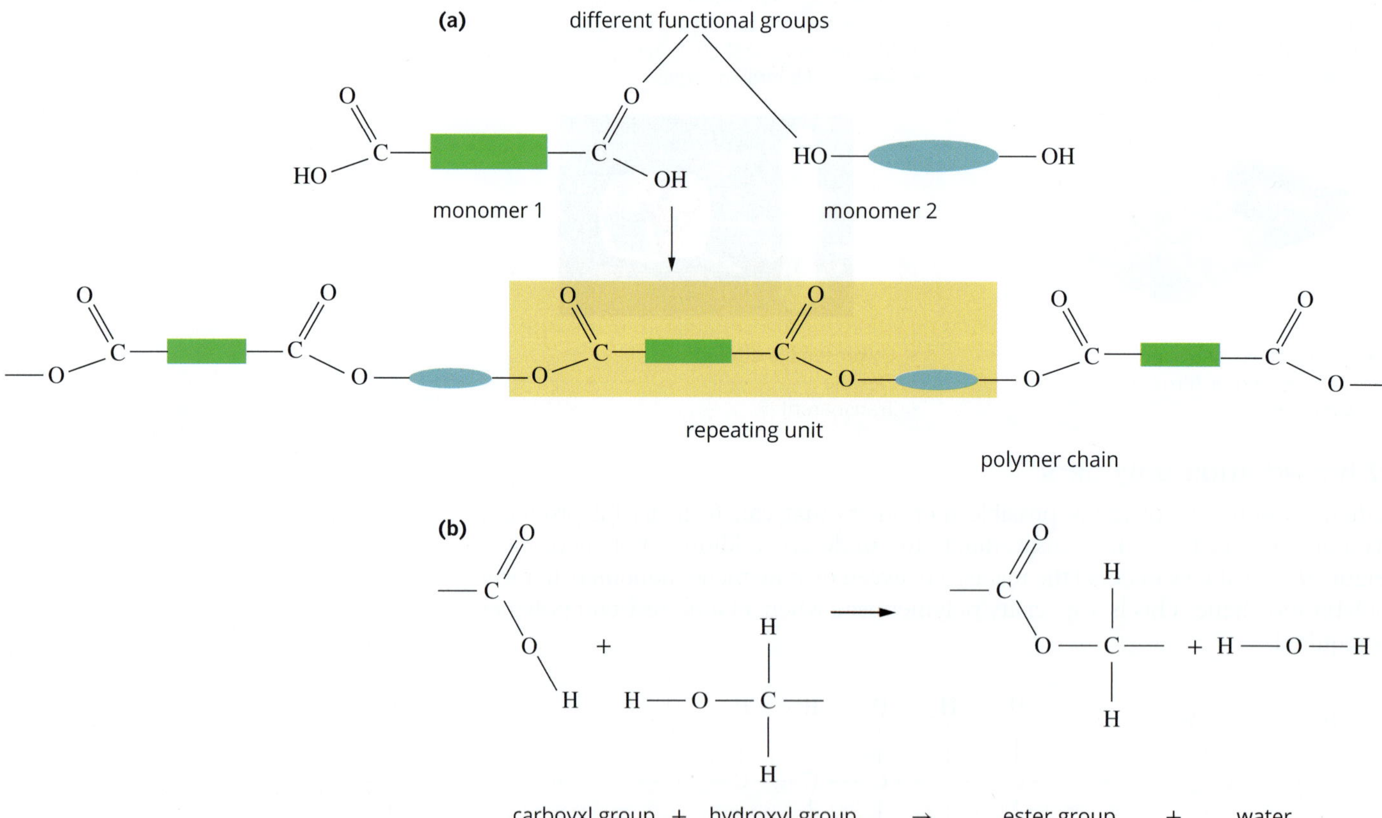

FIGURE 9.1.13 (a) Condensation polymerisation reactions require that the monomers have functional groups at each end of the molecule. These functional groups react together to form the polymer chain. (b) A condensation reaction occurs each time two monomers combine. An ester group is formed from the reaction between the carboxyl and hydroxyl groups and a water molecule is also formed.

Figure 9.1.14 shows the structure of one of the world's most-used condensation polymers, PET (polyethene terephthalate). Note the presence of either a carboxyl or hydroxyl functional group on the end of each monomer. PET might be familiar to you in the form of polyester fabric (Figure 9.1.15a) or bottled water containers (Figure 9.1.15b). Any plastic item with recycling code 1, as in Figure 9.1.15c, is made from PET.

1,4-benzenedicarboxylic acid + ethane-1,2-diol

+ water

monomer unit

ester groups

FIGURE 9.1.14 PET is formed when benzene-1,4-dioic acid reacts with ethane-1,2-diol. The PET is linked by ester groups, which is why it is part of the polyester family of polymers.

The formation of a condensation polymer does not necessarily require two different monomers. When polylactic acid is formed the only monomer required is lactic acid. It has a different functional group on each end of the monomer. This polymerisation reaction is shown in Figure 9.1.16. PLA is becoming increasingly popular for plastic products due to its ease of biodegradability.

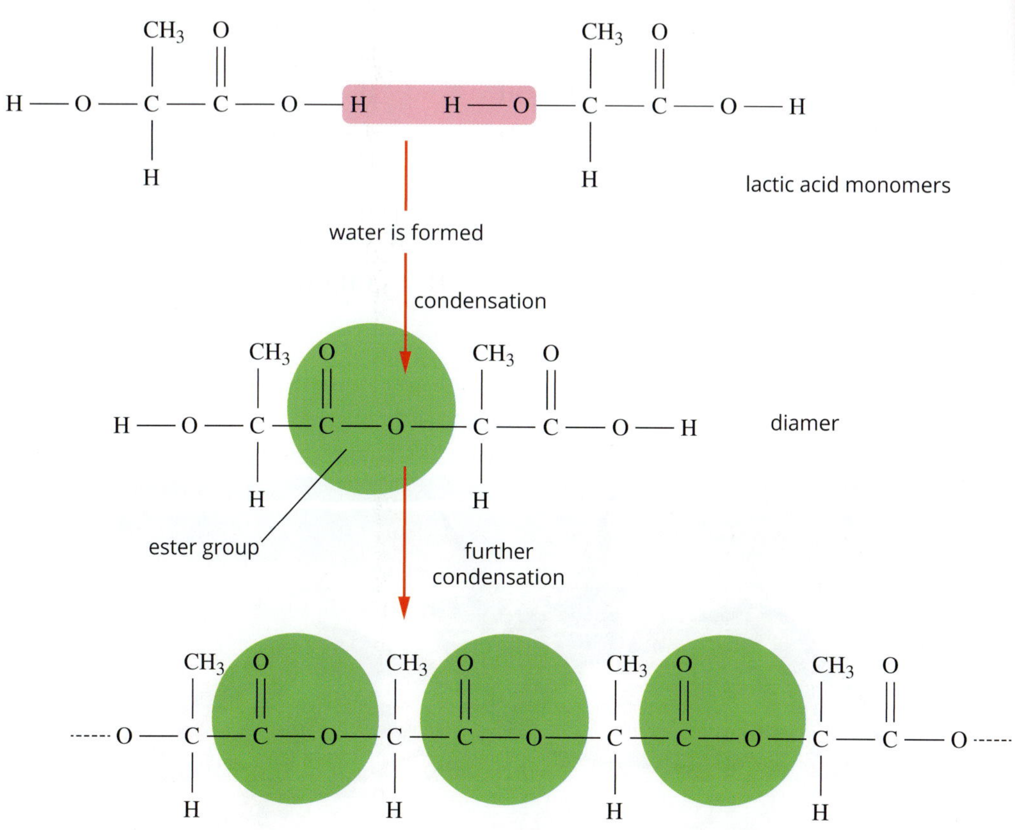

FIGURE 9.1.16 Polylactic acid is a condensation polymer made from one kind of monomer.

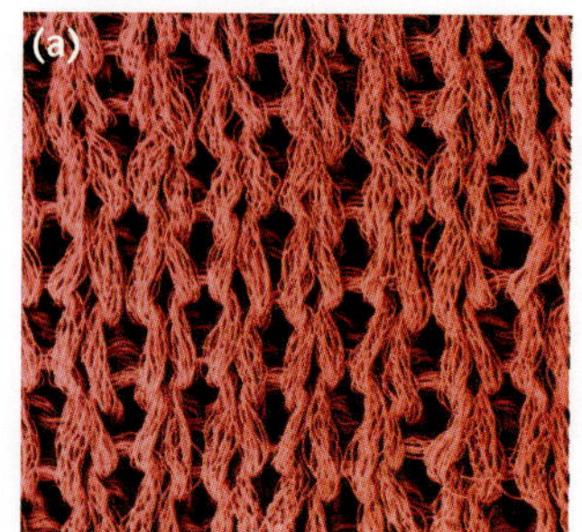

FIGURE 9.1.15 PET has many uses. (a) A scanning electron micrograph showing knitted polyester fibres. (b) PET is commonly used to make plastic water bottles. (c) The recycling code for bottles and other items made of PET is 1.

Table 9.1.4 contrasts the monomers used for addition polymerisation with those used for condensation polymerisation.

TABLE 9.1.4 Comparison of condensation and addition polymerisation

Addition polymerisation	Condensation polymerisation
Monomer has a carbon–carbon double bond. Double bond breaks during polymerisation. Polymer chain contains single C–C only. No other product formed.	Functional group on each end of the monomer. Functional groups react. Polymer chain will contain atoms other than C. Smaller molecules, such as water, formed.

NATURAL POLYMERS

While humans set up large factories to synthesise polymer molecules, a variety of living organisms continuously create natural polymers. Figure 9.1.17a shows the chemical structure of chitin, a polymer that is in the cell walls of fungi and is the main component of the exoskeletons of crustaceans (Figure 9.1.17b) and insects. Chitin is built from a monomer like all polymers. The polymerisation reaction occurs within the organism itself.

Proteins, silk, cellulose and starch are other examples of natural polymers. They are all condensation polymers.

(a)

chitin

FIGURE 9.1.17 (a) The chemical structure of chitin. The condensation polymerisation reaction occurs within the organism. (b) Chitin is the main component of the exoskeletons of crustaceans and insects.

9.1 Review

OA ✓✓

SUMMARY

- Polymers are long molecules formed by the reaction of thousands of monomer units.
- In general, polymers are durable and have relatively low densities. They are non-conductors of electricity and have relatively low melting points.
- Addition polymers are formed from the reactions of monomers containing carbon–carbon double bonds.
- The most common addition polymer used is polyethene. It can be manufactured in two different ways to make two different products: high-density polyethene (HDPE) and low-density polyethene (LDPE).
- Condensation polymers are formed when there are functional groups on each end of a monomer. A small molecule is also produced in this process.
- A number of polymers, such as cellulose, silk, chitin, protein and starch, exist in nature (natural polymers). Their structures are often complex, but they are examples of condensation polymers.

KEY QUESTIONS

Knowledge and understanding

1 **a** What is an addition reaction?
  b What is an addition polymerisation reaction?

2 **a** Draw a molecule of chloroethane and a molecule of chloroethene.
  b Which one of these molecules can form a polymer? Justify your answer.

3 Explain why a molecule containing 18 repeating units from the monomer ethene would not be considered a polymer.

4 Draw a section of the polymer that will form from the monomer shown.

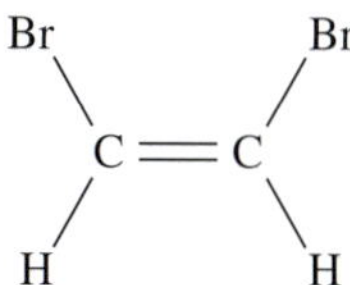

5 **a** What is a condensation reaction?
  b What is a condensation polymerisation reaction?

Analysis

6 Draw a section of the polymer that will form from the monomer shown.

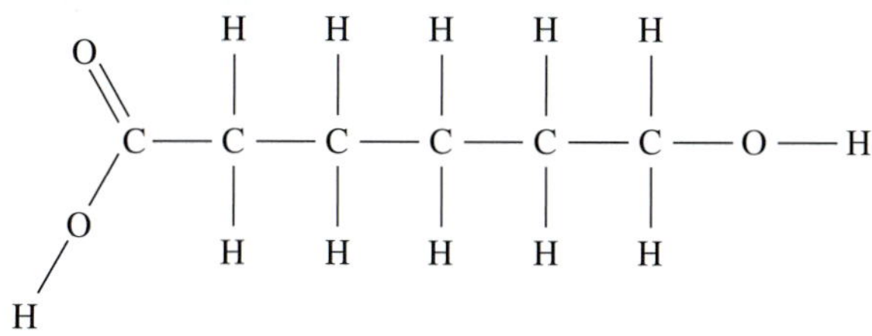

7 How many repeating monomer units does the following polymer segment contain?

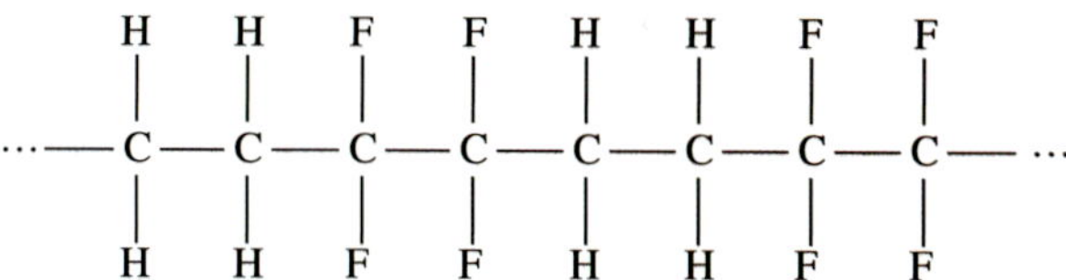

8 **a** In terms of their structures, explain the difference in properties between HDPE and LDPE.
  b Which of these two forms would be a suitable material for the following products?
    i a soft, flexible plastic wrap
    ii a 2-litre drink container
    iii wrapping material for frozen food

9 Is the melting point of polyethene manufactured by different companies likely to be exactly the same temperature? Justify your answer.

9.2 Thermoplastic and thermosetting polymers

Polymers can be classified into two groups on the basis of their behaviour when heated:

- **thermoplastic** polymers
- **thermosetting** (or thermoset) polymers.

THERMOPLASTIC POLYMERS

Thermoplastic polymers soften when heated, which means they can be remoulded or recycled. Polymers are only thermoplastic if the bonds between the long polymer chains are hydrogen bonds, dipole–dipole attractions or weak dispersion forces (Figure 9.2.1). When heated, the molecules in thermoplastic materials have enough energy to overcome the intermolecular forces and become free to move and slip past one another. If the polymer can be remoulded, it can probably be recycled easily, a desirable property in modern society. Most addition polymers are thermoplastic.

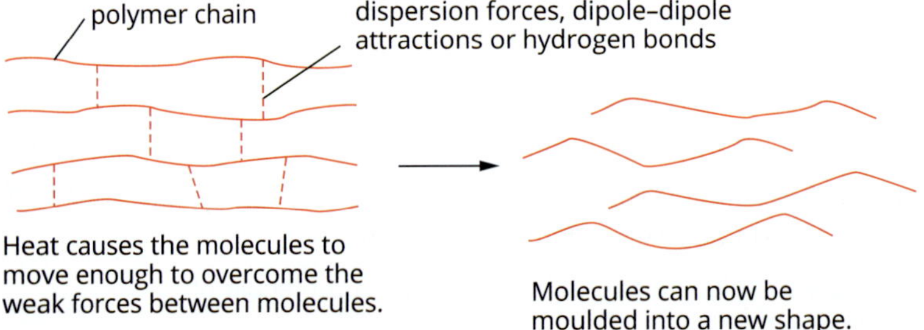

FIGURE 9.2.1 Heat overcomes the forces between molecules in a thermoplastic polymer.

THERMOSETTING POLYMERS

The monomers in some condensation reactions contain more than two functional groups. When polymerisation occurs a three-dimensional structure can form, rather than a linear polymer. Urea-formaldehyde, shown in Figure 9.2.2a and b, is an example. Notice that there are strong covalent bonds throughout the structure.

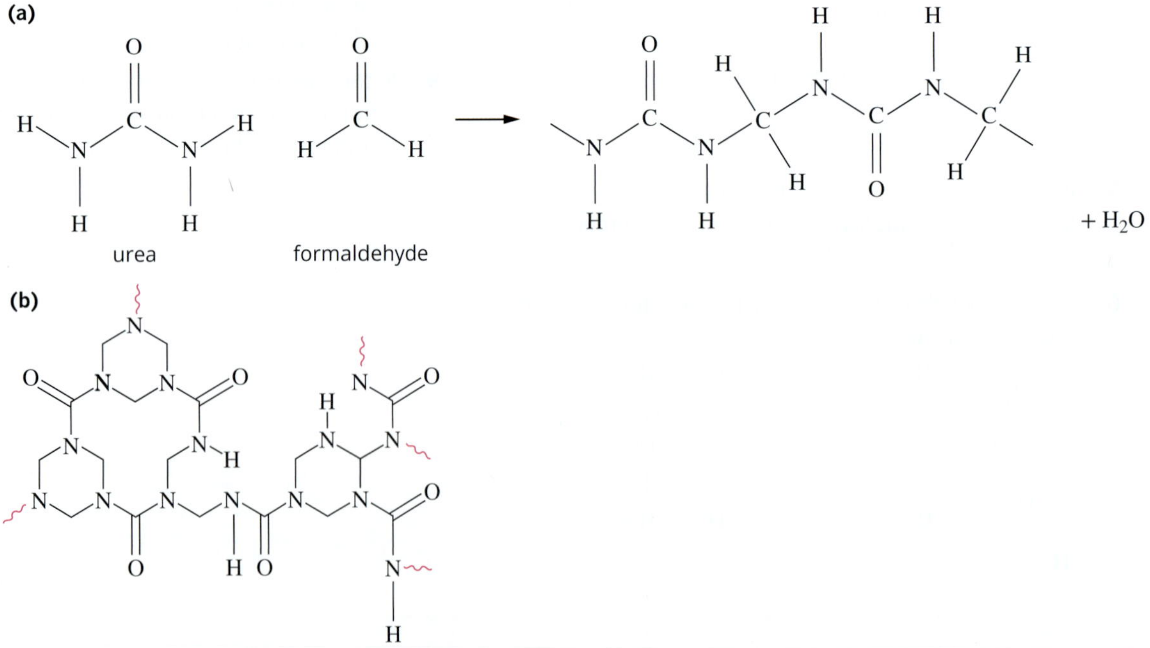

(a)

urea formaldehyde $+ H_2O$

(b)

FIGURE 9.2.2 (a) Segment of urea-formaldehyde condensation polymer. (b) Polymerisation of urea and formaldehyde continues to make a 3D structure.

Covalent bonds between polymer molecules are called **cross-links**. Cross-links limit movement between the polymer molecules and make the polymer rigid, hard and heat resistant. Thermosetting polymers are used to make items such as saucepan handles, bowling balls and shatterproof crockery.

Thermosetting polymers decompose or burn when heated. They do not soften because of the presence of the strong covalent bonds between the chains. (Figure 9.2.3). If the temperature becomes high enough to break the covalent bonds, the breaks may occur at any point, causing the polymer to decompose. It is difficult to recycle thermosetting polymers as they cannot be remoulded into new shapes.

> Thermoplastic polymers should be recyclable as they return to a liquid when heated, allowing them to be remoulded.
>
> Thermosetting polymers are difficult to recycle as they decompose or burn when heated.

All bonds in the structure are covalent bonds.
The bonds are heat resistant, but the whole structure collapses if heated strongly.

FIGURE 9.2.3 A thermosetting polymer has strong covalent bonds between chains.

In general, most modern plastic products are made from thermoplastic polymers to allow recycling to occur. Thermosetting polymers are only used when the strength or heat resistance they offer is needed.

ELASTOMERS

Elastomers are an interesting class of polymers that are formed when only occasional cross-links are present. The chains in these polymers can still move past each other when stretched, but the cross-links return the chains to their original positions once the force causing the stretching is released. Elastic bands and other rubber items are made of elastomers.

The cross-links stop elastomers from completely melting when heated, which makes recycling difficult. For example, the sulfur cross-links in the polymer in car tyres (see Figure 9.2.4) make the tyres non-recyclable.

```
~C—C~~C—C~~C—C~
 |       |
 S       S
 |       |
 S       S
 |       |
~C—C~~C—C~~C—C~
 |           |
 S           S
 |           |
 S           S
 |           |
~C—C~~C—C~~C—C~
```

FIGURE 9.2.4 The elastomer chains in rubber car tyres are cross-linked by sulfur atoms.

9.2 Review

OA ✓✓

SUMMARY

- A polymer is thermoplastic if it will soften when heated, allowing it to be reshaped. A thermoplastic polymer can be recycled by moulding it into a new shape. Thermoplastic materials have no strong bonds between polymer chains.
- Some polymers have covalent bonds, or cross-links, between polymer chains. Such thermosetting materials do not melt and they cannot be reshaped. If they cannot be reshaped, recycling is limited.

KEY QUESTIONS

Knowledge and understanding

1 Which of the following items are likely to be made from thermosetting polmers?
   a saucepan handle
   b banknotes
   c heat shield
   d school ruler

2 There are two main types of forces present in a sample of polyethene.
   a Name the types of forces present.
   b Which forces are disrupted when the polymer melts?

Analysis

3 Thermoset items are considered less sustainable than thermoplastic goods. Explain why this is the case.

4 Both thermosetting and thermoplastic polymers contain covalent bonds, yet thermoset polymers generally have better heat resistance. Explain why.

5 A thermoset polymer and a thermoplastic polymer are both heated until visible changes occur. Which sample is likely to produce toxic gases when heated? Explain your answer.

9.3 Designing polymers for a purpose

When it comes to polymers, it is definitely not the case that 'one size fits all'. Intensive research by chemists has enabled them to tailor the properties of a particular polymer to meet the exact requirements of its end use. To do this, chemists vary:

- the length of the polymer chains
- the monomer chosen
- the degree of branching
- other additives in the product, such as foaming agents, **plasticisers** and antioxidants.

FORMS OF POLYETHENE

Polyethene exists in several different forms that provide good examples of the effects of polymer chain length and branching on polymer properties.

Effect of chain length

Ultra-high molecular weight polyethene (UHMWPE) consists of extremely long polymer molecules. As a consequence, dispersion forces between chains are much stronger than in shorter chains of polyethene. Because of this, UHMWPE is such a tough polymer that it can be used to make artificial hip joints, safety helmets and even bulletproof vests.

Copolymerisation with ethene

As you learnt in Section 9.1, the amount of branching in polyethene can be varied by changing the process used for its manufacture. When ethene is copolymerised with a small amount of another alkene, a third form of polyethene is possible, LLDPE or **linear low-density polyethene** (see Alkatane Chemfile). A **copolymer** is a polymer made from at least two different monomers. The form of branching in LLDPE is a compromise between that of HDPE and LDPE—it has branches but they are very short (Figure 9.3.1). LLDPE retains the toughness of HDPE, but at a lower density, and hence lower cost.

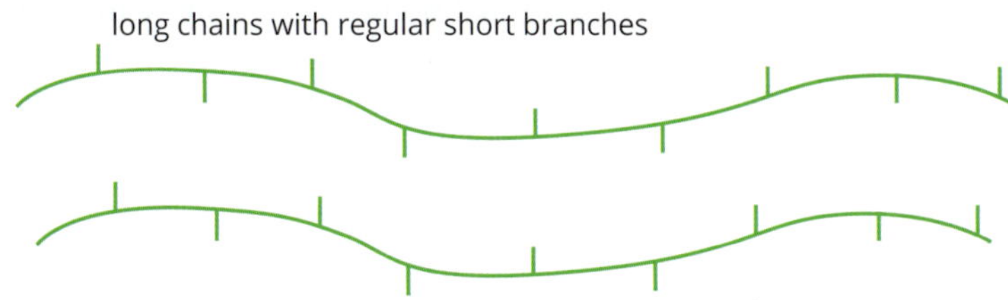

FIGURE 9.3.1 LLDPE has many branches, but they are very short.

CHEMFILE

Alkatane

When the summer Olympics were held in Atlanta in 1996, it was an Australian polymer, Alkatane®, that was used in the production of 54 000 seats in the main stadium. The same polymer is used in Australia to make 2 L milk bottles. Alkatane is manufactured in Altona by the petrochemical company Qenos.

Australian dairy farms produce over 9 billion litres of milk per year. The toughness of Alkatane has allowed the mass of each milk bottle to be reduced significantly, leading to significant savings in material used. Alkatane is formed when ethene is copolymerised with a small percentage of a larger alkene, such as but-1-ene. The use of revolutionary new catalysts known as metallocenes leads to a linear structure with regular small branches. The product is referred to as LLDPE, or linear low-density polyethene. The small branches are formed each time a but-1-ene is included in the chain and they increase the strength of the polymer.

CHOICE OF MONOMER

Many applications require polymers with more specialised properties than polyethene. For example, low flammability is essential for electrical wiring, and a baby's bottle needs a polymer with a higher melting point. One way to improve the properties of polyethene is to replace one or more of the hydrogen atoms on the monomer with more electronegative atoms, or with a larger group of atoms.

As can be seen in Figure 9.3.2, when a chlorine atom replaces a hydrogen atom on ethene, the polymer polyvinyl chloride, PVC, is formed.

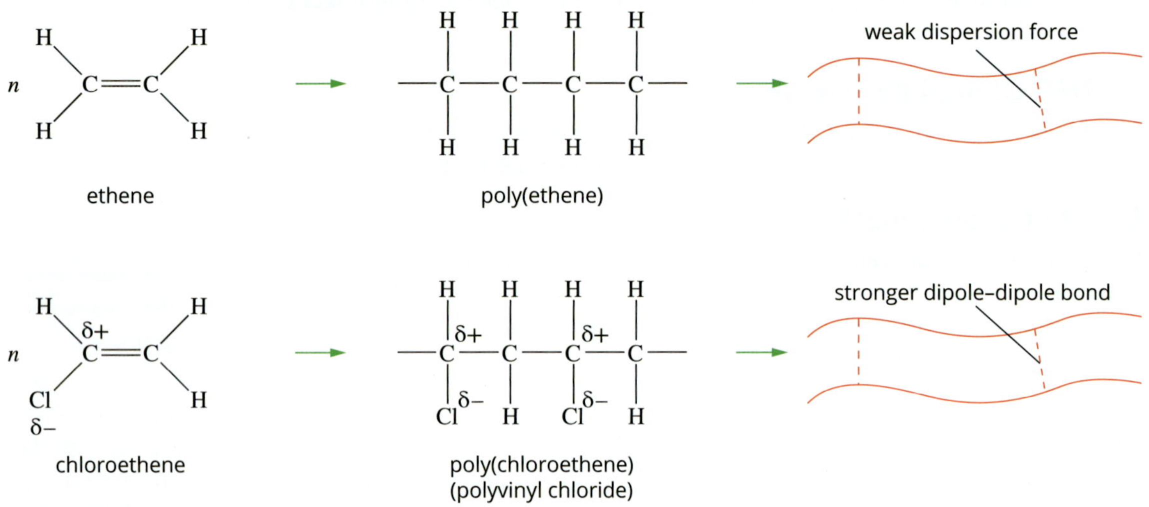

ethene

poly(ethene)

weak dispersion force

chloroethene

poly(chloroethene)
(polyvinyl chloride)

stronger dipole–dipole bond

FIGURE 9.3.2 Polyvinyl chloride is formed from polymerisation of polar chloroethene monomers and has dipole–dipole attractions between chains.

The chlorine atoms introduce dipoles into the long molecules. This increases the strength of the forces between polymer molecules, which leads to a higher melting point. PVC offers another advantage over polyethene: it has a low flammability. A PVC item burning in a flame will extinguish itself when it is removed from the flame. PVC is used in conveyor belts, cordial bottles, water pipes and the covering of electrical wiring.

Tetrafluoroethene ($CF_2{=}CF_2$) (shown in Figure 9.3.3) is formed when all of the hydrogen atoms in ethene are replaced by highly electronegative fluorine atoms.

tetrafluoroethene monomers

polytetrafluoroethene (PTFE) segment

usually represented like this

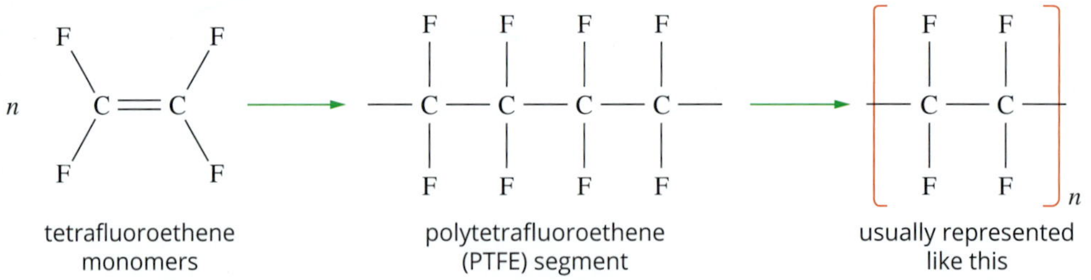

FIGURE 9.3.3 Addition polymerisation of tetrafluoroethene makes the polymer Teflon.

Molecules of tetrafluoroethene react with themselves to form the polymer polytetrafluoroethene, known as Teflon™. Teflon has quite exceptional properties that are very different from those of polyethene. The electronegative fluorine atoms reduce the strength of intermolecular bonds with other substances. The properties of Teflon are summarised in Table 9.3.1.

TABLE 9.3.1 A summary of the properties of Teflon™

Property	Description
non-stick	Teflon repels all other substances, both hydrophobic (oil, fat) and hydrophilic (water).
heat resistant	The melting point of Teflon is 335°C and the upper operating temperature for this polymer is 260°C.
chemical resistant	The polymer is extremely resistant to all known chemicals. It is not attacked by strong acids and bases and is inert towards all organic solvents.
good mechanical properties	Teflon is strong and durable, but not as hard as PVC.
low friction coefficient	Teflon is slippery to the touch. The friction coefficient between two pieces of Teflon is very low.
flame resistant	Teflon is non-flammable.

CHEMFILE

Teflon—a wonder material

American polymer company DuPont first manufactured Teflon in 1938. It was used during World War II as part of the process of isolating uranium for the first atomic bomb. After the war, its uses spread to plumber's tape, non-stick cookware and artificial hips and vocal chords, as manufacturers sought to take advantage of its heat resistance and low coefficient of friction.

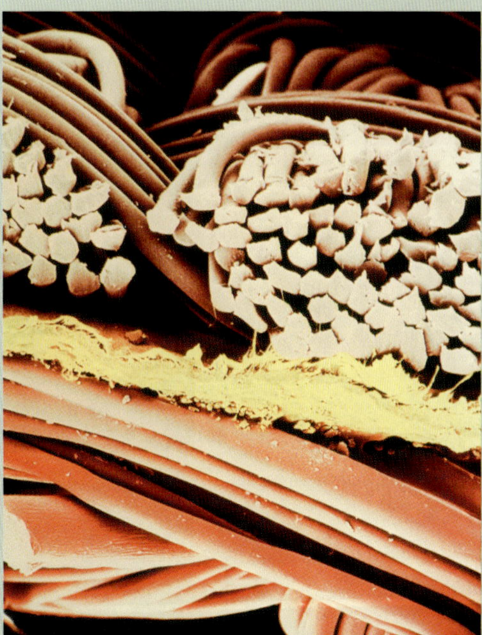

Scanning electron micrograph of Gore-Tex fabric. The small yellow and white particles are Teflon. The small holes between these particles allow vapour to pass through the fabric but not water.

The non-stick nature of Teflon has the disadvantage of making it difficult to apply to a surface such as a clothes iron. The metal surface has to be sand-blasted and the Teflon applied in several layers, starting with a type of primer.

Another innovative application of Teflon is in Gore-Tex® (shown below). Gore-Tex is a fabric that 'breathes'. Liquid water from rain cannot penetrate a Gore-Tex raincoat, but water vapour from sweat can escape through it. High-quality camping clothing is often made from Gore-Tex.

Typical weatherproof Gore-Tex jacket

There are thousands of commercial addition polymers. Table 9.3.2 shows a number of other polymers that may be familiar to you. Each of these polymers offers a unique property or properties that make them of commercial interest.

TABLE 9.3.2 Commercial addition polymers

Monomer	Poylmer	Properties	Examples	Application
$CH_3(H)C{=}C(H)H$ propene	polypropene (PP, polypropylene)	durable, cheap	artificial grass, dishwasher-safe plastic, ice-cream containers, rope	
$F_2C{=}CF_2$ tetrafluoroethene	polytetrafluoroethene (PTFE, teflon)	non-stick, high melting point	frying pan and iron coatings, plumber's tape, Gore-Tex fabric	
$Cl_2C{=}CH_2$ dichloroethene	polyvinylidene chloride (PVDC)	sticks to self, transparent, stretchy	food wrap	
$H_2C{=}C(H)CN$ propenenitrile	polypropenenitrile (acrylic)	strong, able to form fibres	acrylic fibres, fabrics	
$H_2C{=}C(H)C_6H_5$ phenylethene (styrene)	polyphenylethene (PS, polystyrene)	hard, brittle, low melting point	toys, packaging, expanded foams	
$H_2C{=}C(CN)C(=O)OCH_3$ methylcyanoacrylate	polymethylcyanoacrylate	polymerises on contact with water	super glue	
$H_2C{=}C(CH_3)C(=O)OCH_3$ methyl 2-methylpropenoate	polymethyl methacrylate (perspex)	transparent, strong	perspex (a glass substitute)	

OTHER MODIFICATIONS

Even after a monomer has been selected, chemists have developed further techniques to increase the diversity of the uses of plastics.

Bulky side groups

Bulky **side groups** in polymer chains make it difficult for the chains to slide over one another or stack closely together. This prevents the formation of **crystalline regions** that refract light. As a result, an amorphous material is produced that is often transparent, making it a useful substitute for applications where glass was once traditionally used.

Polystyrene

The side group in styrene is a flat ring of six carbon atoms called a phenyl group, shown in Figure 9.3.4.

Styrene polymerises, as shown in Figure 9.3.5, where the C_6H_5 side groups are covalently bonded to every second carbon atom in the polymer chain. This causes polystyrene (PS) to be a hard but quite brittle plastic with a low density. It is used to make food containers, picnic sets, refrigerator parts, and CD and DVD cases.

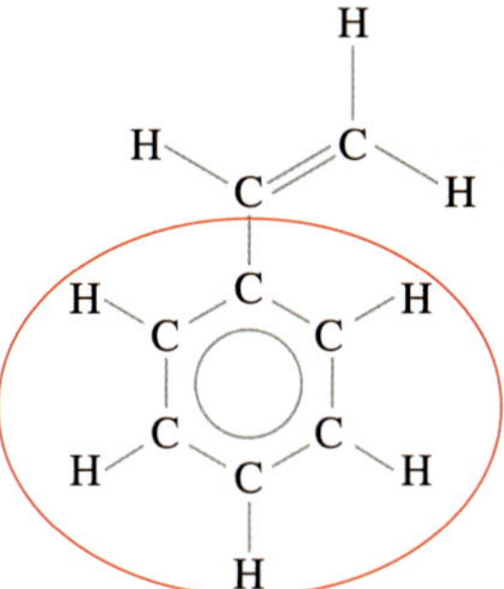

FIGURE 9.3.4 The chemical structure of the styrene monomer. The bulky side group is circled in red.

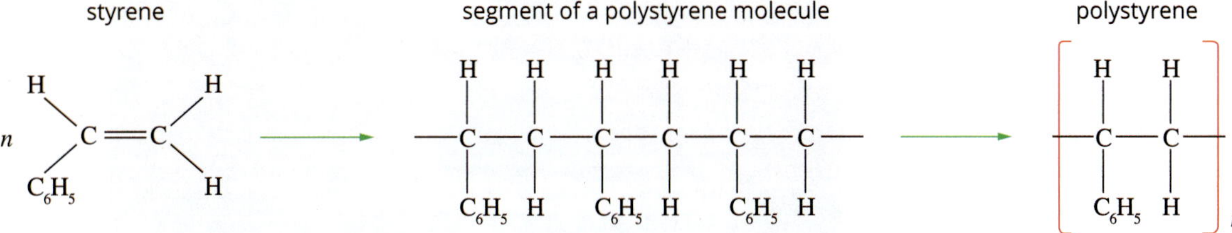

FIGURE 9.3.5 The polymerisation of styrene to form polystyrene

Foamed polymers

Foamed polymers are formed by blowing a gas through melted polymer materials. Foaming can drastically change the physical properties of a polymer material, as shown by the two examples of polystyrene in Figure 9.3.6. Polystyrene foam is produced by introducing pentane into melted polystyrene beads. The beads swell up to produce the lightweight, insulating, shock-absorbing foam that is commonly used for take-away hot drink containers, beanbag beans and packaging materials.

FIGURE 9.3.6 A model plane made of rigid polystyrene sitting on a block of foamed polystyrene

Specialty copolymers

You saw earlier that copolymerisation could be used to improved the properties of polyethene. The same principle of mixing monomers can be used to create other specialty polymers.

The Water Cube Stadium built for the 2008 Beijing Olympic Games (Figure 9.3.7) was made with a copolymer of ethene ($CH_2=CH_2$) and tetrafluoroethene ($CF_2=CF_2$) monomers. The copolymer is known as ethene tetrafluoroethene (ETFE). The stadium has over 100 000 m^2 of very thin ETFE 'bubble walls'. The walls allow more light to penetrate than traditional glass does, lowering energy costs. ETFE weighs only 1% of the weight of glass and is a better thermal insulator. It was designed by a consortium including two Australian companies: PTW Architects and Arup.

FIGURE 9.3.7 The outer shell of the Water Cube Stadium in Beijing, China, is made of ETFE, a new copolymer building material.

Another copolymer called styrene–butadiene rubber (SBR) is formed from styrene ($CH_2=CH(C_6H_5)$) and butadiene ($CH_2=CHCH=CH_2$). Its properties can be varied simply by altering the ratio of the monomers. For example, an elastomer similar to natural rubber is produced when the two monomers are present in close to equal amounts. The styrene monomers increase the abrasion resistance of the polymer and make it suited for use in car tyres, its main application.

Addition of a third monomer, acrylonitrile ($CH_2=CHC\equiv N$), produces the polymer acrylonitrile–butadiene–styrene (ABS), which is used to make Lego® blocks. This polymer is rigid and strong, but can be melted easily. These properties have made this thermoplastic polymer very popular for use in 3D printing.

In Figure 9.3.8, a 3D printer melts a thin cord of ABS, called a filament. The molten ABS is built up in layers, where it sets to make the solid object.

FIGURE 9.3.8 This 3D printer uses an ABS copolymer filament to make an object.

Advantages and disadvantages of polymers

Polymers have become the dominant material used in our society. There are many reasons why this has occurred and it is wise to be aware of potential drawbacks that arise from their use. Some of these advantages and disadvantages are shown in Table 9.3.3.

TABLE 9.3.3 Some advantages and disadvantages of using polymers

Advantages	Disadvantages
available in an enormous variety of different forms, with distinctive properties	many are derived from petroleum, a non-renewable resource
generally biologically inert, chemical resistant, corrosion resistant	microorganisms cannot break down most synthetic polymers (they are not biodegradable)
easy to process by moulding	thermoplastic polymers have a limited thermal stability
low density	some plastic products crack, scratch or break easily
good mechanical strength	many plastics produce toxic gases, such as hydrogen chloride, hydrogen cyanide and dioxins, when burnt
properties can be modified, e.g. by foaming or adding plasticisers	some plasticisers can leach out of containers or wraps and pose a health risk
many can be recycled	thermosetting polymers are currently difficult to recycle

9.3 Review

OA ✓✓

SUMMARY

- Materials composed of polymers offer an almost limitless variety of properties, enabling them to be used for many different purposes.
- Polymers can be designed for a particular purpose by selecting suitable monomers, reaction conditions or additives.
- Factors that affect the physical properties of polymers include the:
  - polarity of side groups in the polymer
  - use of more than one monomer (copolymers)
  - polymer chain length
  - extent of branching of polymer chains, e.g. LDPE and HDPE
  - arrangement of side groups in the polymer chain.
- A copolymer is a polymer that is made from more than one monomer.

KEY QUESTIONS

Knowledge and understanding

1 Refer to Table 9.3.2 on page 304 and draw equations to represent the formation of:

a polypropene (CH_2CHCH_3)

b Teflon (CF_2CF_2)

c polypropenenitrile (CH_2CHCN)

2 Draw a segment of a copolymer that might be formed between propene and chloroethene.

3 a How many repeating monomer units does the polymer segment below contain?

b Name the monomer used to make the polymer.

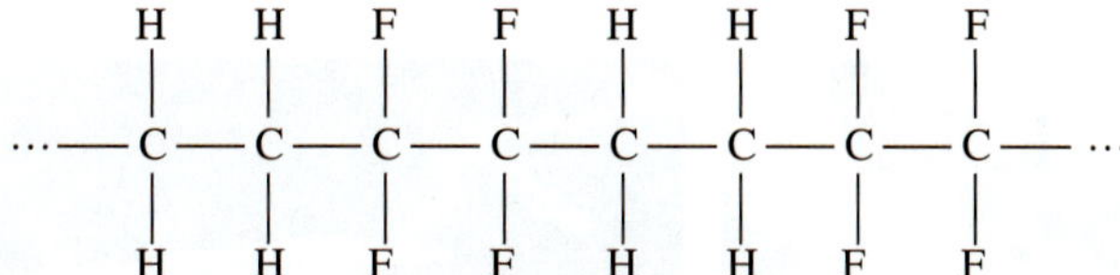

4 The following side groups are found in different polymers. Sort the groups from smallest to bulkiest.
$-F$ (in Teflon); $-NC_{12}H_8$ (in polyvinyl carbazole);
$-H$ (in polyethene); $-C_6H_5$ (in polystyrene);
$-Cl$ (in polyvinyl chloride)

Analysis

5 Identify the strongest type of intermolecular force present between polymer molecules produced from the following monomers.

a ethene ($CH_2{=}CH_2$)

b vinyl chloride ($CH_2{=}CHCl$)

c styrene ($CH_2{=}CHC_6H_5$)

d propene ($CH_2{=}CHCH_3$)

e acrylonitrile ($CH_2{=}CHCN$)

6 Unbranched polymers containing chains of equal length are made from each of these monomers: chloroethene, propene and tetrafluroethene.

a What is the likely order of the polymers, ranked from lowest melting point to highest melting point.

b Why did part **a** of this question state 'unbranched polymers containing chains of equal length'?

7 Many small eskies are made from polystyrene foam.

a How does polystyrene foam differ from polystyrene?

b List four properties of polystyrene foam that make it suitable for use as an esky.

8 The density of polystyrene is around $0.95\,g\,cm^{-3}$. When polystyrene is expanded to form a foam the density drops to around $0.05\,g\,cm^{-3}$.

a Calculate the mass of polystyrene required to manufacture a cube of side length 10 cm from both forms of polystyrene.

b Your answer to part **a** highlights one of the advantages of using a foam. What is this advantage?

9.4 Recycling plastics

FIGURE 9.4.1 Plastic persists for a very long time in the environment.

FIGURE 9.4.2 Young boy removing plastic bags from the mouth of a dead dolphin

Australians consume more than 1.5 million tonnes (1.5 billion kilograms) of polymer materials every year, which includes many different plastics. The disposal of the waste polymer material is a serious issue in our society.

Plastics are durable, chemically resistant and lightweight. These properties make plastics very useful, but they also have a serious environmental impact. The biodegradation of plastics is very slow. Once discarded, plastics persist for a very long time, possibly hundreds of years. Because synthetic polymers have low density, waste plastic takes up more volume than other kinds of waste. They occupy the limited space available in landfills and litter the environment (Figure 9.4.1). Figure 9.4.2 is an all-too-familiar image of the harm plastic items can do to marine species.

Visible plastic pollution is one problem; however, there is another emerging environmental issue, **microplastics**. Often, plastic items do not actually degrade, they just break up into ever smaller pieces, until they are smaller than 5 mm wide. These often invisible particles are referred to as microplastics and scientists are now researching their impact on marine creatures.

Ideally, we should create a lot less waste from plastic. This sounds simple in theory, but unfortunately it is more complex than it seems.

PRODUCTS MADE FROM RECYCLED PLASTICS

The problems related to disposal have resulted in more waste plastic being collected from users to be recycled. Some recyclers shred mixed plastics into pellets, then remould them into general purpose products, such as garden furniture and retaining walls. Figure 9.4.3a shows the shredded pellets and 9.4.3b shows a picnic table and bench seats made from these pellets.

Other recyclers separate plastics to their individual polymers to remould them into higher quality items. For instance, artificial clothing fibres can be spun from PET (polyethene terephthalate) bottles. Recycling can use only thermoplastic polymers. The recycling examples in this section are referred to as examples of **mechanical recycling**, where the polymer structure is unchanged but it is moulded into a different product.

Mechanical recycling is when a polymer is remoulded into a new product, but its chemical structure is unchanged.

FIGURE 9.4.3 (a) Reprocessing plastic by shredding and moulding new products (b) Bench seat and table produced by Replas—a company that recycles plastics into new products

IDENTIFICATION OF RECYCLABLE PLASTICS

A numbering scale is used to identify plastics for recycling (Figure 9.4.4). The items listed under each code number are made from that plastic. Most are recyclable, but some, such as plastic wrap and expanded polystyrene foam, are not practical to recycle. If the product is recyclable, the recycle code is usually printed on it.

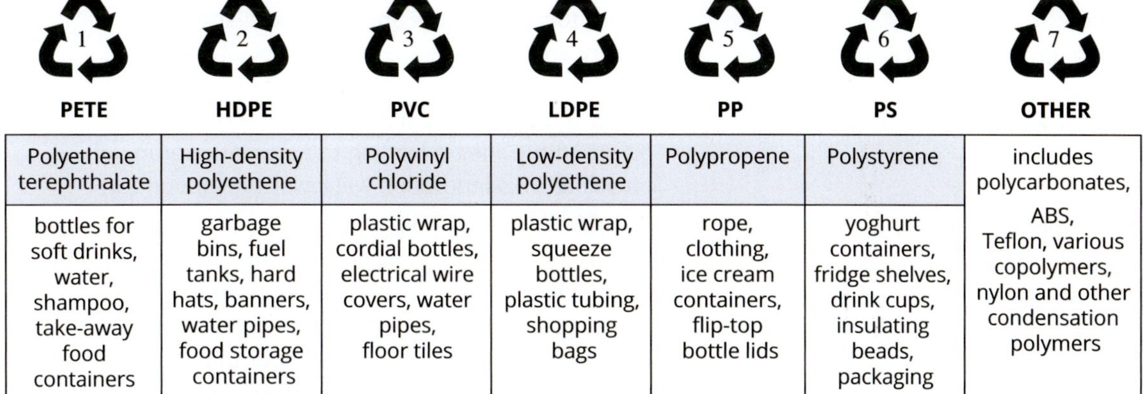

1 PETE	2 HDPE	3 PVC	4 LDPE	5 PP	6 PS	7 OTHER
Polyethene terephthalate	High-density polyethene	Polyvinyl chloride	Low-density polyethene	Polypropene	Polystyrene	includes polycarbonates, ABS, Teflon, various copolymers, nylon and other condensation polymers
bottles for soft drinks, water, shampoo, take-away food containers	garbage bins, fuel tanks, hard hats, banners, water pipes, food storage containers	plastic wrap, cordial bottles, electrical wire covers, water pipes, floor tiles	plastic wrap, squeeze bottles, plastic tubing, shopping bags	rope, clothing, ice cream containers, flip-top bottle lids	yoghurt containers, fridge shelves, drink cups, insulating beads, packaging	

FIGURE 9.4.4 International number code used to identify recyclable plastics

The success of recycling varies depending on the country and the polymer. Statistics from 2019 show that Norway recycled an impressive 97% of its PET plastic compared to Australia with 21%. Norway, however, recycles less than 40% of its polyethene. This reflects the ease of identifying and separating PET drink bottles and of remoulding the polymer into products of the same quality as the original. Government and industry are seeking new ways to improve the rates of plastic recycling. Soft plastics such as plastic wrap are not permitted in recycling bins. The collapse of recycling organisation REDcycle in 2022 illustrates that there are still many challenges to be worked through before we have the capability to recycle all plastic. Before the collapse, REDcycle had partnered with Coles and Woolworths to collect soft plastic at supermarkets. Other recycling companies such as Replas were then able to use the soft plastic recovered, mixed with other recycled plastics, to make outdoor products such as furniture and Polyrok (Figure 9.4.5a), an additive to concrete that replaces the usual minerals used (Figure 9.4.5b).

FIGURE 9.4.5 Soft plastic can be recycled to make (a) Polyrok, an additive to (b) footpath and road surfaces.

CASE STUDY ANALYSIS

Some choices are smarter than others

Typical take-away food packages, like those shown in Figure 9.4.6, might include:

- a biodegradable coffee cup lid that will remain unchanged after 5 years in landfill
- a starch-based container made from potatoes that will dissolve back to liquid starch after 1 minute of being soaked in water
- a biopolyethene cup made from ethanol sourced from sugar waste that will not degrade in landfill
- a compostable spoon made from crude oil that will degrade in 2–3 months
- a PET drink container that will not degrade, but can be reformed into polyester fibre
- a fork made from crude oil that does not have a recycle number on it.

FIGURE 9.4.6 Cross-section of the variety of packaging options used in modern society

The figure above highlights our role as consumers in the success or otherwise of addressing issues relating to polymer sustainability. Plastics disposal will not be resolved by recycling alone. The CSIRO advises that for real progress to occur, each of the following strategies needs to be pursued.

- Revolutionising packaging and waste systems. There are many examples emerging of smart packaging—yoghurt containers that are suitable for seedlings, biscuit packaging that dissolves in water, and a return to paper straws and bamboo disposable cutlery are just a few examples. If society packages less, it is able to reduce the amount of material that needs to be disposed of. Similarly, the use of lasers on assembly lines is leading to automatic sorting of recycling. This innovation will lower the labour cost of recycling significantly.
- Behaviour change and incentives. The South Australian container refund system is an example of a successful incentive scheme. Another positive change is the increase in the number of bins Victorian households are supplied with. Each new change will need to be accompanied by advertising to inform the public of how the scheme works.
- Waste innovation. The use of recycled materials such as Polyrok, and the use of technologies that can generate new polymers from waste and produce oil from waste plastic, are examples of waste innovation.
- Supporting best practice and standards. Standards need to be in place for industry to follow; for example, in relation to the quality of polymers that will be in contact with food and the necessary durability of a product.
- Information for decision making. It seems at times that each industry is having to solve waste issues on its own. Central government bodies can be used to inform industries of possible solutions.

Analysis

1. What are the advantages and disadvantages of each packaging item listed above?
2. Consider the environmental merit of the packaging items listed above.
   - **a** What factors would you take into account?
   - **b** What would your overall ranking be?
3. Is sustainability a relevant issue to you when you purchase items? Explain.
4. Research the regulations regarding labelling of plastics in each Australian state. Are they consistent in all states and territories?

9.4 Review

OA ✓✓

SUMMARY

- Australia, like most countries, is consuming high volumes of plastic material each year. We need to improve the percentage of the items that we are recycling. Two of the important factors that will help lift recycling rates are public education about the recycling categories of plastics, and innovation from industry in the recycling processes and technologies they use. Most thermoplastic polymers should be able to be recycled in some way.
- Recycling may involve remoulding mixed polymers to lower grade items or it may involve more careful sorting to produce higher quality items.
- Common thermoplastics are given a recycle code from 1 to 7 to inform consumers and industry of their recycling potential.

KEY QUESTIONS

Knowledge and understanding

1 **a** How do you know if a plastic item can be placed in our recycling bins?

b Which category of polymer is easier to recycle, thermoplastic or thermoset?

2 Label each of the following as suitable or not suitable for Victorian recycle bins:
2 L milk cartons; plastic wrap; polystyrene foam; 600 mL plastic water bottles; Teflon tape

3 Usage of common plastic items is changing as a response to sustainability issues. List two examples of this change, e.g. banning plastic straws.

Analysis

4 PET is the most recycled polymer in the world. List reasons for the high rate of PET recycling.

5 Many city councils are purchasing high technology sorting machines to use at their recycling centres. Explain how these devices offer potential gains in achieving a circular economy.

6 Explain why the production of PET jackets from recycled material is considered a more advanced form of recycling than the remoulding of recycled plastic into garden furniture.

9.5 Innovations in polymer manufacture

Discussions about sustainability of polymer use centre on two main aspects: the use of scarce resources such as crude oil to make the polymer and the high volume of waste created by plastics once they have been used. Scientists and industry are finding new solutions to these problems, offering varying degrees of sustainability. This section describes a number of these innovations, grouping them based on the source of the raw materials (fossil fuel or biobased) and on whether the polymer that is produced can biodegrade.

WASTE TERMINOLOGY

The following terms will be helpful to your understanding of sustainability:

- **Biobased**: The material or product is at least partly derived from biomass (plants). Sugarcane, corn, starch, potatoes and fruit are examples of sources of **bioplastics**.
- **Biodegradable**: Biodegradation is a chemical process during which micro-organisms that are available in the environment convert materials into natural substances such as water, carbon dioxide and compost. The problem with the term biodegradable is the lack of a time frame. In an oxygen-scarce landfill site, many plastics labelled as biodegradable simply do not degrade.
- **Compostable**: A product that is capable of disintegrating into natural elements in a compost environment, leaving no toxicity in the soil. The term compostable is being used to identify materials that really do degrade under normal conditions.

Compostable products are preferred over products that are simply biodegradable. Compostable products will biodegrade naturally within 90 days and leave no harmful substances in the soil.

FOSSIL FUEL SOURCED, COMPOSTABLE POLYMERS

The polyvinyl acetate from which the bag shown in Figure 9.5.1a is made, can be converted to polyvinyl alcohol (PVA). Both polyvinyl acetate and PVA are made from fossil fuels. The structure of PVA is shown in Figure 9.5.1b. PVA contains –OH groups that can form hydrogen bonds with water. PVA degrades by dissolving in water and has many medical and cleaning uses, where its ability to dissolve in the body or water is put to use.

The use of water-soluble plastics such as PVA still depletes scarce petroleum resources but does not produce landfill or toxic products.

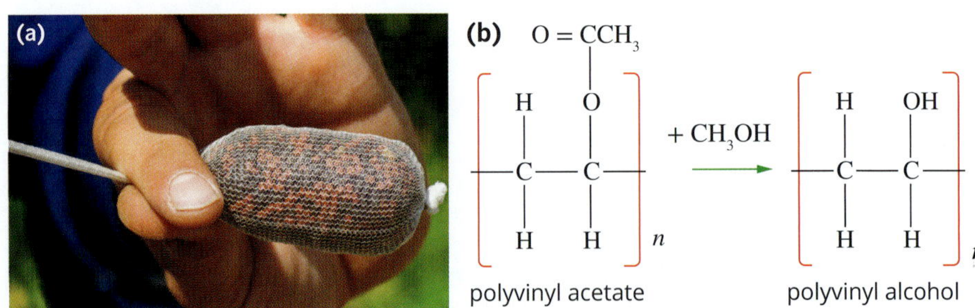

FIGURE 9.5.1 (a) Polyvinyl alcohol (PVA) bags that dissolve in water. (b) The formation of PVA from polyvinyl acetate. PVA is water-soluble because it can form hydrogen bonds with water.

The example above shows that the water solubility of PVA can be used to solve the problem of PVA adding to the volume of landfill and ruining habitats. A different strategy to solve this problem is to improve the compostability of a plastic. PET would normally take over 400 years to degrade naturally, but Japanese scientists discovered bacteria in 2016 in a particular landfill site that were breaking down plastic products at six times the normal rate. They have isolated an enzyme they named PETase. Figure 9.5.2 shows the enzyme attacking the covalent bonds linking the monomers (the ester groups) in PET polymer items. This process leads to the monomers re-forming, offering the possibility of reusing the monomer to build the plastic into a new product.

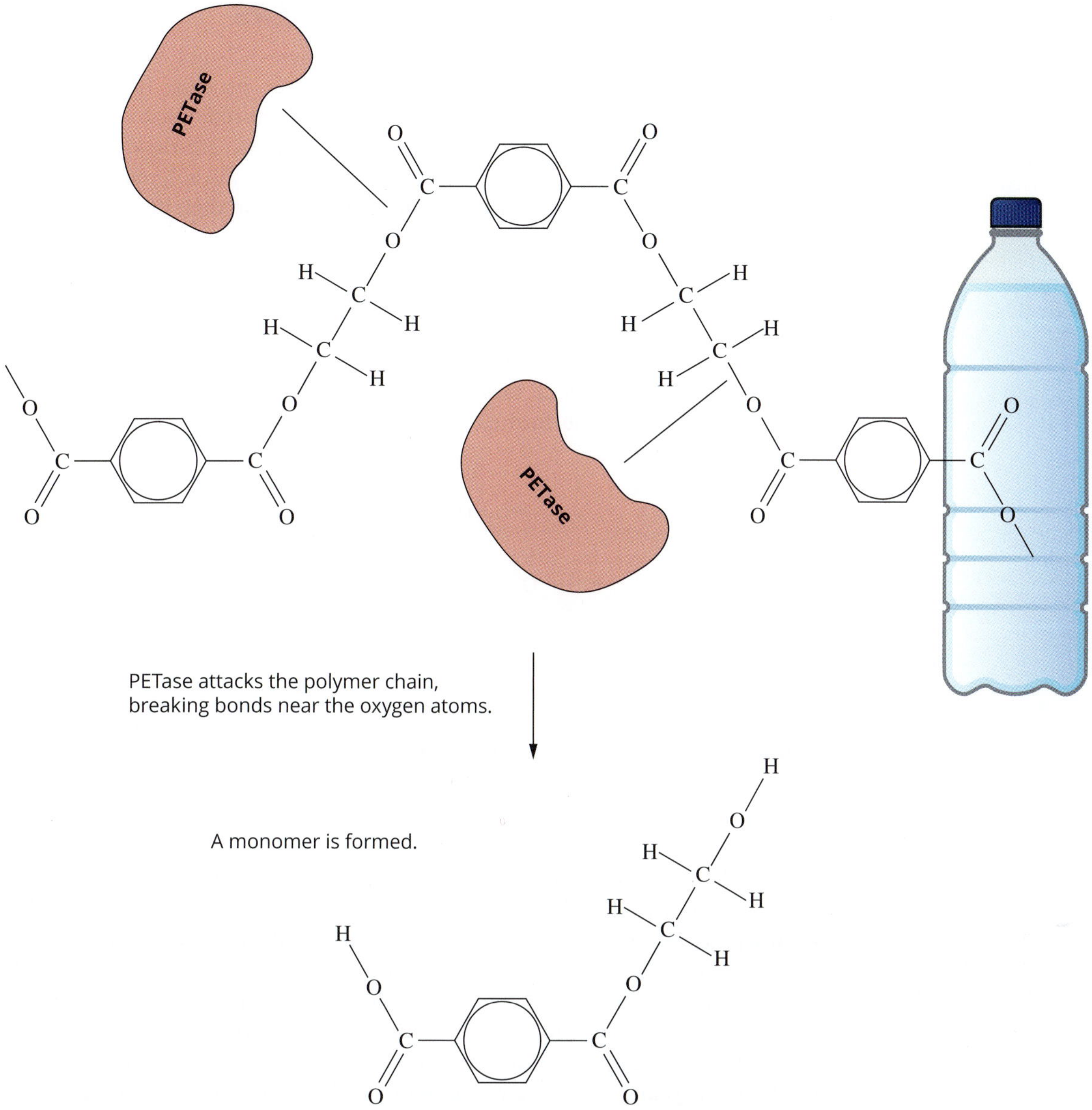

PETase attacks the polymer chain, breaking bonds near the oxygen atoms.

A monomer is formed.

FIGURE 9.5.2 PETase enzyme can break the bonds that join the monomers in PET polymer. The enzyme has the potential to solve any problems with PET disposal contributing to landfill.

Meanwhile, Austrian researchers have found that bacteria from a cow's stomach is capable of breaking the bonds between monomer units in some polymers, enabling the starting monomers to be re-formed. The use of bacteria to break down polymer structures is an example of **organic recycling**. Processes that harness microorganisms such as bacteria are described as **microbial**.

Condensation polymers such as PET are more likely to be compostable than addition polymers due to the presence of oxygen in the molecule chain. The polarity of the C=O bond makes it susceptible to attack from microorganisms. The breaking of bonds in a condensation polymer chain is referred to as **hydrolysis** (Figure 9.5.3), as water molecules released when the polymer is formed are consumed when the polymer breaks down.

> Organic recycling uses microorganisms to significantly reduce the time it takes for polymers to decompose to useful monomers or non-toxic smaller molecules.

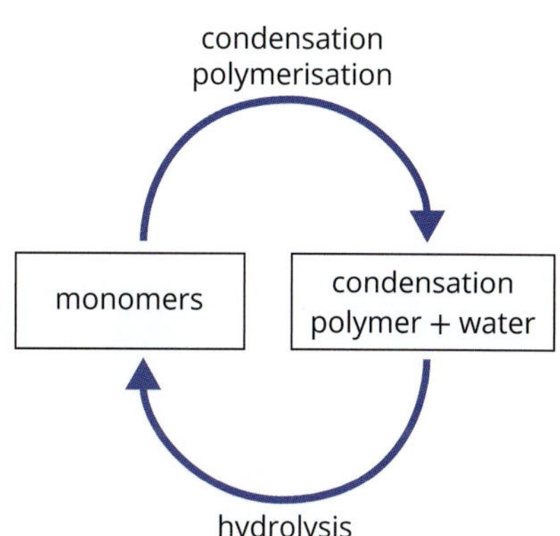

FIGURE 9.5.3 Condensation reactions can form polymers, while hydrolysis reactions break the polymers back to the monomer.

BIOBASED, NOT COMPOSTABLE POLYMERS

> The use of bioethanol to produce monomers reduces world demand for scarce petroleum, but a non-compostable product is formed.

One of the monomers used to make PET is ethane-1,2-diol (Figure 9.5.4). Traditionally, this is sourced from crude oil. A newer process, however, uses fermentation of plant waste to produce the monomer required. Yeast is added to a solution of carbohydrates and the carbohydates in the waste are slowly converted to bioethanol. The bioethanol is then converted to ethane-1,2-diol, making this an example of a biobased monomer (**bio-monomer**). This can then be used to make products such as drink bottles. The plastic in such bottles is referred to **bio-PET**.

Alternatively, the bioethanol can be converted to ethene to make what industry calls **bio-polyethene (bio-PE)**, or to propene to make **bio-polypropene (bio-PP)**. The first commercial biopolyethene plant was built in Brazil in 2010, not a surprise as Brazil is the world's leading producer of bioethanol. The use of biomass to produce the ethene lowers the demand on non-renewable petroleum reserves, but the polyethene formed has the same lack of compostability as any other form of polyethene.

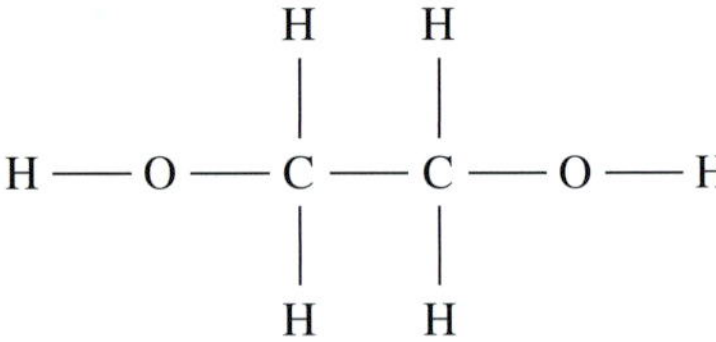

FIGURE 9.5.4 One of the monomers in PET, ethane-1,2-diol

Chemical breakdown of polymers

Recycled plastic is often used for items such as outdoor furniture and detergent bottles, where a high level of purity is not required. The KitKat wrapper shown in Figure 9.5.5, although just a prototype, represented an Australian 'first'—the first time recycled plastic had been used to manufacture a food-safe, flexible polymer.

> Thermal energy and catalysts can be used to break polymer molecules down to synthetic oil. This is an example of chemical **cracking**.

The company Licella is running a trial plant in New South Wales that is using plastic waste as its feedstock. Using a combination of high temperature steam and catalysts, Licella is breaking down plastics to form a synthetic, renewable oil with similar properties and uses to crude oil. Monomers such as propene can then be formed from this oil and used to make plastics such as the KitKat chocolate wrapper. This innovation produces oil in 20 to 30 minutes, rather than millions of years, and uses waste items to do so. The Licella process is an example of **chemical recycling**, where the polymer structure is changed in the recycling process.

FIGURE 9.5.5 A food wrapper manufactured from oil formed from 30% recycled plastic

BIOBASED AND COMPOSTABLE POLYMERS

Several companies sell products made mainly from starch. Figure 9.5.6a shows a range of these products. The starch can be obtained as a by-product from food industries, where it is used for processing potato and corn (Figure 9.5.6b). The products will compost easily to form either the glucose monomer or CO_2.

FIGURE 9.5.6 (a) A range of starch-based food containers and utensils. (b) The formation of starch polymer from glucose.

Polylactic acid, PLA, is another emerging biobased polymer (Figure 9.5.7a). The action of bacteria on biomass can produce lactic acid. Waste fruit from a fruit-canning plant or waste from a dairy plant are examples of suitable sources of biomass for this process. The lactic acid is then polymerised to PLA. The PLA composts easily to form natural products, as shown in Figure 9.5.7b. Table 9.5.1 on the following page sums up the sustainability promise of widespread use of PLA in place of PET.

The use of compostable biobased polymers such as PLA is a genuine example of a circular economy. The monomer is formed from waste and the polymer composts quickly to harmless products.

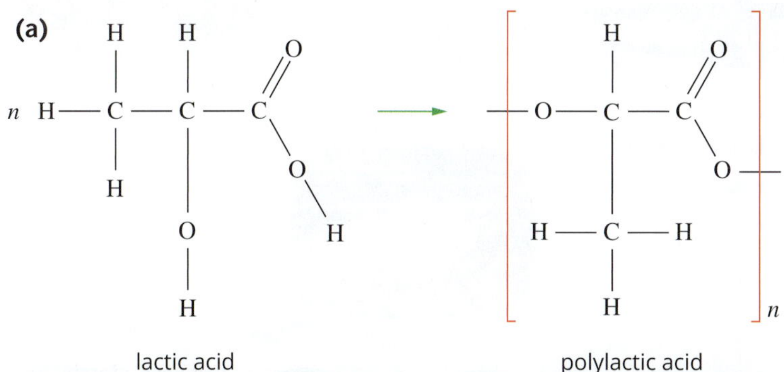

FIGURE 9.5.7 (a) Polymerisation of lactic acid and (b) degradation of a PLA bottle

TABLE 9.5.1 Comparison of the sustainability of PET and PLA

Property	PET	PLA
Made from	oil	plant waste
Biodegrade?	no	yes
Recycle?	yes	yes
Toxins released if incinerated?	yes	no
Concerns of toxins leaching?	yes	no

Figure 9.5.8 shows the movement towards a circular economy as new technologies are adopted. The most sustainable options are those that make the monomer from waste and produce a polymer that is fully compostable. The emphasis is on the design of better packaging (or its elimination), increasing recycling rates, and introducing new types of polymers that are derived from natural, renewable sources.

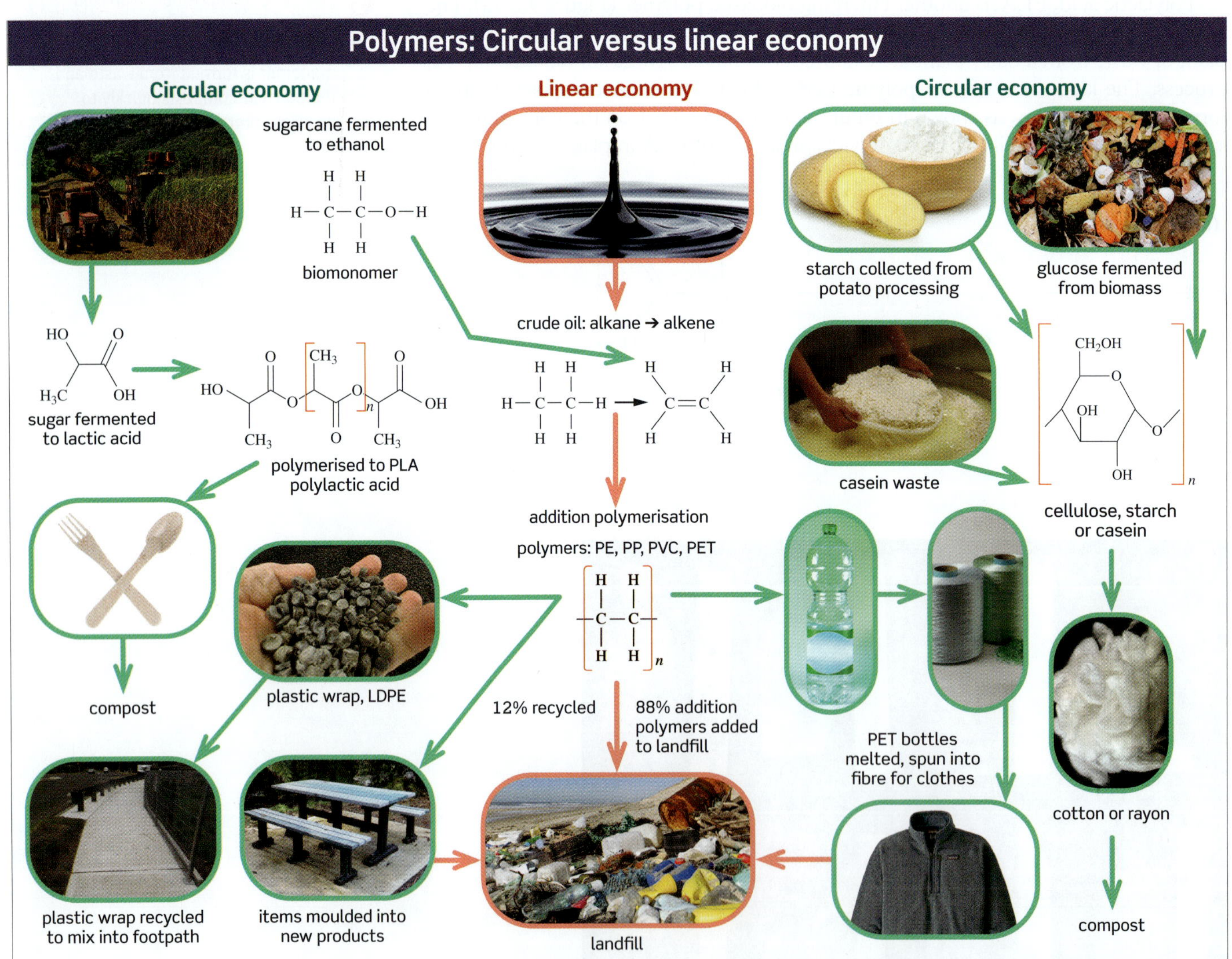

FIGURE 9.5.8 Classifying the sustainability of polymers. Processes employed in a linear economy are shown with red arrows, those associated with a circular economy have green arrows.

The above discussion can be used to loosely classify polymers according to their origin and sustainability (Table 9.5.2). Some polymers can appear in more than one category depending upon how they are made.

TABLE 9.5.2 Classifying the sustainability of polymers

Category	Characteristic	Examples
fossil fuel-based polymers	monomers derived from crude oil Can be recycled but not composted	HDPE, PVC, LDPE, PP, PS, PET
bio-monomers	monomers produced from biomass the polymer made from the monomer may or may not be recyclable or compostable.	bio-ethanol and monomers derived from it, like ethene, used to make bio-PE, bio-PP or bio-PET
bio-plastics	monomers made from biomass compostable	starch, PLA, nanollose

CHEMFILE

Nanollose

Rayon, or viscose, is a popular fabric made from cotton or timber. The farming of cotton or growing of forest plantations both involve considerable amounts of water and resources. A Perth company has patented a process to manufacture rayon from waste. The company uses bacteria such as acetobacter to convert waste biomass into glucose. Different bacteria then convert the glucose to cellulose. The rayon produced is called nanollose (Figure 9.5.9). The steps in the process are outlined in Figure 9.5.10.

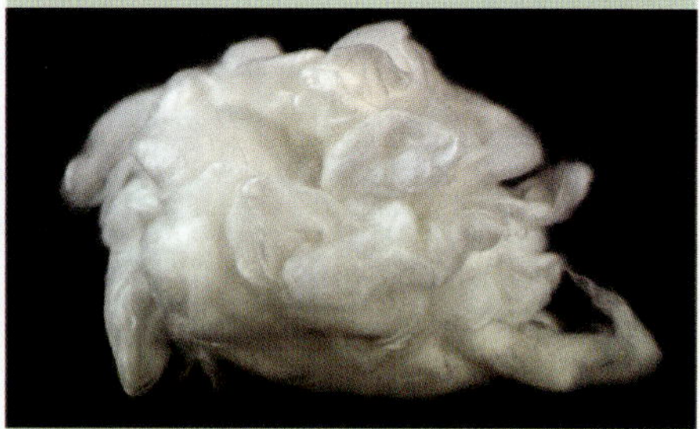

FIGURE 9.5.9 This nanollose fibre was manufactured from coconut waste using bacteria.

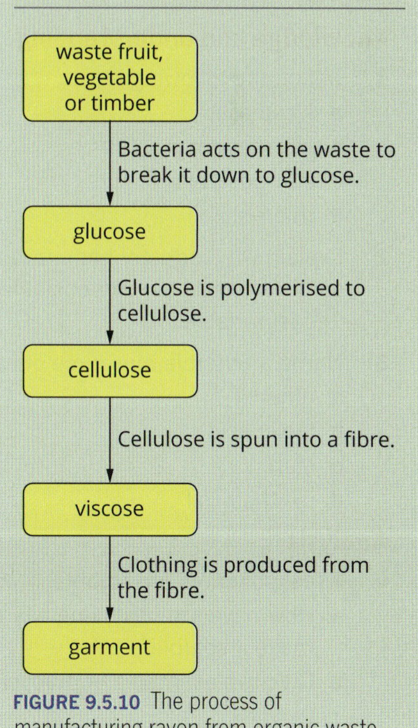

FIGURE 9.5.10 The process of manufacturing rayon from organic waste

9.5 Review

OA ✓✓

SUMMARY

- Scientists are researching many different strategies to progress the plastics industry towards a circular economy.
  Strategies include:
  - sourcing the monomers from plant matter (biobased monomers)
  - producing monomers from bioethanol
  - improving the compostability of degradable polymers
  - incorporating bacteria in plastic production
  - heat-treating plastic waste to produce synthetic oil.
- Polymers are classified as bio-plastics if the source material or the product is at least partially derived from biomass such as corn or starch. Polylactic acid (PLA), bio-polyethene (bio-PE), bio-PET and bio-polypropene (bio-PP) are examples of bioplastics, whereas HDPE, PVC, LDPE, polypropene (PP) and polystyrene (PS) are examples of fossil fuel based plastics.
- The term compostable has been introduced to identify polymers that are biodegradable in a reasonable time frame.
- Plastic recycling or composting methods usually fit into one of three categories

Recycling category	Description	Example
mechanical	remould the same polymer into a new shape	garden furniture
organic	microorganisms break the polymer structure	PLA composting
chemical	temperature and catalysts used to break polymer structure	Licella process

KEY QUESTIONS

Knowledge and understanding

1 Define the following terms.
- **a** bio-plastic
- **b** bio-monomer
- **c** compostable
- **d** microbial

2 Explain why industry has moved to labelling some products as 'compostable' in preference to 'biodegradable'.

3 Name a possible biosource of each of the following:
- **a** ethanol
- **b** lactic acid
- **c** starch
- **d** synthetic oil

Analysis

4 A few polymers are soluble in water.
- **a** How have scientists engineered the polymer to make it soluble?
- **b** Give two examples of practical uses for soluble plastics.

5 Landfill problems are usually eased if a polymer is compostable. The composting of some polymers, however, is still of concern. Explain why this is the case.

6
- **a** Explain how ethanol is made from biomass.
- **b** Explain how the use of biomass can improve sustainability of polyethene.

7 Polyethene can be made and disposed of several different ways. Consider the alternatives in the following table.

Option	Process for manufacture and disposal
A	crude oil → ethene → polyethene → garden hose → landfill
B	potato starch → ethanol → ethene → polyethene → garden hose → landfill
C	crude oil → ethene → copolymer with lactic acid → garden hose → compost
D	crude oil → ethene → polyethene → garden hose → bench seat

- **a** Discuss the relative merits of each option.
- **b** Is there an option that stands out as being the most sustainable?

8 PET is a condensation polymer where each monomer is joined with a type of link called an ester group. Use PET as an example to explain the difference between the following recycling techniques:
- **a** mechanical
- **b** organic
- **c** chemical

Chapter review

09

KEY TERMS

addition polymer
addition polymerisation
amorphous
biobased
biodegradable
bio-monomer
bio-polyethene (bio-PE)
bio-polyethylene terephthalate (bio-PET)
bio-plastic
bio-polypropene (bio-PP)
branches
catalyst
chemical recycling
compostable
condensation polymer
copolymer
cracking
cross-links
crystalline region
elastomer
ester
high-density polyethene (HDPE)
hydrolysis
linear low-density polyethene (LLDPE)
low-density polyethene (LDPE)
mechanical recycling
microbial
microplastics
monomer
natural polymer
organic recycling
plastic
plasticiser
polymer
polymerisation
side group
thermoplastic
thermosetting
unsaturated

REVIEW QUESTIONS

Knowledge and understanding

1 Which of the following changes to a polymer is likely to lead to an increase in the melting point?

A increasing the degree of branching
B increasing the length of the molecule chains
C decreasing the length of the molecule chains
D converting the polymer to a foam

2 The main reason PVC will have a higher melting point than polyethene is:

A the bonding within the molecule is stronger.
B the length of the molecules must be greater.
C that PVC will contain cross-links between chains.
D the dipoles between polymer molecules are stronger.

3 Select the alternative that best compares the properties of HDPE with LDPE.

A LDPE is denser, more transparent and has a higher melting point.
B LDPE is stronger, more crystalline and has a lower melting point.
C HDPE is denser, harder and less transparent.
D HDPE is denser, more flexible and more transparent.

4 Which one or more of the following molecules can act as monomers in addition polymerisation?

A propene
B propane
C chloroethene
D $CH_2{=}CHF$

5 A segment of a PVC polymer is shown below.

$$—CH_2—\underset{\displaystyle Cl}{\underset{|}{CH}}—CH_2—\underset{\displaystyle Cl}{\underset{|}{CH}}—CH_2—\underset{\displaystyle Cl}{\underset{|}{CH}}—CH_2—\underset{\displaystyle Cl}{\underset{|}{CH}}—CH_2—\underset{\displaystyle Cl}{\underset{|}{CH}}—CH_2—\underset{\displaystyle Cl}{\underset{|}{CH}}—$$

a Draw the structure of the monomer used to make PVC.
b How many repeating monomer units are shown in the section of polymer?
c What is the strongest type of bonding between PVC polymer chains?

6 Define the following terms.

a monomer
b thermoplastic
c thermosetting
d cross-link

7 State whether each of the following statements about polymers is true or false.

a Each chain in a polymer is the same length.
b The chains in thermosetting polymers are held together by dipole–dipole bonds.
c HDPE has no branches.
d The properties of a polymer are different from the properties of the monomer.

8 Ethene (C_2H_4) is the smallest alkene.

a Why is it described as unsaturated?
b Draw the structural formula of ethene.
c Could ethane (C_2H_6) act as a monomer? Explain.

9 Draw a section of the polymer made from each of these monomers in an addition polymerisation process.

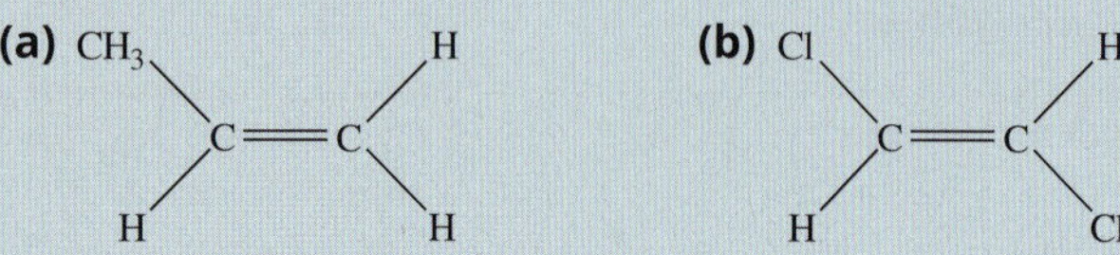

10 The polymer chains of a certain material can be cross-linked to varying extents.

- **a** Sketch the polymer chains when there is:
  - **i** no cross-linking
  - **ii** a little cross-linking
  - **iii** a large degree of cross-linking.
- **b** Use diagrams to show the effect of stretching each of these materials.

11 **a** How does the strength of the inter-chain bonding differ between thermosetting and thermoplastic polymers?

- **b** Why do thermosetting polymers decompose rather than melt when heated strongly?

12 Would a thermoplastic or thermosetting polymer be the most suitable material for the following purposes?

- **a** handle of a kettle
- **b** 'squeeze' container for shampoo
- **c** knob of a saucepan lid
- **d** shopping bag
- **e** rope

13 Elastic bands, golf balls and saucepan handles are made from polymers with some cross-linking.

- **a** Which material has the greatest degree of cross-linking?
- **b** Describe the properties of the material you chose in part **a** to support your answer.

14 How do you decide if a plastic item can be recycled in Victoria?

15 **a** What is a copolymer?

- **b** Explain why copolymerisation is used.

16 Is a plastic that can be recycled more likely to be thermoplastic or thermosetting?

17 The following terms are related but different. Explain each term.

- **a** biobased
- **b** biodegradable
- **c** compostable

Application and analysis

18 A typical section of a certain polymer molecule is shown below.

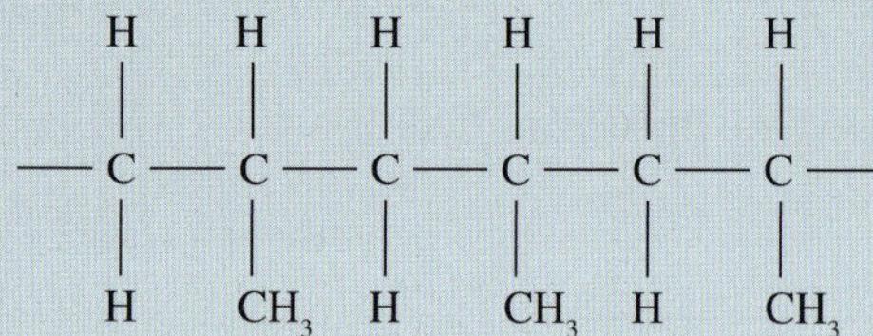

- **a** Draw the monomer from which this polymer is produced.
- **b** Identify how many repeating monomer units are shown.
- **c** Would you expect the melting point of this polymer to be higher or lower than that of polyethene?
- **d** How would you suggest items made from this polymer would be best disposed of?

19 PVA (or white) adhesive contains polyvinyl acetate in water. A section of the polymer is shown below.

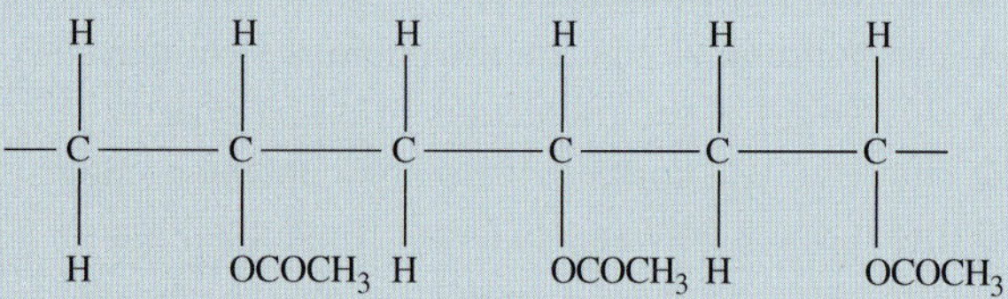

- **a** Give the structural formula for its monomer, vinyl acetate.
- **b** PVA is relatively soluble in water. Explain why it is soluble when most polymers are not.
- **c** What are the advantages and disadvantages of a polymer being water soluble?

20 But-2-ene can act as a monomer. Draw a segment of the polymer that it can form.

21 Part of the copolymer acrylonitrile–butadiene–styrene (ABS) that Lego® is made from is shown here. Identify the molecular formula of the monomers responsible for the sections in each box.

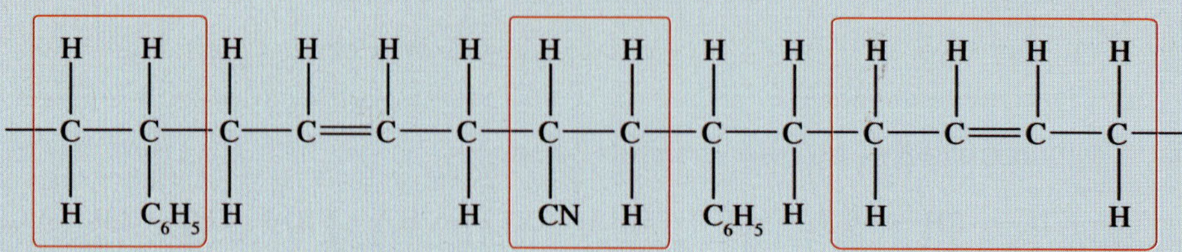

22 Explain how the following polymer production techniques will affect the properties of a polymer.
   a production conditions that favour longer polymer molecules
   b incorporation of side groups with benzene rings ($-C_6H_5$)
   c addition of a foaming agent to a polymer

23 A polymer chemist is investigating the properties of polymers by increasing the length of chains in a controlled way. The chemist identified the effect on the relative molecular mass, melting point, strength of inter-chain forces and electrical conductivity. What would you expect the results for each property test to be?

24 Polyethene (PE) can be made from crude oil, while bio-PE can be made from corn starch.
   a Research each of these processes and draw a flowchart for the steps involved in each.
   b Explain why all polyethene required is not manufactured as bio-PE.

25 The molecule below is 4-hydroxybutanoic acid. It can undergo condensation polymerisation to form a compostable polymer.

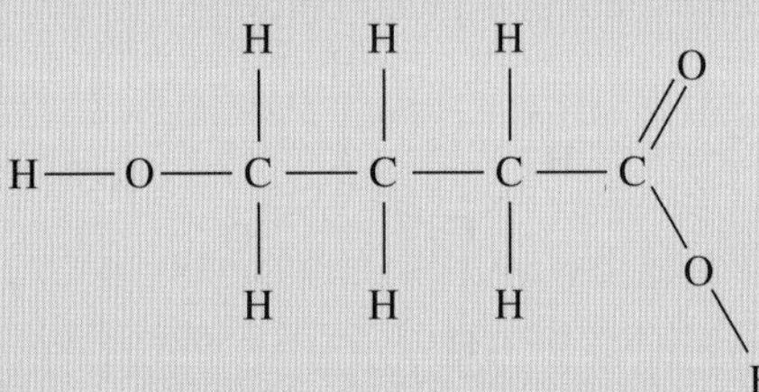

   a Draw a segment of the polymer formed.
   b Use this segment to explain how a condensation polymer differs from an addition polymer.
   c Use this segment to explain why condensation polymers are more susceptible to the action of microorganisms than addition polymers.

26 The production of polylactic acid, PLA, is described in Section 9.5. PLA is one of the best examples of the use of polymers in a circular economy. Explain why this is the case.

27 World production of polyethene is higher than any other polymer. Research and discuss the following sustainability aspects of the use of polyethene (your answer should include the terms in bold).
   a 'The use of over 80% of plastics such as polyethene is an example of the **linear economy**.' Explain.
   b A small percentage of the polyethene marketed contains a **biomonomer**. What is a biomonomer and how is this monomer formed?
   c Explain what **mechanical recycling** is and how recycled polyethene is used.
   d Licella's trial **chemical recycling** plant accepts polyethene items. What is chemical recycling and how does it use polyethene?
   e Polyethene is not suitable for **organic recycling**. Explain why.

REVIEW QUESTIONS

How are materials quantified and classified?

Multiple-choice questions

1 The empirical formula of a molecular compound is best defined as the:

A actual number and type of each element present
B simplest whole number mole ratio of each ion present
C simplest whole number mole ratio of each element present
D simplest whole number mass ratio of each element present.

2 The hydrocarbon with the formula C_3H_8 is:

A propane
B prop-1-ane
C prop-1-ene
D propene

3 The total number of atoms in 3.5 mol of pentan-1-ol is:

A 63
B 3.5×10^{23}
C 2.1×10^{24}
D 3.8×10^{25}

4 Which one of the following alternatives lists only compounds that are likely to be found in crude oil?

A C_2H_4, C_3H_7OH, C_8H_{16}, C_8H_{18}
B C_2H_6, C_4H_{10}, C_8H_{18}, $C_{16}H_{34}$
C C_2H_4, C_2H_6, C_2H_5OH, CH_3COOH
D C_2H_4, C_2H_5COOH, C_6H_{12}, C_8H_{16}

5 One of the four hydrocarbons listed below is analysed. It was found that 0.235 mol of the hydrocarbon has a mass of 16.0 g. The hydrocarbon is:

A CH_4
B C_3H_6
C C_5H_8
D C_6H_{14}

6 A hydrocarbon is found to contain 80.0% carbon. The molecular formula of this hydrocarbon could be:

A C_4H
B CH_3
C CH_4
D C_2H_6

7 How many different alkenes are there with the molecular formula C_4H_8?

A 1
B 2
C 3
D 4

8 Which one of the following alternatives lists the compounds in order of increasing boiling points?

A ethane, propane, ethanol, propan-1-ol
B ethane, ethanol, propane, propan-1-ol
C ethanol, propan-1-ol, ethane, propane
D ethanol, ethane, propan-1-ol, propane

9 The relative atomic mass of molybdenum is listed as 96.0. This means that:

A each molybdenum atom weighs 96 times more than each carbon-12 atom
B ten atoms of molybdenum will weigh eight times the mass of ten atoms of carbon-12
C molybdenum atoms have eight times the number of protons that carbon-12 atoms have
D eight atoms of carbon-12 will have the same number of protons as one atom of molybdenum.

10 Polyvinyl alcohol is a polymer often used as a water soluble film in packaging. The monomer used to form polyvinyl alcohol is given below.

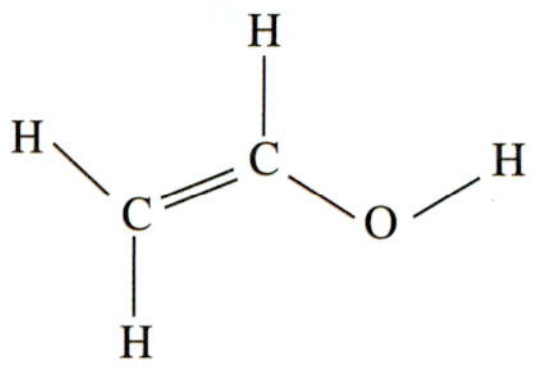

H
H C=C O H
H

Which one of the following structures shows a possible segment of the polymer?

A
H H H H
—C—C—C—C—
H OH OH H

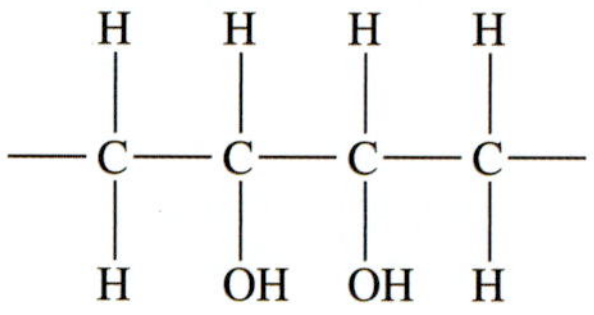

B
H H H H
—C=C C=C
H O H O

C
H H H H
—C=C—C=C—
H OH H OH

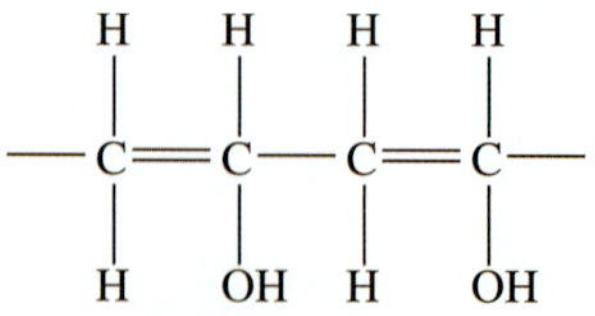

D
H H H H
—C=C C—C
H O H O

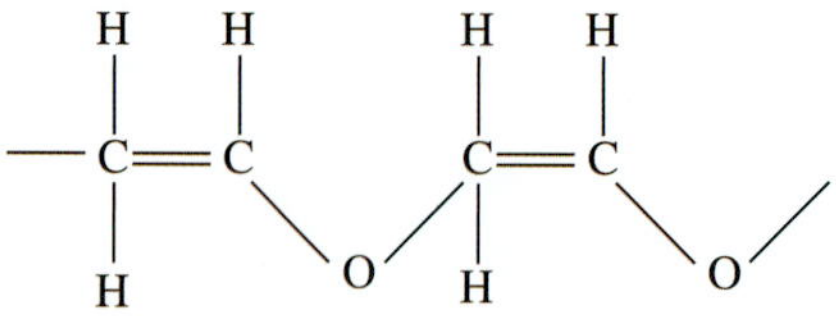

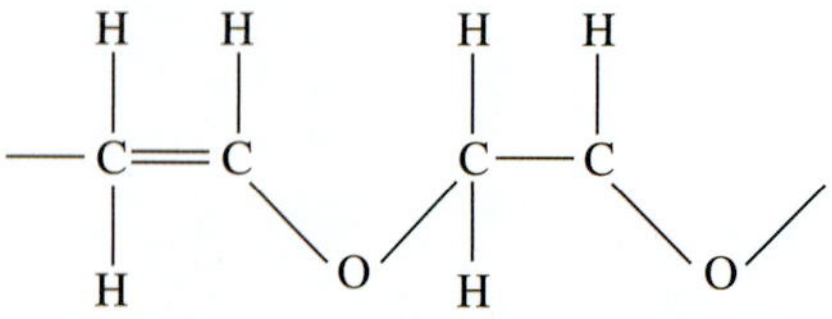

11 A hydrocarbon has a molar mass of 56 g mol^{-1}. It reacts with bromine to form a symmetrical molecule. The hydrocarbon is:

A butane
B 2-methylprop-1-ene
C but-2-ene
D but-1-ene

12 Polyethene is a polymer that has a wide range of uses. It can be produced as a high-density product (HDPE) or a lower density form (LDPE) that is softer and more flexible.

Compared to LDPE, HDPE has:

A higher softening temperature due to a greater degree of branching of the polymer chain
B higher softening temperature due to a smaller degree of branching of the polymer chain
C lower softening temperature due to a greater degree of branching of the polymer chain
D lower softening temperature due to a smaller degree of branching of the polymer chain.

Short-answer questions

13 Crude oil is an important resource to our society.

a What is the origin of crude oil?
b Crude oil is not a pure substance but a mixture.
  i Briefly describe the composition of crude oil.
  ii Give the molecular formula of two hydrocarbons present in crude oil.

14 The mass spectrum for magnesium is shown below:

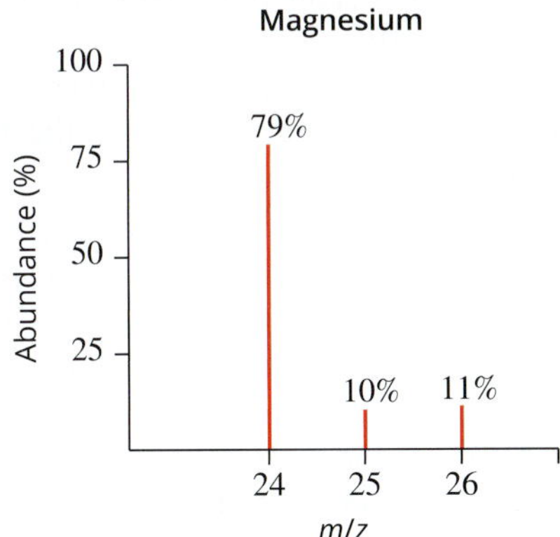

a Describe what this spectrum tells you about magnesium.
b Use the data provided to determine the relative atomic mass of magnesium.

15 The molecular formula of phosphorus pentoxide is P_4O_{10}.
Given a 5.00 mol sample of P_4O_{10}:

a How many mol of phosphorus atoms are in the sample?
b How many mol of atoms are in the sample?
c How many atoms are present?

16 Give the name of each of the following compounds.

a $CH_3CH{=}CHCH_3$
b $CH_3CH_2CH_2CH_2CH_2CH_2CH_3$
c

```
          H                 H
          |                 |
       H--C--H           H--C--H
    H     |                 |     H     H
    |     |                 |     |     |
 H--C-----C-----------------C-----C-----C--H
    |     |                 |     |     |
    H     |                 H     H     H
       H--C--H
          |
          H
```

d

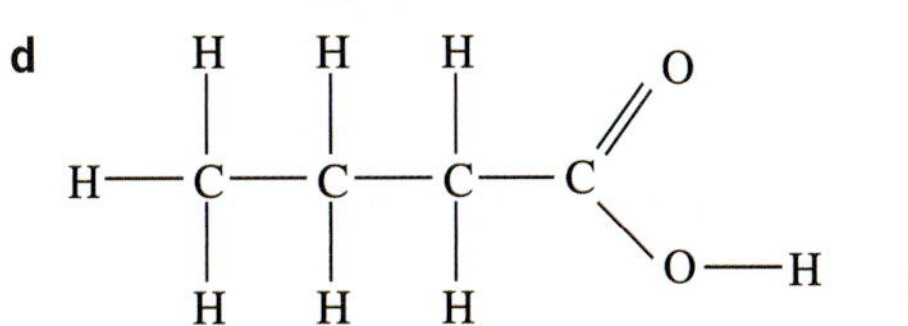

e
```
CH3CHCH2OH
   |
   CH3
```

17 Draw the structural formula of each of the following compounds.

a hex-1-ene
b propanoic acid
c 2,3-dichloropentane
d 2-methylpropan-2-ol

18 Give a condensed structural formula (a semi-structural formula) for each of the following compounds.

a hexanoic acid
b 4-ethyl-2,2-dimethylheptane
c butan-2-ol
d 2,4-dimethylhex-1-ene

19 Polymers are very large covalent molecular substances.

a The following is a representation of a section of an addition polymer.

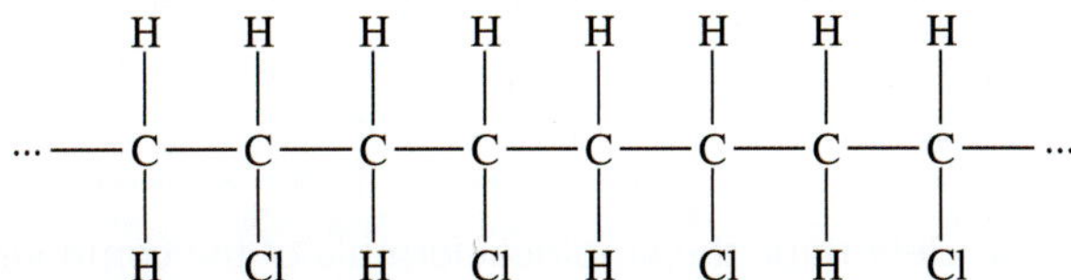

Draw the structure of the monomer from which this polymer was formed.

b A small section of a thermosetting polymer is heated over a flame.
  i Describe what observations you would expect to make.
  ii Describe the structure of thermosetting polymers and use it to explain your answer to part **bi**.

20 Using a suitable example where possible, clearly explain the difference between the following pairs of terms.

- **a** a monomer and a polymer
- **b** a thermoplastic polymer and a thermosetting polymer
- **c** a branched polymer and a cross-linked polymer
- **d** an addition polymer and a condensation polymer.

21 Element X forms a compound with oxygen in which the ratio of the number of atoms of X to the number of atoms of oxygen is 2:3. X makes up 59.7% of the mass of the compound.

Calculate the molar mass of X and use it to identify this element.

22 Nitrogen can form at least five different compounds when it reacts with oxygen:

$$NO, NO_2, N_2O, N_2O_4, N_2O_5$$

- **a** Which of these oxides contains the highest percentage mass of nitrogen?
- **b** Identify two oxides with the same percentage mass of nitrogen.
- **c** How many atoms are there in 1.00 mol of N_2O_5?

23 Hexane is used by the food industry when a very non-polar solvent is required.

- **a** Use a diagram to explain why hexane is non-polar.
- **b** There are less polar solvents than hexane but they are not suitable for the food industry. Give two reasons for some non-polar hydrocarbons being unsuitable for the food industry.

24 A compound consists of carbon, hydrogen and oxygen only. It contains 54.54% carbon and 36.36% oxygen.

- **a** Determine the empirical formula of this compound.
- **b** 0.350 mol of this compound has a mass of 30.8 g. Calculate the relative molecular mass of the compound.
- **c** Determine the molecular formula of the compound.
- **d** There are several compounds with this molecular formula. Draw the structural formula and give the name of two compounds with this molecular formula.

25 A compound consists of carbon, chlorine and hydrogen. A sample of 4.738 g of the compound contains 0.476 g of carbon and 4.22 g of chlorine.

- **a** Determine the empirical formula of the compound.
- **b** Given that an amount of 0.326 mol of the compound has mass of 39.0 g, calculate the molar mass and then deduce the molecular formula of the compound.
- **c** Draw the structure of the compound.
- **d** This compound is a liquid at room temperature. What kind of interactions are present between molecules of this compound?
- **e** What mass of this compound contains 7.2×10^{23} atoms of chlorine?

26 Ammonium nitrate is a common fertiliser with a formula NH_4NO_3. A particular brand of fertiliser contains 85.0% ammonium nitrate by mass.

- **a** Calculate the percentage nitrogen in ammonium nitrate.
- **b** Calculate the mass of ammonium nitrate in a 5.00 kg bag.
- **c** Determine the mass of nitrogen, present as ammonium nitrate, in the bag.
- **d** Calculate the number of mole of nitrogen, present as ammonium nitrate, in the bag.

27 Nylon is a commercial name for a group of polymers formed from monomers like the one drawn below. This monomer can react with itself to form a condensation polymer.

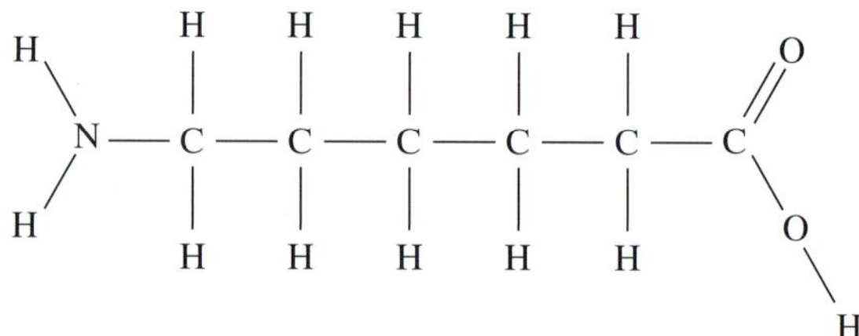

- **a** You are not familiar with one of the functional groups on this monomer, but given that water is also formed in the reaction, you should be able to draw several monomers end to end, then deduce what the structure of the polymer will look like. Draw a segment of the polymer that will form from this monomer.
- **b** If 3000 monomers join, how many molecules of water are formed?
- **c** This polymer can be degraded by microorganisms. Explain why it is more susceptible to this form of degradation that polyethene is.

28 If acid is added to milk, the milk curdles. The main component of the curds is the protein, casein, a natural polymer. Before the introduction of synthetic polymers, an industry existed where the curds were moulded into plastic objects, such as buttons or combs. Casein items were durable if used indoors. If used externally they degraded quickly in wet conditions.

- **a** Do you think casein would be an addition or condensation polymer? Justify your answer.
- **b** Discuss the sustainability of casein polymer.

29 Ethanol and ethanoic acid (also known as acetic acid) are used in modern society in high volumes. Both products can be synthesised from crude oil or by fermentation of sugars from biomass. Ethanoic acid is the main ingredient in vinegar.

- **a** Draw the structures of ethanol and ethanoic acid.
- **b** **i** Calculate the percentage oxygen in ethanol.
  - **ii** Calculate the percentage oxygen in ethanoic acid.

c Which will have the higher boiling point, ethanol or ethanoic acid? Explain your answer.

d Fermentation processes usually occur at temperatures around 35°C, whereas synthesis from crude oil may involve temperatures over 300°C. Suggest a reason for this difference.

e Discuss the relative sustainability of the formation of these products from crude oil compared to the production from sugars.

30 A class is asked to determine the empirical formula of magnesium oxide. The method used is outlined below.

1 Clean then weigh a small piece of magnesium.
2 Weigh a crucible and lid. Add the magnesium to the crucible.
3 Heat the crucible until the magnesium starts to burn (react with oxygen).
4 Lift the lid of the crucible just enough to allow the reaction to continue, but to not allow the magnesium oxide ash to escape.
5 Allow to cool once the reaction is complete.
6 Weigh the crucible, lid and ash.

The students' results are presented in the table below.

Student	Mass mg (g)	Mass ash (g)
A	0.414	0.663
B	0.532	0.853
C	0.588	0.959
D	0.622	0.859
E	0.668	1.084
F	0.723	1.168
G	0.788	1.288

Graph of results:

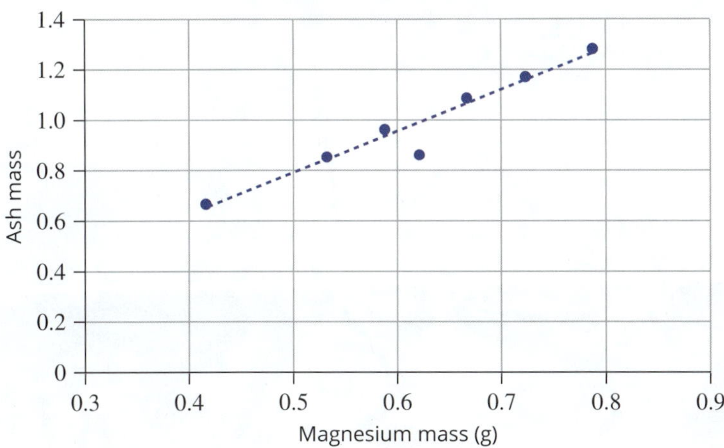

a i Identify the independent variable.

ii Identify the dependent variable.

b Comment on the precision of the class data.

c i Use the graph provided to find the mass of ash expected if a mass of 0.500 g of magnesium is used.

ii Use this value to determine the empirical formula of the magnesium oxide.

iii The mole ratio of magnesium to oxygen is not an exact value. What might it suggest and give a possible reason for the difference?

d What would you expect the formula of magnesium oxide to be? Justify your answer.

WS 22 EQ U102

UNIT 2 How do chemical reactions shape the natural world?

To achieve the outcomes in Unit 2, you will draw on key knowledge outlined in each area of study and the related key science skills on pages 11 and 12 of the Study Design. The key science skills are discussed in Chapter 1 of this book.

AREA OF STUDY 1

How do chemicals interact with water?

Outcome 1: On completion of this unit the student should be able to explain the properties of water in terms of structure and bonding, and experimentally investigate and analyse applications of acid–base and redox reactions in society.

AREA OF STUDY 2

How are chemicals measured and analysed?

Outcome 2: On completion of this unit the student should be able to calculate solution concentrations and predict solubilities, use volumetric analysis and instrumental techniques to analyse for acids, bases and salts, and apply stoichiometry to calculate chemical quantities.

AREA OF STUDY 3

How do quantitative scientific investigations develop our understanding of chemical reactions?

Outcome 3: On completion of this unit the student should be able to draw an evidence-based conclusion from primary data generated from a student-adapted or student-designed scientific investigation related to the production of gases, acid–base or redox reactions, or the analysis of substances in water.

CHAPTER

10 Water as a unique chemical

The water you see and use every day is a unique chemical with unusual properties. Water falls from the sky and fills rivers and oceans. You drink it, cook with it, bathe in it and wash your clothes in it. Reactions in your body take place between chemicals that are dissolved in it. Without water there is no life.

In this chapter, you will learn about where water exists on Earth and how much of that water is available for us to drink. You will learn about some of the unusual properties of water and how they support life on Earth. You will also learn how each property can be explained using knowledge and understanding of the structure and bonding present in water.

Key knowledge

- the existence of water in all three states at Earth's surface, including the distribution and proportion of available drinking water **10.1**
- explanation of the anomalous properties of H_2O (ice and water) with reference to hydrogen bonding:
  - trends in the boiling points of Group 16 hydrides **10.1**
  - the density of solid ice compared with liquid water at low temperatures **10.1**
  - specific heat capacity of water, including units and symbols **10.2**
- the relatively high latent heat of vaporisation of water and its impact on the regulation of the temperature of the oceans and aquatic life. **10.3**

10.1 Essential water

Water is a special chemical. It moderates our weather, shapes our lands and is essential for the existence of life. Water is the most abundant liquid on Earth, covering more than 70% of our planet. The total water supply on Earth is estimated to be more than 1.3 billion cubic kilometres and is in continuous movement between land, ocean, rivers and creeks, and the atmosphere.

THE AVAILABILITY OF DRINKING WATER

FIGURE 10.1.1 Icebergs are solid water, the ocean is liquid water and gaseous water is present in the atmosphere.

In Australia, we take for granted that clean drinking water will come out of the taps in our homes, schools and workplaces, but this is not always in the case in other parts of the world. About 780 million people worldwide lack basic drinking water access. While there is a very large volume of water on Earth, the ability of water to dissolve many substances means many water sources do not provide fresh water.

Water sources on Earth

Water exists naturally on Earth in the states of solid, liquid and gas, as seen in Figure 10.1.1.

Table 10.1.1 shows the distribution of water on Earth in each of these states.

TABLE 10.1.1 The distribution of water on Earth

Location of water	State of matter	Volume (km^3)	Per cent of total water (%)
oceans	liquid	1 300 000 000	96.54
ice caps and glaciers	solid	24 000 000	1.74
groundwater	liquid	23 000 000	1.69
ground ice and permafrost	solid	300 000	0.022
lakes	liquid	180 000	0.013
soil moisture	liquid	17 000	0.001
atmosphere as water vapour	gas	13 000	0.001
rivers	liquid	2100	0.0002

Despite the large quantities of water, there is a limited supply of fresh water on our planet. Figure 10.1.2 shows that only 2.5% of the water on Earth is fresh and therefore drinkable, and that most of this water is not accessible as it is locked up in icecaps, glaciers or the soil.

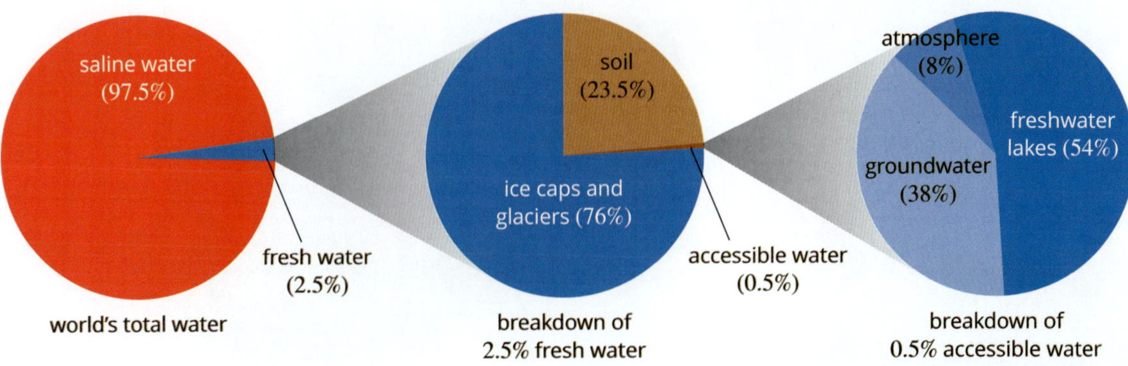

FIGURE 10.1.2 Distribution of water on Earth. Only 2.5% of the water on Earth is potentially drinkable.

Australia is the driest inhabited continent and has extremely variable rainfall. Since Australia is also very flat and hot, much of the rain that falls evaporates before it can enter rivers and reservoirs.

Obtaining clean drinking water

Obtaining clean drinking water, or **potable water**, has been a challenge for people living in communities for thousands of years. While water is abundant on Earth, a large infrastructure is required to supply a town or city with high-quality drinking water from the water that is naturally occurring.

Major sources of drinking water

Drinking water in Australia is obtained from a variety of sources:

- reservoirs filled by run-off from rivers and streams
- water obtained directly from rivers and lakes
- groundwater
- **recycled water**
- **desalinated seawater**.

In most of Australia's major cities, water comes from reservoirs built on rivers in protected areas. These reservoirs are able to provide the population with safe, clean water. Water from such sources remains very clean and needs only minor treatment before being released for consumption. Reservoirs like the one shown in Figure 10.1.3 are surrounded by protected land and are often located in national parks and forests with limited access to ensure the quality of the water remains very high.

However, in some parts of inland Australia, and in some other countries in the Asia–Pacific region, water comes from sources other than protected catchments. In some cases, water is taken directly from rivers and lakes that may be subject to contamination from run-off from farms and urban areas. Drinking water may also be obtained from groundwater, which is often referred to as **bore water** in Australia. Such sources may contain levels of contamination that require a more complex purification process than for water from protected catchments.

Table 10.1.2 shows the major sources of drinking water in some Australian towns and cities.

TABLE 10.1.2 Sources of drinking water in Australia

City or town	Main water source
Alice Springs, inland Northern Territory	groundwater
Adelaide, South Australia	Torrens and Murray rivers
Bourke, inland New South Wales	Darling River
Broome, coastal Western Australia	groundwater
Melbourne, Victoria	Reservoirs
Mildura, inland Victoria	Murray River
Birdsville, inland Queensland	groundwater
Hobart, Tasmania	Derwent River

Drinking water in Victoria

Most Victorians obtain their drinking water from a mains water supply. The water is piped from a reservoir and is controlled by a local water authority. Where a piped water supply is not available, drinking water may be obtained from:

- rainwater tanks
- bores
- dams
- rivers and creeks.

FIGURE 10.1.3 The Silvan Reservoir stores water from the Upper Yarra, Thomson and O'Shannassy reservoirs for Melbourne's water supply. Note the surrounding protected land.

CHEMFILE

Great Artesian Basin

The Great Artesian Basin lies under one-fifth of the Australian land mass and provides a reliable source of water for irrigation, stock and domestic use for a large part of inland Australia. An artesian basin is an underground water supply. The Great Artesian Basin is the largest artesian basin in the world and has the following key features.

- It covers an area of over 1 700 000 km^2 underlying nearly one-fifth of the Australian continent (see map below).
- In some places, the basin is up to 3000 m deep.
- The temperature of the water in the basin has been recorded to be anywhere from 30°C to 100°C.

Traditionally, the water could be readily accessed through bore holes as it flowed close to the surface. However, in the last century, government bodies have set up initiatives to manage the basin and try to maintain stores of water within the basin.

The Great Artesian Basin provides water for stock and the human population for a large area of inland Australia.

The water from the mains supply is tested rigorously and is of the highest quality and poses the lowest risk of contamination. The risk of contamination from different water sources can be seen in Figure 10.1.4.

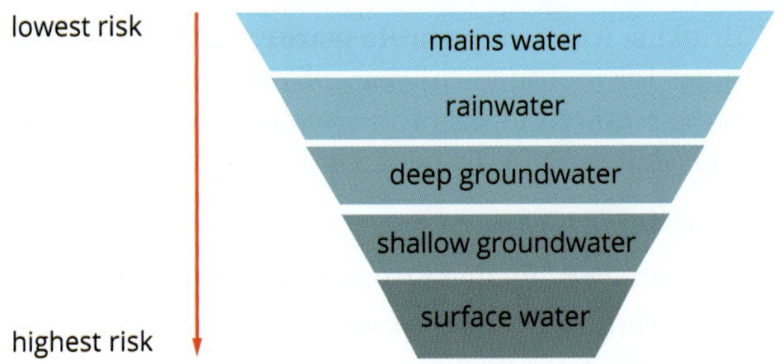

FIGURE 10.1.4 Different sources of drinking water present different risks of contamination.

SPECIAL PROPERTIES OF WATER

Water has a number of special properties that allow it to support life. Water and life are so strongly linked that space scientists search the universe for water in their quest to discover possible life beyond our planet.

The special properties of water can be explained through underlying knowledge of the structure of a water molecule and the **hydrogen bonding** that occurs between the water molecules.

Structure and bonding of water

Water has the chemical formula H_2O, which means that each water molecule contains one oxygen atom covalently bonded to two hydrogen atoms.

The covalent bonds in a water molecule are **polar** due to the oxygen atom having a higher electronegativity than the hydrogen atoms. This means that the shared pair of electrons in the O–H covalent bond are more attracted to the oxygen atom than the hydrogen atom. The molecule is polar overall.

Hydrogen bonding

Each water molecule has the potential to form four hydrogen bonds with surrounding water molecules. There are two partially charged hydrogen atoms and two lone pairs of electrons on the oxygen atom in each molecule, so all the hydrogen atoms and lone pairs in the molecule can be involved in hydrogen bonding.

The main type of **intermolecular force** existing between molecules in water is a hydrogen bond. The hydrogen bonds are formed by an **electrostatic attraction** between the partial positive charge on a hydrogen atom on one water molecule and a **non-bonding pair (lone pair)** of electrons on the oxygen atom of a neighbouring water molecule.

Figure 10.1.5 shows the hydrogen bonding between molecules of water. The two hydrogen atoms (white) in each water molecule have a slight positive charge, while the central oxygen atom (red) has a slight negative charge. The black dashed lines represent the hydrogen bonds that form between water molecules. (For more details on hydrogen bonding, see Section 3.4 on page 119.)

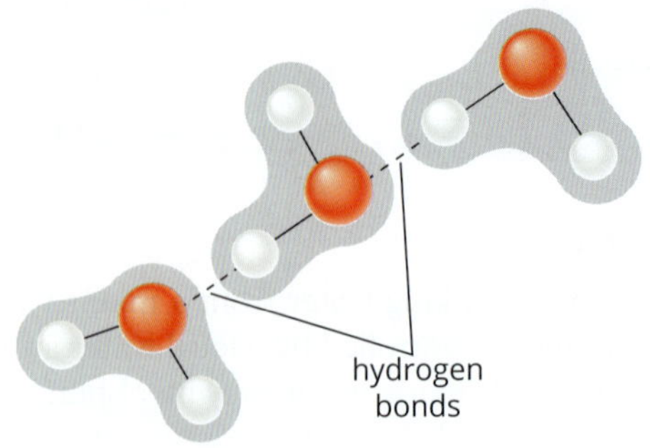

FIGURE 10.1.5 The hydrogen bonding between molecules of water

Because there is a large electronegativity difference between oxygen and hydrogen atoms, the partial charges on the atoms in a water molecule are relatively large. The electrostatic attraction between these opposite partial charges makes the hydrogen bonds between water molecules relatively strong.

As Figure 10.1.6 shows, each water molecule is capable of forming up to four hydrogen bonds with four other water molecules.

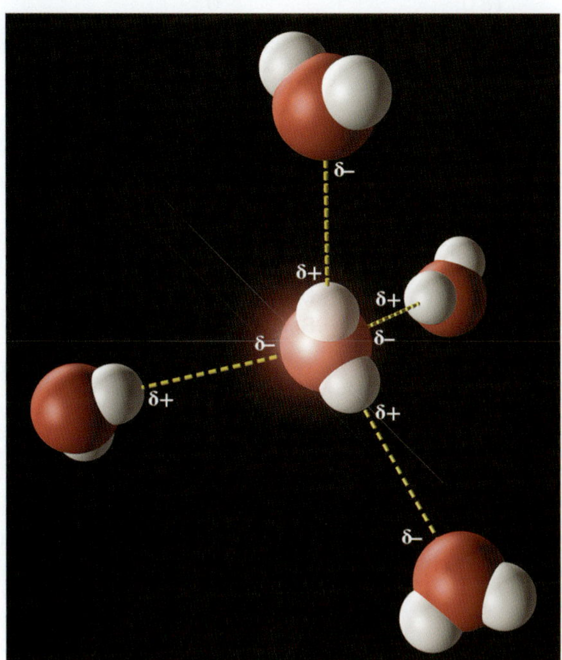

FIGURE 10.1.6 Each water molecule is capable of forming hydrogen bonds to up to four other water molecules.

Relatively high melting and boiling points of water

Compared to other molecules of a similar size, water has high boiling and melting points. This can most easily be seen in the observed trends in these properties for the group 16 hydrides.

The group 16 elements include oxygen (O), sulfur (S), selenium (Se), tellurium (Te) and polonium (Po). Each of these elements can bond with hydrogen to form a compound known as a **hydride**. Water can be classified as one of the group 16 hydrides (and could have the name hydrogen oxide). The other group 16 hydrides are listed in Table 10.1.3.

TABLE 10.1.3 Names and formulas of the group 16 hydrides

Group 16 element	Name of hydride	Formula of hydride
O	water	H_2O
S	hydrogen sulfide	H_2S
Se	hydrogen selenide	H_2Se
Te	hydrogen telluride	H_2Te
Po	hydrogen polonide	H_2Po

The group 16 hydrides are all molecular compounds. Their melting and boiling points reflect the size of the forces between their molecules. The higher the melting and boiling points, the stronger the intermolecular forces must be because more energy is required to overcome the forces and allow the molecules to move apart from each other to melt or evaporate.

The melting and boiling points of some of the group 16 hydrides are shown in Table 10.1.4, with the trend in melting and boiling points evident in Figure 10.1.7. Generally, as the molar mass of the hydrides increases, so do the melting and boiling points.

TABLE 10.1.4 Melting and boiling points of the group 16 hydrides

Hydride	Molar mass (g mol^{-1})	Melting point (°C)	Boiling point (°C)
H_2O	18.0	0	100
H_2S	34.1	–82	–60.7
H_2Se	81.0	–66	–41.5
H_2Te	129.6	–49	–2.2
H_2Po	212.0	–35	36.1

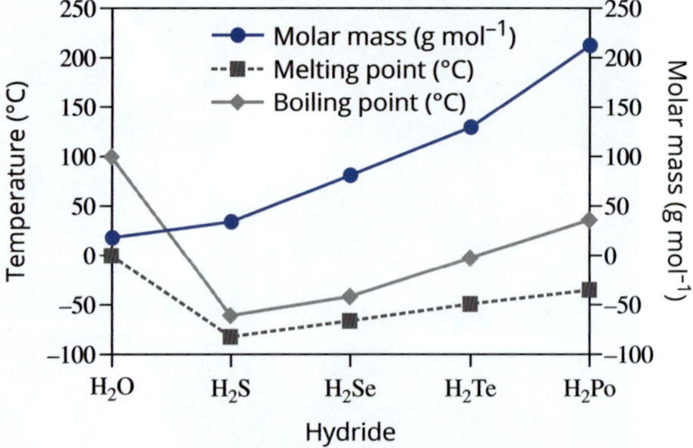

FIGURE 10.1.7 Water is an exception to the trend that melting and boiling points of group 16 hydrides increase as the mass increases.

You can see from Figure 10.1.7 that the boiling and melting points of water are relatively high. Apart from water, the melting and boiling points both increase going down the group of elements. The intermolecular forces are getting stronger down the group as the atom bonding with hydrogen has increasing mass. H_2Po has a larger mass than H_2S, H_2Se and H_2Te, hence it has stronger dispersion forces and higher melting and boiling points. However, H_2O, which has the smallest mass, has the highest melting and boiling points.

The exceptional melting and boiling points of water

Water has a melting point of 0°C and a boiling point of 100°C. Not only are these values significantly higher than those of the larger group 16 hydrides, they are also significantly higher than those of other molecular substances of a similar size, as can be seen in Figure 10.1.8.

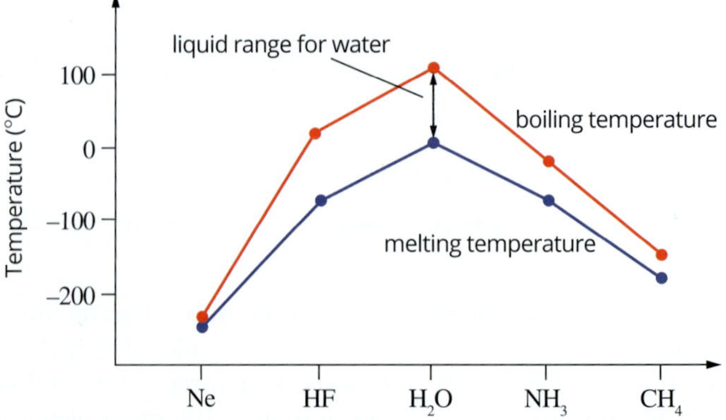

FIGURE 10.1.8 The melting and boiling points of water and other molecules of similar size

It is the strong hydrogen bonds between its molecules that give water its relatively high melting and boiling points. Of the group 16 hydrides, only water has hydrogen bonding as the strongest intermolecular force. A significant amount of energy is needed to disrupt the hydrogen bonds between water molecules, resulting in the higher melting points and boiling points observed for water.

The melting and boiling points of water do not follow the trend otherwise observed in the group 16 hydrides. This is due to water's hydrogen bonds.

Density of liquid water and solid ice

Another unusual property of water is that solid ice floats in liquid water because solid ice has a lower density. For most other substances, the solid state is denser than the liquid, meaning the solid will sink in the liquid. **Density** is a measurement of the mass of a unit volume of a substance. It can be calculated using the formula

$$\text{density (g cm}^{-3}) = \frac{\text{mass (g)}}{\text{volume (cm}^3)}$$

In solid ice there is less mass present in the same volume than in liquid water. The water molecules are arranged in ice in a way that is more spread out than in the liquid state. Solid ice has a density of 0.917 g mL^{-3}, whereas the density of water varies with temperature, but is 0.997 g mL^{-3} at 25°C.

Density is a measurement of the mass of a unit volume of a substance. Solid ice has a lower density than liquid water.

Expansion of water on freezing

Figure 10.1.9 shows a measuring cylinder filled with water, which is then frozen. It can clearly be seen that the volume occupied by the water molecules when they are in solid ice is greater than when the same amount of water is in the liquid state.

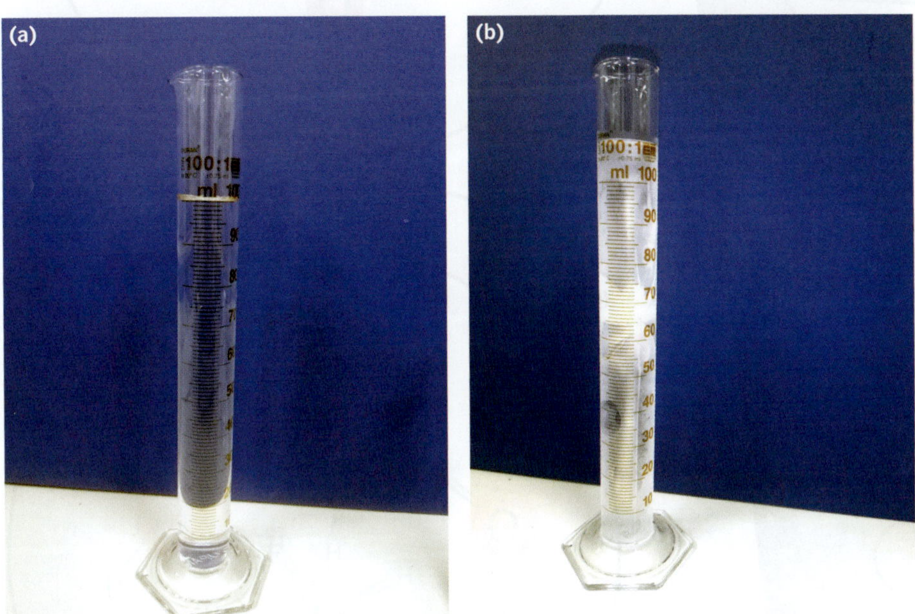

FIGURE 10.1.9 (a) A sample of liquid water occupies 100 mL in a measuring cylinder. (b) When the water was frozen, it expanded to approximately 110 mL. The expanding water also caused the measuring cylinder to crack.

CHEMFILE

The deep blue sea

When light passes through water, the vibration of O–H bonds in water molecules, combined with the hydrogen bonds between molecules, causes a small amount of light to be absorbed in the red part of the visible colour spectrum. This leaves the remaining complementary colour, which is blue. This colour cannot be seen in small volumes of water. However, when light travels through several metres of water or ice, the blue colour becomes visible (see figure below).

Large volumes of water or ice appear blue.

As liquid water is cooled, the water molecules move more slowly. Upon approaching the freezing temperature of water, the molecules arrange in such a way that each water molecule forms four hydrogen bonds to four neighbouring water molecules. This creates a very open arrangement of molecules, as shown in Figure 10.1.10, meaning that the water molecules are more widely spaced than in liquid water. Therefore, ice is less dense than liquid water and floats. When ice melts, the water molecules move more freely and become closer together.

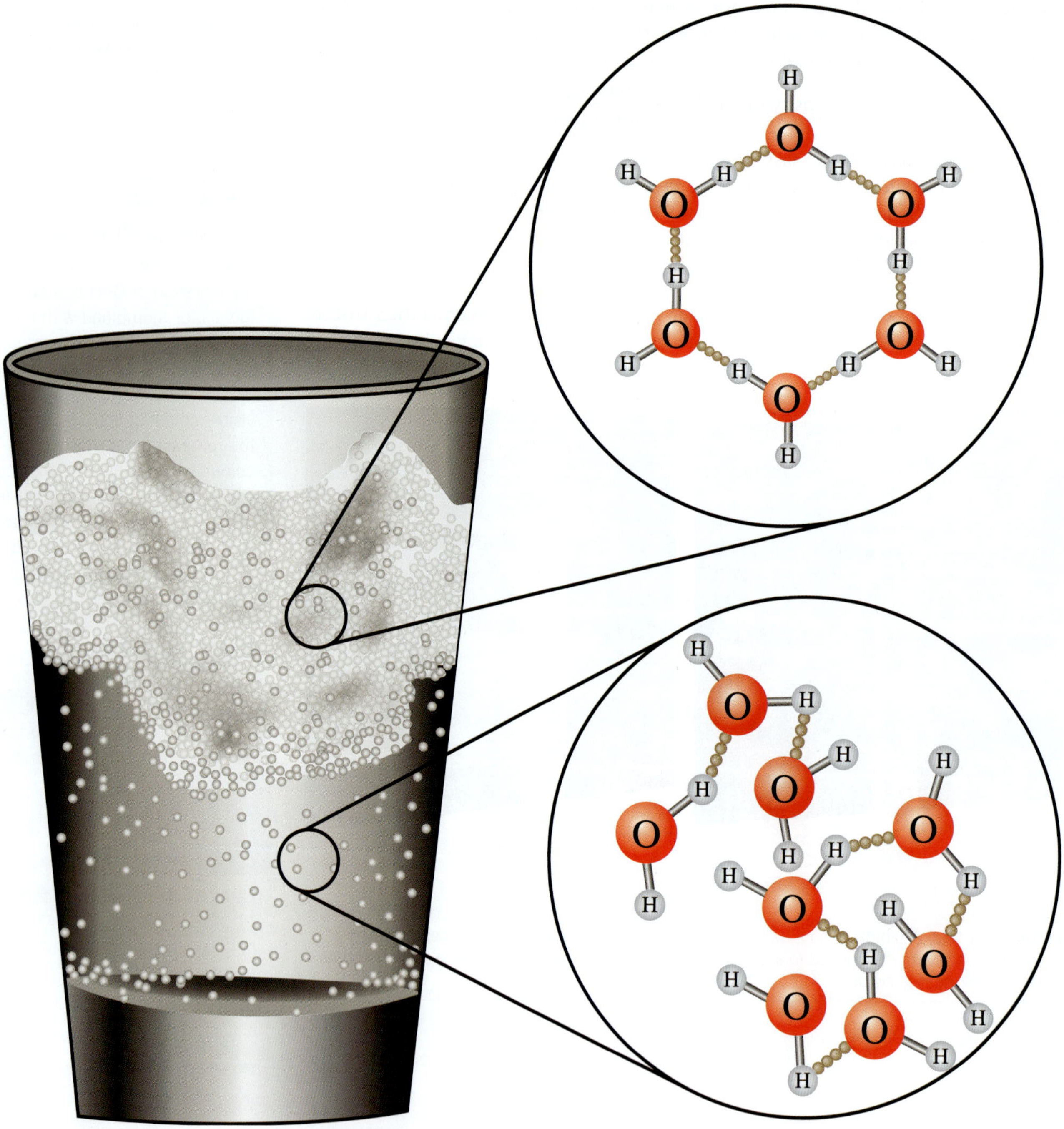

FIGURE 10.1.10 Ice floats because the water molecules form a crystal lattice in which the molecules are spaced more widely apart than in liquid water. This arrangement means ice is less dense than liquid water.

CASE STUDY **ANALYSIS**

The importance of floating ice

In countries where the temperature drops lower than it does in Australia, such as parts of North America, the United Kingdom and Europe, lakes and rivers completely freeze over in winter. Icebreaker ships are used to help ships continue to navigate rivers and lakes, as seen in Figure 10.1.11, by breaking through the ice that floats on the surface. While inconvenient for ships, the low density of solid ice is essential for the survival of living things within river and lake systems. A layer of ice over the surface of bodies of water forms an insulating layer that separates the warmer water below from the cold air temperatures. If ice were denser than water and sank to the bottom, new water would become exposed to cold air temperatures and also sink. Eventually the lake or river could be frozen completely solid from the bottom up and living organisms would not be able to survive. Water below the ice of lakes maintains a temperature of about 4°C, which is very cold but above freezing.

FIGURE 10.1.11 Icebreakers working on a river in the state of Michigan in the USA

The density of liquid water does vary a little with temperature, as seen in the graph in Figure 10.1.12.

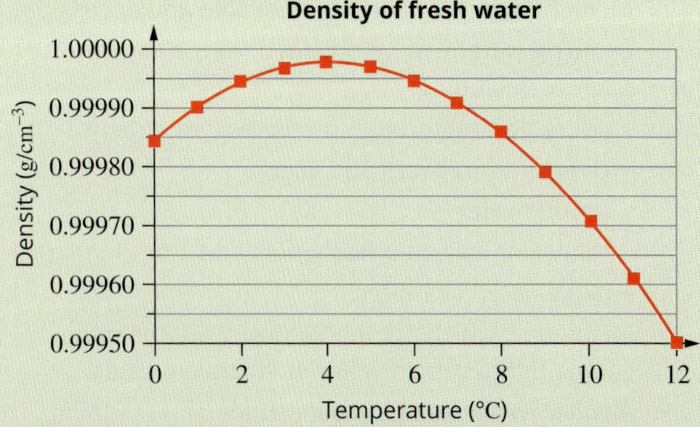

FIGURE 10.1.12 The density of water varies with changing temperatures.

Analysis

1. At what temperature does liquid water have the highest density?
2. At what temperatures does liquid water have a density above 0.99995 g cm^{-1}?
3. Solid ice has a density of 0.9168 g cm^{-3}. With reference to the graph, explain why ice floats on water.
4. Calculate the mass of 5.000 L of water at:
   - **a** 2°C
   - **b** 10°C
5. Explain why ice being less dense than liquid water allows life to exist in the waters surrounding Antarctica.

10.1 Review

OA ✓✓

SUMMARY

- Most of the water on Earth is in liquid form and is contained in the oceans.
- 2.5% of all the water on Earth is fresh, but just 0.5% of all the water on Earth is drinkable and accessible.
- Drinking water can be obtained from a variety of sources, including:
  - rivers flowing through protected catchments
  - directly from rivers and lakes
  - groundwater
  - rainwater collected from roofs and stored in tanks
  - desalinated seawater.
- Reservoirs, which are fed by rivers, are the main source of household water in Australian cities.
- A water molecule has a bent shape and contains polar covalent bonds.
- The forces that attract one water molecule to another are relatively strong hydrogen bonds.
- Each water molecule can form up to four hydrogen bonds with other water molecules.
- The strength of water's hydrogen bonds means relatively large amounts of energy are required to disrupt the bonds and separate the molecules from each other. This gives water relatively high melting and boiling points.
- Water has significantly higher melting and boiling points than the other group 16 hydrides.
- With the exception of water, the melting and boiling points of the group 16 hydrides increase going down the group. This is due to the increasing strength of dispersion forces.
- Ice is less dense than liquid water because of its unique geometric arrangement of water molecules as a result of hydrogen bonding.

KEY QUESTIONS

Knowledge and understanding

1 Which source of water has the highest risk of contamination? Explain why.

2 **a** List some physical properties that are unusual to water.
  b Explain the significance of polarity and hydrogen bonding in relation to these properties of water.

3 The overall structure of a water molecule is important in explaining its unique properties. Draw the structural formula of water, indicating any partial charges and dipoles.

Analysis

4 Describe why only 0.5% of water on Earth is available for drinking.

5 Explain why Australia needs to store more water per person than any other country.

6 Explain why a water molecule can form up to four hydrogen bonds with other water molecules.

7 With the exception of water, the group 16 hydrides exhibit a trend in boiling and melting points.
  a Sort the group 16 hydrides (H_2O, H_2Po, H_2S, H_2Se, H_2Te) from lowest to highest boiling point.
  b Why is it that water does not follow the trend of the other group 16 hydrides?

10.2 Heat capacity

Several of the ways in which water supports life on Earth are based on how water interacts with heat energy. Water has a high **heat capacity**, meaning water can absorb a relatively large quantity of heat energy for very small changes in temperature.

When you walk on the beach on a hot summer's day, the dry sand can be burning hot on your feet while the shallow water at the edge of the sea is cool (Figure 10.2.1). Both surfaces are receiving and absorbing similar amounts of heat energy from the Sun, but the temperature of each is markedly different, based on their differing heat capacities.

FIGURE 10.2.1 On a hot day, the sand at the beach can be hot to touch while the water remains pleasantly cool.

COMPARING THE HEAT CAPACITY OF DIFFERENT SUBSTANCES

The heat capacity of a substance is a measure of the substance's capacity to absorb and store heat energy. When the same quantity of heat energy is applied to two substances with different heat capacities, they will undergo different temperature changes.

This can be seen in a laboratory by comparing water and ethanediol, a liquid used as a coolant in car radiators. Water has a higher heat capacity than ethanediol. When equal amounts of the two liquids are heated with the same amount of heat energy, the temperature of the two liquids will rise by different amounts. As shown in Figure 10.2.2, when 100 g of water and 100 g of ethanediol are heated with the same amount of energy, the temperature of the beaker containing water will increase by a smaller amount than the beaker containing ethanediol. Water has a higher heat capacity than ethanediol.

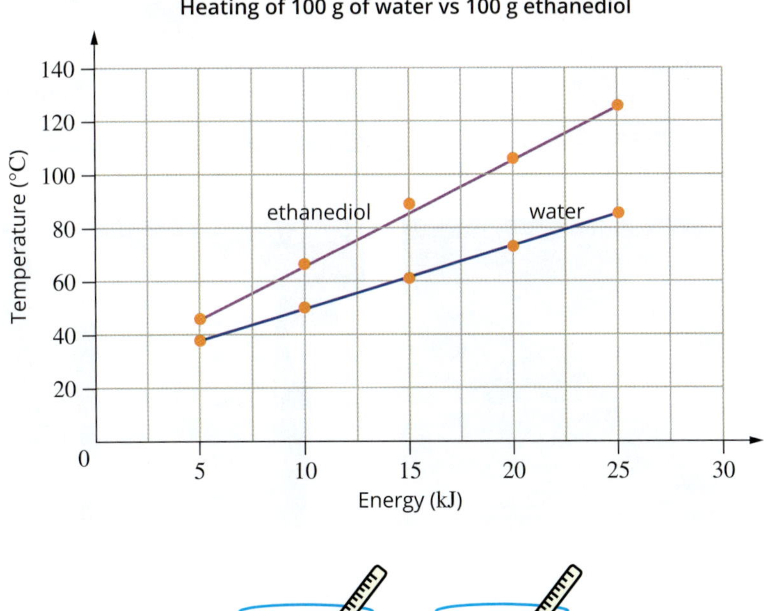

FIGURE 10.2.2 When 100 g of water and 100 g of ethanediol are heated with the same amount of energy, the temperature of the beaker containing water increases less than the temperature of the ethanediol.

Specific heat capacity (c) is a measure of the energy required to raise 1 g of substance by 1°C. It is reported in units of J g^{-1} °C^{-1}.

The specific heat capacity of a substance reflects the type of bonding in that substance. For molecular substances, the heat capacity will depend on the strength of the intermolecular forces between molecules.

SPECIFIC HEAT CAPACITY

The **specific heat capacity** of a substance measures the amount of energy (in joules) needed to increase the temperature of a certain amount (usually 1 gram) of that substance by 1°C.

Specific heat capacity is given the symbol *c* and is expressed in joules per gram per degree Celsius, i.e. J g^{-1} °C^{-1}.

The specific heat capacities of some common substances are shown in Figure 10.2.3. You can see that the value for water is relatively high. It takes more energy to increase the temperature of water compared to other substances. Metals, covalent molecular substances and composite substances exhibit a variety of specific heat capacities. The top row illustrates metals, the middle row of substances shows some small molecular substances, and the third row shows some composite materials.

The specific heat capacity of a substance is a reflection of the type of bonds holding the molecules, ions or atoms together in the substance. The relatively high specific heat capacity of water is related to the number and strength of the hydrogen bonds between water molecules. Hydrogen bonds are stronger than other intermolecular forces and they are able to absorb and store large amounts of heat energy before they are disrupted.

iron: 0.45 J g^{-1} °C^{-1}

copper: 0.39 J g^{-1} °C^{-1}

aluminium: 0.90 J g^{-1} °C^{-1}

water: 4.18 J g^{-1} °C^{-1}

ethanediol (antifreeze): 2.42 J g^{-1} °C^{-1}

chlorofluorocarbon (CCl_2F_2): 0.60 J g^{-1} °C^{-1}

wood: 1.3–2.0 J g^{-1} °C^{-1}

concrete: 0.8 J g^{-1} °C^{-1}

glass: 0.84 J g^{-1} °C^{-1}

FIGURE 10.2.3 Heat capacities of some selected substances

Specific heat capacity of water

Water has a specific heat capacity of 4.18 J g^{-1} °C^{-1}. This means that 4.18 joules of heat energy are needed to increase the temperature of 1 gram of water by 1°C.

As water is a liquid, it is often convenient to measure a quantity of water as a volume, in mL or L, rather than as a mass measured in grams. 1 mL of water has a mass of approximately 1 g, so it is easy to convert a given volume of water to an approximate mass:

- 1 mL of water has a mass of approximately 1 g.
- 100 mL of water has a mass of approximately 100 g.
- 250 mL of water has a mass of approximately 250 g.

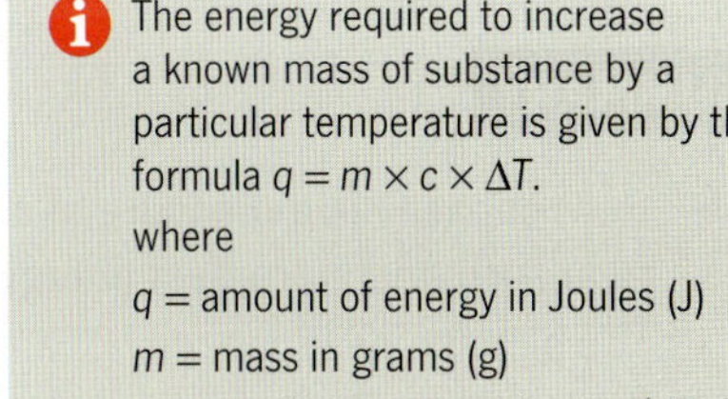

The energy required to increase a known mass of substance by a particular temperature is given by the formula $q = m \times c \times \Delta T$.

where

q = amount of energy in Joules (J)

m = mass in grams (g)

c = specific heat capacity (J g^{-1} °C^{-1})

ΔT = temperature change (°C)

Calculations using specific heat capacity

Heat energy is given the symbol q.

A useful equation can be written to calculate the heat energy needed to change the temperature of a mass of substance:

Heat energy = mass × specific heat capacity × temperature change

Using symbols, the equation can be written:

$$q = m \times c \times \Delta T$$

where q is the amount of heat energy (J), m is the mass (g), c is the specific heat capacity (J g^{-1} °C^{-1}) and ΔT is the temperature change (°C).

Worked example 10.2.1

CALCULATING THE AMOUNT OF ENERGY REQUIRED TO HEAT A SPECIFIED MASS OF A SUBSTANCE USING SPECIFIC HEAT CAPACITY

Calculate the heat energy, in kJ, needed to increase the temperature of 200 g of water by 15.0°C.

Thinking	Working
Find the specific heat capacity (c) of the substance from Figure 10.2.3.	The specific heat capacity of water is 4.18 J g^{-1} °C^{-1}.
Calculate the quantity of heat energy in joules, using the formula: $q = m \times c \times \Delta T$	$q = 200 \times 4.18 \times 15.0$ $= 1.25 \times 10^4$ J
Express the quantity of energy in kJ. Remember that to convert from J to kJ, you multiply by 10^{-3}.	$q = 1.25 \times 10^4 \times 10^{-3}$ = 12.5 kJ

Worked example: Try yourself 10.2.1

CALCULATING THE AMOUNT OF ENERGY REQUIRED TO HEAT A SPECIFIED MASS OF A SUBSTANCE USING SPECIFIC HEAT CAPACITY

Calculate the heat energy, in kJ, needed to increase the temperature of 375 g of water by 45.0°C.

CHEMFILE

Removing excess heat

Car drivers are often reminded to check the level of radiator fluid in their engine, especially before going for a long drive. The water that is part of this fluid is needed to keep the car's engine cool. The combustion of fuel in the engine is a highly exothermic reaction, giving off a large amount of heat to its surroundings. If the engine overheats, it can be permanently damaged. Excess heat is removed from the engine by water. Water is cycled through the engine by a series of pipes and then returns to the radiator where the heat energy is transferred from the water to the metal of the radiator and then to the air passing through the grill (see figure below). This heat is also used to heat the passenger compartment of the vehicle. This is why in the past, if a car engine was heating on very hot days, drivers might turn on the heater in order to divert heat away from the engine.

Water is an effective coolant because of its high heat capacity.

10.2 Review

OA ✓✓

SUMMARY

- Heat capacity is a measure of a substance's capacity to absorb and store heat energy.
- The specific heat capacity of a substance measures the quantity of energy (in joules) needed to increase the temperature of a certain amount (usually 1 gram) of that substance by 1°C.
- When compared to other substances, water has a relatively high specific heat capacity of 4.18 J g^{-1} $°C^{-1}$.
- The specific heat capacity of water is relatively high due to the ability of the hydrogen bonds between water molecules to absorb and store heat energy.
- The heat energy required to increase a given mass of substance by a particular temperature can be calculated using the equation:
  $q = m \times c \times \Delta T$
  where q is heat energy (J), m is mass of the substance (g), c is the specific heat capacity (J g^{-1} $°C^{-1}$), and ΔT is the temperature change (°C).
- The mass of a sample of liquid water in grams (g) at 25°C is approximately equal to its volume in millilitres (mL).

KEY QUESTIONS

Knowledge and understanding

1 What is the definition of specific heat capacity?

2 Figure 10.2.3, on page 340, shows the specific heat capacities of some substances. Order the substances from the one that absorbs the least amount of energy to the one that absorbs the most for the same increase in temperature.

Analysis

3 Calculate the heat energy, in J, needed to increase the temperature of 5.0 g of sand by 12°C. The specific heat capacity of sand is 0.48 J g^{-1} $°C^{-1}$.

4 Calculate the heat energy, in kJ, needed to increase the temperature of 1.5 kg of water by 15°C.

5 **a** How much energy, in kJ, will be required to heat 600 mL of water in a kettle from 21°C to 100°C?

b Given that the specific heat capacity of lead is 0.13 J g^{-1} $°C^{-1}$, how much energy would it take to raise the same mass of lead by the same number of degrees as the water described in part **a**?

6 A 250 mL beaker of water at a temperature of 22°C is heated with the addition of 10 kJ of energy. Calculate the temperature reached by the beaker of water.

7 A 500 g sample of an unknown substance was heated with the addition of 9.75 kJ of energy. The temperature is noted to rise from 25°C to 75°C. Calculate the specific heat capacity and use Figure 10.2.3 to identify the unknown substance.

10.3 Latent heat

When you perspire (sweat), your body uses a very effective method to cool down. The water released onto your skin as sweat sits on the surface of the skin and absorbs relatively large amounts of heat energy from the body before it evaporates as a gas (Figure 10.3.1).

In combination with its high specific heat capacity, it is the **latent heat** of water that makes it such an effective coolant. In this section you will learn about latent heat and its significance to ocean temperature regulation and aquatic organisms.

When a solid is heated at constant pressure, the temperature of the solid increases until it reaches its melting point. The temperature then remains constant as the solid melts, even though further energy is being absorbed. Similarly, when a liquid reaches its boiling point, its temperature remains constant as boiling occurs until all of the liquid has evaporated. The gaseous water that forms can increase in temperature, but the liquid will not.

Think about a pot of boiling water. The temperature remains at 100°C until all of the water in the pot has evaporated. Figure 10.3.2 shows the temperature of water as it is heated over time. No matter how long water is heated, the temperature never rises above 100°C until it has become gaseous. This is because the heat energy being supplied is being used to change the state of the substance.

FIGURE 10.3.1 One of the signs of an increasing body temperature on a hot day or during exercise is when you start to perspire. Your body cools itself by releasing water onto the surface of your skin.

> Latent heat is the energy required to change a fixed amount of substance, usually 1 mole, from either a solid to a liquid at melting temperature or a liquid to a gas at boiling point. Over the period of time that latent heat is being absorbed, the temperature of a substance will not change.

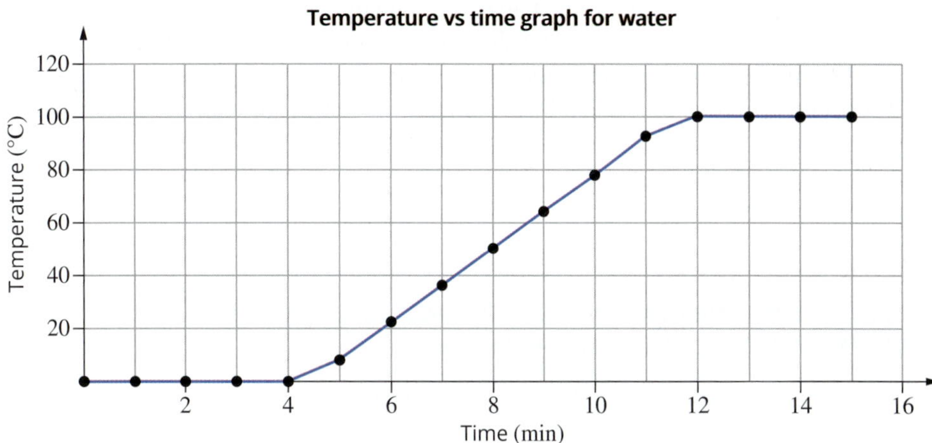

FIGURE 10.3.2 A temperature versus time graph for water, showing the plateaus at 0°C and 100°C

Even though the temperature does not change while a phase change is taking place, the substance is still absorbing energy. This energy is referred to as latent heat and is defined as the energy absorbed by a fixed amount of substance as it changes state from a solid to a liquid or a liquid to a gas at its melting point or boiling point respectively.

VALUES OF LATENT HEAT

Latent heat values are a measure of the quantity of heat energy required to melt or boil a given amount of a solid or liquid. Latent heat values are given the symbol L and have the unit kilojoules per mole, kJ mol^{-1}.

The **latent heat of fusion** of a substance is the energy needed to change 1 mole of the substance from a solid to a liquid at its melting point.

The **latent heat of vaporisation** of a substance is the energy needed to change one mole of the substance from a liquid to a gas at its boiling point.

LATENT HEAT OF WATER

The latent heat of fusion of water is 6.0 kJ mol^{-1}. This means that 6.0 kJ of energy is needed to change 1 mole of water from a solid to a liquid at 0°C. This energy is needed to disrupt the ice lattice by breaking some of the hydrogen bonds between water molecules.

CHEMFILE

Importance of latent heat values for water storage

Much of Australia's fresh water supplies are stored in open reservoirs, such as the Maroondah Reservoir in Melbourne (see figure below). Although evaporation losses from storage facilities can potentially be large, they would be far greater if water had a lower latent heat of vaporisation as energy from the Sun would cause a far greater rate of evaporation.

The Maroondah Reservoir in Melbourne. The water level would be depleted by evaporation much more quickly if water had a lower latent heat of vaporisation.

The heat energy required for a state change is given by the formula $q = n \times L$. Always check if the change is from liquid to gas or solid to liquid and substitute the appropriate latent heat value.

The latent heat of vaporisation of water is 40.7 kJ mol^{-1}. This means that 40.7 kJ is needed to change the state of 1 mole of water from a liquid to a gas at 100°C. This relatively large quantity of energy is required to completely break the hydrogen bonds between the water molecules so they can separate and form a gas.

The latent heat values of water and some other substances are listed in Table 10.3.1. Much like specific heat capacity, latent heat depends on the strength of the intermolecular forces between molecules of the substance.

TABLE 10.3.1 Latent heat values for some common molecular substances

Substance	Latent heat of fusion (kJ mol^{-1})	Latent heat of vaporisation (kJ mol^{-1})
water	6.0	40.7
hydrogen	0.06	0.45
oxygen	0.22	3.4

As you can see, the latent heat values of water are relatively high. This is due to the strength and number of water's hydrogen bonds relative to its molecular size.

CALCULATIONS INVOLVING LATENT HEAT

The amount of heat required to change the state of a given amount of a substance can be calculated if the latent heat values for the substance are known.

The amount of heat energy required for a change in state is equal to the amount of substance, in mol, multiplied by the latent heat value. This can be written as:

$$q = n \times L$$

where q is the heat energy (kJ), n is the amount (mol) of the substance changing state and L is the latent heat value of fusion or vaporisation (kJ mol^{-1}).

Worked example 10.3.1

CALCULATING THE HEAT ENERGY REQUIRED TO EVAPORATE A GIVEN MASS OF WATER AT ITS BOILING TEMPERATURE

Calculate the heat energy, in kJ, required to evaporate 200 g of water at 100°C.

Thinking	Working
Determine the amount, in mol, of the substance using the formula: $n = \frac{m}{M}$	$n = \frac{200}{18.0}$ $= 11.1$ mol
Find the relevant latent heat value of the substance.	Water is being evaporated, so the latent heat of vaporisation of water is required. $L = 40.7$ kJ mol^{-1}
Calculate the heat energy, in kJ, using the formula: $q = n \times L$	$q = 11.1 \times 40.7$ $= 452$ kJ

Worked example: Try yourself 10.3.1

CALCULATING THE HEAT ENERGY REQUIRED TO EVAPORATE A GIVEN MASS OF WATER AT ITS BOILING TEMPERATURE

Calculate the heat energy, in kJ, required to evaporate 75.0 g of water at 100°C.

CASE STUDY

Regulation of ocean temperatures

Ocean temperatures vary throughout ocean waters across the Earth. The ocean is generally warmer near the equator and cooler near the North and South Pole.

Figure 10.3.3 is a satellite image of global ocean currents. The colours indicate ocean temperatures, showing the warm water (red) and cooler water (green).

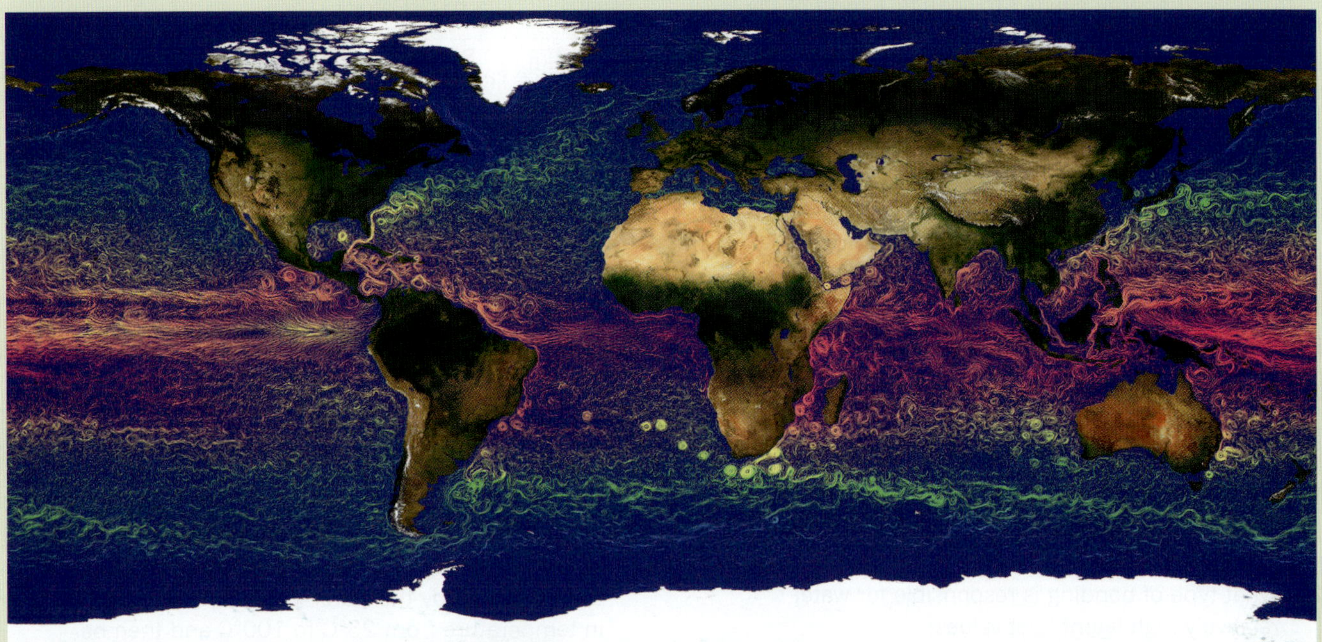

FIGURE 10.3.3 Global ocean currents. Red colours indicate warm water and green colours are cooler.

The average temperature of Earth's water at the ocean surface is about 17°C. The Indian Ocean has the warmest temperatures, ranging from 19° to 30°C, while the Arctic Ocean is the coldest at around −2.2°C (the salt dissolved in ocean water lowers the water's melting temperature). Although there is variation in ocean temperatures around the globe, the temperature of each individual region of the ocean stays relatively stable. This is important for aquatic marine life, such as the coral and fish seen in Figure 10.3.4, as large fluctuations in temperature can cause them harm. It is water's high latent heat value of vaporisation and its high heat capacity that aid temperature regulation. High heat capacity means high levels of sunlight do not result in large increases in water temperature; high latent heat values mean high levels of sunlight do not result in excessive evaporation.

FIGURE 10.3.4 Fish and coral in the sea can survive due to the relatively consistent temperature of the sea water.

10.3 Review

OA ✓✓

SUMMARY

- Latent heat is the energy absorbed by a fixed amount of substance as it changes state from a solid to a liquid or a liquid to a gas.
- The latent heat of vaporisation of a substance is the heat energy needed to change 1 mole of the substance from a liquid to a gas at its boiling point.
- The value of the latent heat of vaporisation of water is a relatively high 40.7 kJ mol^{-1}. This has significance for cooling of organisms and preservation of water supplies.
- The latent heat of fusion of a substance is the heat energy needed to change 1 mole of the substance from a solid to a liquid at its melting point.
- The value of the latent heat of fusion of water is a relatively high 6.0 kJ mol^{-1}.
- The heat energy required to change the state of a substance can be determined using the equation $q = n \times L$.

KEY QUESTIONS

Knowledge and understanding

1 A heating curve for a substance is produced by plotting the temperature change against the energy input. When this is done from the melting point to the boiling point of a substance, two flat regions are observed. Explain what the flat regions of the graph represent.

2 What type of bonding is responsible for water's relatively high latent heat values?

3 Explain why water's latent heat of vaporisation is much higher than its latent heat of fusion.

Analysis

4 Calculate the heat energy, in kJ, required to evaporate 2.50 mol of water at 100°C.

5 A student places a beaker containing 100 mL of water at 25°C over a lit Bunsen burner, but gets distracted and does not notice that the beaker boils dry. Calculate the amount of heat energy, in kJ, that was absorbed by the water in the beaker to increase in temperature from 25°C to 100°C and then be completely evaporated at 100°C.

6 Calculate the heat energy, in kJ, required to melt 300 g of ice at 0°C.

Chapter review

10

KEY TERMS

bore water
density
desalinated seawater
electrostatic attraction
heat capacity
hydride
hydrogen bond
intermolecular force
latent heat
latent heat of fusion
latent heat of vaporisation
lone pair
non-bonding pair
polar
potable water
recycled water
specific heat capacity

REVIEW QUESTIONS

Knowledge and understanding

1 Copy and complete the following sentences about water and its properties by selecting the correct term from the options in italics.

Water is a *polar/non-polar molecule*. Within a single molecule, hydrogen and oxygen atoms are held together by strong *covalent/hydrogen* bonds. Between different molecules, the most significant forces are *hydrogen bonds/dispersion forces*.

It is the relatively *low/high* strength of the intermolecular forces that gives water its unique properties of relatively:

- *low/high* boiling point, __________°C
- *low/high* latent heat values, __________ kJ mol^{-1} and __________ kJ mol^{-1}
- *low/high* specific heat capacity, __________ J g^{-1} °C^{-1}.

2 The image below shows the arrangement of water molecules in ice. Identify the parts of the diagram labelled **a–d.**

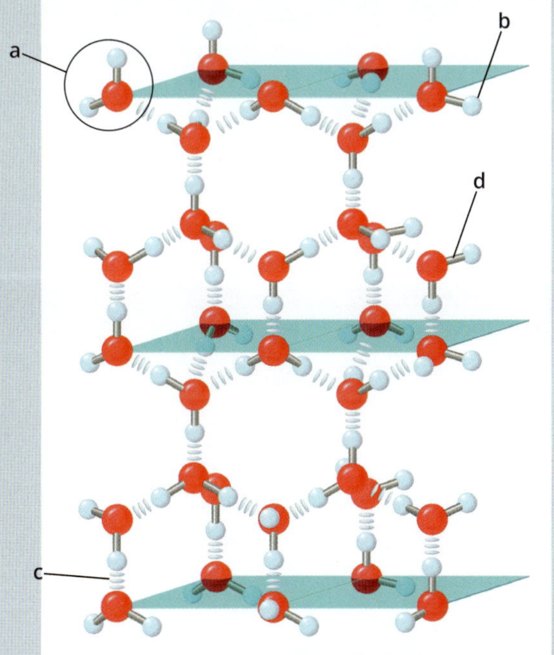

3 Where is most of the drinkable water found on Earth?

A Ice
B Rivers
C Groundwater
D Bore water

4 What is the source of drinking water for most Australians and why?

5 Arrange the following sources of water on Earth in order from highest volume to lowest volume:

oceans; groundwater; rivers; soil moisture; ground ice and permafrost; lakes; ice caps and glaciers; atmosphere as water vapour

6 Water boils at 100°C. However, a much higher temperature (1000°C) is needed to decompose (break down) water molecules into hydrogen gas and oxygen gas. Explain, with reference to bonding types, the reasons for the difference in temperature.

7 Hydrogen sulfide and water are both group 16 hydrides. Explain why water exhibits a much higher boiling point than hydrogen sulfide.

8 In your own words, discuss the significance of polarity and hydrogen bonding in relation to the relatively high boiling point of water.

9 When heat energy is applied to a substance, what determines the temperature increase of the substance? Explain.

10 Match the following terms with their definitions: latent heat of fusion; specific heat capacity; latent heat of vaporisation; boiling point

a the temperature at which a liquid boils to form a gas
b the heat energy required to melt a solid to a liquid at its melting point
c the amount of heat energy required to increase 1 g of a substance by 1°C
d the heat energy required to evaporate a liquid to a gas at its boiling point

Application and analysis

11 An evaporative air conditioner is used to cool the air inside some buildings. The air conditioner lowers the air temperature by using heat from the air to evaporate water. What property of water is most important in its use in an evaporative air conditioner? Explain.

12 A student boils water to make a cup of coffee. Calculate the energy required, in kJ, to raise 250 mL of water from 18°C to 100°C. Express your answer to three significant figures.

13 Calculate the quantity of energy, in kJ, required to raise the temperature of 1.50 L of water from 23°C to 90°C.

14 Calculate the heat energy required to evaporate 250 g of water at its boiling temperature.

15 Calculate the heat energy, in kJ, needed to increase the temperature of 1.0 kg of ethanol by 4.0°C. The specific heat capacity of ethanol is 2.4 J g^{-1} $°C^{-1}$.

16 Calculate the mass of water that could be evaporated at 100°C if 1000 kJ of heat were absorbed by the water.

17 Calculate the difference, to the nearest kJ, in the amount of heat energy required to melt 500 g of ice at 0°C and to evaporate 500 g of water at 100°C.

18 A student was asked to record the temperature changes as a sample of ice was heated. The ice was placed in a beaker and heated with a Bunsen burner for 20 minutes. The graph below shows the temperature, in degrees Celsius, recorded at 1 minute intervals.

Temperature vs time graph of sample of ice being heated

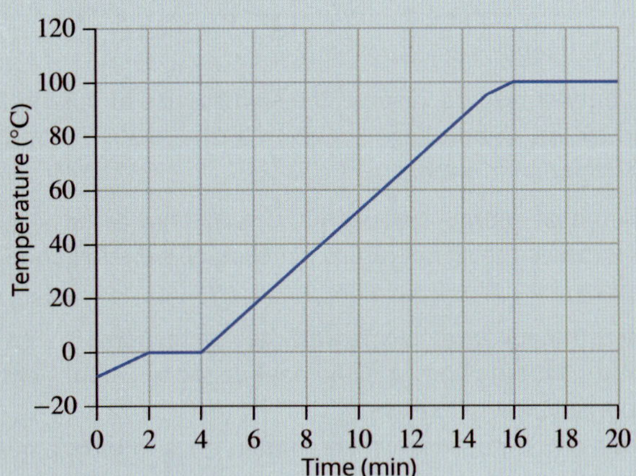

a Explain what is happening, at a molecular level, between 2.0 and 4.0 minutes on the graph.

b Even though heating is continued for 20 minutes, no further temperature rise is observed after 16 minutes. What happens to the added heat between 16 and 20 minutes?

19 A copper pan of mass 750 g at a temperature of 25°C is placed on a stovetop and heated with the addition of 12.5 kJ of heat energy. Given that the heat capacity of copper is 0.39 J g^{-1} $°C^{-1}$, calculate the final temperature of the pan.

20 A rock weighing 15 g is placed next to a campfire and absorbs 50 J of energy, increasing in temperature by 8.0°C. Calculate the specific heat capacity of the rock.

21 200 g of water is placed in one beaker and 200 g of ethanol in another. Initially, both are the same temperature. The specific heat capacity of ethanol is 2.4 J g^{-1} $°C^{-1}$ and that of water is 4.18 J g^{-1} $°C^{-1}$. Each beaker is heated by the addition of 5 kJ of heat energy. The molecular structures of water and ethanol are shown in the diagram below.

a How would you expect the temperature of the ethanol to compare to that of the water after heating?

b Explain your answer to part **a**, making reference to the specific heat capacities and the strength of the bonding that would exist between molecules.

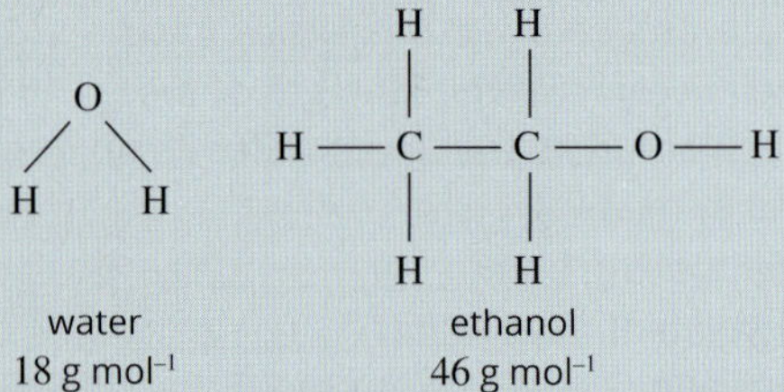

22 A 25 g piece of an unknown metal alloy at 150°C is dropped into an insulated container with 200 g of ice. When the temperature of the metal stops changing, 9.0 g of ice has melted. Calculate the specific heat capacity of the metal.

OA ✓✓

CHAPTER

11 Acid–base reactions

Acids and bases are found in homes and are used extensively in industry and agriculture. They are also the reactants and products of many chemical reactions that take place in environmental and biological systems.

In this chapter, you will study a theory that explains the characteristic reactions of acids and bases. You will learn to represent the ionisation of acids and bases in water using ionic equations and investigate the distinction between strong and weak acids and bases, and concentrated and dilute solutions of acids and bases. Studying the use of the logarithmic pH scale will enable you to quantitatively determine the concentration of acidic and basic solutions. You will also learn how to measure the pH of acidic and basic solutions using methods which vary in their accuracy and precision.

Finally, you will apply your understanding of acid–base chemistry to the contemporary environmental issue of increasing ocean acidity.

Key knowledge

- the Brønsted–Lowry theory of acids and bases, including polyprotic acids and amphiprotic species, and the writing of balanced ionic and full equations, with states, for their reactions in water **11.1**
- the distinction between strong and weak acids and strong and weak bases, and between concentrated and dilute acids and bases, including common examples **11.2**
- neutralisation reactions to produce salts:
  - reactions of acids with metals, carbonates and hydroxides, including balanced full and ionic equations, with states **11.3**
  - types of antacids and their use in the neutralisation of stomach acid **11.3**
- use of the logarithmic pH scale to rank solutions from most acidic to most basic; calculation of pH for strong acid and strong base solutions of known concentration using the ionic product of water (K_w at a given temperature) **11.4**
- accuracy and precision in measurement by the comparison of natural indicators, commercial indicators and pH meters to determine the relative strengths of acidic and basic solutions **11.5**
- applications of acid–base reactions in society; for example, natural acidity of rain due to dissolved CO_2 and the distinction between the natural acidity of rain and acid rain, or the action of CO_2 forming a weak acid in oceans and the consequences for shell growth in marine invertebrates. **11.6**

OA
✓✓

11.1 Acids and bases

Acids and **bases** make up some of the household products in your kitchen and laundry (Figure 11.1.1). In this section, you will be introduced to the Brønsted–Lowry theory that explains the chemical properties of acids and bases. You will learn to write balanced ionic and full equations that show the reactions of acids and bases with water, and also learn that some species may act as either an acid or a base, depending on the other reactants present.

FIGURE 11.1.1 Common products that contain acids, bases or salts

CHEMFILE

Saving the Nobel Prize gold medal

George de Hevesy (1885–1966) worked for the Niels Bohr Institute in Denmark during World War II. The institute was looking after a number of valuable gold medals that had been awarded to recipients of the Nobel Prize.

When Germany invaded Denmark at the beginning of World War II, de Hevesy was concerned the gold medals would be confiscated. He dissolved the gold medals in aqua regia, which is a mixture of concentrated hydrochloric and nitric acids, and hid the bottle containing the gold solution among the hundreds of other bottles on his laboratory shelves. It was never discovered and after the war he retrieved the precious bottle and precipitated the gold out of solution. The gold was sent to the Nobel foundation where the medals were recast into duplicates of the originals and returned to their rightful owners.

Recipients of the Nobel Prize are given a sum of money and a gold medal weighing about 175 g.

PROPERTIES OF ACIDS AND BASES

Acids are used in our homes, in agriculture and in industry. They also have an important role in our bodies. Table 11.1.1 gives the names, chemical formulas and uses of some common acids.

TABLE 11.1.1 Common acids and their everyday uses

Name	Formula	Uses
hydrochloric acid	HCl	present in stomach acid to help break down proteins; used as a cleaning agent for brickwork
sulfuric acid	H_2SO_4	one of the most common chemicals manufactured; used in car batteries and in the manufacture of fertilisers and detergents
nitric acid	HNO_3	used in the manufacture of fertilisers, dyes and explosives
ethanoic acid (acetic acid)	CH_3COOH	found in vinegar; used as a preservative
carbonic acid	H_2CO_3	found in carbonated soft drinks and beer
phosphoric acid	H_3PO_4	used in some soft drinks and in the manufacture of fertilisers
citric acid	$C_6H_8O_7$	found in citrus fruits
ascorbic acid	$C_6H_8O_6$	found in citrus fruits (vitamin C)

Many cleaning agents used in the home, such as washing powders and oven cleaners, contain bases. Solutions of ammonia are used as floor cleaners, and sodium hydroxide is the major active ingredient in oven cleaner and dishwasher powder. Bases are effective cleaners because they react with fats or oils to produce water-soluble soaps. Soluble bases, such as NH_3 and $NaOH$, are referred to as **alkalis**. An example of an insoluble base is calcium carbonate, which reacts with acids; however, because it is insoluble, it is not considered an alkali.

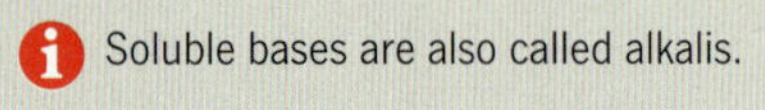

Soluble bases are also called alkalis.

Table 11.1.2 gives the names, chemical formulas and uses of some common bases.

TABLE 11.1.2 Common bases and their uses

Name	Formula	Uses
sodium hydroxide (caustic soda)	$NaOH$	used in drain and oven cleaners, and soap making
ammonia	NH_3	used in household cleaners, fertilisers and explosives
calcium hydroxide	$Ca(OH)_2$	found in cement and mortar; used in garden lime to adjust soil pH
magnesium hydroxide	$Mg(OH)_2$	key ingredient in some antacids, such as milk of magnesia, to overcome indigestion
sodium carbonate	Na_2CO_3	used in the manufacture of washing powder and glass

All acids have some properties in common. Similarly, bases have common properties. The properties of acids and bases are summarised in Table 11.1.3.

TABLE 11.1.3 Properties of acids and bases

Properties of acids	Properties of bases
turn litmus indicator red	turn litmus indicator blue
tend to be corrosive*	are corrosive, caustic** and slippery
taste sour	taste bitter
react with bases	react with acids
solutions have a relatively low pH	solutions have a relatively high pH
solutions conduct an electric current	solutions conduct an electric current

*corrosive – able to dissolve the structure of an object.

**caustic – able to burn or corrode organic tissue by chemical action

CHEMFILE

Handle strong bases with care

Bases feel slippery to the touch because they react with fats in our skin to produce soap. Strong bases should therefore be handled with care. Oven cleaners contain about 4% of the strong base sodium hydroxide. A common name of sodium hydroxide is caustic soda.

The following figure shows the safety instructions on a container of oven cleaner.

Oven cleaner removes fatty deposits by turning them into soap. Note the safety instructions.

CHANGING IDEAS ABOUT THE NATURE OF ACIDS AND BASES

Over the years, there have been many attempts to define acids and bases. At first, acids and bases were explained in terms of their *observed* properties, such as their taste and effect on other substances.

For example, in the seventeenth century, British scientist Robert Boyle described the properties of acids in terms of taste, their action as solvents, and how they changed the colour of certain vegetable extracts. He also noticed that alkalis (soluble bases) could reverse the effect that acids had on these extracts.

It wasn't until the late eighteenth century that attempts were made to define acids and bases on the basis of the nature of their constituent elements. Antoine Lavoisier, a French chemist, thought that acidic properties were due to the presence of oxygen. While this explanation applied to sulfuric acid (H_2SO_4), nitric acid (HNO_3) and phosphoric acid (H_3PO_4), it did not explain why hydrochloric acid (HCl) was an acid.

In about 1810, Humphrey Davy suggested that the acid properties of substances were associated with hydrogen and not oxygen. He came to this conclusion after producing hydrogen gas by reacting acids with metals. Davey also suggested that acids react with bases to form compounds, called **salts**, and water.

The Swedish scientist Svante Arrhenius further developed this theory. In 1887, he defined acids and bases as follows.

- Acids are substances that **dissociate** (break apart) and **ionise** (form ions) in water to produce hydrogen ions (H^+).
- Bases dissociate in water to produce **hydroxide ions** (OH^-).

In 1923, Danish physical chemist Johannes Nicolaus Brønsted and English chemist Thomas Martin Lowry were working independently on the behaviour of acids and bases. They each developed the theory which now bears both of their names. The Brønsted–Lowry theory is more general than the one proposed by Arrhenius and provides an explanation for some observed acid–base behaviours that cannot be explained by the earlier theories.

THE BRØNSTED–LOWRY THEORY OF ACIDS AND BASES

According to the **Brønsted–Lowry theory**, a substance behaves as an acid when it donates a proton, i.e. H^+, to a base. A substance behaves as a base when it accepts a proton from an acid.

In summary:

- acids are proton donors
- bases are proton acceptors
- an **acid–base reaction** involves an exchange of protons from an acid to a base.

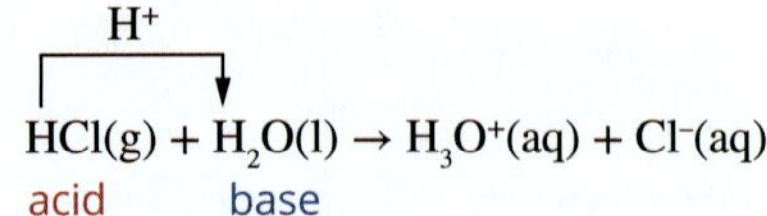

FIGURE 11.1.2 The reaction between hydrochloric acid and water is an example of an acid–base reaction, which involves a transfer of protons from an acid to a base.

For example, hydrogen chloride (HCl) is a gaseous molecular compound that is very soluble in water. The molecules dissociate and ionise in water according to the following reaction:

$$HCl(g) + H_2O(l) \longrightarrow H_3O^+(aq) + Cl^-(aq)$$

In an aqueous solution of hydrogen chloride, nearly all the hydrogen chloride is present as ions—virtually no molecules of hydrogen chloride remain. This solution is known as hydrochloric acid.

> Remember that the symbol $H^+(aq)$ is often used interchangeably with $H_3O^+(aq)$ to represent the hydronium ion.

In this reaction, each HCl molecule has donated a proton to a water molecule, forming the **hydronium ion**, $H_3O^+(aq)$. According to the Brønsted–Lowry theory, the HCl has acted as an acid. The water molecule has accepted a proton from the HCl molecule, so the water has acted as a base. This is shown in Figure 11.1.2.

The hydronium ion can be represented as either $H_3O^+(aq)$ or $H^+(aq)$. The reaction of HCl(g) with water can be written as either:

$$HCl(g) + H_2O(l) \longrightarrow H_3O^+(aq) + Cl^-(aq)$$

or

$$HCl(g) \longrightarrow H^+(aq) + Cl^-(aq)$$

However, writing the hydronium ion as $H^+(aq)$ in an equation makes it harder to see that a proton transfer has occurred. The hydronium ion is therefore usually written as $H_3O^+(aq)$ in this chapter.

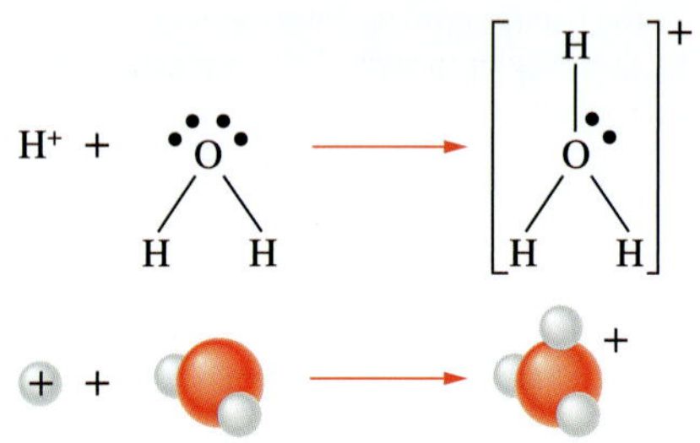

FIGURE 11.1.3 Formation of the hydronium ion

The structural formulas of the hydronium ion and the water molecule are shown in Figure 11.1.3.

Advantages of the Brønsted–Lowry model

Unlike earlier models, the Brønsted–Lowry model for acid–base reactions is not restricted to reactions that occur in aqueous solutions. A reaction between two gases can be an acid–base reaction. For example, the salt ammonium chloride can be formed by a reaction between:

- solutions of hydrochloric acid and ammonia:

$$HCl(aq) + NH_3(aq) \longrightarrow NH_4^+(aq) + Cl^-(aq)$$

- gaseous hydrogen chloride and gaseous ammonia:

$$HCl(g) + NH_3(g) \longrightarrow NH_4Cl(s)$$

The Brønsted–Lowry model classifies both of these reactions as acid–base reactions because in each case the acid donates a proton to the base.

CONJUGATE ACID–BASE PAIRS

A solution of hydrochloric acid is produced when hydrogen chloride ionises in water:

$$HCl(aq) + H_2O(l) \rightarrow H_3O^+(aq) + Cl^-(aq)$$

Because Cl^- can be formed from HCl by the loss of a single proton, it is called the **conjugate base** of HCl. Similarly, HCl is described as the **conjugate acid** of Cl^-. HCl and Cl^- are called a **conjugate acid–base pair**. H_3O^+ and H_2O are also a conjugate pair.

The relationship between acid–base conjugate pairs is represented in Figure 11.1.4.

Some common acid–base pairs are listed in Figure 11.1.5.

In the reaction between NH_3 and H_2O (Figure 11.1.6), the conjugate acid–base pairs are NH_4^+/NH_3 and H_2O/OH^- because each acid differs from its corresponding conjugate base by one proton. By convention, we usually write the acid form of the conjugate pair first.

$$NH_3(aq) + H_2O(l) \longrightarrow NH_4^+(aq) + OH^-(aq)$$

base | acid | conjugate acid of NH_3 | conjugate base of H_2O

FIGURE 11.1.6 The reaction between ammonia and water, showing the conjugate acid–base pairs

When acids react with water, hydronium (H_3O^+) ions are produced. When bases react with water, hydroxide (OH^-) ions are produced.

The reactions of acids or bases with water are called **ionisation reactions** because ions are produced.

$$HCl(g) + H_2O(l) \longrightarrow H_3O^+(aq) + Cl^-(aq)$$

acid | base | conjugate acid of H_2O | conjugate base of HCl

FIGURE 11.1.4 Conjugate acid–base pairs are formed when an acid donates a proton to a base.

A conjugate acid–base pair is two species whose formulas differ by a proton.

Acids ... donate a proton to form:

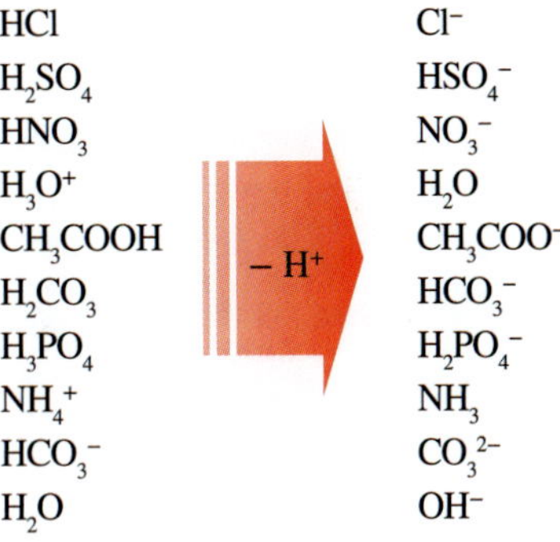

Bases ... accept a proton to form:

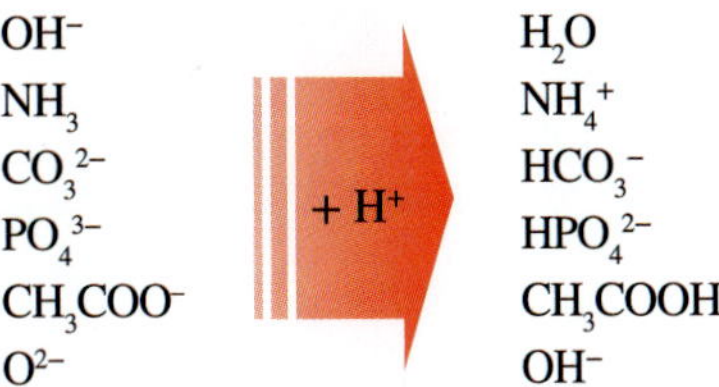

FIGURE 11.1.5 Some common acid–base conjugate pairs

WRITING EQUATIONS FOR ACID-BASE REACTIONS

You can see that balanced equations can be used to represent the reactions of acids and bases. When the species that exist as ions are shown in the equation, these equations are called ionic equations. If water is the solvent used, the ions will be in an aqueous state and the state symbol (aq) is used in the equations.

The use of arrows

The use of the arrow symbol '$\rightarrow$' in an equation usually indicates the reaction is 'complete', which means at least one of the reactants is consumed completely. For example, the equation:

$$HCl(g) + H_2O(l) \rightarrow H_3O^+(aq) + Cl^-(aq)$$

indicates that if sufficient water is present all the HCl molecules will react.

A reversible (or double) arrow symbol '$\rightleftharpoons$' is used to indicate that only a small proportion of reactant molecules is consumed, so the reaction mixture will have some reactant and product molecules present. Many of the equations that you will encounter in this chapter are of this type.

AMPHIPROTIC SUBSTANCES

Some substances can either donate or accept protons, depending on what they are reacting with. Therefore, they can behave as either acids or bases. Such substances are said to be **amphiprotic**.

For example, in the following reaction, water gains a proton from HCl and acts as a base.

$$HCl(g) + H_2O(l) \rightarrow Cl^-(aq) + H_3O^+(aq)$$

However, in the reaction below, water donates a proton to NH_3 and acts as an acid.

$$NH_3(aq) + H_2O(l) \rightarrow NH_4^+(aq) + OH^-(aq)$$

These reactions are represented in Figure 11.1.7.

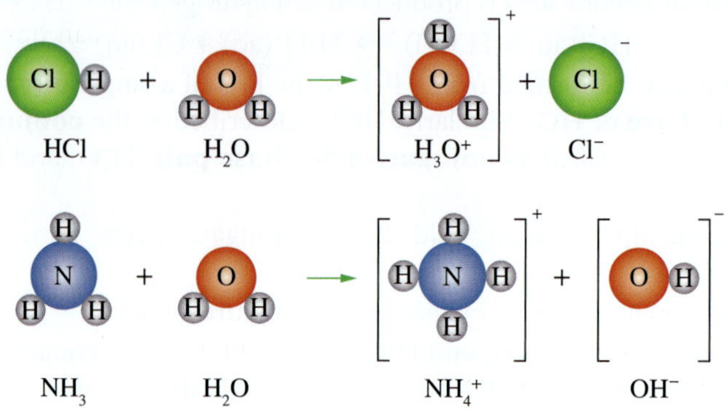

$HCl + H_2O \rightarrow [H_3O^+] + Cl^-$

$NH_3 + H_2O \rightarrow [NH_4^+] + [OH^-]$

FIGURE 11.1.7 The amphiprotic nature of water is demonstrated by its reactions with HCl and NH_3.

It is evident that water can act as either an acid or a base, depending on the other reactant present. If this reactant is a stronger acid than water, then water will react as a base. If it is a stronger base than water, then water will react as an acid.

Some common amphiprotic substances are listed in Figure 11.1.8.

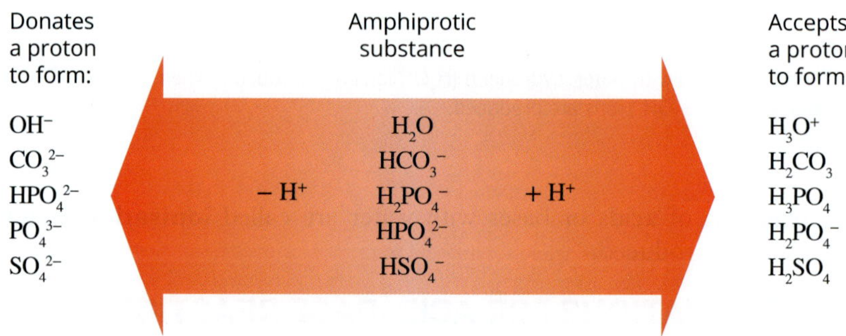

FIGURE 11.1.8 Substances that are amphiprotic can act as either an acid or a base.

When an amphiprotic substance is placed in water, it reacts as both an acid and a base. For example, the hydrogen carbonate ion (HCO_3^-) reacts according to the equations:

as an acid: $HCO_3^-(aq) + H_2O(l) \rightleftharpoons CO_3^{2-}(aq) + H_3O^+(aq)$

as a base: $HCO_3^-(aq) + H_2O(l) \rightleftharpoons H_2CO_3(aq) + OH^-(aq)$

Since HCO_3^- can act as both an acid and a base, it is amphiprotic. Note the use of $\rightleftharpoons$ arrows, indicating that $HCO_3^-(aq)$ does not react completely as an acid nor as a base.

Although both reactions are possible for all amphiprotic substances in water, generally one of these reactions dominates. The dominant reaction can be identified by measuring the pH (a measure of the amount of hydronium ion in solution). (You will look more closely at pH in Section 11.4.)

MONOPROTIC ACIDS

Monoprotic acids can donate only one proton. These acids include hydrochloric acid (HCl), hydrofluoric acid (HF), nitric acid (HNO_3) and ethanoic acid (CH_3COOH).

While ethanoic acid contains four hydrogen atoms, each molecule can donate only one proton to produce an ethanoate ion (CH_3COO^-), and is therefore a monoprotic acid. Only the hydrogen that is bonded to the electronegative oxygen atom and is part of the highly polar O–H bond is donated. This hydrogen atom is called the **acidic proton** (Figure 11.1.9).

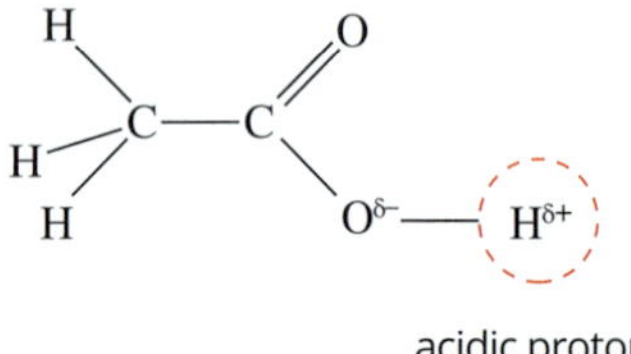

FIGURE 11.1.9 The structure of ethanoic acid. Each molecule can donate only one proton to a water molecule when ethanoic acid reacts with water.

POLYPROTIC ACIDS

Some acids can donate more than one proton from each molecule and are said to be **polyprotic acids**. The number of hydrogen ions an acid can donate depends on the structure of the acid. Polyprotic acids do not donate all their protons at once, but instead do so in steps when reacting with a base.

Diprotic acids

Diprotic acids, such as sulfuric acid (H_2SO_4) and carbonic acid (H_2CO_3), can donate two protons.

A diprotic acid ionises in two stages. These stages are described below using sulfuric acid as an example.

Stage 1:

$$H_2SO_4(l) + H_2O(l) \longrightarrow HSO_4^-(aq) + H_3O^+(aq)$$

Sulfuric acid is described as a **strong acid** in water because it readily donates a proton and so this stage occurs almost to completion. Virtually no H_2SO_4 molecules are found in an aqueous solution, so the symbol '$\longrightarrow$' is used as the arrow in the equation.

Stage 2:

The HSO_4^- ion formed can also act as an acid. In a 1.0 M solution, only a small proportion of those ions react further to produce H_3O^+ ions and SO_4^{2-} ions.

$$HSO_4^-(aq) + H_2O(l) \rightleftharpoons SO_4^{2-}(aq) + H_3O^+(aq)$$

HSO_4^- is described as a **weak acid** because it is only partially ionised. The reversible arrow symbol '$\rightleftharpoons$' indicates that an incomplete reaction occurs. (You will learn more about strong and weak acids later in this chapter.) Therefore, a solution of sulfuric acid contains hydrogen ions, hydrogen sulfate ions and sulfate ions, as well as billions of water molecules.

Triprotic acids

Triprotic acids can donate three protons. These include phosphoric acid (H_3PO_4) and arsenic acid (H_3AsO_4).

A triprotic acid, such as phosphoric acid, ionises in three stages.

Stage 1: $H_3PO_4(aq) + H_2O(l) \rightleftharpoons H_2PO_4^-(aq) + H_3O^+(aq)$

Stage 2: $H_2PO_4^-(aq) + H_2O(l) \rightleftharpoons HPO_4^{2-}(aq) + H_3O^+(aq)$

Stage 3: $HPO_4^{2-}(aq) + H_2O(l) \rightleftharpoons PO_4^{3-}(aq) + H_3O^+(aq)$

Phosphoric acid is a moderately weak acid in water. Therefore, in a 1.0 M solution of phosphoric acid, only a small proportion of the protons is donated at each **ionisation** stage. The extent of the ionisation decreases progressively from stage 1 to stage 3, because a positive proton is being removed at each step from an already negatively charged species. This becomes a more difficult process as each ionisation steps proceeds.

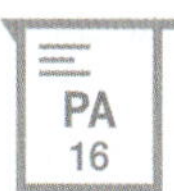

11.1 Review

OA ✓✓

SUMMARY

- The Brønsted–Lowry model describes acid–base properties in terms of proton transfer. A Brønsted–Lowry acid is a proton donor and a base is a proton acceptor.
- When an acid donates a proton, it forms its conjugate base. When a base accepts a proton, it forms its conjugate acid.
- Conjugate acid–base pairs are molecules and/or ions whose formulas differ from each other by an H^+ ion.
- Ionic and full equations can be written to show the reactions of acids and of bases with water. They are often called ionisation reactions and use (aq) to denote the states of matter of the ions in solution.
- The use of complete arrows ($\rightarrow$) and reversible arrows ($\rightleftharpoons$) indicates which reactions are complete and which only partially occur.
- A proton or hydrogen ion in solution can be represented by the hydronium ion, $H_3O^+(aq)$, or simply as $H^+(aq)$.
- Amphiprotic substances can act as either acids or bases, depending on the substance with which they are reacting.
- A polyprotic acid can donate more than one proton to a base.
- The first ionisation of a diprotic acid occurs to a greater extent than the second ionisation. In a triprotic acid, the third ionisation occurs to the least extent.

KEY QUESTIONS

Knowledge and understanding

1 Complete the following statements about the Brønsted–Lowry theory by using words from the following list:

acceptors; amphiprotic; proton; donors; base; H^+; acid; gain; lose

Using the Brønsted–Lowry theory of acids and bases, an acid–base reaction is a reaction in which a ________ transfer occurs. This theory states that:

- Acids are proton ____________.
- Bases are proton ____________.
- Hence acids and bases must always act together.
- Acids and bases react with each other and form their respective conjugates. Acids ____________ a proton and form their conjugate ____________. Bases ____________ a proton and form their conjugate ____________.
- The formulas of conjugate pairs differ by a ____________.
- An ____________ substance can act as an acid or a base depending on what the other reactant is.

2 An acidic solution is formed when hydrogen chloride gas (HCl) is mixed with water (H_2O). There are virtually no HCl molecules in the solution after the reaction.

a Write a balanced ionic equation for this reaction.

b List the conjugate pairs.

3 What are the two acid–base conjugate pairs in the reaction below?

$$H_2SO_4(l) + HNO_3(l) \rightarrow HSO_4^-(l) + H_2NO_3^+(l)$$

4 For each equation below:

i list the reactant acting as an acid and the one acting as a base

ii identify the conjugate pairs.

a $HF(aq) + OH^-(aq) \rightarrow H_2O(l) + F^-(aq)$

b $HCOOH(aq) + H_2O(l) \rightarrow H_3O^+(aq) + HCOO^-(aq)$

c $CH_3NH_2(aq) + HCl(aq) \rightarrow CH_3NH_3^+(aq) + Cl^-(aq)$

5 What is the conjugate acid of the following bases?

a NH_3

b CH_3COO^-

c HPO_4^{2-}

d CO_3^{2-}

Analysis

6 Write equations for each of the following amphiprotic substances when they react with water, to show how each acts as an acid and as a base.

a HCO_3^-

b HPO_4^{2-}

c HSO_4^-

d H_2O

7 **a** Write equations for the ionisation reactions that occur in solutions of the diprotic acid, H_2CO_3.

b Write balanced equations for the three ionisation stages of arsenic acid (H_3AsO_4), which is a weak acid.

c Identify the amphiprotic species in parts **a** and **b**.

11.2 Strength of acids and bases

The acid solutions in the two beakers shown in Figure 11.2.1 are of equal **concentration**, yet the acid in the beaker on the left reacts more vigorously with zinc than the acid on the right. The acid on the left is described as a stronger acid than the one in the beaker on the right.

FIGURE 11.2.1 Zinc reacts more vigorously with a strong acid (left) than with a weak acid (right). The acid solutions are of equal concentration and volume.

In Section 11.1, you learnt that the Brønsted–Lowry theory defines acids as proton donors and bases as proton acceptors. In this section, you will investigate the differences between:

- strong and weak acids
- strong and weak bases.

As you have seen previously, some acids do not ionise to the same extent as others. Experiments have shown that different acids of the same concentration do not all have the same H_3O^+ concentrations and therefore will not have the same pH (a measure of the **acidity** of the solution).

Some acids can donate a proton more readily than others. The Brønsted–Lowry theory describes the strength of an acid as its ability to donate H^+ ions to a base. The strength of a base is a measure of its ability to accept hydrogen ions from an acid.

> ℹ The strength of an acid is determined by how readily it can donate a proton.

Since aqueous solutions of acids and bases are most commonly used, it is convenient to use an acid's tendency to donate a proton to water, or a base's tendency to accept a proton from water, as a measure of its strength.

Table 11.2.1 gives the names and chemical formulas of some strong and weak acids and bases.

TABLE 11.2.1 Examples of common strong and weak acids and bases

Strong acids	Weak acids	Strong bases	Weak bases
hydrochloric acid, HCl	ethanoic acid, CH_3COOH	sodium hydroxide, NaOH	ammonia, NH_3
sulfuric acid, H_2SO_4	carbonic acid, H_2CO_3	potassium hydroxide, KOH	
nitric acid, HNO_3	phosphoric acid, H_3PO_4	calcium hydroxide, $Ca(OH)_2$	

STRONG ACIDS

As you saw previously, when hydrogen chloride gas (HCl) is bubbled through water, it ionises completely—virtually no HCl molecules remain in the solution (Figure 11.2.2a). Similarly, pure HNO_3 and H_2SO_4 are covalent molecular compounds, which also ionise completely in water:

$$HCl(g) + H_2O(l) \rightarrow H_3O^+(aq) + Cl^-(aq)$$

$$HNO_3(l) + H_2O(l) \rightarrow H_3O^+(aq) + NO_3^-(aq)$$

$$H_2SO_4(l) + H_2O(l) \rightarrow H_3O^+(aq) + HSO_4^-(aq)$$

The single reaction arrow '$\rightarrow$' in each equation above indicates that the ionisation reaction is complete.

Strong acids ionise completely in water. They donate protons very readily.

Acids that readily donate a proton are called strong acids. Strong acids donate protons easily. Therefore, solutions of strong acids contain ions, with virtually no unreacted acid molecules remaining. Hydrochloric acid, nitric acid and sulfuric acid are the most common strong acids.

WEAK ACIDS

Vinegar is a solution of ethanoic acid. Pure ethanoic acid is a polar covalent molecular compound that ionises in water to produce hydronium ions and ethanoate ions. In a 1.0 M solution of ethanoic acid (CH_3COOH), only a small proportion of ethanoic acid molecules are ionised at any one time (Figure 11.2.2b). As a result, a 1.0 M solution of ethanoic acid contains a high proportion of CH_3COOH molecules and only some hydronium ions and ethanoate ions. At 25°C, in a 1.0 M solution of ethanoic acid, the concentrations of $CH_3COO^-(aq)$ and H_3O^+ are only about 0.004 M.

Weak acids partially ionise in water. Only a small proportion of the acid molecules donate a proton to the water molecules.

The partial ionisation of a weak acid is shown in an equation using reversible (double) arrows '$\rightleftharpoons$':

$$\underset{\text{acid}}{CH_3COOH(aq)} + \underset{\text{base}}{H_2O(l)} \rightleftharpoons CH_3COO^-(aq) + H_3O^+(aq)$$

Therefore, ethanoic acid is described as a weak acid in water.

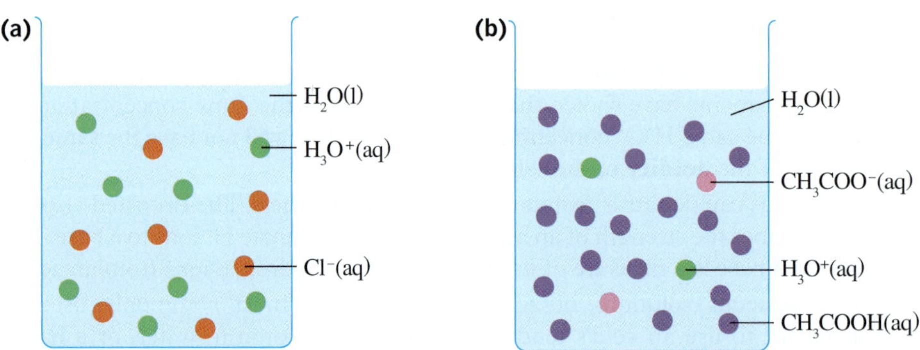

FIGURE 11.2.2 (a) In a 1.0 M solution of a hydrochloric acid the acid, molecules are completely ionised in water. (b) However, in a 1.0 M solution of ethanoic acid, only a small proportion of the ethanoic acid molecules is ionised.

STRONG BASES

The ionic compound sodium oxide (Na_2O) dissociates in water, releasing sodium ions (Na^+) and oxide ions (O^{2-}). The oxide ions react completely with the water, accepting a proton to form hydroxide ions (OH^-):

$$\underset{\text{base}}{O^{2-}(aq)} + \underset{\text{acid}}{H_2O(l)} \rightarrow OH^-(aq) + OH^-(aq)$$

The oxide ion is an example of a **strong base**. Strong bases accept protons easily.

Strong bases accept protons very readily from acid molecules.

Sodium hydroxide (NaOH) is often referred to as a strong base. However, according to the Brønsted–Lowry definition of acids and bases, it is more correct to say that sodium hydroxide is a soluble ionic compound that is a source of the strong base OH^-.

WEAK BASES

Ammonia is a covalent molecular compound that ionises in water by accepting a proton. This ionisation can be represented by the equation:

$$\underset{\text{base}}{NH_3(aq)} + \underset{\text{acid}}{H_2O(l)} \rightleftharpoons NH_4^+(aq) + OH^-(aq)$$

Ammonia behaves as a base because it gains a proton. Water has donated a proton and so it behaves as an acid.

Only a small proportion of ammonia molecules are ionised at any instant, so a 1.0 M solution of ammonia contains mostly ammonia molecules, together with a smaller number of ammonium ions and hydroxide ions. The double arrow in the equation indicates that this is a partial reaction. Ammonia is a **weak base** in water.

At any instant only a small proportion of molecules of weak bases accept a proton from an acid.

RELATIVE STRENGTH OF CONJUGATE ACID–BASE PAIRS

You will recall from Section 11.1 that conjugate acids and bases differ by one proton (H^+). In the reaction represented by the equation:

$$HF(aq) + OH^-(aq) \rightleftharpoons H_2O(l) + F^-(aq)$$

HF is the conjugate acid of F^- and OH^- is the conjugate base of H_2O. HF and F^- are a conjugate acid–base pair. H_2O and OH^- are another conjugate acid–base pair in this reaction.

The stronger an acid is, the weaker is its conjugate base. Similarly, the stronger a base is, the weaker is its conjugate acid. The relative strength of some conjugate acid–base pairs is shown in Figure 11.2.3.

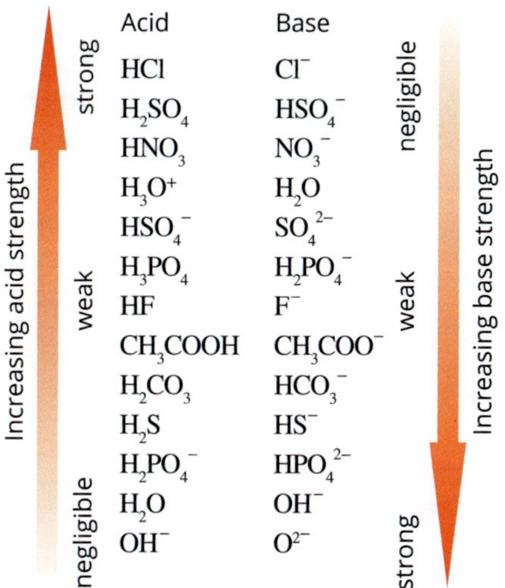

FIGURE 11.2.3 The relative strength of some conjugate acid–base pairs

STRENGTH VERSUS CONCENTRATION

When referring to solutions of acids and bases, it is important not to confuse the terms 'strong' and 'weak' with 'concentrated' and 'dilute'. Concentrated and dilute describe the amount of acid or base dissolved in a given volume of solution. Hydrochloric acid is a strong acid because it readily donates protons to a base. A **concentrated solution** of hydrochloric acid can be prepared by bubbling a large amount of hydrogen chloride into a given volume of water. By using only a small amount of hydrogen chloride, a dilute solution of hydrochloric acid would be produced. However, in both cases, the hydrogen chloride is completely ionised—it is a strong acid.

Similarly, a solution of ethanoic acid may be concentrated or dilute. However, as it is only partially ionised, it is a weak acid (Figure 11.2.4).

Terms such as 'weak' or 'strong' acids, or solutions classified as 'dilute' or 'concentrated', are qualitative (or descriptive) terms. Solutions can be given accurate, quantitative descriptions by stating concentrations in mol L^{-1} or g L^{-1}.

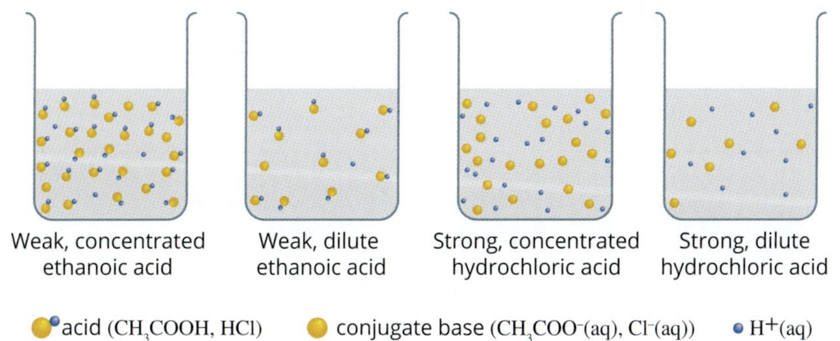

FIGURE 11.2.4 The concentration of ions in an acid solution depends on both the concentration and strength of the acid.

CHEMFILE

Super acids

Fluorosulfuric acid (HSO_3F) is one of the strongest acids known. It has a similar geometry to that of the sulfuric acid molecule (see figure below). The highly electronegative fluorine atom causes the oxygen–hydrogen bond in fluorosulfuric acid to be significantly more polarised than the oxygen–hydrogen bond in sulfuric acid. Hence the acidic proton is significantly more easily transferred to a base.

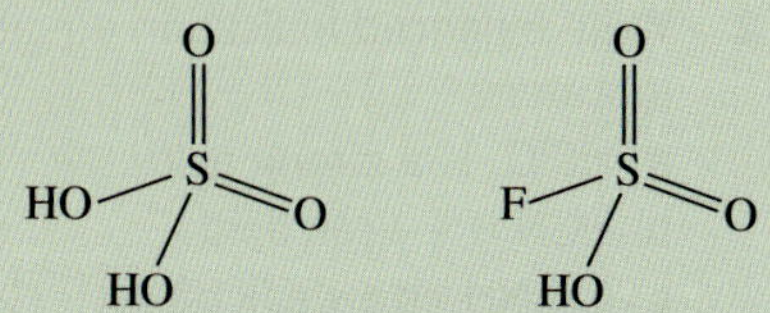

Structure of sulfuric acid (left) and fluorosulfuric acid (right) molecules

Fluorosulfuric acid is classified as a super acid, which are defined as acids that have acidity greater than the acidity of pure sulfuric acid.

Super acids such as fluorosulfuric acid and triflic acid (CF_3SO_3H) are about 1000 times stronger than sulfuric acid. Carborane acid ($H(CHB_{11}Cl_{11})$) is 1 million times stronger than sulfuric acid. The strongest known super acid is fluoroantimonic acid (H_2FSbF_6), which is 10^{16} times stronger than 100% sulfuric acid.

11.2 Review

OA ✓✓

SUMMARY

- A concentrated acid or base contains more moles of solute per litre of solution than a dilute acid or base.
- In the context of acids and bases, the terms 'strong' and 'weak' refer to the relative tendency to accept or donate protons.
  - A strong acid donates a proton more readily than a weak acid.
  - A strong base accepts a proton more readily than a weak base.
- The stronger an acid, the weaker is its conjugate base. The stronger a base, the weaker is its conjugate acid.

KEY QUESTIONS

Knowledge and understanding

1 Complete the spaces in the following description of strength and concentration in acids and bases using words from the following list:

extent; concentrated; litre; ionise; dilute; strong; weak

A ____________ solution has a larger amount of solute dissolved in a specific volume of solvent, whereas a ____________ solution has a smaller amount of solute dissolved in the same specific volume of solvent. In chemistry, these terms should not be confused with the meaning of strong and weak when applied to acids and bases. All the molecules of a ____________ acid ionise, whereas at any instant, a small proportion of the molecules of a ____________ acid ionise. This is true for bases as well. Concentrated and dilute refer to moles per____________. Strong and weak refer to what ____________ the molecules ____________ in water.

2 Write balanced ionisation equations to show that, in water:

a $HClO_4$ is a strong acid

b HCN is a weak acid

c CH_3NH_2 is a weak base.

3 For each of the following equations **a** to **d** in water, indicate which represents the reaction of:

- a strong acid
- a strong base
- a weak acid
- a weak base.

a $HNO_3(aq) + H_2O(l) \rightarrow H_3O^+(aq) + NO_3^-(aq)$

b $HF(aq) + H_2O(l) \rightleftharpoons H_3O^+(aq) + F^-(aq)$

c $O_2^-(aq) + H_2O(l) \rightarrow 2OH^-(aq)$

d $NH_3(aq) + H_2O(l) \rightleftharpoons NH_4^+(aq) + OH^-(aq)$

4 Explain the difference between a concentrated solution of a weak acid, such as CH_3COOH, and a dilute solution of a strong acid, such as HCl.

Analysis

5 Perchloric acid is a stronger acid than ethanoic acid. If you had a 1 M solution of each, which acidic solution would you expect to be the better conductor of electricity? Explain why.

11.3 Reactions of acids and bases

The chemicals we describe as acids were originally grouped together because of their similar chemical behaviour. Acids and bases react readily with many other chemicals, and some of the early definitions of acids and bases were derived from these reactions.

In this section, you will learn to use the patterns in the reactions of acids and bases to predict the products that are formed.

GENERAL REACTIONS OF ACIDS AND BASES

Acids and bases react in many ways. However, it is possible to categorise some reactions according to the similarity of the reactants involved and products formed. While the identification of products should be based on observation of reactions, these groups, or reaction types, can be useful. The reaction types you will study in this section are the reaction of acids with:

- metal hydroxides
- metal carbonates and hydrogen carbonates
- reactive metals.

Acids and metal hydroxides

Soluble metal hydroxides, such as NaOH, dissociate in water to form metal cations and hydroxide ions, $OH^-(aq)$. The products of a reaction of an acid with a metal hydroxide are an ionic compound, called a salt, and water. These reactions are known as **neutralisation reactions**.

The general reaction between acids and metal hydroxides can be expressed as:

$$\text{acid} + \text{metal hydroxide} \longrightarrow \text{salt} + \text{water}$$

The general equation for the reaction between acids and metal hydroxides is: acid + metal hydroxide ⟶ salt + water.

For example, solutions of sulfuric acid and sodium hydroxide react to form sodium sulfate and water. This can be represented by the full (or overall) equation:

$$H_2SO_4(aq) + 2NaOH(aq) \longrightarrow Na_2SO_4(aq) + 2H_2O(l)$$

The salt formed in the reaction between sulfuric acid and sodium hydroxide is sodium sulfate. If water were evaporated from the reaction mixture, solid sodium sulfate would be left behind. Salts consist of the positive ion or cation from the base and the negative ion or anion from the acid.

Table 11.3.1 lists the names of salts formed from some neutralisation reactions of acids with metal hydroxides. Note that the name of the positive ion is listed first and the name of the negative ion from the acid is listed second.

TABLE 11.3.1 Salts formed from some common neutralisation reactions

Reactants (acid + metal hydroxide)	Name of salt formed	Formulas of ions present in the salt solution
hydrochloric acid + potassium hydroxide	potassium chloride	$K^+(aq) + Cl^-(aq)$
hydrochloric acid + magnesium hydroxide	magnesium chloride	$Mg^{2+}(aq) + Cl^-(aq)$
nitric acid + sodium hydroxide	sodium nitrate	$Na^+(aq) + NO_3^-(aq)$
sulfuric acid + zinc hydroxide	zinc sulfate	$Zn^{2+}(aq) + SO_4^{2-}(aq)$
phosphoric acid + potassium hydroxide	potassium phosphate	$K^+(aq) + PO_4^{3-}(aq)$
ethanoic acid + calcium hydroxide	calcium ethanoate	$Ca^{2+}(aq) + CH_3COO^-(aq)$

The hydroxide ions from metal hydroxides, such as sodium hydroxide (NaOH), calcium hydroxide ($Ca(OH)_2$) and magnesium hydroxide ($Mg(OH)_2$), react readily with the hydrogen ion ($H^+(aq)$) from acids.

The reaction between an acid and a metal hydroxide can be represented by an **ionic equation** as well as by an overall equation. You were introduced to ionic equations when studying precipitation reactions in Chapter 5, page 177. Remember that **spectator ions** are not shown in a ionic equation.

Figure 11.3.1 is a diagrammatic representation of the ions in solutions of HCl and NaOH when mixed in a neutralisation reaction. If a solution of a metal hydroxide is added to a solution of an acid, the hydroxide ions will react with the hydronium ions. The acid and base are said to have been neutralised at the point when all the hydroxide ions have reacted with all the hydronium ions, forming water (H_2O).

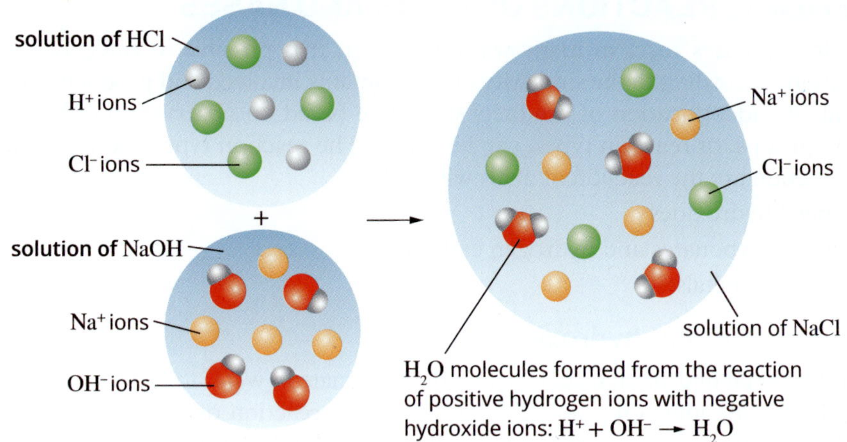

FIGURE 11.3.1 A representation of the reaction between $H^+(aq)$ and $OH^-(aq)$ ions that occurs when solutions of HCl and NaOH are mixed

Worked example 11.3.1 indicates the steps to follow when writing ionic equations for the neutralisation reaction of acids and metal hydroxides.

Worked example 11.3.1

WRITING AN IONIC EQUATION FOR AN ACID–BASE REACTION

Write an ionic equation for the reaction that occurs when hydrochloric acid is added to sodium hydroxide solution. (A representation of this reaction is shown in Figure 11.3.1.)

Thinking	Working
What is the general reaction? Identify the products formed.	acid + metal hydroxide → salt + water A solution of sodium chloride and water is formed.
Identify the reactants and products. Indicate the state of each, i.e. (aq), (s), (l) or (g).	Reactants: HCl(aq) is ionised in solution, forming $H^+(aq)$ and $Cl^-(aq)$. NaOH(aq) is dissociated in solution, forming $Na^+(aq)$ and $OH^-(aq)$. Products: Sodium chloride is dissociated and exists as $Na^+(aq)$ and $Cl^-(aq)$. Water is a molecular compound and its formula is $H_2O(l)$.
Write the equation showing all reactants and products, in ionised form where possible. (There is no need to balance the equation yet.)	$H^+(aq) + Cl^-(aq) + Na^+(aq) + OH^-(aq) \rightarrow Na^+(aq) + Cl^-(aq) + H_2O(l)$
Identify the spectator ions: the ions that have an (aq) state, both as a reactant and as a product.	$Na^+(aq)$ and $Cl^-(aq)$
Rewrite the equation without the spectator ions. Balance the equation with respect to the number of atoms of each element and charge.	$H^+(aq) + OH^-(aq) \rightarrow H_2O(l)$ Note that if hydronium ions are represented as $H_3O^+(aq)$, rather than as $H^+(aq)$, this reaction would be written as: $H_3O^+(aq) + OH^-(aq) \rightarrow 2H_2O(l)$

Worked example: Try yourself 11.3.1

WRITING AN IONIC EQUATION FOR AN ACID–BASE REACTION

Write an ionic equation for the reaction that occurs when sulfuric acid is added to potassium hydroxide solution.

CASE STUDY ANALYSIS

Benefits of neutralisation

As you have seen, acids react with bases and form their conjugate base or acid respectively. Since acids can be neutralised by bases, these reactions can be used to adjust how much acid is present in a solution. If excess acid is harmful, it can be neutralised by adding a base. Conversely, an environment that is alkaline (contains a soluble base) can be neutralised by adding an acid. Some common examples of the use of neutralisation reactions are the reduction of pain of insect bites and treatments for toothache.

- Insect bites
  - Methanoic acid, HCOOH, is released from the sting of ants, bees and nettles (Figure 11.3.2). If affected skin is rinsed with water and then a dilute solution of ammonia or sodium bicarbonate solution the acid becomes neutralised and is no longer painful to the person affected.
  - The venom from wasps is alkaline. A common treatment is to rinse the site of the wasp bite with water and then a dilute solution of vinegar, as this neutralises the base within the venom.

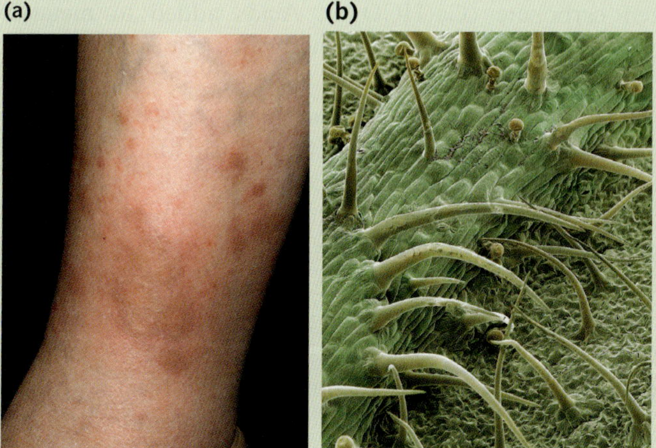

FIGURE 11.3.2 A base can be used to neutralise an acidic ant bite or the acidic solution from a nettle sting. (a) Irritation on the leg of a person bitten by ants. (b) Scanning electron microscope image of a nettle (Urtica sp.) surface showing stinging, hair-like structures (colourised).

- Toothache
  - The bacteria on tooth enamel feed on the sugars present in food. The products of their metabolism are acids that attack the enamel, which leads to tooth decay. This is why toothpastes (Figure 11.3.3) are weak bases.

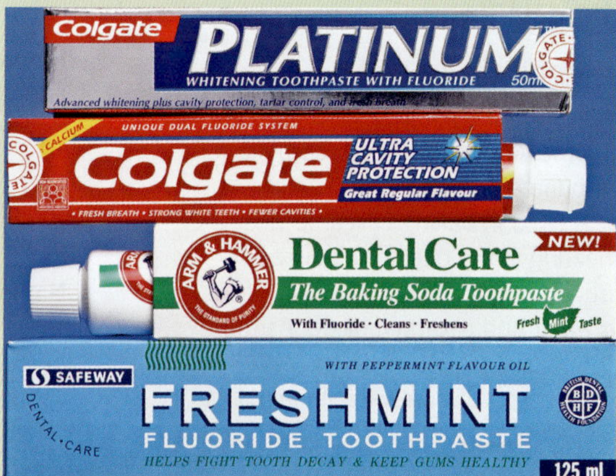

FIGURE 11.3.3 Weakly basic toothpastes neutralise food acids on your teeth. These products are advertised as providing protection against gum disease and tooth decay.

Analysis

1 Write balanced chemical equations for the following reactions when treating ant bite.

   (a) Methanoic acid in the sting (a monoprotic acid with formula HCOOH) and ammonia solution

   (b) Methanoic acid in the sting (a monoprotic acid with formula HCOOH) and sodium bicarbonate solution

2 Ammonia is a weak base and ethanoic acid is a weak acid. Why are they used instead of sodium hydroxide or hydrochloric acid, to neutralise the effects of an ant or wasp bite?

Acids and metal carbonates

FIGURE 11.3.4 This statue has been heavily eroded by acid rain, which reacts with carbonate salts in limestone. Acid rain is formed when gases, such as sulfur dioxide and nitrogen oxides, dissolve in water to form acidic solutions.

The weathering of buildings and statues (Figure 11.3.4) is due in part to the reaction between acid rain and the carbonate minerals in the stone.

Acids reacting with metal carbonates and metal hydrogen carbonates produce carbon dioxide gas, together with a salt and water. Metal carbonates include sodium carbonate (Na_2CO_3), magnesium carbonate ($MgCO_3$) and calcium carbonate ($CaCO_3$).

The general reaction for metal carbonates with acids can be summarised as:

$$\text{acid} + \text{metal carbonate} \rightarrow \text{salt} + \text{water} + \text{carbon dioxide}$$

For example, a solution of hydrochloric acid reacting with sodium carbonate solution produces a solution of sodium chloride, water and carbon dioxide gas. The reaction is represented by the equation:

$$2HCl(aq) + Na_2CO_3(aq) \rightarrow 2NaCl(aq) + CO_2(g) + H_2O(l)$$

The reaction between hydrochloric acid and sodium carbonate is represented in Figure 11.3.5.

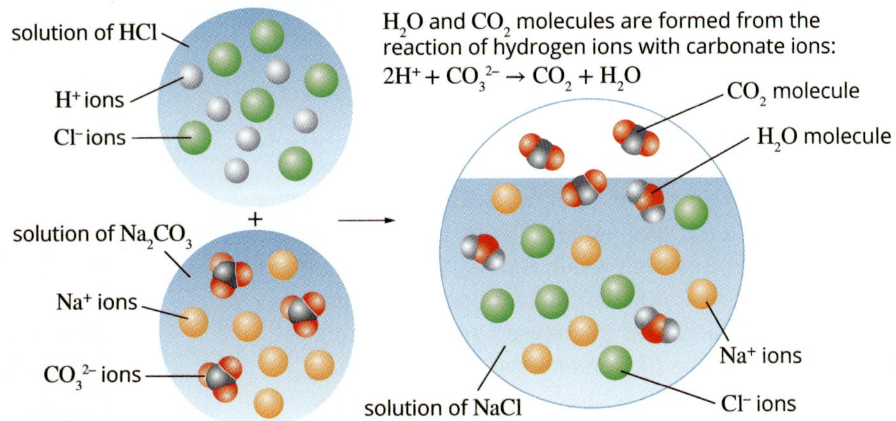

FIGURE 11.3.5 The reaction between solutions of sodium carbonate and hydrochloric acid

Metal hydrogen carbonates (also known as bicarbonates) include sodium hydrogen carbonate ($NaHCO_3$), potassium hydrogen carbonate ($KHCO_3$) and calcium hydrogen carbonate ($Ca(HCO_3)_2$). Acids added to metal hydrogen carbonates also produce carbon dioxide together with a salt and water. The general equation is:

$$\text{acid} + \text{metal hydrogen carbonate} \rightarrow \text{salt} + \text{water} + \text{carbon dioxide}$$

For example, when solutions of hydrochloric acid and sodium hydrogen carbonate are mixed, the following reaction occurs:

$$HCl(aq) + NaHCO_3(s) \rightarrow NaCl(aq) + H_2O(l) + CO_2(g)$$

The reactions between acids and metal carbonates, and reactions between acids and metal hydrogen carbonates, can also be represented as ionic equations by following steps similar to the steps for writing reactions between acids and metal hydroxides (Worked example 11.3.2).

The use of **antacids** is an everyday example of the reaction between acids and carbonates (Figure 11.3.6). Excess hydrochloric acid in the stomach can be the cause of indigestion, heartburn and an upset stomach. This condition can be treated with bases that neutralise the excess acid. Antacids such as calcium carbonate and sodium hydrogen carbonate are quick-acting and very effective, but they should not be used for long periods of time due to the possibility of adverse side effects. Other antacids contain aluminium hydroxide or magnesium hydroxide and dissolve more slowly.

CHEMFILE

Bicarbonate of soda

Self-raising flour contains tartaric acid and some sodium hydrogen carbonate (bicarbonate of soda). This type of flour is used in baking cakes because on heating in the oven, the acid and hydrogen carbonate react to produce a salt, water and carbon dioxide. As the carbon dioxide is released, the cake mixture rises. Alternatively, if plain flour is substituted, tartaric acid and sodium hydrogen carbonate would need to be added.

Bicarbonate of soda (sodium hydrogen carbonate) leads to the production of carbon dioxide causing these muffins to rise.

The general equation for the reaction between acids and carbonates is acid + carbonate → salt + water + carbon dioxide.

Antacids contain bases such as carbonates, hydrogen carbonates and hydroxides that neutralise stomach acidity.

(a)

Drug Facts

Active ingredients (in each tablet)	***Purpose***
Anhydrous citric acid 1000 mg................................	Antacid
Aspirin 325 mg (NSAID)*.......................................	Analgesic
Sodium bicarbonate (heat-treated) 1916 mg.........	Antacid

*nonsteroidal anti-inflammatory drug

Used for the temporary relief of:
- heartburn, acid indigestion and sour stomach when accompanied with headache or body aches and pains
- upset stomach with headache from overindulgence in food or drink
- headache, body aches and pain alone

(b)

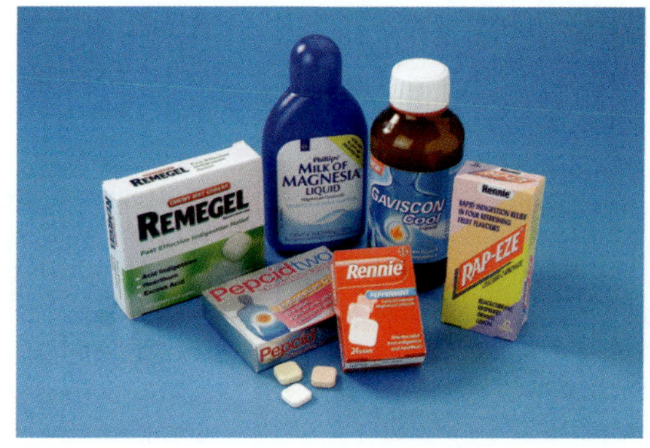

FIGURE 11.3.6 A commercial antacid. The aspirin is a pain reliever and anti-inflammatory agent and the sodium bicarbonate is an antacid. The citric acid reacts with some of the sodium bicarbonate to create bubbles of carbon dioxide when the tablets dissolve.

Worked example 11.3.2

WRITING IONIC EQUATIONS FOR REACTIONS BETWEEN ACIDS AND METAL CARBONATES

What products are formed when a dilute solution of nitric acid is added to solid magnesium carbonate? Write an ionic equation for this reaction.

Thinking	Working
What is the general reaction? Identify the products.	acid + metal carbonate ⟶ salt + water + carbon dioxide Products of this reaction are magnesium nitrate in solution, water and carbon dioxide gas.
Identify the reactants and products. Indicate the state of each, i.e. (aq), (s), (l) or (g).	Reactants: Nitric acid is ionised in solution, forming $H^+(aq)$ and $NO_3^-(aq)$ ions. Magnesium carbonate is an ionic solid, $MgCO_3(s)$. Products: Magnesium nitrate is dissociated into $Mg^{2+}(aq)$ and $NO_3^-(aq)$ ions. Water has the formula $H_2O(l)$. Carbon dioxide has the formula $CO_2(g)$.
Write the equation showing all reactants and products. (There is no need to balance the equation yet.)	$H^+(aq) + NO_3^-(aq) + MgCO_3(s) \longrightarrow Mg^{2+}(aq) + NO_3^-(aq) + H_2O(l) + CO_2(g)$
Identify the spectator ions.	$NO_3^-(aq)$
Rewrite the equation without the spectator ions. Balance the equation with respect to the number of atoms of each element and charge.	The equation with spectator ions omitted is: $H^+(aq) + MgCO_3(s) \longrightarrow Mg^{2+}(aq) + H_2O(l) + CO_2(g)$ The balanced equation is: $2H^+(aq) + MgCO_3(s) \longrightarrow Mg^{2+}(aq) + H_2O(l) + CO_2(g)$ Note that if hydronium ions are represented as $H_3O^+(aq)$, rather than as $H^+(aq)$, this reaction would be written as: $2H_3O^+(aq) + MgCO_3(s) \longrightarrow Mg^{2+}(aq) + 3H_2O(l) + CO_2(g)$

Worked example: Try yourself 11.3.2

WRITING IONIC EQUATIONS FOR REACTIONS BETWEEN ACIDS AND METAL CARBONATES

What products are formed when a solution of hydrochloric acid is added to a solution of sodium hydrogen carbonate? Write an ionic equation for this reaction.

A neutralisation reaction is a reaction between an acid and a base, which can be a metal hydroxide or metal carbonate.

Acids and reactive metals

When dilute acids are added to reactive metals, bubbles of hydrogen gas are released and a salt is formed. In general, the equation for the reaction is:

$$\text{acid} + \text{reactive metal} \longrightarrow \text{salt} + \text{hydrogen}$$

Reactive metals include calcium, magnesium, iron and zinc, but *not* copper, silver or gold. For example, the reaction between dilute hydrochloric acid and zinc metal can be seen in Figure 11.3.7 and is represented by the equation:

$$2HCl(aq) + Zn(s) \longrightarrow ZnCl_2(aq) + H_2(g)$$

This reaction can also be represented by an ionic equation. In an aqueous solution, the hydrochloric acid is ionised and the salt, zinc chloride (a soluble ionic compound), is dissociated. The ionic equation can be determined as shown in Worked example 11.3.3.

The general equation for the reaction between an acid and a reactive metal is:

acid + reactive metal $\longrightarrow$ salt + hydrogen gas.

Worked example 11.3.3

WRITING IONIC EQUATIONS FOR REACTIONS BETWEEN ACIDS AND REACTIVE METALS

Write an ionic equation for the reaction that occurs when dilute hydrochloric acid is added to a sample of zinc metal.

Thinking	Working
What is the general reaction? Identify the products formed.	acid + reactive metal $\longrightarrow$ salt + hydrogen Hydrogen gas and zinc chloride solution are produced.
Identify the reactants and products. Indicate the state of each, i.e. (aq), (s), (l) or (g).	Reactants: zinc is a solid, $Zn(s)$. Hydrochloric acid is ionised, forming $H^+(aq)$ and $Cl^-(aq)$ ions. Products: hydrogen gas, $H_2(g)$. Zinc chloride is dissociated into $Zn^{2+}(aq)$ and $Cl^-(aq)$ ions.
Write the equation showing all reactants and products. (There is no need to balance the equation yet.)	$H^+(aq) + Cl^-(aq) + Zn(s) \longrightarrow Zn^{2+}(aq) + Cl^-(aq) + H_2(g)$
Identify the spectator ions.	$Cl^-(aq)$
Rewrite the equation without the spectator ions. Balance the equation with respect to number of atoms of each element and charge.	$2H^+(aq) + Zn(s) \longrightarrow Zn^{2+}(aq) + H_2(g)$

Worked example: Try yourself 11.3.3

WRITING IONIC EQUATIONS FOR REACTIONS BETWEEN ACIDS AND REACTIVE METALS

Write an ionic equation for the reaction that occurs when aluminium is added to a dilute solution of hydrochloric acid.

CHEMFILE

Testing for carbonate salts

Acids can be used to detect the presence of carbonate salts. Carbon dioxide gas is produced when an acid is added to a carbonate. So if an acid is added to an unknown sample and carbon dioxide is produced, the sample is a carbonate salt.

The **limewater test** is a simple laboratory test used to confirm the presence of carbon dioxide gas. Limewater is a saturated solution of calcium hydroxide ($Ca(OH)_2$). When carbon dioxide is bubbled through this solution, it will turn 'milky' or 'cloudy' in appearance (see figure) due to the precipitation of calcium carbonate:

$$Ca(OH)_2(aq) + CO_2(g) \longrightarrow CaCO_3(s) + H_2O(l)$$

Limewater test. Carbon dioxide is bubbled through limewater, turning the limewater cloudy.

FIGURE 11.3.7 Hydrogen gas is produced from the reaction of zinc with dilute hydrochloric acid.

11.3 Review

OA ✓✓

SUMMARY

- Generalisations can be made about the probable products of reactions involving acids and bases:
  acid + metal hydroxide ⟶ salt + water
  acid + metal carbonate ⟶ salt + water + carbon dioxide
  acid + metal hydrogen carbonate ⟶ salt + water + carbon dioxide
  acid + reactive metal ⟶ salt + hydrogen
- Each of these reactions can be represented by full and ionic equations.
- An ionic equation only shows those ions, atoms or molecules that take part in the reaction.
- Spectator ions (ions that do not take part in the reaction) are not included in ionic equations.
- Ionic equations are balanced with respect to both the number of atoms of each element and charge.
- Antacids, which contain bases such as carbonates, hydrogen carbonates and hydroxides, can be used to neutralise stomach acidity.

KEY QUESTIONS

Knowledge and understanding

1 Write full and ionic chemical equations for the reactions between:
  - a magnesium and sulfuric acid
  - b calcium and hydrochloric acid
  - c zinc and ethanoic acid
  - d aluminium and nitric acid.

2 Name the salt produced in each of the reactions in Question **1**.

3 For each of the following reactions, write:
  - i a full chemical equation to represent the reaction (remember to include states)
  - ii an ionic equation.
  - a solid zinc carbonate and sulfuric acid
  - b solid calcium and nitric acid
  - c solid copper(II) hydroxide and nitric acid
  - d solid magnesium hydrogen carbonate and hydrochloric acid

Analysis

4 Predict the products of the following reactions and write full and ionic chemical equations for each.
  - a A solution of sulfuric acid is added to a solution of potassium hydroxide.
  - b Nitric acid solution is mixed with sodium hydroxide solution.
  - c Hydrochloric acid solution is poured onto some solid magnesium hydroxide.
  - d Blue copper(II) carbonate powder is added to dilute sulfuric acid.
  - e Dilute hydrofluoric acid is mixed with a solution of potassium hydrogen carbonate.
  - f Dilute nitric acid is added to a spoon coated with solid zinc.
  - g Hydrochloric acid solution is added to some marble chips (calcium carbonate).
  - h Solid bicarbonate of soda (sodium hydrogen carbonate) is mixed with vinegar (a dilute solution of ethanoic acid).

5 Students were given four unlabelled bottles containing 0.1 M solutions of different acids. Each bottle contained a clear liquid and they all looked identical. They were also given several similar-sized pieces of zinc, a stopwatch and as many test tubes as necessary.
  - a Design an experiment that could qualitatively demonstrate the strengths of the acids.
  - b Name the products you would expect from mixing the zinc with an acid.
  - c Explain how you would determine the strength of the acid and then rank the acids in order of strength.

  The results one student obtained are shown in the table.

Acid added to the zinc strip	Observations	Time taken for reaction to go to completion (s)
A		342
B		22
C		65
D		178

  - d Predict the strongest acid, based on the results above.
  - e Complete the table with the observations you might expect to see from this experiment.
  - f Rank the acids in order from weakest to strongest acid.
  - g Propose a list of possible acids that could have been used in this experiment.

11.4 pH: a measure of acidity

In the previous sections, you were introduced to the Brønsted–Lowry theory, which defines an acid as a proton donor and a base as a proton acceptor. You also learnt that acids and bases can be classified as strong or weak, depending on how readily they donate or accept protons.

In this section, you will learn about the pH scale, which is a measure of the acidity of solutions. You will also learn that water, being an amphiprotic species, can react with itself to a small extent and produce hydronium and hydroxide ions. From the relationship that exists between hydronium and hydroxide ions in different aqueous solutions, you can determine their concentrations and **pH**.

IONIC PRODUCT OF WATER

Water molecules can react with each other as represented by the equation:

$$H_2O(l) + H_2O(l) \rightleftharpoons H_3O^+(aq) + OH^-(aq)$$

The production of the H_3O^+ ion and OH^- ion within this reaction can be seen in Figure 11.4.1.

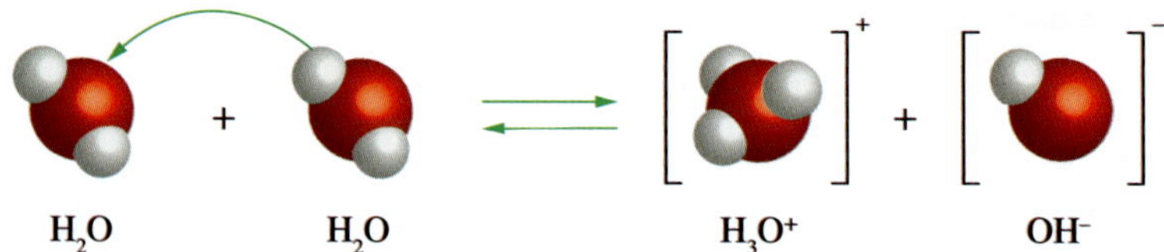

FIGURE 11.4.1 The ionisation of water molecules

Pure water undergoes this **self-ionisation** to a very small extent. In this reaction, water behaves as both a very weak acid and as a very weak base, producing one hydronium ion (H_3O^+) for every one hydroxide ion (OH^-). Water is displaying amphiprotic properties.

The concentration of hydronium and hydroxide ions in water is very low. In pure water at 25°C, the concentrations of H_3O^+ and of OH^- are 1×10^{-7} M, i.e. $[H_3O^+] = [OH^-] = 1 \times 10^{-7}$ M in pure water, where [ ] represents the concentration of the ion in moles per litre (mol L^{-1}), also known as the molar concentration (M). For each H_3O^+ ion (or OH^- ion) present in a glass of water, there are 560 million H_2O molecules!

Experimental evidence shows that all aqueous solutions contain both H_3O^+ and OH^- ions, and that the product of their molar concentrations, $[H_3O^+] \times [OH^-]$, is always 1.00×10^{-14} at 25°C. If either $[H_3O^+]$ or $[OH^-]$ in an aqueous solution increases, then the concentration of the other must decrease proportionally.

> ℹ The ionic product $K_w = [H_3O^+][OH^-] = 1.00 \times 10^{-14}$ M^2 at 25°C, where the square brackets [ ] represent molar concentration.

Remember that $[H_3O^+]$ represents the concentration of hydrogen ions in mol L^{-1}, or M, and $[OH^-]$ represents the concentration of hydroxide ions in mol L^{-1}, or M. The expression $[H_3O^+][OH^-]$ is known as the **ionic product (or ionisation constant) of water** and is represented by the symbol K_w:

$$K_w = [H_3O^+][OH^-] = 1.00 \times 10^{-14} \text{ M}^2 \text{ at } 25°C$$

Acidic and basic solutions

In solutions of acidic substances, H_3O^+ ions are formed by reaction of the acid with water, as well as from self-ionisation of water. So, the concentration of H_3O^+ ions will be greater than 10^{-7} M at 25°C. Since the product $[H_3O^+][OH^-]$ remains constant, the concentration of OH^- ions in an **acidic solution** at this temperature must be less than 10^{-7} M.

The opposite is true for **basic solutions**. The concentration of OH^- ions in a basic solution is greater than 10^{-7} M and that of H_3O^+ ions is less than 10^{-7} M.

In summary, at 25°C:

- pure water and **neutral solutions**: $[H_3O^+] = [OH^-] = 10^{-7}$ M
- acidic solutions: $[H_3O^+] > 10^{-7}$ M and $[OH^-] < 10^{-7}$ M
- basic solutions: $[H_3O^+] < 10^{-7}$ M and $[OH^-] > 10^{-7}$ M

> ℹ The higher the concentration of H_3O^+ ions in a solution, the more acidic the solution is.

Calculating the concentration of H_3O^+ in aqueous solutions

The expression for K_w can be used to determine the concentrations of hydronium and hydroxide ions in solution, knowing that the value of K_w in solutions at 25°C is 1.00×10^{-14} M^2.

Worked example 11.4.1

CALCULATING THE CONCENTRATION OF HYDRONIUM AND HYDROXIDE IONS IN AN AQUEOUS SOLUTION

For a 0.10 M HCl solution at 25°C, calculate the concentration of:
- H_3O^+ ions
- OH^- ions.

Thinking	Working
Find the concentration of the hydronium (H_3O^+) ions.	HCl is a strong acid, so it will ionise completely in solution. Each molecule of HCl donates one proton to water to form one H_3O^+ ion: $HCl(aq) + H_2O(l) \longrightarrow H_3O^+(aq) + Cl^-(aq)$ Because HCl is completely ionised in water, a 0.10 M HCl solution has a concentration of H_3O^+ ions of 0.10 M: i.e. $[H_3O^+] = 0.10$ M
Use the expression for the ionic product of water to calculate the concentration of OH^- ions.	$K_w = [H_3O^+][OH^-] = 1.00 \times 10^{-14}$ M^2 $[OH^-] = \frac{1.00 \times 10^{-14}}{[H_3O^+]}$ Since $[H_3O^+] = 0.10$ M $[OH^-] = \frac{1.00 \times 10^{-14}}{0.10}$ $= 1.0 \times 10^{-13}$ M

Worked example: Try yourself 11.4.1

CALCULATING THE CONCENTRATION OF HYDRONIUM AND HYDROXIDE IONS IN AN AQUEOUS SOLUTION

For a 5.6×10^{-6} M HNO_3 solution at 25°C, calculate the concentration of:
- H_3O^+
- OH^-

DEFINITION OF pH

The range of H_3O^+ concentrations in solutions is so great that a convenient scale, called the **pH scale**, has been developed to measure acidity. This is a logarithmic scale, which means that extremely large differences in concentration can be represented on a simple scale. The pH scale was first proposed by the Danish scientist Sören Sörenson in 1909, as a way of expressing levels of acidity. The pH of a solution is defined as:

$$pH = -\log_{10}[H_3O^+]$$

Alternatively, this expression can be rearranged to give:

$$[H_3O^+] = 10^{-pH}$$

The pH scale eliminates the need to write cumbersome powers of 10 when we describe hydrogen ion concentration, simplifying the measurement and calculation of acidity. Since the scale is based upon the *negative logarithm* of the hydrogen ion concentration, the pH of a solution *decreases* as the concentration of hydrogen ions *increases*.

$pH = -\log_{10}[H_3O^+]$; The more acidic a solution, the higher its $[H_3O^+]$, the lower its pH.

pH OF ACIDIC AND BASIC SOLUTIONS

On the pH scale, the most acidic solutions have pH values slightly less than 0 and the most basic solutions have values of about 14. Acidic, basic and neutral solutions can be defined in terms of their pH at 25°C.

- Neutral solutions have a pH equal to 7.
- Acidic solutions have a pH less than 7.
- Basic solutions have a pH greater than 7.

The pH values of some common substances, ranked from most acidic to most basic, are listed in Table 11.4.1. In the case of blood, the normal pH range is 7.35–7.45, and even small deviations from this range for any length of time can lead to serious illness and death. The body has mechanisms for controlling blood pH, which guard against sudden shifts in acidity and alkalinity.

TABLE 11.4.1 pH values of some common substances at 25°C

Solution	pH	$[H_3O^+]$ (M)	$[OH^-]$ (M)	$[H_3O^+] \times [OH^-]$ (M^2)
1.0 M HCl	0.0	1	10^{-14}	10^{-14}
lemon juice	3.0	10^{-3}	10^{-11}	10^{-14}
vinegar	4.0	10^{-4}	10^{-10}	10^{-14}
tomatoes	5.0	10^{-5}	10^{-9}	10^{-14}
rain water	6.0	10^{-6}	10^{-8}	10^{-14}
pure water	7.0	10^{-7}	10^{-7}	10^{-14}
blood	7.4	4×10^{-8}	2.5×10^{-7}	10^{-14}
seawater	8.0	10^{-8}	10^{-6}	10^{-14}
soap	9.0	10^{-9}	10^{-5}	10^{-14}
oven cleaner	13.0	10^{-13}	10^{-1}	10^{-14}
1.0 M NaOH	14.0	10^{-14}	10^{-0}	10^{-14}

A solution with pH 2 has 10 times the concentration of hydronium ions as one of pH 3 and one-tenth the concentration of hydroxide ions as one of pH 1.

CALCULATIONS INVOLVING pH

pH can be calculated using the formula $pH = -\log_{10}[H_3O^+]$. Your scientific calculator has a logarithm function that will simplify pH calculations.

Worked example 11.4.2

CALCULATING THE pH OF AN AQUEOUS SOLUTION FROM $[H_3O^+]$

What is the pH of a solution in which the concentration of $[H_3O^+]$ is 0.14 M? Express your answer to two decimal places.

Thinking	Working
Write down the concentration of H_3O^+ ions in the solution.	$[H_3O^+] = 0.14$ M
Substitute the value of $[H_3O^+]$ into: $pH = -\log_{10}[H_3O^+]$ Use the logarithm function on your calculator to determine the answer.	$pH = -\log_{10}[H_3O^+]$ $= -\log_{10}(0.14)$ (use your calculator) $= 0.85$

Worked example: Try yourself 11.4.2

CALCULATING THE pH OF AN AQUEOUS SOLUTION FROM $[H_3O^+]$

What is the pH of a solution in which the concentration of $[H_3O^+]$ is 6×10^{-9} M? Express your answer to two significant figures.

Worked example 11.4.3

CALCULATING pH IN A SOLUTION OF A BASE

What is the pH of a 0.005 M solution of $Ba(OH)_2$ at 25°C?

Thinking	Working
Write down the reaction in which $Ba(OH)_2$ dissociates.	In water, each mole of $Ba(OH)_2$ completely dissociates to release 2 moles of OH^- ions. $Ba(OH)_2(aq) \longrightarrow Ba^{2+}(aq) + 2OH^-(aq)$
Determine the concentration of OH^- ions.	$[OH^-] = 2 \times [Ba(OH)_2]$ $= 2 \times 0.005$ M $= 0.01$ M
Determine the $[H_3O^+]$ in the diluted solution by substituting the $[OH^-]$ into the ionic product of water: $K_w = [H_3O^+][OH^-] = 1.00 \times 10^{-14}$ M^2	$K_w = [H_3O^+][OH^-] = 1.00 \times 10^{-14}$ M^2 $[H_3O^+] = \frac{K_w}{[OH^-]}$ $= \frac{1.00 \times 10^{-14}}{0.01}$ $= 1 \times 10^{-12}$ M
Substitute the value of $[H_3O^+]$ into: $pH = -\log_{10}[H_3O^+]$ Use the logarithm function on your calculator to determine the answer.	$pH = -\log_{10}[H_3O^+]$ $= -\log_{10}(1 \times 10^{-12})$ (use your calculator) $= 12.0$

Worked example: Try yourself 11.4.3

CALCULATING pH IN A SOLUTION OF A BASE

What is the pH of a 0.01 M solution of $Ba(OH)_2$ at 25°C?

Worked example 11.4.4

CALCULATING $[H_3O^+]$ IN A SOLUTION OF A GIVEN pH

Calculate the $[H_3O^+]$ in a solution of pH 5.0 at 25°C.

Thinking	Working
Decide which form of the relationship between pH and $[H_3O^+]$ should be used: $pH = -\log_{10}[H_3O^+]$ or $[H_3O^+] = 10^{-pH}$	As you have the pH and are calculating $[H_3O^+]$, use: $[H_3O^+] = 10^{-pH}$
Substitute the value of pH into the relationship expression and use a calculator to determine the answer.	$[H_3O^+] = 10^{-pH}$ $= 10^{-5.0}$ $= 1 \times 10^{-5}$ M (or 0.000 01 M)

Worked example: Try yourself 11.4.4

CALCULATING $[H_3O^+]$ IN A SOLUTION OF A GIVEN pH

Calculate the $[H_3O^+]$ in a solution of pH 10.4 at 25°C.

CHEMFILE

Effect of temperature on pH

Earlier, we defined the ionic product of water as:

$$K_w = [H_3O^+]\,[OH^-] = 1.00 \times 10^{-14}\text{ M}^2 \text{ at } 25^\circ\text{C}$$

You can use this relationship to calculate either $[H_3O^+]$ or $[OH^-]$ at 25°C in different solutions. But what happens if the temperature is not 25°C?

From experimental data we know that the value of K_w increases as the temperature increases.

While the pH of pure water is 7.00 at 25°C, at 0°C the pH is 7.47 and at 55°C the pH is 6.57°C. However, at all temperatures pure water is still described as neutral, because the concentrations of H_3O^+ and OH^- ions are equal, even though the pH may not be equal to 7.00.

Worked example 11.4.5

CALCULATING pH IN A SOLUTION WHERE THE SOLUTE CONCENTRATION IS NOT GIVEN

What is the pH of a solution at 25°C, which contains 1.0 g NaOH in 100 mL of solution?

Thinking	Working
Determine the number of moles of NaOH.	$n(\text{NaOH}) = \frac{m}{M}$ $= \frac{1.0}{40.0}$ $= 0.025$ mol
Write the equation for dissociation of NaOH.	$NaOH(aq) \rightarrow Na^+(aq) + OH^-(aq)$ NaOH is completely dissociated in water.
Determine the number of moles of OH^- based on the dissociation equation.	$n(OH^-) = n(\text{NaOH})$ $= 0.025$ mol
Use the formula for determining concentration given number of moles and volume: $c = \frac{n}{V}$	$n = 0.025$ mol $V = 0.100$ L $c = \frac{0.025}{0.100}$ $= 0.25$ M
Determine the $[H_3O^+]$ in the diluted solution by substituting the $[OH^-]$ into the ionic product of water: $K_w = [H_3O^+][OH^-] = 1.00 \times 10^{-14}$ M^2	$K_w = [H_3O^+][OH^-] = 1.00 \times 10^{-14}$ M^2 $[H_3O^+] = \frac{K_w}{[OH^-]}$ $= \frac{1.00 \times 10^{-14}}{0.025}$ $= 4.0 \times 10^{-14}$ M
Substitute the value of $[H_3O^+]$ into: $pH = -\log_{10}[H_3O^+]$ Use the logarithm function on your calculator to determine the answer.	$pH = -\log_{10}[H_3O^+]$ $= -\log_{10}(4.0 \times 10^{-14})$ (use your calculator) $= 13.40$

Worked example: Try yourself 11.4.5

CALCULATING pH IN A SOLUTION WHERE THE SOLUTE CONCENTRATION IS NOT GIVEN

What is the pH of a solution at 25°C that contains 0.50 g KOH in 500 mL of solution?

11.4 Review

OA ✓✓

SUMMARY

- Water self-ionises according to the equation:
  $H_2O(l) + H_2O(l) \rightleftharpoons H_3O^+(aq) + OH^-(aq)$
- The ionic product for water is:
  $K_w = [H_3O^+][OH^-] = 1.00 \times 10^{-14}\ M^2$ at 25°C
- The concentration of H_3O^+ is measured using the pH scale, which can be written:
  $pH = -\log_{10}[H_3O^+]$ or $[H_3O^+] = 10^{-pH}$
- At 25°C the pH of a neutral solution is 7. The pH of an acidic solution is less than 7 and the pH of a basic solution is greater than 7.
- The more acidic a solution, the higher its $[H_3O^+]$ and the lower its pH. The more basic a solution, the higher its $[OH^-]$ and the higher its pH.
- The pH scale can be used to rank solutions from most acidic to most basic.
- The ionic product of water can be used to calculate the pH for strong acid and strong base solutions of known concentration.

KEY QUESTIONS

Knowledge and understanding

1 Calculate the concentration of OH^- ions at 25°C in an aqueous solution with a concentration of H_3O^+ ions of 0.01 M.

2 What is the concentration of OH^- ions in a solution at 25°C which has a concentration of H_3O^+ ions of 5.70×10^{-10} M?

3 Calculate $[H_3O^+]$ at 25°C in an aqueous solution in which $[OH^-] = 1.0 \times 10^{-4}$ M.

4 What is the pH of a solution in which H_3O^+ concentration is 0.0001 M?

5 Calculate the pH of a 0.01 M solution of nitric acid (HNO_3).

6 The pH of water in a lake is 6.0. Calculate $[H_3O^+]$ in the lake.

Analysis

7 Determine the pH of a 200 mL solution that contains 0.365 g of dissolved HCl.

8 Consider 0.01 M solutions of the following acids and bases: HCl, CH_3COOH, NaOH, NH_3.

  a State whether they are weak or strong acids or bases.

  b For the strong acids and strong bases, calculate the $[H_3O^+]$ and $[OH^-]$.

  c Calculate the pH of the strong acids and strong bases.

9 If a 0.1 M acid solution has a pH of 4.3, determine whether it is a strong or weak acid. Explain your answer.

11.5 Measuring pH

An indicator is a weak acid with a colour that is different to the colour of its weak conjugate base.

Indicators are weak acids that have a different colour in acidic solutions to their colour in basic solutions. They are used to distinguish between acids and bases. The conjugate acid form of the indicator is one colour and its conjugate base form is a different colour. Indicators can be used to determine the pH of a solution and they are also used for acid–base analyses.

NATURAL INDICATORS

One of the characteristic properties of acids and bases is their ability to change the colour of certain plant extracts, such as rose petals, blackberries and red cabbage. Such plant extracts are a form of **natural indicator**.

Litmus

Litmus is one of the oldest natural indicators in use and is a purple water-soluble mixture of dyes obtained from lichen. In the presence of acids, litmus turns red and it turns blue in basic solutions.

FIGURE 11.5.1 Natural acid–base indicators are found in plants such as red cabbage. Red cabbage extract turns a different colour in (from left to right) strong acid (pH 1–3), weak acid (pH 4–6), neutral solution (pH 7), weak base (pH 8–10) and strong base (pH 11–14).

Red cabbage

Figure 11.5.1 shows the variation in colour that can be achieved using red cabbage as an indicator. The colour of juice extracted from the leaves of red cabbage is dependent upon the pH of the solution, so red cabbage juice can be used to measure the pH of solutions with reasonable **accuracy**.

COMMERCIAL INDICATORS

Today, most indicators used in laboratories and industry are commercially produced. These indicators include bromothymol blue, methyl orange and phenolphthalein. Some common indicators and their pH ranges can be seen in Figure 11.5.2. These indicators can be used to determine the pH change over a small range of values only and are useful in acid–base analyses.

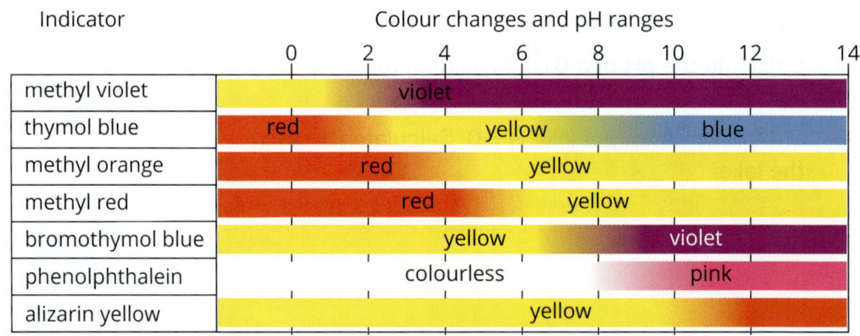

FIGURE 11.5.2 Common indicators and their pH ranges

Bromothymol blue

Bromothymol blue is a widely used indicator, which is yellow in acidic solutions and blue in basic solutions. In a neutral solution of pH 7.0, the indicator is green, midway between yellow and blue.

A chart showing the variation in colour of bromothymol blue from pH 0 to pH 14 is shown in Figure 11.5.3.

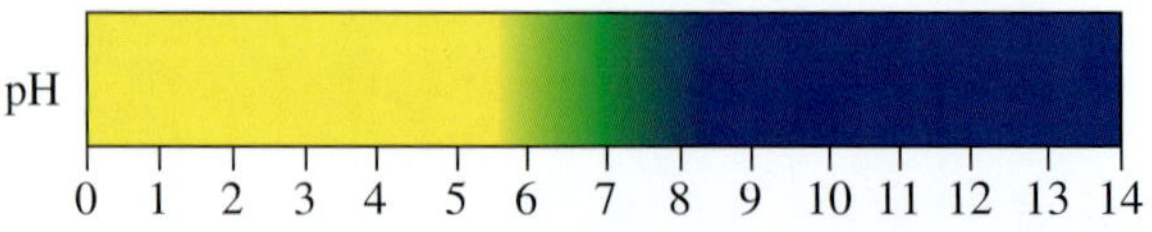

FIGURE 11.5.3 Bromothymol blue is yellow in acidic solutions and blue in basic solutions. The indicator changes colour over the pH range 6.0 to 7.6.

Methyl orange

Methyl orange is another synthetic indicator often used in the analysis of weak acids. It is also used as a textile dye. Methyl orange is red in acidic solutions and yellow in basic solutions. The indicator changes colour between pH 3.1 and pH 4.4. Between these pH values the indicator has an orange colour.

A colour chart for methyl orange indicator is shown in Figure 11.5.4.

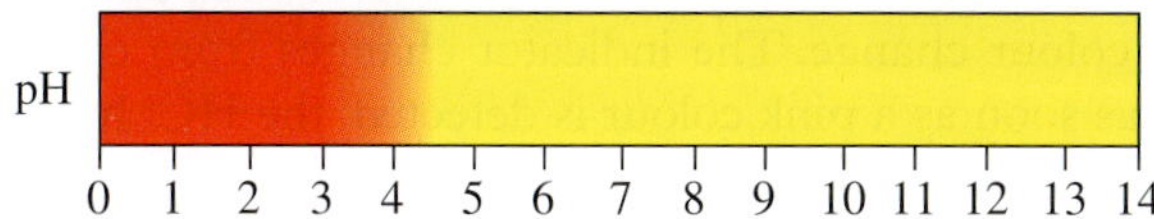

FIGURE 11.5.4 Methyl orange is red in acidic solutions and yellow in basic solutions. The indicator changes colour over the pH range 3.1 to 4.4.

Phenolphthalein

Phenolphthalein is a synthetic indicator used in the analysis of acids and bases. In the past it was also used as the active ingredient in some laxatives. In acidic solutions phenolphthalein is colourless, whereas in basic solutions it has a pink to magenta colour. Phenolphthalein changes colour over the pH range 8.3 to 10.0. A colour chart for phenolphthalein indicator is shown in Figure 11.5.5.

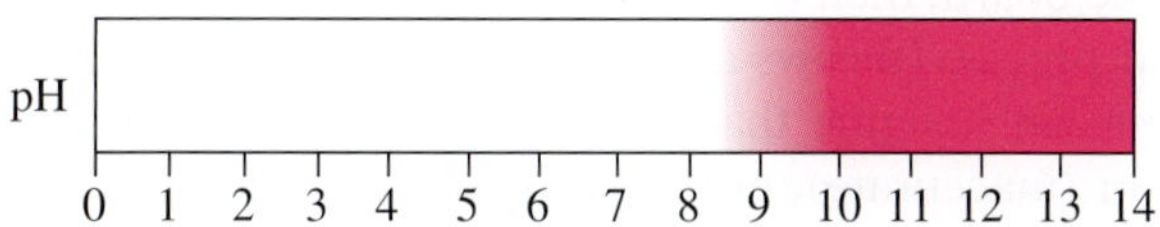

FIGURE 11.5.5 Phenolphthalein is colourless in solutions when the pH is below 8.3 and pink when the pH is greater than 10. The indicator changes colour over the pH range of 8.3 to 10.

Universal indicator

Universal indicator (Figure 11.5.6) is widely used to estimate the pH of a solution. It is a mixture of several indicators and displays a range of colours, from red through yellow, green and blue, to violet. Each pH value between 0 and 14 is a specific colour. Universal indicator can provide the chemist with a more accurate determination of the pH of a sample, than if the other indicators described above were used.

FIGURE 11.5.6 Universal indicator pH scale. When universal indicator is added to a solution, it changes colour depending on the solution's pH. The tubes contain solutions of pH 0 to 14 from left to right. The tube in the centre with a green-coloured solution is neutral, pH 7.

When an even more accurate measurement of the pH of a solution is required, a pH meter can be used (Figure 11.5.7).

CHEMFILE

Indicators in gardening

The hydrangeas shown below are different colours because of the pH of the soil. In acidic soil, hydrangeas are blue, but in alkaline soil they are pink. To change their colour, a gardener can determine the pH of the soil and add either a weak acid or weak base to the area around the plant. By making the soil more acidic using aluminium sulfate, which is a weak acid, the colour of the flowers will become more blue. More alkaline conditions can be achieved by adding lime (calcium carbonate), which is a weak base, to make the colour more pink.

To measure the pH of soil, a sample without any organic matter in it is mixed with water and the suspension allowed to settle. The water above the sediment is then tested with an indicator, such as universal indicator, to determine the pH. For more accurate results, a pH meter is used.

The different colours displayed by these hydrangeas are caused by the acidity or basicity of the soil. Indicators can be used to determine the soil pH and therefore the chemicals that can be added to the soil to change their colour.

FIGURE 11.5.7 A pH meter can be used to measure the acidity of a solution.

ACCURACY AND PRECISION IN pH MEASUREMENT

Accurate measurements are close to the true value. Precise measurements are close to each other and are reproducible.

We have seen that different indicators can be used to measure pH over different ranges and for different purposes.

For example, suppose you wished to exactly neutralise a solution of HCl using NaOH solution. Phenolphthalein is a suitable indicator for this purpose. If you add a few drops of phenolphthalein to the acid and then slowly add the base, the exact point when equal amounts of acid and base have been added can be detected by watching the colour change. The indicator changes from colourless in acid to pink in base. So as soon as a pink colour is detected, the HCl has been completely neutralised.

On the other hand, indicators such as phenolphthalein, methyl orange and bromothymol blue, which change colour over narrow pH ranges, would not be suitable indicators to use if the pH of an unknown sample is to be measured. To measure the pH of a sample that could have any value between 0 and 14, universal indicator or a pH meter (or even red cabbage juice) could be used. A pH meter would be the best choice if you want the measurement to be as accurate and precise as possible. However, for some purposes it may be sufficient to use universal indicator to obtain a less accurate and less precise reading.

The difference between accuracy and **precision** can be understood by considering a dart game. If a player throws darts that are all close to the bullseye in the centre of the board, then these throws are accurate (close to the true value—the bullseye) and precise (throws are very close to each other and can be reliably reproduced). You can imagine that another player might miss the bullseye by a significant margin (inaccurate), but all of their throws may be very close to each other (precise).

Table 11.5.1 compares the accuracy and precision of three methods for determining pH.

Red cabbage juice gives less accurate and less precise measurements of pH than universal indicator, whereas universal indicator is less accurate and less precise than a pH meter.

TABLE 11.5.1 Accuracy and precision of three methods of determining the pH of a sample

	Using red cabbage juice	Using universal indicator	Using a pH meter
Accuracy	sample value can be determined to within 2 pH units of the true value	sample value can be determined to within 1 pH unit of the true value	sample value can be determined to within 0.01 pH units of the true value
Precision	generally reproducible to within 2 pH units	generally reproducible to within 1 pH unit	generally reproducible to within 0.01 pH units of the true value
Overall conclusion	Red cabbage juice is both less accurate and less precise than universal indicator. When accuracy and precision are required to within one-hundredth of a pH value, a pH meter is used.		

11.5 Review

OA ✓✓

SUMMARY

- Indicators are used to determine the pH of a solution.
- The colour of an indicator in acidic solution is distinctly different from its colour in basic solution.
- Indicators are solutions of a weak acid or base in solution.
- Different indicators change colours at different pH values.
- Indicators such as phenolphthalein, methyl orange and bromothymol blue can be used to detect when the neutralisation of an acid and base occurs.
- Indicators such as red cabbage juice, universal indicator and pH meters can be used to measure the pH of a sample.
- Accurate values are those that are close to the true or well accepted value.
- Precise readings are close to each other, but not necessarily close to the true value.
- A pH meter provides the most accurate and precise measurement of pH.

KEY QUESTIONS

Knowledge and understanding

1 Using the charts above, determine what colour the following indicators would be in pure water.

- **a** phenolphthalein
- **b** methyl orange
- **c** universal indicator

2 Explain why universal indicator shows a range of colours across the pH scale.

3 Bromothymol blue is added to a solution of pH 5.

- **a** Determine the colour of the solution.
- **b** Explain what happens when a solution of NaOH is then added to the solution.

4 Referring to Figure 11.5.2, on page 374 identify at what pH range you would expect the following to happen.

- **a** methyl violet to appear yellow
- **b** phenolphthalein to appear pink
- **c** methyl red to transition from red to yellow

Analysis

5 Congo red is an acid–base indicator. It is blue–violet at pH 3.0 and red at pH 5.0. Predict the colour of the solution at the following pH values:

- **a** pH 4
- **b** pH 8
- **c** pH 2

6 Which indicator in Figure 11.5.2 (page 374) should be used if you need to see the pH change according to the following values?

- **a** From 2 to 6
- **b** From 0 to 4

7 Red cabbage juice was used to measure the pH of several solutions with concentrations of 1 M. Use Figure 11.5.1 on page 374 to determine whether the solutions are neutral or contain strong or weak acids, or strong or weak bases, if the following colours were observed.

- **a** green
- **b** pink
- **c** purple

8 The pH readings below were determined at 25°C using either red cabbage juice, universal indicator or a pH meter. Explain which method for measuring pH was most likely to have been used. Use the terms accuracy and precision when explaining the reasons for your choice.

- **a** pure water pH readings: 6.99, 6.98, 7.01, 7.02
- **b** acidic solution believed to have a pH of 5: 5, 4, 5, 4
- **c** basic solution believed to contain a weak base: 8, 7, 6, 7

11.6 Acid–base reactions in the environment

CHEMFILE

'Ocean acidity' does not mean a pH below 7

Seawater is slightly alkaline, with an average pH of 8.14. In the last 200 years, the pH has dropped by 0.11. Since pH is a logarithmic scale, this is about a 30% decline in the concentration of $H^+(aq)$ and if this continues, by 2100 it is estimated that seawater will become 100% more acidic, with a pH of about 7.85. Notice that the seawater will still be alkaline, but more acidic than it has ever been.

Since the start of the Industrial Revolution in the 1760s, there has been a dramatic increase in the combustion of fossil fuels. This has resulted in significantly higher levels of acidic gases, including carbon dioxide, sulfur dioxide and nitrogen dioxide in the atmosphere. As well as contributing to an increase in global warming, the increase in atmospheric carbon dioxide is responsible for changes in **ocean acidity**, which has impacts on marine organisms.

In this section, you will apply your understanding of acid–base reactions to study the chemistry and effects of increasing ocean acidity. These are significant environmental, social and economic issues that result from human activity.

CARBON DIOXIDE IN NATURE AND OCEAN ACIDITY

Carbon dioxide is essential to life on Earth. Through the carbon cycle shown in Figure 11.6.1, carbon is exchanged within the biosphere (the global sum of all ecosystems on Earth).

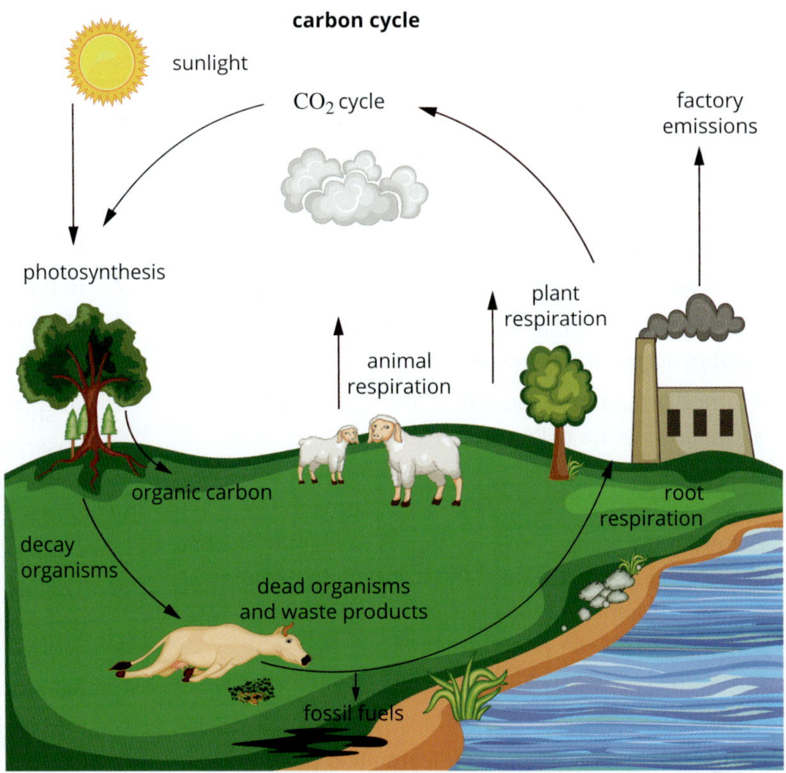

FIGURE 11.6.1 The element carbon cycles between land, atmosphere and oceans.

Plants take carbon dioxide from the atmosphere and, via **photosynthesis**, use it to make glucose ($C_6H_{12}O_6$), which acts as an energy source for the plant. The glucose can then be converted into larger molecules, which store energy or make up structural parts within the plant. Photosynthesis can only occur in the presence of sunlight and the pigment chlorophyll that is found in the green parts of the plant.

The process of photosynthesis can be summarised by the equation:

$$6CO_2(g) + 6H_2O(l) \xrightarrow[\text{chlorophyll}]{\text{sunlight}} C_6H_{12}O_6(aq) + 6O_2(g)$$

Carbon dioxide in the atmosphere

Studies have shown that the levels of atmospheric CO_2 have varied naturally over the last several thousand years. Atmospheric carbon dioxide and other greenhouse gases are critical in maintaining the average surface temperature of the Earth by trapping energy and re-radiating it in all directions, returning about half to Earth and half to outer space. Scientists have used ratios of carbon isotopes to distinguish between natural and human-caused contributions. In recent times, the human contributions to CO_2 levels have exceeded the natural fluctuations, as can be seen in Figure 11.6.2.

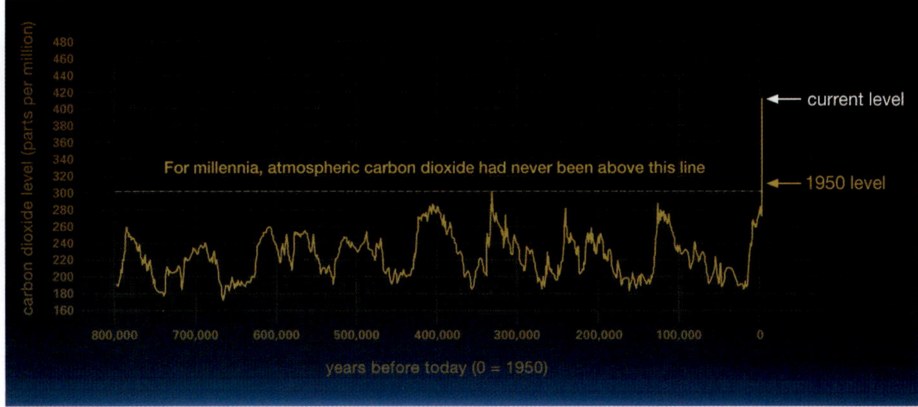

FIGURE 11.6.2 Concentrations of atmospheric carbon dioxide for the last 800 000 years. The data was determined from measurements of CO_2 concentrations in ice cores from Antarctica.

It is thought that these changes in atmospheric carbon dioxide levels are influencing the acidity of rain, surface land temperatures and causing global warming. Global weather patterns have altered as a consequence and chemical processes in the oceans are also changing.

Carbon dioxide in the oceans

As levels of carbon dioxide in the atmosphere increase, more carbon dioxide dissolves in the ocean. This increases the concentration of a weak acid, carbonic acid, which in turn increases the concentration of $H_3O^+(aq)$ ions, and therefore decreases pH. The overall result is an increase in ocean acidity.

The pattern of changes in pH and dissolved carbon dioxide expected over time can be seen in Figure 11.6.3. Combined with increasing surface temperatures, there is an impact on the complex chemical systems in the oceans. These systems involve huge amounts of soluble metal salts, such as calcium and sodium salts, together with carbonate ions, organic matter and gases.

> Ocean acidity is a decrease in pH that is occurring due to the absorption of carbon dioxide from the atmosphere.

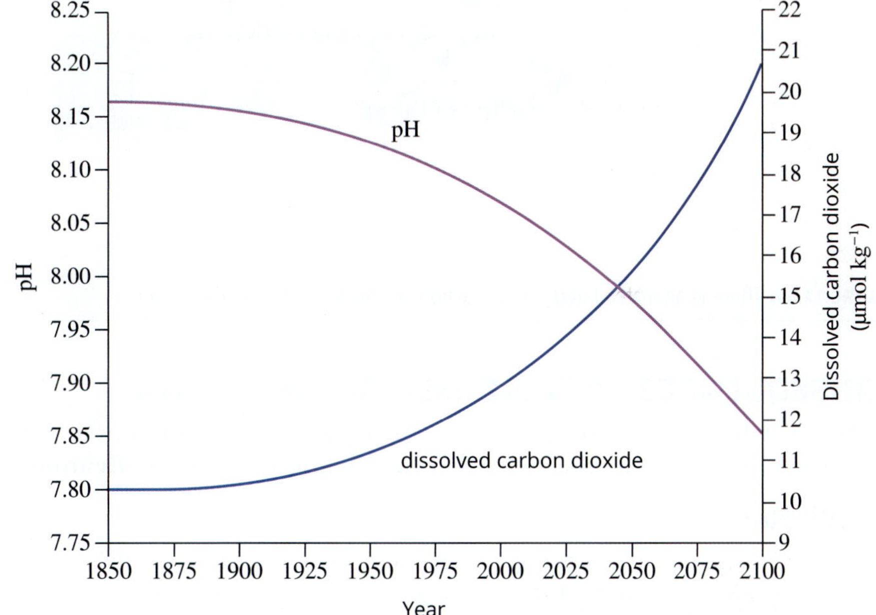

FIGURE 11.6.3 Historical (from 1850) and projected (to 2100) pH and dissolved CO_2 levels in the ocean

Scientists calculate that between one-third and one-half of the carbon dioxide emissions from human activity are absorbed by the oceans. Although this reduces the **greenhouse effect**, which is mainly due to carbon dioxide in the atmosphere, the pH of the oceans has decreased by 0.11 units since the Industrial Revolution. This represents a 30% increase in the hydronium ion concentration in seawater.

The chemistry of dissolved carbon dioxide

Carbon enters the ocean mainly by the dissolution of carbon dioxide gas, which is slightly soluble in water. The equation for carbon dioxide dissolving in water can be written as:

$$CO_2(g) \rightleftharpoons CO_2(aq)$$

Most of the carbon dissolved in seawater is present as $CO_2(aq)$, but some reacts further to form carbonic acid:

$$CO_2(aq) + H_2O(l) \rightleftharpoons H_2CO_3(aq)$$

Carbonic acid is a weak diprotic acid, ionising in two steps to form hydrogen carbonate ions (HCO_3^-) and carbonate ions (CO_3^{2-}):

$$H_2CO_3(aq) + H_2O(l) \rightleftharpoons H_3O^+(aq) + HCO_3^-(aq)$$

$$HCO_3^-(aq) + H_2O(l) \rightleftharpoons H_3O^+(aq) + CO_3^{2-}(aq)$$

There are an interrelated series of reactions in seawater that involve dissolved carbon dioxide gas, hydrogen carbonate ions, HCO_3^-, and carbonate ions, CO_3^{2-}.

This process can be seen in Figure 11.6.4. The double arrow $\rightleftharpoons$ indicates that reactions can go in either direction, depending on which species are in excess. (You will learn more about this in Year 12.) For example, if the concentration of H_3O^+ increases, which occurs as the ocean becomes more acidic, the lower reaction above will go backwards, causing H_3O^+ to react with CO_3^{2-}. These processes that occur in the ocean can be summarised as follows:

- As carbon dioxide in the atmosphere increases, the ocean becomes more acidic (i.e. increased $[H_3O^+]$).
- The increased amount of H_3O^+ reacts with dissolved carbonate ions (CO_3^{2-}).
- Some of the solid $CaCO_3$ in shells, exoskeletons of marine animals and corals, then dissolves to produce more carbonate ions, CO_3^{2-}.
- The increase in dissolved carbonate ions has consequences for shell growth.

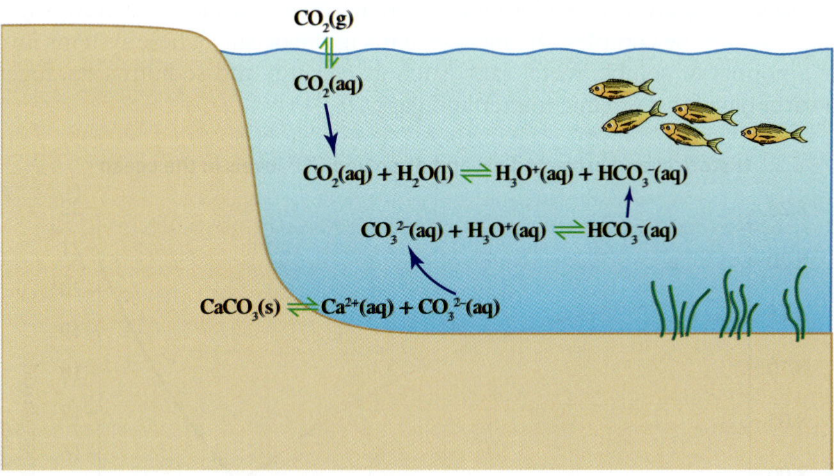

FIGURE 11.6.4 There is an interrelated set of reactions involving carbon in the ocean.

CONSEQUENCES OF INCREASED OCEAN ACIDITY

As the pH of the oceans decreases, the amount of available carbonate ions decreases and this has an impact on processes called **calcification** and **decalcification**.

Calcification

Many aquatic organisms, including marine invertebrates such as shellfish, starfish, coral, sea snails, crabs and lobsters have a protective covering made of calcium carbonate ($CaCO_3$).

These organisms absorb calcium ions and carbonate ions from seawater to build and maintain the calcium carbonate structure essential for their survival, as represented in the following equation:

$$Ca^{2+}(aq) + CO_3^{2-}(aq) \rightleftharpoons CaCO_3(s)$$

This process is called calcification.

Calcium carbonate is virtually insoluble in water, and the oceans can be regarded as **saturated solutions** of calcium and carbonate ions. Once formed, solid calcium carbonate is usually quite stable. The health and growth of these animals depends critically on the concentration of carbonate ions and therefore carbon dioxide in the oceans.

> Calcification involves the precipitation of dissolved Ca^{2+} ions and CO_3^{2-} ions as solid $CaCO_3$ in shells and coral.

Decalcification

The increased acidity of the oceans causes some of the additional hydronium ions to react with carbonate ions via the following equation:

$$H_3O^+(aq) + CO_3^{2-}(aq) \rightleftharpoons HCO_3^-(aq) + H_2O(l)$$

This reaction has the effect of reducing the concentration of free CO_3^{2-} ions in seawater, making it more difficult for marine creatures to build or maintain their protective structures.

This process is called decalcification. It is estimated that the pH of the ocean will fall from 8.14 to 7.90 over the next 50 years, decreasing the rate of calcification and increasing the rate of decalcification, putting coral reefs and other marine organisms at risk.

> As the level of dissolved CO_2 in the oceans increases, $[H_3O^+]$ increases, pH decreases and solid $CaCO_3$ starts to dissolve.

Figure 11.6.5 shows the effects of decalcification on sea snails. These small, free-floating snails are at the base of the ocean food web. While the healthy specimen has a glass-like shell with smooth edges, the shell of the specimen affected by increased ocean acidity is starting to dissolve. Weak spots in the shell have an opaque, cloudy appearance and the shell edges are more ragged.

FIGURE 11.6.5 These tiny free-swimming sea snails are an important food source for other marine animals. The healthy specimen has a glass-like shell with smooth edges, the unhealthy specimen has been affected by increased ocean acidity and is starting to dissolve.

Figure 11.6.6 shows a food web that indicates the relationships between tiny ocean species with shells of $CaCO_3$ and other creatures. As the pH of the oceans falls and the concentrations of essential chemicals, such as CO_3^{2-} ions, needed for the development and survival of the organisms, decrease, it can be seen that ocean ecosystems are at risk.

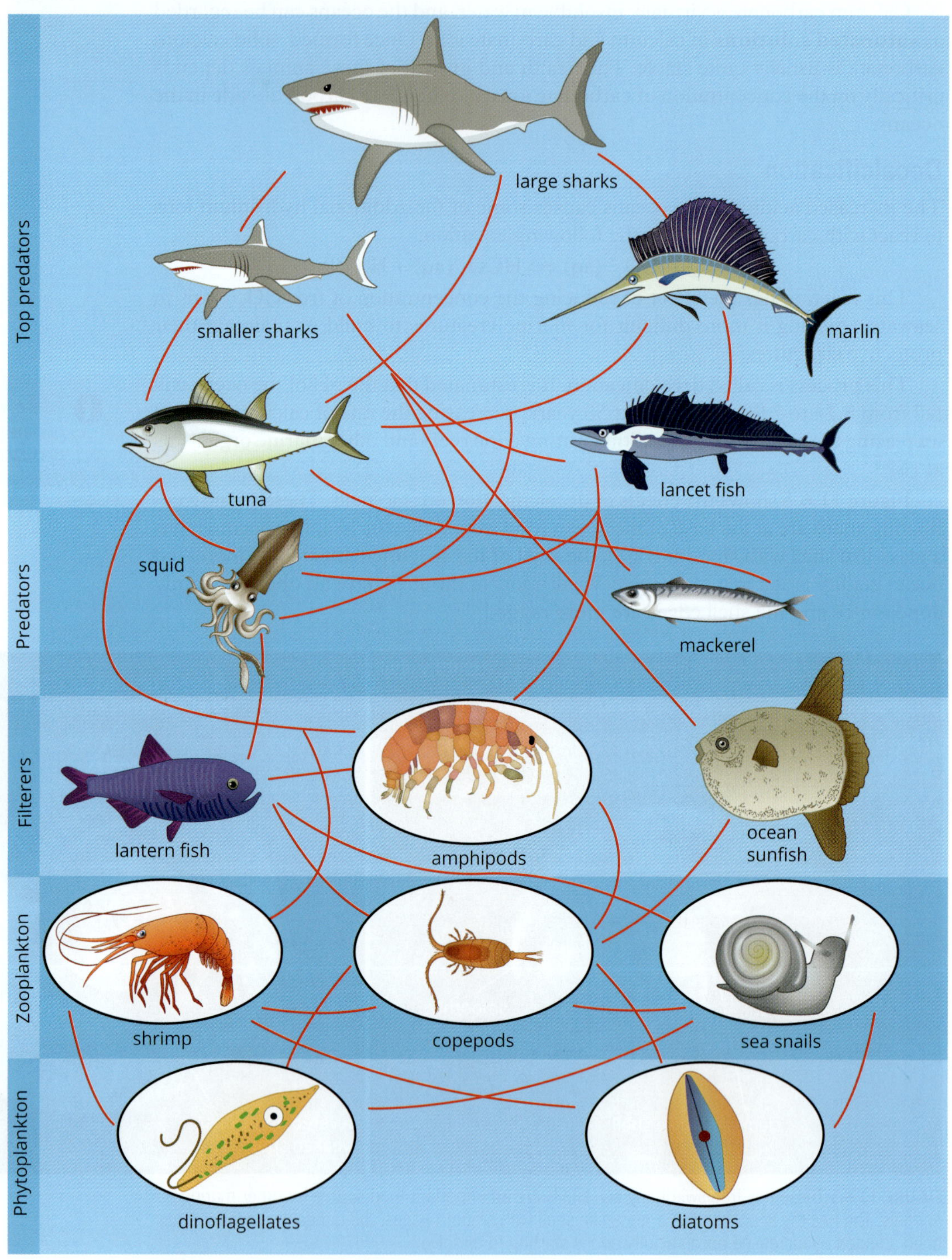

FIGURE 11.6.6 Ocean food web showing the importance of animals such as sea snails (pteropods) and shrimp

CASE STUDY **ANALYSIS**

Other impacts of ocean acidity

Ocean acidity is a global issue involving all the oceans. It affects all marine species, not just those which use $CaCO_3$ for protective coatings. For example, it has a detrimental effect on cold ocean organisms, such as plankton and krill (Figure 11.6.7). Diatoms are single-celled algae that are an important component of plankton, and form the foundation of marine and freshwater food chains. Krill feed on plankton and can be found in swarms that are kilometres wide. They are a major food source for many marine organisms, from small fish, such as sardines, to huge mammals, such as whales.

FIGURE 11.6.7 (a) Diatoms are single-celled algae that are an important component of plankton. (b) Krill feed on plankton.

Krill eggs do not hatch successfully at a lower pH. Therefore, an increase in ocean acidity is predicted to have an impact on the species of plankton and krill upon which other species depend for their survival. A collapse of the krill population coupled with ocean warming would have a disastrous effect on the ecosystem of the ocean. Because oceans provide a diverse range of food sources for human consumption, increased ocean acidity would affect us all.

The reduction in the free CO_3^{2-} ions in seawater is also considered an important threat to coral reef ecosystems, making it harder for corals to build their skeletons. Coral reefs provide protection for coastal communities from storms and erosion, and, because of their colour and diversity, they are popular tourist attractions. The deterioration and destruction of these reefs would affect the economies of these communities and perhaps eventually destroy them.

Analysis

1 Krill and plankton populations are expected to decrease with increasing ocean acidity.
   - **a** What is the direct effect of ocean acidity on krill?
   - **b** Why is the decrease of krill and plankton population numbers a concern?

2 Briefly describe two other effects of increased ocean acidity.

11.6 Review

OA ✓✓

SUMMARY

- An increasing concentration of atmospheric carbon dioxide is responsible for increasing ocean acidity.
- Carbonic acid is a weak diprotic acid that reacts with water to produce hydronium ions, hydrogen carbonate ions and carbonate ions.
- The chemistry of ocean acidity can be summarised as follows:
  $CO_2(g) \rightleftharpoons CO_2(aq)$
  $CO_2(aq) + H_2O(l) \rightleftharpoons H_2CO_3(aq)$
  $H_2CO_3(aq) + H_2O(l) \rightleftharpoons H_3O^+(aq) + HCO_3^-(aq)$
  $HCO_3^-(aq) + H_2O(l) \rightleftharpoons H_3O^+(aq) + CO_3^{2-}(aq)$
- Some marine organisms use calcium ions and carbonate ions to protect themselves with layers of calcium carbonate. The equation for the reaction is: $Ca^{2+}(aq) + CO_3^{2-}(aq) \rightleftharpoons CaCO_3(s)$
  This process is called calcification.
- Absorption of increased amounts of carbon dioxide has increased the concentration of hydronium ions in seawater and lowered its pH.
- Hydronium ions can react with carbonate ions, reducing the concentration of free carbonate ions, causing decalcification.
- Reduced carbonate ion concentration impairs the ability of some marine organisms to build and maintain calcium carbonate structures.
- Increased ocean acidity has environmental, economic and social impacts.

KEY QUESTIONS

Knowledge and understanding

1 Write equations to show the following processes.
  a the dissolution of atmospheric carbon dioxide
  b the formation of carbonic acid
  c the two steps involved in the ionisation of carbonic acid.

2 Many marine animals have a protective skeleton that is seriously affected by ocean acidity.
  a What is the name of the compound that forms a protective layer for many marine organisms?
  b What is the name of the process for the formation of this layer?
  c Write an equation for this process.

3 Increased ocean acidity causes a decrease in the concentration of carbonate ions. Write an equation for the process that causes this decrease and name this process.

Analysis

4 The following three reaction systems are present in the ocean.
  (1) $CO_2(g) \rightarrow CO_2(aq)$
  (2) $CO_2(g) + H_2O(l) \rightarrow H_2CO_3(aq)$
  (3) $H_2CO_3(aq)\ ) + H_2O(l) \rightarrow HCO_3^-(aq) + H_3O^+(aq)$
  Using these equations, explain how increased carbon dioxide concentration in the atmosphere reduces the pH of the oceans.

Chapter review

OA ✓✓

11

KEY TERMS

accuracy
acid
acid–base reaction
acidic proton
acidic solution
acidity
alkali
amphiprotic
antacids
base
basic solution
Brønsted–Lowry theory
calcification
concentrated solution
concentration
conjugate acid
conjugate acid–base pair
conjugate base
decalcification
diprotic acid
dissociate
greenhouse effect
hydronium ion
hydroxide ion
indicator
ionic equation
ionic product of water
ionisation
ionisation constant of water
ionisation reaction
ionise
limewater test
monoprotic acid
natural indicator
neutral solution
neutralisation reaction
ocean acidity
pH
pH scale
photosynthesis
polyprotic acid
precision
salt
saturated solution
self-ionisation
spectator ion
strong acid
strong base
triprotic acid
weak acid
weak base

REVIEW QUESTIONS

Knowledge and understanding

1 Which one of the following statements is true about the reaction represented by the equation below?
$SO_4^{2-}(aq) + H_3O^+(aq) \longrightarrow HSO_4^-(aq) + H_2O(l)$
- **A** SO_4^{2-} is acting as a base and its conjugate acid is HSO_4^-.
- **B** H_3O^+ is acting as a base and its conjugate acid is H_2O.
- **C** H_2O is the conjugate acid of H_3O^+.
- **D** H_3O^+ and HSO_4^- are both bases.

2 Identify the reactant that acts as an acid in each of the following equations.
- **a** $NH_4^+(aq) + H_2O(l) \longrightarrow NH_3(aq) + H_3O^+(aq)$
- **b** $NH_3(g) + HCl(g) \longrightarrow NH_4Cl(s)$
- **c** $HCO_3^-(aq) + OH^-(aq) \longrightarrow H_2O(l) + CO_3^{2-}(aq)$
- **d** $CO_3^{2-}(aq) + CH_3COOH(aq) \longrightarrow HCO_3^-(aq) + CH_3COO^-(aq)$

3 Write balanced equations to show that in water:
- **a** PO_4^{3-} acts as a base
- **b** $H_2PO_4^-$ acts as an amphiprotic substance
- **c** H_2S acts as an acid.

4 Identify which one of the following reactions is a Brønsted–Lowry acid–base reaction.
- **A** $HCl(aq) + KOH(aq) \longrightarrow KCl(aq) + H_2O(l)$
- **B** $2HNO_3(aq) + Mg(s) \longrightarrow Mg(NO_3)_2(aq) + H_2(g)$
- **C** $AgNO_3(aq) + NaCl(aq) \longrightarrow AgCl(s) + NaNO_3(aq)$
- **D** $Cu(s) + H_2SO_4(aq) \longrightarrow CuSO_4(aq) + H_2(g)$

5 Which one of the following is a correct acid–base conjugate pair?
- **A** HCl / H_2O
- **B** NH_4^+ / NH_3
- **C** H_3O^+ / OH^-
- **D** CH_3COOH / $HCOO^-$

6 Write the formula for the conjugate base of each of the following.
- **a** HCl
- **b** H_2O
- **c** OH^-
- **d** HSO_4^-

7 Write concise definitions for the following.
- **a** Brønsted–Lowry acid
- **b** strong base
- **c** conjugate acid

8 Using suitable examples, distinguish between:
- **a** a diprotic acid and an amphiprotic substance
- **b** a strong acid and a concentrated acid.

9 Construct a concept map that demonstrates your understanding of the links between the following terms.
Use the terms: Brønsted–Lowry acid–base theory, acid, base, protons, HCl, Cl^- ions
Use the links: proton donor is an, proton acceptor is a, contains, an example is, is a conjugate pair.

10 The CH_3COOH molecule is a weak acid. Which one of the following species is most likely to be the most abundant in a 1 M CH_3COOH solution?
A CH_3COOH
B CH_3COO^-
C H_3O^+
D OH^-

11 Write an equation to show the following reactions.
a perchloric acid ($HClO_4$) acts as a strong acid in water
b hypochlorous acid ($HClO_3$) acts as a weak acid in water
c ammonia (NH_3) acts as a weak base in water
d $H_2PO_4^-$ ion acting as a weak base in water.

12 pH can be measured using either universal indicator or a pH meter.
a Suggest a situation when universal indicator might be used in preference to pH meter.
b Suggest a situation a pH meter might be used in preference to universal indicator.
c 'A pH meter can give a more accurate and more precise measurement of pH than universal indicator.' Explain the meaning of this statement.

13 The decrease in the pH of the oceans has a consequential effect on the processes of calcification and decalcification and on the hard, protective outer layer of some marine organisms.
a Define these processes.
b Write the appropriate ionic equation for each process.

Application and analysis

14 Complete, and balance, the following chemical equations.
a $HNO_3(aq) + KOH(aq) \longrightarrow$
b $H_2SO_4(aq) + K_2CO_3(aq) \longrightarrow$
c $H_3PO_4(aq) + Ca(HCO_3)_2(s) \longrightarrow$
d $HF(aq) + Zn(OH)_2(s) \longrightarrow$

15 Hydrogen is produced when dilute sulfuric acid reacts with aluminium metal.
a Write a balanced full equation for this reaction.
b Write a balanced ionic equation for this reaction.

16 An antacid composed of $MgCO_3$ is relatively insoluble in water.
a Write a balanced ionic equation for the reaction between an acid HA and the antacid, $MgCO_3$.
b Explain the purpose of antacids.

17 Calculate the concentration of OH^- ions in aqueous solutions at 25°C with H_3O^+ ion concentrations equal to:
a 0.001 M
b 10^{-5} M
c 5.7×10^{-9} M
d 3.4×10^{-12} M
e 6.5×10^{-2} M
f 2.23×10^{-13} M

18 Human blood has a pH of 7.4.
a Is blood acidic, basic or neutral?
b Calculate the [H+] of this sample of human blood.

19 A solution is yellow in both bromothymol blue and methyl orange. Select which one of the following is the most accurate approximation of the pH of this solution.
A between 3.1 and 6.0
B between 3.1 and 7.6
C between 4.4 and 7.6
D between 5.0 and 6.0

20 What is the concentration of these ions in solutions, at 25°C, with the following pH values?
i hydronium ions
ii hydroxide ions
a 1
b 3
c 7
d 11.7

21 The pH of a cola drink is 3 and of black coffee is 5. How many more times acidic is the cola than black coffee?

22 The pH of tomato juice is 5.4 at 25°C. What is the concentration of OH^- ions in tomato juice?

23 Calculate the concentration of H^+ and OH^- ions in solutions, at 25°C, with the following pH values.
a 3.0
b 10.0
c 8.5
d 5.8
e 9.6
f 13.5

24 The increased amount of carbon dioxide in the atmosphere causes more to dissolve in the oceans.
a List all the species, apart from H_2O, that exist in an aqueous solution of carbon dioxide.
b Write the two steps in which carbonic acid ionises to form hydrogen carbonate ions and carbonate ions.

25 A laboratory technician forgot to label 0.1 M solutions of sodium hydroxide (NaOH), hydrochloric acid (HCl), glucose ($C_6H_{12}O_6$), ammonia (NH_3) and ethanoic acid (CH_3COOH). To identify them, temporary labels A–E were placed on the bottles and the electrical conductivity and pH of each solution was measured. The results are shown in the table below. Identify each solution and briefly explain your reasoning.

Solution	Electrical conductivity	pH
A	poor	11
B	zero	7
C	good	13
D	good	1
E	poor	3

CHAPTER

12 Redox reactions

Some of the most colourful and energy-releasing reactions are classified as redox reactions. This group of reactions also includes some that are vitally important to our existence.

In this chapter, you will learn how redox reactions can be defined in terms of the loss and gain of electrons. You will understand how to write balanced half-equations that describe the transfer of electrons and then combine these half-equations to create an overall equation for the reaction. You will become familiar with redox reactions such as metal displacement reactions, and you will be able to predict when these reactions will occur.

Finally, you will see how these redox reactions have important applications in society, including galvanic cells, which are portable generators of electricity, and corrosion, which can cause the failure of structures.

Key knowledge

- oxidising and reducing agents, and redox reactions, including writing of balanced half and overall redox equations (including in acidic conditions), with states **12.1**
- the reactivity series of metals and metal displacement reactions, including balanced redox equations with states **12.2**
- applications of redox reactions in society: for example, corrosion or the use of simple primary cells in the production of electrical energy from chemical energy. **12.3**

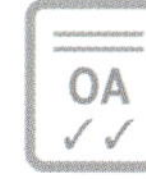

12.1 Introducing redox reactions

FIGURE 12.1.1 Potassium burning in chlorine gas—a spectacular example of a redox reaction

Your everyday life depends on a large number of chemical reactions. Many of these are **redox reactions** (a term that combines the names of two types of reaction, *red*-uction and *ox*-idation, which are explained below). From the respiration reactions that enable your cells to produce energy and the combustion reactions that warm your home, to the reactions in the batteries that keep your mobile phone working, redox reactions are occurring within you and around you all the time. The spectacular reaction between potassium and chlorine gas, shown in Figure 12.1.1, is also an example of a redox reaction.

In this section, you will learn how redox reactions are defined in terms of electron transfer and how to represent this transfer of electrons using half-equations.

EARLY UNDERSTANDINGS OF REDOX REACTIONS

When chemistry evolved from the ancient study of alchemy, many of the reactions known to early chemists involved air. French chemist Antoine Lavoisier identified the reactive component of air and named it oxygen. As a result, reactions in which oxygen was a reactant were described as oxidation reactions. In air, the **combustion** of an element such as carbon, sulfur or iron produces an oxide:

$$C(s) + O_2(g) \rightarrow CO_2(g)$$

$$S(s) + O_2(g) \rightarrow SO_2(g)$$

$$4Fe(s) + 3O_2(g) \rightarrow 2Fe_2O_3(s)$$

Because elemental iron reacts readily with oxygen, iron is generally found in nature in ores containing minerals, such as haematite (Fe_2O_3) and magnetite (Fe_3O_4). The iron metal used extensively for construction has been extracted from iron ore in a blast furnace. Pouring the molten iron from the blast furnace is shown in Figure 12.1.2. (You learnt about this process in Chapter 4, page 148.)

The extraction of iron from iron ore in a blast furnace can be represented by the equation:

$$Fe_2O_3(s) + 3CO(g) \rightarrow 2Fe(l) + 3CO_2(g)$$

In this reaction, the iron(III) oxide (Fe_2O_3) has lost oxygen and the carbon monoxide has gained oxygen. The iron(III) oxide is described as having been **reduced** and the carbon monoxide is described as having been **oxidised**.

Oxidation and reduction always occur simultaneously, hence the term 'redox reaction'.

FIGURE 12.1.2 Molten iron from a blast furnace being poured into a bucket

CHEMFILE

Origins of the words 'oxidation' and 'reduction'

Scientists first used the term 'oxidation' in the late 18th century after the work of Antoine Lavoisier. Lavoisier showed that the 'burning' of metals, such as mercury, involved a combination with oxygen.

The term 'reduction' was used long before this to describe the process of extracting metals from their ores. The word 'reduction' comes from the Latin *reduco*, meaning to restore. The process of metal extraction was seen as restoring the metal from its compounds, such as iron from iron(III) oxide or copper from copper(II) oxide. The reduction of copper(II) oxide to form copper powder occurs when copper(II) oxide is heated in the presence of hydrogen or methane gas, as shown in the figure on the right. Some fine particles of copper escape with the gas, causing the green flame.

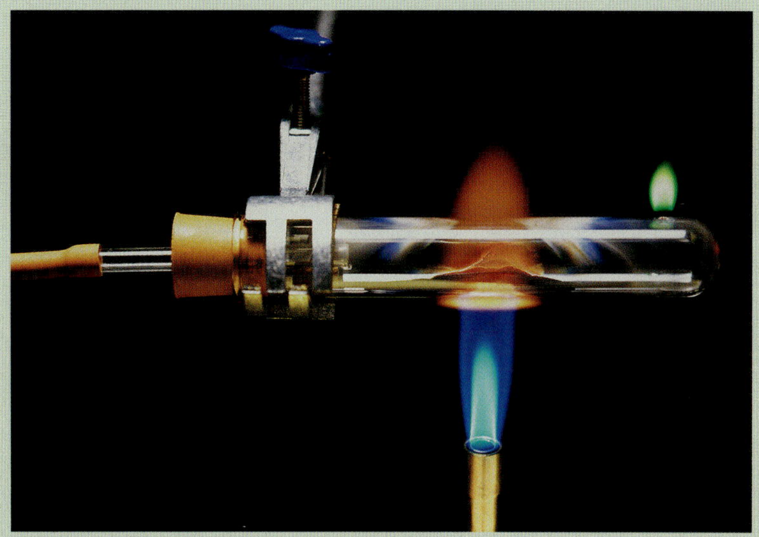

The reduction of copper(II) oxide to form copper powder occurs when it is heated in the presence of hydrogen or methane gas.

TRANSFER OF ELECTRONS

If you have heated a piece of magnesium ribbon in an experiment, as shown in Figure 12.1.3, you will remember that it burns with a brilliant white flame. Magnesium oxide powder is formed.

This reaction can be represented by the equation:

$$2Mg(s) + O_2(g) \rightarrow 2MgO(s)$$

The reaction involves a loss and gain of electrons by the reactants, which can be represented by two **half-equations**.

Each magnesium atom loses two electrons to form a magnesium ion (Mg^{2+}). The half-equation for this part of the overall reaction is written as shown below. The electronic configurations are also shown:

$$Mg(s) \rightarrow Mg^{2+}(s) + 2e^-$$

2,8,2 2,8

At the same time, each oxygen atom in the oxygen molecule (O_2) gains two electrons (i.e. four electrons per oxygen molecule):

$$O_2(g) + 4e^- \rightarrow 2O^{2-}(s)$$

2,6 2,8

Notice that when electrons are gained, they appear as reactants in the half-equation. The electrons that are gained by the oxygen have come from the magnesium atoms.

The burning of magnesium involves the transfer of electrons from magnesium atoms to oxygen atoms. Atoms also lose and gain electrons in many other reactions. This transfer of electrons is the basis of the following widely used definitions of oxidation and reduction.

Oxidation is defined as the loss of electrons.

Reduction is defined as the gain of electrons.

FIGURE 12.1.3 Magnesium ribbon burns brightly when heated in air to form a white powder, magnesium oxide.

Notice that when electrons are lost, they appear as products in the half-equation.

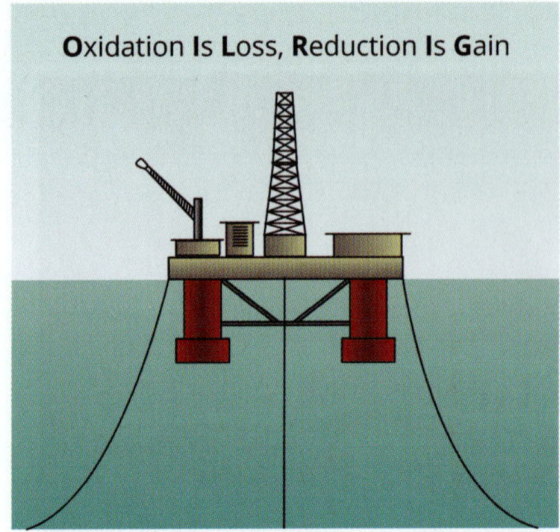

FIGURE 12.1.4 The mnemonic OIL RIG is a useful way to remember that **O**xidation **I**s the **L**oss of electrons and **R**eduction **I**s the **G**ain of electrons.

The definitions of oxidation and reduction can be recalled using the mnemonic or memory aid, OIL RIG (Figure 12.1.4).

Note that there is no overall loss of electrons, but a transfer of electrons from one atom to another. If an atom loses electrons, there must be another atom that gains electrons. Therefore, oxidation and reduction always occur simultaneously.

Other examples of redox reactions

Many redox reactions do not involve a reaction with oxygen. The reaction between potassium and chlorine shown in Figure 12.1.1 is an example:

$$2K(s) + Cl_2(g) \rightarrow 2KCl(s)$$

Oxidation half-equation: $K(s) \rightarrow K^+(s) + e^-$

Reduction half-equation: $Cl_2(g) + 2e^- \rightarrow 2Cl^-(s)$

In the previous chapter, you saw that some metals react with the $H^+(aq)$ ions in solutions of acids to produce a salt and hydrogen gas. For example, zinc metal reacts with dilute hydrochloric acid according to the equation:

$$Zn(s) + 2H^+(aq) \rightarrow Zn^{2+}(aq) + H_2(g)$$

Oxidation half-equation: $Zn(s) \rightarrow Zn^{2+}(aq) + 2e^-$

Reduction half-equation: $2H^+(aq) + 2e^- \rightarrow H_2(g)$

Worked example 12.1.1

IDENTIFYING OXIDATION AND REDUCTION

Write the oxidation and reduction half-equations for the reaction with the overall equation:

$$Mg(s) + 2H^+(aq) \rightarrow Mg^{2+}(aq) + H_2(g)$$

Thinking	Working
Write the half-equation for the reactant that loses electrons (undergoes oxidation) and balance the equation with electrons.	$Mg(s) \rightarrow Mg^{2+}(aq) + 2e^-$
Write the half-equation for the reactant that gains electrons (undergoes reduction) and balance the equation with electrons.	$2H^+(aq) + 2e^- \rightarrow H_2(g)$

Worked example: Try yourself 12.1.1

IDENTIFYING OXIDATION AND REDUCTION

Write the oxidation and reduction half-equations for the reaction with the overall equation:

$$Fe(s) + Sn^{2+}(aq) \rightarrow Fe^{2+}(aq) + Sn(s)$$

OXIDISING AGENTS AND REDUCING AGENTS

Just as an employment agent enables a client to become employed, an **oxidising agent** or **oxidant** enables or causes another chemical to be oxidised. Similarly, a **reducing agent**, or **reductant**, enables or causes another chemical to be reduced. Redox reactions always involve an oxidising agent and a reducing agent that react together.

Reducing agents cause another reactant to be reduced. In the reaction, they are oxidised.
Oxidising agents cause another reactant to be oxidised. In the reaction, they are reduced.

In the reaction between magnesium and oxygen represented in Figure 12.1.5, magnesium is being oxidised by oxygen. So, oxygen is the oxidising agent. In turn, oxygen is gaining electrons from magnesium. It is being reduced by the magnesium, so magnesium is the reducing agent.

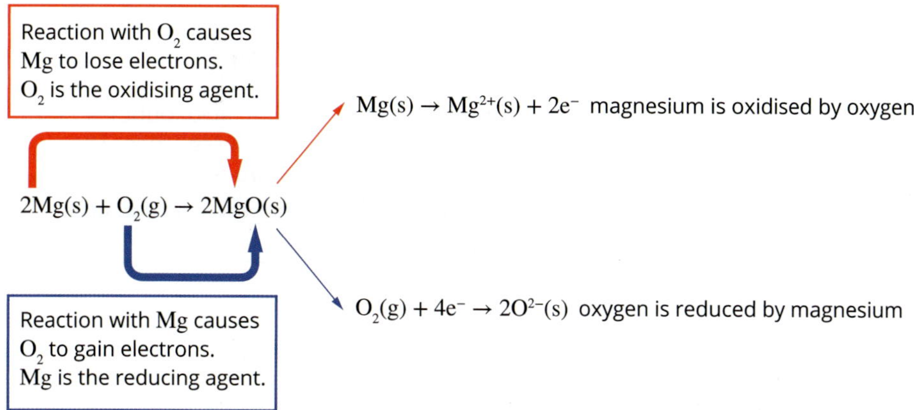

FIGURE 12.1.5 In the reaction between magnesium and oxygen, magnesium is the reducing agent and oxygen is the oxidising agent.

Since metals tend to lose electrons, they often act as reducing agents. Figure 12.1.6 summarises the list of redox terms introduced in this section.

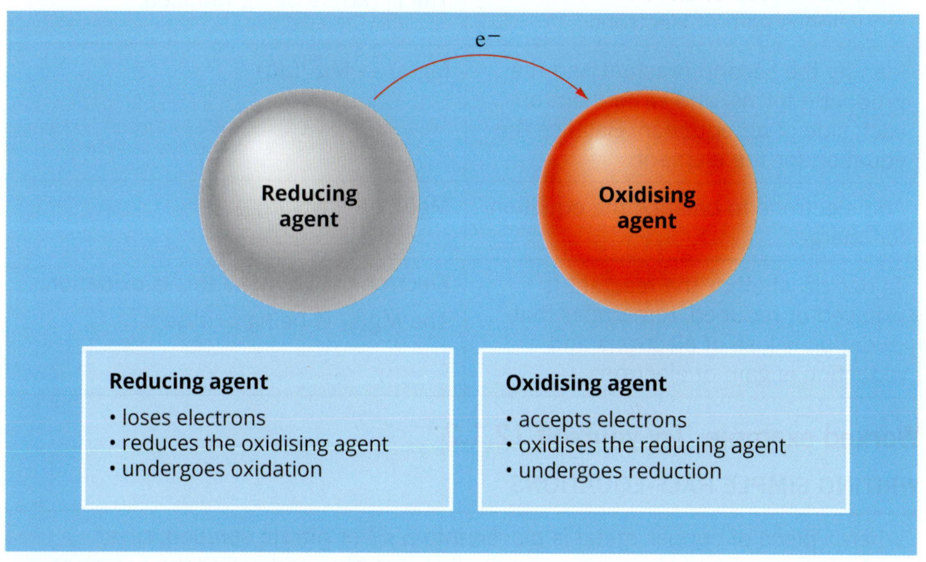

FIGURE 12.1.6 Summary of redox reaction terms

WRITING REDOX EQUATIONS

When writing equations for redox reactions, the two half-equations are normally written first and then added together to obtain an overall equation. The steps involved in writing half-equations and combining them to obtain the overall equation are described below.

Writing simple half-equations

Half-equations enable you to see the detail of what is happening in a redox reaction. Like other chemical equations, half-equations must be balanced, so there is the same number of atoms of each element on each side of the arrow. Similarly, charge must also be balanced. Half- and full equations should indicate the states of all the species in the reaction as well.

Worked example 12.1.2

WRITING SIMPLE HALF-EQUATIONS

When a strip of magnesium metal is placed in dilute sulfuric acid, bubbles of hydrogen gas are formed and the metal 'dissolves' to form Mg^{2+} ions. The oxidation and reduction reactions can be represented by two half-equations.

Write these half-equations and identify the substances that are oxidised and reduced.

Thinking	Working
Identify one reactant and the product it forms and write them on each side of an equation. Balance the equation for the element.	$H^+(aq)$ ions in the acid react to form H_2 gas: $2H^+(aq) \rightarrow H_2(g)$
Add electrons to balance the equation for charge.	$2H^+(aq) + 2e^- \rightarrow H_2(g)$
To decide whether the reactant is oxidised or reduced, remember that oxidation is loss of electrons and reduction is gain of electrons.	Electrons are gained, so this is reduction. The $H^+(aq)$ is being reduced.
Identify the second reactant and the product it forms, and write them on each side of an equation. Balance the equation for the element.	$Mg(s) \rightarrow Mg^{2+}(aq)$
Add electrons to balance the equation for charge.	$Mg(s) \rightarrow Mg^{2+}(aq) + 2e^-$
To decide whether the reactant is oxidised or reduced, remember that oxidation is loss of electrons and reduction is gain of electrons.	Electrons are lost, so this is oxidation. The Mg(s) is being oxidised.

Worked example: Try yourself 12.1.2

WRITING SIMPLE HALF-EQUATIONS

When a piece of copper metal is placed into a silver nitrate solution, silver metal is formed and the solution gradually turns blue, indicating the presence of copper(II) ions in the solution. The oxidation and reduction reactions can be represented by two half-equations.

Write these half-equations and identify the substances that are oxidised and reduced.

Writing an overall redox equation

When writing equations for redox reactions, the two half-equations are normally written first and then added together to obtain an overall equation.

An overall equation does not show any electrons transferred; all the electrons lost in the oxidation reaction are gained in the reduction reaction. One, or perhaps both, of the half-equations may need to be multiplied by a factor to ensure that the electrons balance and can be cancelled out in the overall equation.

Consider the reaction of sodium and water to produce hydrogen gas and a solution of sodium hydroxide (Figure 12.1.7). In this reaction, each Na atom is oxidised and loses one electron. Each water molecule is reduced and gains two electrons to form hydrogen and hydroxide ions, $OH^-(aq)$:

$$Na(s) \rightarrow Na^+(aq) + e^-$$

$$2H_2O(l) + 2e^- \rightarrow H_2(g) + 2OH^-(aq)$$

FIGURE 12.1.7 Sodium metal reacting with water. The hydrogen gas produced by the reaction burns and a solution of sodium hydroxide is also produced.

Two Na atoms must therefore be oxidised to provide the electrons required by each water molecule that is reduced. To write an overall equation for this reaction, the half-equation involving the oxidation of Na is multiplied by a factor of two before combining it with the half-equation for reduction of H_2O:

$$2Na(s) \rightarrow 2Na^+(aq) + 2e^-$$

You can now write the two half-equations and add them to find the overall equation:

$$2Na(s) \rightarrow 2Na^+ + 2e^-$$
$$2H_2O(l) + 2e^- \rightarrow H_2(g) + 2OH^-(aq)$$

Since the number of electrons in each half-equation is the same, they may be cancelled (as there is no overall change). The overall equation is:

$$2Na(s) + 2H_2O(l) \rightarrow 2Na^+(aq) + H_2(g) + 2OH^-(aq)$$

When writing an overall redox equation from two half-equations, you need to balance the number of electrons in each half equation.

Worked example 12.1.3

WRITING OVERALL REDOX EQUATIONS FROM HALF-EQUATIONS

The reaction between silver nitrate solution and zinc metal forms silver metal and zinc nitrate solution.

Write the half-equations for the reaction and then write the balanced overall equation.

Thinking	Working
Identify one reactant and the product(s) it forms, and write them on each side of the equation. Balance the equation for the element.	$Ag^+(aq)$ ions in silver nitrate have formed $Ag(s)$. $Ag^+(aq) \rightarrow Ag(s)$
Add electrons to balance the equation for charge.	$Ag^+(aq) + e^- \rightarrow Ag(s)$
Identify the second reactant and the product it forms, and write them on each side of the equation. Balance the equation for the element.	$Zn(s)$ has formed $Zn^{2+}(aq)$. $Zn(s) \rightarrow Zn^{2+}(aq)$
Add electrons to balance the equation for charge.	$Zn(s) \rightarrow Zn^{2+}(aq) + 2e^-$
Multiply one equation by a suitable factor to ensure that the number of electrons is balanced.	$[Ag^+(aq) + e^- \rightarrow Ag(s)] \times 2$ $2Ag^+(aq) + 2e^- \rightarrow 2Ag(s)$
Add the oxidation and the reduction half-equations together, cancelling out electrons so that none are in the final equation. If necessary, combine ions to create the formula of the product.	$2Ag^+(aq) + 2e^- \rightarrow 2Ag(s)$ $Zn(s) \rightarrow Zn^{2+}(aq) + 2e^-$ When the electrons have been cancelled, the overall equation is: $2Ag^+(aq) + Zn(s) \rightarrow 2Ag(s) + Zn^{2+}(aq)$

Worked example: Try yourself 12.1.3

WRITING OVERALL REDOX EQUATIONS FROM HALF-EQUATIONS

When potassium metal reacts with water, the water undergoes reduction to form hydrogen gas and hydroxide ions (see above) and the metal is oxidised to $K^+(aq)$ ions.

Write the half-equations for the reaction and then write the balanced overall equation.

CHEMFILE

Photochromic sunglasses

Photochromic glasses (see figure below) avoid the need for standard sunglasses. On exposure to sunlight, chloride ions in tiny crystals of silver chloride incorporated in the lens are oxidised:

$$Cl^- \rightarrow Cl + e^-$$

Electrons produced from this reaction cause Ag^+ ions to be reduced to metallic silver:

$$Ag^+ + e^- \rightarrow Ag$$

The metal causes light to be reflected and the lenses to darken. To prevent the metal and chlorine atoms re-forming AgCl immediately, copper(I) chloride is also added to the lens. It reacts with the chlorine atoms, reducing the rate at which AgCl can be re-formed:

$$Cl + Cu^+ \rightarrow Cl^- + Cu^{2+}$$

The darkening process must be reversible for the glasses to be effective. In the absence of strong sunlight, silver ions are re-formed by a redox reaction involving the silver metal and Cu^{2+} ions:

$$Cu^{2+} + Ag \rightarrow Cu^+ + Ag^+$$

As a consequence, the lenses of the glasses recover their transparency.

The lenses of these photochromic glasses darken in sunlight as a result of redox reactions involving silver chloride.

An oxidising agent and its corresponding reducing agent are known as a conjugate redox pair.

Conjugate redox pairs

When a half-equation is written for an oxidation reaction, the reactant, a reducing agent, loses electrons. The product is the oxidised form of the reactant and it is called the conjugate oxidising agent. The reactant and the product that it forms are known as a **conjugate redox pair**.

Consider the half-equation in which zinc is oxidised:

$$Zn(s) \rightarrow Zn^{2+}(aq) + 2e^-$$

The zinc metal is a reducing agent because it would cause another reactant to gain electrons. In the half-equation, $Zn^{2+}(aq)$ is formed. This is called the conjugate oxidising agent. $Zn(s)$ and $Zn^{2+}(aq)$ are a conjugate redox pair.

As another example, consider the half-equation in which $Cu^{2+}(aq)$ is reduced:

$$Cu^{2+}(aq) + 2e^- \rightarrow Cu(s)$$

The copper(II) ion is the oxidising agent, because it would cause another reactant to lose electrons. In the half-equation, $Cu(s)$ is formed. This is called the conjugate reducing agent. $Cu(s)$ and $Cu^{2+}(aq)$ are a conjugate redox pair.

There are two conjugate redox pairs in a redox reaction, usually written in the format oxidising agent/reducing agent, as shown in the two examples in Table 12.1.1.

TABLE 12.1.1 Identifying conjugate redox pairs in a redox reaction

	Example 1	Example 2
Overall equation	$Fe(s) + Sn^{2+}(aq) \rightarrow Fe^{2+}(aq) + Sn(s)$	$Cu(s) + 2Ag^+(aq) \rightarrow Cu^{2+}(aq) + 2Ag(s)$
Oxidation half-equation	$Fe(s) \rightarrow Fe^{2+}(aq) + 2e^-$	$Cu(s) \rightarrow Cu^{2+}(aq) + 2e^-$
Reduction half-equation	$Sn^{2+}(aq) + 2e^- \rightarrow Sn(s)$	$Ag^+(aq) + e^- \rightarrow Ag(s)$
Conjugate redox pairs	$Fe^{2+}(aq)/Fe(s)$ $Sn^{2+}(aq)/Sn(s)$	$Cu^{2+}(aq)/Cu(s)$ $Ag^+(aq)/Ag(s)$

HALF-EQUATIONS FOR COMPLEX REDOX REACTIONS

Most laboratories have strong oxidising agents that can be used when a substance needs to be oxidised. For safety reasons, these are often stored together and well away from flammable materials, since they could cause a fire. This class of strong oxidising agents includes potassium permanganate ($KMnO_4$), potassium dichromate ($K_2Cr_2O_7$) and potassium chromate ($KCrO_4$). To write the more complex redox half-equations for these oxidising agents in acidic solutions, a specific method involving the use of $H^+(aq)$ and $H_2O(l)$ has to be followed.

For example, consider the equation for the redox reaction that occurs when potassium permanganate ($KMnO_4$) reacts with iron(II) sulfate ($FeSO_4$) in acidified solution:

$$MnO_4^-(aq) + 8H^+(aq) + 5Fe^{2+}(aq) \rightarrow Mn^{2+}(aq) + 4H_2O(l) + 5Fe^{3+}(aq)$$

The potassium and sulfate ions are spectator ions and do not appear in the overall equation. You can describe this reaction in words as follows: the permanganate ions oxidise iron(II) ions to iron(III) ions. The permanganate ions are reduced to manganese(II) ions.

To write the overall equation for a redox reaction such as this, begin with the oxidation of iron(II) ions to iron(III) ions. It can be represented by the simple half-equation:

$$Fe^{2+}(aq) \rightarrow Fe^{3+}(aq) + e^-$$

The conversion of MnO_4^- to Mn^{2+} involves reduction. The steps to write a half equation for this reduction reaction involve balancing oxygens with H_2O, hydrogens with H^+ and charge with electrons. These steps are shown in the worked example below, which also shows you how to write the overall equation for the reaction.

In complex redox half equations, balance oxygens with H_2O, hydrogens with H^+ and charge with electrons.

Worked example 12.1.4

WRITING EQUATIONS FOR COMPLEX REDOX REACTIONS

Permanganate ions (MnO_4^-) undergo reduction to Mn^{2+} in acidic solution.

Write:

a the half-equation for the reduction of MnO_4^- to Mn^{2+}.

b an overall ionic equation for the reaction in which MnO_4^- oxidises Fe^{2+} to Fe^{3+} in acidic solution.

Thinking	Working
Balance all atoms in the half-equation except oxygen and hydrogen.	$MnO_4^- \rightarrow Mn^{2+}$
Balance the oxygen atoms by adding water. (As the reaction is in aqueous solution, oxygen is present as water.)	$MnO_4^- \rightarrow Mn^{2+} + 4H_2O$
Balance the hydrogen atoms by adding H^+ ions (which are present in the acidic solution).	$MnO_4^- + 8H^+ \rightarrow Mn^{2+} + 4H_2O$
Balance the charges on both sides of the equation by adding electrons to the more positive side. Add states.	$MnO_4^-(aq) + 8H^+(aq) + 5e^- \rightarrow Mn^{2+}(aq) + 4H_2O(l)$
Add the oxidation half-equation to the reduction half-equation to obtain the overall equation, making sure that the number of electrons used in reduction equals the number of electrons released during oxidation.	$MnO_4^-(aq) + 8H^+(aq) + 5e^- \rightarrow Mn^{2+}(aq) + 4H_2O(l)$ $5 \times [Fe^{2+}(aq) \rightarrow Fe^{3+}(aq) + e^-]$ $MnO_4^-(aq) + 8H^+(aq) + 5Fe^{2+}(aq) \rightarrow Mn^{2+}(aq) + 4H_2O(l) + 5Fe^{3+}(aq)$

Worked example: Try yourself 12.1.4

WRITING EQUATIONS FOR COMPLEX REDOX REACTIONS

Potassium dichromate ($K_2Cr_2O_7$) reacts with potassium iodide (KI) in acidified solution. The iodide ion (I^-) is oxidised to I_2 according to the half-equation:

$$2I^-(aq) \rightarrow I_2(aq) + 2e^-$$

Dichromate ions ($Cr_2O_7^{2-}$) undergo reduction to Cr^{3+}.

Write:

a the half-equation for the reduction of $Cr_2O_7^{2-}(aq)$ to $Cr^{3+}(aq)$ in acidic solution.

b an overall ionic equation for the reaction.

(The potassium ions are spectator ions and do not appear in the overall equation.)

12.1 Review

OA ✓✓

SUMMARY

- Redox (*red*-uction; *ox*-idation) reactions involve the transfer of electrons from one species to another.
- Oxidation and reduction always occur at the same time.
- Half-equations are used to represent oxidation and reduction.
- Oxidation is defined as the loss of electrons, e.g. $Mg(s) \rightarrow Mg^{2+}(aq) + 2e^-$.
- Reduction is defined as the gain of electrons, e.g. $Br_2(aq) + 2e^- \rightarrow 2Br^-(aq)$.
- The reducing agent donates electrons to another substance, causing that substance to be reduced. The reducing agent is itself oxidised.
- Metals can act as reducing agents.
- The oxidising agent accepts electrons from another substance, causing that substance to be oxidised. The oxidising agent is itself reduced.
- Half-equations are added together to determine the overall redox equation. It may be necessary to multiply one or both half-equations by a factor to balance the electrons.
- In a redox reaction, a conjugate redox pair is made up of a reducing agent and the oxidising agent that is formed, or an oxidising agent and the reducing agent that is formed. The conjugate redox pair is usually written as oxidising agent/reducing agent.
- More complex redox half-equations involving acidic solutions of strong oxidising agents, such as $KMnO_4$ and $K_2Cr_2O_7$, can be written by following a series of steps.

KEY QUESTIONS

Knowledge and understanding

1 Identify each of the following half-equations as involving either oxidation or reduction.
  - **a** $Sn^{2+}(aq) + 2e^- \rightarrow Sn(s)$
  - **b** $O_2(g) + 4e^- \rightarrow 2O^{2-}(s)$
  - **c** $S^{2-}(aq) \rightarrow S(s) + 2e^-$
  - **d** $Zn(s) \rightarrow Zn^{2+}(aq) + 2e^-$

2 Balance the following half-equations and then identify each as an oxidation or a reduction reaction.
  - **a** $Fe(s) \rightarrow Fe^{2+}(aq)$
  - **b** $K(s) \rightarrow K^+(aq)$
  - **c** $F_2(g) \rightarrow F^-(aq)$
  - **d** $O_2(g) \rightarrow O^{2-}(s)$

3 Write half-equations to represent the following reactions in acidic solution:
  - **a** reduction of $SO_4^{2-}(aq)$ to $SO_2(g)$
  - **b** oxidation of $H_2O_2(aq)$ to $O_2(g)$
  - **c** oxidation of $H_2S(g)$ to $S(s)$
  - **d** reduction of $MnO_4^-(aq)$ to $MnO_2(s)$

Analysis

4 Iron reacts with hydrochloric acid according to the ionic equation:

$$Fe(s) + 2H^+(aq) \rightarrow Fe^{2+}(aq) + H_2(g)$$

  - **a** What has been oxidised in this reaction? What is the product?
  - **b** Write a half-equation for the oxidation reaction.
  - **c** Identify the oxidising agent.
  - **d** What has been reduced in this reaction? What is the product?
  - **e** Write a half-equation for the reduction reaction.
  - **f** Identify the reducing agent.
  - **g** Identify the two conjugate redox pairs in this reaction.

5 When a strip of magnesium metal is placed in a blue solution containing copper(II) ions ($Cu^{2+}(aq)$), crystals of copper metal appear and the solution becomes paler in colour.
  - **a** Show that this reaction is a redox reaction by identifying the substance that is oxidised and the one that is reduced.
  - **b** Write a half-equation for the oxidation reaction.
  - **c** Write a half-equation for the reduction reaction.
  - **d** Write an overall redox equation.
  - **e** Identify the oxidising agent and the reducing agent.
  - **f** Explain why the solution loses some of its blue colour as a result of the reaction.

6 Some ions, such as the Sn^{2+} ion, can be either oxidised or reduced. Tin is commonly found as Sn metal, and the ions Sn^{2+} and Sn^{4+}.

   a Write the formula for the product of the oxidation of the Sn^{2+} ion.

   b Write the formula for the product of the reduction of the Sn^{2+} ion.

7 Potassium metal that is exposed to the air forms an oxide coating.

   a What is the formula of potassium oxide?

   b What has been oxidised in this reaction?

   c Write a half-equation for the oxidation reaction.

   d What has been reduced in this reaction?

   e Write a balanced half-equation for the reduction reaction.

   f Write an overall equation for this redox reaction.

   g Copy the following statement and fill in the blank spaces with the appropriate words.

   Potassium has been ____________ by ____________ to form potassium ions. The ____________ has gained electrons from the ____________. The oxygen has been ____________ by ____________ to form oxide ions. The ______________ has lost electrons to the ____________.

12.2 Metal displacement reactions

In Chapter 4, you learnt about the reactivity of metals with water, oxygen and acids. In these reactions, the water, oxygen and acids are all behaving as oxidising agents. The redox reaction between a metal and an oxidising agent depends on the ability of the metal to lose electrons and that of the oxidising agent to gain electrons. Some metals react quickly and vigorously, while the reaction between other metals with oxygen is quite slow. Copper pipes oxidise slowly in the air, producing a brown-black copper oxide coating, while the reaction of magnesium ribbon in acid is more vigorous.

This section will examine the reactivity of different metals and describe how to write equations for reactions between metals and solutions of another type of oxidising agent, metal ions.

REACTIVITY OF METALS

Sodium, magnesium and iron are metals that are relatively easily oxidised. Sodium is oxidised so readily that it has to be stored under paraffin oil to prevent it from reacting with oxygen in the atmosphere. The oxidation of iron, which can eventually result in the formation of rust, can be an expensive problem.

Other metals do not corrode as readily. For example, platinum and gold are sufficiently inert to be found as pure elements in nature.

In Chapter 4 you saw that the metals can be ranked in a list according to their reactivity (or ability to act as reducing agents). Figure 12.2.1 shows such a ranking, which is also known as a **reactivity series of metals**. The series shows the reduction half-equations for the metal cations as each cation gains electrons to form the corresponding metal.

The metals are listed on the right-hand side of the series, from the least reactive (Au) at the top to the most reactive (K and Li) at the bottom. The lower down the table a metal is placed, the more reactive it is. Remember that the most reactive metals are those that are oxidised most easily.

Metals, with their small number of valence electrons, generally act as reducing agents. A relatively small amount of energy is required to remove these valence electrons. In general, the lower the amount of energy required to remove the valence electrons, the more readily a metal will act as a reducing agent.

strongest oxidising agent

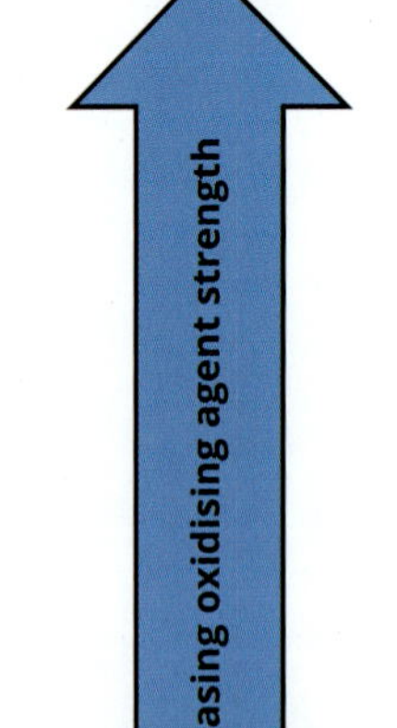

weakest oxidising agent

$Au^{3+}(aq) + 3e^- \rightleftharpoons Au(s)$
$Ag^+(aq) + e^- \rightleftharpoons Ag(s)$
$Cu^{2+}(aq) + 2e^- \rightleftharpoons Cu(s)$
$Pb^{2+}(aq) + 2e^- \rightleftharpoons Pb(s)$
$Sn^{2+}(aq) + 2e^- \rightleftharpoons Sn(s)$
$Ni^{2+}(aq) + 2e^- \rightleftharpoons Ni(s)$
$Co^{2+}(aq) + 2e^- \rightleftharpoons Co(s)$
$Fe^{2+}(aq) + 2e^- \rightleftharpoons Fe(s)$
$Cr^{3+}(aq) + 3e^- \rightleftharpoons Cr(s)$
$Zn^{2+}(aq) + 2e^- \rightleftharpoons Zn(s)$
$Mn^{2+}(aq) + 2e^- \rightleftharpoons Mn(s)$
$Al^{3+}(aq) + 3e^- \rightleftharpoons Al(s)$
$Mg^{2+}(aq) + 2e^- \rightleftharpoons Mg(s)$
$Na^+(aq) + e^- \rightleftharpoons Na(s)$
$Ca^{2+}(aq) + 2e^- \rightleftharpoons Ca(s)$
$K^+(aq) + e^- \rightleftharpoons K(s)$
$Li^+(aq) + e^- \rightleftharpoons Li(s)$

weakest reducing agent

increasing reducing agent strength

strongest reducing agent

FIGURE 12.2.1 Reactivity series of metals

CHEMFILE

Use of gold in communication satellites

Gold is one of the most unreactive of all the metals. This lack of reactivity makes it ideal for use in the advanced technologies used by the aerospace and satellite telecommunications industries. Gold-coated plastic sheets are used as a radiation shield in space satellites (see figure below) and in astronauts' space suits. Gold reflects both infrared and ultraviolet radiation, which can damage the delicate instruments in a satellite. Because gold is softer and more malleable than most other metals, it is easier to work with. It also requires less maintenance, because it will not react with oxygen. Microprocessors in the satellite are made of gold because it is an excellent conductor of electricity. An ultra-thin layer of gold is sprayed onto the mirrors used in space telescopes to improve the mirror's reflective properties.

A gold-coated radiation shield is installed on a section of a satellite.

As you go down the reactivity series of metals in Figure 12.2.1:

- the metals, which are on the right-hand side, become more reactive. This means the metals lower in the series lose electrons more easily and are therefore stronger reducing agents
- the metal cations, which are on the left-hand side, become increasingly harder to reduce and are therefore weaker oxidising agents.

> The strongest reducing agents are found at the bottom on the right-hand side of the reactivity series.
> The strongest oxidising agents are found at the top on the left-hand side of the reactivity series.

CASE STUDY

Discovery of metals through the ages

Since the dawn of civilisation, our progress has been dependent on the discovery and development of materials. The epochs of civilisation are commonly described in terms of metals, e.g. the Bronze Age and the Iron Age.

The first metals discovered were low in reactivity and were found as pure elements, such as gold, in the environment. In later times, samples of copper and tin, which had been reduced from their ores, were found in the ashes of campfires. The mixture of copper and tin that produced the hard **alloy**, bronze, initiated the Bronze Age. Bronze could be beaten into shapes for tools and weapons that held a sharper edge than stone and were easily resharpened once blunt. The Trojan War was conducted by bronze-shielded warriors throwing bronze-tipped spears.

Lead and iron, which are more reactive metals, required higher temperatures to extract them from their ores, and efficient production methods were invented much later. Wood was inadequate as a fuel to attain high enough temperatures, so it was the availability of manufactured charcoal which led to the cheap manufacture of iron and weapons that could be produced at a fraction of the cost of those made from bronze.

FIGURE 12.2.2 A 20 franc aluminium coin from 1857

FIGURE 12.2.3 A row of electrolytic cells used for the modern production of aluminium

Highly reactive metals, such as aluminium and sodium, are so easily oxidised and have such stable cations that the invention of electricity in the late 1800s was needed to extract them from their mineral ores. It was so difficult to extract aluminium metal that Napoleon III proudly displayed a small bar of the metal with his crown jewels in 1855. He also had a rattle made of aluminium for his son, and in 1857, 20 franc coins were made of aluminium as well as from gold (see Figure 12.2.2).

Modern production methods use vast quantities of electricity to produce these highly reactive metals from molten minerals. Consequently, these metals tend to be expensive to produce. Aluminium is produced using devices called electrolytic cells, shown in Figure 12.2.3. The great expense of producing aluminium from its ore, alumina, makes recycling of any aluminium product an absolute necessity.

A metal ion higher in the reactivity series (an oxidising agent) will react with a metal lower in the reactivity series (a reducing agent).

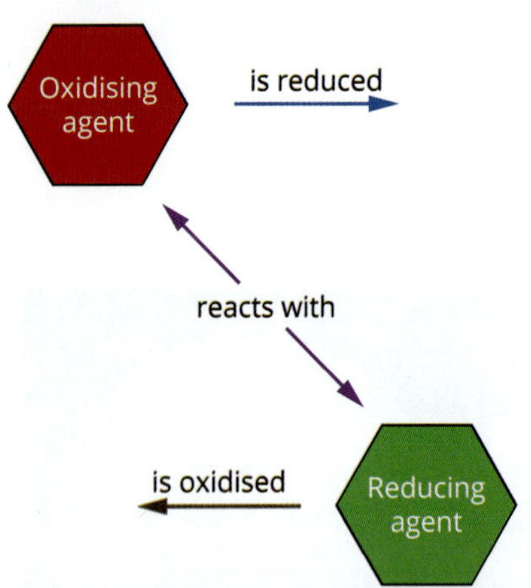

FIGURE 12.2.4 Predicting the reaction between an oxidising agent and a reducing agent

FIGURE 12.2.5 When a strip of copper wire is suspended in a solution of silver nitrate in the flask on the left, long crystals of silver metal start to form. In the flask on the right, the copper has displaced the silver from the solution.

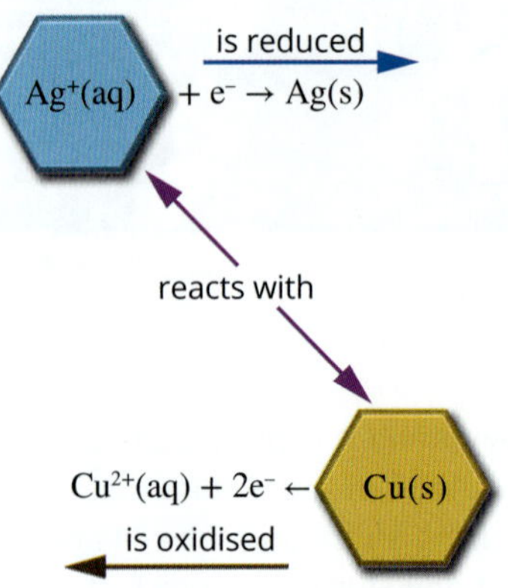

FIGURE 12.2.6 A spontaneous redox reaction occurs when copper is added to a solution of silver nitrate. Silver ions oxidise copper atoms.

METAL DISPLACEMENT REACTIONS

The order in which metals and their metal ions appear in the reactivity series enables us to predict which metals will **displace** other metals from solutions of their ions. Such reactions are known as **metal displacement reactions**.

A more reactive metal will be oxidised by, and donate its electrons to, the cation of a less reactive metal. The cation receives the electrons and is reduced. In other words, for a naturally occurring **spontaneous redox reaction** to occur, the metal ions of one metal must be above the other metal in the reactivity series, as shown in Figure 12.2.4. The more reactive metal acts as the reducing agent, and the metal ions of the other metal act as the oxidising agent.

When a strip of copper wire is placed in a solution of silver nitrate, as shown in Figure 12.2.5, silver ions are reduced to silver atoms by copper atoms. The silver atoms are deposited as silver crystals. The copper atoms are oxidised to form a blue solution containing copper(II) ions. As a result of the reaction, copper(II) ions have displaced silver ions from the solution.

Silver ions have oxidised copper atoms, consistent with their order in the reactivity series, as shown in Figure 12.2.6. The overall redox reaction can be represented by the equation:

$$Cu(s) + 2Ag^+(aq) \rightarrow Cu^{2+}(aq) + 2Ag(s)$$

According to the reactivity series, a metal displacement reaction is predicted to occur when zinc is added to copper(II) sulfate solution, as shown in Figure 12.2.7.

FIGURE 12.2.7 A brown deposit of copper metal is observed forming on the zinc and the blue copper(II) sulfate solution gradually becomes colourless as the concentration of Cu^{2+} ions decreases.

Worked example 12.2.1

PREDICTING METAL DISPLACEMENT REACTIONS

Using the reactivity series in Figure 12.2.1, on page 398, predict whether zinc will displace copper from a solution containing copper(II) ions and, if appropriate, write the overall equation for the reaction.

Thinking	Working
Locate the metal and the metal ions in the reactivity series.	Metals (reducing agents) are found on the right-hand side of the reactivity series and metal ions (oxidising agents) are on the left-hand side of the reactivity series. $Cu^{2+}(aq) + 2e^- \rightleftharpoons Cu(s)$ $Pb^{2+}(aq) + 2e^- \rightleftharpoons Pb(s)$ $Sn^{2+}(aq) + 2e^- \rightleftharpoons Sn(s)$ $Ni^{2+}(aq) + 2e^- \rightleftharpoons Ni(s)$ $Co^{2+}(aq) + 2e^- \rightleftharpoons Co(s)$ $Fe^{2+}(aq) + 2e^- \rightleftharpoons Fe(s)$ $Cr^{3+}(aq) + 3e^- \rightleftharpoons Cr(s)$ $Zn^{2+}(aq) + 2e^- \rightleftharpoons Zn(s)$
Determine whether the metal is below (and to the right of) the metal ion in Figure 12.2.1. If this is the case, there will be a reaction.	You can see from the reactivity series that Zn is on the right-hand side because it is a reducing agent and it is below Cu^{2+}, so there will be a reaction, as shown in Figure 12.2.7.
Write the reduction reaction for the metal ion directly as it is written in the reactivity series, including states, but with the arrow now only one way.	$Cu^{2+}(aq) + 2e^- \rightarrow Cu(s)$
Write the oxidation reaction for the metal from the reactivity series, writing the metal on the left-hand side of the arrow (as a reactant). Include states and draw a one-way arrow.	$Zn(s) \rightarrow Zn^{2+}(aq) + 2e^-$
Combine the two half-equations, balancing electrons, to give the overall equation for the reaction.	$Zn(s) \rightarrow Zn^{2+}(aq) + 2e^-$ $Cu^{2+}(aq) + 2e^- \rightarrow Cu(s)$ $Zn(s) + Cu^{2+}(aq) \rightarrow Zn^{2+}(aq) + Cu(s)$

Worked example: Try yourself 12.2.1

PREDICTING METAL DISPLACEMENT REACTIONS

Using the reactivity series in Figure 12.2.1, on page 398, predict whether cobalt will displace copper from a solution containing copper(II) ions and, if appropriate, write the overall equation for the reaction.

12.2 Review

OA ✓✓

SUMMARY

- The reactivity series lists half-equations involving metals and their corresponding cations.
- The half-equations involving stronger oxidising agents (the ones more easily reduced) appear higher in the reactivity series than those involving weaker oxidising agents.
- The half-equations involving stronger reducing agents (the ones more easily oxidised) appear lower in the reactivity series.
- The reactivity series can be used to predict whether a redox reaction is likely to occur.
- Metal displacement reactions involve the transfer of electrons from a more reactive metal to the positive ions of a less reactive metal in solution.

KEY QUESTIONS

Knowledge and understanding

1 In each of the following groups of metals, use the reactivity series in Figure 12.2.1 (page 398) to identify the strongest reducing agent.

a Cu, Al, Cr
b Zn, K, Pb
c Mg, Ni, Mn

2 In each of the following groups of metal ions, use the reactivity series in Figure 12.2.1 (page 398) to identify the strongest oxidising agent.

a Na^+, Ni^{2+}, Mn^{2+}
b Co^{2+}, Cu^{2+}, Ca^{2+}
c Fe^{2+}, Pb^{2+}, Ag^+

3 Use the reactivity series to:

a classify each of the following species as an oxidising agent or a reducing agent
b order the groups of oxidising agents and reducing agents from weakest to strongest.
Cr^{3+}, Zn, Mg, Al^{3+}, Sn^{2+}, Ag, Cu, Na^+, Pb^{2+}, Ni, Fe^{2+}, Au^+, Mn

Analysis

4 Refer to the reactivity series and predict whether the following reactions will spontaneously occur.

a Tin metal is placed in a copper(II) nitrate solution.
b A strip of aluminium is placed in a sodium chloride solution.
c Magnesium is added to a solution of iron(II) sulfate.
d The element zinc is placed in a tin(II) sulfate solution.
e A piece of tin is placed in a silver nitrate solution.

5 Use the reactivity series to predict whether a reaction will occur in each of the following situations. Write an overall equation for each reaction that you predict will occur.

a Copper(II) sulfate solution is stored in an aluminium container.
b Sodium chloride solution is stored in a copper container.
c Silver nitrate solution is stored in a zinc container.

6 Solutions of zinc nitrate and iron(II) nitrate have been prepared in a laboratory, but have accidentally been left unlabelled. Name one metal that could be used to identify which solution is which and explain why that metal could be used.

12.3 Redox reactions in society

Redox reactions are very important to our everyday lives. Many of us are highly reliant on our mobile phones, laptop computers, calculators and other mobile electronic devices. Without the redox reactions in rechargeable batteries, which generate electricity for these devices, we would be doomed to stay plugged into the mains electricity to make these devices work. In this section, you will see how the redox reactions you have already learnt about are used to generate electricity in simple **galvanic cells**.

SIMPLE GALVANIC CELLS

Heat energy is released when a spontaneous reaction occurs between a metal and a non-metal. The surrounding solution becomes warm and sometimes even hot. This type of reaction, called an **exothermic reaction**, is typical of spontaneous redox reactions. Up to this point, redox reactions have been described as electron transfer reactions, without concrete evidence of the movement of electrons from the reducing agent to the oxidising agent. In this section, you will see how the flow of electrons between the two reactants in a redox reaction can be used to operate electrical devices, thus providing evidence of the movement of electrons.

Evidence for electron transfer

When zinc metal is placed in a solution of copper(II) sulfate, the zinc is oxidised and a brown deposit of copper metal forms:

Oxidation of zinc	$Zn(s) \rightarrow Zn^{2+}(aq) + 2e^-$
Reduction of copper(II) ions	$Cu^{2+}(aq) + 2e^- \rightarrow Cu(s)$
Overall equation	$Zn(s) + Cu^{2+}(aq) \rightarrow Zn^{2+}(aq) + Cu(s)$

Electrons flow from the zinc atoms to the copper(II) ions as these collide. In the direct reaction, heat is released. If the reactants are separated using the equipment shown in Figure 12.3.1, we can gain evidence for the flow of electrons.

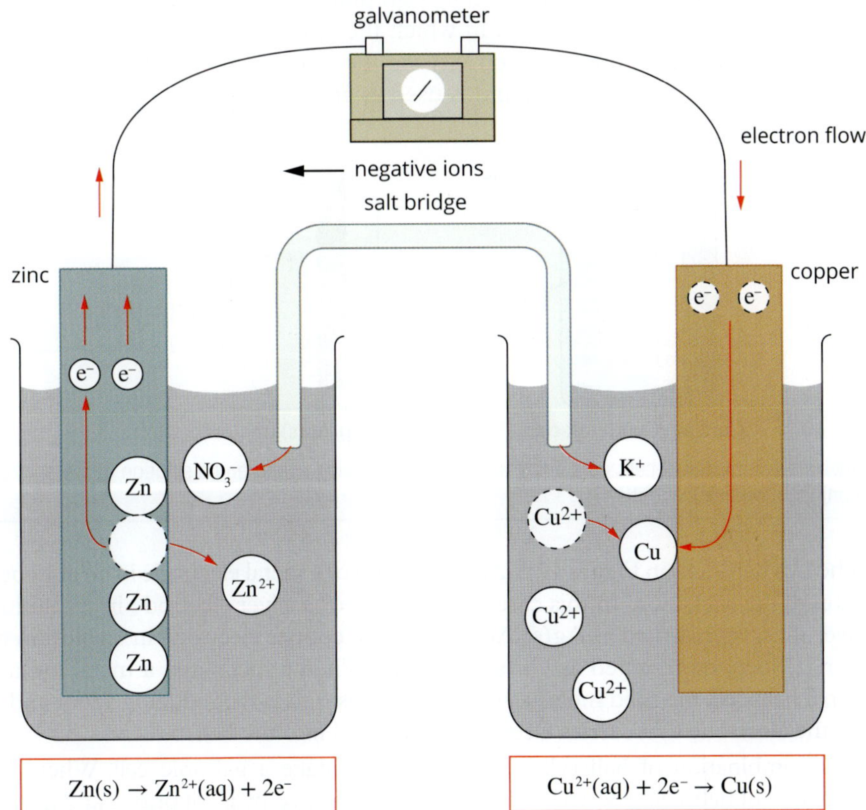

FIGURE 12.3.1 The equipment used to demonstrate electron flow during the reaction between zinc and copper(II) sulfate

The zinc metal and the solution containing copper(II) ions are in separate beakers. This prevents a spontaneous, heat-releasing reaction from occurring. Instead, the electrons are forced to travel through the **external circuit**, the wire, in order to reach the oxidising agent. A copper strip, dipping in the solution of copper(II) ions, is connected to the zinc strip and a **galvanometer**—a meter for detecting a flow of electrons—is also in the circuit. The solutions in the two beakers are connected by a **salt bridge**. The salt bridge contains a solution of an ionic compound, such as potassium nitrate, KNO_3, which will not react with either solution. The salt bridge may be as simple as a piece of filter paper soaked in the potassium nitrate solution. The flow of charge, in the form of electrons and ions can be seen in Figure 12.3.1. The electrons move through the wire from the zinc electrode to the copper electrode. The ions move through the solutions, with negative ions travelling in the same direction as the electrons, in this case the clockwise direction, and positive ions travelling in the opposite direction.

The salt bridge completes the circuit in a galvanic cell by allowing ions to move from one half-cell to the other.

This apparatus is known as a galvanic cell, also known as a **voltaic cell,** which is a type of **electrochemical cell**. A galvanic cell converts chemical energy into electrical energy and is one of the most important devices in modern life. It is a simple form of what we call a battery. When the galvanic cell in Figure 12.3.1 operates, the positive reading on the galvanometer indicates that electrons are flowing from the zinc strip to the copper strip.

This observation provides evidence that the same oxidation and reduction reactions are occurring when the two reactants (zinc and copper(II) ions) are separated as when they are put together in a beaker. It also supports the idea that in redox reactions there is a transfer of electrons.

Oxidation and reduction in galvanic cells

All galvanic cells are composed of two **half-cells**. Two typical half-cells are shown in Figure 12.3.2. Each half cell contains a reducing agent and its conjugate oxidising agent. Oxidation occurs in one half-cell and reduction occurs in the other.

A half-cell must contain an **electrode** and an **electrolyte**. The electrode is a solid electrical conductor, such as a metal or graphite rod. It provides a link between the external circuit and the solution in which the reaction occurs. The electrolyte is a solution which can conduct electricity via the movement of charged particles, ions. A solution of an ionic compound is used as an electrolyte.

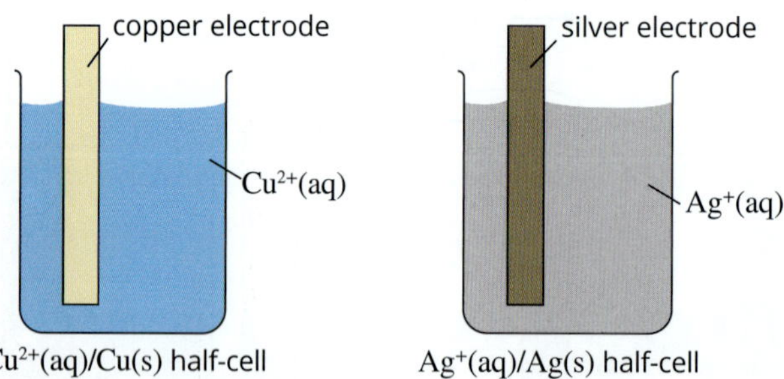

FIGURE 12.3.2 Typical half-cells containing a metal (the reducing agent) and a solution of metal ions (the conjugate oxidising agent)

The two half-cells in Figure 12.3.2 are made up of a metal electrode in an aqueous solution of the metal ion. In this case, they consist of a silver electrode in a solution of silver ions, represented as $Ag(s)/Ag^{+}(aq)$, and a copper electrode in a solution of copper(II) ions, represented as $Cu(s)/Cu^{2+}(aq)$. When a metal and a metal ion are used in a half-cell, the solid metal serves two purposes. It is the reducing agent and it is also the electrode that carries electrons out of and into the half-cell.

Any combination of half-cells can be used to make a galvanic cell. When the combination of half-cells shown in Figure 12.3.2 are joined by a salt bridge and a wire, which makes the external circuit, the galvanic cell in Figure 12.3.3 is created. In this case, the external circuit leads to a light globe, which the flow of electrons can light up.

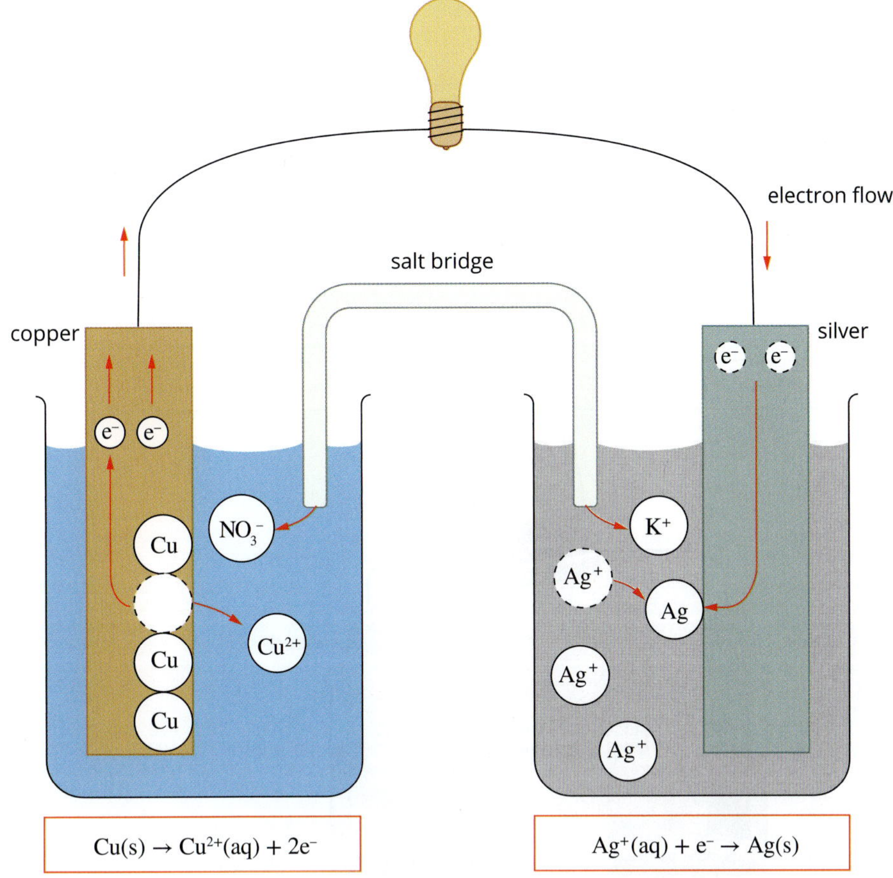

FIGURE 12.3.3 A galvanic cell made up of a $Cu(s)/Cu^{2+}(aq)$ half-cell and a $Ag(s)/Ag^+(aq)$ half-cell

Using the reactivity series in Figure 12.2.1 on page 398 you can see that copper is a stronger reducing agent than silver. This means that copper will be oxidised in the half-cell on the left-hand side. The silver ions are a stronger oxidising agent than the copper(II) ions, so $Ag^+(aq)$ is reduced in the half-cell on the right-hand side. The equations for the half-cell reactions occurring in the two half-cells are:

$$Cu(s) \rightarrow Cu^{2+}(aq) + 2e^-$$

and

$$Ag^+(aq) + e^- \rightarrow Ag(s)$$

Because electrons are leaving the copper electrode to move into the external circuit, the copper electrode is known as the negative electrode—the source of electrons. Electrons then flow through the wire from the negative electrode to the positive electrode, the silver electrode. The silver metal is not able to gain electrons, but it does conduct the electrons to the Ag^+ solution, where reduction of the $Ag^+(aq)$ occurs.

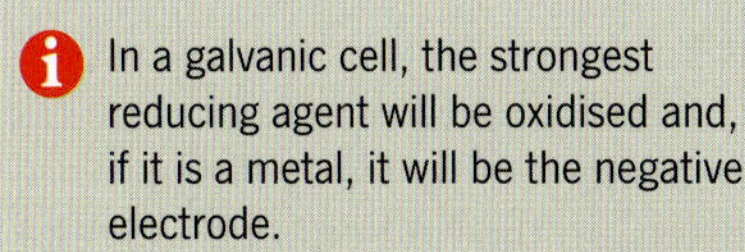

> In a galvanic cell, the strongest reducing agent will be oxidised and, if it is a metal, it will be the negative electrode.

The electrode at which oxidation occurs is called the **anode**, while the electrode at which reduction occurs is called the **cathode** (see Figure 12.3.4). It is important to remember that the names anode and cathode relate to the type of reaction occurring at each electrode, rather than the charge on the electrode. However, you will see that in a galvanic cell, the anode is negative and the cathode is positive. In Figure 12.3.3, the copper electrode is the anode and the silver electrode is the cathode.

> Oxidation occurs at the anode and reduction occurs at the cathode.

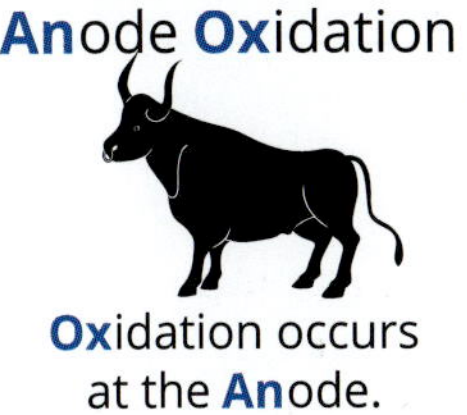

Reduction Cathode

Reduction occurs
at the Cathode.

FIGURE 12.3.4 A useful memory aide to help you remember that oxidation occurs at the anode and reduction occurs at the cathode

Worked example 12.3.1

LABELLING A SIMPLE GALVANIC CELL

A galvanic cell is set up using a $Zn(s)/Zn^{2+}(aq)$ half-cell and a $Pb(s)/Pb^{2+}(aq)$ half-cell. Sketch this galvanic cell and use the reactivity series in Figure 12.2.1 on page 398 to predict:

a the reducing agent and the oxidising agent in the galvanic cell

b the reactions occurring at each half-cell

c the direction of flow of electrons through the external circuit

d the negative and positive electrodes

e the anode and the cathode

Thinking	Working
When sketching a galvanic cell, always draw two half-cells, each containing an electrode and an ionic solution, with a wire between the two electrodes and a salt bridge between the two beakers. A galvanometer or light globe can be drawn in the external circuit.	zinc lead salt bridge $Zn^{2+}(aq)$ $Pb^{2+}(aq)$
Locate the two metals on the right-hand side of the reactivity series and determine which one is the stronger reducing agent (closer to the bottom).	$Cu^{2+}(aq) + 2e^- \rightleftharpoons Cu(s)$ $Pb^{2+}(aq) + 2e^- \rightleftharpoons Pb(s)$ $Sn^{2+}(aq) + 2e^- \rightleftharpoons Sn(s)$ $Ni^{2+}(aq) + 2e^- \rightleftharpoons Ni(s)$ $Co^{2+}(aq) + 2e^- \rightleftharpoons Co(s)$ $Fe^{2+}(aq) + 2e^- \rightleftharpoons Fe(s)$ $Cr^{3+}(aq) + 3e^- \rightleftharpoons Cr(s)$ $Zn^{2+}(aq) + 2e^- \rightleftharpoons Zn(s)$ The two circles show that zinc, Zn, is closer to the bottom of the reactivity series, so Zn is the stronger reducing agent.
The stronger reducing agent is oxidised, so the half-equation can be copied from the reactivity series, but must be written with the reactant on the left-hand side of the equation.	$Zn^{2+}(aq) + 2e^- \rightleftharpoons Zn(s)$ is reversed to give the oxidation half-equation: $Zn(s) \rightarrow Zn^{2+}(aq) + 2e^-$ (note the one-way arrow now)
The oxidising agent for the galvanic cell will be the positive metal ion in the other half-cell, so the reduction half-equation can be copied exactly from the reactivity series.	Pb^{2+} is the oxidising agent (conjugate oxidising agent in the other half-cell) Reduction half-equation: $Pb^{2+}(aq) + 2e^- \rightarrow Pb(s)$

The electrons will flow from the electrode, which is the reducing agent, through the external circuit, to the other electrode. Add an arrow to the diagram with the label 'electron flow' to show this.	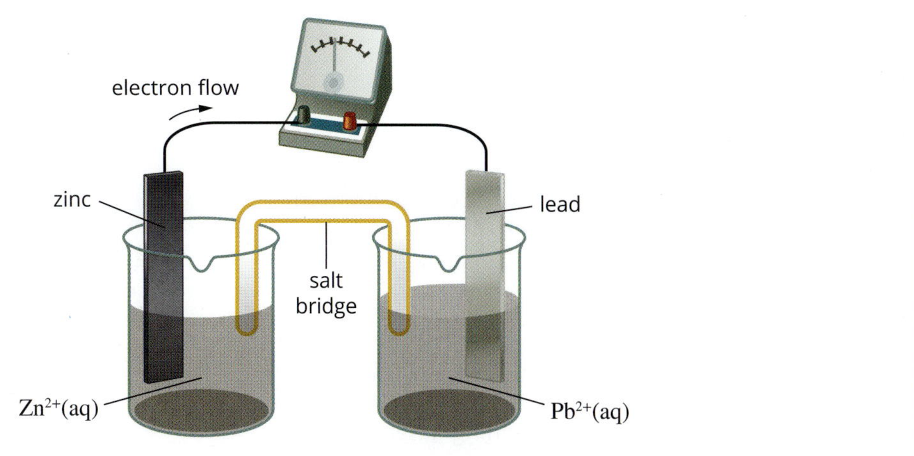
Add the labels 'negative electrode' to the electrode from which the electrons are leaving and 'positive electrode' to the other electrode.	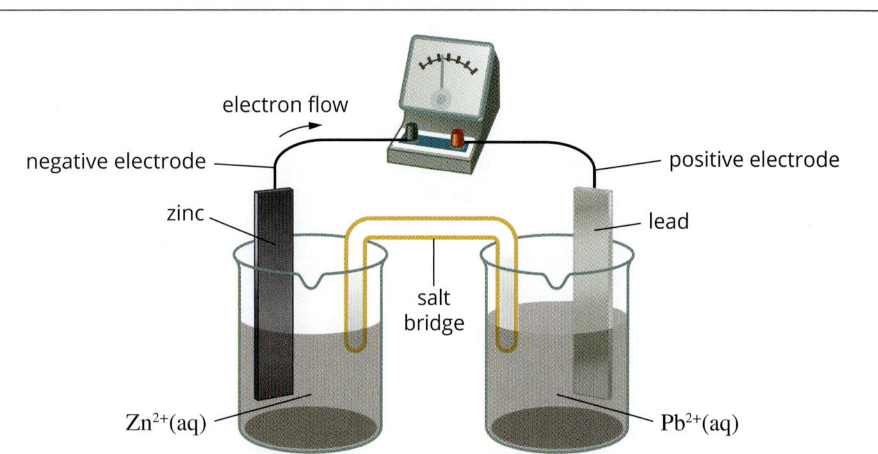
Add the label 'anode' to the electrode where oxidation is occurring and add the label 'cathode' to the electrode where reduction is occurring.	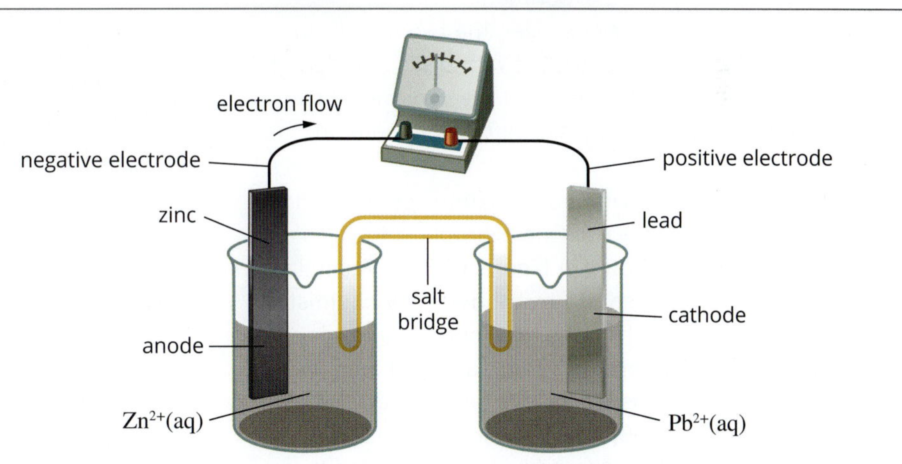

Worked example: Try yourself 12.3.1

LABELLING A SIMPLE GALVANIC CELL

A galvanic cell is set up using a $Cu(s)/Cu^{2+}(aq)$ half-cell and a $Ni(s)/Ni^{2+}(aq)$ half-cell. Sketch this galvanic cell and use the reactivity series in Figure 12.2.1 on page 398 to predict:

a the reducing agent and the oxidising agent in the galvanic cell

b the reactions occurring at each half-cell

c the direction of flow of electrons through the external circuit

d the negative and positive electrodes

e the anode and the cathode.

CASE STUDY **ANALYSIS**

The earliest batteries

Luigi Galvani (1737–1798) was Professor of Anatomy at the University of Bologna, Italy. His interest in the anatomy of frogs and the effects of electricity on muscles led to the discovery of a flow of electricity between two different metals. Galvani had been experimenting extensively with frogs under various conditions. He is credited with making the muscles of a skinned frog contract by touching a nerve with a pair of scissors during an electrical storm. Most famously, he observed the twitching of a frog's legs when he pressed a copper hook into the frog's spinal cord while it was hanging on an iron railing. In 1791, when he published his findings about the movement of electricity through the frog, he concluded that the animal tissue contained a new form of electricity called 'animal electricity', which activated the nerves and muscles when metal probes were introduced on either side of the muscles.

While Galvani's findings were accepted by many, Alessandro Volta (1745–1827) who was Professor of Physics at the University of Pavia, Italy, did not accept the idea of 'animal electricity'. Instead, he proposed that the frog in Galvani's experiments was just conducting the electricity between the two metals. In 1792, Volta began to experiment with disks of different metals, and no frogs! He detected the very weak flow of electricity between the disks by placing them on his tongue.

Volta experimented with combinations of electrodes connected in various ways. One example is the 'crown of cups', as shown in Figure 12.3.5b. A similar arrangement is often set up by modern-day chemistry students to increase the voltage of simple galvanic cells made in the laboratory. Volta favoured zinc and copper, or silver, as his electrodes, and the electrolyte was saltwater. By 1800, he had built the first battery by alternating disks of zinc and silver, separated by paper or cloth soaked in saltwater. This was known as a voltaic pile. Volta received great accolades for his work from Napoleon, I as well as from the Austrian emperor, Francis I. Most notable of these honours was that the unit of electromotive force, the volt, was named in his honour in 1881.

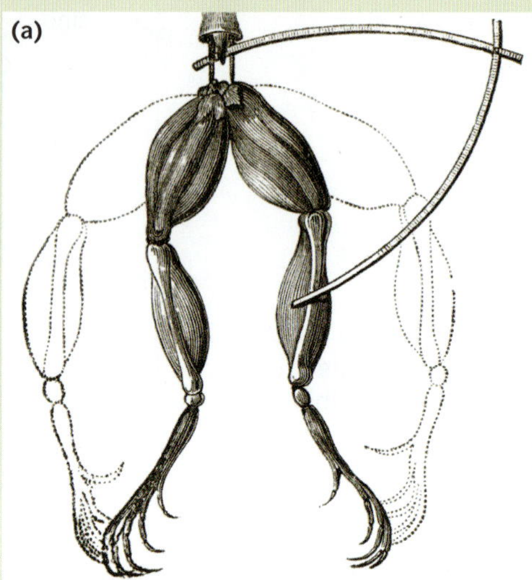

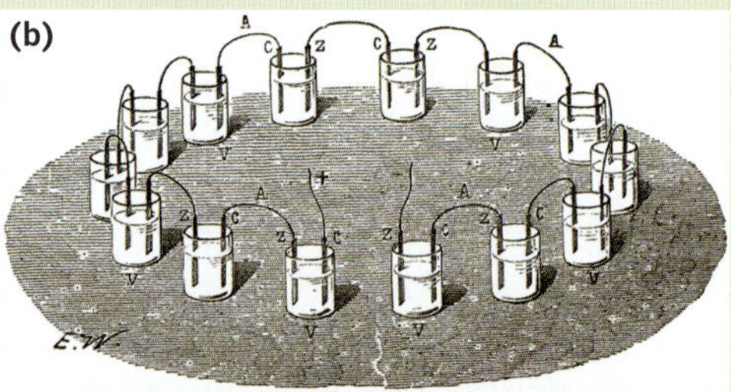

FIGURE 12.3.5 (a) In 1791 Luigi Galvani observed that a frog's legs twitched due to the flow of electricity between copper and iron. (b) A 'crown of cups' battery constructed by Alessandro Volta in 1800. This is a series of containers (half-cells) holding saltwater and electrodes, consisting of plates of zinc (Z) and copper (C). Metal hoops (A) connect the cells in series.

Analysis

1 When Galvani joined the copper hook to the iron rail with the skinned frog, electricity passed through the frog. Predict which metal would be the anode and which would be the cathode in this situation.

2 The historical reports do not give much detail about the twitching of the frog's legs in Galvani's experiment. Suggest a method that could be used to measure the twitching of the frog's legs and so use this as a dependent variable.

3 If you were repeating Galvani's frog on the rail experiment as an investigation, with your answer to question **2** as the dependent variable, identify:
  a the independent variable
  b three controlled variables.

Galvanic cells in society

The portable power source that is essential for our everyday modern lifestyle, the **battery**, is based on a simple galvanic cell. While a battery is strictly defined as a series of cells connected together, like Volta's 'crown of cups' in the case study above, many portable power sources, which we may refer to as batteries, are actually cells. They are made up of two electrodes and an electrolyte. A cell that is disposable and designed not to be recharged is called a **primary cell** (see Figure 12.3.6).

FIGURE 12.3.6 Primary cells are used as a portable power source. Once the chemicals in the cell have reacted fully, the cell goes flat and is discarded.

The simplest primary cell which became widely available during the 20th century was known as a **dry cell**. It was based on a cell developed in 1866 by French scientist George Leclanché. If you use a primary cell in a device such as a wireless computer mouse or a child's toy, you are more likely to be using an improvement on the dry cell, called an **alkaline cell**. The difference between a dry cell and an alkaline cell relates to the electrolyte in the cell. While the electrolyte in a dry cell is acidic, the electrolyte in an alkaline cell is, predictably, alkaline.

Another improvement on the dry cell is the ability for these cells to be recharged. A rechargeable cell that is designed to be reused many times is called a **secondary cell**. Many of the latest developments in technology, such as electric vehicles and state-of-the-art power storage, are based on the improvement in battery technology that has occurred over recent years. Most of this has been focused on secondary cells. Batteries, in one form or another, are widely seen as the key to transitioning away from fossil fuel dependence towards making better use of renewable energy sources (see Figure 12.3.7). Secondary cells and their wide applications are discussed more fully in the Unit 3 Chemistry course.

FIGURE 12.3.7 New battery technologies play a major part in making better use of renewable energy sources.

CHEMFILE

Why you should recycle batteries

In July 2019 the Victorian government banned all e-waste from landfill, meaning that used batteries should not be thrown into general waste or your home recycling bin. While not breaking the law can be a strong motive for some people, many would like a more compelling reason to go to the trouble of recycling batteries.

So, what are the reasons for recycling batteries? First of all, they contain many different metals, all of which are non-renewable resources. Furthermore, if the batteries break apart in landfill, there is the chance that toxic metals, such as lead and mercury, will leach into the soil or waterways.

Some of the metals used in rechargeable batteries, such as cobalt, are mined in parts of the world where poverty makes people desperate, and corruption is common. These metals must be recycled to help minimise the human suffering created by the mining of these metals.

Separate bins for recycling batteries and other electronic devices can be found at your local tip.

12.3 Review

OA ✓✓

SUMMARY

- A simple galvanic cell is constructed from two half-cells, each containing a reducing agent and its conjugate oxidising agent.
- In a galvanic cell, a spontaneous redox reaction occurs, but the reactants are separated so the electrons travel through the external circuit and this electric current can be used to power portable electronic devices.
- The negative electrode in a galvanic cell is the source of electrons. It is where the oxidation reaction is occurring, so it is also called the anode.
- The positive electrode in a galvanic cell is where the electrons travel to. It is where the reduction reaction is occurring, so it is also called the cathode.
- Redox reactions play a major part in society in their application to cells and batteries, which provide portable sources of electricity.

KEY QUESTIONS

Knowledge and understanding

1 Explain why the electrodes in a half-cell must be made of metal or graphite.

2 Describe the purpose of a salt bridge in a galvanic cell.

3 Draw a labelled diagram of a galvanic cell constructed from an iron, Fe, electrode in a solution of Fe^{2+}(aq) and a lead, Pb, electrode in a solution of Pb^{2+}(aq). The labels should include:
   - **a** the half-equations for the oxidation and the reduction reactions
   - **b** the negative and the positive electrodes
   - **c** the direction of flow of electrons through the external circuit
   - **d** the anode and the cathode
   - **e** the direction of flow of positive and negative ions through the salt bridge.

4 Describe the difference between a cell and a battery.

Analysis

5 Use the reactivity series in Figure 12.2.1 on page 398 to identify the negative electrode in each of the following galvanic cells.
   - **a** Zn(s)/Zn^{2+}(aq) connected to Fe(s)/Fe^{2+}(aq)
   - **b** Cu(s)/Cu^{2+}(aq) connected to Cr(s)/Cr^{3+}(aq)
   - **c** Ag(s)/Ag^{+}(aq) connected to Fe(s)/Fe^{2+}(aq)

6 Explain how primary cells are useful to society and why recycling plays a major part in our ongoing use of these cells.

7 The overall equation for the redox reaction between silver ions and nickel metal is:

$$2Ag^{+}(aq) + Ni(s) \rightarrow 2Ag(s) + Ni^{2+}(aq)$$

Sketch a suitable galvanic cell to demonstrate that there is a flow of electrons between the reactants. Fully label the cell and show the direction of electron flow through the external circuit.

Chapter review

12

OA ✓✓

KEY TERMS

alkaline cell
alloy
anode
battery
cathode
combustion
conjugate redox pair
displace
dry cell
electrochemical cell
electrode
electrolyte
exothermic reaction
external circuit
galvanic cells
galvanometer
half-cell
half-equation
metal displacement reaction
oxidant
oxidation
oxidised
oxidising agent
primary cell
reactivity series of metals
redox reaction
reduced
reducing agent
reductant
reduction
salt bridge
secondary cell
spontaneous redox reaction
voltaic cell

REVIEW QUESTIONS

Knowledge and understanding

1 Define oxidation and reduction in terms of the transfer of:
  a oxygen
  b electrons

2 Copy and complete the following statements relating to the reactivity series.
  The reactivity series arranges metals (and other elements) in order of their ability to react with ________ agents. The most reactive metals are found at the ________ of the reactivity series, on the ________ -hand side. For a redox reaction to occur, an oxidising agent must react with a ________ agent. In this reaction, the oxidising agent must be ________ in the reactivity series than the conjugate ________ agent of the reducing agent in the reaction. During the reaction, the oxidising agent ________ electrons and the reducing agent ________ electrons.

3 Identify the following half-equations as involving either oxidation or reduction.
  a $Ni(s) \rightarrow Ni^{2+}(aq) + 2e^-$
  b $Fe^{2+}(aq) + 2e^- \rightarrow Fe(s)$
  c $Ag^+(aq) + e^- \rightarrow Ag(s)$
  d $Cu(s) \rightarrow Cu^{2+}(aq) + 2e^-$

4 Which one of the following alternatives describes what happens when zinc and oxygen react?
  A Each zinc atom gains two electrons.
  B Each zinc atom loses two electrons.
  C Each oxygen atom gains one electron.
  D Each oxygen atom loses one electron.

5 Balance these half-equations:
  a $Ag^+(aq) \rightarrow Ag(s)$
  b $Cu(s) \rightarrow Cu^{2+}(aq)$
  c $Zn(s) \rightarrow Zn^{2+}(aq)$

6 Using the reactivity series in Figure 12.2.1 on page 398, order the following metals in order of decreasing reactivity (highest to lowest): magnesium, iron, zinc, tin, aluminium, silver.

7 Which one of the following redox equations has not been balanced correctly?
  A $Ni^{2+}(aq) + Zn(s) \rightarrow Ni(s) + Zn^{2+}(aq)$
  B $Fe^{2+}(aq) + Mg(s) \rightarrow Fe(s) + Mg^{2+}(aq)$
  C $Ag^+(aq) + Ni(s) \rightarrow Ag(s) + Ni^{2+}(aq)$
  D $Cu^{2+}(aq) + Mg(s) \rightarrow Cu(s) + Mg^{2+}(aq)$

8 Magnesium reacts with copper(II) ions according to the following equation:
$$Mg(s) + Cu^{2+}(aq) \rightarrow Mg^{2+}(aq) + Cu(s)$$
  Which one of the following is the correct set of conjugate redox pairs for this reaction?
  A $Mg^{2+}(aq)/Mg(s)$ and $Cu^{2+}(aq)/Cu(s)$
  B $Cu(s)/Mg^{2+}(aq)$ and $Cu^{2+}(aq)/Mg(s)$
  C $Mg(s)/Cu(s)$ and $Cu^{2+}(aq)/Mg^{2+}(aq)$
  D $Mg(s)/Cu^{2+}(aq)$ and $Cu(s)/Mg^{2+}(aq)$

9 Draw a labelled diagram of a galvanic cell constructed from a magnesium electrode in a solution of $Mg^{2+}(aq)$ and a nickel electrode in a solution of $Ni^{2+}(aq)$. The labels should include:
  a the half-equations for the oxidation and the reduction reactions
  b the negative and the positive electrodes
  c the flow of electrons through the external circuit and the ions through the salt bridge
  d the anode and the cathode.

Application and analysis

10 Consider the following half-equations and the overall equation for the reaction between metallic lead and a solution of silver ions.

$$Pb(s) \rightarrow Pb^{2+}(aq) + 2e^-$$
$$Ag^+(aq) + e^- \rightarrow Ag(s)$$
$$\overline{Pb(s) + 2Ag^+(aq) \rightarrow Pb^{2+}(aq) + 2Ag(s)}$$

Which one of the following sets of statements correctly describes this redox reaction?

- **A** Pb(s) is the oxidising agent, $Ag^+(aq)$ is the reducing agent, lead metal is reduced.
- **B** Pb(s) is the reducing agent, $Ag^+(aq)$ is the oxidising agent, lead metal is reduced.
- **C** $Pb^{2+}(aq)$ is the oxidising agent, Ag(s) is the reducing agent, silver metal is oxidised.
- **D** Pb(s) is the reducing agent, $Ag^+(aq)$ is the oxidising agent, lead metal is oxidised.

11 Construct a concept map that shows the links between the following terms:
electrons, anode, cathode, oxidation, reduction, galvanic cell, half-cell, negative, positive

12 The following equations form part of the reactivity series. In the reactivity series, they are ranked in the order shown.

$Ag^+(aq) + e^- \rightleftharpoons Ag(s)$
$Pb^{2+}(aq) + 2e^- \rightleftharpoons Pb(s)$
$Fe^{2+}(aq) + 2e^- \rightleftharpoons Fe(s)$
$Zn^{2+}(aq) + 2e^- \rightleftharpoons Zn(s)$
$Mg^{2+}(aq) + 2e^- \rightleftharpoons Mg(s)$

- **a** Which species is the strongest oxidising agent and which is the weakest oxidising agent?
- **b** Which species is the strongest reducing agent and which is the weakest reducing agent?
- **c** Lead rods are placed in solutions of silver nitrate, iron(II) sulfate and magnesium chloride. In which solutions would you expect to see a coating of another metal form on the lead rod? Explain.
- **d** Which of the metals silver, zinc or magnesium might be coated with lead when immersed in a solution of lead(II) nitrate?

13 Predict whether the following mixtures would result in spontaneous reactions. Write an overall equation for each reaction that you predict will occur.

- **a** Zinc metal is added to a solution of silver nitrate.
- **b** Copper metal is placed in an aluminium chloride solution.
- **c** Tin(II) sulfate is placed in a copper container.
- **d** Magnesium metal is added to a solution of lead nitrate.
- **e** Silver metal is added to nickel chloride solution.
- **f** copper(II) nitrate solution is added to a iron container.

14 Iron nails are placed into 1 M solutions of $CuSO_4$, $MgCl_2$, $Pb(NO_3)_2$ and $ZnCl_2$. In which solutions would you expect a coating of another metal to appear on the nail? Explain your answer.

15 An unknown metal is placed in solutions of aluminium nitrate and iron(II) sulfate. After a period of time, the metal is found to have reacted with the iron(II) sulfate solution, but not the aluminium nitrate solution. Suggest the name of the unknown metal.

16 For each of the following uses, state which type of cell or battery might be the most commonly used in society currently and explain your answer. The types of cells or batteries to choose from are: alkaline dry cell, rechargeable cell, rechargeable battery

- **a** a mobile phone
- **b** a wireless mouse
- **c** an electric car
- **d** an energy storage unit for a home solar power system

17 Write half-equations to represent the following reactions in acidic solution:

- **a** reduction of $Ta_2O_5(s)$ to Ta(s)
- **b** oxidation of $SO_3^{2-}(aq)$ to $SO_4^{2-}(aq)$
- **c** reduction of $IO_3^-(aq)$ to $I^-(aq)$.

18 Balance the following redox equations by separating them into two half-equations, balancing each equation, and then combining the pair into a balanced overall equation.

- **a** $H_2O_2(aq) + PbS(s) \rightarrow PbSO_4(s) + H_2O(l)$
- **b** $I_2(aq) + H_2S(g) \rightarrow I^-(aq) + S(s) + H^+(aq)$
- **c** $SO_3^{2-}(aq) + MnO_4^-(aq) + H^+(aq) \rightarrow SO_4^{2-}(aq) + Mn^{2+}(aq) + H_2O(l)$
- **d** $NO(g) + Cr_2O_7^{2-}(aq) + H^+(aq) \rightarrow NO_3^-(aq) + Cr^{3+}(aq) + H_2O(l)$

19 You are given three colourless solutions (A, B and C) known to be potassium nitrate, silver nitrate and tin(II) nitrate, but not necessarily in this order. You also have some pieces of magnesium ribbon and copper sheet. Describe the experiment that you would carry out to identify each of the solutions using only the chemicals supplied.

20 Consider the galvanic cell in the figure below. Explain why it will not generate electricity and describe how it could be changed so that electricity is produced.

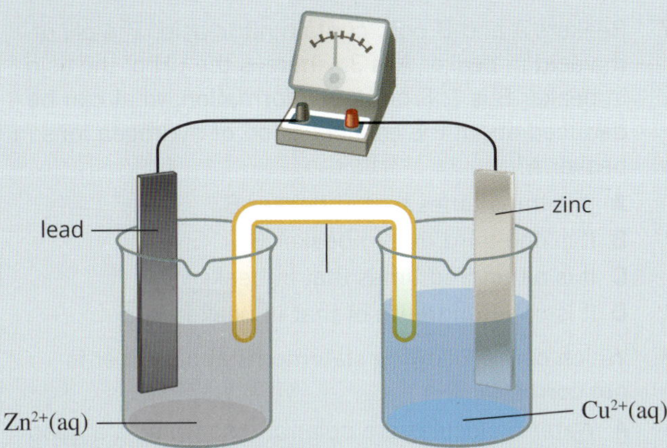

21 Consider the following information relating to metals, W, X, Y and Z, and solutions of their salts, $W(NO_3)_2$, $X(NO_3)_2$, $Y(NO_3)_2$ and $Z(NO_3)_2$.
Four experiments are carried out with these metals to determine their order of reactivity. Equations representing what happened in these experiments are listed below.
Experiment 1: $W(s) + X(NO_3)_2(aq) \rightarrow W(NO_3)_2(aq) + X(s)$
Experiment 2: $Z(NO_3)_2(aq) + X(s) \rightarrow$ no reaction
Experiment 3: $W(s) + Z(NO_3)_2(aq) \rightarrow W(NO_3)_2(aq) + Z(s)$
Experiment 4: $Y(s) + W(NO_3)_2(aq) \rightarrow Y(NO_3)_2(aq) + W(s)$
Determine the order of reactivity of the four metals, from least reactive to most reactive.

22 Cobalt is a metal that is a major component of electric vehicle batteries. It has been predicted that the demand for cobalt will exceed its supply by 2030. Most of this metal is mined in the Democratic Republic of Congo.
Recycling, becomes even more important than ever. Consider the figure below, which shows the contrast in the demand for cobalt and lithium with and without recycling, and answer the following questions.

a If as much as possible is recycled, which metal would have its supplies prolonged by the greater amount?

b A single lithium-ion electric vehicle (EV) battery pack contains more than 14 kg of cobalt. The scale on the figure below indicates that 20 million tonnes (20×10^9 kg) of cobalt would be saved by using future battery technology, while recycling as much as possible. Calculate how many EV battery packs are represented by this saving.

c Suggest ways in which students at your school could be encouraged to recycle batteries and mobile phones.

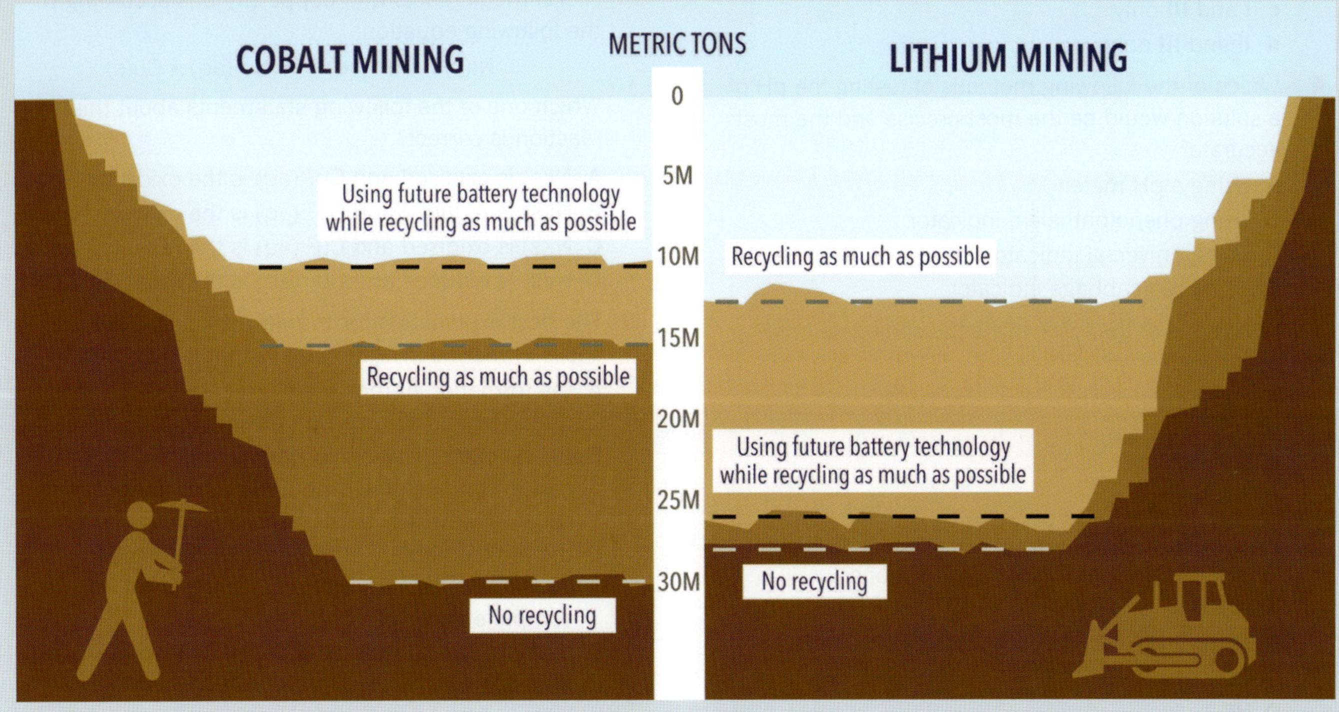

Mined metals needed for batteries by 2050

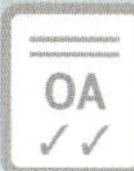

UNIT 2 • Area of Study 1

REVIEW QUESTIONS

How do chemicals interact with water?

Multiple-choice questions

1 Which one of the following represents a concentrated solution of a weak acid?

A 6.0 M CH_3COOH
B 0.01 M CH_3COOH
C 6.0 M HCl
D 0.01 M HCl

2 Which of the following acids can be classified as polyprotic in water?

I CH_3COOH
II H_2SO_4
III NH_4^+

a **I** only
b **II** only
c **I** and **II** only
d **I**, **II** and **III**

3 Which of the following statements about a reducing agent is/are correct?

I A reducing agent is reduced in a reaction
II A reducing agent causes another substance to be reduced
III A reducing agent donates electrons to another substance

a **I** only
b **II** only
c **I** and **III** only
d **II** and **III** only

4 Which of the following methods of testing the pH of a solution would be the most precise and the most accurate?

A Using a pH meter
B Using phenolphthalein indicator
C Using universal indicator
D Using red cabbage indicator

5 Which one of the following is the conjugate base of OH^-?

A H_2O
B O^{2-}
C H_3O^+
D NaOH

6 Which one of the following can act as both a Brønsted–Lowry acid or base in aqueous solutions?

A HS^-
B NH_4^+
C CO_3^{2-}
D BF_3

7 Beakers A and B both contain nitric acid. The pH of the acid in beaker A is 3, whereas the pH of the acid in beaker B is 1. From this information, what can be deduced about the concentration of hydrogen ions in beaker A?

A It is three times that in beaker B
B It is one-third that in beaker B
C It is a hundred times that in beaker B
D It is one-hundredth of that in beaker B

8 Which of the following statements about water is not correct?

A Water molecules are polar and so water is a good solvent.
B Water has a high latent heat value and so it is a good coolant.
C Water expands on freezing and so lakes are covered with a layer of ice in very cold climates.
D Water has a low specific heat capacity and so large bodies of water moderate temperatures on Earth.

9 Which of the species below is the strongest reducing agent?

A Ag
B Ag^+
C Mg
D Zn

10 Nickel metal reacts with copper(II) ions according to the following equation:

$$Ni(s) + Cu^{2+}(aq) \longrightarrow Ni^{2+}(aq) + Cu(s)$$

Which one of the following statements about this reaction is correct?

A Ni(s) is oxidised and Cu^{2+}(aq) is the oxidising agent.
B Ni(s) is reduced and Cu^{2+}(aq) is the oxidising agent.
C Ni(s) is oxidised and Cu^{2+}(aq) is the reducing agent.
D Ni(s) is reduced and Cu^{2+}(aq) is the reducing agent.

11 The best explanation for the decalcification of the shells of marine invertebrates is that it is caused by:

A the formation of more calcium carbonate due to the dissolving of more carbon dioxide in the oceans
B the reaction between carbonate ions and increased concentrations of hydronium ions in the oceans causing the shells to dissolve
C the reduced ability of marine invertebrates to digest calcium carbonate and successfully create a complete shell
D the reduced number of sharks in the oceans due to over-fishing

12 Which one of the following is not a redox reaction?

A $Fe(s) + Sn^{2+}(aq) \longrightarrow Fe^{2+}(aq) + Sn(s)$

B $Mg(s) + O_2(g) \longrightarrow 2MgO(s)$

C $2Na(s) + 2H_2O(l) \longrightarrow H_2(g) + 2NaOH(aq)$

D $NaOH(aq) + HCl(aq) \longrightarrow NaCl(aq) + H_2O(l)$

13 When the following equation for the reaction between permanganate ions, MnO_4^- and silver metal, Ag, is correctly balanced, what is the coefficient for Ag(s)?

$$\_MnO_4^-(aq) + \_H^+(aq) + \_Ag(s) \longrightarrow \_Mn^{2+}(aq) + \_H_2O(l) + \_Ag^+(aq)$$

A 1

B 2

C 4

D 5

14 Which of the following statements would correctly describe a simple galvanic cell constructed from a $Zn(s)/Zn^{2+}(aq)$ half cell and a $Ag(s)/Ag^+(aq)$ half cell?

A The anode would be silver and the cathode would be zinc.

B The negative electrode would be zinc and it would be the cathode.

C The positive electrode would be silver and this is where oxidation would occur.

D The anode would be zinc and this is where oxidation would occur.

15 Metal X was added to a solution of lead nitrate ($Pb(NO_3)_2$). A reaction occurred and a precipitate of lead was produced. On the basis of this result, which one of the following can be deduced?

A Lead metal must be more reactive than metal X.

B A solution containing ions of metal X will not react with lead metal.

C Metal X must be able to react with a solution of $Mg(NO_3)_2$ to produce a precipitate of Mg.

D Lead metal must be able to react with a solution of $X(NO_3)_2$ to produce a precipitate of X.

Short-answer questions

16 a Rearrange the following list in increasing order of pH (lowest to highest).

distilled water; lemon juice; 1.0 M hydrochloric acid; 1.0 M sodium hydroxide

b Explain your choice of the substance with the lowest pH and the one with the highest pH (i.e. the first and last substances in your list).

17 The following graph shows the boiling points of the group 16 hydrides.

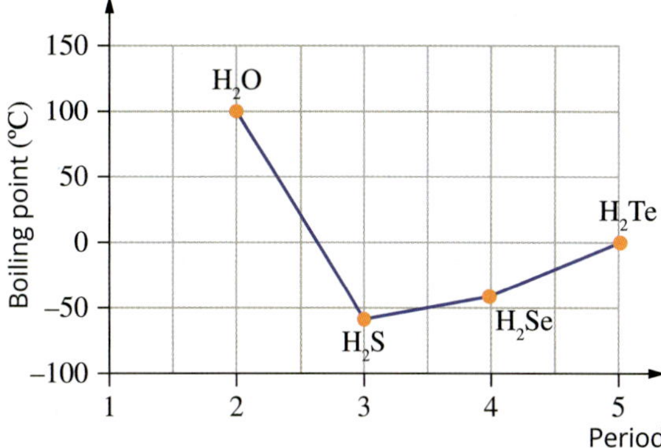

Explain the following.

a The boiling point of H_2S is much lower than that of H_2O.

b The boiling point of H_2S is also lower than that of H_2Se.

18 Water is present on Earth in three states—solid, liquid and gas.

a Explain why a large amount of energy is needed to change liquid water into gaseous water, using the correct term for this energy in your answer.

b Explain why the large amount of energy needed to change liquid water into gaseous water is useful for the regulation of the temperature of the oceans around the world.

19 The specific heat capacity of some common substances are listed in the table below.

Substance	Specific heat capacity (J g^{-1} °C^{-1})
water	4.18
ethanol	2.4
copper	0.39

a Calculate the amount of heat energy, in J, required to increase the temperature of 250.0 g of water from 18°C to 100°C.

b How much energy would be required to raise the temperature of an equal mass of copper by the same amount?

c If the same amount of energy as in part **b** is used to heat 250.0 g ethanol, what would the temperature rise of the ethanol be?

d Explain why water has a high specific heat capacity.

e Give one use for water based on its high specific heat capacity.

20 Since the mid-1700s, the pH of the world's oceans has changed from 8.2 to 8.1. Explain why this change has occurred, including in your answer an identification of which substance has caused this change.

21 **a** Write balanced full chemical equations to represent each of the following reactions.
- **i** a solution of hydrochloric acid reacts with zinc powder
- **ii** a solution of nitric acid reacts with a solution of calcium hydroxide
- **iii** a solution of sulfuric acid reacts with solid sodium carbonate
- **iv** a piece of solid copper reacts with a solution of silver nitrate

b Write ionic equations for the reaction between:
- **i** a solution of sulfuric acid and zinc metal
- **ii** a solution of sodium hydroxide and a solution of nitric acid
- **iii** a solution of hydrochloric acid and a solution of magnesium carbonate
- **iv** magnesium metal and copper(II) nitrate solution

22 Calculate the pH of each of the following at 25°C.
- **a** 0.0010 M HCl
- **b** 0.0025 M NaOH
- **c** 0.00075 M HNO_3
- **d** 0.015 M $Ca(OH)_2$

23 **a** Define the term 'strong acid'.

b Give the formula of a substance that is polyprotic.

c HCO_3^- is an amphiprotic ion. Write chemical equations to show it acting in water as:
- **i** a base
- **ii** an acid

d Give the formula of:
- **i** the conjugate acid of H_2O
- **ii** the conjugate base of H_2O

24 Justify the statement 'seawater with a pH of 7.85 is 100% more acidic than seawater with a pH of 8.15' by calculating the H_3O^+ concentration of both solutions and comparing them

25 Consider the following statements, which compare 20.00 mL of a 0.10 mol L^{-1} solution of hydrochloric acid with 20.00 mL of a 0.10 mol L^{-1} solution of ethanoic acid.
- **I** Both solutions are of the same strength.
- **II** The pH of the hydrochloric acid solution will be higher.
- **III** The ethanoic acid will react less vigorously with solid calcium carbonate than will the hydrochloric acid.
- **IV** Both solutions will require the same volume of 0.10 mol L^{-1} NaOH for neutralisation.

a Explain why statements **III** and **IV** are correct.

b Explain why statements **I** and **II** are incorrect.

26 As the concentration of ions in a solution increases, the electrical conductivity of the solution increases.

An experiment compared the conductivity of four solutions, A, B, C and D. Each solution had the same concentration and the pH was measured using a pH meter. The results are shown in the table below.

	pH	Litmus paper test	Conductivity (relative units)
A	?	turns blue litmus paper red	3
B	1.5	turns blue litmus paper red	6
C	7.0	does not change the colour of red or blue litmus	0
D	10.5	turns red litmus paper blue	3

a When considering the pH measurements and litmus paper observations given for solutions A to D, which results are likely to be the least precise?

b Which solution (A, B, C or D) is most likely to be an aqueous sodium hydroxide solution? Give a reason for your choice.

c Which solution (A, B, C or D) contains the strongest acid? Explain your answer.

d Give an estimation for the pH of solution A. Give a reason for your choice.

e Determine the hydrogen ion concentration in:
- **i** solution B
- **ii** solution D

f Explain why solution C does not conduct electricity.

27 A galvanic cell was constructed from an $Fe(s)/Fe^{2+}(aq)$ half-cell connected to a $Zn(s)/Zn^{2+}(aq)$ half-cell. Sketch this galvanic cell and use the reactivity series in Figure 12.2.1 on page 398 to predict:
- **a** the reducing agent and the oxidising agent in the galvanic cell
- **b** the reactions occurring at each half-cell
- **c** the direction of flow of electrons through the external circuit
- **d** the negative and positive electrodes
- **e** the anode and the cathode

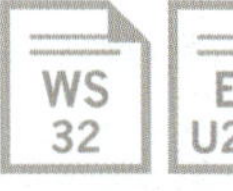

CHAPTER

13 Measuring solubility and concentration

Aqueous solutions are found all around you. The Earth's oceans, rivers and lakes are aqueous solutions. They contain dissolved minerals and gases. The plasma of human blood and the sap of plants are aqueous solutions carrying dissolved nutrients and wastes. Each body cell contains aqueous solutions. Even rain contains small quantities of dissolved gases and other materials.

In Chapter 6 you learnt how different types of compounds dissolve in water in different ways. You also learnt that some kinds of compounds are very soluble in water and others are insoluble.

Solubility is the term we use to describe how much of a compound dissolves in a particular amount of solvent.

The focus of this chapter is on the solubility of compounds in water and on the units we use to measure it.

Key knowledge

- solution concentration as a measure of the quantity of solute dissolved in a given mass or volume of solution (mol L^{-1}, g L^{-1}, %(m/v), %(v/v), ppm), including unit conversions **13.2**
- the use of solubility tables and solubility graphs to predict experimental determination of ionic compound solubility; the effect of temperature on the solubility of a given solid, liquid or gas in water **13.1**
- the use of precipitation reactions to remove impurities from water. **13.1**

WS 33

OA ✓✓

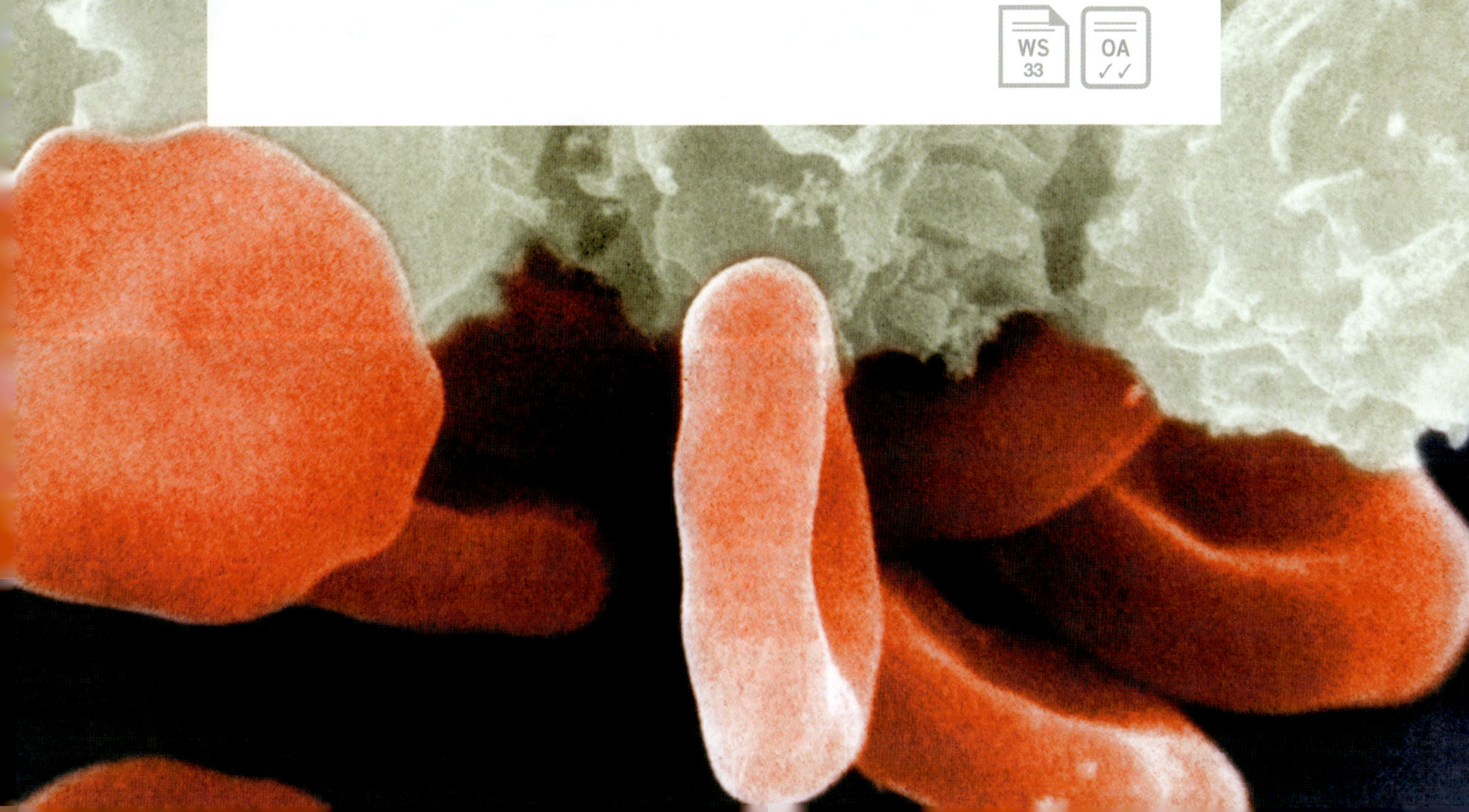

13.1 Measuring solubility

In fields such as medicine, pharmaceutical manufacturing, and even food preparation, it is vital to know how much of a compound is present in a solution.

In this section, you will learn how to determine how much of a substance will dissolve in a given amount of water at a particular temperature. You will also learn how to calculate the amount of a compound dissolved in a given solution.

DEFINITION OF SOLUBILITY

The **solubility** of a substance is a measure of how much of the substance will dissolve in a given amount of a solvent. For example, glucose is very soluble in water and is used by your body as a readily available energy source. Most rocks are made of minerals that are insoluble in water, but **limestone** (which contains calcium carbonate) is slightly soluble and so caves form in areas of limestone rock over long periods. One consequence can be the formation of sinkholes, like that shown in Figure 13.1.1.

FIGURE 13.1.1 The Umpherston sinkhole in Mount Gambier, South Australia, was formed when the roof of a limestone cave collapsed.

In chemistry, the term solubility has a specific meaning. It refers to the maximum amount of a solute that can be dissolved in a given quantity of a solvent at a certain temperature (the common quantity often used is 100 g). Table 13.1.1 gives the solubility of some common substances in 100 g of water at 18°C.

TABLE 13.1.1 Solubility of solutes at 18°C

Solute	Solubility (g per 100 g of water)
sugar (sucrose)	200
salt (sodium chloride)	35
limestone (calcium carbonate)	0.0013

> The solubility of a compound refers to the maximum amount of solute that can be dissolved in a given quantity of solvent at a particular temperature.

DIFFERENT KINDS OF SOLUTIONS

There are three terms that can be used to describe the different solutions that result from dissolving solutes in solvents. They are:

- saturated
- unsaturated
- supersaturated.

A **saturated solution** is one in which no more solute can be dissolved at a particular temperature.

An **unsaturated solution** contains less solute than is needed to make the solution saturated. Unsaturated solutions can dissolve more solute.

A **supersaturated solution** is an unstable solution that contains more dissolved solute than a saturated solution. If this type of solution is disturbed, some of the solute will separate from the solvent as a solid.

Figure 13.1.2 shows a supersaturated solution of sodium ethanoate. Supersaturated sodium ethanoate is prepared by cooling a saturated solution very carefully so that solid crystals do not form. Adding a small **seed crystal** to the supersaturated solution causes the solute to **crystallise** (form solid crystals) so that a saturated solution remains.

FIGURE 13.1.2 Crystals of sodium ethanoate form after a seed crystal is added to a supersaturated solution of the compound.

SOLUBILITY CURVES

The solubility of many substances changes significantly as the temperature changes. For example, you can dissolve more chocolate powder in hot milk than in cold milk. For most solids, increasing the temperature increases the solubility in a liquid. This is because at higher temperatures, both the solute and solvent have more energy to overcome the forces of attraction that hold the particles together in the solid.

The relationship between solubility and temperature can be represented by a **solubility curve**, as shown in Figure 13.1.3.

Solubility curves show the solubility of a substance as a function of temperature. For the solubility curves featured in Figure 13.1.3, note that:

- each point on a curve represents the maximum amount of the solute that can be dissolved in 100 g of water at a particular temperature. Therefore, each point on the curve represents a saturated solution
- any point below a curve represents an unsaturated solution
- any point above a curve represents a supersaturated solution
- for most solids, as temperature increases, the solubility increases.

ℹ For the solubility curves shown in Figure 13.1.3 each point on the curve represents the maximum amount of solute that can be dissolved in 100 g of water at a particular temperature.

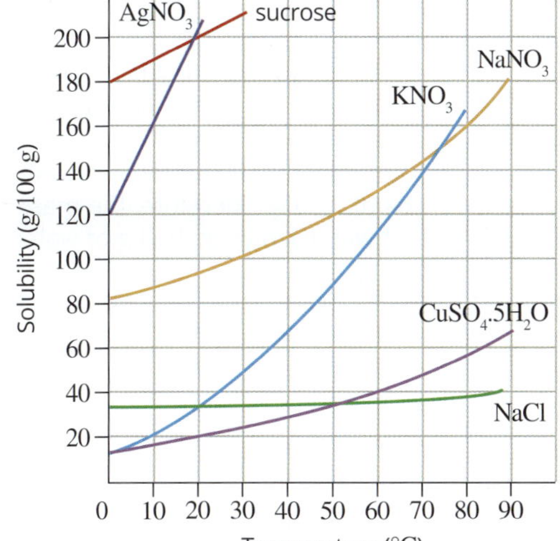

FIGURE 13.1.3 Solubility curves for some common substances

ℹ For most solids, solubility in water increases as the temperature increases.

Worked example 13.1.1

SOLUBILITY CURVE CALCULATIONS

Use Figure 13.1.3 on page 419 to find how many grams of sodium nitrate ($NaNO_3$) will dissolve in 100 g of water at 50°C.

Thinking	Working
Draw a vertical line from the required temperature on the horizontal axis to intersect with the appropriate solubility curve.	Draw a vertical line from 50°C on the horizontal axis to intersect with the solubility curve for $NaNO_3$. 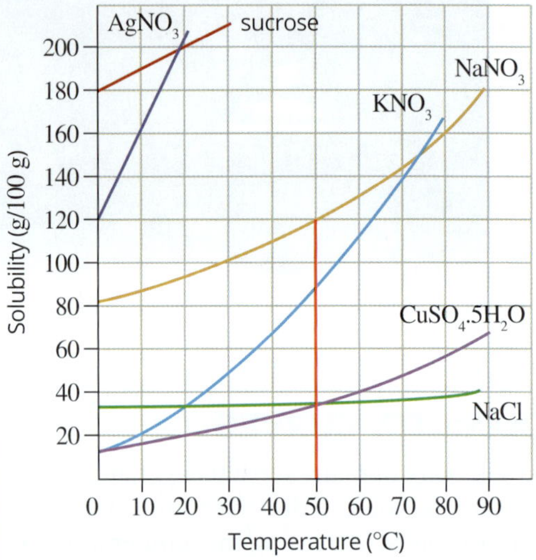
Draw a horizontal line from the intersection point of the vertical line drawn in the previous step. The point where this horizontal line intersects with the vertical axis will give the mass of dissolved substance.	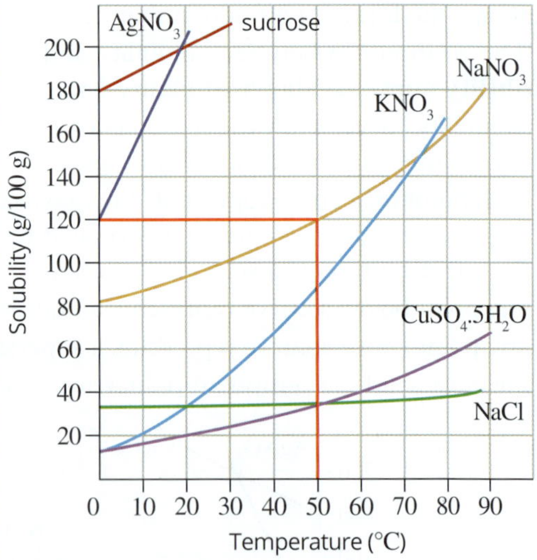
Read the solubility from the graph.	The horizontal line intersects with the vertical axis at 120 g. Therefore, the mass of $NaNO_3$ that will dissolve in 100 g of water at 50°C is 120 g.

Worked example: Try yourself 13.1.1

SOLUBILITY CURVE CALCULATIONS

Use the solubility curve below to find how many grams of potassium nitrate (KNO_3) will dissolve in 100 g of water at 70°C.

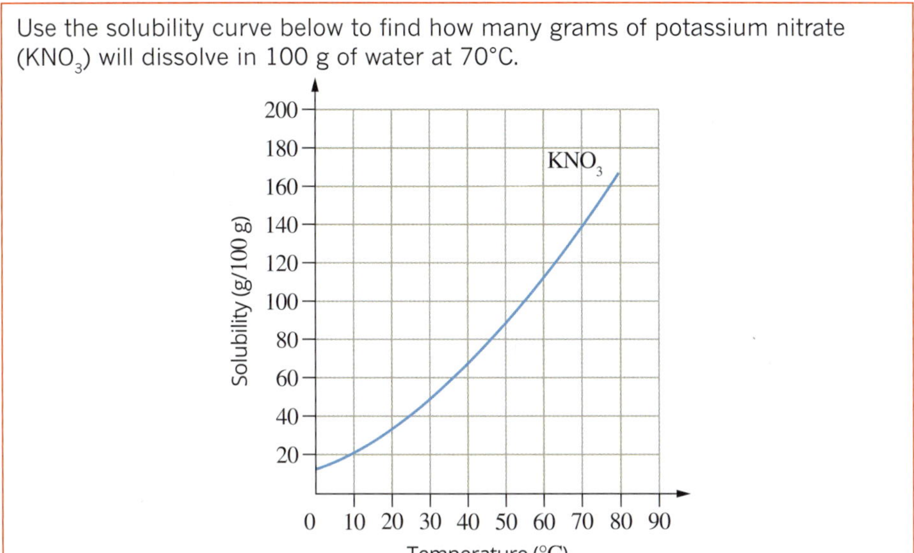

MORE CALCULATIONS USING SOLUBILITY CURVES

Solubility curves allow you to directly calculate the mass of a compound that will dissolve in 100 g water at different temperatures. It is also possible to use the curves to calculate solubilities of compounds in quantities of water other than 100 g. Worked example 13.1.2 shows you how.

Worked example 13.1.2

SOLUBILITY CURVES

An 80 g sample of sodium nitrate ($NaNO_3$) is added to 200 mL of H_2O at 20°C. Use Figure 13.1.3 on page 419 to calculate how much more $NaNO_3$ must be added to make the solution saturated with $NaNO_3$ at 20°C.

Thinking	Working
Use the solubility curve to find the mass of solute in a saturated solution of 100 g of H_2O at the required temperature.	Draw a line from 20°C on the horizontal axis to the solubility curve for $NaNO_3$ and find the corresponding value on the vertical axis. The value is 92 g.

Use the amount of solute that will dissolve in 100 g of H_2O to find the mass of solute to make a saturated solution in the mass of H_2O for this question.	The density of water is 1.0 g mL^{-1}, so 200 mL of water will weigh 200 g. So, twice the mass of solute can dissolve in 200 g of water as in 100 g. $m(NaNO_3) = 2 \times 92$ $= 184$ g
To find out how much extra solute you need to add, find the difference between the mass of solute needed to make a saturated solution and how much has already been added.	80 g of $NaNO_3$ has already been added to 200 g H_2O. So, the extra mass of $NaNO_3$ needed: $= 184 - 80$ $= 104$ g

Worked example: Try yourself 13.1.2

SOLUBILITY CURVES

A 120 g sample of sodium nitrate ($NaNO_3$) is added to 300 mL of H_2O at 40°C.
Use Figure 13.1.3 on page 419 to calculate how much more $NaNO_3$ must be added to make the solution saturated with $NaNO_3$ at 40°C.

Making your own solubility curve

Solubility curves can be developed relatively easily in a school laboratory. For example, a solubility curve for potassium nitrate in water can be derived in a group activity by carrying out the steps shown in Figure 13.1.4.

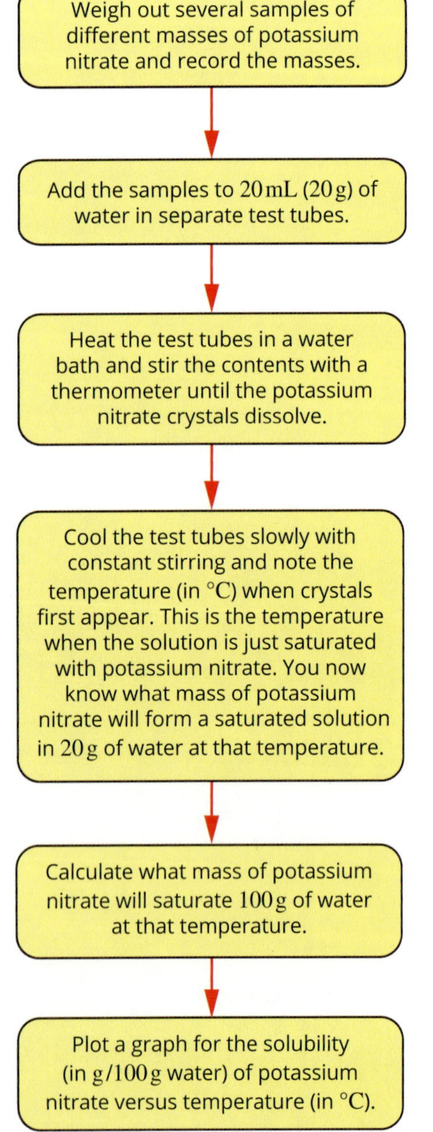

FIGURE 13.1.4 Process for obtaining a solubility curve

CRYSTALLISATION

Crystallisation from solutions occurs when an unsaturated solution becomes saturated and crystals form.

Cooling a solution may reduce the solubility of a dissolved solute to the point where not all of the substance present is soluble. An example of this occurs when copper(II) sulfate crystals form after a hot solution of the compound is cooled (Figure 13.1.5). At lower temperatures, less of the compound can dissolve in the water and a point will be reached when the solution becomes saturated. Further cooling will result in crystals being formed.

Calculations involving crystallisation

You can predict how much of a compound will crystallise out of a solution as temperature decreases using a solubility curve. Worked example 13.1.3 shows how this can be done.

Worked example 13.1.3

CRYSTALLISATION

50 g of potassium nitrate (KNO_3) is dissolved in 100 g water at 40°C. What mass of KNO_3 crystals will form if the temperature is reduced to 20°C? You will need to refer to Figure 13.1.3 on page 419 to complete this question.

Thinking	Working
Identify the mass of solute dissolved in the original solution.	Mass of KNO_3 in solution = 50 g
Find the maximum mass of solute that will remain dissolved in 100 g water at the final temperature.	The solubility curve shows that the maximum mass that will dissolve at 20°C is 32 g.
Calculate the mass of solute crystals that will form in the solution at the final temperature.	Mass of crystals formed = original mass − remaining $= 50 - 32$ $= 18$ g

Worked example: Try yourself 13.1.3

CRYSTALLISATION

200 g of sucrose is dissolved in 100 g water at 20°C. What mass of sucrose crystals will form if the temperature is reduced to 10°C?

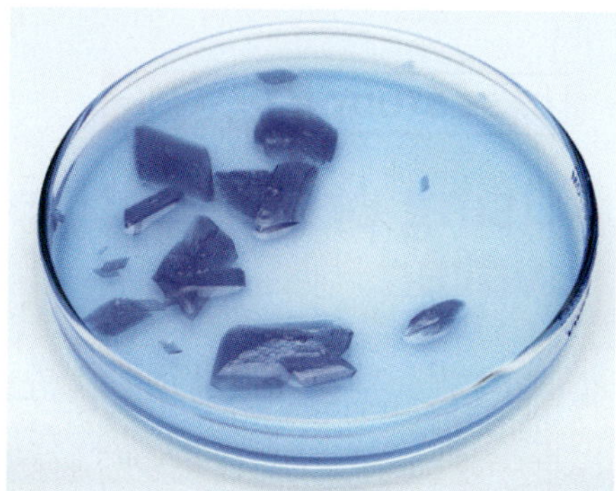

FIGURE 13.1.5 Copper sulfate crystals form when a solution cools to below a temperature where the solution becomes saturated.

REMOVING IMPURITIES FROM WATER

Precipitation reactions are often used as a means of removing or reducing the levels of impurities in drinking water and wastewater from domestic or industrial sources.

Each of these processes involves converting soluble ionic compounds into insoluble ones. This is done by adding solutions containing ions, which will combine with the soluble ions in the impurity to form a precipitate. The precipitate can then be separated from the water by filtering or settling processes. Three examples of these procedures are discussed below.

Removing phosphates from domestic sewage

In many inland towns of Australia treated sewage is released into rivers and lakes. Phosphate and nitrates in sewage can result in eutrophication (growth of algae due to excess nutrients) of waterways, so control of the levels of these compounds is important, especially when river flow rates are low. (You will learn more about eutrophication in Chapter 16.) Phosphate ions can be precipitated from sewage water by the addition of solutions of compounds, such as aluminium sulfate or iron(III) chloride. Ions from these compounds will react with phosphate ions to form precipitates of aluminium phosphate and iron(III) phosphate respectively. A simple ionic equation for the reaction between iron(III) ions and phosphate ions is:

$$Fe^{3+}(aq) + PO_4^{\ 3-}(aq) \rightarrow FePO_4(s)$$

Note that it is not possible to remove nitrates from sewage water by precipitation reactions because these compounds are soluble in water. Nitrate levels are reduced in sewage waste using biological methods.

Removing heavy metals from industrial wastewater

Many industrial processes involve the use of heavy metals and, as a result, waste streams from these industries can contain dissolved heavy metal compounds. The metals are present in wastewater in the form of their cations.

In Melbourne, wastewaters end up in Port Phillip Bay or Bass Strait, so it is important that heavy metals such as lead, copper, chromium, nickel and zinc are removed, preferably at their source. A common and effective method of removing heavy metal ions from solution is by reacting them with suitable anions to form insoluble compounds. The insoluble compounds then come out of solution as precipitates.

In Chapter 5, Table 5.3.2 on page 178 lists anions, which form mainly insoluble compounds when combined with cations. Hydroxide, carbonate, sulfide and phosphate ions are included in this list.

Calcium hydroxide is a relatively inexpensive chemical that is commonly used in this way. For example, the hydroxide ions from a solution of calcium hydroxide will react with lead ions in wastewater to form a precipitate of lead(II) hydroxide.

$$Pb^{2+}(aq) + 2OH^-(aq) \rightarrow Pb(OH)_2(s)$$

The treatment process allows for the precipitate to settle to form a sludge, which is then separated from the purified wastewater.

Soluble carbonate and sulfide compounds have also been successfully used in the removal of heavy metal ions from wastewater. A problem with using sulfides is that at low pH levels, the toxic gas hydrogen sulfide can be formed.

CASE STUDY ANALYSIS

Using precipitation reactions in the purification of Melbourne's water

Melbourne's water is some of the cleanest in the world. This is because most of the reservoirs around Melbourne have protected catchments (see Figure 13.1.6), which means that water flowing into them is not contaminated by pollutants such as chemical fertilisers and pesticides.

An exception to this is Sugarloaf Reservoir, which takes some of its water from the Yarra River. The Yarra flows through farmland and so is prone to contamination.

The different stages of the treatment of water at the Winneke treatment plant at Sugarloaf Reservoir are shown in Figure 13.1.7.

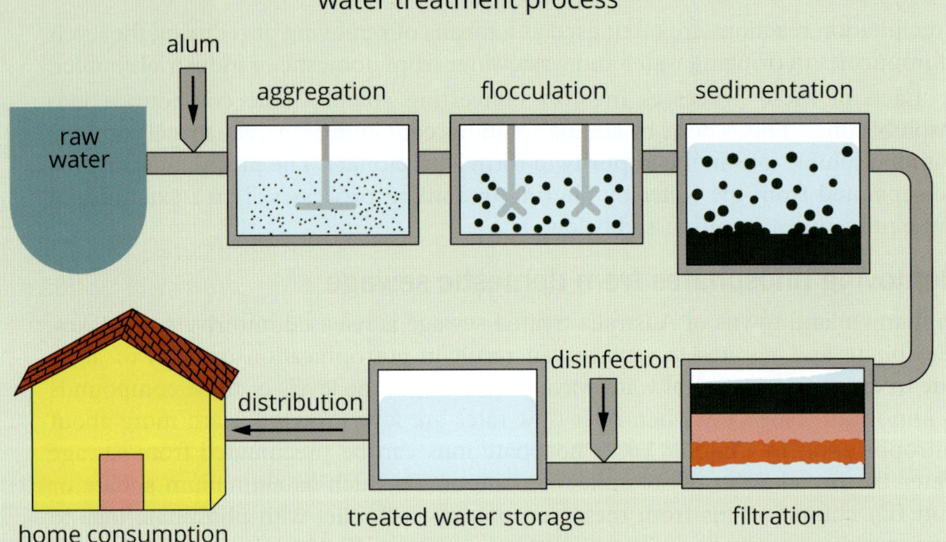

FIGURE 13.1.7 Stages in the purification of drinking water at the Winneke treatment plant, Sugarloaf Reservoir

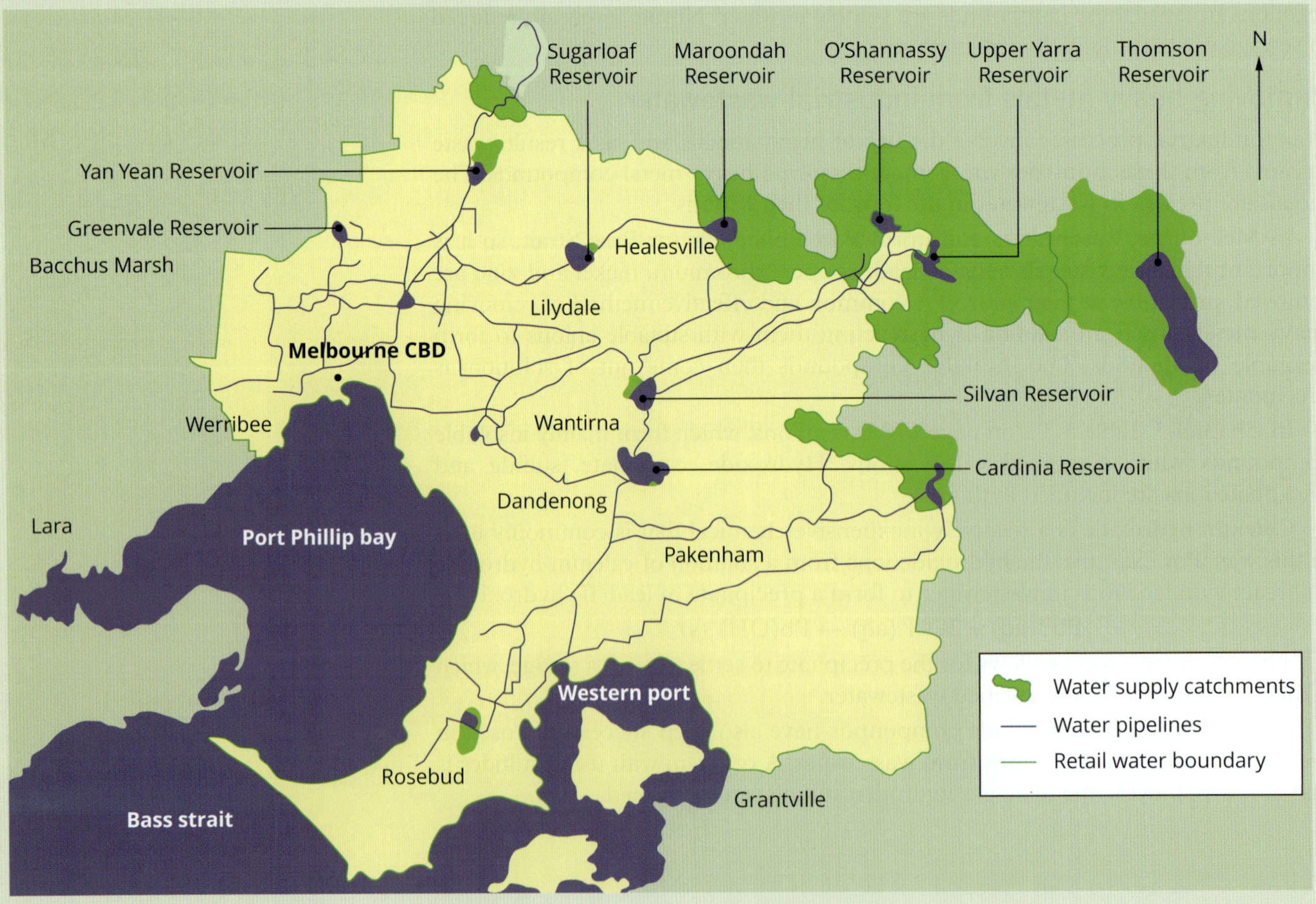

Sugarloaf Reservoir
Maroondah Reservoir
O'Shannassy Reservoir
Upper Yarra Reservoir
Thomson Reservoir
N
Yan Yean Reservoir
Greenvale Reservoir
Bacchus Marsh
Healesville
Lilydale
Melbourne CBD
Werribee
Wantirna
Silvan Reservoir
Cardinia Reservoir
Dandenong
Lara
Port Phillip bay
Pakenham
Western port
Rosebud
Bass strait
Grantville
Water supply catchments
Water pipelines
Retail water boundary

FIGURE 13.1.6 Melbourne's reservoirs

In the **aggregation** stage, alum (aluminium sulfate) is added. Aluminium ions from the dissolved alum combine with hydroxide ions in the water to form a precipitate of aluminium hydroxide.

$$Al^{3+}(aq) + 3OH^{-}(aq) \rightarrow Al(OH)_3(s)$$

The precipitate traps fine particles and some micro-organisms within the precipitate particles. The addition of a positively charged polymer results in the fine particles accumulating to form larger particles, in a process called **flocculation**. The larger particles settle to form a sediment and the clear water is filtered. The filtered water is treated with a small amount of chlorine to remove micro-organisms, and fluoride is also added. The treated water is then distributed to holding reservoirs (Figure 13.1.8) across the city. From these elevated locations, water is gravity-fed to local homes, schools and industries.

FIGURE 13.1.8 Water tower in the Melbourne suburb of Surrey Hills

Analysis

1 If the pH of water to be treated is too low, reservoir water can be treated with lime (calcium oxide) to provide hydroxide ions before alum is added.
   - **a** Write a balanced full equation for calcium oxide reacting with water to produce calcium hydroxide.
   - **b** Write a balanced ionic equation to show the dissociation of dissolved calcium hydroxide in water to produce hydroxide ions.

2 Iron(III) chloride has sometimes been used in place of alum to treat water. Write an ionic equation for the reaction between this compound and calcium hydroxide dissolved in a sample of water.

SOLUBILITY OF LIQUIDS AND GASES IN WATER

It is generally true that as temperature increases, the solubility of solids in solution increases. However, the solubility of liquids in other liquids does not show clear trends in solubility as the temperature changes.

Gases generally become less soluble as the temperature increases. The graph in Figure 13.1.9 shows the solubility of different gases in water at varying temperatures.

Table 13.1.2 shows the solubility of some common gases in water at different temperatures. For each gas, it can be seen that as temperature increases, less gas is able to dissolve in water.

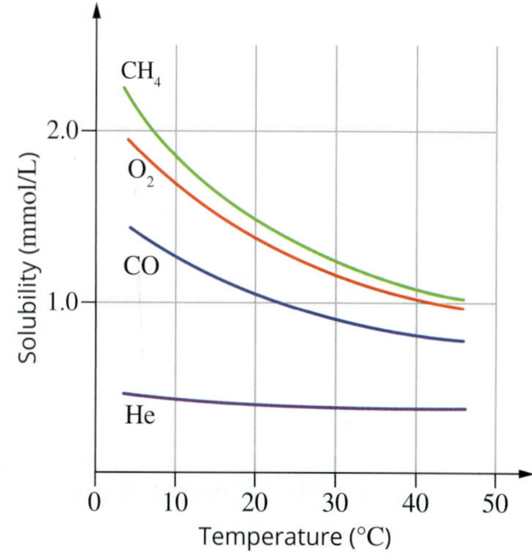

FIGURE 13.1.9 Solubility curves for some gases

TABLE 13.1.2 Solubility of some gases in water at different temperatures

Gas	Solubility (g of gas per kg of water) at 0°C	Solubility (g of gas per kg of water) at 20°C	Solubility (g of gas per kg of water) at 60°C
oxygen	0.069	0.043	0.023
carbon dioxide	3.4	1.7	0.58
nitrogen	0.029	0.019	0.011
methane	0.040	0.023	0.011
ammonia	897	529	168

FIGURE 13.1.10 Bubbles of air appear as water is heated. This is because solubility of gases decreases at higher temperatures.

Figure 13.1.10 shows the bubbles that appear when water is first heated. The bubbles are not steam (water vapour), but bubbles of air. Some of the air that was dissolved in the water comes out of solution as the temperature increases and the gas becomes less soluble.

You may have noticed that soft drinks will go 'flat' more quickly if they are left standing in the sun on a hot day than if they are in a cold refrigerator (Figure 13.1.11). The dissolved carbon dioxide in the drink comes out of solution as the drink heats up.

The effect of temperature on the solubility of a gas can have environmental implications. If the temperature of the water in rivers, lakes and oceans increases even slightly, it will contain less dissolved oxygen. This can have serious consequences for oxygen-breathing aquatic organisms, such as the Australian Murray River cod in Figure 13.1.12.

FIGURE 13.1.11 Soft drinks left standing in the sun go 'flat' quickly.

FIGURE 13.1.12 Fish, such as this native Australian Murray River cod, are susceptible to lowered levels of oxygen in natural waters.

Therefore, hot water from power stations and other industries must be cooled before it can be discharged into waterways. Even a small increase in water temperature can cause the oxygen concentration in the water to drop below levels that are necessary for aquatic life to survive.

CHEMFILE

Green sea turtles

Climate change has had a significant impact on large areas of Australia's Great Barrier Reef. Increased water temperatures have had a direct effect, for example, on corals, and parts of the reef have experienced bleaching.

An associated effect of increasing ocean temperatures has been decreased levels of dissolved oxygen in the reef waters. Figure 13.1.9 demonstrates that increasing temperatures result in a decrease in solubility for most gases, including oxygen.

Decreased levels of dissolved oxygen in the oceans around reefs can have an impact on small organisms forming part of the reef ecosystem. This, in turn, can affect the food sources of larger reef animals, such as the green sea turtle (see figure on the right). Although adult sea turtles feed mainly on sea grass, young turtles are omnivorous and their diet includes sea sponges, molluscs and small crustaceans.

Depleted oxygen levels in oceans due to climate change can have an impact on ecosystems where green sea turtles are found.

13.1 Review

OA ✓✓

SUMMARY

- Solubility is a measure of how much solute will dissolve in a given amount of solvent at a specified temperature.
- Solubility curves show how the solubility of a compound changes with temperature.
- Solubility curves can be used to calculate the amount of a substance that will dissolve in a given amount of solvent at a specified temperature.
- Precipitation reactions can be used to remove impurities from water.
- As the temperature of a solution increases, the solubility of solids generally increases and the solubility of gases decreases. Liquids show no overall trend in solubility with temperature.

KEY QUESTIONS

Knowledge and understanding

1 Select from the list of terms to fill in the gaps in the following sentences. Some terms may be used more than once, some not at all.

unsaturated; saturated; supersaturated; temperature; solute; solvent; solution

The solubility of a substance can be measured by how much ________ will dissolve in a given quantity of ________ at a given ________. An ________ solution is one which is able to dissolve more solute at a given temperature. A ________ solution is one in which there is more solute dissolved at a given temperature than is necessary to make the solution saturated. A ________ solution is one which contains the maximum amount of dissolved solute at a given temperature.

Refer to Figure 13.1.3, on page 419, to help you answer the questions below.

2 Solubility curves represent the maximum amount of a solute that can dissolve in 100 g of water at different temperatures. Use the list of terms provided to fill in the gaps in the sentences below.

a saturated solution; an unsaturated solution; a supersaturated solution

- Any point below a solubility curve represents ________
- Any point above a solubility curve represents ________
- Any point on a solubility curve represents ________

3 How many grams of each of the following compounds would be required to make a saturated solution in 100 g of water at the temperatures shown?

a $NaNO_3$ at 50°C

b KNO_3 at 30°C

c $CuSO_4{\cdot}5H_2O$ at 60°C

Analysis

4 What is the maximum mass of sodium nitrate, $NaNO_3$, that would dissolve at 50°C in the masses of water shown?

a 100 g

b 20 g

c 300 g

5 What is the maximum mass of solute that would dissolve in each of the following situations?

a $CuSO_4{\cdot}5H_2O$ in 200 g of water at 60°C

b $AgNO_3$ in 20 g of water at 9°C

c Sucrose in 40 g of water at 20°C

6 Would the following solutions be saturated, unsaturated or supersaturated?

a 60 g of NaCl dissolved in 300 g of water at 80°C

b 16 g of $NaNO_3$ dissolved in 10 g of water at 80°C

c 90 g of sucrose dissolved in 50 g of water at 10°C

d 1 kg of KNO_3 dissolved in 1 kg of water at 50°C

e 0.018 g N_2 gas dissolved in 2 kg water at 60°C. (Use Table 13.1.2, on page 425, to help you with this part of the question.)

7 Calculate the mass of crystals that will form when a solution containing 50 g of $CuSO_4{\cdot}5H_2O$ in 100 g of water is cooled from 80°C to 20°C.

8 Wastewater from an industrial plant contains dissolved chromium(III) and nickel(II) ions.

a Would it be possible to remove these metal ions from solution using either sodium sulfate or sodium sulfide solutions? (Refer to the solubility tables, Appendix 3 on page 534 for help with this question.)

b Write an ionic equation for the reaction of chromium(III) ions with carbonate ions.

c List some compounds which, when dissolved in water, could be used to form a precipitate with dissolved nickel(II) ions.

13.2 Calculating concentration

The concentration of a solution describes the relative amount of solute and solvent present in the solution. A solution in which the ratio of solute to solvent is high is said to be concentrated. Cordial that has not had any water added to it is an example of a **concentrated solution**. A solution in which the ratio of solute to solvent is low is said to be a **dilute solution**. A quarter of a teaspoon of sugar dissolved in a litre of water will produce a dilute solution that tastes slightly sweet.

'Concentrated' and 'dilute' are general terms. However, sometimes you need to know the actual concentration of a solution—the exact ratio of solute to solvent. The use of exact solution concentrations is important in the prescription of medicines, chemical manufacturing and chemical analysis (Figure 13.2.1).

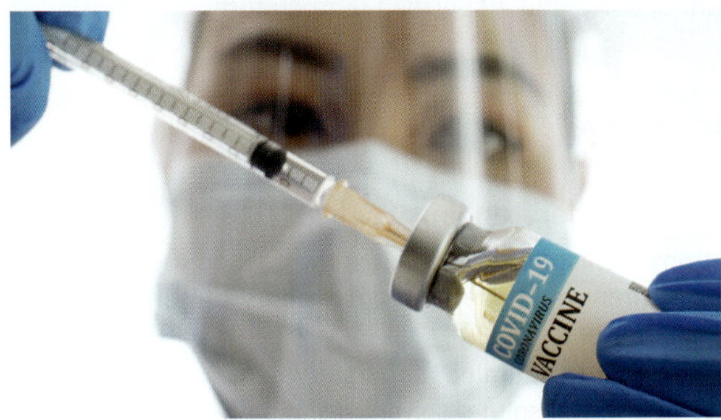

FIGURE 13.2.1 It is vitally important that the concentration of pharmaceuticals in prescribed medications is exactly as indicated on the label of the container.

UNITS OF CONCENTRATION

Chemists use different measures of concentration depending on the situation. The most common measures describe the amount of solute in a given amount of solution.

Some examples of units of concentration are:

- mass of solute per litre of solution (grams per litre, g L^{-1})
- moles of solute per litre of solution (moles per litre, mol L^{-1})

Units of concentration measured in this way have two parts.

- The first part gives information about how much solute there is.
- The second part gives information about how much solution there is.

For example, if a solution contains sodium chloride (NaCl) with a concentration of 17 g L^{-1}, then (first part) 17 g of NaCl is dissolved in every (second part) 1 L of the solution.

You will now look at how to perform calculations of concentration using different units.

> Concentration units have two parts. The first part provides information about the amount of solute. The second part provides information about the amount of solution.

Concentration in grams per litre (g L^{-1})

The concentration of a solution in grams per litre (g L^{-1}) indicates the mass, in grams, of solute dissolved in 1 litre of the solution. For example, if the concentration of sodium chloride in seawater is 20 g L^{-1}, this means that in 1 L of seawater there is 20 g of dissolved sodium chloride. A formula used to calculate the concentration in g L^{-1} is:

$$\text{concentration (g L}^{-1}\text{)} = \frac{\text{mass of solute (in g)}}{\text{volume of solution (in L)}}$$

Worked example 13.2.1

CALCULATING CONCENTRATION IN GRAMS PER LITRE ($g\ L^{-1}$)

What is the concentration, in $g\ L^{-1}$, of a solution containing 8.00 g of sodium chloride in 500 mL of solution?

Thinking	Working
Change the volume of solution so it is expressed in litres.	$500\ mL = \frac{500}{1000}$ $= 0.500\ L$
Calculate the concentration in $g\ L^{-1}$.	$c = \frac{\text{mass of solute (in g)}}{\text{volume of solution (in L)}}$ $= \frac{8.00}{0.500}$ $= 16.0\ g\ L^{-1}$

> Concentration in grams per litre can be calculated using the formula:
> $$\text{concentration } (g\ L^{-1}) = \frac{\text{mass of solute (in g)}}{\text{volume of solution (in L)}}$$

Worked example: Try yourself 13.2.1

CALCULATING CONCENTRATION IN GRAMS PER LITRE ($g\ L^{-1}$)

What is the concentration, in $g\ L^{-1}$, of a solution containing 5.00 g of glucose in 250 mL of solution?

Concentration in parts per million (ppm)

When very small quantities of solute are dissolved to form a solution, the concentration is often measured in **parts per million (ppm)**. For example, the concentration of mercury in fish that is suitable for consumption is usually expressed in parts per million. The maximum concentration allowed for sale in Australia is 1 ppm (Figure 13.2.2).

FIGURE 13.2.2 Fish sold in Australia must have no more than 1 ppm of mercury (Australia New Zealand Food Standards Code).

In simple terms, the concentration in parts per million can be thought of as the mass of solute, in grams, dissolved in 1 000 000 g of solution. But a million grams of a solution is a large and unwieldy quantity. A simpler, alternative way of thinking about concentrations in ppm is to consider how many milligrams of solute are dissolved in 1 kilogram of solution, because there are 1 million milligrams in 1 kilogram. A formula used to calculate the concentration of a solution in ppm is therefore:

$$\text{concentration (ppm)} = \frac{\text{mass of solute (in mg)}}{\text{mass of solution (in kg)}}$$

For example, a solution of sodium chloride that contains 154 mg of sodium chloride dissolved in 1 kg of solution has a concentration of 154 ppm.

> Concentration in parts per million can be calculated using the formula:
> $$\text{Concentration (ppm)} = \frac{\text{mass of solute (in mg)}}{\text{mass of solution (in kg)}}$$

Worked example 13.2.2

CALCULATING CONCENTRATION IN PARTS PER MILLION (ppm)

A saturated solution of calcium carbonate was found to contain 0.0198 g of calcium carbonate dissolved in 2000 g of solution. Calculate the concentration, in ppm, of calcium carbonate in the saturated solution.

Thinking	Working
Calculate the mass of solute in mg. Remember: mass (in mg) = mass (in g) × 1000	Mass of solute (calcium carbonate) in mg: $= 0.0198 \times 1000$ $= 19.8$ mg
Calculate the mass of solution in kg. Remember: $\text{mass (in kg)} = \frac{\text{mass (in g)}}{1000}$	$\text{Mass of solution in kg} = \frac{2000}{1000}$ $= 2.000$ kg
Calculate the concentration of the solution in mg kg^{-1}. This is the same as concentration in ppm.	Concentration of calcium carbonate in ppm: $= \frac{\text{mass of solute (in mg)}}{\text{mass of solution (in kg)}}$ $= \frac{19.8}{2.000}$ $= 9.90$ mg kg^{-1} $= 9.90$ ppm

Worked example: Try yourself 13.2.2

CALCULATING CONCENTRATION IN PARTS PER MILLION (ppm)

A sample of tap water was found to contain 0.0537 g of NaCl per 250.0 g of solution. Calculate the concentration of NaCl in parts per million (ppm).

Other units of concentration

You might have noticed symbols such as w/v or v/v on the labels of some foods, drinks and pharmaceuticals. These symbols represent other concentration units based on masses and volumes of solutes and solutions. These are useful in practical situations because people are familiar with these quantities. (Figure 13.2.3).

FIGURE 13.2.3 Consumer products from hardware stores and supermarkets show a wide range of concentration units on their labels.

CHEMFILE

Threshold limit values of common substances

The threshold limit value (TLV) of a substance is the level to which it is believed a worker or consumer can be exposed day after day without adverse health effects. Three examples of TLVs are:

- sulfur dioxide (SO_2) at 5 ppm. Sulphur dioxide is produced when coal and other fossil fuels are burnt. It is also used as a preservative in foods.
- ozone (O_3) at 0.1 ppm. Ozone is produced in car engines. The figure below shows an example of an ozone warning sign.
- octane at 30 ppm. Octane is a component of petrol. Workers in garages and service stations are exposed to petrol fumes.

An ozone exposure warning

Percentage by volume: % (v/v)

The abbreviation v/v indicates that the percentage is based on volumes of both solute and solution. The same units must be used to record both volumes. Percentage by volume is a convenient unit to use when the solute is a liquid.

Percentage by volume is frequently expressed as volume per 100 mL of solution. For example, the wine label in Figure 13.2.4 shows 14.5% alc./vol. This means the wine contains 14.5% alcohol (ethanol) by volume (14.5%(v/v)). There will be 14.5 mL of alcohol in 100 mL of the wine.

For example, if a 200 mL glass of champagne contains 28 mL of alcohol, the concentration as %(v/v) of alcohol in this solution can be calculated as follows:

$$\begin{aligned}\text{concentration \% (v/v)} &= \frac{\text{volume of solute (in mL)}}{\text{volume of solution (in mL)}} \times 100\% \\ &= \frac{28}{200} \times 100\% \\ &= 14\%\text{(v/v)}\end{aligned}$$

FIGURE 13.2.4 The label from a bottle of wine that shows the alcohol concentration in units of %(v/v).

Percentage mass/volume: % (m/v)

Percentage mass/volume describes the mass of solute, measured in grams, present in 100 mL of the solution.

For example, if a solution of plant food contains a particular potassium compound at a concentration of 3%(m/v), this indicates that there is 3 g of potassium in 100 mL of solution. This is sometimes written as %(w/v) on product labels, to represent percentage weight/volume.

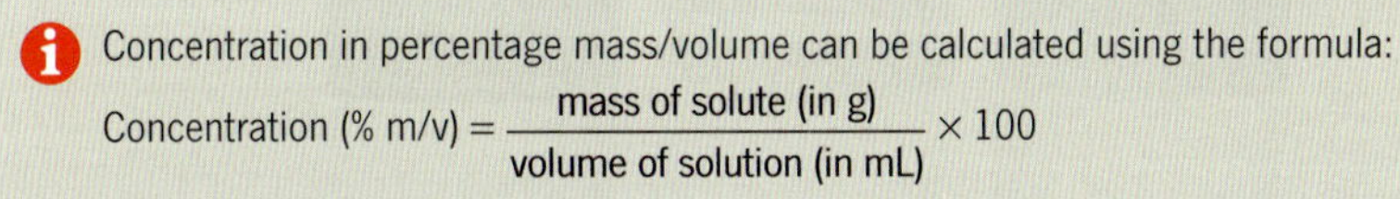

Concentration in percentage mass/volume can be calculated using the formula:

$$\text{Concentration (\% m/v)} = \frac{\text{mass of solute (in g)}}{\text{volume of solution (in mL)}} \times 100$$

CALCULATING CONCENTRATION IN MOLES PER LITRE

We have seen that units of concentration, such as g L^{-1}, %(m/v) and %(v/v), are commonly used on the labels of consumer products found in supermarkets, pharmacies and hardware stores. These units are rarely used in school chemistry laboratories and in chemistry research laboratories generally. This is because quantitative calculations using these units are cumbersome. The unit of concentration favoured by chemists is mol L^{-1}. However, unless people have had some training in chemistry, they will not have an understanding of the chemical terms 'mole' and mol L^{-1}. It is unlikely, then, that you will see this unit used on product labels in your local supermarket.

Concentrations expressed in moles per litre (mol L^{-1}) allow chemists to compare the amount, in moles, of atoms, molecules or ions present in a given volume of solution.

The concentration, in mol L^{-1}, of a solution can be calculated as follows:

$$\text{concentration (mol L}^{-1}\text{)} = \frac{\text{amount of solute (in mol)}}{\text{volume of solution (in L)}}$$

This can be written using variables:

$$c = \frac{n}{V}$$

Where:

c is the concentration (mol L^{-1})

n is the amount (mol)

V is the volume (L).

CHEMFILE

Saline drip

A 'saline drip' is sometimes used during medical procedures to replenish body fluids in patients (see figure below). The solution used contains sodium chloride, commonly with a concentration of 0.9%(m/v). This means that 100 mL of the solution contains 0.9 g of dissolved sodium chloride.

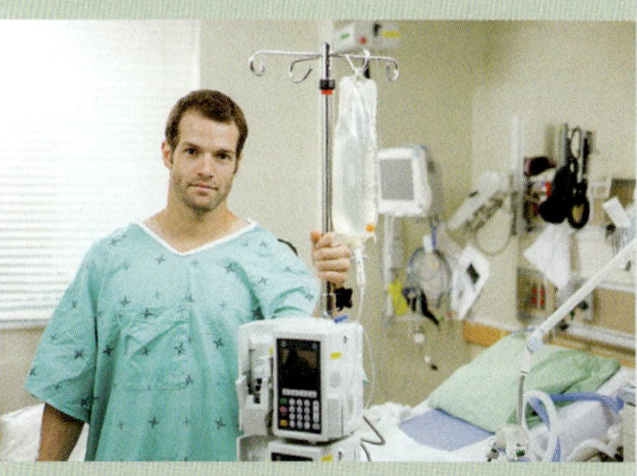

A saline drip is a solution of sodium chloride.

Concentration in moles per litre can be calculated using the formula:

$$\text{concentration (mol L}^{-1}) = \frac{\text{amount of solute (in mol)}}{\text{volume of solution (in L)}}$$

The symbol M was used earlier to represent molar mass. Take care not to confuse the two different uses; one is a quantity symbol, the other a unit symbol.

The concentration of a solution in moles per litre is often referred to as the **molarity**, or molar concentration of the solution, and is given the unit M.

A solution containing 1 mole of solute dissolved in 1 litre of solution can therefore be described in several different ways. You can say that the solution:

- has a concentration of 1 mole per litre
- has a concentration of 1 mol L^{-1}
- is 1 molar
- is 1 M
- has a molarity of 1 M.

For example, a 0.1 M solution of hydrochloric acid would contain 0.1 moles of HCl in 1 L of the solution. (Bottles of acids of this concentration are shown in Figure 13.2.5.)

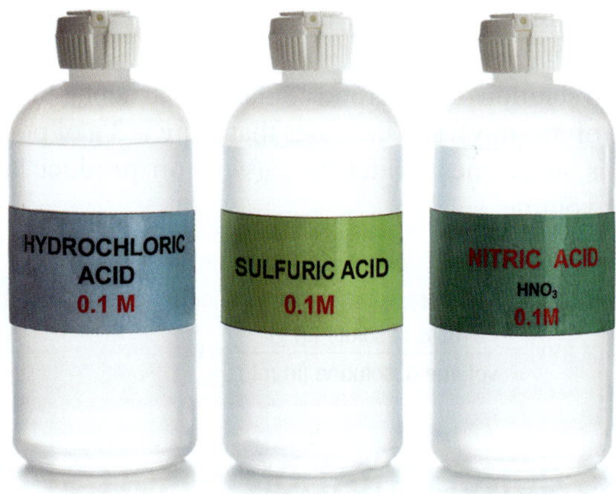

FIGURE 13.2.5 These containers hold solutions with concentrations shown in units of mol L^{-1}.

Worked example 13.2.3

CALCULATING MOLAR CONCENTRATIONS (MOLARITY)

Calculate the molar concentration of a solution that contains 0.105 mol of potassium nitrate dissolved in 200 mL of solution.

Thinking	Working
Convert the given volume of the solution to litres.	$V(KNO_3) = \frac{200}{1000}$ $= 0.200$ L
Calculate the molar concentration using the formula: $c = \frac{n}{V}$	$c(KNO_3) = \frac{n}{V}$ $= \frac{0.105}{0.200}$ $= 0.525$ mol L^{-1} or 0.525 M

Worked example: Try yourself 13.2.3

CALCULATING MOLAR CONCENTRATIONS (MOLARITY)

Calculate the molar concentration of a solution that contains 0.24 mol of glucose dissolved in 500 mL of solution.

Calculating molarity given the mass of solute

Sometimes you will know the mass of a solute and volume of solution and want to calculate the molarity. Two main calculations are involved.

1 Calculate the number of moles of solute from its mass.

2 Calculate the concentration using the number of moles and the volume (in litres).

Worked example 13.2.4

CALCULATING MOLARITY GIVEN THE MASS OF SOLUTE

Calculate the concentration, in mol L^{-1}, of a solution that contains 16.8 mg of silver nitrate ($AgNO_3$) dissolved in 150 mL of solution.

Thinking	Working
Convert the volume of the solution to litres.	$V(AgNO_3) = \frac{150}{1000}$ $= 0.150$ L
Convert the mass to grams.	$m(AgNO_3) = \frac{16.8}{1000}$ $= 0.0168$ g
Calculate the molar mass of the solute. To do this, add up the atomic masses of all the atoms in the compound.	$M(AgNO_3) = 107.9 + 14.0 + (3 \times 16.0)$ $= 169.9$ g mol^{-1}
Calculate the number of moles of solute using the formula: $n = \frac{m}{M}$	$n(AgNO_3) = \frac{m}{M}$ $= \frac{0.0168}{169.9}$ $= 9.89 \times 10^{-5}$ mol
Calculate the molar concentration using the formula: $c = \frac{n}{V}$	$c(AgNO_3) = \frac{n}{V}$ $= \frac{9.89 \times 10^{-5}}{0.150}$ $= 6.59 \times 10^{-4}$ mol L^{-1} or 6.59×10^{-4} M

Worked example: Try yourself 13.2.4

CALCULATING MOLARITY GIVEN THE MASS OF SOLUTE

Calculate the concentration, in mol L^{-1}, of a solution that contains 4000 mg of ethanoic acid (CH_3COOH) dissolved in 100 mL of solution.

Calculating the number of moles of solute in a solution

The formula used to calculate the molarity of a solution is $c = \frac{n}{V}$. If you rearrange the formula, you can use it to calculate the number of moles of solute in a solution of given concentration and volume:

$$n = c \times V$$

where n is the amount (mol), c is the concentration (mol L^{-1}) and V is the volume (L).

Worked example 13.2.5

CALCULATING THE NUMBER OF MOLES OF SOLUTE IN A SOLUTION

Calculate the amount, in moles, of ammonia (NH_3) in 25.0 mL of a 0.3277 M ammonia solution.

Thinking	Working
Convert the given volume of the solution to litres.	$V(NH_3) = \frac{25.0}{1000}$ $= 0.0250$ L
Calculate the amount of compound, in moles, using the formula: $n = c \times V$	$n(NH_3) = c \times V$ $= 0.3277 \times 0.0250$ $= 8.19 \times 10^{-3}$ mol

The number of moles of a compound in solution can be calculated using the formula $n = c \times V$, where:
n = number of moles of compound
c = concentration of the solution in mol L^{-1}
V = volume of the solution in litres.

Worked example: Try yourself 13.2.5

CALCULATING THE NUMBER OF MOLES OF SOLUTE IN A SOLUTION

Calculate the amount, in moles, of potassium permanganate ($KMnO_4$), in 100 mL of a 0.0250 M solution of the compound.

DILUTION

Many commercially available domestic and industrial products come in the form of concentrated solutions. Examples are pesticides (Figure 13.2.6), fertilisers, detergents, fruit juices, acids and other chemicals. A major reason for using concentrates is to save on transportation costs. Diluted solutions contain a lot of water and that extra mass has to be transported, which increases costs. It is also more convenient to buy concentrated products and dilute them with water, whether at home or in the workplace.

Everyday examples of dilution are:

- adding water to cordial
- a laboratory technician making a 1 M solution of hydrochloric acid from a bottle of concentrated hydrochloric acid
- a home gardener diluting fertiliser concentrate to spray on the lawn
- a farmer diluting weedkiller concentrate to spray on a wheat crop
- an assistant in a commercial kitchen diluting concentrated detergent solution before using it to wash dishes.

In this section, you will learn how to calculate the concentrations of solutions after they have been diluted.

FIGURE 13.2.6 Pesticide solutions used in aerial spraying are prepared by diluting a concentrated solution of the pesticide compound.

CALCULATING CONCENTRATIONS WHEN SOLUTIONS ARE DILUTED

The process of adding more solvent to a solution is known as **dilution**. When a solution is diluted, its concentration is decreased.

For example, if 50 mL of water is added to 50 mL of 0.10 mol L^{-1} sugar solution, the amount of sugar remains unchanged, but the volume of the solution in which it is dissolved doubles. As Figure 13.2.7 shows, this means the sugar molecules are spread further apart during the dilution process, and so the concentration of the sugar solution is decreased (it will become 0.050 mol L^{-1} in this instance).

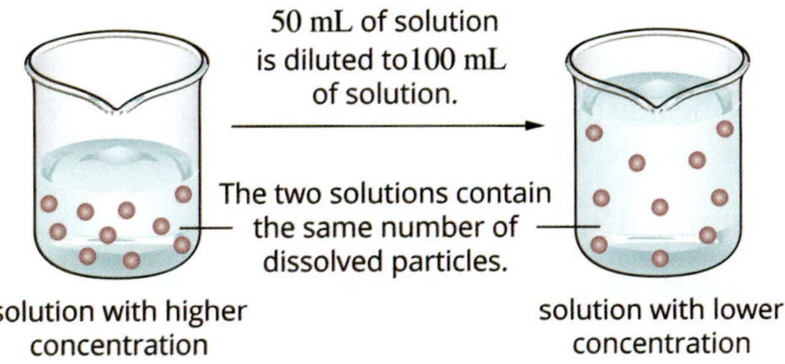

FIGURE 13.2.7 Dilution does not change the number of solute particles, but the concentration of the solute decreases.

It is important to recognise that diluting a solution (by adding more solvent) does not change the amount of solute present.

Suppose you had V_1 litres of a solution and the concentration was c_1 mol L^{-1}. The amount of solute, in moles, is given by:

$$n_1 = c_1 \times V_1$$

Suppose water was added to make a new volume, V_2, and change the concentration to c_2. The amount of solute, n_2, in this diluted solution is given by:

$$n_2 = c_2 \times V_2$$

Since the number of moles of solute has not changed, $n_1 = n_2$; therefore:

$$c_1V_1 = c_2V_2$$

This formula is useful when solving problems involving diluted solutions. (Note that this formula can also be used with different concentration and volume units, as long as the units are the same on both sides of the formula.)

> The formula $c_1V_1 = c_2V_2$ can be used when solving problems involving the dilution of solutions. Units of c and V must be the same on both sides of the formula.

Worked example 13.2.6

QUESTIONS INVOLVING DILUTION

Calculate the concentration of the solution formed when 10.0 mL of water is added to 5.00 mL of 1.2 M HCl.

Thinking	Working
Write down the value of c_1 and V_1. Note: c_1 and V_1 refer to the original solution, before water was added.	$c_1 = 1.2$ M $V_1 = 5.00$ mL
Write down the value of V_2. Note: V_2 is the total volume of the original solution plus the added water.	$V_2 = 10.0 + 5.00$ $= 15.0$ mL
Transpose the equation $c_1V_1 = c_2V_2$ to allow calculation of the concentration, c_2, of the new solution.	$c_1V_1 = c_2V_2$ $c_2 = \frac{c_1V_1}{V_2}$
Calculate the concentration of the diluted solution.	$c_2 = \frac{1.2 \times 5.00}{15.0}$ $= 0.40$ M

Worked example: Try yourself 13.2.6

QUESTIONS INVOLVING DILUTION

Calculate the concentration of the solution formed when 95.0 mL of water is added to 5.00 mL of 0.500 M HCl.

CHANGING THE UNITS OF CONCENTRATION

At times it is useful to change, or 'convert', concentration values from one unit to another unit. One way to do this is in a series of simple steps. Worked example 13.2.7 shows how this can be done.

Worked example 13.2.7

CONCENTRATION UNIT CONVERSIONS

What is the concentration, in %(m/v), of a 0.200 M solution of NaCl?

Thinking	Working
Calculate the number of moles of solute in 1.00 L of the solution.	$n(\text{NaCl}) = c \times V$ $= 0.200 \times 1.00$ $= 0.200$ mol
Calculate the mass, in grams, of solute in 1.00 L of the solution.	$M(\text{NaCl}) = 35.5 + 23.0$ $= 58.5$ g mol^{-1} $m(\text{NaCl}) = n \times M$ $= 0.200 \times 58.5$ $= 11.7$ g
Calculate the mass, in g, of solute in 100 mL of the solution.	$m(\text{NaCl}) = 11.7 \times \frac{100}{1000}$ $= 1.17$ g
Express the concentration of the solute as %(m/v).	c(NaCl) = 1.17%(m/v)

Worked example: Try yourself 13.2.7

CONCENTRATION UNIT CONVERSIONS

What is the concentration, in %(m/v), of a 0.100 M solution of NaOH?

CASE STUDY **ANALYSIS**

Diluting strong acids

When schools purchase acid supplies for their science departments, they usually choose concentrated solutions. If the solution is concentrated, the volume of acid to transport is smaller. The school can prepare solutions of varying concentration by diluting this acid.

A laboratory assistant diluting the strong acid needs to be aware of the following basic principles.

- Dilutions should be conducted in a fume cupboard.
- A spill plan must be ready.
- The laboratory assistant must wear goggles, gloves and a lab coat.
- It is easier to manage the release of heat energy if small amounts of acid are added gradually, with stirring.
- All equipment must be rinsed thoroughly after use.
- The solutions must be labelled and stored appropriately.
- Acid is added to water; water is not added to acid (Figure 13.2.8).

When a concentrated acid is added to water it produces a lot of heat. The process is exothermic. During the dilution process, it is important that acid is added slowly to water and with constant stirring. This distributes heat evenly throughout the mixture.

Concentrated acid is denser than water. The reason that the acid is added to water rather than the other way around is so that the acid will sink into the water and mix naturally. However, it is important that stirring is also carried out to aid the mixing process. If water is added to concentrated acid, the heat generated will be localised where the two liquid layers meet. The resulting large temperature increase can cause the liquid to boil and be ejected from the container.

Analysis

A teacher prepared 9 bottles of 0.50 M sulfuric acid, H_2SO_4 for her chemistry class of 18 students. The volume of acid in each of the bottles was 50 mL. To make sure there was an adequate supply of acid, the teacher prepared 600 mL of the 0.50 M acid. She did this by diluting concentrated sulfuric acid from the chemistry store. The concentration of this acid was 18.4 M.

1. What volume of the 18.4 M sulfuric acid was required to prepare the 600 mL of diluted acid?
2. What mass of sulfuric acid would there be in each of the 9 bottles containing the diluted acid?

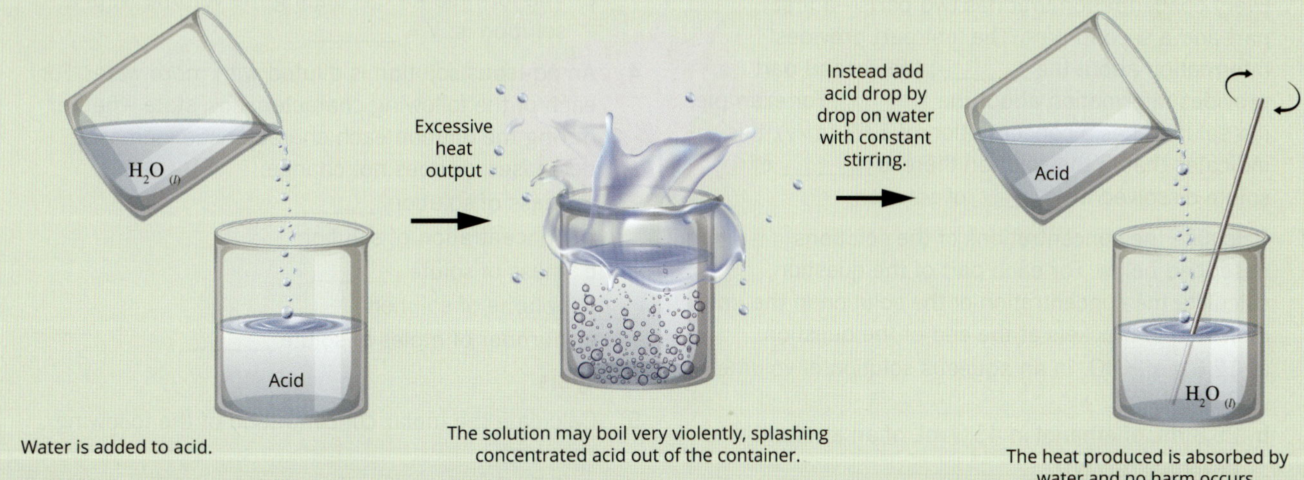

FIGURE 13.2.8 When preparing solutions of a particular concentration, laboratory technicians must often dilute a stock solution.

13.2 Review

OA ✓✓

SUMMARY

- The concentration of a solution is defined as the quantity of solute dissolved in a quantity of solution.
- Concentration units have two parts. The first part provides information about the quantity of solute. The second part provides information about the quantity of solution.
- The concentrations of solutions can be expressed in different units, including g L^{-1}, %(m/v), %(v/v) and ppm.
- Molarity is defined as the number of moles of solute per litre of solution.
- Molarity is calculated using the formula $c = \frac{n}{V}$, where c is the concentration in mol L^{-1}, n is the amount in moles and V is the volume in litres.
- The formula $n = c \times V$ can be used to calculate the number of moles of solute in a given volume of solution.
- Numerical problems involving dilution can be solved using the formula:
  $c_1V_1 = c_2V_2$
  where c is concentration and V is volume.
  When using this formula, c_1 and c_2 must be in the same units of concentration, and V_1 and V_2 must be in the same units of volume.
- One type of concentration unit can be converted to another unit in simple steps. For example, units of %(m/v) can be converted to units of mol L^{-1}.

KEY QUESTIONS

Knowledge and understanding

1 Select from the following terms to fill in the gaps in the following sentences. Some terms may be used more than once, some not at all.

1.6 g; 100 g; quantity of solution;
100 mL; quantity of solute

Every concentration unit has two parts—a first part and a second part. The first part provides information about the _______; the second part provides information about the _______. For example, if a solution has a concentration of 1.6%(m/v), this indicates that in the solution there is _______ of solute dissolved in _______ of solution.

2 Calculate the concentrations of the solutions described below. For each part of the question, calculate the concentration of the solution in the units indicated in brackets at the end of the question.

a 12.6 g of KNO_3 in an aqueous solution of volume 260 mL (g L^{-1})

b 25.8 mL of ethanol in 45.0 mL of an alcoholic beverage(%(v/v))

c 0.066 g of a pesticide in 5.0 L of a drum of solution (ppm). Assume that 1 litre of solution has a mass of 1 kg.

3 When dealing with calculation questions involving molar concentration, the following formula is used:

$$n = c \times V$$

Where:

V = volume in litres
c = concentration in mol L^{-1}
n = number of moles.

Rearrange the formula for use in the following situations.

a The formula for calculating the molar concentration, c, of a solution is: c = _______

b The formula for calculating the number of moles, n, of solute in a solution is: n = _______

c The formula for calculating the volume, V, of a solution is: V = _______

4 An aqueous solution is diluted with more water. For each of the following characteristics, state whether during the dilution each characteristic increases, decreases, or does not change:

a mass of solution _______

b concentration of solution _______

c mass of solute _______

d volume of solution _______

e number of moles of solute _______

Analysis

5 Calculate the molar concentration of the following solutions.

a A 50.0 mL solution that contains 0.17 mol of KNO_3

b A 3.5 L solution that contains 2.6 mol of Na_2CO_3

c A 2.4×10^3 L solution that contains 85 mol of NaCl

d A 250 mL solution that contains 5.6 g $Ca(CH_3COO)_2$

e A 1.25 L solution that contains 4.6 g NH_4Br

6 a Calculate the amount, in mol, of solute in each of the following solutions.
   i 0.50 L of a 0.40 M $CaCl_2$ solution
   ii 60.0 mL of a 1.3 M $Mg(NO_3)_2$ solution
   iii 22.5 mL of a 0.025 M KCl solution
 b Calculate the mass, in grams, of solute in each of the following solutions.
   i 1.3 L of a 0.025 M $FeCl_3$ solution
   ii 25.0 mL of a 1.75 M NH_4Cl solution

7 a What volume, in mL, of a 0.20 M NaCl solution would be required to prepare 80.0 mL of a 0.050 M solution of the compound?
 b 25.0 mL of a glucose solution was diluted with water to produce 200 mL of a 0.80 M solution. What was the molar concentration of the initial glucose solution?
 c When 5.6 mL of a 0.16 M solution of KNO_3 was diluted, the concentration of the diluted solution was 0.024 M. What was the volume of water, in mL, that was added to the original solution?
 d 40.0 mL of a sodium chloride solution, with a concentration of 75.0 g L^{-1} was diluted with water to produce 200 mL of solution. What was the concentration, in g L^{-1}, of the diluted solution?

8 The concentration of ethanol in a bottle of wine is 9.5%(m/v).
 a Calculate the concentration of the ethanol in
   i g L^{-1}
   ii mol L^{-1}
 b The density of ethanol is 0.789 g mL^{-1}. Calculate the concentration of the wine in %(v/v).

Chapter review

KEY TERMS

aggregation
concentrated solution
crystallise
dilute solution
dilution
flocculation
limestone
molarity
parts per million (ppm)
saturated solution
seed crystal
solubility
solubility curve
supersaturated solution
unsaturated solution

REVIEW QUESTIONS

Knowledge and understanding

1 Which one of the following statements about solubility curves is correct?
- **A** Supersaturated solutions are represented by points below the curve.
- **B** Saturated solutions are represented by points on the curve.
- **C** Unsaturated solutions are represented by points above the curve.
- **D** Supersaturated solutions are represented by points on the curve.

2 In the diagram below, the pink dot could represent:
- **A** a supersaturated solution of KNO_3
- **B** a saturated solution of $NaNO_3$
- **C** a supersaturated solution of NaCl
- **D** an unsaturated solution of $CuSO_4{\cdot}5H_2O$

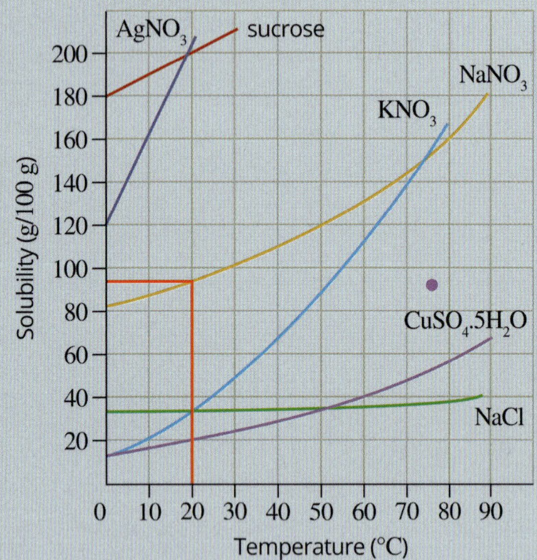

3 10 g of a compound was the minimum amount required to form a saturated solution in 20 g of water at a temperature of 30°C. The compound was:
- **A** KNO_3
- **B** NaCl
- **C** $CuSO_4{\cdot}5H_2O$
- **D** $NaNO_3$

4 What volume of 3.00 M NaOH would be required to prepare 200 mL of a 0.400 M solution?
- **A** 173 mL
- **B** 150 mL
- **C** 120 mL
- **D** 26.7 mL

5 Determine what mass of the following compounds will dissolve in 200 g of water at 30°C to form a saturated solution. Refer to the graph in Figure 13.1.3 (page 419) to help you answer this question.
- **a** $CuSO_4{\cdot}5H_2O$
- **b** $NaNO_3$
- **c** NaCl

6 What is the minimum temperature required to dissolve the following compounds? Refer to the graph in Figure 13.1.3 (page 419) to help you answer this question.
- **a** 138 g of KNO_3 in 100 g of water.
- **b** 81 g of $AgNO_3$ in 50 g of water
- **c** 100 g of $CuSO_4{\cdot}5H_2O$ in 250 g of water

7 What is the molar concentration (molarity) of the following?
- **a** A solution containing 0.60 mol NaOH dissolved in 2.0 L of solution.
- **b** A solution containing 1.78 mol of NH_3 dissolved in 250 mL of solution.
- **c** A solution containing 0.40 mol of HNO_3 dissolved in 600 mL of solution.

8
- **a** How many moles of solute are present in the following solutions?
  - **i** 35 mL of a 0.50 M solution of CH_3COOH
  - **ii** 2.5 L of a 0.030 M solution of $K_2Cr_2O_7$
  - **iii** 15.5 mL of a 2.55 M solution of H_2SO_4
- **b** What mass of solute is there in each of solutions **i–iii**?

9 What is the volume, in litres, of the following solutions?
- **a** A 2.80 M solution of NaOH that contains 0.30 mol of the compound.
- **b** A 0.075 M solution of NaCl that contains 1.5 mol of the compound.

10 Calculate the molar concentration of the diluted solution for the following.

a 21 mL of a 0.40 M solution of NaOH is diluted with water to a final volume of 3.5 L.

b 60 mL of a 0.150 M solution of $KMnO_4$ is diluted with water to a final volume of 80 mL.

11 Calculate the final volume of a solution for the following.

a 55 mL of a 0.60 M solution of Na_2SO_4 is diluted to produce a 0.17 M solution.

b 3.6 L of a 1.5 M solution of KOH is diluted to produce a 0.55 M solution.

Application and analysis

12 Copy and complete the diagram in the figure below. Insert the missing quantities in the yellow boxes or the missing processes in the white boxes.

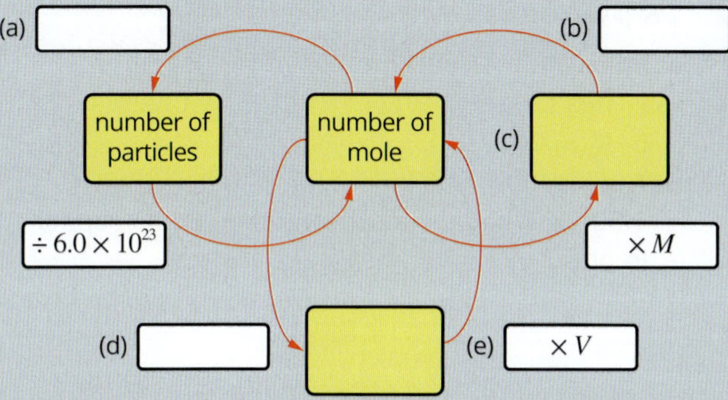

13 Is the following statement true or false? Explain your answer.

A supersaturated solution of $NaNO_3$ in 200 g water at 30°C will contain less dissolved solute than a saturated solution of KNO_3 in 400 g water at 20°C.

14 Use Table 13.1.2 from page 425 to help you answer this question.

a Using data from the table, make a general statement about the relationship between the amount of a gas dissolved in water as the temperature increases.

b When a saucepan of cold water is heated on a stove, bubbles of gas appear on the bottom of the saucepan long before the water boils. Explain this observation.

c A litre of water weighs 1 kg. A bucket of water at 20°C contains 0.24 g of dissolved oxygen. What is the volume of water in the bucket?

15 160 g of KNO_3 was dissolved in 200 g water at 60°C. What mass of crystals would form if the temperature were decreased to 30°C?

16 Use Table 13.1.2 on page 425 to help you to answer this question.

Two students were carrying out an investigation which involved measuring the amount of dissolved oxygen in two tanks of water. One tank held 350 litres of water and was kept at a temperature of 60°C. The other tank held 200 litres of water and was kept at a temperature of 20°C.

Assuming that the water in each tank was saturated with oxygen and that 1 litre of water weighs 1 kg, which tank held the greater mass of oxygen?

17 A pair of chemistry students investigated the solubility of potassium nitrate, KNO_3, in water. They weighed out 25 g of water into a test tube and then heated the test tube and contents to 60°C in a beaker of hot water. They then weighed out 13.0 g of potassium nitrate and dissolved it, while stirring it, in the water in the test tube. The students then took the test tube out of the hot water and allowed it to cool, stirring the solution gently with a thermometer. Small crystals of potassium nitrate first appeared when the temperature of the solution had cooled to 35°C.

According to the students' results, what was the solubility of potassium nitrate in g/100 g water at 35°C?

18 Precipitation reactions are used to remove impurities from wastewater. Write ionic equations for the following precipitation reactions.

a Aluminium sulfate is added to domestic wastewater to remove phosphate ions.

b Calcium hydroxide is added to industrial waste-water to remove cadmium(II) ions.

c Sodium sulfide is added to industrial wastewater to remove mercury(II) ions.

19 A sample of industrial wastewater was known to contain quantities of both lead(II) and chromium(III) ions. It was proposed to remove these ions from the wastewater by precipitation reactions followed by filtration. Aqueous solutions of three different compounds were proposed for the purpose of producing a precipitate when added to the wastewater. The three compounds were aluminium sulfate, calcium hydroxide and magnesium nitrate.

Write ionic equations for any precipitation reactions that would occur if the three aqueous solutions were added to the wastewater. Give reasons for your answers.

20 A sample of NaOH was dissolved in a beaker of water to make 40.0 mL of a solution with a concentration of 26.5 g L^{-1}. This solution was then diluted to make a solution with a concentration of 0.340 g L^{-1}.

How much water was added to the original solution of NaOH?

21 8.50 g of aluminium sulfate, $Al_2(SO_4)_3$, was dissolved in water to produce 320 mL of solution.

a Write a balanced ionic equation for aluminium sulfate dissolving in water

b i Calculate the concentration of aluminium sulfate in the solution in mol L^{-1}.

ii Calculate the concentration of aluminium ions in the solution in mol L^{-1}.

iii Calculate the concentration of sulfate ions in the solution in mol L^{-1}.

c Calculate the concentration of sulfate ions in the solution in %(m/v).

22 25.0 mL of a 0.30 M potassium sulfate solution and 40.0 mL of a 0.40 M potassium sulfate solution were added together and mixed thoroughly in an empty beaker.

a What was the final concentration of potassium sulfate in the new beaker?

b What was the concentration of potassium ions in the new beaker?

23 Calculate the mass, in g, of solute in each of the following solutions.

a 35 mL of a 0.70 M solution of H_2SO_4

b 75 ml of a solution of NaCl with concentration 1.9%(m/v)

c 2.5 kg of a solution of a lead compound in which the concentration of lead is 15.5 ppm

d 4.5 L of a solution of $CaCl_2$ with a concentration of 3.6 g L^{-1}

24 In the boxes labelled a–d of the table below, provide the unknown concentrations of the solutions as shown. Assume that 1 litre of all solutions has a mass of 1 kg.

	Concentration		
Compound/ion	**% (m/v)**	**mol L^{-1}**	**ppm**
NaOH	**a**	0.100	**b**
Cd^{2+}	**c**	**d**	3.00

25 A solution of volume 7.45 L contains 255 g of dissolved NaOH. Calculate the concentration of NaOH in each of these units:

a %(m/v)

b mol L^{-1}

c g L^{-1}

26 The concentration of copper in a wastewater sample is 1.9×10^{-3} g L^{-1}

a Calculate the concentration of the copper in:

i mol L^{-1}

ii %(m/v)

b Assume that 1.00 litre of the solution weighs 1.00 kg. What is the concentration of the copper in ppm?

OA ✓✓

CHAPTER

14 Analysis for acids and bases

There are many situations in which a chemist may want to find out the amount of a substance in solution. For example, the acidity of polluted water, the composition of antacid tablets or the analysis of consumer products in which the active ingredient is an acid or a base. In this chapter, you will learn about volumetric analysis, a method used for the analysis of acids and bases in water. By the end of this chapter, you will have a greater understanding of how specialised pieces of glassware are used to measure accurate volumes of solutions. You will also learn how measurements are used to calculate the concentration of an acid or a base in an aqueous solution.

Key knowledge

- volume–volume stoichiometry (solutions only) and application of volumetric analysis, including the use of indicators, calculations related to the preparation of standard solutions, dilution of solutions, and use of acid–base titrations (excluding back titrations) to determine the concentration of an acid or a base in a water sample. **14.1**, **14.2**

14.1 Principles of volumetric analysis

In Chapter 11, the properties of acids and bases were discussed in detail. As a general rule, acids are compounds that can donate a proton (a hydrogen ion) and bases are proton acceptors.

CHEMFILE

Borax

Sodium borate (sodium tetraborate), or borax, can be used as a primary standard since it does not decompose under normal storage conditions. Borax is a naturally occurring alkaline compound used in the manufacture of boric acid. Boric acid has many uses: preservative, antiseptic and fungicide, to manufacture glazes and enamels, and to fireproof textiles and wood. Domestically, boric acid and borates in tablet or powder form are used to kill insects and used in the manufacture of soaps and detergents (see figure below). A solution of sodium borate and borax is used to preserve and maintain the pH of eyedrops.

(a)

(b)

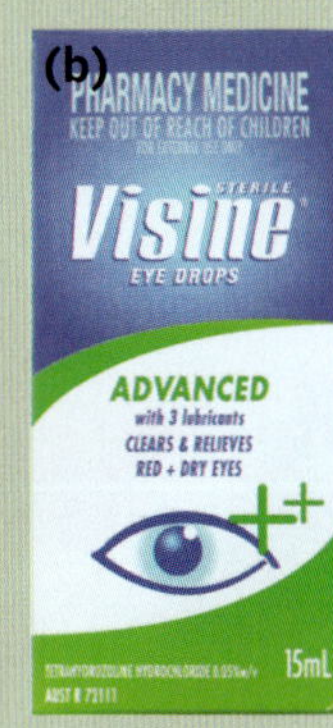

Borax powder can be used in the manufacture of (a) detergents and (b) eye drops.

Volumetric analysis is an analytical technique commonly employed by chemists. Volumetric analysis can be used to accurately determine the amount or concentration of a dissolved substance in a solution. The technique uses the reaction between a solution of unknown concentration and a solution of accurately known concentration. Although more sensitive and faster forms of instrumental analysis are now available, volumetric analysis is still used because it is simple and cheap.

In this section, you will learn about the principles of volumetric analysis with reference to standard solutions, indicator selection and the issues of accuracy of measurements and sources of error. You will then examine how volumetric analysis can be used to analyse for acids and bases in aqueous solutions.

PREPARING STANDARD SOLUTIONS

Commercial brick cleaners contain hydrochloric acid (HCl), a strong, monoprotic acid. To find the concentration of hydrochloric acid in a brick cleaner sample, the brick cleaner can be reacted with a standard solution of a weak base, such as sodium carbonate (Na_2CO_3). A standard solution is a solution with an accurately known concentration.

Before discussing the procedure of volumetric analysis further, it is important to understand how standard solutions are prepared.

Primary standards

Pure substances are widely used in the laboratory to prepare solutions of accurately known concentrations.

Substances that are so pure that the amount of substance, in moles, can be calculated accurately from their mass, are called **primary standards**.

A primary standard should:

- be readily obtainable in a pure form
- have a known chemical formula
- be easy to store without deteriorating or reacting with the atmosphere
- have a high molar mass to minimise the effect of errors in weighing
- be inexpensive.

Examples of primary standards are:

- acids: hydrated oxalic acid ($H_2C_2O_4{\cdot}2H_2O$) and potassium hydrogen phthalate ($KH(C_8H_4O_4)$).
- bases: sodium borate ($Na_2B_4O_7{\cdot}10H_2O$) and anhydrous sodium carbonate (Na_2CO_3).

The term '**anhydrous**' indicates there is no water present in the compound. For example, a sample of sodium carbonate may contain water molecules incorporated into the crystal lattice structure. Anhydrous sodium carbonate is obtained by heating a sample of sodium carbonate to above 100°C to remove any water. Storage in a desiccator prevents further absorption of water from the atmosphere.

Standard solutions

Standard solutions are prepared by dissolving an accurately measured mass of a primary standard in water to make an accurately measured volume of solution.

Digital balances are used in analytical laboratories to accurately weigh primary standards. A top-loading balance can weigh to an accuracy of between 0.1 g and 0.001 g, depending on the model. Analytical balances can weigh to an accuracy of between 0.0001 g and 0.000 01 g. The two types of balances commonly used can be seen in Figure 14.1.1.

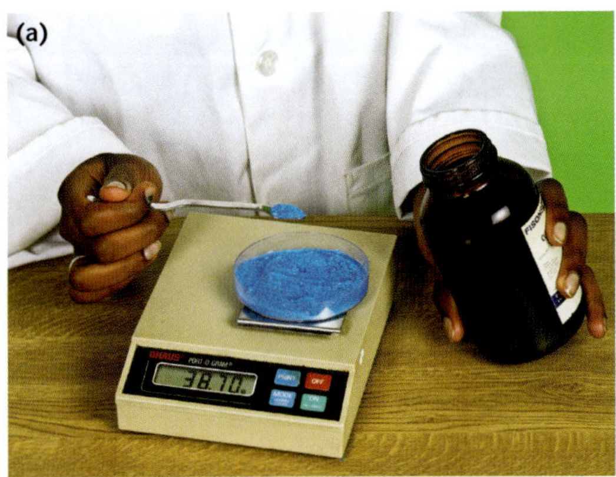
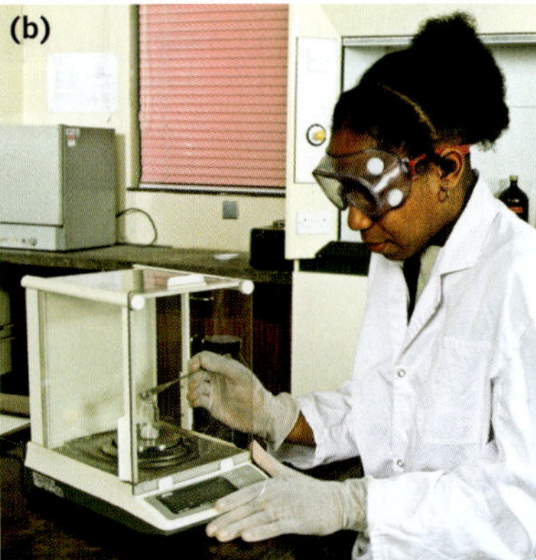

FIGURE 14.1.1 (a) A top-loading balance and (b) an analytical balance can be used to accurately weigh substances used in chemical analysis.

A **volumetric flask**, or standard flask (Figure 14.1.2), is used to prepare a solution that has an accurately known volume. Volumetric flasks of 100.00 mL, 250.0 mL and 1000.0 mL are frequently used in the laboratory.

FIGURE 14.1.2 Volumetric flasks of various sizes are used to prepare standard solutions.

FIGURE 14.1.3 This close-up view of the neck of a standard flask shows the bottom of the meniscus level with the graduation line.

A volumetric flask is filled so that the bottom of the meniscus (the curved surface of the solution) is level with the graduation line on the neck of the flask (Figure 14.1.3). Your eye should be level with the line to avoid **parallax errors** (errors caused by the shift in the apparent position of an object due to the viewing angle).

To prepare a standard solution from a primary standard, you need to dissolve an accurately known amount of the substance in **deionised water** to produce a solution of known volume. Deionised water is water that has had all ions removed. With its high level of purity, deionised water is used for cleaning glassware and preparing solutions. The steps in the preparation of standard solutions are shown in Figure 14.1.4.

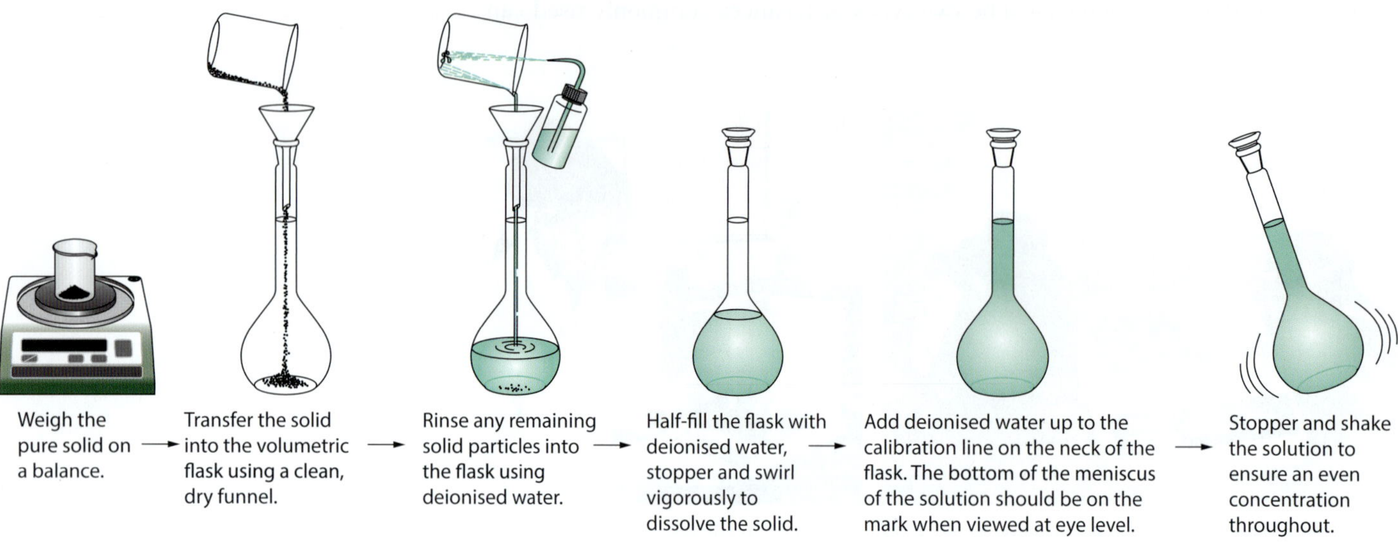

FIGURE 14.1.4 The steps taken to prepare a standard solution from a primary standard

In practice, making a standard solution directly from a primary standard is only possible for a few of the chemicals encountered in the laboratory. Many chemicals are impure because they decompose or react with chemicals in the atmosphere. For example:

- strong bases, such as sodium hydroxide (NaOH) and potassium hydroxide (KOH), absorb water and react with carbon dioxide in the air
- many hydrated salts, such as hydrated sodium carbonate ($Na_2CO_3{\cdot}10H_2O$), lose water to the atmosphere over time.

In addition, the concentrations of commercial supplies of strong acids (HCl, H_2SO_4 and HNO_3) cannot be accurately specified.

Solutions such as HCl(aq), H_2SO_4(aq), NaOH(aq) and KOH(aq) must be **standardised** to determine their concentration. An accurately measured volume of the solution is reacted with a known amount of a standard solution, such as Na_2CO_3(aq) or $KH(C_8H_4O_4)$(aq).

Many standard solutions, such as sodium hydroxide, are prepared as **stock solutions**. To obtain a convenient concentration to use in volumetric analysis, these stock solutions must first be diluted.

A stock solution is a large volume of a common reagent with a known concentration.

Concentration of standard solutions

In Chapter 13 you learnt how to calculate the concentration of a solution and determine how much solute is required to prepare a solution of a specific concentration.

The molar concentration of a standard solution is found using the following formulas:

$$\text{amount in mol, } n = \frac{\text{mass of solute (in g)}}{\text{molar mass (in g mol}^{-1}\text{)}} = \frac{m}{M}$$

$$\text{concentration, } c = \frac{\text{amount of solute (in mol)}}{\text{volume of solution (in L)}} = \frac{n}{V} = \frac{m}{M \times V}$$

Figure 14.1.5 summarises the steps involved in calculating the concentration of a standard solution from the accurately measured mass of a primary standard.

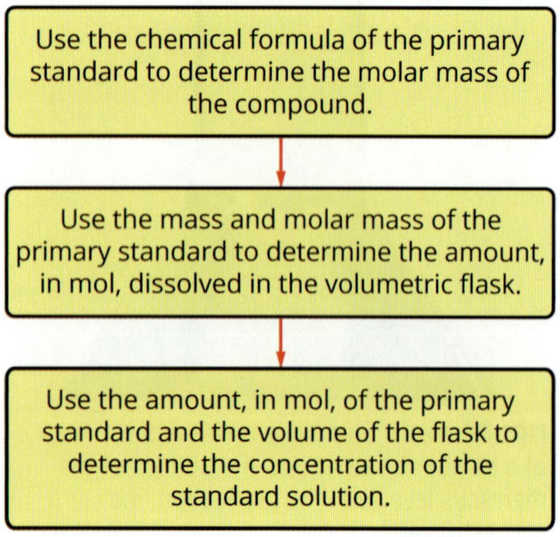

Use the chemical formula of the primary standard to determine the molar mass of the compound.

↓

Use the mass and molar mass of the primary standard to determine the amount, in mol, dissolved in the volumetric flask.

↓

Use the amount, in mol, of the primary standard and the volume of the flask to determine the concentration of the standard solution.

FIGURE 14.1.5 A summary of the steps used in the calculation of the concentration of a standard solution prepared from a primary standard

Worked example 14.1.1

CALCULATING THE CONCENTRATION OF A STANDARD SOLUTION PREPARED FROM A PRIMARY STANDARD

Calculate the concentration of a standard solution prepared from 58.36 g of hydrated oxalic acid ($H_2C_2O_4.2H_2O$) dissolved in a 1.00 L volumetric flask.

Thinking	Working
Use the chemical formula to determine the molar mass (*M*) of the compound.	The molar mass (*M*) of $H_2C_2O_4.2H_2O$ is 2.0 + 24.0 + 64.0 + 36.0 = 126.0 g mol^{-1}
Use the mass (*m*) and molar mass (*M*) of the compound and the formula $n = \frac{m}{M}$ to determine the amount, in mol.	$n = \frac{58.36}{126.0}$ = 0.4632 mol
Use the amount, in mol, to determine the concentration of the solution using the formula $c = \frac{n}{V}$. Express your answer to the appropriate number of significant figures.	$c = \frac{0.4632}{1.00}$ = 0.463 M The final result is rounded to three significant figures, corresponding to the smallest number of significant figures in the original data. Four significant figures are used in the earlier steps of the calculation to avoid rounding errors.

Worked example: Try yourself 14.1.1

CALCULATING THE CONCENTRATION OF A STANDARD SOLUTION PREPARED FROM A PRIMARY STANDARD

Calculate the concentration of a standard solution prepared from 65.03 g of hydrated oxalic acid ($H_2C_2O_4{\cdot}2H_2O$) dissolved in a 500 mL volumetric flask.

CONDUCTING VOLUMETRIC ANALYSES

Now that you know how to prepare a standard solution, you will learn how to perform a volumetric analysis. This technique is an experimental method using specialised glassware, and it is widely used to analyse solutions of acids and bases.

Titration

Volumetric analysis involves reacting a measured volume of a standard solution with a measured volume of the solution of unknown concentration in a process known as a **titration**.

The solutions are mixed until they have just reacted completely in the mole ratio indicated by the balanced chemical equation. This is called the **equivalence point**.

The equivalence point of a titration occurs when the reactants have reacted in the mole ratio indicated by the balanced chemical equation.

The number of moles of solute in the standard solution can be calculated from its concentration and volume. By using the mole ratio from the equation for the reaction, the number of moles of the solute in the solution of unknown concentration can be determined. The unknown solution concentration can then be calculated from the number of moles and the volume of solution. You will learn how to complete these calculations in Section 14.2.

Once the concentration of a solution has been determined from a titration, the solution can be described as a standard solution, and it is said to be standardised.

For example, if you want to determine the concentration of a hydrochloric acid solution, you can titrate a measured volume of the hydrochloric acid with a standard solution of sodium carbonate.

The reaction between hydrochloric acid and sodium carbonate solution is represented by the equation:

$$2HCl(aq) + Na_2CO_3(aq) \rightarrow 2NaCl(aq) + CO_2(g) + H_2O(l)$$

Because these reactants are in a 2:1 ratio, the equivalence point occurs when the amount of HCl added is exactly twice the amount of Na_2CO_3 present in the conical flask.

Table 14.1.1 describes the glassware used to carry out this analysis.

TABLE 14.1.1 Laboratory equipment used to determine the concentration of a solution of hydrochloric acid by volumetric analysis

Equipment	Diagram	Use
pipette		A **pipette** is used to accurately measure a specific volume, or **aliquot**, of the solution of hydrochloric acid.
conical flask		The hydrochloric acid aliquot from the pipette is added to a **conical flask** for analysis.
burette		A **burette** delivers a **titre** of the sodium carbonate standard solution. A titre is an accurately known volume of solution measured by the burette during a titration.

The reaction is complete when the equivalence point is reached. The concentration and volume of sodium carbonate (from the burette) is now known, as is the volume of the aliquot of hydrochloric acid used. The concentration of the acid can then be calculated.

As acid and base solutions are often colourless, an **acid–base indicator** is added to determine when the reaction is complete. (The indicator chosen should change colour when the solutions are neutralised.) The **end point** is the point during the titration at which the indicator changes colour. For an accurate analysis, the end point should be very close to the equivalence point.

The end point occurs when the indicator changes colour.

Figure 14.1.6 shows the equipment used to complete a titration for the reaction between hydrochloric acid and sodium hydroxide.

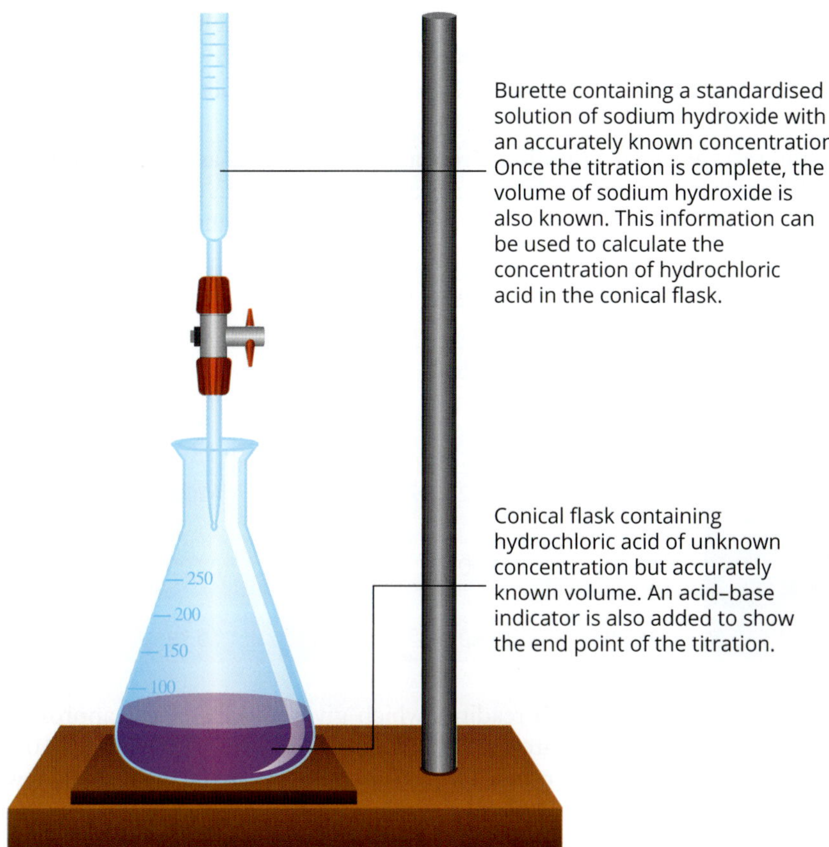

FIGURE 14.1.6 The equipment used in the titration of HCl (in the conical flask) with NaOH (in the burette)

The steps involved in an acid–base titration are as follows.

1 A known volume, or aliquot, of one of the solutions is measured using a pipette and transferred into a conical flask (see Figure 14.1.7).

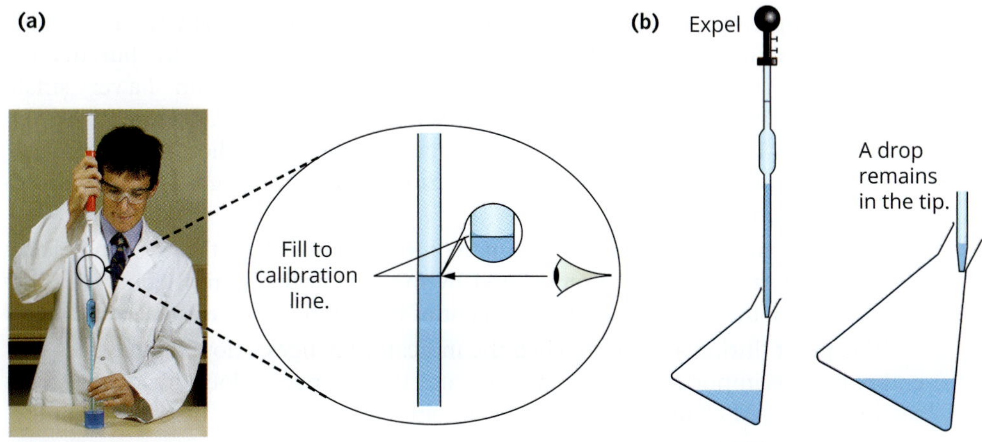

FIGURE 14.1.7 Taking an aliquot of a solution using a pipette. (a) Fill the pipette with the solution so that the bottom of the meniscus (curved upper surface of the solution) is level with the calibration line. (b) Expel the solution; the pipette has been designed so that a drop remains in the tip after an aliquot has been delivered.

2 A few drops of an appropriate acid–base indicator are added, so that a colour change signals the point at which the titration should stop.

3 The other solution is dispensed slowly into the conical flask from a burette until the indicator changes colour permanently (see Figure 14.1.8).

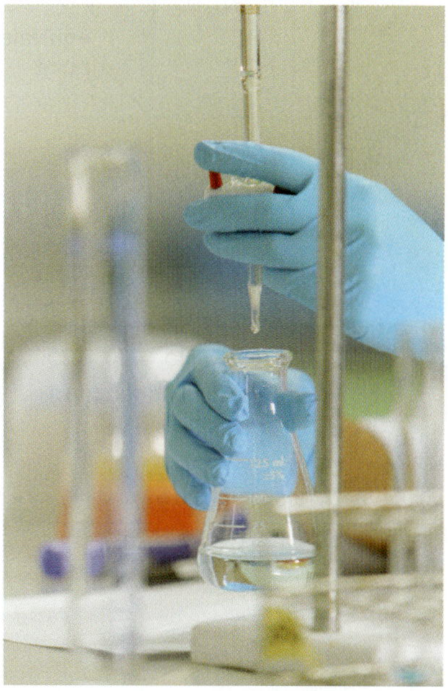

FIGURE 14.1.8 Titration from the burette into the conical flask containing the aliquot of the first solution

FIGURE 14.1.9 The burette is read from the bottom of the meniscus and the volume is estimated to the second decimal place. The volume measurement in this case is 22.18 mL.

Reading a burette scale

Burettes are usually calibrated in intervals of 0.10 mL. The volume of solution in a burette is measured at the bottom of the meniscus of the solution. The reading is estimated to the nearest 0.01 mL, as shown in Figure 14.1.9.

The titre delivered from the burette is calculated by subtracting the initial burette reading from the final burette reading.

To minimise errors, the titration is repeated several times and the **average titre** is found. Usually three **concordant titres** are used to find this average. Concordant titres are titres that are within a range of 0.10 mL from highest to lowest of each other.

Consider the titration data represented in Table 14.1.2.

TABLE 14.1.2 Titration data collected over five trials

Titration number	1	2	3	4	5
Final burette reading	20.20	40.82	20.64	41.78	21.86
Initial burette reading	0.00	21.00	1.00	22.00	2.00
Titre (mL)	20.20	19.82	19.64	19.78	19.86

The first reading is a rough reading, which gives an idea of the approximate end point. Titres 2, 4 and 5 are the concordant titres: they are within 0.10 mL from highest to lowest of each other. The difference between the highest and lowest readings is 19.86 − 19.78 = 0.08 mL, which is within the acceptable range for concordant results. The average (mean) titre is:

$$\frac{19.82+19.78+19.86}{3}=19.82 \text{ mL}$$

Concordant titres are titres that are within a range of 0.10 mL from highest to lowest of each other.

Selecting an indicator

During a titration, the pH of the solution in the conical flask changes as the solution is delivered from the burette. A graph showing this change in pH is called a **titration curve** or **pH curve**.

An example of a pH curve is shown in Figure 14.1.10. Initially the pH of the sodium hydroxide in the conical flask is 14. As HCl is added from the burette, the pH decreases slowly at first. Near the equivalence point, the addition of a very small volume of hydrochloric acid produces a large change in pH. In this titration, the pH changes from 10 to 4 with just one drop of acid. By using an indicator that changes colour within this pH range, one drop will cause a colour change. This is referred to as a sharp end point.

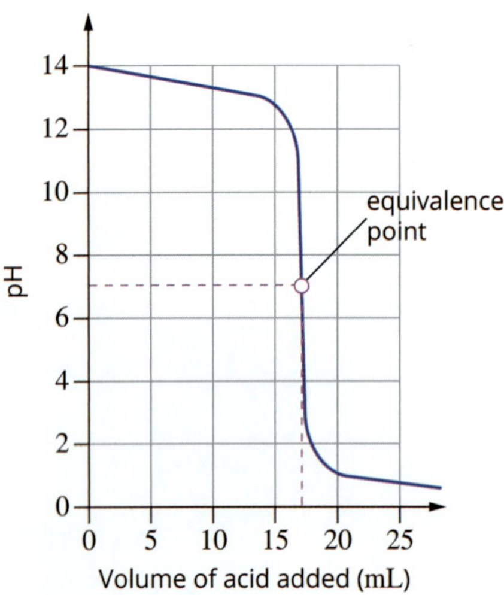

FIGURE 14.1.10 Change in pH during a titration between a strong base (sodium hydroxide solution) and a strong acid (hydrochloric acid)

The equivalence point is the point in the titration when the two chemicals have reacted in the mole ratio indicated by the balanced chemical equation. The equivalence point occurs when the gradient of the titration curve is steepest.

The point during a titration when the indicator changes colour is known as the end point. It is important to select an indicator that changes colour during the steep section of the pH change, so that the end point and equivalence point occur at the same time during the titration.

The equivalence point occurs when the gradient of the pH curve is steepest. The end point occurs when the indicator changes colour.

The colours of common acid–base indicators and the pH range over which they change colour are listed in Table 14.1.3 and shown in Figure 14.1.11.

TABLE 14.1.3 pH range of some common indicators

Indicator	pH range	Colour change from lower pH to higher pH in range
methyl orange	3.1–4.4	red → yellow
bromophenol blue	3.0–4.6	yellow → blue
methyl red	4.4–6.2	red → yellow
bromothymol blue	6.0–7.6	yellow → blue
phenol red	6.8–8.4	yellow → red
phenolphthalein	8.3–10.0	colourless → pink

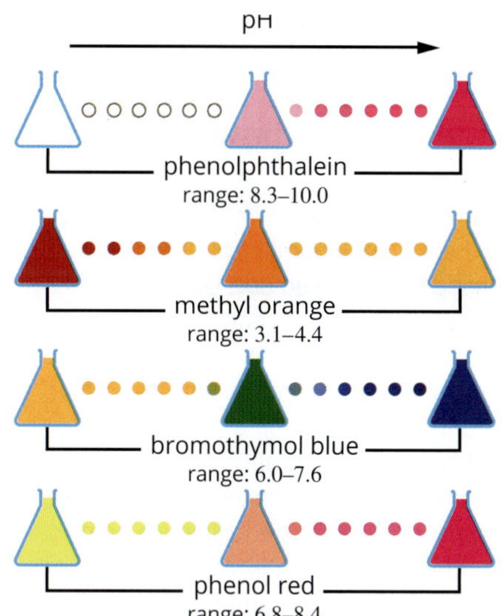

FIGURE 14.1.11 To identify the equivalence point of a titration, you must use a suitable indicator. Although all of these indicators display different colours at low pH and high pH, the pH range in which they change colour is different.

Figure 14.1.12 compares the titration curves obtained when a strong acid is added to a strong base and when a strong acid is added to a weak base. Note that there is a sharp drop in pH in both graphs at the equivalence point. However, because this drop occurs over different ranges, the choice of indicator is important in detecting the equivalence point. Phenolphthalein can be used successfully for the first titration, but not for the second.

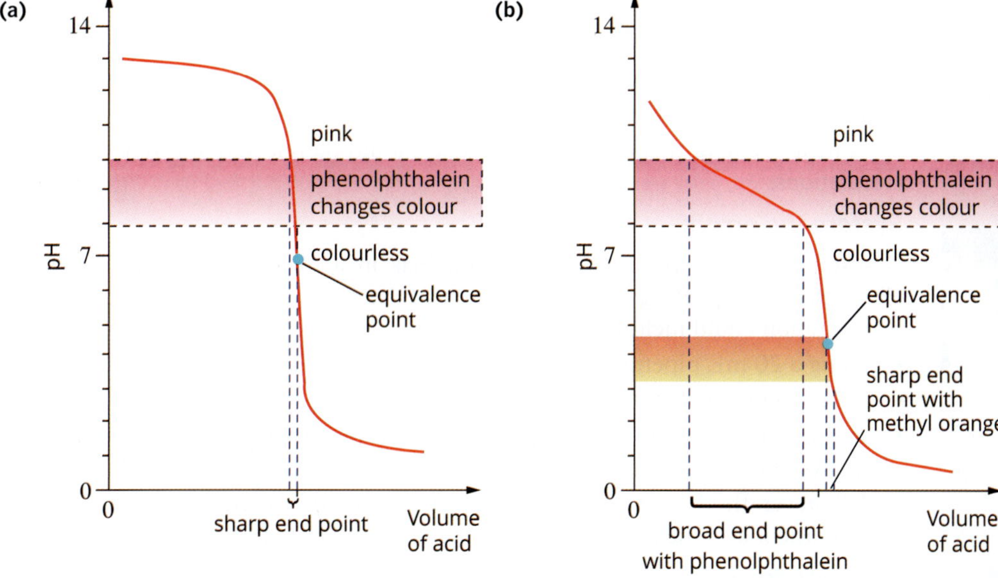

FIGURE 14.1.12 Titration curves showing the pH during a titration of (a) a strong base with a strong acid, and (b) a weak base with a strong acid. Phenolphthalein, which changes colour in the pH range 8.3–10.0, gives a sharp end point in (a) but a broad end point in (b). Methyl orange, which changes colour between pH 3.1 and 4.4, would be a more suitable indicator for the second titration.

Commonly used indicators for different types of acid–base titrations are shown in Table 14.1.4. No indicator is suitable when a weak acid is titrated with a weak base, because there is no sharp change in pH at the end point.

TABLE 14.1.4 Indicators for different acid base titrations

Titration type	Typical pH change near the equivalence point	Commonly used indicator
strong acid – strong base	4–10	bromothymol blue or phenolphthalein or methyl orange
strong acid – weak base	3–7	methyl orange
weak acid – strong base	7–11	phenolphthalein

CHEMFILE

Automated titrators

Automated titrators, such as the one shown below, are used in laboratories where a large number of titrations need to be carried out. During the titration, a pH or conductivity probe is lowered into a beaker containing the solution to be analysed. A sensor controls the addition of the solution of known concentration that is being added from an electronic burette. The sensor detects the equivalence point, and a microcomputer calculates the volume of solution added from the burette. The microcomputer calculates the concentration of the solution being analysed using the pH and/or conductivity data obtained.

Different formulas can be programmed into the microcomputer, depending on the nature of the substance being analysed. The titrator shown has an electronic burette that is refilled automatically. As the circular platform rotates, the next beaker containing a new sample to be analysed moves under the sensor.

Automated titrators can rapidly analyse a large number of samples.

For more accurate volumetric analyses, a pH meter or pH probe can be used instead of an indicator. The data is used to plot a titration curve, and the equivalence point can be identified from the steepest section of the curve.

PRECISION AND ACCURACY IN VOLUMETRIC ANALYSIS

In Chapter 1 you learnt how to evaluate investigation methods and suggest ways to improve accuracy (how close a measured value is to the true value) and precision (how close measured values are to each other). Methods used for accurate and precise **quantitative analysis** should be designed to minimise errors. Where errors cannot be avoided, any discussion of results should refer to the level of inaccuracy that may have accumulated. This requires an understanding of the different types of experimental errors.

Types of error

Poor accuracy in measurements is usually associated with a systematic error; poor precision is associated with random errors. It is important to distinguish mistakes from experimental errors. Mistakes can be avoided, whereas experimental errors can be minimised, but not entirely avoided because they are part of the process of measurement.

Mistakes

Mistakes are avoidable errors. Mistakes made during acid–base titrations could include:

- misreading the numbers on a scale
- using a pipette of incorrect volume
- spilling a portion of a sample
- incorrect rinsing of glassware.

A measurement that involves a mistake should be rejected and not included in calculations.

Systematic errors

A systematic error produces a constant bias in a measurement that cannot be eliminated by repeating the measurement. Systematic errors that affect an acid–base titration could include:

- using a 20 mL pipette that delivers 20.2 mL
- using an unsuitable indicator
- the presence of extra components in the sample other than the chemical being analysed that react during the titration.

Whatever the cause, the resulting error is in the same direction for every measurement and the average is either higher or lower than the true value.

Random errors

Random errors follow no regular pattern. The measurement is sometimes too large and sometimes too small. Random errors in volumetric analysis include:

- mass uncertainties from an analytical balance, usually ±1 in the last decimal place of the reading.
- changes in the volume of a titre due to fluctuations of temperature or minor variations in concentration
- difficulty in judging the fraction between two 0.1 mL scale markings on a burette.

The effects of random errors can be reduced by taking the average of several observations. In volumetric analysis, the average of three concordant titres is used to reduce random error.

Rinsing volumetric glassware

To ensure that glassware is completely clean, it is rinsed before a volumetric analysis is conducted. This removes any trace chemicals from the glassware, and makes the analytical results more precise and accurate.

Table 14.1.5 describes how the glassware used in volumetric analysis should be rinsed. Rinsing with the wrong liquid can cause errors in the analysis.

TABLE 14.1.5 Techniques for rinsing equipment for volumetric analysis

Glassware	Correct	Incorrect
burette pipette	The final rinse should be with the acid or base to be transferred by them.	Rinsing with water will dilute the acid or base solution.
volumetric flask conical flask	Only rinse these with deionised water.	Rinsing with acidic or basic solutions will introduce unmeasured amounts of acids or bases into the flask which can react and affect the results.

CASE STUDY

Native Australian fruits

Native Australian fruits have been part of the diet of Indigenous Australians for tens of thousands of years. Fruits are eaten not only for their taste, but also for their nutritional and medicinal values. The taste of Australian fruits, as with introduced fruits, varies from sweet to sour depending on the relative amounts of sugars and acids in the fruit. Native Australian fruits such as the riberry (Figure 14.1.13) have a sweet taste because they contain a high proportion of sugars such as glucose, fructose and sucrose.

FIGURE 14.1.13 Riberries are valued by Indigenous Australians for their sweet taste, nutritional value and medicinal properties.

Fruits that contain a high proportion of acids, such as the Kakadu plum (Figure 14.1.14a), Davidson plum (Figure 14.1.14b) and quandong (Figure 14.1.14c), have a sour or tart taste. Acids found in native and introduced fruits include citric acid, tartaric acid, malic acid and ascorbic acid (vitamin C). A quandong has twice the vitamin C content of an orange, and a Davidson plum has 100 times the vitamin C content of an orange.

Ascorbic acid also acts as an antioxidant, interrupting the continued formation of free radicals associated with illnesses such as cancer, heart disease, diabetes and Alzheimer's disease. Free radicals are molecules or atoms with unpaired outer-shell electrons. They are unstable and highly reactive, causing damage to biologically important molecules such as lipids (fats), proteins and nucleic acids (including DNA). Damage caused by free radicals may be reduced by eating fruits that are rich in antioxidants, such as berries, black plums and red grapes.

FIGURE 14.1.14 Native Australian fruits such as (a) the Kakadu plum, (b) the Davidson plum and (c) the quandong have a tart or sour taste because of their high acid content. These fruits each have a higher vitamin C content than an orange.

The acid content of fruits can be determined using the volumetric techniques described in this section. Analysis has shown that the Kakadu plum has significantly higher antioxidant levels than blueberries, which are considered to have the highest antioxidant content of the introduced fruits. The Kakadu plum has been recorded as having a vitamin C content as high as 7000 mg per 100 g of fruit, the highest of any known fruit; in comparison, an orange has a typical vitamin C content of about 53 mg per 100 g. The Kakadu plum is also rich in other compounds that have antioxidant properties.

14.1 Review

OA ✓✓

SUMMARY

- Volumetric analysis is an analytical technique for determining the concentration of a solution by titrating it against a solution of known concentration and volume.
- A solution of accurately known concentration is referred to as a standard solution.
- Standard solutions can be prepared from primary standards or by titrating an existing solution with another standard solution to determine its concentration.
- A substance is suitable for use as a primary standard if it:
  - is readily obtainable in a pure form
  - has a known chemical formula
  - is easy to store without deteriorating or reacting with the atmosphere
  - has a high molar mass to minimise the effect of errors in weighing
  - is inexpensive.
- The concentration, in mol L^{-1}, of a prepared standard solution can be determined by measuring the mass of solid dissolved and the volume of solution prepared.
- Volumetric flasks, pipettes and burettes are accurately calibrated pieces of glassware used in volumetric analysis.
- In a titration, a measured volume of a standard solution is reacted with a measured volume of the solution whose concentration is to be determined.
- The equivalence point of a reaction occurs when the reactants have been mixed in the mole ratio shown by the reaction equation.
- The end point is the point during the titration when the indicator changes colour.
- Indicators for a titration should be selected so that the end point occurs when the equivalence point has been reached.
- Three concordant titres are usually obtained during a titration. Concordant titres vary within narrowly specified limits, usually within a range of 0.10 mL from the highest to the lowest.
- When rinsing glassware before a titration, it is important to ensure that:
  - conical and volumetric flasks are rinsed with deionised water
  - the burette and pipette are rinsed with the acid or base to be transferred by them.

KEY QUESTIONS

Knowledge and understanding

1 Hydrated oxalic acid ($H_2C_2O_4{\cdot}2H_2O$) is used as a primary standard in volumetric analysis. Some properties of $H_2C_2O_4{\cdot}2H_2O$ are listed below. Which one of these properties is *not* important in its use as a primary standard?

A It is a soft, white crystalline solid.

B It has a molar mass of 126 g mol^{-1}.

C It is highly soluble in water.

D Its purity is greater than 99.5%.

2 Select from the following list of words to complete the summary paragraph about a titration:

pipette; measuring cylinder; beaker; volumetric flask; primary standard; indicator; base; standard solution; burette; indicator; titre; aliquot

A sample of anhydrous sodium carbonate of approximately 2 g is weighed accurately. (The solid must be dry if it is to be used as a ____________.) The solid is tipped into a ____________ and shaken with about 50 mL of deionised water until the solid dissolves. More water is added to make the solution to a volume of exactly 100.0 mL. A 20.00 mL ____________ of the solution is taken by using a ____________ and placed in a conical flask. A few drops of methyl orange ____________ are added and the mixture is titrated against dilute hydrochloric acid.

3 Explain the difference between:

a a standard solution and a primary standard

b the equivalence point and the end point

c a burette and a pipette

d an aliquot and a titre.

4 A student weighed out the required mass of a primary standard, placed it in a 500 mL volumetric flask, then added 500 mL of deionised water to the flask. Explain why this procedure would not produce a standard solution of accurately known concentration.

5 The following titres were obtained when a solution of sulfuric acid was titrated against a standard solution of potassium carbonate.

Trial	1	2	3	4	5
Titre (mL)	26.28	25.46	25.38	25.62	25.42

Calculate the average of the concordant titres of sulfuric acid.

6 **a** Use Table 14.1.3 (page 451) to match the equivalence points with the following pH values with an appropriate indicator.

i 3.6

ii 9.9

iii 7.8

iv 5.2

b If a solution was pink in phenolphthalein indicator, what colour would the same solution be in methyl orange indicator?

7 Potassium hydrogen phthalate ($KH(C_8H_4O_4)$) is used as a primary standard for the analysis of bases. Calculate the concentration of a standard solution prepared in a 100.00 mL volumetric flask by dissolving 3.527 g of potassium hydrogen phthalate in deionised water. The molar mass of $KH(C_8H_4O_4)$ is 204.1 g mol^{-1}.

Analysis

8 **a** Calculate the mass of anhydrous potassium carbonate (K_2CO_3) required to prepare 250.0 mL of a 0.400 M standard solution.

b Calculate the mass of hydrated oxalic acid ($H_2C_2O_4.2H_2O$) required to prepare 400.0 mL of a 0.095 M standard solution.

9 A student uses aliquots of a standardised solution of sodium hydroxide to determine the concentration of an ethanoic acid (CH_3COOH) solution by volumetric analysis. State the solution the student should use for the final rinsing of each piece of glassware used in the titration.

14.2 Stoichiometry

Measuring and predicting quantities is a very important part of chemistry. In Chapter 7, you were introduced to the mole concept. Chemists can apply the mole concept to chemical reactions to determine the quantities of reactants involved or products formed. These types of calculations are called stoichiometric calculations. In this section, you will use data from a titration and stoichiometric calculations to calculate the concentration of the acid or base being analysed.

Stoichiometry is the study of ratios of moles of substances and is based on the law of conservation of mass. In a chemical reaction, the total mass of all products is equal to the total mass of all reactants. Another way of expressing this is that in a chemical reaction, atoms are neither created nor destroyed. Consequently, given the **amount**, or mole, of one substance involved in a chemical reaction and a balanced equation for the reaction, you can calculate the amounts of all other substances involved.

MOLE RATIOS

Consider the reaction between solutions of sodium carbonate and hydrochloric acid:

$$Na_2CO_3(aq) + 2HCl(aq) \rightarrow 2NaCl(aq) + CO_2(g) + H_2O(l)$$

The **coefficients**, or whole numbers used to balance the equation, show the **mole ratio** between the reactants and products involved in the reaction. The equation indicates that 1 mole of $Na_2CO_3(aq)$ reacts with 2 moles of HCl(aq) to form 2 moles of NaCl(aq), 1 mole of $CO_2(g)$ and 2 moles of $H_2O(l)$. Examples of what this means in more general terms are:

- the amount of hydrochloric acid required to react with the sodium carbonate will be double the amount of sodium carbonate reacted.
- The amount of carbon dioxide produced will be the same as the amount of sodium carbonate reacted.

These ratios can be expressed in formulas:

$$\frac{n(HCl)}{n(Na_2CO_3)} = \frac{2}{1} \text{ and } \frac{n(CO_2)}{n(Na_2CO_3)} = \frac{1}{1}$$

In general, for stoichiometric calculations, you will be told, or will be able to work out, the number of moles of one chemical in the reaction (called the 'known chemical'). This can then be used with the mole ratio to determine the number of moles of one of the other reactants or products involved in the reaction (called the 'unknown chemical').

The mole ratio shows the relationship between the unknown and known chemicals using ratios:

$$\frac{n(\text{unknown chemical})}{n(\text{known chemical})} = \frac{\text{coefficient of unknown chemical}}{\text{coefficient of known chemical}}$$

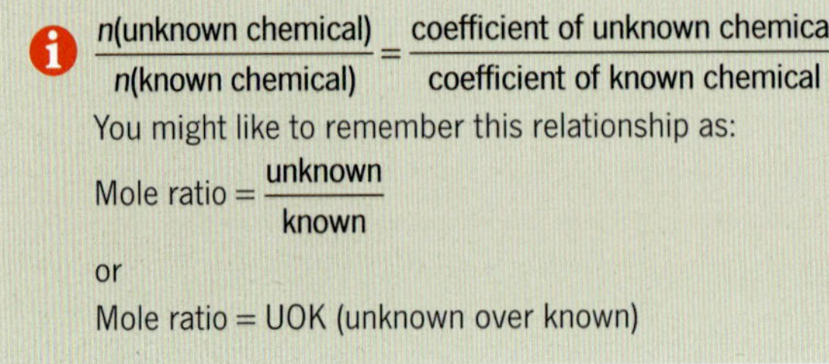

$$\frac{n(\text{unknown chemical})}{n(\text{known chemical})} = \frac{\text{coefficient of unknown chemical}}{\text{coefficient of known chemical}}$$

You might like to remember this relationship as:

$$\text{Mole ratio} = \frac{\text{unknown}}{\text{known}}$$

or

Mole ratio = UOK (unknown over known)

Worked example 14.2.1

USING MOLE RATIOS

Calculate the amount of hydrochloric acid (HCl), in mol, required to completely react with 2.4 mol of sodium carbonate (Na_2CO_3).

Thinking	Working
Write a balanced equation for the reaction.	$Na_2CO_3(aq) + 2HCl(aq) \rightarrow 2NaCl(aq) + CO_2(g) + H_2O(l)$
Note the number of moles of the known substance.	$n(Na_2CO_3) = 2.4$ mol
Identify the mole ratio using: $\frac{n(\text{unknown chemical})}{n(\text{known chemical})} = \frac{\text{coefficient of unknown chemical}}{\text{coefficient of known chemical}}$	$\frac{n(HCl)}{n(Na_2CO_3)} = \frac{2}{1}$
Calculate the number of moles of the unknown substance using: $n(\text{unknown}) = \text{mole ratio} \times n(\text{known})$	$n(HCl) = \frac{2}{1} \times 2.4$ $= 4.8$ mol

Worked example: Try yourself 14.2.1

USING MOLE RATIOS

Calculate the amount of nitric acid (HNO_3), in mol, required to completely react with 0.50 mol of potassium carbonate (K_2CO_3).

When carrying out any stoichiometric calculations, you must always clearly state the mole ratio for the reaction you are working with.

SOLUTION VOLUME–VOLUME STOICHIOMETRY

When a titration is carried out in the laboratory, the calculations using the data from the titration involve determining the number of moles of the reactants. Such calculations are referred to as examples of **solution volume–volume stoichiometry**.

Calculating the concentration of an acid or base in aqueous solution

Volume–volume stoichiometry can be combined with your knowledge of acid–base reactions to calculate the concentration of an acid or base in an aqueous solution.

There are several steps involved in volume–volume stoichiometry calculations:

1 Write a balanced equation for the reaction.

2 Calculate the amount, in mol, of the substance with known volume and concentration using $n = c \times V$.

3 Use the mole ratio from the equation to calculate the amount, in mol, of the unknown substance.

4 Calculate the required volume or concentration using $c = \frac{n}{V}$ or $V = \frac{n}{c}$.

The steps in a stoichiometric calculation can be summarised as shown in the flow chart in Figure 14.2.1.

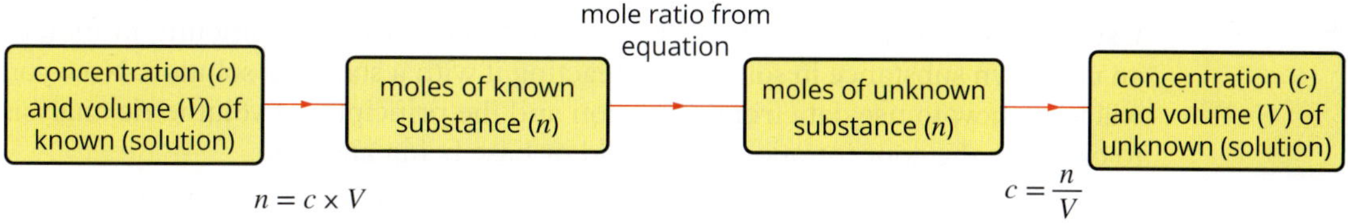

FIGURE 14.2.1 Flow chart for solution volume–volume stoichiometry calculations

CHEMFILE

Phosphate mining

The major use of sulfuric acid is in the production of fertilisers, such as superphosphate of lime and ammonium sulfate. The South Pacific island of Nauru was once a major source of rock phosphate for the Australian superphosphate industry. The phosphate was formed over thousands of years from the excretions and bodies of sea birds. Over 80% of the island has been mined, resulting in significant environmental damage and changes to the islander's lifestyle and culture. The island's phosphate deposits were virtually depleted by the year 2000. In 1997 the Australian Government paid Nauru $75 million as compensation for the environmental damage.

Worked out phosphate fields on the island of Nauru in the South Pacific

Worked example 14.2.2

A SOLUTION VOLUME–VOLUME CALCULATION

What volume of 0.100 M sulfuric acid reacts completely with 17.8 mL of 0.15 M potassium hydroxide?

Thinking	Working
Write a balanced full equation for the reaction.	$2KOH(aq) + H_2SO_4(aq) \rightarrow K_2SO_4(aq) + 2H_2O(l)$
Calculate the amount, in mol, of the substance with known volume and concentration.	The volume and concentration of potassium hydroxide solution are given, so you use $n = c \times V$ (remember that volume must be expressed in litres). $n(KOH) = c \times V$ $= 0.15 \times 0.0178$ $= 0.00267$ mol
Use the mole ratio from the equation to calculate the amount, in mol, of the required substance.	The balanced equation shows that 1 mol of sulfuric acid reacts with 2 mol of potassium hydroxide. $\frac{n(H_2SO_4)}{n(KOH)} = \frac{1}{2}$ $n(H_2SO_4) = \frac{1}{2} \times n(KOH)$ $= \frac{1}{2} \times 0.00267$ $= 0.00134$ mol
Calculate the volume or concentration required. Express your answer to the appropriate number of significant figures.	The volume of H_2SO_4 is found by using $n = cV$. $V(H_2SO_4) = \frac{n}{c}$ $= \frac{0.00134}{0.100}$ $= 0.0134$ L $= 13.4$ mL So 13 mL of 0.100 M H_2SO_4 will react completely with 17.8 mL of 0.150 M KOH solution. The final result is rounded off to two significant figures, corresponding to the smallest number of significant figures in the original data. Three significant figures are used in the earlier steps of the calculation to avoid rounding off errors.

Worked example: Try yourself 14.2.2

A SOLUTION VOLUME–VOLUME CALCULATION

What volume of 0.500 M hydrochloric acid (HCl) reacts completely with 25.0 mL of 0.100 M calcium hydroxide ($Ca(OH)_2$) solution? The salt formed in this acid–base reaction is calcium chloride.

CALCULATIONS IN VOLUMETRIC ANALYSIS

Using volumetric analysis, chemists are able to determine the amount, in mol, of an unknown substance in solution by reacting it with a standard solution. Now you will learn how to use data from a titration and the principles of volumetric analysis to calculate the concentration of the acid or base being analysed.

Consider the data gained in an acid–base titration in which the concentration of a dilute solution of hydrochloric acid was determined by titration with a standard solution of sodium carbonate. The data gained from the titration is summarised in Table 14.2.1.

TABLE 14.2.1 Sample data obtained from a volumetric analysis

volume of aliquot of HCl	25.00 mL
concentration of standard Na_2CO_3 solution	1.00 M
titre volumes of Na_2CO_3	25.05 mL, 22.10 mL, 22.00 mL, 22.05 mL

The concentration of hydrochloric acid is calculated by following a number of steps. These steps are summarised in the flow chart shown in Figure 14.2.2.

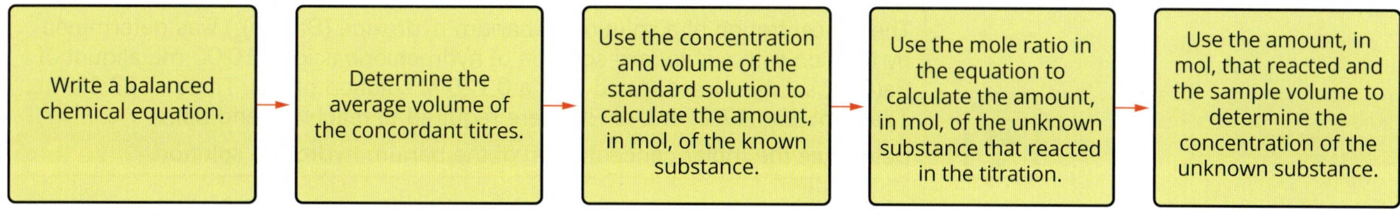

FIGURE 14.2.2 Flow chart summarising the steps in the calculation of the concentration of an unknown substance using data from a titration

Worked example 14.2.3 shows how the concentration of an acid can be determined using data from an acid–base titration.

Worked example 14.2.3

SIMPLE TITRATION CALCULATION

The concentration of hydrochloric acid was determined by titration with a standard solution of sodium hydroxide.

A 25.00 mL aliquot of HCl was titrated with a 1.00 M solution of sodium hydroxide. Titres of 25.05 mL, 22.10 mL, 22.05 mL and 22.00 mL were required to reach the end point. Determine the molar concentration of the hydrochloric acid solution.

Thinking	Working
Write a balanced chemical equation for the reaction.	An acid is reacting with a metal hydroxide, or base, so the products will be a salt and water. $HCl(aq) + NaOH(aq) \rightarrow NaCl(aq) + H_2O(l)$
Determine the average volume of the concordant titres.	The titre of 25.05 mL is discarded as it is not concordant (i.e. it is not within a range of 0.10 mL from highest to lowest of the other titre volumes). $\text{average titre} = \frac{22.10 + 22.05 + 22.00}{3}$ $= 22.05 \text{ mL}$
Calculate the amount, in mol, of the standard solution that was required to reach the end point.	$n(NaOH) = cV$ $= 1.00 \times 0.02205$ $= 0.02205 \text{ mol}$
Use the mole ratio in the equation to calculate the amount, in mol, of the unknown substance that would have reacted with the given amount, in mol, of the standard solution.	$\text{mole ratio} = \frac{n(HCl)}{n(NaOH)} = \frac{1}{1}$ $n(HCl) = n(NaOH)$ $= 0.02205 \text{ mol}$
Determine the concentration of the unknown substance.	$c(HCl) = \frac{n}{V}$ $= \frac{0.02205}{0.02500}$ $= 0.882 \text{ M}$
Express your answer to the appropriate number of significant figures.	Concentration of HCl = 0.882 M The final result is rounded off to three significant figures, corresponding to the smallest number of significant figures in the original data. Four significant figures are used in the earlier steps of the calculation to avoid rounding off errors.

Worked example: Try yourself 14.2.3

SIMPLE TITRATION CALCULATION

The concentration of a solution of barium hydroxide ($Ba(OH)_2$) was determined by titration with a standard solution of hydrochloric acid. A 10.00 mL aliquot of $Ba(OH)_2$ solution was titrated with a 0.125 M solution of HCl. Titres of 17.23 mL, 17.28 mL and 17.21 mL of HCl were required to reach the end point.

Determine the molar concentration of the barium hydroxide solution.

TABLE 14.2.2 Data from a titration involving diluted concrete cleaner

volume of undiluted concrete cleaner	25.00 mL
volume of diluted concrete cleaner	250.0 mL
volume of titre of diluted concrete cleaner	19.84 mL
concentration of standard Na_2CO_3 solution	0.4480 M
volume of aliquot of Na_2CO_3 solution	20.00 mL

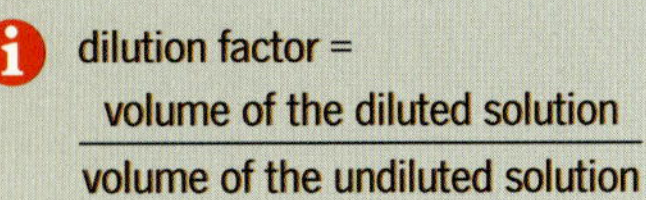

Titrations that involve dilution

It is often necessary to dilute a solution before carrying out a titration in order to obtain concentrations that are more convenient to use. This will result in titres that are within the volume range of the burette.

Suppose you want to perform an acid–base titration to find the concentration of hydrochloric acid in concrete cleaner. The concrete cleaner is so concentrated that it has to be accurately diluted before the titration. The following additional data is recorded:

- the volume of the aliquot of undiluted concrete cleaner
- the volume of diluted solution that is prepared.

The data obtained from such a titration is summarised in Table 14.2.2.

In this titration, 25.00 mL of concrete cleaner was diluted to 250.0 mL in a volumetric flask prior to taking a sample for titration. This means the **dilution factor** is $\frac{250.0}{25.00} = 10.00$. The undiluted concrete cleaner will be 10.00 times more concentrated than the concentration of the aliquot. This will be taken into account in the calculations.

The steps required to calculate the concentration of undiluted concrete cleaner are summarised in the flow chart shown in Figure 14.2.3.

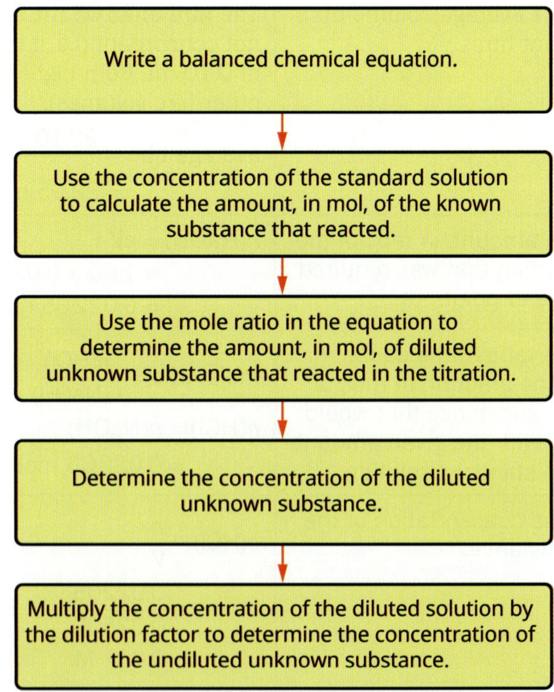

FIGURE 14.2.3 This flow chart shows the steps used in the calculation of the concentration of an unknown substance that has been diluted for use in a titration.

Worked example 14.2.4

A TITRATION CALCULATION THAT INVOLVES DILUTION

A commercial concrete cleaner contains hydrochloric acid. A 25.00 mL volume of cleaner was diluted to 250.0 mL in a volumetric flask. A 20.00 mL aliquot of 0.448 M sodium carbonate solution was placed in a conical flask. Methyl orange indicator was added and the solution was titrated with the diluted cleaner. The indicator changed permanently from yellow to pink when 19.84 mL of the cleaner was added.

Calculate the molar concentration of hydrochloric acid in the concrete cleaner.

Thinking	Working
Write a balanced chemical equation.	A dilute acid is reacting with a metal carbonate so the products will be a salt, water and carbon dioxide gas. $2HCl(aq) + Na_2CO_3(aq) \rightarrow 2NaCl(aq) + H_2O(l) + CO_2(g)$
Using the concentration of the standard solution, calculate the amount, in mol, of the known substance that reacted in the titration. Remember that volume must be expressed in litres.	$n(Na_2CO_3) = cV$ $= 0.448 \times 0.02000$ $= 0.008960$ mol
Use the mole ratio in the equation to calculate the amount, in mol, of the diluted unknown solution that reacted in the titration.	$\frac{n(HCl)}{n(Na_2CO_3)} = \frac{2}{1}$ $n(HCl) = \frac{2}{1} \times n(Na_2CO_3)$ $= \frac{2}{1} \times 0.008960$ $= 0.01792$ mol
Calculate the concentration of diluted unknown solution reacting in the titration.	V(diluted HCl) = 0.01984 L $c(HCl) = \frac{n}{V}$ $= \frac{0.01792}{0.01984}$ $= 0.9032$ M
Multiply by the dilution factor to determine the concentration of undiluted concrete cleaner. Express your answer to the appropriate number of significant figures.	Dilution factor $= \frac{250.0}{25.00} = 10.00$ So undiluted c(HCl) = diluted c(HCl) × 10.00 $= 0.9032 \times 10.00$ $= 9.03$ M The final result is rounded off to three significant figures, corresponding to the smallest number of significant figures in the original data. Four significant figures are used in the earlier steps of the calculation to avoid rounding off errors.

Worked example: Try yourself 14.2.4

A TITRATION CALCULATION THAT INVOLVES DILUTION

A commercial concrete cleaner contains hydrochloric acid. A 20.00 mL volume of cleaner was diluted to 250.0 mL in a volumetric flask. A 20.00 mL aliquot of 0.241 M sodium carbonate solution was placed in a conical flask. Methyl orange indicator was added and the solution was titrated with the diluted cleaner. The indicator changed permanently from yellow to pink when 18.68 mL of the cleaner was added.

Calculate the molar concentration of hydrochloric acid in the concrete cleaner.

CASE STUDY **ANALYSIS**

Determination of the ammonia content of window cleaner

In April 2021, paramedics attended a Melbourne childcare centre in response to an emergency call reporting multiple children suffering a suspected reaction to an unknown substance. Twelve children were treated at the centre after accidently drinking cleaning chemicals. The cause of the reaction was determined to be detergent mistakenly added to a drink given to the children.

In Australia, young children aged 0–4 years have the highest rate of accidental poisoning hospitalisation. Approximately 30% of child poisoning admissions relate to the ingestion of household chemicals such as bleach, disinfectants, detergents, cleaning agents, cosmetics and vinegar.

Ammonia (NH_3) is the active ingredient in a number of household cleaning products. Due to its volatility, ammonia readily escapes from open containers. Mixing bleach with ammonia cleaners generates highly toxic gases. Generating these toxic gases in limited, often poorly ventilated bathroom spaces can be extremely hazardous.

A strong irritant to skin, eyes, throat and lungs, ammonia can not only damage mucous membranes and airways, but may cause kidney and liver damage. Hospitalisation for ammonia poisoning can result from inhalation exposure or ingestion. Exposure to ammonia at concentrations above 5% increases the risk of adverse health consequences, particularly for those with underlying health conditions, such as asthma.

Manufacturers of household ammonia products, such as the window cleaners shown in Figure 14.2.4, frequently claim an ammonia content of 5% by mass. Volumetric analysis is one technique that can be used to confirm ammonia content before distribution to consumers.

FIGURE 14.2.4 Many commercial window cleaning products contain ammonia.

Analysis

A student designed and performed an experiment to determine the ammonia content of a window-cleaning solution. The experiment involved titrating aliquots of diluted window cleaner with a standardised hydrochloric acid solution. Part of the report presented by the student is shown below.

Aim

To determine the concentration of ammonia in a commercial window cleaner and compare the experimental value to the manufacturer's stated value.

Equation for the reaction: $NH_3(aq) + HCl(aq) \rightarrow NH_4Cl(aq)$

Method

1 Dilute the window cleaner by pipetting 25.00 mL of the cleaner into a 250.0 mL volumetric flask. Add 250 mL of deionised water, stopper the flask and invert several times so that the solution is thoroughly mixed.

2 Fill a burette with the standardised hydrochloric acid solution.

3 Place a 20.00 mL aliquot of the diluted window cleaner solution into a conical flask. Add 3 drops of methyl orange indicator.

4 Titrate the hydrochloric acid with the diluted window cleaner aliquot until the indicator just shows a permanent colour change from yellow to orange.

5 Repeat the titration until 3 concordant titres are obtained.

Results

Volume of window cleaner aliquot = 25.00 mL

Volume of diluted window cleaner = 250.0 mL

Volume of diluted window cleaner aliquots = 20.00 mL

Concentration of the standardised hydrochloric acid solution = 0.187 M

Trial	1	2	3	4	5
Titre (mL)	24.57	24.34	24.28	24.16	24.29

Average titre hydrochloric acid = 24.24 mL

1 Why is the hydrochloric acid referred to as a 'standardised' solution?

2 Is the average titre calculated by the student appropriate for use in a volumetric titration calculation? Explain your answer.

3 Using the student's data, calculate the molar concentration of ammonia in the window cleaner.

4 Identify one mistake in the student's procedure. Explain how the procedure should be changed to eliminate this mistake.

5 Suggest one possible systematic error that may affect the accuracy of the calculated ammonia concentration.

6 The calculated concentration of ammonia was considerably less than the value given on the label of the window cleaner. Which one or more of the following errors would account for the lower than expected calculated concentration? Justify your answers.
   i The 20.00 mL pipette was rinsed only with deionised water prior to its use.
   ii The burette was rinsed only with deionised water prior to its use.
   iii The volumetric flask was rinsed only with deionised water to its use.
   iv Phenolphthalein indicator was used during the titration instead of the specified methyl orange indicator. You can refer to Table 14.1.3, page 451, for information about indicator colour changes.

7 The safety data sheet (SDS) for ammonia solution includes the following information:
   - Causes skin irritation and serious eye irritation
   - May cause respiratory irritation

   Apart from a laboratory coat, what personal protective equipment (PPE) should be used by the student during the experiment?

14.2 Review

OA ✓✓

SUMMARY

- The coefficients in a balanced equation show the ratio of the amount, in moles, of reactants and products involved in the reaction.
- Solution volume–volume stoichiometric calculations follow the following steps:
  1 Write a balanced equation for the reaction.
  2 Calculate the amount, in mol, of the known substance using $n = c \times V$.
  3 Use the mole ratio from the balanced chemical equation to calculate the amount, in mole, of the unknown substance.
  4 Convert the amount, in mole, of the unknown substance to the quantity required in the question using $c = \frac{n}{V}$ or $V = \frac{n}{c}$.
- The equivalence point of a titration is the point at which chemically equivalent amounts of reactants have been mixed. The end point of the titration is the point at which the indicator changes colour to show that the reaction is complete.
- The concentration of acidic or basic solutions can be determined by volumetric analysis.
- Dilution of the unknown solution is sometimes required to obtain manageable titre volumes.

KEY QUESTIONS

Knowledge and understanding

1 Aqueous solutions of nitric acid and sodium carbonate react according to the equation:

$$2HNO_3(aq) + Na_2CO_3(aq) \rightarrow 2NaNO_3(aq) + CO_2(g) + H_2O(l)$$

Complete the expressions for the following mole ratios:

a $\frac{n(HNO_3)}{n(CO_2)} =$

b $\frac{n(HNO_3)}{n(Na_2CO_3)} =$

c $\frac{n(Na_2CO_3)}{n(H_2O)} =$

2 Create a flow chart for completing solution volume–volume stoichiometric calculations by placing the steps in an appropriate order.
- Identify the known and unknown substances in the question.
- Use mole ratios from the equation to calculate the amount of the unknown substance.
- Calculate the volume of the unknown substance using $V = \frac{n}{c}$.
- Write a balanced equation for the reaction.
- Calculate the amount, in mol, of the known substance using $n = c \times V$.

continued over page

14.2 Review *continued*

3 Calculate the amount of sulfuric acid (H_2SO_4), in mol, required to completely react with 1.37 mol of sodium hydroxide (NaOH).

4 18.26 mL of dilute nitric acid reacts completely with 20.00 mL of 0.099 27 M potassium hydroxide solution.

- **a** Write a balanced chemical equation for the reaction between nitric acid and potassium hydroxide.
- **b** Calculate the amount, in mol, of potassium hydroxide consumed in this reaction.
- **c** Calculate the amount, in mol, of nitric acid that reacted with the potassium hydroxide.
- **d** Calculate the concentration of the nitric acid.

5 The reaction between solutions of hydrochloric acid and sodium carbonate is represented by this equation:

$$Na_2CO_3(aq) + 2HCl(aq) \rightarrow 2NaCl(aq) + CO_2(g) + H_2O(l)$$

- **a** Calculate the volume of 0.250 M HCl required to react with 20.00 mL of 0.200 M Na_2CO_3.
- **b** Calculate the concentration of a Na_2CO_3 solution if 21.25 mL of Na_2CO_3 reacts completely with 18.75 mL of 0.520 M HCl solution.

6 20.00 mL aliquots of 0.354 M sulfuric acid solution (H_2SO_4) were titrated against a potassium hydroxide solution of unknown concentration. The equation for the reaction is:

$$2KOH(aq) + H_2SO_4(aq) \rightarrow K_2SO_4(aq) + 2H_2O(l)$$

Given the average titre of potassium hydroxide was 19.84 mL, calculate the molar concentration of the potassium hydroxide solution.

Analysis

7 A commercial concrete cleaner contains hydrochloric acid. A 25.00 mL volume of cleaner was diluted to 250.0 mL in a volumetric flask. A 25.00 mL aliquot of 0.508 M sodium carbonate solution was placed in a conical flask. Methyl orange indicator was added and the solution was titrated with the diluted cleaner. The indicator changed permanently from yellow to pink when 23.92 mL of the diluted cleaner had been added from the burette.

- **a** Write a balanced equation for the reaction between hydrochloric acid and sodium carbonate.
- **b** Calculate the amount, in mol, of Na_2CO_3 in the conical flask.
- **c** Calculate the amount, in mol, of hydrochloric acid that reacted with the Na_2CO_3.
- **d** Calculate the molar concentration of the hydrochloric acid in the diluted concrete cleaner.
- **e** Calculate the concentration of hydrochloric acid in the commercial concrete cleaner.

8 A student is required to determine the accurate concentration of a solution of hydrochloric acid by titration with 20.00 mL aliquots of a sodium carbonate standard solution using methyl orange indicator to identify the end point of the reaction. The sodium carbonate solution is prepared by dissolving 1.247 g of anhydrous sodium carbonate in deionised water and making the solution up to 250.0 mL in a volumetric flask. The titres recorded were 21.23 mL, 20.95 mL, 21.02 mL and 20.98 mL.

- **a** Calculate the molar concentration of the sodium carbonate solution.
- **b** Write a balanced equation for the reaction between hydrochloric acid and sodium carbonate.
- **c** Identify the colour change at the end point.
- **d** Calculate the average of the concordant titres of hydrochloric acid.
- **e** Calculate the molar concentration of the hydrochloric acid solution.

Chapter review

KEY TERMS

acid–base indicator
aliquot
amount
anhydrous
average titre
burette
coefficient
concordant titres
conical flask
deionised water
dilution factor
end point
equivalence point
mole ratio
parallax error
pH curve
pipette
primary standard
quantitative analysis
solution volume–volume stoichiometry
standard solution
standardised
stock solution
stoichiometry
titration
titration curve
titre
volumetric analysis
volumetric flask

REVIEW QUESTIONS

Knowledge and understanding

1 Calculate the concentration of a standard solution of sodium borate ($Na_2B_4O_7.10H_2O$) prepared by dissolving 25.21 g of sodium borate in 250.0 mL of deionised water.

2 Anhydrous sodium carbonate is used as a primary standard in determining the concentration of hydrochloric acid by volumetric analysis.

- **a** List the criteria that are used to determine whether or not a substance is suitable for use as a primary standard.
- **b** Explain how you would prepare a standard solution of anhydrous sodium carbonate.

3 Oxalic acid ($C_2H_2O_4$) is used in commercial laundries for removing rust stains or yellowing of laundry caused by iron deposition from the water supply.
To determine the concentration of oxalic acid in a particular laundry solution, a chemist titrates the oxalic acid with a sodium hydroxide solution that had been previously standardised by titrating it with hydrochloric acid of a known concentration.
Explain why the chemist cannot prepare a standard solution of sodium hydroxide directly from solid sodium hydroxide.

4 A volumetric analysis was performed and the following five titres were obtained: 24.22, 25.02, 24.20, 24.16 and 25.13 mL.

- **a** Which of these titres would be considered concordant titres?
- **b** Calculate the average volume based on the concordant titres.

5 A student obtained titration curves for two different acid–base reactions. The range in which two indicators change colour was then superimposed on each graph, as shown below. Determine which indicator is best used for each titration.

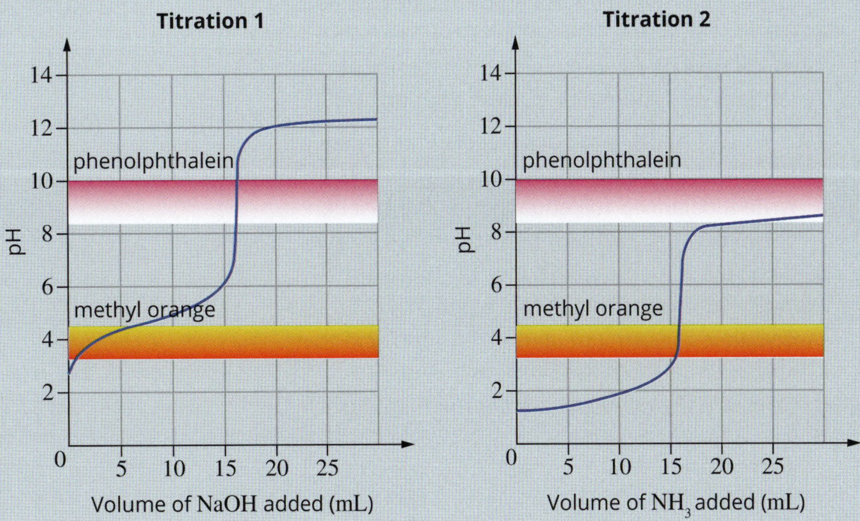

6 Sodium hydroxide cannot be used as a primary standard. Before use in analysis, a sodium hydroxide solution must be standardised. The concentration of a solution of sodium hydroxide was found by titration with a standard solution of potassium hydrogen phthalate. The following steps (not in the correct order) were used when carrying out the titration:

- i The burette was filled with potassium hydrogen phthalate solution.
- ii The conical flask was rinsed with deionised water.
- iii The burette was rinsed with potassium hydrogen phthalate solution.
- iv An aliquot of sodium hydroxide was placed in the conical flask.
- v The initial volume of potassium hydrogen phthalate solution in the burette was read.
- vi The final volume of potassium hydrogen phthalate solution in the burette was read.
- vii A pipette was rinsed with sodium hydroxide solution.
- viii Potassium hydrogen phthalate solution was added from the burette until the end point was reached.
- ix An indicator was added to the sodium hydroxide solution.

The titration was repeated several times and the following titres recorded: 25.70 mL, 25.12 mL, 25.10 mL, 25.14 mL.

- a Explain why the potassium hydrogen phthalate solution is described as a 'standard solution'.
- b From the list above, list the steps required to complete a titration in an appropriate order.
- c Titrations are repeated until concordant results are obtained. Explain what is meant by the term 'concordant'.
- d Using the concordant titres, calculate the average titre of potassium hydrogen phthalate solution.
- e For the titration values 25.12, 25.10 and 25.14 mL, which type of error explains the differences—random or systematic? Explain your answer.

7 19.37 mL of dilute nitric acid reacts completely with 20.00 mL of 0.098 55 M potassium hydroxide solution.

- a Write a balanced full equation for the reaction between nitric acid and potassium hydroxide.
- b Calculate the amount, in mol, of potassium hydroxide consumed in this reaction.
- c Calculate the amount, in mol, of nitric acid that reacted with the potassium hydroxide.
- d Calculate the concentration of the nitric acid.

8 A 0.104 M H_2SO_4 solution is neutralised with 10.0 mL of a solution of 0.315 M KOH.

- a Write a balanced equation for this reaction.
- b Calculate the volume of sulfuric acid neutralised by the potassium hydroxide.

9 17.2 mL of a nitric acid solution is required to react completely with 10.0 mL of a 0.0995 M $Ba(OH)_2$ solution.

- a Write a balanced equation for this reaction.
- b Calculate the concentration of the nitric acid solution.

10 The concentration of ethanoic acid (CH_3COOH) in a vinegar solution was determined by titration with standard sodium hydroxide solution using phenolphthalein indicator. The equation for the reaction is:

$$CH_3COOH(aq) + NaOH(aq) \rightarrow CH_3COONa(aq) + H_2O(l)$$

A 25.00 mL aliquot of vinegar was titrated with a 1.08 M solution of sodium hydroxide. Titres of 24.05, 22.10, 22.05 and 22.00 mL were required to reach the end point. Determine the molar concentration of ethanoic acid in the vinegar.

Application and analysis

11 When an aliquot of sodium hydroxide solution was titrated with ethanoic acid, the indicator changed from pink to colourless after the addition of 20 mL of ethanoic acid from the burette.

- a Determine the most likely indicator used.
- b Explain whether the pH is progressively increasing or decreasing during the titration.
- c When the equivalence point has been reached, will the pH of the solution be less than 7, greater than 7 or equal to 7? Explain your answer. You may wish to refer to Table 14.1.4, page 451.

12 The titration curve in in the graph below shows the change in pH as a solution of 0.10 M sodium hydroxide is added to 20 mL of a 0.10 M nitric acid solution.

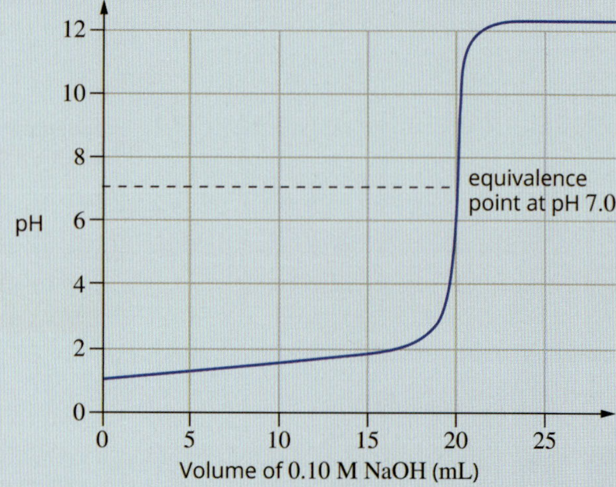

The titration is repeated using 0.20 M sodium hydroxide solution. Which one of the following statements about the second titration is correct?

A The equivalence point occurs when 10 mL of NaOH is added and the pH is then 7.
B The equivalence point occurs when 10 mL of NaOH is added and the pH is then greater than 7.
C The equivalence point occurs when 40 mL of NaOH is added and the pH is then 7.
D The equivalence point occurs when 40 mL of NaOH is added and the pH is then greater than 7.

13 Methanoic acid (HCOOH) is found in stinging nettles. In an analysis of the concentration of methanoic acid in a solution extracted from nettles, the solution was titrated with a standard solution of potassium hydroxide. Aliquots of the methanoic acid were transferred to a conical flask and titrated against the potassium hydroxide solution in a burette.
Determine which one of the following would cause the calculated concentration of methanoic acid to be higher than it actually is.

A The conical flask was rinsed with deionised water, but not dried, directly before use.
B The conical flask was rinsed with potassium hydroxide solution.
C The burette was rinsed with deionised water, but not dried, directly before use.
D The pipette was rinsed with deionised water, but not dried, directly before use.

14 A standard solution of potassium carbonate is made by adding 1.227 g of K_2CO_3 to a 250.0 mL volumetric flask and filling to the mark with water. 20.00 mL aliquots are taken and titrated against a sulfuric acid solution, using methyl orange indicator. The average titre was 22.56 mL of sulfuric acid.

a Write the equation for the reaction.
b Calculate the molar concentration of the K_2CO_3 solution.
c Calculate the molar concentration of the sulfuric acid solution.

15 1.104 g of sodium carbonate (Na_2CO_3) is dissolved in water in a 250.0 mL volumetric flask. 20.00 mL aliquots of this solution were titrated with nitric acid. The average titre of nitric acid was 23.47 mL.

a Calculate the molar concentration of the Na_2CO_3 solution.
b Calculate the molar concentration of the nitric acid.

16 A solution of sodium hydroxide (NaOH) was standardised by titrating it with a solution of potassium hydrogen phthalate. The concentration of ethanoic acid (CH_3COOH) in vinegar was then found by diluting a 10.00 mL aliquot of vinegar in a volumetric flask. Aliquots of the diluted vinegar were transferred from the volumetric flask to a conical flask and titrated with the standardised NaOH solution using phenolphthalein indicator.

a Name the solution that would have been made directly from a primary standard.
b Determine the solution that should be used to rinse the following pieces of equipment used for the titration of the diluted vinegar.
  i 10.00 mL pipette used to dilute the vinegar
  ii volumetric flask used to dilute the vinegar
  iii burette
c The pipette used to transfer the diluted vinegar to the conical flask was rinsed with deionised water immediately before use. What would be the effect on the calculated concentration of ethanoic acid in the vinegar?
d State the colour change observed at the end point.

17 A 20.00 mL solution of potassium hydroxide is standardised in a titration with a solution of 0.0515 M malonic acid ($C_3H_4O_4$). An average titre of 16.48 mL of malonic acid was needed to neutralise the potassium hydroxide solution.

a Calculate the amount, in mol, of malonic acid that reacted with the potassium hydroxide.
b The equation for the reaction between malonic acid and potassium hydroxide is:
$$C_3H_4O_4(aq) + 2KOH(aq) \rightarrow K_2C_3H_2O_4(aq) + 2H_2O(l)$$
Calculate the amount, in mol, of KOH in the 20.00 mL sample.
c Determine the molar concentration of the potassium hydroxide solution.
d Calculate the concentration of potassium hydroxide in g L^{-1}.

18 During a practical class, the concentration of an ammonia solution was determined by titration with a standard solution of hydrochloric acid using methyl orange indicator. A 25.00 mL aliquot of the ammonia solution was titrated with a 0.187 M solution of HCl. Titres of 24.76, 24.37, 24.32 and 24.38 mL of HCl were required to reach the end point. The equation for the reaction is:
$$NH_3(aq) + HCl(aq) \rightarrow NH_4Cl(aq)$$

a Calculate the molar concentration of the ammonia solution.
b The calculated value in **a** was considerably lower than the actual value. For each of the possible errors that could have been made during the analysis, comment on whether it would account for the lower than expected calculated value.
  i The 25.00 mL pipette was rinsed only with deionised water before its use.
  ii The burette was rinsed only with deionised water before its use.

iii The conical flask was rinsed only with deionised water before its use.

iv Phenolphthalein indicator was used during the titration instead of the specified methyl orange.

19 Oxalic acid is a diprotic acid ($C_2H_4O_2$) that can be used as a rust remover. The following results were obtained during an analysis of a rust remover solution. The rust solution was diluted and then titrated with a sodium hydroxide solution using phenolphthalein indicator.

Volume of rust remover used: 5.00 mL

Volume of diluted rust remover solution: 500.0 mL

Volume of diluted rust remover aliquot: 10.00 mL

Titre of 0.0212 M NaOH solution required: 13.45 mL

The equation for the reaction is:

$C_2H_2O_4(aq) + 2NaOH(aq) \rightarrow Na_2C_2O_4(aq) + 2H_2O(l)$

a State the colour change observed at the end point.

b Calculate the oxalic acid concentration in the rust remover in % m/v.

20 A manufacturer wants to know the exact concentration of hydrochloric acid in the concrete cleaner it produces.

a What substance would you use to make a standard solution for use in this titration and why have you chosen this substance?

b How would you prepare this standard solution for volumetric analysis?

c How should each piece of glassware used in a titration be rinsed to ensure you obtain accurate and precise results?

d A 25.00 mL sample of the concrete cleaner was diluted to 250.0 mL in a volumetric flask. A 25.00 mL aliquot of 0.505 M potassium carbonate solution was placed in a conical flask. Methyl orange indicator was added and the solution was titrated with the diluted cleaner. The indicator changed permanently from yellow to pink when 18.44 mL of the diluted cleaner had been added. Calculate the molar concentration of hydrochloric acid in the concrete cleaner.

21 The concentration of ethanoic acid (CH_3COOH) in vinegar was determined by titrating vinegar against a 20.00 mL aliquot of 0.10 M sodium hydroxide. The pH was measured throughout the titration. The titration curve is shown below.

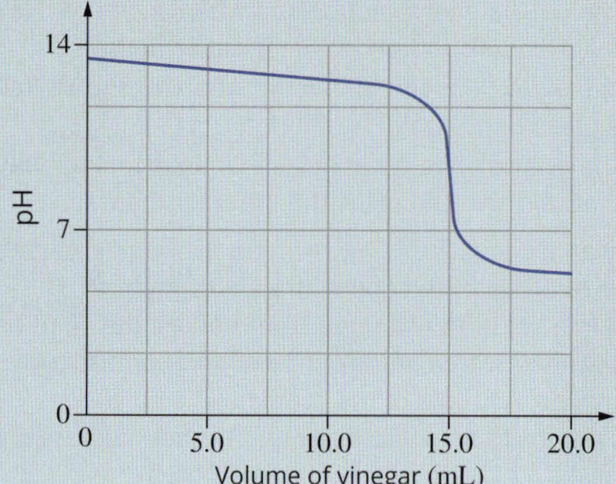

The equation for the reaction is:

$CH_3COOH(aq) + NaOH(aq) \rightarrow CH_3COONa(aq) + H_2O(l)$

Calculate the molar concentration of ethanoic acid in the vinegar.

22 Propanoic acid is a preservative in animal feed products. The propanoic acid content of a commercial animal feed solution was determined by titration with a standard sodium hydroxide solution using phenolphthalein indicator. The animal feed solution was tested by taking a 20.00 mL sample and making it up to 250.0 mL in a volumetric flask. 25.00 mL aliquots of this diluted solution were then titrated against 0.250 M sodium hydroxide solution. The average titre was 32.10 mL. The equation for the reaction between propanoic acid and sodium hydroxide is:

$CH_3CH_2COOH(aq) + NaOH(aq) \rightarrow$

$CH_3CH_2COONa(aq) + H_2O(l)$

Calculate the concentration of propanoic acid in the sample of feed solution in units of:

a mol L^{-1}

b %(m/v).

OA ✓✓

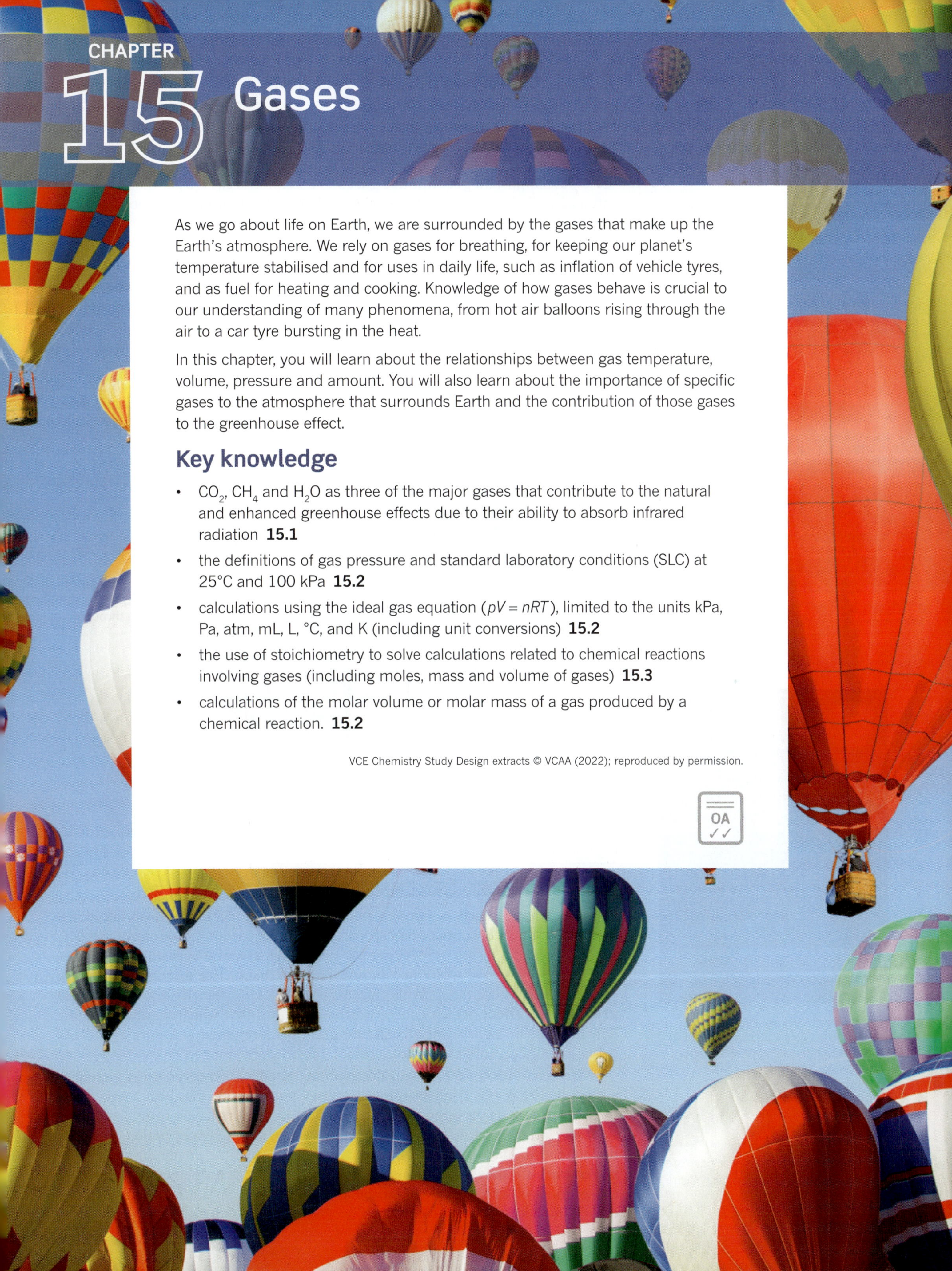

CHAPTER

15 Gases

As we go about life on Earth, we are surrounded by the gases that make up the Earth's atmosphere. We rely on gases for breathing, for keeping our planet's temperature stabilised and for uses in daily life, such as inflation of vehicle tyres, and as fuel for heating and cooking. Knowledge of how gases behave is crucial to our understanding of many phenomena, from hot air balloons rising through the air to a car tyre bursting in the heat.

In this chapter, you will learn about the relationships between gas temperature, volume, pressure and amount. You will also learn about the importance of specific gases to the atmosphere that surrounds Earth and the contribution of those gases to the greenhouse effect.

Key knowledge

- CO_2, CH_4 and H_2O as three of the major gases that contribute to the natural and enhanced greenhouse effects due to their ability to absorb infrared radiation **15.1**
- the definitions of gas pressure and standard laboratory conditions (SLC) at 25°C and 100 kPa **15.2**
- calculations using the ideal gas equation ($pV = nRT$), limited to the units kPa, Pa, atm, mL, L, °C, and K (including unit conversions) **15.2**
- the use of stoichiometry to solve calculations related to chemical reactions involving gases (including moles, mass and volume of gases) **15.3**
- calculations of the molar volume or molar mass of a gas produced by a chemical reaction. **15.2**

OA
✓✓

15.1 Greenhouse gases

The atmosphere is a mixture of gases that surround Earth and help make life possible. One of the ways the atmosphere supports life is through temperature regulation via the greenhouse effect. Carbon dioxide (CO_2), methane (CH_4) and water vapour (H_2O) are important **greenhouse gases**. A greenhouse gas is a gas that can absorb infrared radiation. Other greenhouse gases include chlorofluorocarbons (CFCs) and nitrous oxide.

THE NATURAL GREENHOUSE EFFECT

The **greenhouse effect** is a natural process that keeps the Earth's surface warm enough to support life. When energy from the Sun reaches the Earth's atmosphere, some of the energy is reflected back into space, but the majority of the Sun's energy is absorbed by the Earth, causing the Earth to warm up. Heat then radiates from the Earth back towards space in the form of infrared radiation. Greenhouse gases in the atmosphere trap some of that infrared radiation, keeping the surface temperature relatively warm and stable, and suitable for human habitation. The greenhouse gases act like the outer layer of glass in a garden greenhouse, as seen in Figure 15.1.1.

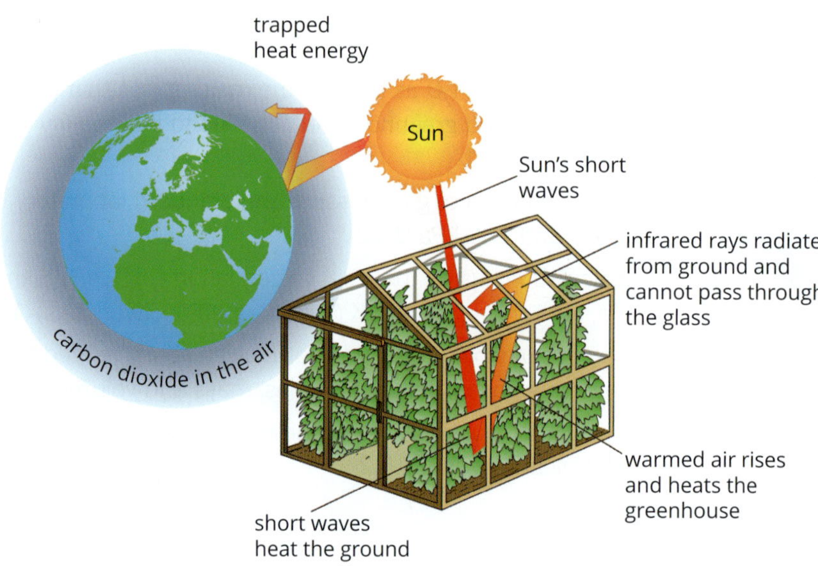

FIGURE 15.1.1 Greenhouse gases in the Earth's atmosphere behave like the glass of a greenhouse. Solar radiation (arrows) passes through and re-radiates. Some of this returns into space, but the build-up of greenhouse gases such as methane, carbon dioxide and water, trap the energy, producing a warming effect.

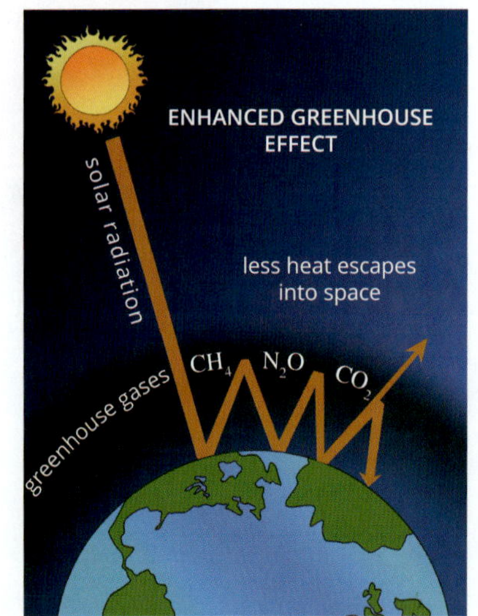

FIGURE 15.1.2 Human activities are increasing the amount of greenhouse gases in the atmosphere, leading to an enhanced greenhouse effect.

> The greenhouse effect is a natural process that is crucial for life. The enhanced greenhouse effect increases surface warming beyond that which occurs naturally.

THE ENHANCED GREENHOUSE EFFECT

Human activities, such as burning fossil fuels for energy, have greatly increased the amount of water vapour and carbon dioxide released into the atmosphere. The production and transport of coal, natural gas and oil, as well as agricultural practices and decay of organic waste in landfills, emit methane. The increasing presence of these greenhouse gases in the atmosphere has contributed to an **enhanced greenhouse effect**, which causes warming beyond that which naturally occurs. The increase in the layer of greenhouse gases is represented in Figure 15.1.2. As the amount of greenhouse gases increases, more heat is trapped.

Water accounts for 60–80% of the greenhouse effect. However, water vapour is a short-lived gas as it condenses into clouds. Clouds briefly trap heat below them, but they also create the opposite effect by reflecting solar radiation from their upper surface. Carbon dioxide remains in the atmosphere for a longer period and does not condense. Although it accounts for only 10–25% of the greenhouse effect, an increase in atmospheric CO_2 causes sustained warming of the planet. This warming causes more water vapour to enter the atmosphere, further adding to the effect of carbon dioxide.

While methane makes up only 0.000 17% of the atmosphere, it is more than 25 times as potent as carbon dioxide as a greenhouse gas. Methane traps significant quantities of heat energy, so human activities that lead to increased methane emissions also make a significant contribution to global warming.

Impacts of the enhanced greenhouse effect

The enhanced greenhouse effect, which is also referred to as climate change or global warming, is causing observed changes to the Earth, including those seen in Figure 15.1.3. Climate change is associated with increases in temperature, frequency and intensity of heat waves, hazardous fire weather and drought conditions, as well as extreme weather events that lead to flooding. Ocean warming is connected with sea level rise, which can have enormous impact on coastal habitats and dwellings.

FIGURE 15.1.3 Increasing (a) bushfires and (b) extreme weather events that lead to flooding are all a result of the climate change caused by the enhanced greenhouse effect.

CHEMFILE

Planet without an atmosphere

The possibility of establishing a human colony on the planet Mars, as seen in the figure, has long captivated the imagination of scientists and science fiction writers alike. One of many challenges to overcome to establish such a colony is the lack of a stable temperature at the surface of Mars. Mars displays very little greenhouse effect as the planet has almost no atmosphere that could trap radiated energy from the Sun. This causes extreme temperature contrasts. For example, a summer's day may reach 20°C near Mars' equator, but at night the temperature can drop to −73°C. Living things, including humans and the plants to be grown for food and resources, could not survive the low temperatures.

The planet Mars does not have an appreciable atmosphere and experiences extreme surface temperature fluctuations as a result.

CHEMFILE

Capturing carbon released during combustion

The release of carbon dioxide into the atmosphere from the combustion of fossil fuels is of significant international concern due to its contribution to the enhanced greenhouse effect. Many scientists are focused on finding ways to reduce carbon dioxide emissions to the atmosphere. One way in which carbon dioxide reduction is being investigated is through carbon sequestration. Carbon sequestration is a process whereby atmospheric carbon dioxide is captured and stored in 'carbon sinks', such as forests and wetlands (figure a). Another method of capture is the proposed CarbonNet Project in Gippsland, where carbon dioxide captured from a range of industries based in Victoria's Latrobe Valley will be delivered via an underground pipeline to offshore storage sites in Bass Strait (figure b)

(a) Scientists return from mangroves with samples collected to assess the health and carbon sequestration capacity of the ecosystem. (b) Proposed carbon capture and storage projects include collecting CO_2 produced from industries and delivering it in pipes to storage underground or under oceans.

15.1 Review

OA
✓✓

SUMMARY

- Greenhouse gases are an important part of the Earth's atmosphere that stabilise our climate.
- Carbon dioxide (CO_2), water (H_2O) and methane (CH_4) are important greenhouse gases. Methane is present in the atmosphere in the lowest quantities, but is the most potent at trapping heat.
- Human activities, including sourcing and combustion of fossil fuels, land clearing and agriculture, have led to increased emissions of carbon dioxide, methane and water.
- Increased emissions of greenhouse gases have contributed to an enhanced greenhouse effect, which is also referred to as climate change or global warming.
- Climate change has significant environmental, social and economic impacts.

KEY QUESTIONS

Knowledge and understanding

1 Name three major greenhouse gases.

2 Explain the difference between the natural and enhanced greenhouse effects.

3 Explain why methane emissions are a concern, even though there are only small amounts of methane in the atmosphere.

4 Describe how an increase in human activities that use energy have contributed to the enhanced greenhouse effect.

Analysis

5 A municipal landfill is exploring how the methane produced by decomposing organic matter might be collected and used for electricity production. One proposal involves burning methane to produce carbon dioxide and water. Discuss the impact this would have on the enhanced greenhouse effect.

15.2 Introducing properties of gases

The gases that make up the Earth's atmosphere have great significance in maintaining a suitable climate for life through the greenhouse effect. An understanding of the properties of gases and ways that gases can be measured is necessary for scientists working on reducing the production and/or impact of greenhouse gases.

Gases differ significantly from liquids and solids in terms of volume, shape, compressibility and ability to mix, as outlined in Table 15.2.1.

TABLE 15.2.1 A comparison of properties of the three states of matter

Property	Gases	Liquids	Solids
Density	low	high	high
Volume and shape	fill all of the space available, because particles move independently of one another	fixed volume, adopt shape of container, because particles are affected by attractive forces	fixed volume and shape, because particles are affected by attractive forces
Compressibility	compress easily	almost incompressible	almost incompressible
Ability to mix	mix together rapidly	mix together slowly unless stirred	do not mix unless finely divided

Some of these differences in properties are evident in gas behavior, examples of which can be seen in Figure 15.2.1. Such examples tell you a great deal about the physical properties of gases—those properties that can be observed and measured without changing the nature of the gas itself.

FIGURE 15.2.1 (a) Air is used to inflate vehicle tyres. Air is a mixture of gases and is easily compressed. When the car goes over a bump in the road, the air compresses slightly and absorbs the impact of the bump. (b) The gases that cause the smell of a freshly brewed cup of coffee rapidly fill an entire room. Gases mix readily and, unlike solids and liquids, occupy all available space. (c) This weather balloon is only partially inflated when released. Its volume increases because of pressure changes as it ascends into the atmosphere, where it will collect data.

KINETIC MOLECULAR THEORY

Scientists have developed a model to explain the properties of a gas based on the behaviour of the particles of the gas. This model is known as the **kinetic molecular theory** of gases. According to this theory:

- gases are composed of small particles, either atoms or molecules
- the volume of the particles in a gas is very small compared with the volume they occupy. Consequently, most of the volume occupied by a gas is empty space
- gas particles move rapidly in random, straight-line motion
- particles collide with each other and with the walls of the container
- the forces between particles are extremely weak
- **kinetic energy** is the energy of motion. The average kinetic energy of gas particles increases as the temperature of the gas increases.

MEASURABLE QUANTITIES OF GASES

Measurable quantities of gases include amount, volume, pressure and temperature.

Commonly measured quantities used to describe a gas in chemistry include:

- amount
- volume
- pressure
- temperature.

Changing any one of these will have an impact on one or more of the others, as seen in Figures 15.2.2 and 15.2.3.

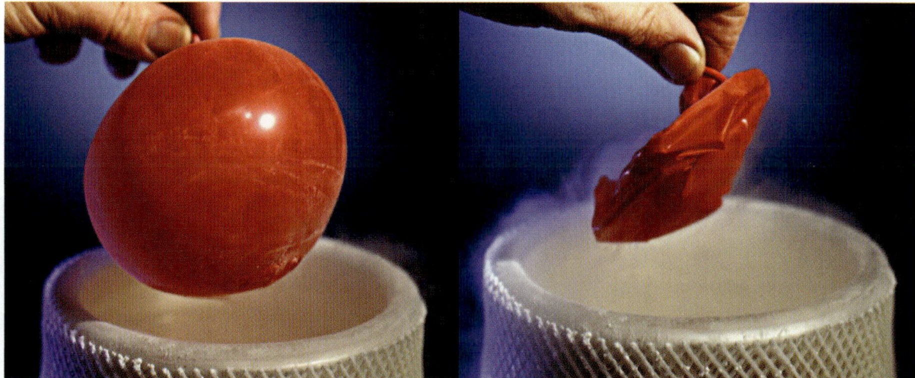

FIGURE 15.2.2 A balloon filled with air at room temperature is dipped into liquid nitrogen at −196°C. The drop in temperature causes the volume of the air inside the balloon to decrease dramatically.

FIGURE 15.2.3 Pumping more air into a tyre increases the pressure in the tyre because more particles are being pumped into a nearly fixed volume.

Amount

As with solids and liquids, the amount of particles of gas present in a sample is measured in terms of number of moles. The amount, in mol, of a gas can be calculated from the mass of the gas using the formula $n = \frac{m}{M}$.

Volume

Volume is the quantity used to describe the space that a gas occupies. Common units used for measuring volume in gases include mL and L.

$$1\ \text{L} = 1000\ \text{mL}$$

Figure 15.2.4 shows how to convert from mL to L, and L to mL.

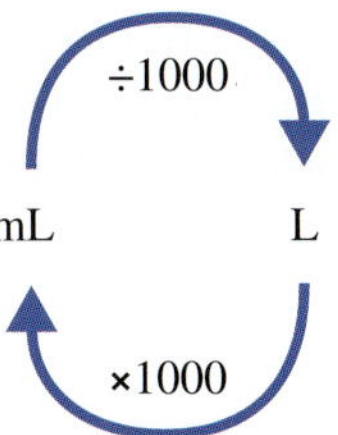

FIGURE 15.2.4 mL and L are common units of volume that can be converted from one to the other.

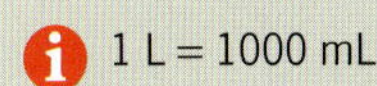
1 L = 1000 mL

Worked example 15.2.1

CONVERTING VOLUME UNITS

A gas has a volume of 255 mL. What is the volume of the gas in litres (L)?

Thinking	Working
Recall the relationship between the volume units.	1 L = 1000 mL Divide volume in mL by 1000 to convert to L.
Apply the relationship to convert the volume to the required unit.	volume in L = $\frac{255}{1000}$ = 0.255 L

Worked example: Try yourself 15.2.1

CONVERTING VOLUME UNITS

A gas has a volume of 700 mL. What is the volume of the gas in litres (L)?

Pressure

The smaller the volume occupied by a gas, the more frequently the gas particles collide with each other and the walls of their container. The increased frequency of collisions with the walls of the container increases the force on the walls of the container, such as the inside of a tyre. The force applied per unit area is described as **pressure**. The relationship can be written as:

$$\text{pressure} = \frac{\text{force}}{\text{area}}$$

The units of pressure depend on the units used to measure force and area. Over the years, scientists in different countries have used different units to measure force and area, so there are a number of different units of pressure:

The Standard International (SI) unit for pressure is newtons per square metre (N m^{-2}). This unit reflects the force, in N, per unit area, in m^2, as per the definition.

One newton per square metre is equivalent to a pressure of one **pascal** (Pa).

Another unit for pressure is the **standard atmosphere** (atm), which is the average atmospheric pressure at sea level. 1 atm = 101.3 kPa.

In 1982, the International Union of Pure and Applied Chemistry (IUPAC) adopted a standard for pressure equivalent to 100 000 Pa or 100 kPa. 100 kPa = 0.987 atm.

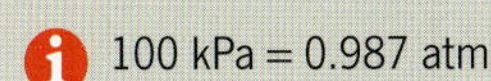
100 kPa = 0.987 atm

Relationships between the common units of gas pressure: pascal, kilopascal and atmosphere, are shown in Table 15.2.2.

TABLE 15.2.2 Relationships between common units of gas pressure

Name of unit	Symbol for unit	Conversions
pascal	Pa	1 Pa = 1 N m^{-2}
kilopascal	kPa	1 kPa = 1×10^3 Pa
atmosphere	atm	0.987 atm = 100 kPa

CHEMFILE

Torricelli's barometer

In the seventeenth century, the Italian physicist Evangelista Torricelli invented the earliest barometer, an instrument used to measure atmospheric pressure. It was a straight glass tube that contained mercury and was closed at one end. The tube was inverted so that the open end was below the surface of a bowl of mercury, as seen in the figure below.

The column of mercury in Torricelli's barometer was supported by the pressure of the gas particles in the atmosphere colliding with the surface of the mercury in the open bowl. At sea level, the top of the column of mercury was about 760 mm above the surface of the mercury in the bowl.

Torricelli found that the height of the mercury column decreased when he took his barometer to higher altitudes in the mountains. At higher altitudes, there are fewer air particles and therefore less frequent collisions on the surface area of mercury. The reduced pressure supports a shorter column of mercury.

Reflecting the work of Torricelli, gas pressure is still sometimes measured in units of mmHg and is the unit commonly used for measuring blood pressures.

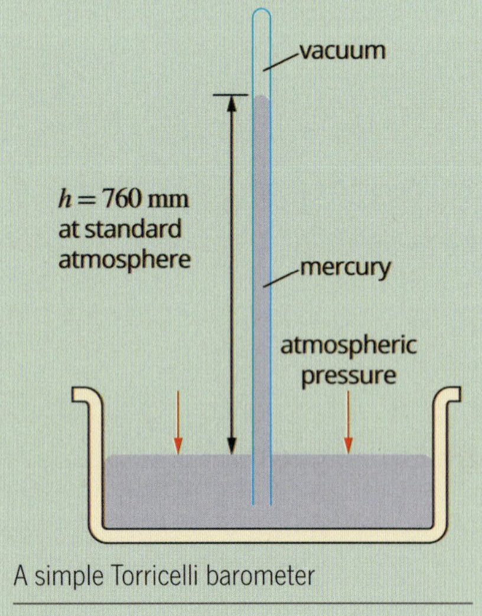

A simple Torricelli barometer

Worked example 15.2.2

CONVERTING PRESSURE UNITS

Mount Everest is the highest mountain on Earth. The atmospheric pressure at the top of Mount Everest is 0.333 atm. What is the pressure in kilopascals (kPa)?

Thinking	Working
Recall the relationship between the two pressure units.	0.987 atm = 100 kPa
Apply the relationship to convert the quantity to the new unit.	$\frac{0.333}{0.987} = \frac{\text{pressure in kPa}}{100}$ pressure in kPa $= \frac{0.333}{0.987} \times 100$ $= 33.7$ kPa

Worked example: Try yourself 15.2.2

CONVERTING PRESSURE UNITS

Cyclone Yasi was one of the biggest cyclones in Australian history. Cyclones are caused by areas of low pressure. If the atmospheric pressure in the eye of Cyclone Yasi was 0.891 atm, what was the pressure in kilopascals (kPa)?

Partial pressure

Air is a mixture of gases, including nitrogen, oxygen, carbon dioxide and argon. In a mixture of gases, each gas collides with the walls of a container, exerting its own **partial pressure**. The measured pressure in a container filled with different gases is the sum of the partial pressures of the individual gases, as seen in Figure 15.2.5.

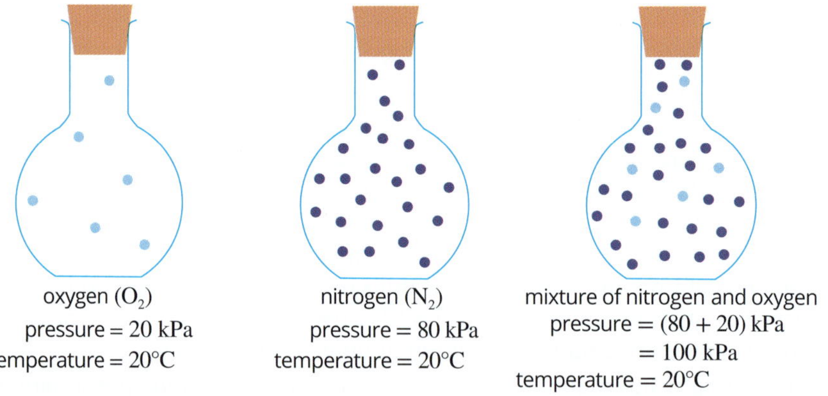

FIGURE 15.2.5 The total pressure of a mixture of gases is the sum of the partial pressures (individual pressures) of each of the gases in the mixture.

Temperature

Kinetic molecular theory describes how increasing the temperature of a gas increases the average kinetic energy of the gas particles. Temperature is therefore an important quantity when describing a gas.

Table 15.2.3 shows the results of an experiment in which the gas in a syringe is heated slowly in an oven. The pressure on the plunger of the syringe is held constant. It can be seen that the volume increases as temperature increases.

TABLE 15.2.3 Variation of volume with temperature

Temperature (°C)	20	40	60	80	100	120	140
Volume (mL)	60.0	64.1	68.2	72.3	76.4	80.5	84.6

The graph of these results is linear, as shown in Figure 15.2.6. When the graph is extrapolated back to a volume of 0 L, it crosses the temperature axis at −273°C. This led scientists to develop a new temperature scale, known as the **kelvin scale** or **absolute temperature scale**. On the kelvin scale, each temperature increment is equal to one temperature increment on the Celsius scale, and 0°C is equal to 273 K.

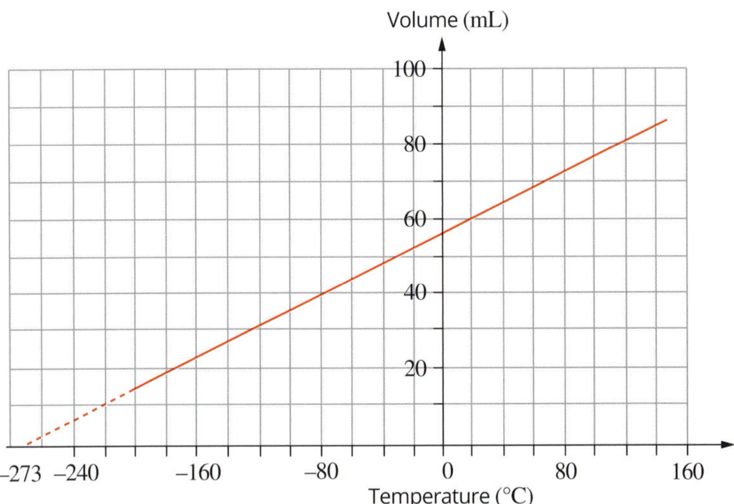

FIGURE 15.2.6 The variation of volume with temperature for fixed amount of gas at constant pressure

 T (in K) = T (in °C) + 273

The relationship between temperature on the Celsius scale and temperature on the kelvin scale is given by the equation:

$$T \text{ (in K)} = T \text{ (in °C)} + 273$$

The temperature 0 K (−273°C) is the lowest temperature theoretically possible. For this reason, 0 K is known as **absolute zero**.

At this temperature, all molecules and atoms have minimum kinetic energy. The kelvin scale has no degrees sign—it is written as just K.

Worked example 15.2.3

CONVERTING TEMPERATURES FROM CELSIUS TO KELVIN

What is 300°C on the kelvin temperature scale?

Thinking	Working
T (in K) = T (in °C) + 273	T (in K) = T (in °C) + 273 = 300 + 273 = 573 K

Worked example: Try yourself 15.2.3

CONVERTING TEMPERATURES FROM CELSIUS TO KELVIN

What is 100°C on the kelvin temperature scale?

Relationship between volume and temperature

We can see in Figure 15.2.6 that the temperature of a gas is directly proportional to volume for a fixed amount of gas at constant pressure.

The mathematical relationship between the temperature in Kelvin, T, of a gas and the volume, V, it occupies can be written as:

$$V \propto T$$

This relationship can also be expressed as $\frac{V}{T} = k$, where k is a constant for a fixed amount of gas at constant pressure.

CHEMFILE

William Thomson—the scientist behind absolute zero

The Kelvin scale is named after Irish physicist and engineer William Thomson (1824–1907), known as Lord Kelvin (see figure). Thomson was born in Belfast and attended Glasgow University from the age of ten. He later attended Cambridge University before becoming the Professor of Natural Philosophy at the University of Glasgow, remaining in the position for 53 years. He is most famous for formulating the second law of thermodynamics and working to install telegraph cables under the Atlantic Ocean. He correctly determined the value of absolute zero (−273°C).

Thomson invented several marine instruments to improve navigation and safety, including a mariner's compass and a deep-sea sounding apparatus, and in 1892 became the first scientist to join the House of Lords in England. Thomson lectured at John Hopkins University on wave theory of light (1884) and was the head of an international commission deciding upon the design of the Niagara Falls power station (1893).

Lord Kelvin (1824–1907) proposed the absolute temperature scale.

MOLAR VOLUME OF A GAS

The **molar volume** of a gas is the volume occupied by one mole of a gas at a particular pressure and temperature. It is given the symbol $V_{m.}$ The molar volume of a gas varies with temperature and pressure. However, molar volume does not vary with the identity of the gas. For example, one mole of neon gas occupies the same volume as one mole of nitrogen gas and one mole of ozone gas at the same temperature and pressure, as seen in Figure 15.2.7.

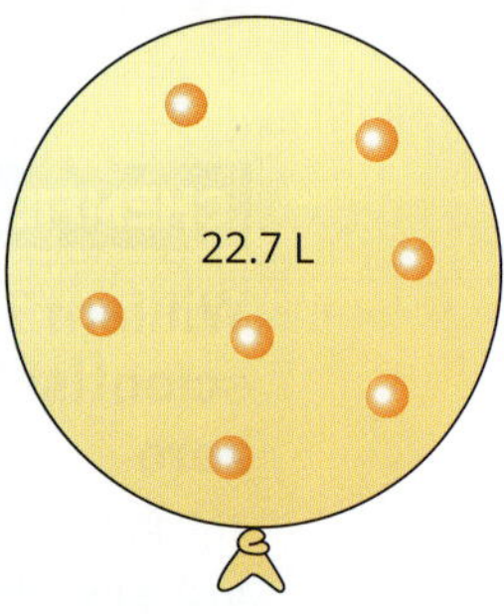

1 mole of neon, Ne
P = 100 kPa
T = 0°C
V = 22.7 L
number of molecules = 6.02×10^{23}
mass = 20.2 g

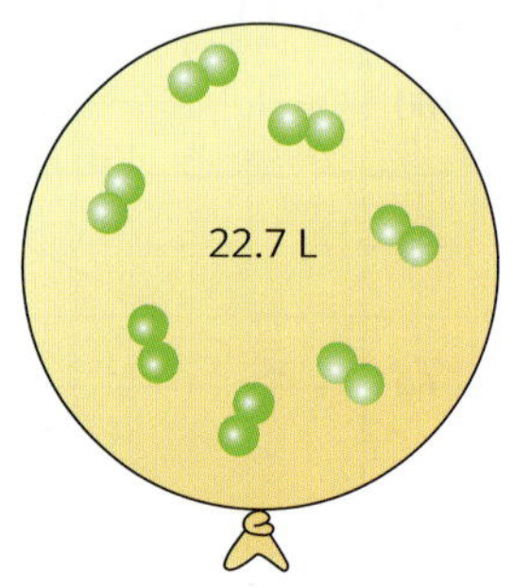

1 mole of nitrogen, N_2
P = 100 kPa
T = 0°C
V = 22.7 L
number of molecules = 6.02×10^{23}
mass = 28.0 g

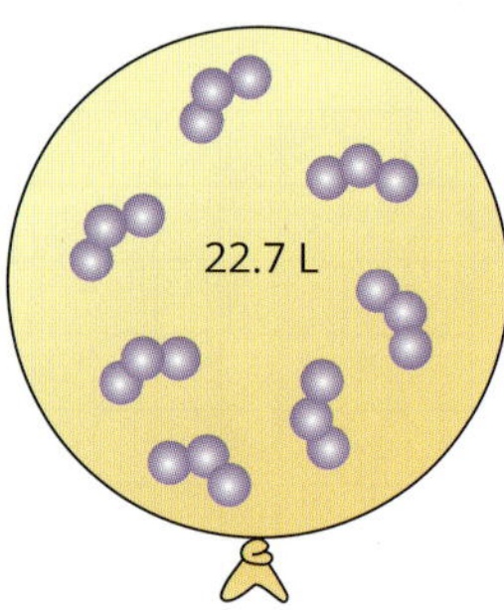

1 mole of ozone, O_3
P = 100 kPa
T = 0°C
V = 22.7 L
number of molecules = 6.02×10^{23}
mass = 48.0 g

FIGURE 15.2.7 One mole of neon, nitrogen and ozone gas each occupy 22.7 L at 100 kPa and 0°C (273 K).

Standard conditions

A temperature of 25°C (298 K) and a pressure of 100 kPa are typical of the conditions you will encounter when working in a laboratory. These conditions are known as **standard laboratory conditions (SLC)**.

An **ideal gas** is a theoretical gas composed of particles that do not interact at all, except during elastic collisions. In reality, no gas is ideal; however, at SLC, most gases behave very like an ideal gas and, therefore, have a molar volume very close to that of an ideal gas (see Table 15.2.4).

It is usual to assume that the molar volume of a gas is 24.8 L mol^{-1} at SLC. From this value, you can calculate the amount, in mol, of a gas given its volume at SLC using the expression:

$$n = \frac{V}{V_m}$$

V_m = 24.8 L mol^{-1} at SLC (25°C or 298 K and 100 kPa)

TABLE 15.2.4 Molar volume at SLC

Gas	Formula	Molar volume at SLC (L mol^{-1})
ideal gas	–	24.79
helium	He	24.83

Worked example 15.2.4

CALCULATING THE VOLUME OF A GAS FROM ITS AMOUNT AT SLC

Calculate the volume, in L, occupied by 0.24 mol of nitrogen gas at SLC.

Thinking	Working
Rearrange $n = \frac{V}{V_m}$ to make volume the subject.	$n = \frac{V}{V_m}$ $V = n \times V_m$
Substitute in the known values where V_m = 24.8 L mol^{-1} (at SLC) and solve.	$V = n \times V_m$ $= 0.24 \times 24.8$ $= 5.952$ L
Consider the units and significant figures. The answer should be given to the smallest number of significant figures present in the data used in the question.	V = 6.0 L

Worked example: Try yourself 15.2.4

CALCULATING THE VOLUME OF A GAS FROM ITS AMOUNT AT SLC

Calculate the volume, in L, occupied by 3.5 mol of oxygen gas at SLC.

THE IDEAL GAS EQUATION

Mathematical relationships that link volume, pressure, temperature and amount were developed over a period of several hundred years by different scientists who performed experiments on gases. Robert Boyle experimentally determined the relationship between the pressure and volume of a gas, and Jacques Charles identified the relationship between the volume and temperature of a gas.

Volume and pressure

Changing the volume of a fixed amount of gas at constant temperature causes a change in the pressure of the gas. For example, pressure of the gas in the syringe shown in Figure 15.2.8 increases as the plunger is pushed in and decreases as the plunger is pulled out.

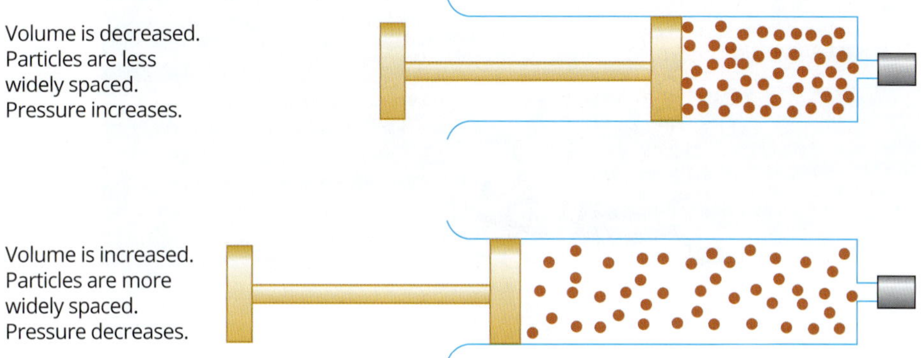

FIGURE 15.2.8 The pressure of the gas in the syringe is affected by a change in volume.

For a given amount of gas at constant temperature, the volume of the gas is inversely proportional to its pressure. This relationship is seen in the changing volume of a weather balloon as it rises to altitudes with much lower pressure than at ground level. A weather balloon filled with helium gas to a volume of 40 L at a pressure of 1 atm increases in volume to 200 L by the time it reaches an altitude with a pressure of 0.2 atm.

The mathematical relationship between the pressure, p, exerted by a gas and the volume, V, it occupies can be written as:

$$p \propto \frac{1}{V}$$

This relationship can also be expressed as $pV = k$, where k is a constant at a given temperature.

This can be rearranged to show that volume is inversely proportional to pressure, i.e. $V \propto \frac{1}{p}$ (for constant T and n).

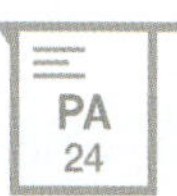

CASE STUDY

Decompression chambers

Scuba diving and water pressure

If you swim at the water's surface, your body experiences a pressure of about 1 atm. Below the surface, your body experiences an additional pressure due to the water. This additional pressure amounts to about 1 atm for every 10 m of depth. Therefore, at 20 m the pressure on your body is about 3 atm.

As the pressure on your body increases, the volume of body cavities, such as lungs and inner ears, decreases. This squeezing effect makes diving well below the water's surface without scuba equipment very uncomfortable. Scuba equipment overcomes this problem by supplying air from tanks to the mouth at the same pressure as that produced by the underwater environment.

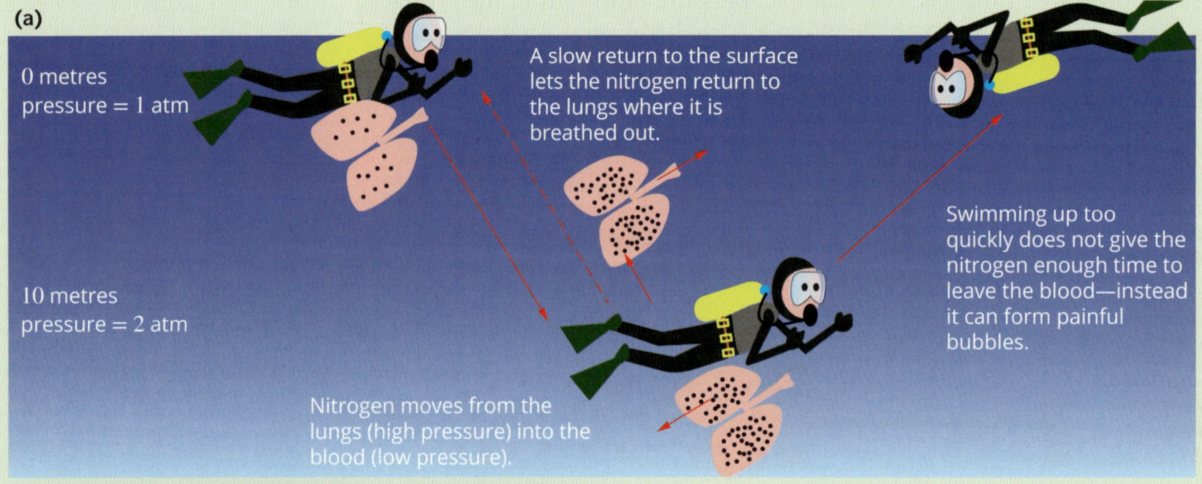

FIGURE 15.2.9 (a) Scuba diving and the bends. (b) A decompression chamber used to treat divers with the bends.

Scuba diving and gas solubility

As the pressure in a diver's lungs increases during a dive, more gas dissolves in the blood. Nitrogen (N_2) is one of these gases. When a diver ascends, the pressure drops, the nitrogen becomes less soluble in the blood and so comes out of solution. If a diver ascends too quickly, the rapid pressure drop causes the nitrogen to come out of the blood as tiny bubbles (see Figure 15.2.9a). This is similar to the bubbles of carbon dioxide you observe when you open a bottle of soft drink.

These bubbles cause pain in joints and muscles. If they form in the spinal cord, brain or lungs, they can cause paralysis or death. Treatment for divers suffering from this effect (commonly known as 'the bends') involves time in a decompression chamber similar to the one shown in Figure 15.2.9b.

The chamber increases the pressure surrounding the diver's body, forcing any nitrogen bubbles to dissolve in the blood, and then slowly reduces the pressure back to 1 atm. The duration of treatment depends on the severity of the symptoms, the dive history and the patient's response to treatment.

Volume and amount of a gas

The volume, V, occupied by a gas also depends directly on the amount of gas, n, in mol.

This relationship is shown in Figure 15.2.10. Both syringes show a gas at a constant temperature and pressure. The volume doubles with twice the number of molecules of gas in the syringe if the pressure is to remain constant.

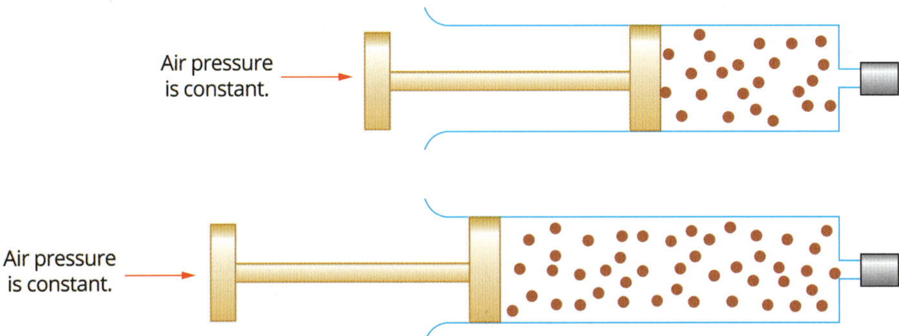

FIGURE 15.2.10 When the amount of gas in the syringe is doubled, the volume doubles, provided the pressure on the plunger and the temperature of the gas remain constant.

The mathematical relationship between the volume, V, occupied by a gas and the amount of gas, n, at constant temperature and pressure can be written as:

$$V \propto n$$

This relationship can also be expressed as $\frac{V}{n} = k$, where k is a constant at a given temperature and pressure.

Combining the relationships

The following three mathematical relationships describe the behaviour of gases under different conditions:

- $V \propto T$ (for constant p and n)
- $V \propto \frac{1}{p}$ (for constant T and n)
- $V \propto n$ (for constant p and T)

These relationships can be combined to show that the volume of a gas is affected by amount, temperature and pressure:

$$V \propto \frac{nT}{p}$$

This relationship can be expressed in the equation $V = \frac{RnT}{p}$, where R is a constant value.

This equation is known as the **ideal gas equation** and is more usually written in the form:

$$pV = nRT$$

where R is called the **ideal gas constant**, or simply, the **gas constant**.

This constant has been determined experimentally by measuring the volume occupied by a known amount of gas at a known temperature and pressure.

The value of R depends on the units of pressure and volume used. It has a value of 8.31 J K^{-1} mol^{-1} when:

- p is measured in kilopascals, kPa
- V is measured in litres, L
- n is measured in moles, mol
- T is measured on the kelvin scale, K.

> The ideal gas equation is usually written in the form:
>
> $$pV = nRT$$
>
> where R is called the ideal gas constant and has a value of 8.31 J K^{-1} mol^{-1}.

Worked example 15.2.5

CALCULATING THE VOLUME OF A GAS USING THE IDEAL GAS EQUATION

Calculate the volume, in L, occupied by 2.24 mol of oxygen gas (O_2) at a pressure of 200 kPa and a temperature of 50°C.

Thinking	Working
Convert units, if necessary. Pressure needs to be in kPa and temperature in K.	p = 200 kPa (no conversion required) $T = 50 + 273$ $= 323$ K
Rearrange the ideal gas equation so that volume, V, is the subject.	$pV = nRT$ $V = \frac{nRT}{P}$
Substitute values for pressure, amount, temperature and the gas constant, R, then solve for V. Express the answer to the correct number of significant figures.	$V = \frac{2.24 \times 8.31 \times 323}{200}$ $= 30.1$ L

Worked example: Try yourself 15.2.5

CALCULATING THE VOLUME OF A GAS USING THE IDEAL GAS EQUATION

Calculate the volume, in L, occupied by 13.0 mol of carbon dioxide gas (CO_2) at a pressure of 250 kPa at 75.0°C.

Calculating molar mass or molar volume of a gas

Experimental data obtained in a laboratory can be used to determine the molar volume or molar mass of a gas produced by a chemical reaction.

Molar volume can be determined experimentally by using a chemical reaction to produce and measure a volume of gas. If the amount, in mol, of gas is known then the molar volume of the gas at that particular temperature and pressure can be calculated using:

$$V_m = \frac{V}{n}$$

Worked example 15.2.6

CALCULATING THE MOLAR VOLUME OF A GAS AT A PARTICULAR TEMPERATURE AND PRESSURE

Calculate the molar volume, in L, of oxygen gas (O_2) at a particular temperature and pressure if 0.500 g of the gas occupies a volume of 360 mL.

Thinking	Working
Determine the amount, in mol, of the gas.	$n(O_2) = \frac{m}{M}$ $= \frac{0.500}{32.0}$ $= 0.0156$ mol
Determine the molar volume of the gas. Remember volume must be converted to L.	$n = \frac{V}{V_m}$ so $V_m = \frac{V}{n}$ $= \frac{0.360}{0.0156}$ $= 23.0$ L mol^{-1}

Worked example: Try yourself 15.2.6

CALCULATING THE MOLAR VOLUME OF A GAS AT A PARTICULAR TEMPERATURE AND PRESSURE

Calculate the molar volume, in L, of carbon dioxide gas (CO_2) at a particular pressure and temperature if 5.00 g of the gas occupies 2.90 L.

Molar mass can also be determined from experimental data of the mass, volume, pressure and temperature of a gas produced in a chemical reaction. The ideal gas equation is used to find the amount, in mol, of gas in the measured volume at the measured temperature and pressure using:

$$n = \frac{PV}{RT}$$

Molar mass, in g mol^{-1}, is then calculated using:

$$M = \frac{m}{n}$$

Worked example 15.2.7

EXPERIMENTAL DETERMINATION OF THE MOLAR MASS OF A GAS

Calculate the molar mass of an unknown gas if 0.0766 g of the gas occupies 32.0 mL at 19°C and 100 kPa.

Thinking	Working
Ensure the volume, temperature and pressure are in the required units.	$p = 100$ kPa $T = 19 + 273 = 292$ K $V = 32.0$ mL $= 0.0320$ L
Calculate the amount, in mol, of the gas using the ideal gas equation $PV = nRT$.	$n = \frac{PV}{RT}$ $= \frac{100 \times 0.0320}{8.31 \times 292}$ $= 0.001\ 32$ mol
Determine molar mass by rearranging the equation $n = \frac{m}{M}$.	$M = \frac{m}{n}$ $= \frac{0.0766}{0.00132}$ $= 58.1$ g mol^{-1}

Worked example: Try yourself 15.2.7

EXPERIMENTAL DETERMINATION OF THE MOLAR MASS OF A GAS

Calculate the molar mass of an unknown gas if 0.2145 g of the gas occupies 120.0 mL at 23 °C and 100 kPa.

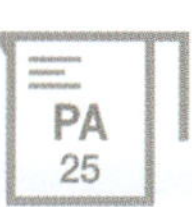

CASE STUDY **ANALYSIS**

Experimental determination of the molar mass of a gas

The ideal gas equation can be used to experimentally determine the molar mass of a gas. An experiment set up for determining the molar mass of butane, C_4H_{10}, is illustrated in Figure 15.2.11. Butane is the main component in lighter fluid because it is a highly flammable, stable and easily compressible gas. In the experiment, a small amount of butane is released from the lighter into the measuring cylinder and the mass lost by the lighter is measured. The volume of gas released is measured by displacement of water. The pressure and temperature of the surrounding environment are also measured.

Sample results for one investigation are provided in Table 15.2.5. The atmospheric pressure was constant and measured to be 0.974 atm and the temperature for all trials was 17°C.

Analysis

1. a Use the ideal gas equation ($pV = nRT$) to determine the amount of butane, in mol, used in each trial.
   b Use the mass of butane in each trial and the equation $n = \frac{m}{M}$ to determine the molar mass of butane for each trial.
   c Calculate the average molar mass of butane.
   d The actual molar mass of butane is 58.0 g mol^{-1}. Compare the experimentally determined value with the actual value and give a possible explanation for any differences.
2. Lighter fluid containers will also contain some oxygen and water vapour, which will displace water and contribute to the overall volume of the gas collected. If a container of lighter fluid at a pressure of 300 kPa contains water vapour at a partial pressure of 7.5 kPa and oxygen at 15 kPa, determine the pressure of the butane gas.

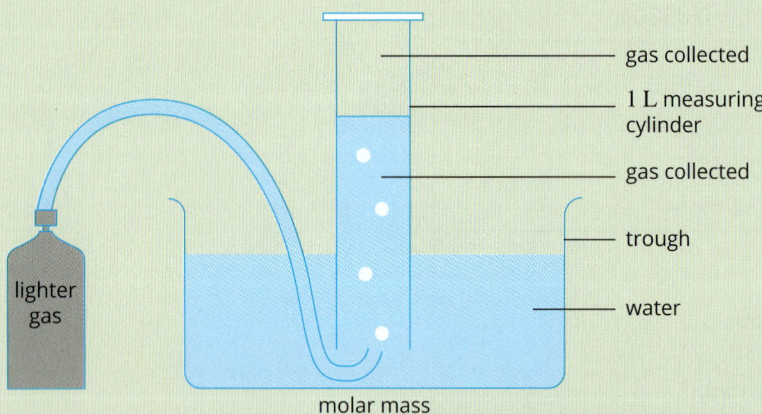

FIGURE 15.2.11 Apparatus used to determine the molar mass of butane, C_4H_{10}

TABLE 15.2.5 Experimental data from the experiment to determine the molar mass of butane

Trial number	Initial mass of lighter (g)	Final mass of lighter (g)	Mass of butane (final – initial mass, g)	Volume of water displaced (mL)
1	15.27	15.05	0.22	91
2	15.02	14.79	0.23	84
3	14.75	14.50	0.25	94
4	14.43	14.20	0.23	82

15.2 Review

OA ✓✓

SUMMARY

- Measurable quantities of a gas include amount, volume, pressure and temperature.
- Volume is the space occupied by a gas. Conversion between different volume units uses the following relationship:
  1 L = 1000 mL
- Pressure is defined as the force per unit area.
  100 kPa = 0.987 atm
- Temperature can be measured on an absolute temperature scale, with units of kelvin.
  $T(\text{in K}) = T(\text{in °C}) + 273$
- Absolute zero is a temperature of −273°C or 0 K. At this temperature, molecules and atoms have minimum kinetic energy.
- The molar volume, V_m, of a gas is the volume occupied by 1 mol of gas at a given temperature and pressure:
  $n = \frac{V}{V_m}$
- Standard laboratory conditions (SLC) refers to a temperature of 25°C (298 K) and a pressure of 100 kPa. The value of V_m at SLC is 24.8 L mol^{-1}.
- $pV = nRT$ is the ideal gas equation. It can be used to calculate one variable (p, V, n or T) when the other three variables are known.
- R is the ideal gas constant and its value is 8.31 J K^{-1} mol^{-1} when pressure is measured in kPa, volume is in L, amount is in mol and temperature is in K.
- Molar volume and molar mass of a gas produced in a chemical reaction can be calculated from experimental data of mass and volume using:
  $V_m = \frac{V}{n}$ or $M = \frac{m}{n}$

KEY QUESTIONS

Knowledge and understanding

1 Convert each of the following pressures to the units specified.
  a 14000 Pa to kPa
  b 4.24 atm to kPa
  c 120 kPa to atm

2 Convert the following Celsius temperatures to absolute temperatures.
  a 120°C
  b −145°C

3 What volume of gas, in litres, is occupied by:
  a 0.20 mol of hydrogen at 115 kPa and 40°C?
  b 12.5 mol of carbon dioxide at 5 atm and 150°C?
  c 8.50 g of hydrogen sulfide (H_2S) at 100 kPa and 27°C?

Analysis

4 Calculate each of the following:
  a pressure in a 5.0 L flask which contains 0.25 mol of nitrogen at a temperature of 5°C.
  b the mass of helium in a balloon if the volume is 100 L at a pressure of 95 000 Pa and a temperature of 0°C.
  c the temperature, in °C, of a sample of nitrogen gas that has a mass 11.3 g and exerts a pressure of 102 kPa in a 10.0 L cylinder.
  d molar volume of a gas at a particular temperature and pressure if 0.525 mol of the gas has a volume of 18.0 L.

5 Which sample of gas contains the greater amount, in mol, of gas: 3.2 L of nitrogen at 25°C and a pressure of 120 kPa or 2.5 L of helium at 23°C and a pressure of 1.2 atm?

6 A helium airship has an envelope of volume 1.0×10^7 L at sea level (100 kPa) and at 20°C. At higher altitudes, the air pressure falls at a rate of 10.0 kPa for every 1000 m increase in altitude.
  a What will the volume of the envelope be at a height of 1500 m, when the airship pilot's thermometer registers a temperature of 5°C?
  b The airship rises through the atmosphere until its volume is 1.05×10^7 L. The temperature is 0°C.
    i What will be the pressure, in kPa, indicated by the pilot's barometer?
    ii Calculate the height of the airship.

7 A student experimentally determined that 0.0148 g of a gas occupied a volume of 184 mL at SLC. Calculate the molar mass of the gas.

15.3 Calculations involving gases

In Chapter 14, you learnt how volume–volume stoichiometry can be used to determine the concentration of an unknown solution. In this section, you will use stoichiometry to carry out calculations involving mole, mass and volume of gases.

MASS-MASS STOICHIOMETRY

When a reaction is carried out in a laboratory, quantities of chemicals are often measured in grams, not moles. For this reason, many stoichiometry-based calculations will require you to start and finish with mass rather than moles of a substance. To calculate the number of moles of an unknown gas from a mass, you can use this relationship:

$$\text{moles } (n) = \frac{\text{mass in g } (m)}{\text{molar mass } (M)}, \text{ which is written as } n = \frac{m}{M}$$

Mass is calculated by rearranging this relationship to:

$$m = n \times M$$

Calculating the mass of a gas produced in a reaction

It can be useful to determine the mass of a gas produced from a specified mass of reactant. For example, in the combustion reactions that produce greenhouse gases, which you learnt about in Section 15.1, a fuel reacts with oxygen to produce carbon dioxide and water. The mass of the greenhouse gases carbon dioxide and water produced can be calculated from the mass of the fuel using stoichiometry.

The steps involved in calculating the mass of a gas produced based on the mass of a reactant are:

1 Calculate the number of moles of the reactant from its mass, using the formula $n = \frac{m}{M}$.

2 Use the mole ratio from the coefficients in the balanced chemical equation to calculate the number of moles of gas produced.

3 Calculate the mass of gas using the formula $m = n \times M$.

Figure 15.3.1 provides a flow chart that summarises this process. Worked example 15.3.1 will help you to understand these steps.

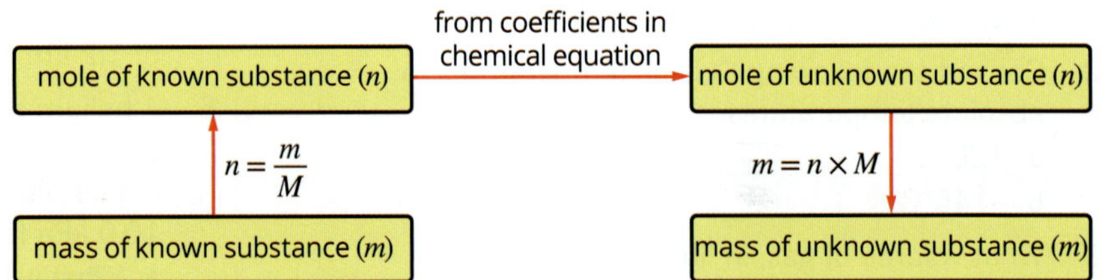

FIGURE 15.3.1 A flow chart for mass–mass stoichiometric calculations is helpful when trying to solve these problems.

Worked example 15.3.1

SOLVING MASS–MASS STOICHIOMETRIC PROBLEMS

Calculate the mass of carbon dioxide, in kg, produced when 540 g of propane (C_3H_8) burns completely in oxygen according to the equation:

$$C_3H_8(g) + 5O_2(g) \longrightarrow 3CO_2(g) + 4H_2O(g)$$

Thinking	Working
Calculate the number of moles of the known substance using: $n = \frac{m}{M}$	$n(C_3H_8) = \frac{540}{44.0}$ $= 12.3$ mol
Find the mole ratio: $\frac{\text{coefficient of unknown}}{\text{coefficient of known}}$	$\frac{n(CO_2)}{n(C_3H_8)} = \frac{3}{1}$
Calculate the number of moles of the unknown substance using: $n(\text{unknown}) = \text{mole ratio} \times n(\text{known})$	$n(CO_2) = \frac{3}{1} \times 12.3$ $= 36.8$ mol
Calculate the mass of the unknown substance using: $m = n \times M$	$m(CO_2) = 36.8 \times 44.0$ $= 1620$ g $= 1.62$ kg

Worked example: Try yourself 15.3.1

SOLVING MASS–MASS STOICHIOMETRIC PROBLEMS

Calculate the mass of carbon dioxide, in kg, produced when 3.60 kg of butane (C_4H_{10}) burns completely in oxygen according to the equation:

$$2C_4H_{10}(g) + 13O_2(g) \longrightarrow 8CO_2(g) + 10H_2O(g)$$

MASS-VOLUME STOICHIOMETRY

Stoichiometric calculations that follow the same general pattern can also be used to calculate the volumes of oxygen required to burn fuels and the volumes of gases produced in any reaction. The number of moles of a 'known' substance is calculated from data that is given to you, the mole ratio from the coefficients in the equation are used to find the number of moles of the 'unknown' substance, and the desired quantity of the unknown substance is then calculated. This is summarised in Figure 15.3.2.

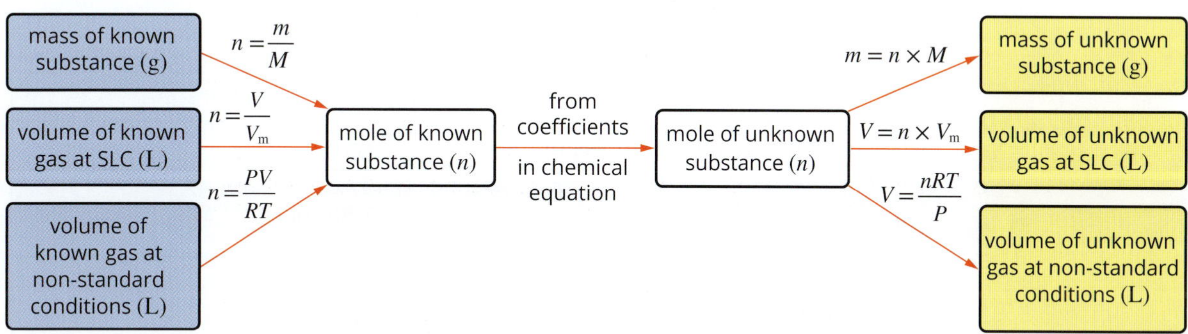

FIGURE 15.3.2 Stoichiometric calculations generally follow the steps shown in the flow chart. Calculating the number of moles and using a mole ratio from a balanced chemical equation are always central to any stoichiometric calculation.

Mass-volume stoichiometric calculations involve the use of the gas equations, which were covered in Section 15.2. Worked examples 15.3.2 and 15.3.3 show how to calculate the volume of gas that is produced from a known mass of reactant at SLC and at non-standard conditions.

Worked example 15.3.2

MASS–VOLUME STOICHIOMETRIC CALCULATIONS AT STANDARD LABORATORY CONDITIONS

Calculate the volume of methane, in L, produced when 1.00 kg of glucose ($C_6H_{12}O_6$) decomposes according to the equation below. The gas volume is measured at SLC.

$$C_6H_{12}O_6(s) \rightarrow 3CO_2(g) + 3CH_4(g)$$

Thinking	Working
Calculate the number of moles of the known substance using: $n = \frac{m}{M}$	$n(C_6H_{12}O_6) = \frac{1000}{180.0} = 5.56$ mol
Find the mole ratio: $\frac{\text{coefficient of unknown}}{\text{coefficient of known}}$	$\frac{n(CH_4)}{n(C_6H_{12}O_6)} = \frac{3}{1}$
Calculate the number of moles of the unknown substance using: $n(\text{unknown}) = \text{mole ratio} \times n(\text{known})$	$n(CH_4) = \frac{3}{1} \times 5.56 = 16.7$ mol
Calculate the volume of the unknown substance using: $V = n \times V_m$	$V(CH_4) = 16.7 \times 24.8 = 413$ L

Worked example 15.3.2: Try yourself

MASS–VOLUME STOICHIOMETRIC CALCULATIONS AT STANDARD LABORATORY CONDITIONS

Calculate the volume of carbon dioxide, in L, produced when 300 g of coal (C) burns in air according to the equation below. The gas volume is measured at SLC.

$$C(s) + O_2(g) \rightarrow CO_2(g)$$

Worked example 15.3.3

MASS–VOLUME STOICHIOMETRIC CALCULATIONS AT NON-STANDARD CONDITIONS

Calculate the volume of hydrogen gas, in L, produced when 50.0 g of aluminium reacts completely with hydrochloric acid according to the equation:

$$2Al(s) + 6HCl(aq) \rightarrow 2AlCl_3(aq) + 3H_2(g)$$

The gas volume is measured at 60°C and 200 kPa.

Thinking	Working
Calculate the number of moles of the known substance using: $n = \frac{m}{M}$	$n(Al) = \frac{50.0}{27.0} = 1.85$ mol
Find the mole ratio: $\frac{\text{coefficient of unknown}}{\text{coefficient of known}}$	$\frac{n(H_2)}{n(Al)} = \frac{3}{2}$
Calculate the number of moles of the unknown substance using: $n(\text{unknown}) = \text{mole ratio} \times n(\text{known})$	$n(H_2) = \frac{3}{2} \times 1.85 = 2.78$ mol
Express the pressure and temperature in required units.	$p = 200$ kPa $T = 60 + 273 = 333$ K

Calculate the volume of the unknown substance using: $V = \frac{nRT}{p}$	$V(H_2) = \frac{2.78 \times 8.31 \times 333}{200}$ $= 38.4$ L

Worked example: Try yourself 15.3.3

MASS–VOLUME STOICHIOMETRIC CALCULATIONS AT NON-STANDARD CONDITIONS

Calculate the volume of hydrogen gas, in L, produced when 120.0 g of magnesium reacts completely with hydrochloric acid according to the equation:

$$Mg(s) + 2HCl(aq) \rightarrow MgCl_2(aq) + H_2(g)$$

The gas volume is measured at 32°C and 150 kPa.

GAS VOLUME–VOLUME CALCULATIONS

For chemical reactions where both the reactants and products are in the gaseous state, it is often convenient to measure volumes, rather than masses.

For example, the reaction between propane gas and oxygen can be represented by the equation:

$$C_3H_8(g) + 5O_2(g) \rightarrow 3CO_2(g) + 4H_2O(g)$$

This equation tells us that when 1 mole of propane reacts with 5 moles of oxygen gas, 3 moles of carbon dioxide and 4 moles of water vapour are produced.

You saw in Section 15.2 that equal amounts, in moles, of different gases occupy equal volumes when measured at the same temperature and pressure. Therefore, the mole ratios of gases will also be volume ratios at the same temperature and pressure. In the above reaction, this means that when 1 litre of propane gas reacts with 5 litres of oxygen gas, 3 litres of carbon dioxide and 4 litres of water vapour are produced.

Worked example 15.3.4

GAS VOLUME–VOLUME CALCULATIONS

Methane gas (CH_4) is burned in a gas stove according to the following equation:

$$CH_4(g) + 2O_2(g) \rightarrow CO_2(g) + 2H_2O(g)$$

If 50 mL of methane is burned, calculate the volume of O_2 gas required for complete combustion of the methane under constant temperature and pressure conditions.

Thinking	Working
Find the mole ratio: $\frac{\text{coefficient of unknown}}{\text{coefficient of known}}$	$\frac{n(O_2)}{n(CH_4)} = \frac{2}{1}$
The temperature and pressure are constant, so volume ratios are the same as mole ratios.	$\frac{V(O_2)}{V(CH_4)} = \frac{2}{1}$
Calculate the volume of the unknown substance using: $V(\text{unknown}) = \text{mole ratio} \times V(\text{known})$	$V(O_2) = \frac{2}{1} \times 50$ $= 100$ mL

Worked example: Try yourself 15.3.4

GAS VOLUME–VOLUME CALCULATIONS

Ethane gas (C_2H_6) is burned in a gas stove according to the following equation:

$$2C_2H_6(g) + 7O_2(g) \rightarrow 4CO_2(g) + 6H_2O(g)$$

If 100 mL of ethane is burned in air, calculate the volume of CO_2 gas produced under constant temperature and pressure conditions.

15.3 Review

OA ✓✓

SUMMARY

- Stoichiometric calculations follow the general steps:
  1. Calculate the amount, in mol, of a known substance from the data given.
     Use $n = \frac{m}{M}$, $n = \frac{V}{V_m}$ or $n = \frac{pV}{RT}$
  2. Use the mole ratio from a balanced chemical equation to determine the amount, in mol, of the unknown substance.
     $$\frac{n(\text{unknown chemical})}{n(\text{known chemical})} = \frac{\text{coefficient of unknown chemical}}{\text{coefficient of known chemical}}$$
  3. Find the desired quantity of the unknown substance from its amount, in mol, using $m = n \times M$, $V = n \times V_m$ or $pV = nRT$
- Stoichiometric calculations can be used to calculate the mass of a gas reacted or produced during a chemical reaction.
- Stoichiometric calculations can be used to calculate the volume a gas reacted or produced during a chemical reaction.
- The mole ratio in a balanced equation is also a volume ratio if all reactants and products are in the gaseous state and the temperature and pressure are kept constant.

KEY QUESTIONS

Knowledge and understanding

1 What volume, in L, of CO_2 is produced when 3.00 L of pentane (C_5H_{12}) undergoes complete combustion in oxygen if the volumes are measured at the same temperature and pressure? The equation for the reaction is:

$$C_5H_{12}(g) + 8O_2(g) \rightarrow 5CO_2(g) + 6H_2O(g)$$

2 Calculate the volume of oxygen needed to completely react with 1.50 L of carbon monoxide according to the following equation. Assume all volumes are measured at the same temperature and pressure.

$$2CO(g) + O_2(g) \rightarrow 2CO_2(g)$$

Analysis

3 Octane (C_8H_{18}) is a component of petrol. 200 g of octane burns in oxygen to produce carbon dioxide and water. The equation for this reaction is:

$$2C_8H_{18}(g) + 25O_2(g) \rightarrow 16CO_2(g) + 18H_2O(g)$$

a Calculate the mass of oxygen required to react.

b Calculate the mass of carbon dioxide produced.

4 A 2.00 g piece of aluminium metal reacts completely with a solution of sulfuric acid to produce aqueous aluminium sulfate and hydrogen gas:

$$2Al(s) + 3H_2SO_4(aq) \rightarrow Al_2(SO_4)_3(aq) + 3H_2(g)$$

a Calculate the mass of hydrogen produced.

b Determine the volume of hydrogen gas produced at SLC.

5 Octane is one of the main constituents of petrol. It burns according to the equation:

$$2C_8H_{18}(g) + 25O_2(g) \rightarrow 16CO_2(g) + 18H_2O(l)$$

What mass of octane must have been used if 50.0 L of carbon dioxide, measured at 120°C and 1.10 atm, was produced?

6 Hydrogen peroxide decomposes according to the equation:

$$2H_2O_2(aq) \rightarrow 2H_2O(l) + O_2(g)$$

Calculate the volume of oxygen, collected at 30.0°C and 91.0 kPa, that is produced when 10.0 g of hydrogen peroxide decomposes.

Chapter review

KEY TERMS

absolute temperature scale
absolute zero
enhanced greenhouse effect
gas constant
greenhouse gas
greenhouse effect
ideal gas
ideal gas constant
ideal gas equation
kelvin scale
kinetic energy
kinetic molecular theory
molar mass
molar volume
partial pressure
pascal
pressure
standard atmosphere
standard laboratory conditions (SLC)
volume

REVIEW QUESTIONS

Knowledge and understanding

1 Which of the following gases is the most potent greenhouse gas?

A CO_2
B H_2O
C O_2
D CH_4

2 Compare the natural greenhouse effect and enhanced greenhouse effect in terms of contributing gases and environmental impacts.

3 Explain why a large focus of world climate change conferences is reduction of the use of fossil fuels such as coal.

4 **a** What is the volume of 0.75 mol of oxygen gas (O_2) at SLC?

b Would the volume of 0.75 mol of carbon dioxide gas (CO_2) at SLC be higher, lower or the same as the oxygen gas? Give a reason for your answer.

5 **a** What is the molar volume of a gas at a particular pressure and temperature if 0.356 mol of the gas has a volume of 10.0 L?

b If the pressure of the gas in part **a** is 100 kPa, is the temperature higher or lower than 25°C? Explain your answer.

6 Use the molar volume of a gas at SLC to find the:

a volume occupied by 8.0 g of oxygen, O_2 at SLC

b the mass of oxygen, O_2, present in a 50.0 L container of oxygen at SLC

c mass of nitrogen dioxide, NO_2, present in 10 L at SLC.

7 If 64.0 g of oxygen gas occupies a volume of 25.0 L when the temperature is 303 K, then the pressure of the gas, in kPa, is closest to:

A 20.0
B 200
C 400
D 6.40×10^3

Application and analysis

8 At what temperature, in °C, will 0.5 g of helium exert a pressure of 150 kPa in a container with a fixed volume of 4.0 L?

9 Propane (C_3H_8) burns in oxygen according to the equation:

$$C_3H_8(g) + 5O_2(g) \rightarrow 3CO_2(g) + 4H_2O(g)$$

6.70 g of propane was burned in excess oxygen.

a What mass of carbon dioxide would be produced?

b What mass of oxygen would be consumed in the reaction?

c What mass of water would be produced?

10 Large quantities of coal are burned in Australia to generate electricity, in the process generating significant amounts of the greenhouse gas carbon dioxide. The equation for this combustion reaction is:

$$C(s) + O_2(g) \rightarrow CO_2(g)$$

Determine the mass of carbon dioxide produced by the combustion of 1.0 tonne (10^6 g) of coal, assuming that the coal is pure carbon.

11 Propane (C_3H_8) burns in oxygen according to the equation:

$$C_3H_8(g) + 5O_2(g) \rightarrow 3CO_2(g) + 4H_2O(g)$$

Calculate the volume of carbon dioxide produced, at SLC, when 5.0 kg of propane reacts completely with excess oxygen.

12 Methane burns in excess oxygen according to the equation:

$$CH_4(g) + 2O_2(g) \rightarrow CO_2(g) + 2H_2O(g)$$

This reaction produces 5 L of carbon dioxide at 200°C and 100 kPa. Assuming all volumes are measured at the same temperature and pressure, calculate the:

a volume of methane used

b volume of oxygen used

c mass of water vapour produced.

13 Consider two containers of equal size; one contains oxygen and the other carbon dioxide. Both containers are at 23°C and at a pressure of 1.0 atm. Answer the following questions about the two gas samples and give a reason for your answers.

a Does one container have more molecules than the other and, if so, which one?

b Which of the two samples of gas contains the greater number of atoms?

c Given that density is defined as the mass per unit volume, which of the two gases has the greater density?

14 A sample of gas of mass 9.68 g occupies a volume of 5.4 L at 27°C and 100×10^3 Pa.

a Calculate the amount, in mol, of gas in the sample.

b Determine the molar mass of the gas.

15 A room has a volume of 220 m^3.

a Calculate the amount, in mol, of air particles in the room at 23°C and a pressure of 100 kPa.

b Assume that 20% of the molecules in the air are oxygen molecules and the remaining molecules are nitrogen. Calculate the mass of air in the room.

16 Carbon dioxide gas is a product of the complete combustion of fuels.

a Calculate the mass of 1.00 mol of carbon dioxide.

b What is the volume occupied by 1.00 mol of carbon dioxide at SLC?

c Given that density is defined as mass/volume, calculate the density of carbon dioxide at SLC in g L^{-1}.

17 There are many scientists investigating possible fuels to replace fossil fuels. A group of Japanese chemists is investigating the following reaction as a source of methane:

$$CaCO_3(s) + 4H_2(g) \longrightarrow CH_4(g) + Ca(OH)_2(s) + H_2O(g)$$

At 400°C, 100 kPa and under suitable reaction conditions calculate the:

a volume of methane produced if 100 L of hydrogen is completely reacted.

b mass of calcium carbonate used in part **a**.

18 An indoor gas heater burns propane (C_3H_8) at a rate of 12.7 g per minute. Calculate the minimum mass of oxygen per minute, in g, that needs to be available for the complete combustion of propane according to the equation:

$$C_3H_8(g) + 5O_2(g) \longrightarrow 3CO_2(g) + 4H_2O(g)$$

19 The Loy Yang A power station in the Latrobe Valley consumes about 60 thousand tonnes (1 tonne = 10^6 g) of coal a day. The coal used in the power station is composed of approximately 25% carbon. Calculate the volume of the greenhouse gas carbon dioxide released each day by the power station at SLC according to the equation:

$$C(s) + O_2(g) \longrightarrow CO_2(g)$$

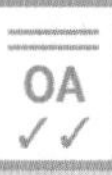

CHAPTER

16 Analysis for salts

As you discovered in Chapter 10, water is a very good solvent for a wide variety of polar molecules and ionic salts. As a result of water's solvent properties, all water systems contain some dissolved salts. Their presence can be attributed to a number of factors, such as natural processes, pollution, farming and industrial activities.

In this chapter, you will examine the ways in which salts can enter the water and soil systems and the methods used by scientists to test water and soil to determine salinity levels. You will also look at a number of methods used to test water and soil samples for the presence of metal contaminants and other ions.

Key knowledge

- sources of salts found in water or soil (which may include minerals, heavy metals, organo-metallic substances) and the use of electrical conductivity to assess the salinity and quality of water or soil samples **16.1**
- quantitative analysis of salts:
  - molar ratio of water of hydration for an ionic compound **16.2**
  - the application of mass–mass stoichiometry to determine the mass present of an ionic compound **16.2**
  - the application of colorimetry and/or UV–visible spectroscopy, including the use of a calibration curve to determine the concentration of ions or complexes in a water or soil sample. **16.3**

16.1 Testing for salts in water

Figure 16.1.1 shows a dry section of Lake Eyre, a very large inland salt lake in South Australia. On the rare occasions when the lake contains water, the salt concentration in the water is very high. When the lake dries up, the salts are deposited on the lake bed. The high concentration of salts in the water in Lake Eyre is an extreme example of **salinity**.

Sodium chloride (NaCl) is commonly referred to as salt; however, in the context of soil and water supplies, the term 'salt' refers to any ionic compounds present. In this section, you will examine ways in which these salts come to be in the water and soil systems. The sources of the salts found in water can include minerals, heavy metals and organometallic substances. This section will also describe one of the most common ways in which the salt concentrations are determined.

FIGURE 16.1.1 Lake Eyre is Australia's lowest natural point at approximately 15 metres below sea level. Salt deposits on the dry bed are a result of the extreme salinity levels in this region.

SALTS FROM MINERALS

Salts are naturally present in water and soil systems. As part of the water cycle, water runs through soil and rocks, dissolving solid **mineral** deposits and transporting these salts into lakes, rivers, creeks and other bodies of water. The centre of Australia was once submerged under the ocean, so the presence of salt in Lake Eyre should not be a surprise.

The region near Buchan in Gippsland, Victoria, was also submerged under the ocean millions of years ago. During that time, the remains of marine organisms containing calcium carbonate ($CaCO_3$) accumulated there. These layers of calcium carbonate deposits eventually formed limestone. The spectacular caves throughout the region are a result of underground rivers cutting through limestone rock. In Figure 16.1.2 you can see the formations have been created by the dissolving and redepositing of limestone by rainwater over time as it passes through the caves.

FIGURE 16.1.2 The action of water dissolving and redepositing minerals has led to these limestone formations and caves.

There are many other examples of regions in Victoria with high mineral concentrations, including:

- Hepburn Springs: Hepburn Springs is famous for its mineral water springs. Tourists bathe in the hot pools which contain high levels of minerals. The ions present in the water include Na^+, K^+, Ca^{2+}, Mg^{2+}, Fe^{2+}, Cl^- and HCO_3^-.
- Wimmera Mineral Sands: The sandy soil in the Wimmera region of north-west Victoria contains significant levels of zirconium and titanium minerals. Iluka Mines operates in this region to extract these valuable minerals.
- Pittong clay mine: Levels of clay minerals in the soil near Pittong in Central Victoria are very high and are mined for commercial use. Pittong clay is a mix of pittongite (a clay mineral containing tungsten) and kaolin (a white clay used in cosmetics and pottery).

SALTS FROM HUMAN ACTIVITY

Human activity can increase salt levels in water and soil. In most cases, the addition of these salts is considered a form of pollution and results from various industries and processes, such as mining, agriculture, domestic sources and sewage treatment plants.

In many countries, governments monitor and regulate the levels of dissolved salts and other contaminants in waterways.

Mining

Mining industries can use large volumes of water to process the materials they are extracting. Some of this water, still containing various ions, may be discharged back into local waterways. Dust from mining sites can be carried by the wind and rain onto land and contaminate soil.

Agriculture

Most farms use fertilisers to improve the yield of crops. When it rains, some of this fertiliser dissolves and may be transported to the waterways through run-off. Ammonium nitrate (NH_4NO_3), ammonium sulfate ($(NH_4)_2SO_4$) and superphosphate ($Ca(H_2PO_4)_2$) are common fertilisers used in Australia that can contribute to the build-up of excess nutrients in waterways such as rivers and lakes. Excessive irrigation can cause the groundwater level to rise to the surface. Salt from the ancient inland sea deep underground dissolves into the groundwater, leading to salty water accumulating on the surface of the soil. The soil then becomes too saline to grow crops.

Domestic sources

Until recently, most detergents contained softening agents made from phosphate compounds. Therefore, the discharge from washing machines and sinks added metal cations and anions, such as phosphate, to the water system. Phosphate is a nutrient for plants and leads to excessive algae growth in waterways, which is known as algal blooms. In Figure 16.1.3, you can see the effect of high levels of phosphates in water. The growth of algal blooms due to excess nutrients leads to a significant problem known as **eutrophication**. When eutrophication occurs in waterways, the oxygen concentration drops below the levels that fish need to survive, leading to mass fish-kills. In addition, blue-green algal blooms, such as *Nodularia*, can cause water to become unusable as they produce highly toxic substances.

FIGURE 16.1.3 Eutrophication due to algal blooms can result from high concentrations of phosphate in the water.

Sewage treatment plants

All cities have treatment plants to process effluent (sewage) and grey water (non-sewage water waste from homes and businesses). Although this water is treated to remove harmful contaminants, the water discharged from the treatment plants may contain a variety of ions similar to those from domestic sources.

Stormwater

In an urban environment, rainfall lands on hard surfaces, such as carparks, rooftops and roads. Any material on these surfaces is transported by the rain into the waterways through the stormwater pipes. Physical rubbish, soil, dust, animal faeces and petrochemicals are some of the many substances that end up in the waterways after rainfall.

Industry

There are many different ways in which human activity contributes to the salt content of waterways and soil. This increase in salt content can have harmful and/or toxic effects on the environment.

Industrial processes commonly produce wastes that contain salts. Unfortunately, these wastes have been inappropriately disposed of in the waterways or soil, especially in the past when the environmental impacts were unknown or ignored. There are many sites around Victoria with contaminated soil, with some dating back to industries in the late 1800s and early 1900s.

HEAVY METAL SALTS

Definitions of **heavy metals** vary, but they are usually described as metals with a high density that have a toxic effect on living organisms. Cadmium, lead, chromium, copper and mercury all fit this description of heavy metals. Some metalloids, including arsenic, are also commonly included in lists of heavy metals due to their high toxicity. Figure 16.1.4 compares the densities of some heavy metals and lighter, non-toxic elements such as aluminium and magnesium. Some scientists regard aluminium ions in solution as toxic.

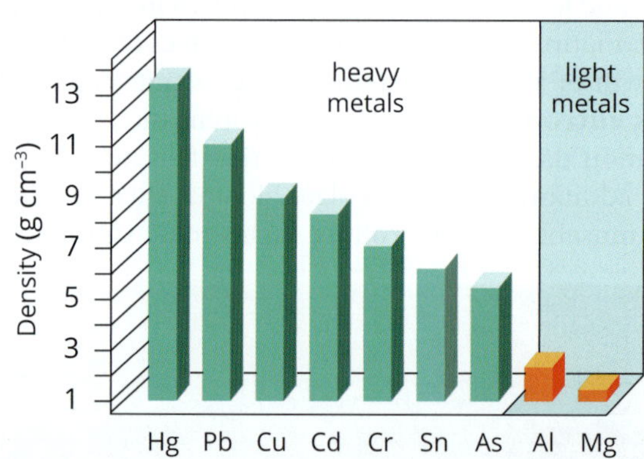

FIGURE 16.1.4 A comparison of the relative densities, in g cm^{-3}, of a number of toxic heavy metal elements with the lighter elements aluminium and magnesium. Tin is considered a heavy metal to a lesser extent due to its lower toxicity levels.

Heavy metals occur naturally within the Earth's crust. Their salts can dissolve into rivers and groundwater and so make their way into drinking water supplies. Usually, the concentrations of heavy metals from these natural sources are very low. However, heavy metals are often used in industry, and various human activities can result in elevated levels of heavy metals in the environment.

Table 16.1.1 lists some heavy metal pollutants, their sources and the effects they can have on human health.

TABLE 16.1.1 Sources of a number of heavy metals and their effects on the human body

Heavy metal pollutant	Source	Essential or non-essential for human metabolism	Health effects through ingestion at toxic levels
Copper	• copper pipes • roofing • coins • algicides	• essential • required for making red blood cells	• anaemia • liver and kidney damage • stomach and intestinal irritation
Lead	• lead pipes • batteries • leaded petrol • paints • ammunition • cosmetics	• non-essential	• negatively affects haemoglobin production • damage to kidneys, gastrointestinal tract, joints and reproductive system • can lower IQ levels in young children
Cadmium	• smelting • improper disposal of rechargeable batteries • metal plating • paints • burning fossil fuels	• non-essential	• kidney failure • liver disease • osteoporosis
Nickel	• power plants • waste incinerators • improper disposal of batteries	• essential • required for hormonal and lipid metabolism	• decreased body weight • damage to the heart and liver
Zinc	• mining • smelting • steel production	• essential • required for many enzymes	• anaemia • damage to nervous system and pancreas
Arsenic	• natural deposits in the ground • industry • agricultural processes • treated pine	• non-essential	• carcinogenic • stomach pain • numbness • blindness
Mercury	• improper disposal of batteries, fluorescent light bulbs, thermometers and barometers • various industrial processes • burning of fossil fuels	• non-essential	• tremors • gingivitis • spontaneous abortion • damage to the brain and central nervous system

Heavy metal ions are released into the environment in two main ways: through human activity directly or as a result of reaction with water in the atmosphere. Heavy metal compounds can be released directly into waterways through waste from industries such as metal processing and mining. Other potential sources of contamination from heavy metals include leachate (water containing toxic substances) from landfill sites and agricultural run-off.

Combustion of fuels and wastes containing heavy metals can release these ions into the atmosphere where they can interact with water molecules. Rain can then take the dissolved salts into soils, rivers and groundwater.

Once released into the environment, heavy metals will persist and accumulate.

Problems caused by heavy metals in the environment can be made worse when they increase in concentration in a food chain. The build-up of heavy metals in higher-order predators is often referred to as **biomagnification**, and is illustrated in Figure 16.1.5.

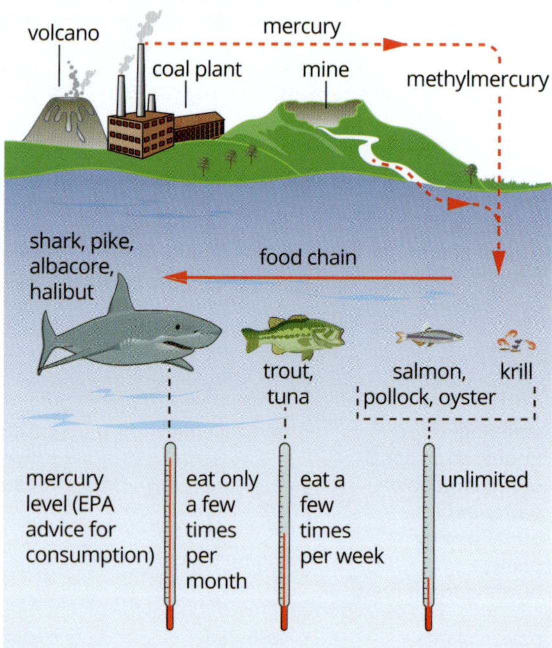

FIGURE 16.1.5 Sources of mercury in the environment are volcanic eruptions, burning fossil fuels, bushfires and mining. As living organisms have no way of removing heavy metals, such as mercury, from their cells, the concentration of heavy metals can be seen to increase in species higher up the food chain.

Heavy metal elements are only toxic when present as cations in water-soluble compounds or as organometallic compounds.

Heavy metals are only toxic to living organisms when present as cations in water-soluble compounds, or as **organometallic compounds**.

ORGANOMETALLIC COMPOUNDS

Organometallic compounds are another source of heavy metals in water and soil systems. Organometallic compounds are substances for which there must be at least one direct bond between a metal atom or ion and a carbon atom. If the metal has a very low electronegativity, this bond may be very polar, almost ionic. These organometallic compounds are usually synthetic substances that are used in industry as catalysts or reagents in chemical processes.

An example of an organometallic compound used in industry is tetraethyl lead ($Pb(C_2H_5)_4$), shown in Figure 16.1.6. You can see this molecule has four bonds between a lead atom and the carbon atom of an ethyl group. Tetraethyl lead was added to petrol in Australia for many years to improve the smoothness of combustion in car engines. It is now banned because lead levels near busy roads were affecting the health of local residents. The lead emissions from the cars deposit on nearby soils and could even enter nearby waterways. The toxicity of tetraethyl lead is a consequence of the toxic lead it contains and the ease with which the molecule enters cells.

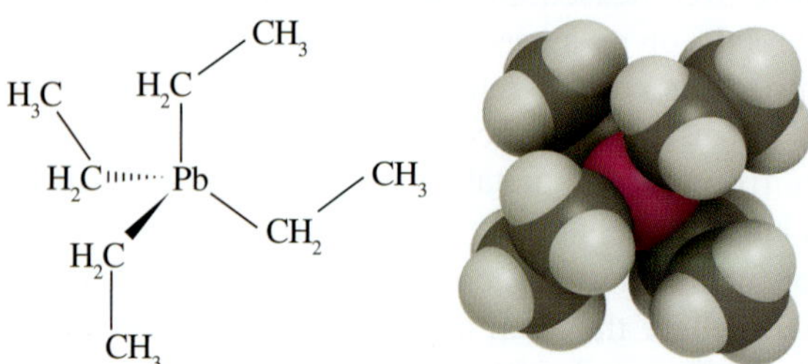

FIGURE 16.1.6 Two different structural representations of tetraethyl lead. This compound has four ethyl groups ($-CH_2CH_3$) bonded to a lead atom in the centre of the molecule.

Another example of an organometallic compound is methylmercury. It is an organometallic cation with the formula CH_3Hg^+. In methylmercury, a methyl group is bonded to a mercury(II) ion. Methylmercury combines with anions to form organometallic salts such as methylmercury chloride and methylmercury hydroxide. Methylmercury compounds are formed in some industrial processes, such as the production of ethyne, but they are also formed when compounds containing mercury are burnt. Methylmercury compounds are more toxic than mercury itself because they can be easily transported around the human body in the same way as proteins.

CASE STUDY

Lasting impact of heavy metals

The levels of heavy metals in waterways are closely monitored because even amounts as small as 24.8 ppm (24.8 mg L^{-1}) can be deadly. The wide-ranging and long-term effects of heavy metal poisoning were clearly shown in Japan in the 1950s. A factory in the small fishing village of Minamata had been discharging toxic waste containing methylmercury into the local bay. The main diet of the people of Minamata consisted of seafood caught in the contaminated bay. As mercury compounds are not easily excreted by the aquatic organisms living in the bay, the toxic mercury compounds built up.

The first indication of a problem was the erratic behaviour of the local cats that were seen to be 'dancing' down the streets before collapsing and dying. This strange behaviour was a direct result of mercury poisoning that causes neurological disorders and eventually death.

Neurological symptoms were also seen in the local population, with many residents suffering irreversible brain and organ damage. Many people died as a result of the high levels of mercury they unknowingly ingested. Originally referred to as 'Minamata disease', the neurological effects were eventually determined to be the direct result of mercury poisoning. The Minamata area residents still struggle with highly toxic levels of mercury to this day. The accumulated mercury levels in the people also led to the development of a number of congenital disorders in children born to parents suffering from Minamata disease. Figure 16.1.7 shows a boy receiving physiotherapy to treat the effects of mercury poisoning.

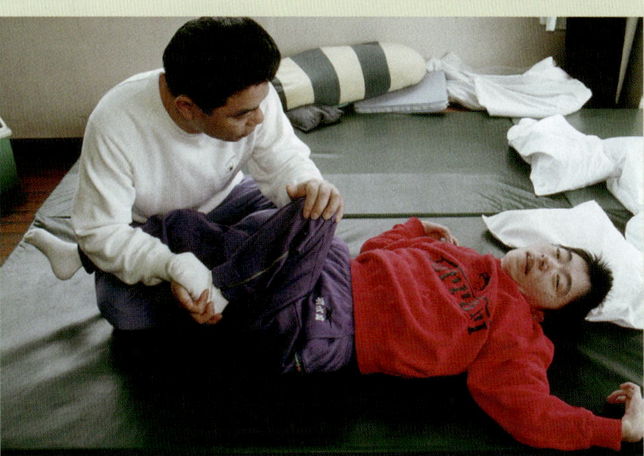

FIGURE 16.1.7 A boy in Japan receives physiotherapy for the ravaging effects of mercury poisoning.

HARD WATER

Hard water is a term used to describe water that requires a lot of soap to obtain a lather or froth. **Hardness** in water is caused by the presence of some metal ions, mainly calcium, magnesium, manganese and iron. These metal ions are due mainly to the presence of dissolved minerals, and they interfere with the washing action of soaps and some detergents. One of the essential ingredients in soap is the compound sodium stearate ($C_{17}H_{35}COONa$). When dissolved in water, soap provides the stearate ion ($C_{17}H_{35}COO^-$), which can act as a dirt remover. The metal ions in hard water react with this ion to produce a precipitate, removing the stearate ions from solution and reducing the amount of lather produced. The reaction can be described using the following equation:

$$2C_{17}H_{35}COO^-(aq) + Ca^{2+}(aq) \rightarrow Ca(C_{17}H_{35}COO)_2(s)$$

These precipitates accumulate on the inside of water pipes. These deposits can lead to the eventual blocking of the pipes (Figure 16.1.8).

FIGURE 16.1.8 Accumulation of limescale in pipes is a result of the precipitation of soap scum due to the calcium, magnesium, iron or manganese ions present in hard water. Eventually, the crystalline deposits can become so severe they block the flow of water.

TESTING FOR SALINITY

FIGURE 16.1.9 A simple circuit for detecting salinity. The brighter the light emitted from the globe, the greater the conductivity of the solution.

Pure water is a poor conductor of electricity. As the concentration of ions in solution increases so does the conductivity.

One of the most common methods for testing salinity levels of water and soil samples is to measure the **electrical conductivity (EC)** of the sample. Electrical conductivity is the degree to which a specified material conducts electricity. If two electrodes in a circuit containing a light globe and a battery are placed into a water sample, as shown in Figure 16.1.9, the intensity of the light emitted from the globe will provide an indication of the salinity level.

Pure water contains very few ions and is a poor conductor of electricity. As soluble salts are added to the water, the ion concentration increases, which in turn increases the conductivity of the solution. It is the flow of ions towards the electrodes that is responsible for the conductivity. Quantitative measurements can be taken if a meter is included in the circuit. Figure 16.1.10 shows how the increase in conductivity with increased ion concentration can be measured.

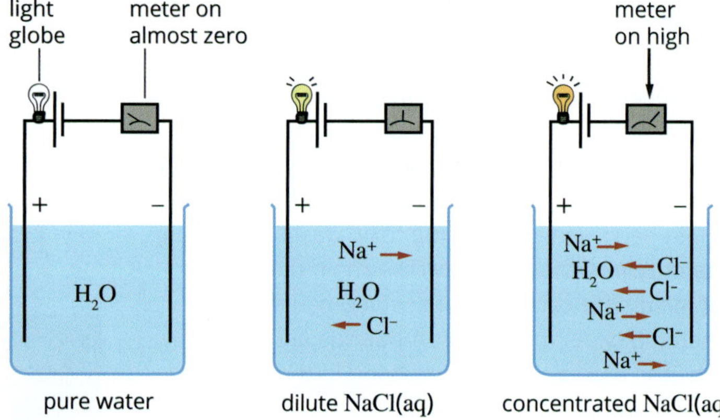

FIGURE 16.1.10 The conductivity of a salt solution depends upon the concentration of salt in the solution. The flow of current increases as the concentration of ions increases.

CHEMFILE

Australian laundry detergents become phosphate-free

Most Australian laundry detergents became phosphate-free by 2014, as major companies agreed to phase out the environmentally damaging ion from their products.

Major supermarkets, such as Coles, Woolworths and Aldi, also agreed to stop selling environmentally damaging detergents.

Zeolites are widely used as a replacement to phosphates. Zeolites are aluminosilicate minerals with large numbers of small pores with diameters smaller than 2 nm. As a consequence, they have extremely high surface areas. The calcium and magnesium in hard water bind to the surface of the zeolites, therefore removing these cations from the water.

Phosphates have been phased out of Australian laundry detergents.

Figure 16.1.11 shows that at low salt concentrations there is a direct relationship between the conductivity and concentration. A graph of conductivity against concentration of salt solutions is linear.

Electrical conductivity is the inverse of electrical resistance. Electrical conductivity is measured in units of micro-Siemens per centimetre, $\mu S\ cm^{-1}$. The electrical conductivity of a solution increases as the temperature of the solution increases, so to ensure consistency, readings for electrical conductivity are generally taken at 25°C and at a constant voltage. In addition, the distance between the electrodes is independent, allowing for the comparison between different brands and types of electrodes.

Typical guidelines for EC of water for different uses are shown in Table 16.1.4.

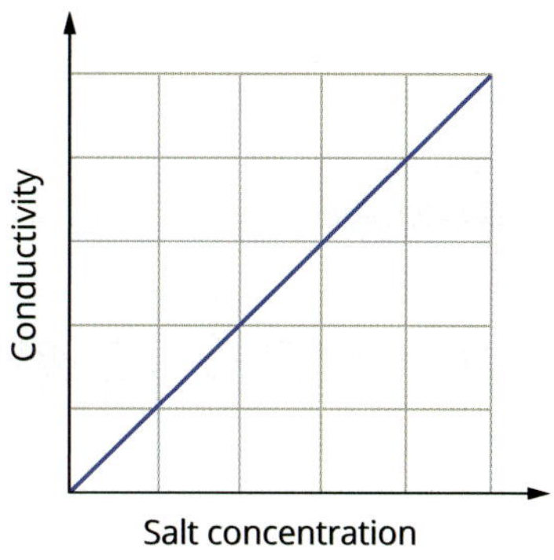

FIGURE 16.1.11 The conductivity of a salt solution increases with concentration. At low concentrations, the relationship is considered to be a linear one.

TABLE 16.1.4 EC guidelines for water use

Electrical conductivity (EC) $\mu S\ cm^{-1}$	Water use
0–800	good drinking water, suitable for irrigation and livestock
800–2500	unpleasant to drink; use in irrigation needs to be managed
2500–10 000	not recommended for human consumption; suitable for salt-tolerant vegetation only
>10 000	not suitable for any plants or animals
40 000–56 000	range measured in ocean waters

Handheld, portable salinity testing probes, such as the one shown in Figure 16.1.12, are widely used by agricultural experts and water authorities who want quick and reliable salinity estimates on the spot. Because these readings are often completed on-site, the probes are calibrated to account for differences in temperature.

Soil salinity is determined by the 1 : 5 weight to volume method. The first step is to mix one part dried soil (in grams) with five parts distilled water (in millilitres). Typically, 20 g of dried soil is mixed with 100 mL of distilled water and stirred well. Once the soil has settled, the salinity of the water ($EC_{1:5}$) is measured using a portable salinity probe. The second step is to determine the soil salinity by multiplying the $EC_{1:5}$ by the conversion value (Table 16.1.5) based on **soil texture**. Soil texture is the percentage of sand, silt and clay in a soil sample. For example, the $EC_{1:5}$ of a clay soil sample will be multiplied by 7 to determine the soil salinity.

TABLE 16.1.5 Soil salinity conversion values based on soil texture

Soil texture	Conversion value
sandy	17
silty sand	14
silty	10
silty clay	9
clay	7

FIGURE 16.1.12 Portable digital salinity meters such as this one usually self-adjust for temperatures other than 25°C. The meter here is measuring the salinity of a water sample as 5 ppt (parts per thousand, or 5000 mg L^{-1}) at 32°C.

16.1 Review

OA ✓✓

SUMMARY

- Pollution resulting from human activity is a source of salts in our water supplies. Activities include run-off from farms and cities, and release of chemicals from industries and mines.
- Heavy metals are metals with a high density, such as lead and mercury, which have a toxic effect on living things. Heavy metals are present in nature, but processes in some industries add to the levels of these metals in water supplies and surrounding soil.
- Organometallic compounds are substances that contain at least one carbon–metal bond. Their use in industry and laboratories can lead to their presence in water and soil systems.
- In some regions of Australia, the high mineral levels in water lead to the water being described as 'hard'. This means it contains high levels of metal ions such as Ca^{2+}, Mg^{2+}, Mn^{2+} and Fe^{2+}. Soaps do not function as effectively in hard water and they cause bathroom sinks, taps and drains to become blocked.
- Electrical conductivity is an effective way to measure the levels of salinity in water and soil samples. The higher the concentration of ions in a solution, the higher the electrical conductivity.
- Electrical conductivity increases as the temperature of a solution increases, so readings are taken at 25°C.

KEY QUESTIONS

Knowledge and understanding

1 List sources of salt in Australian waterways from:
  a natural sources
  b human activities

2 Provide an example of an essential and non-essential heavy metal.

3 List the ways in which toxic heavy metals make their way into waterways.

4 Which of the following are organometallic compounds?
  a $AgCl$
  b CH_3Hg^+
  c $HgCl_2$
  d $PbCr_2O_7$
  e $[Fe(CN)_6]^{3-}$
  f $Pb(C_2H_5)_4$
  g $Zn(CH_3)_2$
  h $Ca(H_2PO_4)_2$

5 The following questions relate to hard water.
  a Define hard water.
  b Name an ion that causes hard water.

Analysis

6 Water samples from households in a country town were taken and analysed for heavy metals. The results were:
arsenic = not detected
cadmium = not detected
lead = 0.52 mg L^{-1}
mercury = not detected
nickel = 0.78 mg L^{-1}
  a Name the main sources of the heavy metals present in the water sample.
  b Using your knowledge of the solubility of ionic compounds and precipitation reactions, suggest a way of removing these heavy metals from the water supply.

7 A farmer wants to know the salinity of their sandy soil. They take 20 grams of dried soil and add 100 mL of distilled water. They stir the mix and measure the $EC_{1:5}$ as 300 $\mu S\ cm^{-1}$. Referring to Table 16.1.5, on page 501, calculate the salinity of their sandy soil.

16.2 Quantitative analysis of salts

In the previous section, you learnt that electrical conductivity can be used to measure the salinity level of a solution. This technique provides a measure for the **total dissolved solids** in a solution, but it does not distinguish between the different salts that have contributed to the conductivity. However, there are many instances in which you may want to know the concentration of a particular salt in a solution. For example, you might wish to determine the concentration of sodium chloride in a particular brand of soup. In this section, you will learn about two simple laboratory techniques to determine the mass of an ionic compound.

WATER OF HYDRATION

Copper(II) sulfate crystals are a brilliant deep blue colour (Figure 16.2.1). Yet, if you heat the crystals, they eventually become a white powder. In this section, you will learn why the physical characteristics of copper(II) sulfate change when it is heated and how to calculate the change.

FIGURE 16.2.1 Copper(II) sulfate forms brilliant blue crystals.

Salts exist mostly as crystals. Crystals are composed of alternating cations and anions in a lattice structure. Often water molecules become trapped in these lattices. The water molecules form weak interactions between the ions as the crystal develops, and so the water molecules become part of the crystal. If you heat the crystal, these water molecules are released from the crystal. The crystal then usually loses its structure, leaving the salt as a powder. The water molecules that were part of the crystal structure are called the **water of hydration**.

 Many common salts exist as hydrated salts under normal laboratory conditions.

The number of water molecules in the salt crystal can be determined using a form of **gravimetric analysis**. Gravimetric analysis is a technique that measures the mass of a reactant or product to determines the quantity of an unknown by stoichiometry. Using the example of copper(II) sulfate above, the copper(II) sulfate crystal would have a set number of water molecules for each cation and anion:

$$CuSO_4{\cdot}xH_2O$$

where x is the number of water molecules per formula unit of $CuSO_4$. The value of x can be determined by heating the crystal.

$$CuSO_4{\cdot}xH_2O(s) \xrightarrow{\text{heat}} CuSO_4(s) + xH_2O(g)$$

The amount of water of hydration within a salt crystal is determined by heating the crystal.

$CuSO_4{\cdot}xH_2O$ is known as **hydrated** copper(II) sulfate, whereas $CuSO_4$ is known as **anhydrous** copper(II) sulfate. You can find the value of x by weighing the salt before and after heating. The difference in the mass is the mass of water evaporated from the crystal. The value of x can then be determined using molar ratios, as described in the following steps.

Worked example 16.2.1

DETERMINING MOLAR RATIO OF WATER OF HYDRATION

5.00 g of blue hydrated copper(II) sulfate crystals was heated until it changed to a whiteish powder. The mass of the anhydrous $CuSO_4$ powder was 3.19 g. Calculate the molar ratio of water of hydration of the original blue crystals.

Thinking	Working
Write a balanced chemical equation for the dehydration reaction.	$CuSO_4{\cdot}xH_2O(s) \xrightarrow{\text{heat}} CuSO_4(s) + xH_2O(g)$
Calculate the amount (in mole) of anhydrous solid using: $n = \frac{m}{M}$	$n(CuSO_4) = \frac{m}{M}$ $= \frac{3.19}{159.6}$ $= 0.0200$ mol
The difference in mass between the hydrated solid and the anhydrous solid is the mass of water lost.	$m(H_2O) = m(CuSO_4.xH_2O) - m(CuSO_4)$ $m(H_2O) = 5.00 - 3.19$ $m(H_2O) = 1.81$ g
Calculate the amount (in mole) of water lost.	$n(H_2O) = \frac{m}{M}$ $= \frac{1.81}{18.0}$ $= 0.101$ mol
Calculate the value of x by using the molar ratio of H_2O to the anhydrous salt calculated above: $\frac{\text{coefficient of } H_2O(x)}{\text{coefficient of anhydrous salt}(1)} = \frac{n(H_2O)}{n(\text{anhydrous salt})}$	$\frac{x}{1} = \frac{n(H_2O)}{n(CuSO_4)}$ $= \frac{0.101}{0.0200}$ $= 5.05$ round to nearest whole number $x = 5$
Write the formula of the hydrated salt using the value of x.	$CuSO_4{\cdot}5H_2O$

Worked example: Try yourself 16.2.1

DETERMINING MOLAR RATIO OF WATER OF HYDRATION

Anhydrous $NiCl_2$ is a yellow solid, whereas $NiCl_2$ hydrate is bright green. 0.251 g of yellow solid was obtained after heating 0.460 g of green hydrated $NiCl_2$ crystals. Calculate the molar ratio of water of hydration of the original green crystal.

CHEMFILE

Why do some salts change colour when dehydrated?

Many transition metals form coloured crystal salts when hydrated. When the crystal is dehydrated (i.e. heated to remove its water molecules), the anhydrous salt is sometimes a different colour. Why is this so? The colour is due to electrons in the cations of the salt absorbing selected wavelengths of light and moving to a higher energy level (see figure below). The colour of the salt that we observe is from wavelengths that are not absorbed. The electrons in the cations in the hydrated salt have different energies to those in the anhydrous salt. Once the water is removed, the wavelength required to excite the electrons is different, and the salt can appear a different colour.

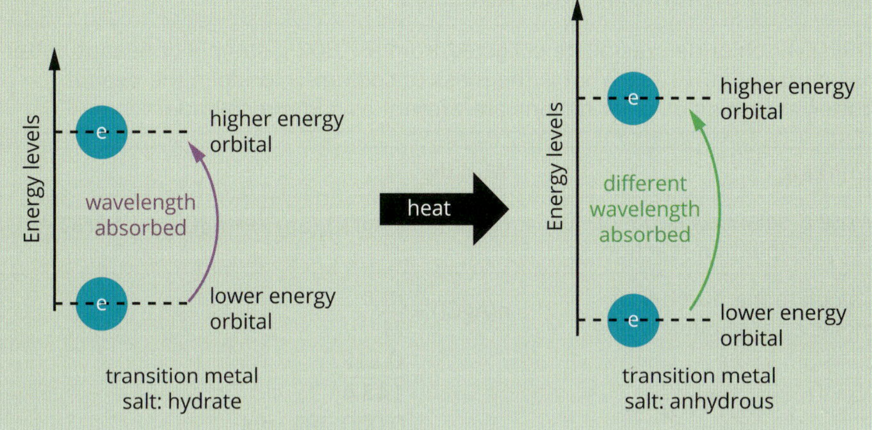

The hydrate salt on the left absorbs a specific wavelength of light. Once the water of hydration molecules are removed, the energy levels change for the anhydrous salt and a different wavelength of light is absorbed.

MASS–MASS STOICHIOMETRY

When you carry out a reaction in the laboratory, you will measure quantities of chemicals in grams, not moles. For this reason, most calculations will require you to start and finish with mass rather than moles of a substance. To calculate the number of moles of the known substance from a mass, you can use the relationship:

$$\text{Number of moles} = \frac{\text{mass (in g)}}{\text{molar mass (in g mol}^{-1}\text{)}}$$

Calculating the mass of a salt in solution from a precipitation reaction

Stoichiometry can be combined with your knowledge of **precipitation reactions** to find the amount of a salt in a solution.

There are several steps involved in calculating the mass of a salt in a water sample, based on the mass of a **precipitate** produced in a precipitation reaction.

1 Write a balanced equation for the reaction.

2 Calculate the number of mole of the precipitate from its mass, using the formula $n = \frac{m}{M}$.

3 Use the mole ratios in the equation to calculate the number of mole of the salt in solution.

4 Calculate the mass of the salt, using $m = n \times M$.

When analysing for salts in samples using precipitation reactions, knowledge of the solubility rules is important

Figure 16.2.2 provides a flow chart that summarises this process.

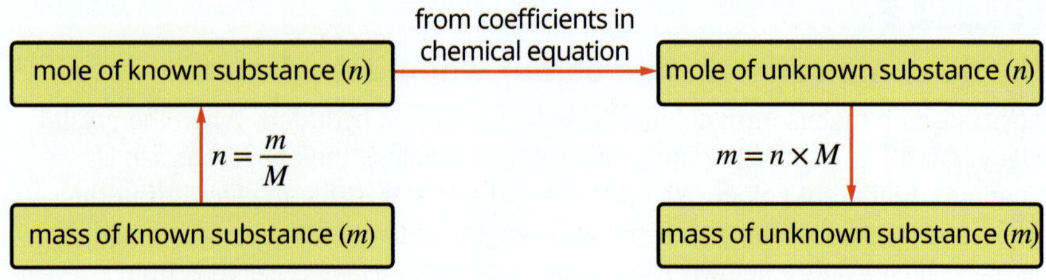

FIGURE 16.2.2 A flow chart for mass–mass stoichiometric calculations

Worked example 16.2.2

MASS–MASS STOICHIOMETRIC PROBLEMS

The silver chloride precipitate collected from a 7.802 g sample of peanut butter has a mass of 0.112 g. What is the mass of sodium chloride in the peanut butter, assuming all chloride ions are present as sodium chloride?

Thinking	Working
Write a balanced equation for the precipitation reaction.	$NaCl(aq) + AgNO_3(aq) \longrightarrow AgCl(s) + NaNO_3(aq)$
Calculate the number of moles of precipitate using: $n = \frac{m}{M}$	$n(AgCl) = \frac{m}{M}$ $= \frac{0.112}{143.4}$ $= 0.000\ 781$ mol
Use the balanced equation to find the mole ratio of the 'known' and 'unknown' substances. The 'known substance' is the one about which you are provided information in the question; the 'unknown substance' is the one whose mass you are required to calculate.	mole ratio $= \frac{n(NaCl)}{n(AgCl)}$ $= \frac{1}{1}$
Calculate the number of moles of unknown substance.	$n(NaCl) = n(AgCl) = 0.000\ 781$ mol
Calculate the mass of unknown substance in the sample.	$m(NaCl) = n \times M$ $= 0.000\ 781 \times 58.5$ $= 0.0457$ g

Worked example: Try yourself 16.2.2

MASS–MASS STOICHIOMETRIC PROBLEMS

Water discharged from a mining plant contains silver ions present as silver nitrate ($AgNO_3$). Excess potassium chromate (K_2CrO_4) solution is added to a 50.0 g sample of water to precipitate the silver as silver chromate. The precipitate is heated to remove any water to produce 1.32 g of silver chromate (Ag_2CrO_4).

Calculate the mass of silver in the water supply. (The molar mass of Ag_2CrO_4 is 331.7 g mol^{-1}.)

Gravimetric analysis

Gravimetric analysis is one of the techniques used by chemists to analyse substances for their composition or concentration. It is inexpensive and can be used for a range of common inorganic substances. Because of its versatility and the relative ease with which it is conducted, gravimetric analysis has been used for analysis in a variety of industries. Gravimetric analysis can be used to determine the salt content of foods, the sulfur content of ores and the level of impurities in water. Since the introduction of modern instruments, gravimetric analysis has usually been replaced by faster, automated techniques. However, it is still used to check the accuracy of analytical instruments.

The aim of gravimetric analysis is to separate one of the ions (either cation or anion) of the salt you want to measure from the other ions present in solution by forming a precipitate. The precipitate is dried and weighed and, using **mass–mass stoichiometry**, the amount of salt you wanted to measure is calculated. For example, the concentration of barium ions in a solution containing barium, sodium and nitrate ions can be determined by adding sodium sulfate (Figure 16.2.3). The sodium and nitrate ions will remain in solution, but a precipitate of barium sulfate is formed, which can be separated from the solution and dried. The mass of the precipitate may then be used to calculate the barium ion concentration in the initial solution.

> A precipitate is usually formed when analysing salts by gravimetric analysis.

Finding the concentration of Ba^{2+} ions in solution.

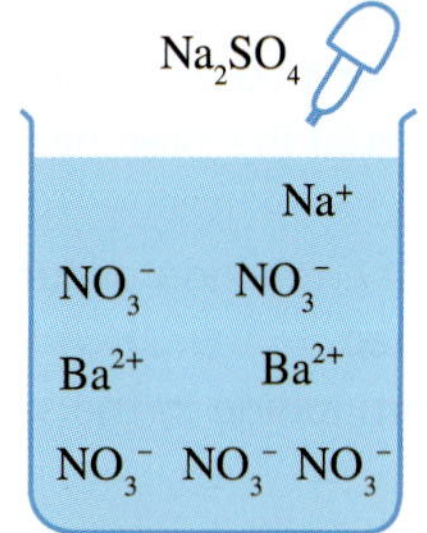

To separate the Ba^{2+} ions from other species present, Na_2SO_4 is added.

Na^+ NO_3^- Na^+ Na^+ Na^+ NO_3^- NO_3^- Na^+ NO_3^- NO_3^-

$BaSO_4$ $BaSO_4$

The Ba^{2+} ions are the only metal ions in the precipitate.

filtered and dried

Measuring the mass of precipitate will allow calculation of the amount of Ba^{2+} ions initially present.

FIGURE 16.2.3 The principle of gravimetric analysis: Ba^{2+} ions in a solution are separated from Na^+ ions and NO_3^- ions by precipitation. The precipitate is then collected and dried, so that the initial mass of Ba^{2+} can be determined using stoichiometry.

> Gravimetric analysis requires knowing the stoichiometry of the precipitation reaction.

PA 26

Regardless of the salt being tested, the steps in the gravimetric analysis method remain the same. Figure 16.2.4 shows the typical laboratory steps in a gravimetric analysis.

Step 1
Accurately weigh the sample.

Step 2
Dissolve sample in suitable solvent.

Step 3
Add an excess of solution that will form a precipitate.

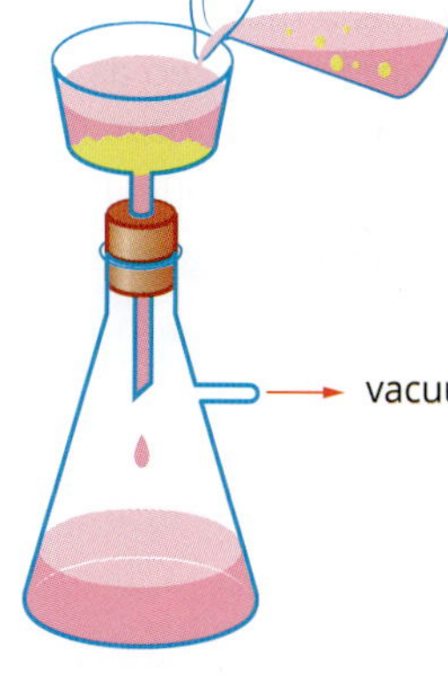

Step 4
Filter the solution to collect precipitate.

Step 5
Dry and weigh the precipitate.

FIGURE 16.2.4 Laboratory procedure for gravimetric analysis

16.2 Review

OA ✓✓

SUMMARY

- Many salts have water molecules within their crystal structures, and these are called water of hydration.
- The water of hydration can be determined by heating a known mass of the salt crystal until all the water has evaporated. The difference in mass is the amount of water present in the hydrate.
- Given the quantity of one of the reactants or products of a chemical reaction, such as a precipitation reaction, the quantity of all other reactants and products can be calculated using stoichiometry
- Gravimetric analysis can be used to determine the amount of salt in food, impurities such as sulfur in ores, or the level of heavy metals in water.

KEY QUESTIONS

Knowledge and understanding

1 Classify the following salts as hydrated or anhydrous:

a $Na_2CO_3{\cdot}10H_2O$

b K_2CrO_4

c $FeCl_3{\cdot}6H_2O$

d $CoCl_2{\cdot}6H_2O$

e $Na_2C_2O_4$

2 You have been provided with some green crystals of $FeSO_4$. Describe the process you will undertake to find its molar ratio of water of hydration.

3 To test for the presence of Cu^{2+} ions in water, a solution of NaOH can be added to the sample.

a Identify the chemical formula of the precipitate formed if Cu^{2+} ions are present.

b Write the ionic equation for the reaction occurring when the precipitate is formed.

Analysis

4 0.187 g of pink $CoCl_2$ crystals was heated in an oven until an anhydrous blue powder was formed. The final mass of the blue powder was 0.102 g. Calculate the formula of the hydrated salt.

5 Borax, $Na_2B_4O_7{\cdot}10H_2O$, is a mineral containing the element boron. It is commonly used as a household cleaner, stain remover and deodoriser. If you heat 10.0 g of borax, what will be the final mass of anhydrous borax, $Na_2B_4O_7$?

6 Some brands of table salt contain low percentages of sodium iodide (NaI). A sample of table salt is dissolved in water. A few drops of $Pb(NO_3)_2$ are added to this salt solution and a yellow precipitate formed. The mass of yellow precipitate formed was 2.63 g.

a Identify the chemical formula of the precipitate formed.

b Write an ionic equation for the reaction occurring when the precipitate is formed.

c Calculate the mass of sodium iodide in the sample of table salt.

7 Students were testing their tap water for copper content by gravimetric analysis. They added an excess of sodium hydroxide to 1 L tap water and collected 0.325 g of solid copper hydroxide.

a Write an ionic equation for the reaction.

b Calculate the minimum amount of sodium hydroxide (by mass) required to react all the copper in the 1 L tap water sample.

8 Soy sauce can have a high salt (NaCl) content. A student wants to measure the amount of salt in soy sauce by gravimetric analysis using $AgNO_3$ solution.

a Write out a full balanced chemical equation for the precipitation reaction of NaCl(aq) with $AgNO_3$(aq).

b The student added an excess amount of $AgNO_3$ solution to 100 mL of soy sauce. 17.82 g of precipitate was collected. Calculate the mass of NaCl in 100 mL of soy sauce.

16.3 Instrumental analysis for salts

In Section 16.2, you learnt how two simple laboratory techniques, combined with knowledge of stoichiometry, can be used to determine the amount of salt in a sample. However, modern chemists have a range of chemical instruments at their disposal that can provide alternative methods of analysis. These methods are often faster, more accurate and can measure very low concentrations that would be difficult to determine with a traditional form of analysis.

This section looks at how instruments called **colorimeters** and **UV–visible spectrophotometers** can be used to determine the concentration of a salt in water and soil samples. These instruments measure the interaction of light with solutions to determine the concentration of a solution. The intensity of the colour of a solution provides an indication of its concentration. Your eye can detect some differences in colour, but the instruments studied in this section can measure their intensity accurately and use this to determine concentration.

SPECTROSCOPY

Light is a form of energy and is a type of **electromagnetic radiation**. Other forms of electromagnetic radiation are radio waves and X-rays. Visible light is only a small part of the range of different forms of electromagnetic radiation. The spread of the different types of radiation arranged according to their relative energies and wavelengths is referred to as the **electromagnetic spectrum** (Figure 16.3.1).

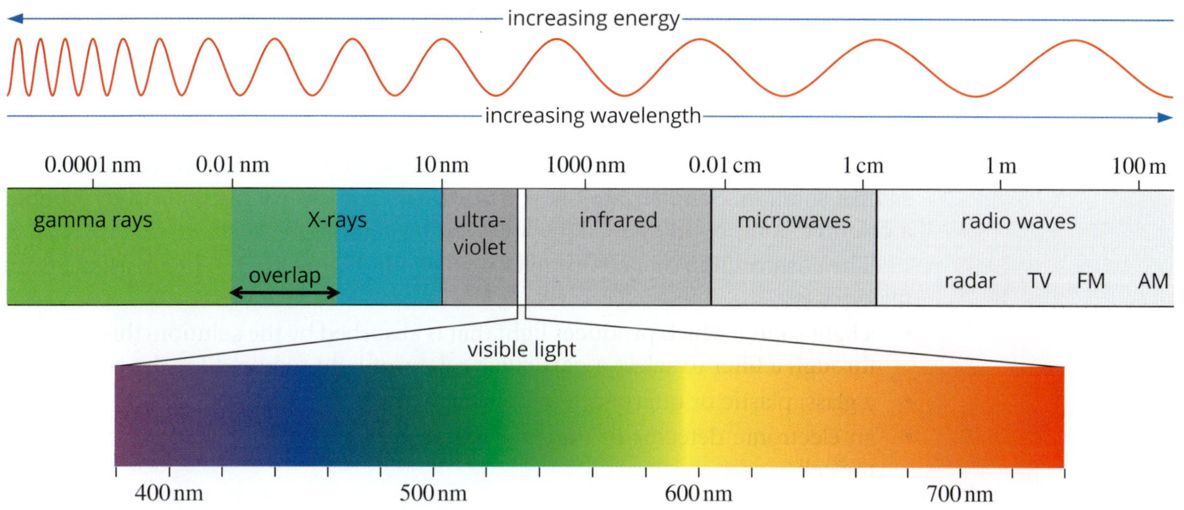

FIGURE 16.3.1 Visible light is only a small part of the electromagnetic spectrum. The spectroscopy techniques detailed in this chapter use radiation within the visible spectrum.

Electromagnetic radiation, such as light, can interact with atoms, and the nature of this interaction depends upon the energy of the electromagnetic radiation. In this section, you will learn about analytical techniques called **spectroscopy**, which use light and other radiation of the electromagnetic spectrum to give us information about the materials around us. The spectroscopic techniques that you will look at in this section will specifically deal with light within the visible region of the electromagnetic spectrum.

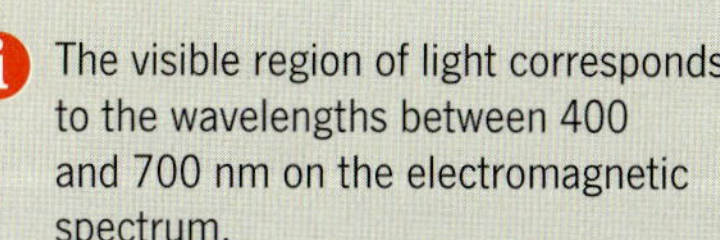

The visible region of light corresponds to the wavelengths between 400 and 700 nm on the electromagnetic spectrum.

When a substance absorbs visible light, it appears coloured. The colour observed is not the same as the colour of the light absorbed. The colour you see is actually due to reflected or transmitted light. For example, plant leaves contain chlorophyll. When you look at the leaves, they appear green because chlorophyll absorbs light in the violet and red ranges of the spectrum. Chlorophyll does not absorb light in the green region of the spectrum, so this is reflected back into your eyes.

The observed colour and the absorbed colour are referred to as **complementary colours**.

Colorimeters and UV–visible spectrophotometers are instruments used to determine the concentration of solutions by measuring their **absorbance** of radiation in the ultraviolet and visible region of the spectrum. The more concentrated the solution, the more radiation it will absorb. Figure 16.3.2 shows solutions of potassium permanganate ($KMnO_4$) at two different concentrations. The solutions appear purple as they absorb light in the yellow–green region of the electromagnetic spectrum (570 nm) and transmit the remaining blue, violet and red light.

The colour of a substance is due to the light that is reflected from its surface (for opaque objects) or transmitted through the substance (for transparent objects and solutions).

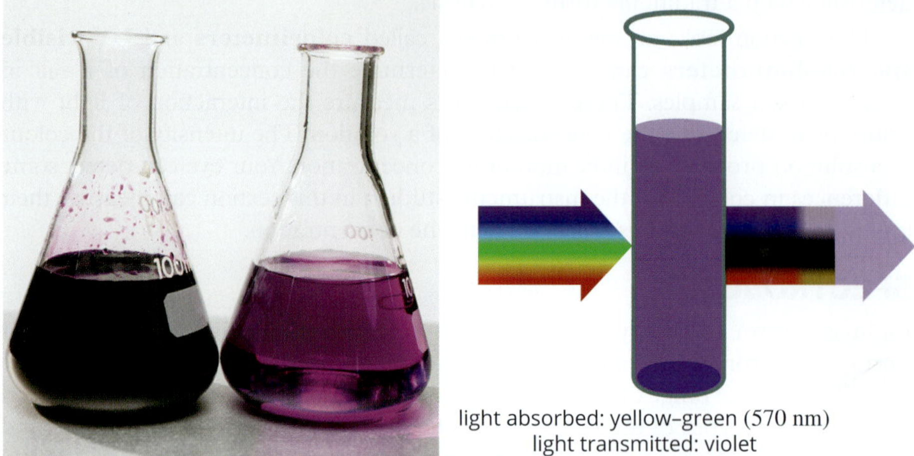

FIGURE 16.3.2 Two solutions of potassium permanganate ($KMnO_4$) appear purple because they absorb yellow–green light. The 0.001 M solution on the right is lighter in colour than the 0.01 M solution because it absorbs less light.

Colorimetry

Colorimetry is a technique that involves measuring the intensity of colour in a sample solution. Samples are often treated with a chemical compound to produce a coloured compound that can be analysed by colorimetry.

The construction of a colorimeter is shown in Figure 16.3.3. It consists of three main parts:

- a light source which produces light that is absorbed by the solution; this is passed through a filter to select a particular colour of light required for the analysis
- a glass, plastic or quartz cell to hold the sample
- an electronic detector to measure the absorbance of light that passes through the cell.

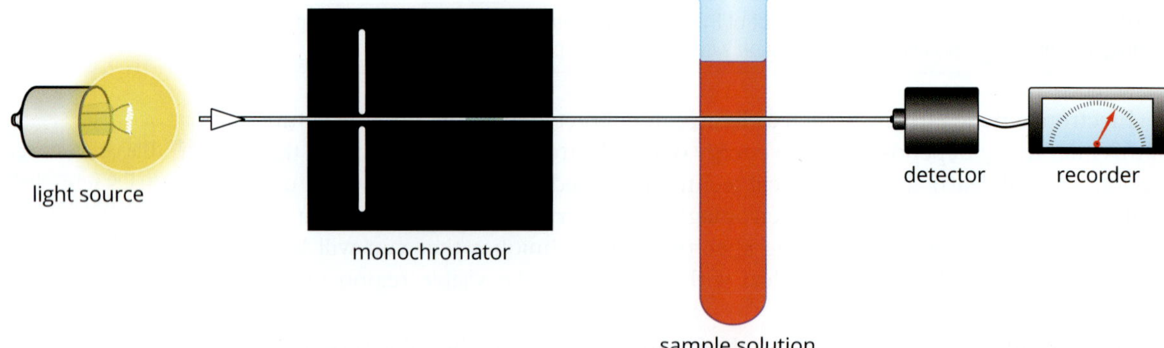

FIGURE 16.3.3 Diagram of the components of a colorimeter. Light of a suitable colour is passed through a sample. The recorder is able to measure the percentage of light absorbed by the sample.

The purpose of the filter is to select light of an appropriate colour that will be strongly absorbed by the sample. For example, since a chlorophyll solution absorbs strongly in the red and blue (purple) regions of the spectrum, a purple filter would be a good choice for chlorophyll analysis. The higher the concentration of chlorophyll, the higher the absorption of the purple light will be.

Table 16.3.1 shows the relationship between the colour of a solution and the selection of a filter for use in a colorimeter. Remember the colour absorbed by the sample is the complementary colour to the colour you observe, as shown in the colour wheel below.

TABLE 16.3.1 Colours of visible light and the complementary colours that are absorbed. The opposite colour on the colour wheel shows the complementary colour.

Wavelength (nm)	Colour absorbed (colour of filter)	Colour observed
380–420	violet	yellow
420–440	violet–blue	yellow–orange
440–470	blue	orange
470–500	blue–green	orange–red
500–520	green	red
520–550	yellow–green	purple (red + violet)
550–580	yellow	violet
580–620	orange	blue
620–680	orange–red	blue–green
680–780	red	green

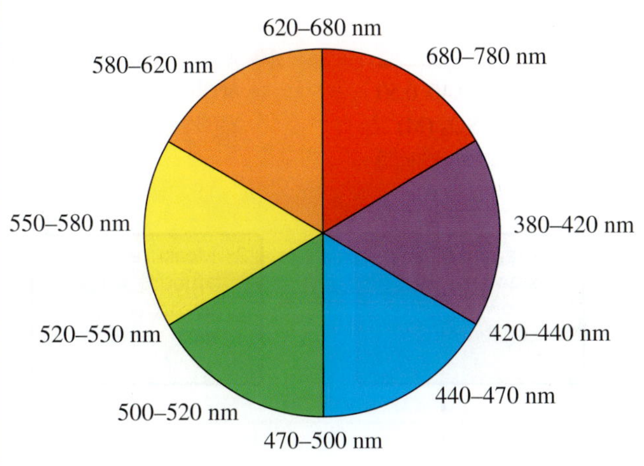

A handy way to remember complementary colours is to write the initials of the main colours in order of their decreasing wavelengths, as ROYGBV (red, orange, yellow, green, blue and violet), and then to write the initials again of the complementary colours directly below, this time starting from green.

R O Y G B V
G B V R O Y

Calibration curves

To determine the concentration of a substance in a solution using colorimetry, a series of **standard solutions** (solutions of accurately known concentration) of the substance must first be prepared and their absorbances measured. The absorbance reading is between 0.00 and 1.00.

For example, if you wished to analyse for nickel(II) sulfate, you would create a series of nickel(II) sulfate solutions of varying concentrations. Depending on the amount of nickel(II) sulfate you suspected to be in your sample, a suitable range of concentrations might be from 0.1 M to 0.5 M. Once your standard solutions are prepared, you can measure their absorbance at the selected wavelength. Table 16.3.2 shows a series of data typical for the absorbances of standard solutions of nickel(II) sulfate.

TABLE 16.3.2 Absorbance of standard solutions of nickel(II) sulfate

Nickel(II) sulfate concentration (M)	Absorbance
0.00	0.00
0.10	0.18
0.20	0.34
0.30	0.49
0.40	0.66
0.50	0.81

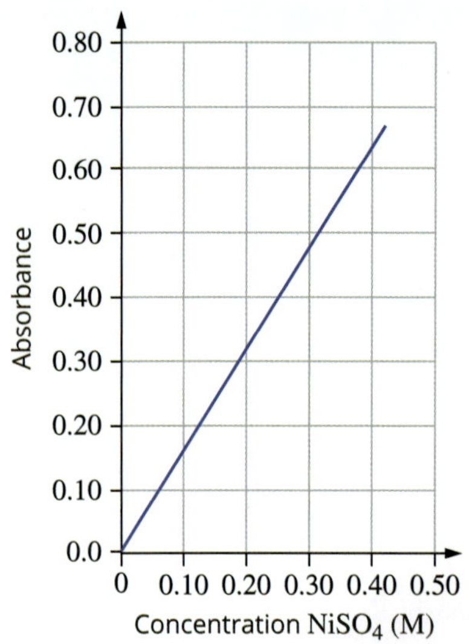

FIGURE 16.3.4 Calibration curve for nickel(II) sulfate

You can then construct a **calibration curve** from the data. A calibration curve is a plot of the absorbances of the standards against their concentration. A calibration curve always has the independent variable (concentration) on the x-axis and the dependent variable (absorbance) on the y-axis. Figure 16.3.4 shows the calibration curve created from the nickel(II) sulfate data in Table 16.3.2.

If the absorbance of a solution of unknown concentration is now measured, the value can be used to determine the concentration from the calibration curve. The unknown absorbance reading must be below the absorbance reading of the highest concentration standard, otherwise the unknown must be diluted.

Figure 16.3.5 outlines the procedure that is followed when analysing a solution using colorimetry.

> A calibration curve is constructed by measuring the absorbance of a series of solutions with accurately known concentrations and then plotting the results on a graph of absorbance versus concentration.

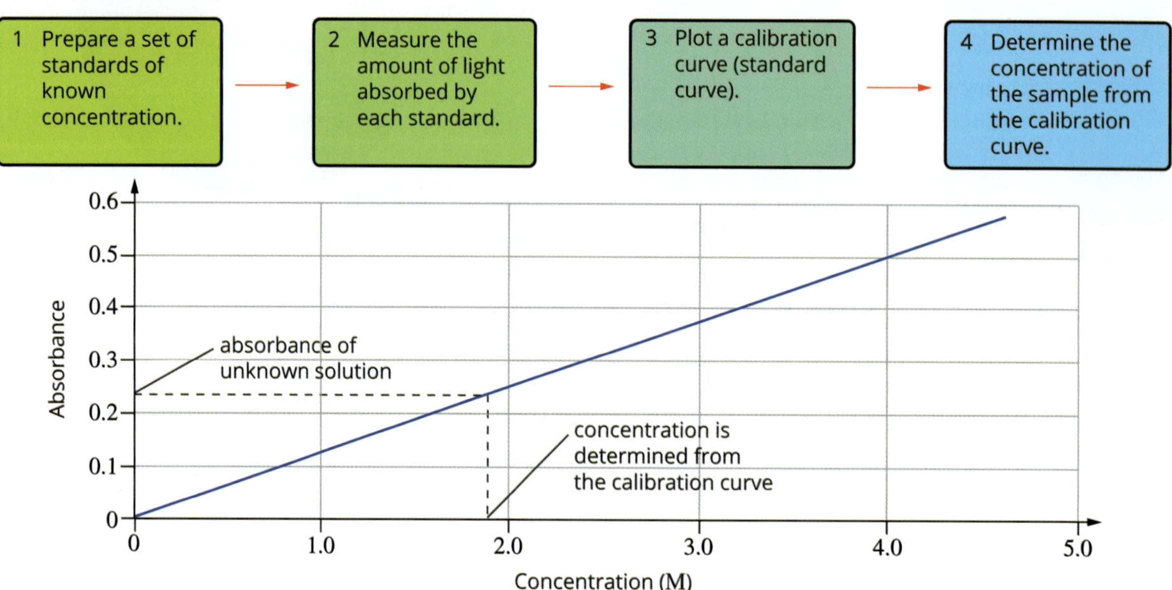

FIGURE 16.3.5 Procedure for the determination of concentration using a colorimeter

Worked example 16.3.1

USING A CALIBRATION CURVE

The concentration of iron in dam water is determined by colorimetry. The absorbances of a series of standard solutions and a sample of dam water are shown in the table below. Determine the concentration of iron (as Fe^{2+}) in the sample.

Concentration of Fe^{2+} (mg mL^{-1})	Absorbance
0.0	0.00
4.0	0.16
8.0	0.31
12.0	0.47
16.0	0.63
sample	0.38

Thinking	Working
Construct a calibration curve from the data above. Concentration will be on the horizontal axis.	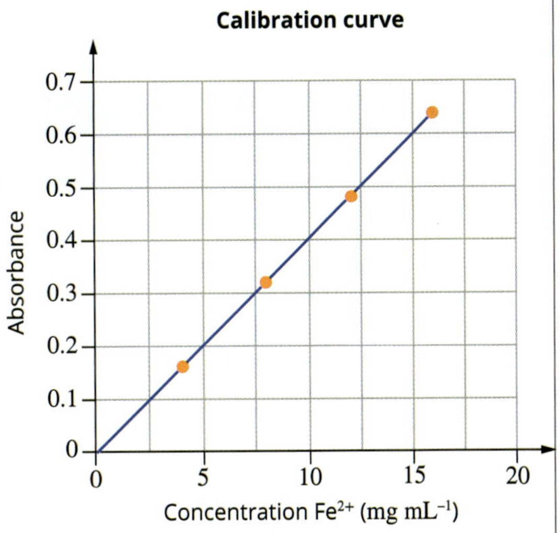 **Calibration curve** Absorbance: 0, 0.1, 0.2, 0.3, 0.4, 0.5, 0.6, 0.7 Concentration Fe^{2+} (mg mL^{-1}): 0, 5, 10, 15, 20
Mark where the absorption of the sample lies on the calibration curve by tracing a horizontal line to the curve.	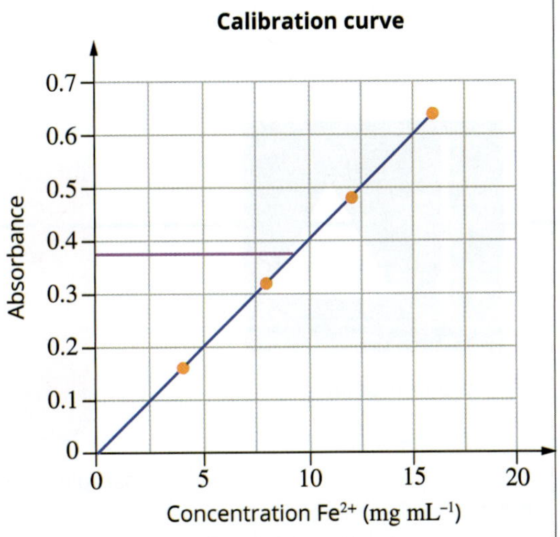 **Calibration curve** Absorbance: 0, 0.1, 0.2, 0.3, 0.4, 0.5, 0.6, 0.7 Concentration Fe^{2+} (mg mL^{-1}): 0, 5, 10, 15, 20 Plot the absorbance value of the unknown solution.
Draw a vertical line from the absorbance value to the *x*-axis.	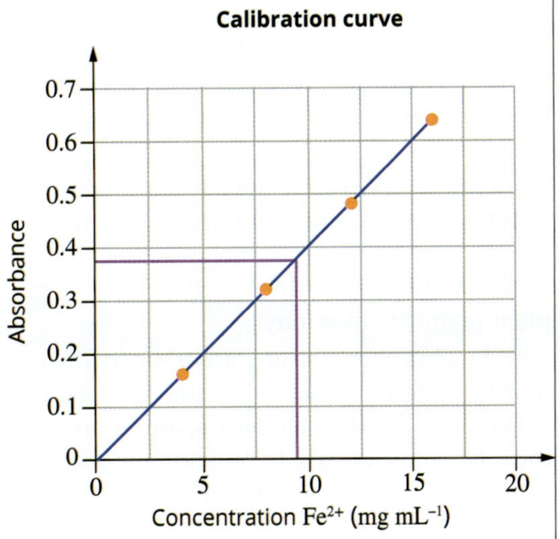 **Calibration curve** Absorbance: 0, 0.1, 0.2, 0.3, 0.4, 0.5, 0.6, 0.7 Concentration Fe^{2+} (mg mL^{-1}): 0, 5, 10, 15, 20
Determine the concentration of the sample by reading the concentration value for the unknown solution from the *x*-axis.	The concentration of iron in the dam water sample can be determined from the graph as 9 mg mL^{-1}.

Worked example: Try yourself 16.3.1

USING A CALIBRATION CURVE

Determine the lead level in a solution using the following colorimetry data.

Concentration of Pb^{2+} (mg L^{-1})	Absorbance
2.5	0.18
5.0	0.35
7.5	0.51
10.0	0.68
sample	0.60

UV-VISIBLE SPECTROSCOPY

A colorimeter is relatively simple and inexpensive, but its accuracy is limited. A UV–visible spectrophotometer, such as the one shown in Figure 16.3.6, is a more sophisticated instrument. A UV–visible spectrophotometer uses a **monochromator** rather than a filter to select light of an exact wavelength to be used in the analysis.

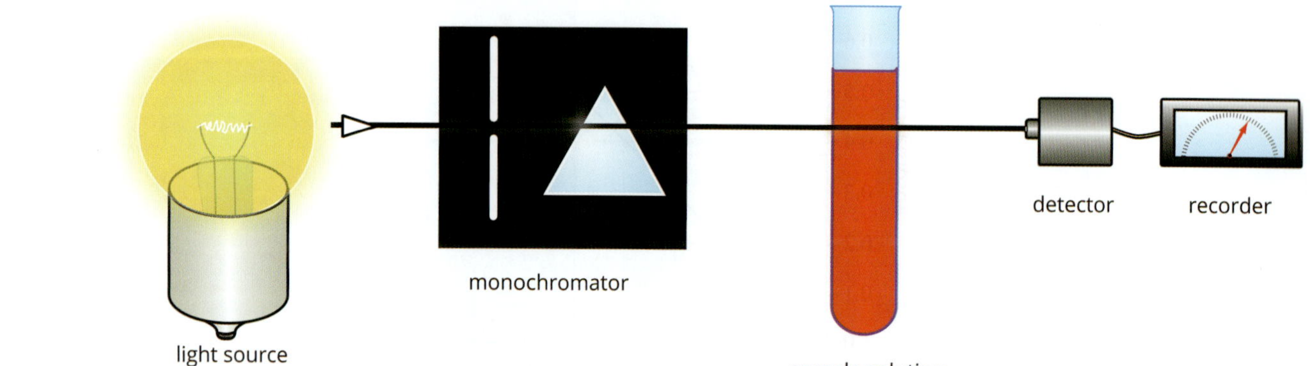

FIGURE 16.3.6 Diagram of the components of a UV–visible spectrophotometer. A monochromator is used to allow selection of specific wavelengths for the analysis of samples.

A monochromator splits white light into individual wavelengths that can be selected using a computer. A UV–visible spectrophotometer can measure in both the ultraviolet and visible wavelengths, whereas colorimeters can only measure visible wavelengths

When a UV–visible spectrophotometer is used, the solution to be tested can first be scanned across multiple wavelengths to select the best wavelength to use. Scanning involves varying the wavelength of light used and checking the absorbance of the sample. Figure 16.3.7 shows an example of a scan for a solution of the green plant pigment chlorophyll.

The scan shows strong absorbance at wavelengths of around 420 and 660 nm, which correspond to violet and red light respectively. The measurements of absorbance for the standard solution and sample would be conducted at one of these two wavelengths. In practice, the wavelength at which other compounds in the solution do not absorb strongly would be chosen.

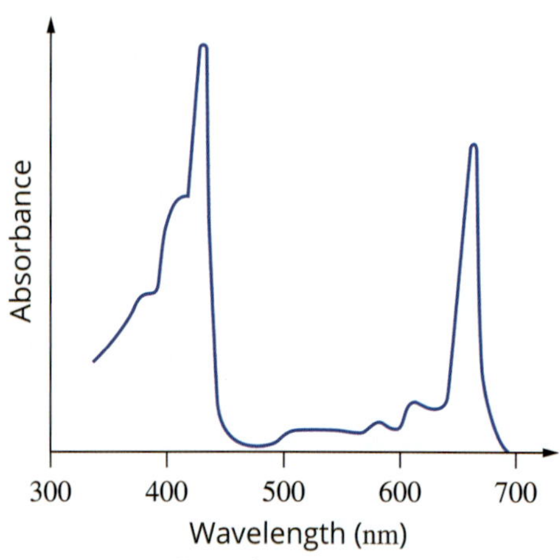

FIGURE 16.3.7 Scan of a green chlorophyll solution showing absorbance maxima at around 420 and 660 nm. Other wavelengths are also absorbed, but not as strongly (e.g. 410 nm and 610 nm).

Metal complexes

Solutions containing Fe^{2+} ions do not absorb very strongly in the ultraviolet or visible part of the spectrum. However, if the Fe^{2+} ions are mixed with an acidic solution of 1,10-phenanthroline (abbreviation to phen), a bright orange solution forms, as seen in Figure 16.3.8. This highly coloured orange solution can then be analysed using either a colorimeter or UV–visible spectrophotometer.

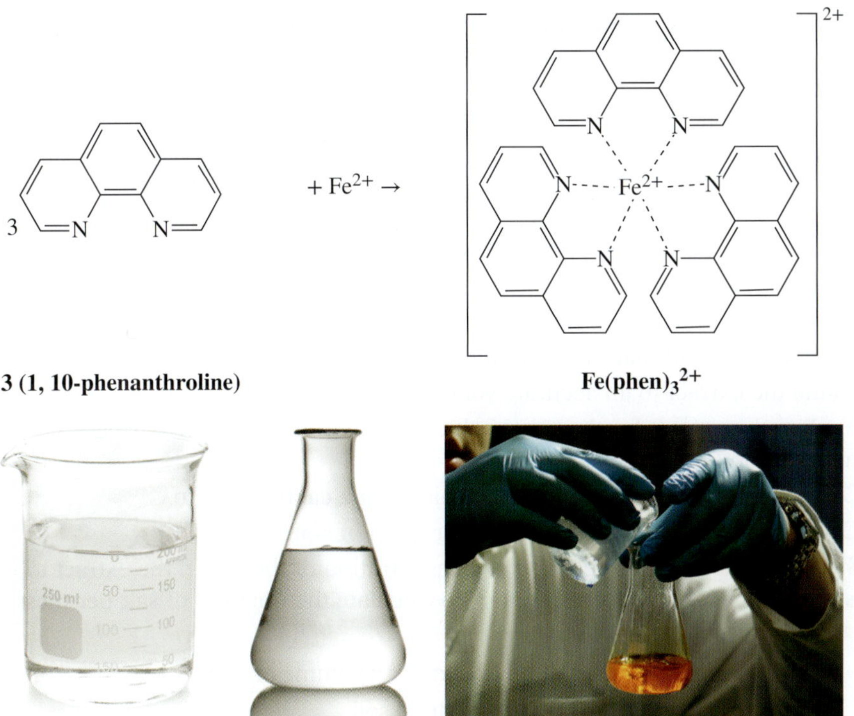

FIGURE 16.3.8 A colourless solution of Fe^{2+} reacts with a colourless solution of 1,10-phenanthroline, forming a bright orange metal complex, that can be analysed by colorimetry or UV–visible spectroscopy.

The reaction shown in Figure 16.3.8 is represented by the equation:

$$Fe^{2+}(aq) + 3phen(aq) \longrightarrow Fe(phen)_3^{2+}(aq)$$

$Fe(phen)_3^{2+}$ is an example of a **metal complex**, in which metal ions (Fe^{2+}) are bound to molecules or anions, known as ligands (phen). Transition metals in particular can form complexes, such as $Fe(phen)_3^{2+}$, many of which are brightly coloured, soluble and suited to analysis with a colorimeter or UV–visible spectrophotometer.

The analysis of a sample using colorimetry or UV–visible spectroscopy can be summarised in the following steps.

1 If the metal ion to be analysed is not strongly coloured, a soluble metal complex may need to be formed.

2 Select the wavelength or filter to be used for the analysis. This will correspond to the wavelength of light absorbed most strongly by the sample.

3 Measure the absorbance of a series of standard solutions of accurately known concentration at the selected wavelength.

4 Plot a calibration curve of absorbance (y-axis) versus concentration (x-axis) for the standard solutions.

5 Measure the absorbance of the sample solution and determine the concentration from reading the corresponding value from the calibration curve.

6 Account for any dilutions that may have been carried out during your sample preparation to calculate the final concentration.

Some metal ions may need to be converted into highly coloured metal complexes to be analysed using UV–visible spectroscopy or colorimetry.

EXTRACTING SALTS FROM SOIL

Soil is made up of a mix of:

- inorganic particles (minerals in the form of sands, silts and clays)
- organic matter (decaying plant or animal)
- living organisms (fungi, bacteria, protozoans, invertebrates)

All these components are known as the **soil matrix**. Salts can be found anywhere within the soil matrix. For example, salts will be within the cells of microbes, bonded to the surface of organic matter, bonded within minerals or dissolved in water. In order to perform an analysis, a liquid, typically water-based, is used to extract the salts from soil. The chemicals used for extracting the salts will depend on the focus of the investigation. For example, in Section 16.1, the extract used to determine soil salinity was distilled water. However, if the scientist wants to know the heavy metal content of a mineral, strong acids are required. The methods of extraction vary depending on the target salt, but will follow these steps:

1 Accurately weigh a dried soil sample.
2 Add a liquid chemical to the dried soil to extract the salt.
3 Remove any solid material from the extract.
4 Dilute the extract to an accurate volume using a volumetric flask.
5 Analyse the extract for the target salt using colorimetry or UV–visible spectroscopy.
6 Determine the concentration of the salt using a calibration curve.
7 Calculate the mass or moles of salt per gram of soil.

Step 7 requires you to convert the concentration of the salt in the extract to the mass or moles of salt per gram of soil. To calculate the quantity of salt per mass of soil, the formula is:

$$\frac{\text{concentration of salt in extract} \times \text{volume of extract}}{\text{mass of dried soil}}$$

To do this calculation, you must have the same volume units for concentration and the volumetric flask used to dilute the extract. For example, suppose HCl was added to 3.405 g of dried soil to determine the amount of magnesium. If the extract was diluted in a 250 mL volumetric flask and the magnesium concentration was determined to be 73 mg L^{-1}, the concentration of magnesium in the soil is calculated as:

$$\frac{73\ \text{mg L}^{-1} \times 0.25\ \text{L}}{3.405\ \text{g}}$$

Therefore, the concentration of magnesium in the soil sample is 5.4 mg g^{-1}. Please note that it does not matter what chemical is used to extract the salt. The key information required is the volume of the extract and the concentration of the salt.

Worked example 16.3.2

AMOUNT OF SALT IN SOIL SAMPLE

A soil sample that was suspected to be contaminated with tin was dried and weighed. The dry weight was 0.298 g. Concentrated nitric acid was added to the soil and the mixture was heated at 90°C for 24 hours. The solids were filtered out and the extract diluted to 50 mL. The extract was analysed and the concentration of tin was recorded to be 2.74 mg L^{-1}. Calculate the concentration of tin in the soil.

Thinking	Working
Check the volume units for both the extract and the ion concentration.	Volume of extract is 50 mL (millilitres). Concentration units are mg L^{-1} (litres)
Convert volume of extract units from millilitres to litres by dividing by 1000.	50 mL = 0.050 L
Multiply the ion concentration by the volume of extract and divide by the mass of soil.	$\frac{2.74 \text{ mg L}^{-1} \times 0.050 \text{ L}}{0.298 \text{ g}}$ $= 0.46 \text{ mg g}^{-1}$ The amount of tin in the soil sample is 0.46 mg g^{-1}

Worked example: Try yourself 16.3.2

AMOUNT OF SALT IN SOIL SAMPLE

An old gold mine had its soil tested for cyanide, as cyanide was once commonly used in the mining process. A dried sample of soil (1.457 g) was mixed with sodium hydroxide solution for 16 hours to extract the cyanide. The solid material was filtered out and the extract diluted to 100 mL. The cyanide concentration was analysed using UV–visible spectroscopy and result was 0.130 M. Calculate the concentration of cyanide in the soil as mmol g^{-1}.

USES OF COLORIMETRY AND UV-VISIBLE SPECTROSCOPY

Colorimeters and UV–visible spectrophotometers are used in many and varied fields. These instruments can determine the concentrations of lead in urine, blood sugar levels, cholesterol levels, levels of haemoglobin in blood and phosphates in water. Portable colorimeters, such as the one shown in Figure 16.3.9a, are now available to make on-site testing easier. UV–visible spectrophotometers (Figure 16.3.9b) are generally found in laboratories due to the sensitivity of the monochromator's moving parts. Portable UV–visible spectrophotometers have set wavelengths and no moving parts.

Further examples of the use of colorimetry or UV–visible spectroscopy are:

- measurement of chromium levels in a workplace: a worker carries a pump and PVC filter unit for a set period of time. The pump samples the air around the worker and solids in the air are collected on the filter paper. Chromium can be extracted from the filter paper and converted to yellow chromate ions (CrO_4^{2-}) before analysis.
- determining phosphate levels in soil: phosphate ions are firstly extracted from a soil sample, then ammonium molybdate and tin(II) chloride are added to the extract, forming a dark-blue complex called phospho-molybdenum blue, which is then analysed using colorimetry or UV–visible spectroscopy.

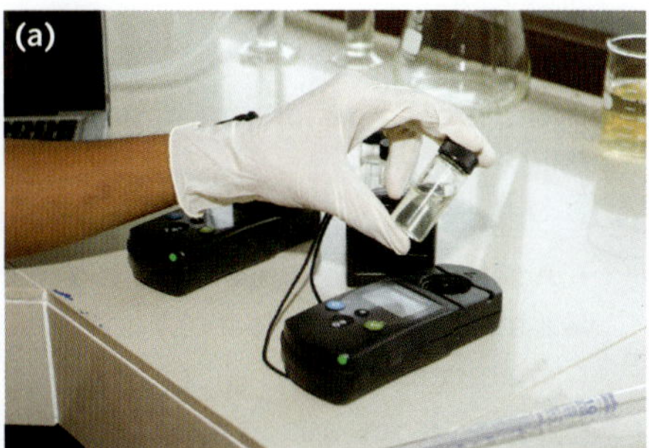

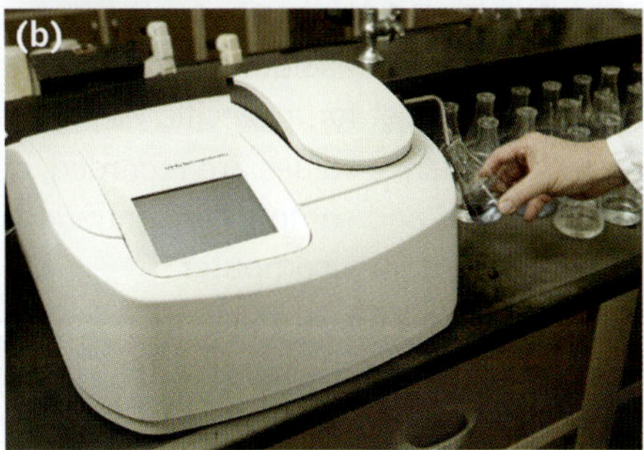

FIGURE 16.3.9 (a) A portable colorimeter. A solution sample is inserted into this unit in a small plastic cell. (b) A laboratory UV–visible spectrophotometer.

CASE STUDY ANALYSIS

Flow injection analysers

Many compounds found in natural waters are detected and analysed using UV–visible spectrophotometers and colorimeters. However, when there are hundreds of samples, multiple compounds required to be analysed or a very small volume of sample provided, some traditional techniques become impractical. Flow injection analysis (FIA) is a technique that is commonly used as it solves many of these practical issues. A flow injection analyser is an instrument that was first designed in 1981 (Figure 16.3.10). It is now commonly used throughout the world for the analysis of many substances, such as nitrate, nitrite, phosphate, ammonia, sulfate, sulfide, fluoride, bromide, chloride, cyanide, aluminium and silica. The instrument mixes specific reagents to the standards and samples and measuring the absorbance at specific wavelengths.

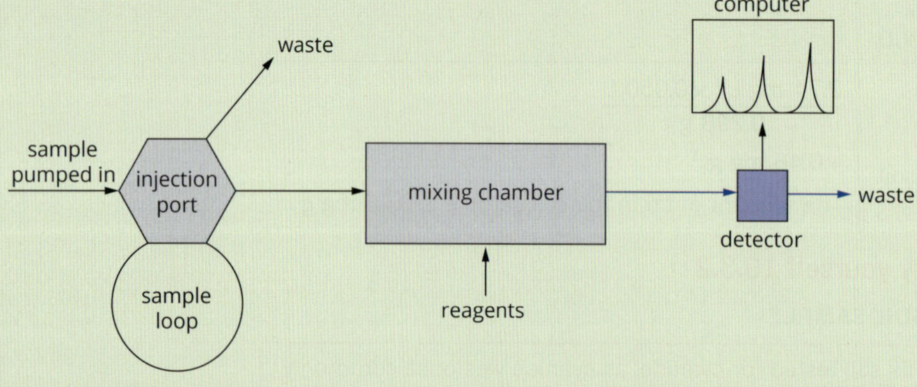

FIGURE 16.3.10 A simplified diagram of an FIA instrument

The procedure for performing an analysis using FIA is as follows.

- Standards and samples are pumped into an injection port.
- A set volume of the standard or sample is collected in a thin tube (sample loop) and injected into the mixing chambers.
- Reagents are added to the standard or sample to create a coloured solution.
- The coloured solution is pumped through the detector which is set for a specific wavelength.
- Absorbance values are recorded, calibration curves are created and sample concentrations are calculated using software on a computer.
- The coloured samples are pumped into waste bottles to be disposed of appropriately.

Portable FIAs provide researchers with flexibility and real-time data when they go out into the field (Figure 16.3.11). The portable instruments are usually powered by car batteries and the researcher would take these, as well as the computer, reagents and standards with them. An example of a typical set of nitrate data is shown in Table 16.3.3. Absorbance measurements of standards and samples from five sites along Dandenong Creek were taken in the field, allowing the researcher to determine the concentrations of nitrate immediately in the creek.

FIGURE 16.3.11 A portable FIA measuring the concentration of nitrate in the water at a local lake. A car battery is used to power the instrument.

TABLE 16.3.3 Nitrate data collected on a field trip at Dandenong Creek

Concentration NO_3^- ($\mu g\ L^{-1}$)	Absorbance measured at 510.0 nm
Standards	
0	0.000
5	0.008
10	0.016
20	0.033
50	0.079
100	0.156
Samples	
blank	0.000
standard check 50 $\mu g\ L^{-1}$	0.080
Site 1	0.042
Site 2	0.026
Site 3	0.140
Site 4	0.048
Site 5	0.031

Analysis

1. Create a calibration curve using the data provided in Table 16.3.3.
2. Determine the blank, standard check and sample concentrations using the calibration curve you have created.
3. Blank samples are used in commercial laboratories to determine any potential contamination. Do your results show potential contamination?
4. Standard checks are also used by commercial laboratories to check if the instrument and chemistry does not change during analysis. What was your measured concentration of the standard check 50 $\mu g\ L^{-1}$? Does it show any issues with the chemistry or instrument?
5. The five sites were in the urban section of Dandenong Creek. Which sites had unusually high concentration of nitrate? Using the information in Section 16.1, suggest some possible sources of nitrate in an urban environment.

16.3 Review

OA ✓✓

SUMMARY

- Some solutions containing metal ions absorb light in the visible and ultraviolet regions of the spectrum. Colorimeters and UV–visible spectrophotometers can be used to determine the concentration of metal ions in these solutions.
- A metal complex consists of a metal ion bonded to molecules or anions. The solutions of many metal complexes are suited to analysis by colorimetry or UV–visible spectroscopy.
- The amount of light absorbed by a solution is related to the concentration of the solution.
- Calibration curves are prepared by measuring the absorbance of a series of standard solutions of known concentration. The absorbance readings are then plotted against concentration. The concentration of a solution can be determined by plotting its absorbance on the calibration curve and reading off the corresponding concentration.
- A colorimeter uses a filter to select the colour of light to be used (Figure 16.3.3 on page 510). The light chosen for the analysis should be complementary to the observed colour of the solution.
- A UV–visible spectrophotometer uses a monochromator in place of a filter (Figure 16.3.6 on page 514). This allows a specific wavelength of light or ultra violet radiation to be chosen.
- A scan across a range of wavelengths is used to determine the wavelength that offers the best absorbance for a particular solution.
- A UV–visible spectrophotometer can usually provide more accurate results than a colorimeter, but it is more expensive. Both instruments offer a means of determining the concentration of salts in a solution.
- The amount of an ion in a soil sample can be determined by first extracting the ion from the soil using a chemical liquid.

KEY QUESTIONS

Knowledge and understanding

1 Why would red light rather than blue light be used in a colorimeter to measure the concentration of a blue-coloured copper(II) sulfate solution?

2 State whether the following sentences are true or false.
  - **a** The complementary colour of blue is yellow.
  - **b** The complementary colour of red is orange.
  - **c** The complementary colour of yellow is violet.
  - **d** The complementary colour of green is red.

3 A colorimeter is used to analyse the concentration of zinc(II) ions (Zn^{2+}) in the rainwater collected from a galvanised roof. The calibration curve below is obtained from measuring the absorbance of a series of standards. The absorbance from the rainwater sample was 0.18. What is the concentration of Zn^{2+} in mol L^{-1} of the Zn^{+} solution analysed in the colorimeter?

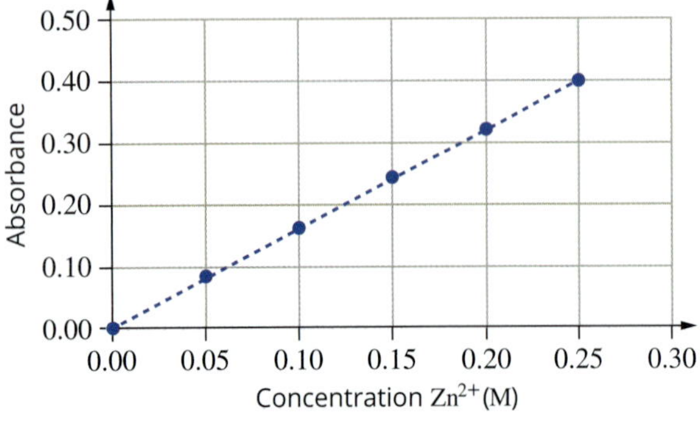

4 Why is a UV–visible spectrophotometer more accurate than a colorimeter for analysis?

5 What are the key characteristics required for a metal complex to be suitable for colorimetric or UV–visible spectrophotometric analysis?

6 Explain why it is important to accurately measure the volume of the extract when analysing an ion in soil.

Analysis

7 The concentration of manganese(II) in groundwater was analysed by colorimetry. The absorbance values from a series of standards and the groundwater are provided below.

Concentration Mn^{2+} (mg L^{-1})	Absorbance
0	0.00
20	0.11
40	0.23
60	0.34
80	0.44
100	0.56
groundwater sample	0.39

a Use the values provided to construct a calibration curve for the analysis of Mn^{2+}.

b Using the calibration curve you created in part **a**, determine the concentration of manganese(II) ions in the groundwater.

8 The absorption spectrum of chlorophyll is shown in Figure 16.3.7 on page 514.

a At what wavelengths is there maximum absorbance of light?

b What wavelength would you select if you were required to determine the concentration of chlorophyll in a leaf extract using UV–visible spectroscopy? Provide an explanation for your answer.

9 A 5.0 mL extract from 0.065 g of lake sediment was collected and diluted to 25 mL to find the concentration of phosphate in the sediment. Ammonium molybdate and $SnCl_2$ were added to a series of standards and the sample to form a dark blue complex. These standards and samples were analysed using a UV–visible spectrophotometer and the following table give the results.

Concentration PO_4^{3-} (µg L^{-1})	Absorbance
0	0.000
5	0.004
10	0.009
15	0.013
20	0.019
25	0.025
lake sediment extract (diluted)	0.010

a Use the values provided to construct a calibration curve for phosphate.

b Determine the phosphate concentration of diluted extract.

c Determine the phosphate concentration of undiluted extract.

d Determine the concentration of phosphate in the sample of lake sediment as µg g^{-1}.

Chapter review

16

OA ✓✓

KEY TERMS

absorbance
anhydrous
biomagnification
calibration curve
colorimeter
complementary colours
electrical conductivity (EC)
electromagnetic radiation
electromagnetic spectrum
eutrophication
gravimetric analysis
hard water
hardness
heavy metals
hydrated
mass–mass stoichiometry
metal complex
mineral
monochromator
organometallic compounds
precipitate
precipitation reaction
salinity
soil matrix
soil texture
spectroscopy
standard solution
total dissolved solids (TDS)
UV–visible spectrophotometer
water of hydration

REVIEW QUESTIONS

Knowledge and understanding

1 Which one of the following reactions is most likely to be part of a gravimetric analysis?

A $H_2SO_4(aq) + 2KOH(aq) \longrightarrow K_2SO_4(aq) + 2H_2O(l)$

B $Na_2S(aq) + HgCl_2(aq) \longrightarrow HgS(s) + 2NaCl(aq)$

C $C_5H_{12}(l) + 8O_2(g) \longrightarrow 5CO_2(g) + 6H_2O(l)$

D $Sn(s) + 2HCl(aq) \longrightarrow SnCl_2(aq) + H_2(g)$

2 Which of the following elements contribute to hard water?

A sodium

B calcium

C phosphorus

D carbon

3 An example of an anhydrous salt is:

A $Fe_2O_3{\cdot}3H_2O$

B $NaCl{\cdot}2H_2O$

C $Na_2CO_3{\cdot}10H_2O$

D $KC_8H_5O_4$

4 Which of the following is not a component of an UV–visible spectrophotometer?

A monochromator

B light source

C filter

D detector

5 You have two salt solutions: solution A has 0.001 M NaCl and solution B has 0.100 M NaCl. Which solution will have the highest electrical conductivity?

6 Provide three examples of human activities that lead to the increase in the amount of salts in waterways.

7 Two samples of copper(II) sulfate solution with concentrations of 0.080 M and 0.30 M were analysed using a spectrophotometer.

a Which sample would allow the most amount of light to pass through to the detector?

b Which sample would show the strongest absorption of light?

8 A scan of a pink solution in a UV–visible spectrophotometer produces the spectrum shown in the graph below.

a At what wavelength does the scan show the strongest absorption?

b What colour of light is the solution absorbing (refer to Table 16.3.1 on page 511)?

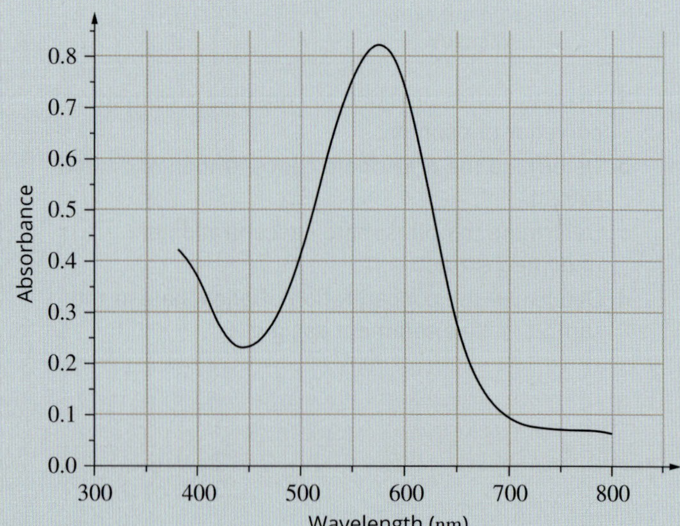

9 Zinc is a heavy metal that is essential for important metabolic processes at low concentrations, but toxic at higher concentrations.

a Suggest a natural source of zinc in waterways.

b What chemical reaction could be used to reduce the concentration of zinc in water?

10 Tetraethyl lead is an organometallic compound. Give another example of an organometallic compound and explain what makes these substances 'organometallic'.

11 Colorimeters and UV–visible spectrophotometers both measure absorbance. List three differences between these two analytical instruments.

12 A colorimeter is used to determine the concentration of copper(II) (Cu^{2+}) in a sample. The calibration curve in the graph below is obtained from measuring the absorbance of a series of standards. The absorbance of the unknown was 0.48. Determine the concentration of copper(II) in the unknown sample.

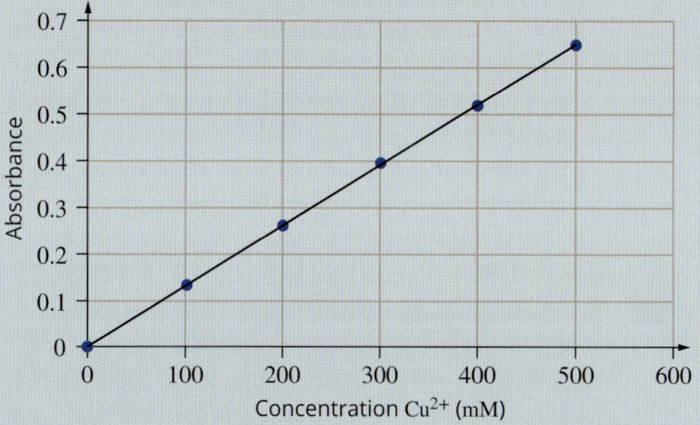

13 The $EC_{1:5}$ of a clay soil sample was measured at 2886 μS cm^{-1}. Calculate the EC of that soil sample by referring to Table 16.1.5 on page 501.

14 A 10 L solution has an unknown amount of manganese(II). You decide to use gravimetric analysis using ammonium sulfide to determine the amount of manganese present, based on the following reaction:

$$(NH_4)_2S(aq) + MnCl_2(aq) \longrightarrow MnS(s) + 2NH_4Cl(aq)$$

The mass of MnS was 85.0 g. Calculate the mass of manganese(II) as manganese chloride in the 10 L solution.

Application and analysis

15 Epsom salts have been used for centuries as a treatment for physical ailments. They are also a common ingredient in bath salts. The salts exist as a hydrate and have the molecular formula: $MgSO_4{\cdot}xH_2O$. Calculate the value of x if 5.00 g of Epsom salts is heated to a final mass of 2.44 g.

16 The amount of ammonia in a water sample can be measured by adding phenol and sodium hypochlorite to form an intense blue compound, which can be measured using by UV–visible spectrophotometry. Absorbance values from a series of standards and samples were determined and the results are given in the following table:

Concentration NH_3 (mg L^{-1})	Absorbance
0	0.000
0.01	0.040
0.02	0.080
0.03	0.120
0.04	0.160
0.05	0.200
sample A	0.058
sample B	0.133
sample C	0.101

a Use the values provided to construct a calibration curve.

b Using the calibration curve you created in part **a**, determine the concentration of ammonia in the three samples.

17 Water pollution can result from phosphate added to washing powders to improve the stability of their suds. The phosphorus in a 2.0 g sample of washing powder is precipitated as $Mg_2P_2O_7$. The precipitate weighs 0.085 g. What is the percentage by mass of phosphorus in the washing powder?

18 The cadmium content of a cadmium alloy is determined by dissolving a sample in sulfuric acid and precipitating the cadmium ions as cadmium(II) chloride. A precipitate of 0.078 g is obtained from a sample with a mass of 0.511 g. Find the percentage by mass of cadmium in the alloy.

19 A chemist determined the salt content of a sausage roll by precipitating chloride ions as silver chloride. If a 8.45 g sample of sausage roll yielded 0.693 g of precipitate, calculate the percentage of salt in the food. Assume that all the chloride is present as sodium chloride.

20 The phosphate content of a detergent may be analysed by UV–visible spectroscopy. In one analysis, a 0.250 g sample of detergent powder was dissolved in water and the solution made up to 250 mL. The solution was treated to convert any phosphate present to a blue-coloured molybdenum phosphorus compound. The absorbance of the solution at a wavelength of 600 nm was measured as 0.17.

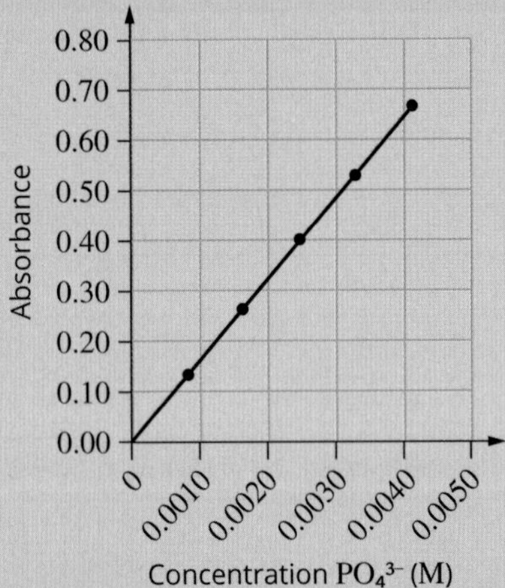

The absorbances of five standard phosphate solutions were measured in a similar fashion and the calibration curve below was obtained.

a What is the concentration of phosphorus, measured in mg L^{-1} in the 250 mL detergent solution?

b Determine the percentage by mass of phosphorus in the detergent powder.

c Why was a wavelength of 600 nm selected for this analysis?

21 A sample of water from Yarra River estuary was analysed using the methods described in this chapter.

a Chloride concentration was analysed using gravimetric analysis. Silver nitrate solution was added to a 100 mL sample until no more precipitate formed. The mass of AgCl solid produced was 3.67g.

i Using the chemical equation: $Ag^+(aq) + Cl^-(aq) \longrightarrow AgCl(s)$, calculate the mass of chloride ions in 100 mL.

ii Calculate the concentration of chloride ions in the units mg L^{-1}.

b Phosphate concentration was measured using colorimetry. A series of standards and a sample were analysed, with the following results.

Phosphate concentration (μmol L^{-1})	Absorbance
0	0.001
5	0.045
10	0.091
20	0.180
40	0.365
estuary sample	0.074

i Plot a calibration curve using the concentration and absorbance values.

ii Using the calibration curve in part **a**, calculate the concentration of phosphate in the estuary.

iii Complete the following statement: High concentration of phosphate in the water is known as ________________.

c Low concentrations of lead, copper and zinc were also found.

i What might be the main sources of these heavy metals in the Yarra River estuary?

ii What are the most toxic forms of these heavy metals?

22 Three soil samples were collected near a historical mining town, as the soil was suspected to be contaminated with mercury and arsenic. Mercury and arsenic were commonly used in the processing of gold from ore, and waste was typically dumped outside the town. A soil scientist added concentrated acids to each sample and heated the sample to extract mercury and arsenic. They diluted each extract to 10 mL after removing all the solid material. Each element was analysed using a UV–visible spectrophotometer using chemicals that formed coloured complexes. The results are shown in the table on the opposite page.

Sample number	Mass of soil (mg)	Volume of extract (mL)	Absorbance mercury	Absorbance arsenic
Standard 0 μg L^{-1}			0.000	0.000
Standard 50 μg L^{-1}			0.025	0.010
Standard 100 μg L^{-1}			0.051	0.019
Standard 150 μg L^{-1}			0.076	0.029
Standard 200 μg L^{-1}			0.102	0.041
Standard 250 μg L^{-1}			0.126	0.051
Soil sample 1	33	10	0.110	0.049
Soil sample 2	42	10	0.030	0.005
Soil sample 3	36	10	0.080	0.045

a Plot two separate calibration curves for mercury and arsenic.

b Calculate the concentrations of mercury and arsenic in μg L^{-1} in each soil extract using your calibration curve in part **a**.

c Calculate the mass in μg of mercury and arsenic per gram of soil

d The Victorian EPA states the soil is contaminated when:

Arsenic > 20 mg kg^{-1}

Mercury > 1 mg kg^{-1}

- **i** Your results are in μg g^{-1}, whereas the Victorian EPA quote their values in mg kg^{-1}. Use your knowledge of units to explain why μg g^{-1} is the same value as mg kg^{-1}.
- **ii** Explain whether your samples are contaminated with arsenic or mercury based on the Victorian EPA guidelines.

UNIT 2 • Area of Study 2

REVIEW QUESTIONS

How are chemicals measured and analysed?

Multiple-choice questions

1 What volume of 3.0 M Na_2CO_3 is required to prepare 750 mL of 0.15 M solution?

A 37.5 mL
B 45.0 mL
C 112.5 mL
D 250 mL

2 What volume of water, in mL, must be added to 20.0 mL of 0.50 M NaCl in order to change its concentration to 0.20 M?

A 30.0
B 50.0
C 200
D 500

3 Cadmium is a metal used in photography, in nickel–cadmium batteries and solar cells, and for metal-plating. Drinking water contaminated with cadmium can have harmful health effects. What is the concentration of cadmium, in ppm, in a lake containing 500 kL (1 kL = 10^3 L) of water if 1.75 kg of cadmium is discharged into it?

A 3.5
B 875
C 1750
D 3000

4 A standard solution of HCl is titrated four times against some NaOH. The following titres are obtained: 21.05 mL, 20.75 mL, 20.65 mL, 20.75 mL.

The average titre is then calculated and used to determine the concentration of the NaOH. The correct value for the average titre is:

A 20.65 mL
B 20.72 mL
C 20.75 mL
D 20.80 mL

5 Identify which one of the following does NOT change when a solution is diluted by the addition of more solvent.

A volume of solvent
B number of moles of solute
C concentration of solution
D mass of solution

The following information relates to Questions 6 and 7.

The absorption spectrum of a coloured compound is shown below.

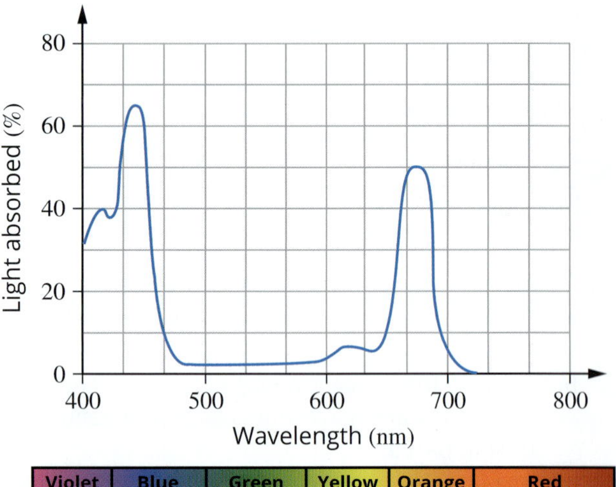

6 What is the best description of the colour of this compound?

a violet
b blue
c green
d orange-red

7 If the concentration of this compound were to be determined by UV–visible spectroscopy, which one of the following wavelengths, in nm, would be most suitable to use?

A 410
B 440
C 550
D 620

8 Dissolved calcium ions are responsible for the hardness of water. The concentration of calcium ions in a dam was determined by adding a dye known as Arsenazo III, which produces a highly coloured solution in the presence of calcium that can be detected with UV–visible spectroscopy. The absorbance of several solutions of known concentration of calcium ions were measured and the results plotted as a calibration curve.

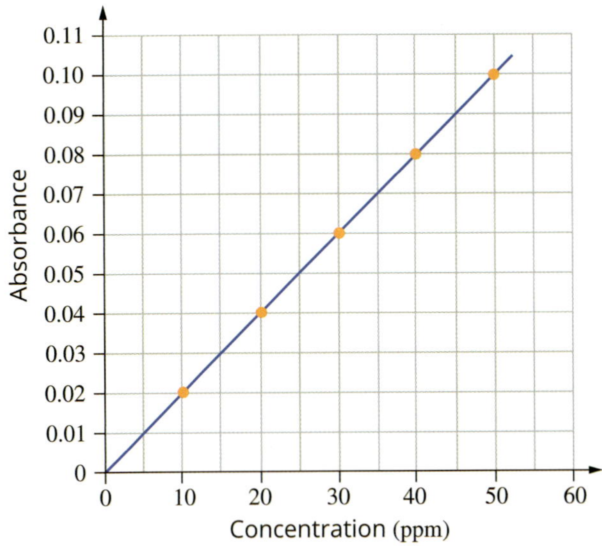

25.0 mL of the dam water is diluted to 100 mL with distilled water. The absorbance of the diluted solution, measured under the same conditions as the standards, is 0.30.

What is the approximate concentration, in ppm, of calcium in the dam water?

A 4
B 16
C 30
D 64

9 Cadmium is a toxic heavy metal which can cause kidney disease. The concentration of cadmium ions in river water was determined by reacting the river water with a suitable dye. The absorbencies of several solutions of known concentration of cadmium ions were measured and the results plotted as a calibration graph.

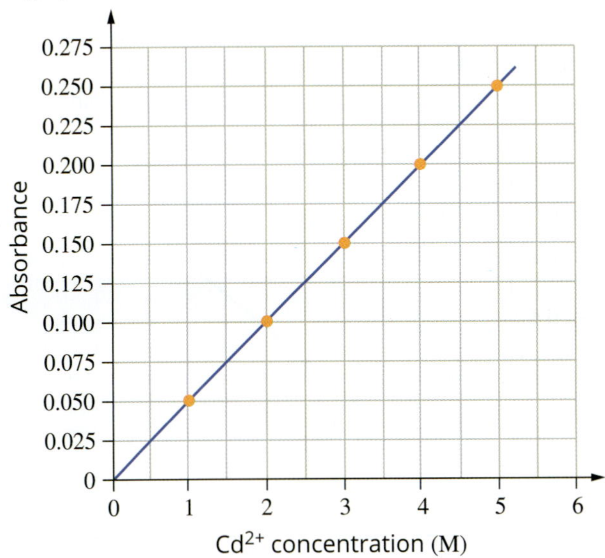

A sample of river water had an absorbance of 0.176. The molar concentration of cadmium ions concentration, in the sample of water is approximately:

A 3.6
B 0.0230
C 5.0
D 3.2×10^{-2}

10 Given that the density, *d*, of a substance is given by the formula $d = \frac{m}{V}$ and that *M* represents molar mass, the density of any gas is given by which formula?

A $d = \frac{R \times T}{P \times M}$

B $d = \frac{P}{R \times T \times M}$

C $d = \frac{R \times T \times M}{P}$

D $d = \frac{P \times M}{R \times T}$

11 A 1.03 kg mass of a gas occupies 520 L at 155°C and 270 kPa pressure. Determine the approximate molar mass of the gas in g mol^{-1}.

A 9
B 26
C 32
D 44

12 Consider the shape of the following solubility curves.

Graph A

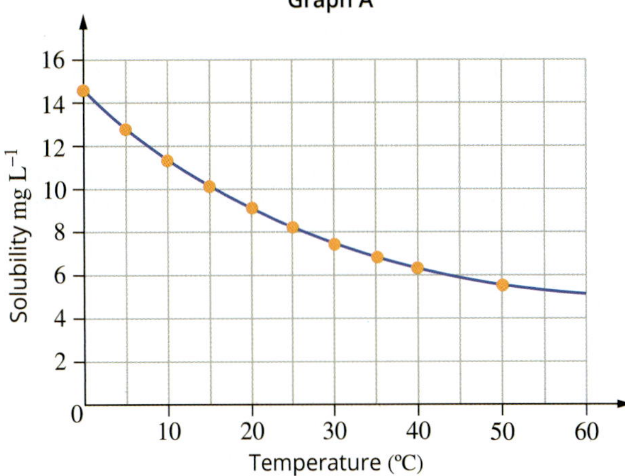

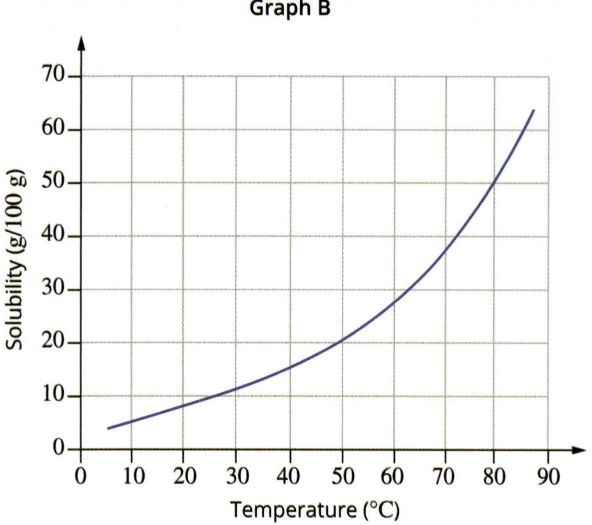

Which one of the following alternatives correctly attributes the most likely shape of the solubility curve for the ionic solid potassium chlorate and the gas oxygen?

	Potassium chlorate	Oxygen
A	Graph A	Graph A
B	Graph A	Graph B
C	Graph B	Graph A
D	Graph B	Graph B

13 The solubility of ammonium chloride at 0°C is 28 g/100 g water and at 70°C it is 85 g/100 g water. 8.0 g of ammonium chloride is dissolved in 25.0 g water at 70°C. The solution is then cooled to 0°C. What mass of ammonium chloride, in grams, crystallises out, assuming a supersaturated solution does not form?

A None will crystallise.
B 1.0
C 7.0
D 8.0

14 Samples of water from a river at two different locations, X and Y, are collected for analysis. The electrical conductivity of the water taken at location X is 23.5 μS cm^{-1}, and that of the water taken at location Y is 21.8 μS cm^{-1}.

Which one of the following alternatives is not a possible cause for this difference in electrical conductivity? Assume the temperature of X and Y is the same in options B, C and D.

A The water temperature at location Y is much lower than that at location X.
B The sample taken at location Y was not stoppered and some of the water had evaporated before testing.
C Location X is much closer than Y to the mouth of the river where the river meets the ocean.
D The river at location X passes through an industrial area and some acidic wastes are dumped in the river at that point.

15 Excess silver nitrate ($AgNO_3$) was added to a 250.0 mL sample of river water to analyse for chloride ions. A precipitate of silver chloride (AgCl) formed, which was dried and weighed. Its mass was found to be 20.37 g. What is the concentration, in g L^{-1}, of chloride ions in the river water?

A 0.142
B 0.569
C 20.2
D 81.5

Short-answer questions

16 A sample of N_2 gas collected at 10.0°C and 0.974 atm pressure occupies 240.0 mL. Calculate the volume it will occupy at SLC.

17 In the following reaction,

$$CaCO_3(s) \longrightarrow CaO(s) + CO_2(g)$$

calculate the mass of:

a $CaCO_3$ that reacts to produce 7.85 g of CO_2.
b CO_2 produced when 52.12 g of CaO reacts.
c CaO produced when 16.02 L of CO_2 reacts at SLC.

18 Deep sea fish can build up high levels of mercury from contaminated water. A 3.0 g sample of fish is ground up and mixed with water. The mixture is made up to 100 mL with water. The concentration of mercury in the solution was analysed by converting the mercury to an orange compound, which was then analysed using UV–visible spectroscopy. An absorbance reading of 0.25 was obtained.

Use the following calibration curve to determine the:

a percentage mass of mercury in the fish
b concentration in ppm of mercury in the fish.

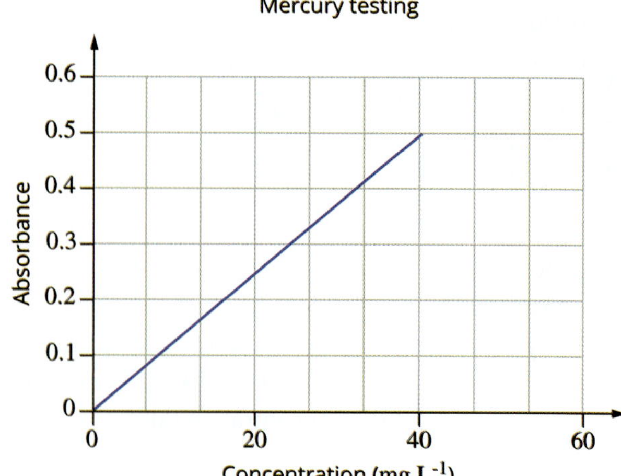

19 The solubilities of three different substances at 20°C and 80°C are given in the table below.

Substance	Solubility in water (g/100 g water)	
	20°C	80°C
potassium sulfate	12.0	21.0
sodium chloride	36	37
graphite	insoluble	insoluble

a If 16.0 g of sodium chloride were mixed with 50 g of water, what mass of sodium chloride would dissolve at 20°C?

b If 9.0 g of potassium sulfate were mixed with 50 g of water, what mass of potassium sulfate would dissolve at:

i 20°C?

ii 80°C?

c If 1.5 g of graphite were mixed with 50 g of water, what mass of graphite would dissolve at 20°C?

d What is the solubility of potassium sulfate at 20°C expressed in mol L^{-1}? The density of water at 20°C is 1.00 g mL^{-1}. Assume that the volume of the solution is equal to the volume of water.

e You are given a mixture that contains 16.0 g of sodium chloride, 9.0 g of potassium sulfate and 1.5 g of graphite. Describe how you could obtain a pure sample of potassium sulfate and a pure sample of graphite from this mixture.

f Use the data provided in the table above to explain why it would be difficult to obtain a pure sample of sodium chloride from the mixture.

20 The concentration of a solution of ammonium chloride was determined by titration against a standard solution of sodium hydroxide. The sodium hydroxide solution was placed in a burette and the ammonium chloride solution was pipetted into a conical flask. Identify the solution or liquid you would use to rinse the following pieces of equipment just prior to use.

Equipment	Rinse with:
burette	
pipette	
conical flask	

21 To analyse the iron content of bore water, all of the iron ions in 1.30 L of bore water were first converted to $Fe^{3+}(aq)$ ions by oxidation. These were then reacted with excess OH^- to form a precipitate $Fe(OH)_3(s)$. The $Fe(OH)_3(s)$ was collected, dried then heated strongly so that it decomposed to produce $Fe_2O_3(s)$ and water.

a Write an ionic equation for the precipitation of $Fe(OH)_3(s)$.

b Write an equation for the production of $Fe_2O_3(s)$ from $Fe(OH)_3(s)$.

c The mass of $Fe_2O_3(s)$ obtained was 1.095 g.

i Calculate the mass of iron in the 1.30 L of water.

ii Calculate the molar concentration of iron in the water.

22 A solution of lead(II) nitrate is prepared by dissolving 9.80 g of solid lead(II) nitrate in water. The total volume of the solution formed was 50.0 mL.

a Calculate the molar concentration of this solution.

b The lead(II) nitrate solution is diluted by adding 30.0 mL of water. What is the new concentration of the solution?

c The lead(II) nitrate solution is mixed with an excess of sodium iodide. A bright yellow precipitate forms, that contains all the lead(II) ions from the solution.

i Write an ionic equation for the formation of the precipitate.

ii Calculate the mass of precipitate that forms.

23 A student makes up a solution of 0.0500 M sodium hydrogen carbonate solution and uses it to determine the concentration of a hydrochloric acid solution.

a Calculate the mass of sodium hydrogen carbonate required to make up 200.0 mL of 0.0500 M solution.

b 20.00 mL of the sodium hydrogen carbonate solution is pipetted into a conical flask and titrated with the hydrochloric acid. An average titre was 35.05 mL of the hydrochloric acid solution.

i Write an equation for the reaction between hydrochloric acid and sodium hydrogen carbonate.

ii Calculate the concentration of the hydrochloric acid.

24 A helium cylinder for the inflation of party balloons holds 25.0 L of gas, and is filled to a pressure of 16 500 kPa at 15°C.

a Express the pressure of helium gas inside the cylinder in units of:

i Pa

ii atm

b What mass of helium does the cylinder contain when full?

c What volume would the helium occupy at standard laboratory conditions (SLC)?

d How many balloons can be inflated from a single cylinder at 30°C if the volume of one balloon is 6.5 L and each needs to be inflated to a pressure of 108 kPa?

25 A solution containing nickel(II) ions is green, so its concentration could be determined using UV–visible spectroscopy.

a Explain how a suitable wavelength is chosen to analyse the solution containing the nickel(II) ions.

b Explain how the concentration of nickel(II) ions in the solution would be experimentally determined.

26 The apparatus shown below was used to determine the molar volume of oxygen. The oxygen was generated in the test tube by heating potassium chlorate and collected by the displacement of water in the measuring cylinder.

$$2KClO_3(s) \rightarrow 2KCl(s) + 3O_2(g)$$

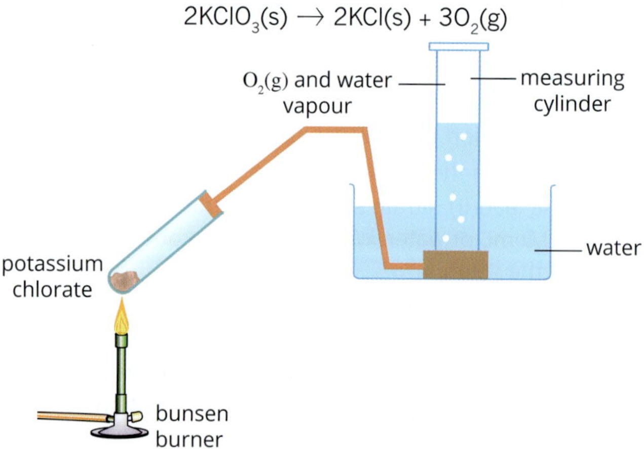

The measurements made during the experiment are provided in the following table.

mass of oxygen produced during the reaction	0.325 g
volume of oxygen collected	242.0 mL
pressure of O_2 in the cylinder	96.66 kPa
room temperature	20.0°C

a Determine the number of moles of oxygen produced.

b Calculate the molar volume of the oxygen at the experimental conditions.

c Use the ideal gas equation to determine the molar volume of a gas under the experimental conditions.

d Calculate the percentage difference between the theoretical and experimental results, where

$$\% \text{ difference} = \frac{\text{theoretical value} - \text{experimental value}}{\text{theoretical value}} \times 100$$

e Comment on some sources of error in the experiment and suggest ways that the experiment could be improved.

27 Consider the following solubility curves.

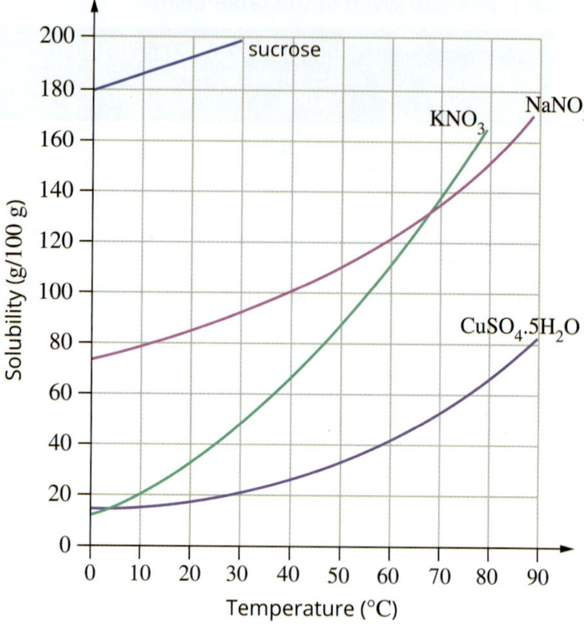

a i What is the minimum mass of $NaNO_3$ that must be dissolved in 100 g of water to produce a saturated solution at 50°C?

ii What would be observed if the solution in part **i** were cooled to 20°C?

iii What mass of water must be added to 110 g of $NaNO_3$ to make a saturated solution at 10°C?

b What is the maximum amount of table sugar (sucrose) that can dissolve in a cup containing 200 g of water at 30°C?

c Three different solutions of KNO_3 are prepared by adding various masses of KNO_3 to 100 g of water. All three solutions are maintained at 50°C.

The masses of KNO_3 contained in the solutions are 30 g, 90 g and 87 g.

i Use this information to clearly explain the difference between a saturated solution, an unsaturated solution and a supersaturated solution.

ii Describe a simple test that you could use to identify which solution is which.

iii Predict what would happen to each solution if it was cooled to 30°C.

d What mass of water must be added to 35 g of $CuSO_2.5H_2O$ to make a saturated solution at 27°C?

28 A chemistry class was asked to prepare a solution of a primary standard to titrate a solution of hydrochloric acid. In response to this task, a student carried out the following procedure.

The student decided to use sodium hydroxide solution as the primary standard, since sodium hydroxide and hydrochloric acid react completely with each other. The student obtained a 500.00 mL volumetric flask, but it looked dirty, so the student rinsed it out with detergent and hot tap water.

The student weighed out 4.321 g of sodium hydroxide pellets on a watch glass and transferred the pellets into the volumetric flask by gently tapping the watch glass. Then, using a 50.00 mL pipette for greater accuracy, added 500.00 mL of tap water into the flask. The student used warm water to help the sodium hydroxide dissolve readily, then shook the flask, preventing the solution from coming out by putting his thumb over the end, until all the solid had dissolved.

Identify five things the student did that would decrease the accuracy of the titration.

29 Victoria has approximately 3.5 million cattle, and each of these animals produces a significant daily amount of methane. Measuring and minimising the amount of methane that each animal produces is a significant challenge for agricultural chemistry.

a Two ways of estimating the methane output from a cow are listed below. For each method, comment on:

- the variables that should be controlled to maximise the validity of each experiment; and
- some steps that must be taken so that the experimental results can be used as a representative estimation for the methane emissions from all of Victoria's cattle.

i The gas can be collected directly, into a 'cow backpack' that is removed and analysed each day.

ii An animal is kept in an enclosed 'methane capture chamber' where the gas produced is constantly captured and measured.

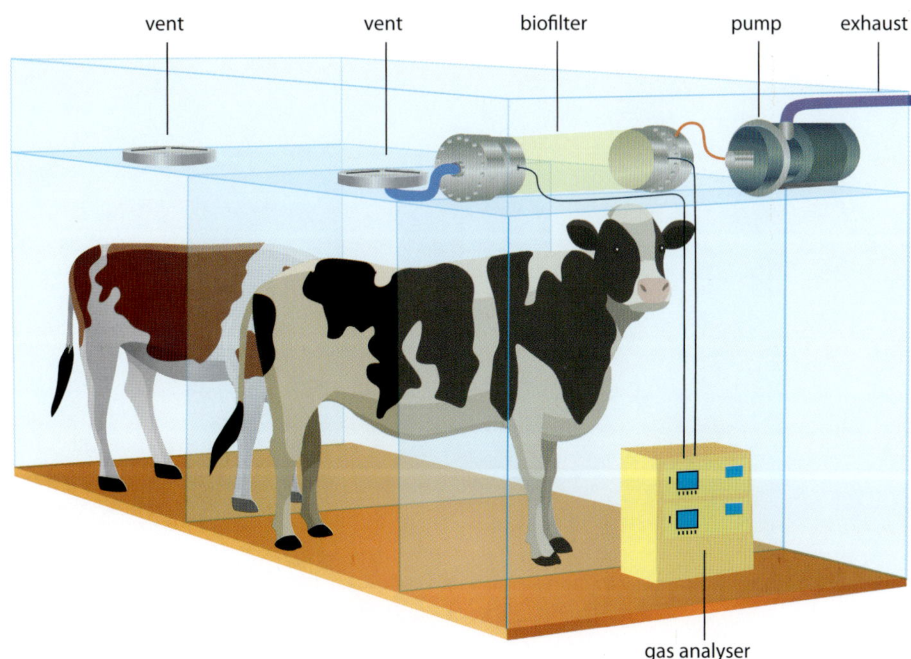

b The cows in one Victorian experiment were measured to emit 146 g of methane per day. Calculate the volume occupied by this mass of gas at SLC.

c Victoria emits approximately 0.63 million tonnes of methane into the environment each year (1 tonne = 10^6 g). Calculate the percentage of these methane emissions that come from the 3.5 million Victorian cows.

APPENDIX 1 Reactivity series of metals

strongest oxidising agent

weakest reducing agent

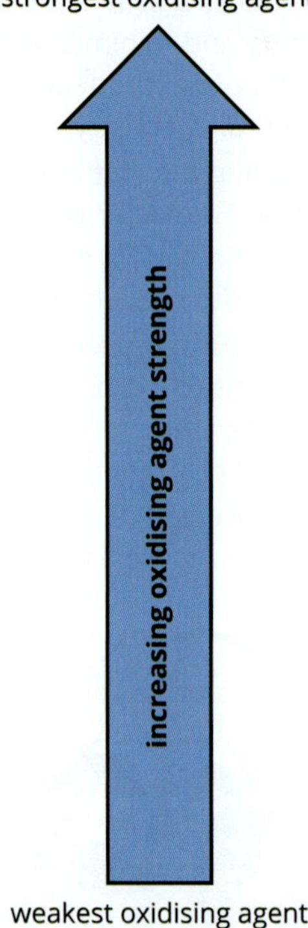

$Au^{3+}(aq) + 3e^- \rightleftharpoons Au(s)$

$Ag^{+}(aq) + e^- \rightleftharpoons Ag(s)$

$Cu^{2+}(aq) + 2e^- \rightleftharpoons Cu(s)$

$Pb^{2+}(aq) + 2e^- \rightleftharpoons Pb(s)$

$Sn^{2+}(aq) + 2e^- \rightleftharpoons Sn(s)$

$Ni^{2+}(aq) + 2e^- \rightleftharpoons Ni(s)$

$Co^{2+}(aq) + 2e^- \rightleftharpoons Co(s)$

$Fe^{2+}(aq) + 2e^- \rightleftharpoons Fe(s)$

$Cr^{3+}(aq) + 3e^- \rightleftharpoons Cr(s)$

$Zn^{2+}(aq) + 2e^- \rightleftharpoons Zn(s)$

$Mn^{2+}(aq) + 2e^- \rightleftharpoons Mn(s)$

$Al^{3+}(aq) + 3e^- \rightleftharpoons Al(s)$

$Mg^{2+}(aq) + 2e^- \rightleftharpoons Mg(s)$

$Na^{+}(aq) + e^- \rightleftharpoons Na(s)$

$Ca^{2+}(aq) + 2e^- \rightleftharpoons Ca(s)$

$K^{+}(aq) + e^- \rightleftharpoons K(s)$

$Li^{+}(aq) + e^- \rightleftharpoons Li(s)$

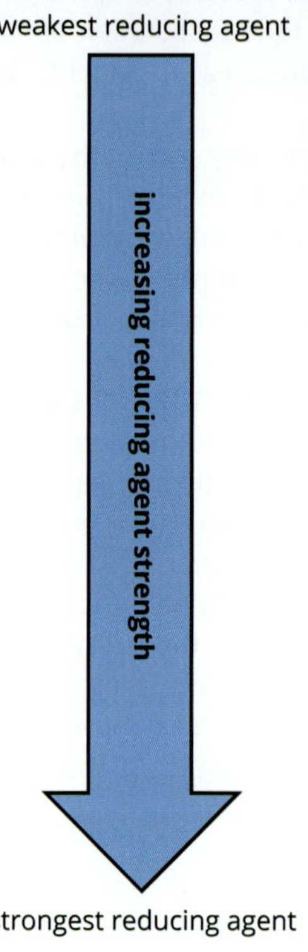

weakest oxidising agent

strongest reducing agent

APPENDIX 2 Chemical relationships, physical constants, unit conversions and indicator data

TABLE 1 Chemical relationships

Name	Formula
number of moles of a substance	$n = \frac{N}{N_A}$; $n = \frac{m}{M}$; $n = c \times V$; $n = \frac{V}{V_m}$
ideal gas equation	$pV = nRT$
pH	$pH = -\log_{10}[H_3O^+]$ $[H_3O^+] = 10^{-pH}$
ionic product of water	$K_w = [H_3O^+][OH^-]$

TABLE 2 Physical constants and standard values

Description	Symbol	Value
Avogadro's constant	N_A	6.02×10^{23} mol^{-1}
gas constant	R	8.31 J K^{-1} mol^{-1}
ionic product for water at 25°C (298 K)	K_w	1.00×10^{-14} mol^2 L^{-2}
molar volume of an ideal gas at 298 K, 100 kPa (standard laboratory conditions – SLC)	V_m	24.8 L mol^{-1}
specific heat capacity of water	c	4.18 J^{-1} g^{-1} K^{-1}
density of water at 25°C (298 K)	d	0.997 g mL^{-1}

TABLE 3 Unit conversions

Measured value	Conversion
0°C	273 K
100 kPa	750 mm Hg or 0.987 atm
1 litre (L)	1 dm^3 or 1×10^{-3} m^3 or 1×10^3 cm^3 or 1×10^3 mL

TABLE 4 Metric (including SI) prefixes

Metric (including SI) prefixes	Scientific notation	Multiplying factor
giga (G)	10^9	1 000 000 000
mega (M)	10^6	1 000 000
kilo (k)	10^3	1000
deci (d)	10^{-1}	0.1
centi (c)	10^{-2}	0.01
milli (m)	10^{-3}	0.001
micro (μ)	10^{-6}	0.000 001
nano (n)	10^{-9}	0.000 000 001
pico (p)	10^{-12}	0.000 000 000 001

TABLE 5 Acid–base indicators

Name	pH range	Colour change from lower pH to higher pH in range
thymol blue (1st change)	1.2–2.8	red → yellow
methyl orange	3.1–4.4	red → yellow
bromophenol blue	3.0–4.6	yellow → blue
methyl red	4.4–6.2	red → yellow
bromothymol blue	6.0–7.6	yellow → blue
phenol red	6.8–8.4	yellow → red
thymol blue (2nd change)	8.0–9.6	yellow → blue
phenolphthalein	8.3–10.0	colourless → pink

APPENDIX 3 Common ions and solubilities of ionic compounds

TABLE 1 The names and formulas of some common positive and negative ions

Positive ions (cations)						Negative ions (anions)			
1+		2+		3+		1−		2−	
caesium	Cs^+	barium	Ba^{2+}	aluminium	Al^{3+}	bromide	Br^-	carbonate	CO_3^{2-}
copper(I)	Cu^+	cadmium(II)	Cd^{2+}	chromium(III)	Cr^{3+}	chloride	Cl^-	chromate	CrO_4^{2-}
gold(I)	Au^+	calcium	Ca^{2+}	gold(III)	Au^{3+}	cyanide	CN^-	dichromate	$Cr_2O_7^{2-}$
lithium	Li^+	cobalt(II)	Co^{2+}	iron(III)	Fe^{3+}	dihydrogen phosphate	$H_2PO_4^-$	hydrogen phosphate	HPO_4^{2-}
potassium	K^+	copper(II)	Cu^{2+}			ethanoate	CH_3COO^-	oxalate	$C_2O_4^{2-}$
rubidium	Rb^+	iron(II)	Fe^{2+}	4+		fluoride	F^-	oxide	O^{2-}
silver	Ag^+	lead(II)	Pb^{2+}	lead(IV)	Pb^{4+}	hydrogen carbonate	HCO_3^-	sulfide	S^{2-}
sodium	Na^+	magnesium	Mg^{2+}	tin(IV)	Sn^{4+}	hydrogen sulfide	HS^-	sulfite	SO_3^{2-}
		manganese(II)	Mn^{2+}			hydrogen sulfite	HSO_3^-	sulfate	SO_4^{2-}
		mercury(II)	Hg^{2+}			hydrogen sulfate	HSO_4^-		
		nickel	Ni^{2+}			hydroxide	OH^-	3−	
		strontium	Sr^{2+}			iodide	I^-	nitride	N^{3-}
		tin(II)	Sn^{2+}			nitrite	NO_2^-	phosphate	PO_4^{3-}
		zinc	Zn^{2+}			nitrate	NO_3^-		
						permanganate	MnO_4^-		

TABLE 2 Solubility of common ionic compounds in water

Soluble ionic compounds		
Soluble in water (> 0.1 mol dissolves per L at 25°C)	**Exceptions: insoluble (< 0.01 mol dissolves per L at 25°C)**	**Exceptions: slightly soluble (0.01–0.1 mol dissolves per L at 25°C)**
most chlorides (Cl^-), bromides (Br^-) and iodides (I^-)	AgCl, AgBr, AgI, PbI_2	$PbCl_2$, $PbBr_2$
all nitrates (NO_3^-)	no exceptions	no exceptions
all ammonium (NH_4^+), salts	no exceptions	no exceptions
all sodium (Na^+) and potassium (K^+), salts	no exceptions	no exceptions
all ethanoates (CH_3COO^-)	no exceptions	no exceptions
most sulfates (SO_4^{2-})	$SrSO_4$, $BaSO_4$, $PbSO_4$	$CaSO_4$, Ag_2SO_4
Insoluble ionic compounds		
Insoluble in water	**Exceptions: soluble**	**Exceptions: slightly soluble**
most hydroxides (OH^-)	NaOH, KOH, $Ba(OH)_2$, NH_4OH*, AgOH**	$Ca(OH)_2$, $Sr(OH)_2$
most carbonates (CO_3^{2-})	Na_2CO_3, K_2CO_3, $(NH_4)_2CO_3$	no exceptions
most phosphates (PO_4^{3-})	Na_3PO_4, K_3PO_4, $(NH_4)_3PO_4$	no exceptions
most sulfides (S^{2-})	Na_2S, K_2S, $(NH_4)_2S$	no exceptions

*NH_4OH does not exist in significant amounts in an ammonia solution. Ammonium and hydroxide ions readily combine to form ammonia and water.
**AgOH readily decomposes to form a precipitate of silver oxide and water.

Answers

Chapter 1 Scientific investigation

1.1 The nature of scientific investigations

1 Student answers will vary.

2 A and D.

3 methodology: description of the investigation's general approach

method: specific steps that must be done, and which can be replicated

4 qualitative: information about what is present

quantitative: numerical information, detailing how many, how much, how often, etc

5

Type of methodology	Type of investigation
case study	in-depth study about a particular chemical process
simulation	using a computer program to look at rotating three-dimensional models of molecules
controlled experiment	designing an experiment with an independent and a dependent variable and keeping everything else constant
literature review	using secondary sources to find information about recycling of metals from second-hand computers

6

Factor	Scientific	Non-scientific
geological assessment of the chosen area	✔	
needs of the residents for employment		economic
local government regulations		political
opinions of local residents about mining		political/ethical
assessment of nearby waterways for potential pollution of ground water	✔	

1.2 Planning investigations

1 a corrosive b toxic c flammable

2 to understand potential hazards in order to identify safety measures that should be taken to reduce or remove risk of harm

3 a a question outlining what is to be investigated

b a prediction of the outcome of an investigation referring to the independent and dependent variables

c the purpose of the investigation

d the variable that is changed on purpose by the investigator

e the variable(s) that are measured in the investigation

f the variables that are kept constant in the investigation in order to see if changing the independent variable causes a change in the dependent variable

4 a observation b theory c hypothesis

5 Student answers will vary.

6

Hazard	Safety measure
flammable liquid	Avoid flames. Use a hot plate instead of a Bunsen burner.
respiratory irritant	Use a fume cupboard.
corrosive solution	Wear gloves and safety goggles, and handle with care.
contamination of wastewater with organic compounds	Dispose of in a labelled organic waste bottle.
toxic solid	Wear gloves and avoid inhalation of powder.

7 a the concentration of sodium chloride solution

b the mass of silver chloride that has been precipitated

c the temperature; the volume of both solutions; the measuring cylinders used to measure the volumes; the method of filtering and drying the precipitate

8 a

	Toxicity	Flammability
acetone	causes serious eye damage/irritation specific target organ toxicity in a single contact	flammable liquid
ethanol	causes serious eye irritation	highly flammable liquid and vapour
cyclohexane	may be fatal if swallowed and enters airways may be harmful in contact with skin causes skin and eye irritation; harmful if inhaled may cause drowsiness or dizziness. very toxic to aquatic life with long-lasting effects	highly flammable liquid and vapour

b Method A: The substitution of ethanol and water for acetone and cyclohexane is an example of substituting safer chemicals (designing for safer chemicals) which achieve their required function while minimising toxicity.

1.3 Data collection and quality

1 a three significant figures b three significant figures

c five significant figures d four significant figures

2 a **Accuracy** is how close a measurement is to the **true value,** whereas **precision** is how closely a set of measurements agree with each other.

b **Validity** refers to whether your results measure what the investigation set out to measure.

c **Repeatability** is the consistency of your results when they are repeated many times as trials under the exact same set of experimental conditions, whereas **reproducibility** is the ability for another experimenter to obtain the same results if they replicate your experiment.

d **Resolution** is the smallest change in the measured quantity that causes a perceptible change in the value shown by the measuring instrument.

3 to ensure that you remain organised and can refer to the data and observations when you are analysing your data and writing your report

4

Information	Quantitative	Qualitative
colour of hydrated salt		✔
colour of dehydrated salt		✔
initial mass of hydrated salt	✔	
texture of dehydrated salt		✔
final mass of dehydrated salt	✔	

5 B

6 a accuracy b resolution c precision d validity e repeatability of data

7 raw data: titres

processed data: the average calculated concordant titre and the results of subsequent calculations

8 a resolution: 0.1°C value: 31.2°C b resolution: 0.5 mL value: 85.0 mL

1.4 Data analysis and presentation

TY 1.4.1 a 2.560×10^3 b 9.71×10^{-5}

TY 1.4.2 $m(CO_2) = 2.76$ g

CSA: Rhodamine B in wastewater

1 1.7 μg L^{-1}

2 a 1.25 μg L^{-1}. b 125 μg L^{-1}

3 Because the sample was tested at 558 nm (yellow-green visible light), the organic solvent is unlikely to absorb that light, so the wastewater calibration curve should still be acceptable to use.

TY 1.4.3 The experimental value is 8.77% more than the theoretical mass.

TY 1.4.4 There has been a 43.95% decrease in the mass.

Key questions

1 a 2×10^{-3} b 2.050×10^3 c 1.234×10^2 d 3.25×10^{-5}

2 A 3 B

4 a pie graph b scatter graph c column graph d line graph

5 The mass has decreased by 51%.

6 The experimental value is 5.60% greater than the theoretical value.

7 A trend is a pattern shown in data. Trends can be positive, negative (inverse), proportional, linear or non-linear (exponential).

8 a Concordant results are from trials 2, 5 and 6: 18.90 mL, 18.90 mL, 18.95 mL.

b mean titre = 18.92 mL

c The result of trial 3, 25.15 mL, is most likely to be due to a mistake.

9 (1) Reverse the axes, as the vertical axis should be the horizontal axis and vice versa.

(2) Add a descriptive title.

(3) Label the y-axis (the vertical axis) 'Absorbance' (no units).

(4) Label the x-axis (the horizontal axis) 'Concentration of Fe^{3+} ions' and include units (mg L^{-1}).

(5) The line should not extend beyond the 0.20 concentration.

10 a

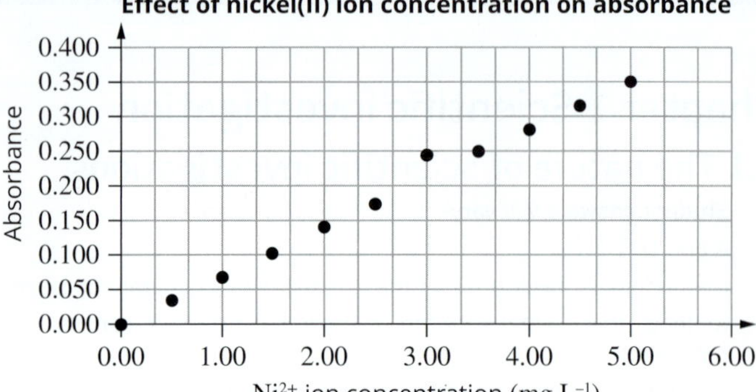

b The data point at 3.00 mg L^{-1} is an outlier.

c

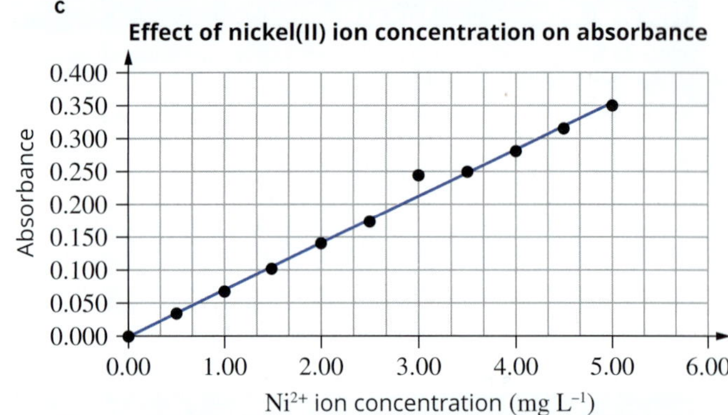

d concentration of nickel ions in solution = 2.17 mg L^{-1}.

1.5 Evaluation and conclusion

1 C

2 a Crystals of a range of metals were grown and their shapes were observed.

b A polymer with unusual properties was observed.

c The physical properties of a covalent network lattice, a covalent layer lattice and a covalent molecular substance were compared.

3 a Random errors will occur—the use of different temperature measuring equipment will introduce variation in the accuracy of temperature results.

b Systematic error will occur—the data logging equipment was not calibrated, so it is unknown whether it was recording true values for conductivity.

c Random errors will occur—the water samples would have changed composition on the different days due to varying weather conditions such as rain.

d Random errors will be reduced—multiple samples were tested at each temperature,and the results can be averaged.

e Systematic errors will be reduced—the same technique was consistently used, and practice improved the skills of the experimenter.

4 The scientist was aiming to study the effect of temperature on electrical conductivity, not pH and electrical conductivity.

5

Limitation in experiment	Effect on the calculated result	Suggested improvement
A measuring cylinder was used to dispense 25 mL of a liquid.	higher uncertainty than using a pipette; higher uncertainty of calculated value	Use a 25.00 mL volumetric pipette.
Only one measurement was obtained.	not clear if the measurement is accurate or precise	Repeat experiment three times and calculate the average and range of results. Where possible, compare a calculated value to a literature value.
Universal indicator was used to measure the end point.	Universal indicator changes colour at many pHs, so the measurement could be completely wrong.	Use an indicator which changes colour at the pH of the end point, or use a calibrated pH meter.

6 **a** Systematic. Practice the titration technique and have a standard colour to aim for. Consider using a pH probe.
b Random. Report the reading to the nearest ± 0.2 s.
c Systematic. Always read a burette at the bottom of the meniscus at eye level.
d Systematic. Insulate the container.
e Systematic. Use a different burette or, if possible, remove the bubble.

1.6 Reporting investigations

1 C **2** A

3 **a** The beaker was observed for 10 minutes, during which time the blue colour was observed to fade and a brown solid appeared.
b The mass of the white solid was observed to decrease every minute for 10 minutes after heating.
c 2.0 g of magnesium was placed in a test tube and 10 mL of 2 M hydrochloric acid was added.

4 To give credit to the work conducted by others and to avoid plagiarism.

5 communication statement: as the central part of the poster
research question: the poster's title
aim, hypothesis, variables, and background information: write concisely, but include all important details
methodology: present as a single, concise sentence
summary flow chart: include dot points within the flow chart
diagram of experimental setup: big enough to read, but not too big
summary results table: set out with mean values of raw data and processed data
graph showing the trend in results: big enough to read, but not too big
brief analysis of results and link to chemical theory: in point form
table of limitations and suggested improvements: in point form
conclusion: in clear, concise language
acknowledgements: in the space at the bottom of the poster

6

Resource type	Information about the reference	Correct format in a bibliography, using APA referencing format
print book	Title of book: Heinemann Chemistry 1 Edition: 6th edition Author: MacEoin, M. et al. Date published: 2022 Section referred to: Chapter 12 page 387–413 Publisher: Pearson Australia	Chan, D., Commons, C., Commons, P., Derry, L., Freer, E., Huddart, E., Lennard, L., MacEoin, M., Moylan., M., O'Shea, P., Ross, B. & Vanderkruk, K. (2022). *Heinemann Chemistry 1* (6th ed.), Chapter 12, pp. 387–413 Pearson Australia.
Journal article	Title of article: Effects of the COVID-19 Pandemic on Student Engagement in a General Chemistry Course Authors: Wu, F and Teets, T Journal name: Journal of Chemical Education Volume: Vol 98, Issue 12 Pages: 3633–3642 Date published: November 2021	Wu, F. & Teets, T. (2021). Effects of the COVID-19 pandemic on student engagement in a general chemistry course. *Journal of Chemical Education,* 98(12): 3633–3642.
internet	Website owner: Royal Society of Chemistry, United Kingdom Name of page: Reactivity of metals video Date posted: no date Website address: https://edu.rsc.org/practical/reactivity-of-metals-practical-videos-14-16-years/4012974.article	United Kingdom, Royal Society of Chemistry (n.d.) Reactivity of metals video. https://edu.rsc.org/practical/reactivity-of-metals-practical-videos-14-16-years/4012974.article

Chapter 2 Elements and the periodic table

2.1 The atomic world

TY 2.1.1 protons: 92, neutrons: 143, electrons: 92

TY 2.1.2 protons: 7, neutrons: 7, electrons: 10

1 10 000–100 000 times larger, depending on the element.

2 Protons and neutrons found in the nucleus.

3 The electrostatic attraction between the protons and electrons; the negative electrons are attracted to the positive protons and are pulled towards them.

4 Mass number

5 28 electrons

6 **a** $^{90}_{39}Y$ **b** 51

7 **a** and **c**

8 It would be easier to separate $^{46}_{20}Ca$ and $^{46}_{22}Ti$ as they are different elements, with different chemical properties. $^{46}_{20}Ca$ and $^{40}_{20}Ca$ would be difficult to separate as they are different types of the same element and have identical chemical properties.

2.2 Emission spectra and the Bohr model

TY 2.2.1 2,8,18,6

1 Each line in an emission spectrum corresponds to a specific amount of energy. This energy is emitted when electrons from higher-energy electron shells transition to a lower-energy shell. Different lines indicate that there are differences in energy between shells. This is evidence that electrons are found in shells with discrete energy levels.

2 Energy is emitted as coloured light or electromagnetic radiation.

3 18: Argon 2,8,8

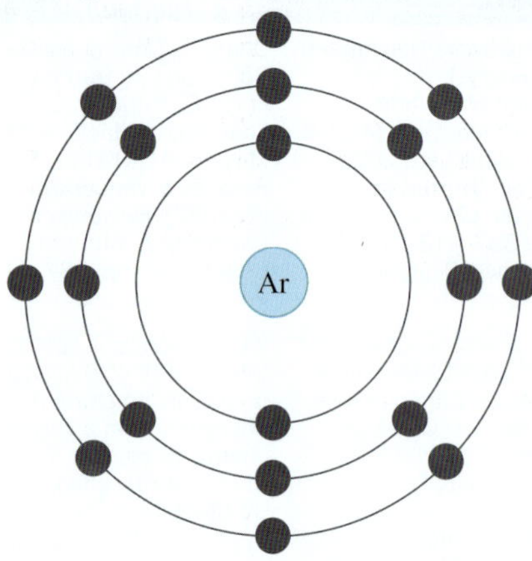

4 50

5 The number of shells corresponds to the row (period) number in the periodic table. The number of valence electrons determines the group number (column).

6 The atom is sulfur because it has 16 electrons. This electronic arrangement is unexpected because the second shell is not filled and electrons have been placed in the third shell. A possible reason for this is that this atom has been excited by an energy source. Two electrons have 'jumped' from the second shell to the third shell.

2.3 The Schrödinger model of the atom

TY 2.3.1 $1s^2 2s^2 2p^6 3s^2 3p^6 3d^3 4s^2$

1

Element (atomic number)	Electronic configuration using the shell model	Electronic configuration using the subshell model
boron (5)	2,3	$1s^2 2s^2 2p^1$
lithium (3)	2,1	$1s^2 2s^1$
chlorine (17)	2,8,7	$1s^2 2s^2 2p^6 3s^2 3p^5$
sodium (11)	2,8,1	$1s^2 2s^2 2p^6 3s^1$
neon (10)	2,8	$1s^2 2s^2 2p^6$
potassium (19)	2,8,8,1	$1s^2 2s^2 2p^6 3s^2 3p^6 4s^1$
scandium (21)	2,8,9,2	$1s^2 2s^2 2p^6 3s^2 3p^6 3d^1 4s^2$
copper (29)	2,8,18,1	$1s^2 2s^2 2p^6 3s^2 3p^6 3d^{10} 4s^1$
bromine (35)	2,8,18,7	$1s^2 2s^2 2p^6 3s^2 3p^6 3d^{10} 4s^2 4p^5$

2 **a** Ga **b** Br **c** Cu **d** Mn **e** Sn
f Sb **g** Xe **h** K

3 The Bohr model was only able to accurately predict the emission spectrum for hydrogen, whereas the Schrödinger model accurately predicts emission spectra for more complex atoms. The Bohr model was unable to explain why the third shell would stop filling after 8 electrons, then would be back filled later. This is explained by the more complex arrangement of subshells and orbitals in the Schrödinger model.

4 The Schrödinger model is a refinement of the Bohr model. The Bohr model proposed that all electrons in the one shell were of equal energy. Evidence from emission spectra indicated that there were different electronic energy levels (called subshells) within a shell. The Schrödinger model includes these subshells.

5 $1s^2 2s^2 2p^6 3s^1 3p^5$ (Other answers are possible.)

2.4 The periodic table
CSA: Naming elements on the periodic table

1 **a** iron (Fe) **b** potassium (K)
c tungsten (W) **d** lead (Pb)
e mercury (Hg)

2 IUPAC guidelines state that any new element must be named after either:

- a mythological concept or character (including an astronomical object),
- a mineral or similar substance
- a place, or geographical region
- a property of the element, or
- a scientist.

Names are proposed by the teams that discovered them, and then the name is chosen and approved by IUPAC.

3 113 – nihonium (Nh): Nihon is one way to say 'Japan'. Named after Japan, where it was discovered. The first element discovered by scientists working in an Asian country.

114 – flerovium (Fl): Honours the Flerov Laboratory of Nuclear Reactions, itself named after Georgiy N. Flerov (1913–1990), a renowned physicist.

115 – moscovium (Mc): Named after the Moscow region, where the discovery experiments were conducted.

116 – livermorium (Lv): Honours the Lawrence Livermore National Laboratory, California.

117– tennessine (Ts): Recognises the laboratories in the Tennessee region of the United States that were involved in the discovery.

118 – oganesson (Og): Named after the Russian nuclear physicist Yuri Oganessian, who is credited with three earlier confirmed elemental discoveries.

Key questions

1 row

2 periods: the horizontal rows in the periodic table

groups: the vertical columns in the periodic table

blocks: a section of the periodic table in which each element has the same type of subshell as their highest energy subshell

3 main group

4 **a** alkali metals **b** alkaline earth metals
c halogens **d** noble gases

5 **a** 2 **b** 13 – 10 = 3
c 15 – 10 = 5 **d** 18 – 10 = 8

6 $1s^2 2s^2 2p^6 3s^2 3p^6 4s^1$

7 **a** **i** Group 13 **ii** Group 17 **iii** Group 1
iv Group 18 **v** Group 14 **vi** Group 14

b **i** Silicon, Si, 2,8,4 or $1s^2 2s^2 2p^6 3s^2 3p^2$
ii Beryllium, Be, 2,2 or $1s^2 2s^2$
iii Argon, Ar, 2,8,8 or $1s^2 2s^2 2p^6 3s^2 3p^6$

c **i** 4 **ii** 2 **iii** 1 **iv** 1
v 7 **vi** 3

8 A critical element is an element heavily relied on by industry and society, which faces some form of supply uncertainty. Examples include endangered elements like osmium and iridium, conflict elements such as tin, tungsten and gold and critical raw materials such as the lanthanides.

2.5 Trends in the periodic table

TY 2.5.1 +7

1 Effective nuclear charge is the pull from the nucleus felt by each valence electron.

Effective nuclear charge of carbon: +4.

2 As effective nuclear charge increases, electronegativity increases.

3 a metal b non-metal c metalloid
d metal e metal f non-metal

4 a i F ii Fr
b i Group 17 ii Group 1

5 N, B, Cl, Ga, Al

6 As you move from left to right across groups 1, 2 and 13–17, the charge on the nucleus increases. Each time the atomic number increases by one, the electrons are attracted to an increasingly more positive nucleus. Within a period, the outer electrons are in the same shell—that is, they have the same number of inner-shell electrons shielding them from the nucleus. Therefore, the additional nuclear charge attracts the electrons more strongly, drawing them closer to the nucleus and so decreasing the size of the atom.

7 K, Na, Ca, Al, P, S, He

8 The reactivity of the alkali metals with water is related to the ease with which the metal ion is ionised. This is based on the first ionisation energy of the elements. Down the group the effective nuclear charge stays constant; however, the number of shells increases. Therefore, the valence electrons are less attracted to the nucleus the further they are from the nucleus. As a result, the energy required to overcome the attraction between the nucleus and the valence electrons is less, and the first ionisation energy decreases down a group. This means that less energy is required to ionise caesium than lithium, and so caesium is more reactive than lithium.

Chapter 2 review

1 D 2 B and C

3 C

4 B 5 C

6 a Atomic number is 24; mass number is 52
b 24 electrons, 24 protons, 28 neutrons

7 magnesium

8 $1s^2 2s^2 2p^6 3s^2 3p^4$

9 a Period 1, *s*-block b Period 2, *p*-block
c Period 3, *p*-block d Period 4, *d*-block
e Period 7, *f*-block

10 a The force of attraction between the nucleus and valence electrons **increases** in a period from left to right.
b Atomic radii of elements **decrease** in a period from left to right.
c Atomic radii of elements **increase** in a group from top to bottom.
d Metallic character of elements **increases** from top to bottom in a group.

11 No. Isotopes have the same number of protons (atomic number) but different numbers of neutrons (and therefore different mass numbers). These atoms have different atomic numbers and different mass numbers.

12 a G and H, D and F b B, C, D, and I c C
d A e 7

13 the third shell

14 a $1s^2 2s^2 2p^6$ b $1s^2 2s^2 2p^6 3s^2 3p^6$
c $1s^2 2s^2 2p^6 3s^2 3p^6 3d^{10}$

15 In the Schrödinger model of the atom, electron shells are divided into subshells, and each subshell can have a different energy level. According to the Schrödinger model, the 4*s*-subshell is lower in energy than the 3*d*-subshell. Therefore, the 4*s*-subshell begins filling after the 3*s*- and 3*p*-subshells but before the 3*d*-subshell.

16 a $1s^2 2s^2 2p^3$ b Period 2, group 15
c 5 d +5

17 a Period 2, group 2 b Period 3, group 14
c Period 4, group 13

18 As you move across period 2 from lithium to fluorine:
a the radius of the atoms decreases as the effective nuclear charge increases.
b there is a trend from metals (lithium, beryllium) to non-metals (boron, carbon, nitrogen, oxygen and fluorine); therefore, metallic character decreases.
c electronegativity increases as the effective nuclear charge increases and size of the atoms decreases.

19 a Chlorine is on the right-hand side of the periodic table and sodium is on the left. Atomic radius decreases across a period because the increasing effective nuclear charge pulls the outer-shell electrons more tightly to the nucleus, causing the volume of the atom to decrease.
b Fluorine is further to the right on the periodic table than lithium, and effective nuclear charge increases from left to right across the periodic table. As effective nuclear charge increases, the electrons are held more tightly to the nucleus and more energy is required to remove the first one.
c Barium and beryllium are in the same group, with beryllium higher than barium. Going down a group the atom size is increasing, meaning the outer-shell electrons are further from the nucleus. The outer electrons of beryllium are, therefore, held more tightly and are less readily released.
d The *s*-block elements have an *s*-subshell as their outer occupied electron subshell. The *s*-subshell can take one or two electrons, so the block is only two groups wide.

20 a Ca b Ar c C d Na or Mg
e Li f N g F

21 Ordered elements in groups:
Group 1 Li, K, Cs, Fr
Group 2 Mg, Ca, Sr, Ba
Group 13 B, Al, Ga
Group 14 C, Ge
Group 15 As, Sb
Group 16 S, Te
Group 17 Br, I, Ts
Group 18 He, Kr, Rn

22 a phosphorus b fluorine

23 K and Cl_2.

24 a A and G both have seven valence electrons; D and F both have two valence electrons.
b A and B are in period 2. D and E are in period 4, C and G are in period 5, F and I are in period 6.
c A d B e F f I
g A, G and I h E

Chapter 3 Covalent substances

3.1 Covalent bonding model

TY 3.1.1

```
   ••
H ×• N •× H
   •×
   H
```

1 A molecule is a discrete group of atoms of known formula, bonded together.

2 a 1 b 3 c 2 d 1

3 a 1 b 2 c 3
d 4 e 1 f 0

4 **a** fluorine (F_2)

F F

b hydrogen fluoride (HF)

H F

c water (H_2O)

H O
H

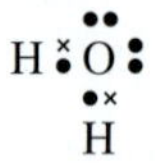

d tetrachloromethane (CCl_4)

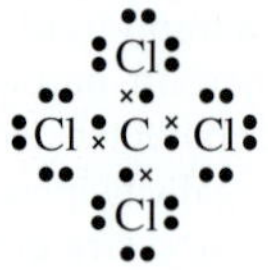

Cl
Cl C Cl
Cl

e phosphine (PH_3)

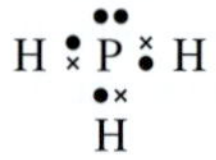

H P H
H

f carbon dioxide (CO_2)

O C O

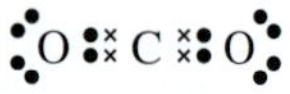

5 To complete its outer shell, the oxygen atom uses two of its valence electrons to form two single bonds or a double bond with other atoms. The remaining four valence electrons are not required for bonding, as the outer shell is now complete, and they arrange themselves as two non-bonding pairs around the oxygen atom.

6 **a** CF_4 **b** PCl_3 **c** CS_2 **d** SiH_4 **e** NBr_3

3.2 Shapes of molecules

TY 3.2.1 To complete its outer shell, the oxygen atom uses two of its valence electrons to form two single bonds or a double bond with other atoms. The remaining four valence electrons are not required for bonding, as the outer shell is now complete, and they arrange themselves as two non-bonding pairs around the oxygen atom.

H S H

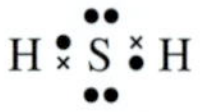

Because there are four electron groups, they will be arranged in a tetrahedral arrangement. The sulfur and hydrogen atoms are arranged in a bent shape.

1 The VSEPR theory is based on the principle that negatively charged electron groups around an atom repel each other. As a consequence, these electron groups are arranged as far away from each other as possible.

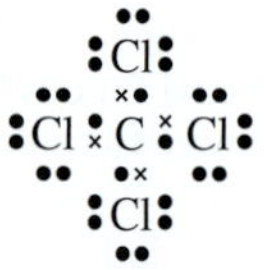

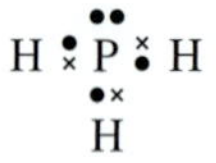

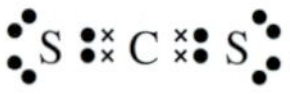

2 four

3 **a** H S
H

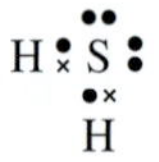

b H I

c Cl
Cl C Cl
Cl

d H P H
H

e S C S

f H
H Si H
H

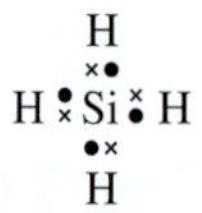

4 **a** Bent **b** Linear **c** Tetrahedral **d** Pyramidal **e** Linear **f** Tetrahedral

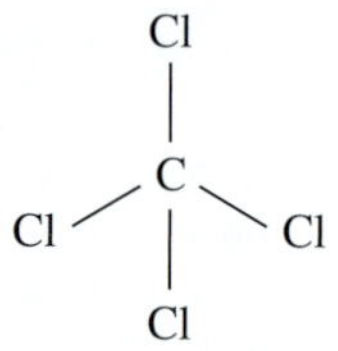

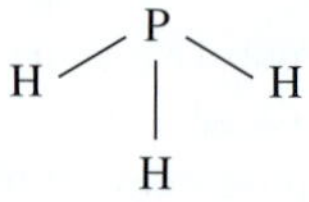

5 **a** S
H H

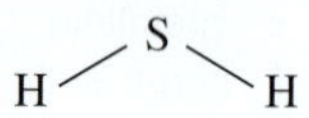

b H—I

c Cl
C
Cl Cl
Cl

d P
H H
H

e S=C=S

f H
Si
H H
H

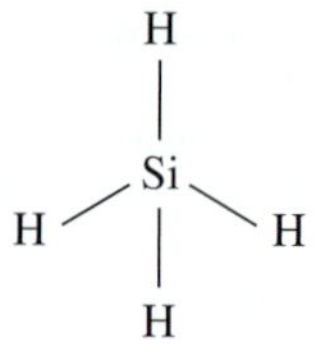

3.3 Polarity in molecules

TY 3.3.1 The bond in HCl is more polar than in NO.

1 **a** Non-polar bond: a covalent bond with an even distribution of bonding electrons. Non-polar bonds occur between atoms with the same electronegativity.

b Polar bond: a covalent bond with an uneven distribution of bonding electrons. Polar bonds occur between atoms of different electronegativity.

2 **a** O **b** C **c** N **d** N **e** F **f** F

3 **a** P–F **b** C–H or N–H

4 In order of increasing polarity: N_2, NO, HBr, HCl

5 δ+
P
δ– F F δ–
δ– F

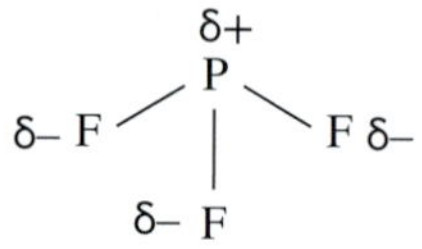

6 **a** non-polar **b** polar **c** polar **d** polar **e** non-polar

7 **a**

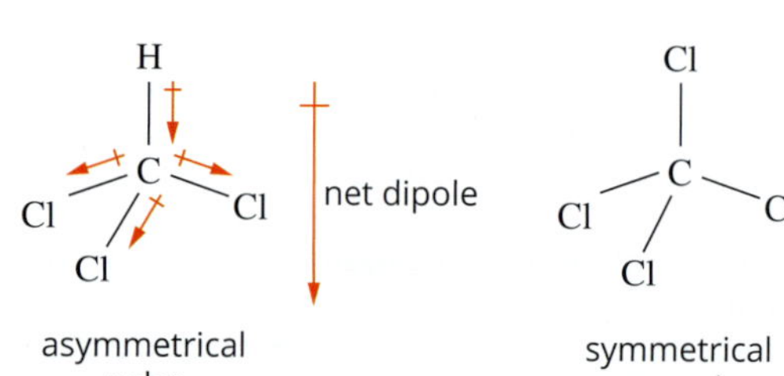

b

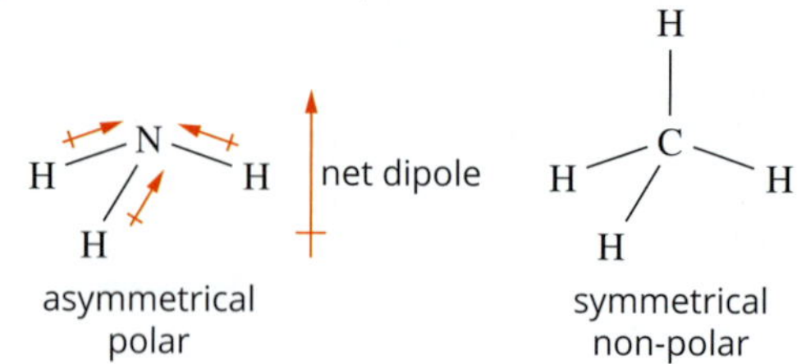

3.4 Intermolecular forces

1 **a** Dipole–dipole attraction or hydrogen bonding, and dispersion forces exist between polar molecules.

b Only dispersion forces exist between non-polar molecules.

2 hydrogen iodide (HI) and fluoromethane (CH_3F)

3 **a**

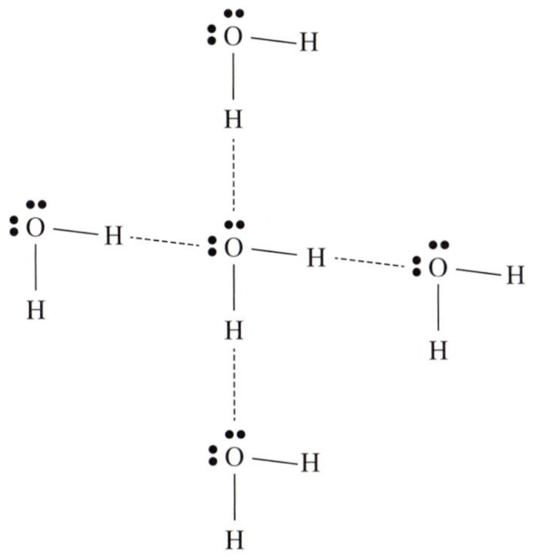

b Hydrogen bonding is the strongest type of intermolecular force in ice.

4

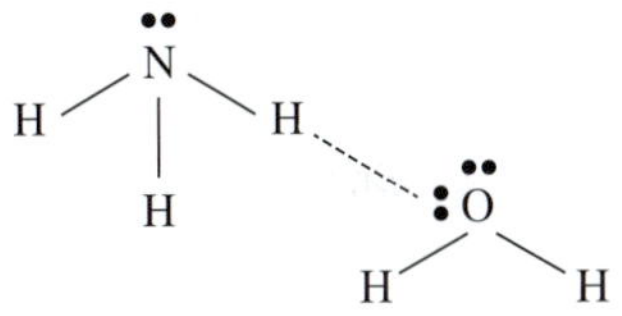

5 Dispersion forces: **e** and **j**; dipole–dipole attraction: **a**, **b**, **c**, **d**, **f**, **g**; hydrogen bonds: **h**, **i**

6 **a** CHF_3 has a higher boiling point. CHF_3 is a polar molecule so dipole–dipole attraction is the strongest force between molecules. CF_4 is a non-polar molecule, so there are only dispersion forces between molecules. Dipole–dipole attraction is stronger than dispersion forces.

b CO_2 has a higher boiling point. O_2 and CO_2 are non-polar molecules, so there are only dispersion forces between molecules. CO_2 molecules have a higher molecular mass so stronger dispersion forces.

c NH_3 has a higher boiling point. NH_3 is a polar molecule capable of forming hydrogen bonds between molecules. CH_4 is a non-polar molecule, so there are only dispersion forces between molecules. Hydrogen bonding is stronger than dispersion forces.

7 When sugar turns to a liquid, it is melting; the intermolecular forces are disrupted. When the liquid turns black and a gas is produced, a chemical reaction is taking place. The intramolecular bonds are broken, allowing new substances to be produced.

3.5 Covalent lattices

CSA: Mined versus synthetic diamonds

1 Similarities: Both contain carbon atoms arranged in a three-dimensional covalent network lattice structure. They are chemically the same.
Differences: Synthetic diamonds are purer with less flaws (inclusion of atoms of other elements in the lattice structure).

2 Environmental: Large open-cut mines have a huge impact on the physical environment. Land is cleared, destroying the habitats of plants and animals. Mines generate pollution that could escape into the local surrounding environment. Mining operations are energy intensive, which generates carbon pollution, contributing to global warming.
Social: In some areas of the world, the mines are located on land belonging to traditional indigenous owners. Traditional owners may become displaced from their lands, interrupting their cultural practices. So called 'blood or conflict diamonds' are sourced from war zones and sold to fund military conflicts.

3 The idea that something is 'real' and 'fake' implies there is a quality difference between the two. However, the chemical and structural composition of natural and synthetic diamonds is identical—they are both made of carbon atoms arranged into a three-dimensional covalent network lattice. Visually, both types of diamonds look identical. There is no significant difference between the two. Experts can only identify minute differences in chemical composition using advanced analytical techniques.

Key questions

1 Diamond: four covalent bonds. Graphite: three covalent bonds.

2 **a** to turn from a solid directly into a gas

b Diamond and graphite contain extended networks of strong covalent bonds, which must be overcome to allow the material to sublime.

3 Any three of: printer toner, ink, reinforcement of rubber, art pencils, charcoal briquettes.

4 **a** Diamond is hard because it has strong covalent bonds throughout the lattice, with all atoms being held in fixed positions.

b Diamond is a non-conductor of electricity because all of its electrons are localised in covalent bonds and are not free to move.

c Diamond is a good heat conductor because the carbon atoms are strongly bonded together in the lattice.

5 **a** Graphite is soft because there are weak dispersion forces between the layers in graphite, so layers can be made to slide over each other easily.

b Graphite is able to conduct electricity because it has delocalised electrons between its layers of carbon atoms.

Chapter 3 review

1 D 2 B 3 A

4 C

5 Intramolecular bonds are the forces that hold the atoms within a molecule together. In ammonia molecules they are the covalent bonds between the nitrogen and hydrogen atoms. Intermolecular forces are between one molecule and its neighbouring molecules. These are much weaker forces. It is the intermolecular forces that are disrupted when ammonia melts, allowing the molecules to move more freely around each other.

6 Neon will not form bonds to other atoms as it has a stable outer shell containing eight electrons.

7 **a** tetrahedral **b** pyramidal
c tetrahedral **d** bent
e pyramidal

8 NBr_3 – pyramidal, H_2O – bent, CH_2F_2 – tetrahedral, HCN – linear

9 F–F, O–Cl, N–O, H–Br, Si–O

10 the O–H bond in water H_2O (answer **b**)

11 If water was a linear molecule, the two polar O–H bonds would cancel each other out and make the molecule non-polar. As water is polar, it cannot be a linear molecule, it is in fact bent.

12 The strength of the intermolecular bonds in pure hydrogen chloride must be relatively weak. Since pure hydrogen chloride exists as a gas at room temperature, it must have a low boiling point, which indicates that not much energy is required to break the intermolecular forces between molecules.

13 a CCl_4

b CH_4 and CCl_4 are both non-polar and so there are only dispersion forces between their molecules. CCl_4 has a higher molecular mass of these two molecules, so the dispersion forces between CCl_4 molecules will be greater than those between CH_4 molecules. As there are stronger dispersion forces between molecules of CCl_4 than for CH_4, CCl_4 has a higher boiling point so it exists as a liquid at room temperature.

14 A permanent molecular dipole is formed if there is asymmetry in the molecule. This causes an asymmetry in the electron distribution around the molecule, causing one end of the molecule to have a partial negative charge while the other end has a partial positive charge. The positive and negative ends of neighbouring molecules attract each other, forming dipole–dipole attractions.

A temporary molecular dipole is caused by random fluctuations in the electron distributions around the molecule. The electrons are constantly moving and can occasionally concentrate at one end of the molecule, causing that end to have a temporary negative charge while the other end has a temporary positive charge. This temporary dipole can then induce dipoles in the neighbouring molecules. The induced dipoles attract each other. Such attractions are known as dispersion forces and are present between all molecules.

15 Carbon exists in different forms with different arrangements of atoms.

16 Diamond has a high sublimation point because it has a covalent network lattice structure. Many strong covalent bonds need to be broken for sublimation to occur.

17 Sublimes at a high temperature—covalent layer structure so many strong covalent bonds need to be broken for sublimation to occur. Conducts electricity—graphite contains delocalised electrons. Lubricant—weak dispersion forces between layers allows the layers to easily slide over each other.

18 Carbon atoms in diamond have a tetrahedral bond geometry; carbon atoms in graphene within graphite have a trigonal planar bond geometry.

19 the molecule will be bent

20 a non-polar **b** polar **c** non-polar
d polar **e** non-polar **f** non-polar

21 a $SiCl_4$: **i** non-polar **ii** dispersion forces
b CF_4: **i** non-polar **ii** dispersion forces
c NF_3: **i** polar **ii** dipole–dipole attraction
d CH_3NH_2: **i** polar **ii** hydrogen bonding

22 a non-polar **b** polar
c non-polar **d** polar

23 a 2 bonding electrons, 6 non-bonding electrons

H — Br

b 8 bonding electrons, 24 non-bonding electrons

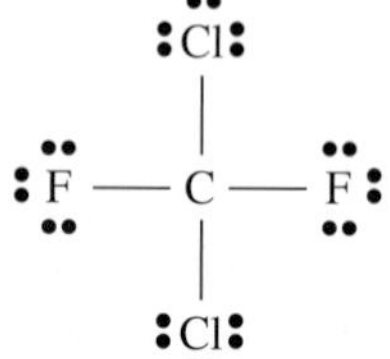

c 14 bonding electrons

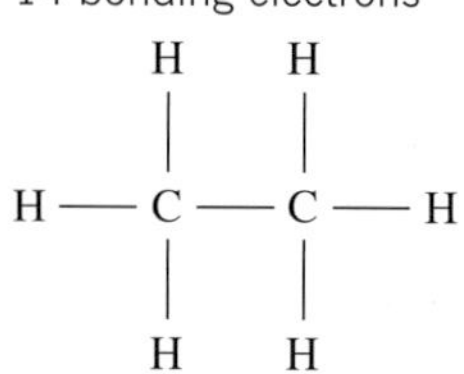

d 6 bonding electrons, 20 non-bonding electrons

F — P — F
F

24 a true **b** true **c** true
d false **e** true

25 Argon has the electronic configuration of 2,8,8 and therefore has eight electrons in its outer shell. The argon atom obeys the octet rule and will therefore exist as stable as single atoms. Chlorine, on the other hand, has the electronic configuration of 2,8,7 and therefore has seven electrons in the outer shell. A single chlorine atom requires another electron to complete its octet. It can do this by bonding to another chlorine atom to form the molecule Cl_2 where both Cl atoms have a complete octet in their valence shell.

26 Melting points increase down the table because the molecules increase in molecular mass. As they are all diatomic non-polar molecules, the strength of the dispersion forces increases.

27 CF_4 has a slightly higher boiling point (–128°C) than OF_2 (–145°C), indicating that the forces between molecules in CF_4 are stronger. OF_2 is slightly polar; CF_4 is non-polar. OF_2 molecules are held together by dipole–dipole attraction and dispersion forces. Although CF_4 molecules are attracted by dispersion forces only, the much larger molecular mass of CF_4 molecules makes the dispersion forces stronger than the sum of the dipole–dipole and dispersion forces between OF_2 molecules.

28 Neon exists as single atoms, with the only forces of attraction being dispersion forces; therefore, neon has a very low boiling point. Hydrogen fluoride molecules, however, are very polar as a hydrogen atom is bonded to the very electronegative fluorine atom. The forces between molecules are hydrogen bonds. These are relatively strong intermolecular forces and hydrogen fluoride, therefore, has a much higher boiling point than neon.

29 a Methane is an example of a covalent molecular substance. It has strong, covalent intramolecular bonds and weak intermolecular dispersion forces. Diamond is an example of a covalent network lattice. It has strong covalent bonds throughout its structure.

b The differences in properties all relate to the fact that methane is a covalent molecular substance whereas diamond is a covalent network lattice. All bonding within diamond is strong, whereas methane has weak intermolecular dispersion forces. The many covalent bonds throughout diamond's lattice structure makes it a very hard substance with a very high sublimation point. The dispersion forces between methane molecules means it has a relatively low boiling point and exists as a gas at room temperature. The covalent bonds in methane are all contained within each individual methane molecule.

30 a Diamond has a much higher hardness value than graphite because diamond has a covalent network lattice structure so there are strong covalent bonds throughout its structure. Graphite has a covalent layer lattice structure with weak dispersion forces between layers which makes it a relatively soft, slippery substance. Graphite has a much higher electrical conductivity value because it contains delocalised electrons. Diamond does not conduct electricity because its valence electrons are all localised in covalent bonds.

b It needs to be clear whether you are referring to thermal conductivity or electrical conductivity as the comparison is different for each.

c The higher the quality of the diamond, the higher the thermal conductivity. High-quality diamonds will cool faster than low-quality ones.

Chapter 4 Metals

4.1 Metallic properties and bonding

TY 4.1.1 Shell configuration of aluminium: 2,8,3. Cation charge: 3+

CSA: Colourful transition metal compounds

1 **a** Co, Al, Cd, Fe, Cr, Pb, Sn
 b Co, Cd, Fe, Cr

2 Different transition metal ions are present in rubies and sapphires. Chromium ions are present in rubies, whereas titanium and iron ions are present in sapphires.

3 Diamonds are composed almost entirely of carbon. Since transition metal ions are not present they are usually colourless.

TY 4.1.2 Mg has 2 electrons in its outer shell. Mg atoms will tend to lose these 2 valence electrons to form a cation with a charge of 2+. The outer-shell electrons become delocalised and form the sea of delocalised electrons within the metal lattice. If the Mg is part of an electric circuit, the delocalised electrons are able to move through the lattice towards a positively charged electrode.

Key questions

1 Any three of: dense, malleable/ductile, good conductors of heat and electricity, lustrous.

2 **a** silver and gold
 b availability, cost, malleability, and ductility

3 ductile: the material is able to be drawn into a wire
malleable: the material can be shaped by beating or rolling

4 **a**

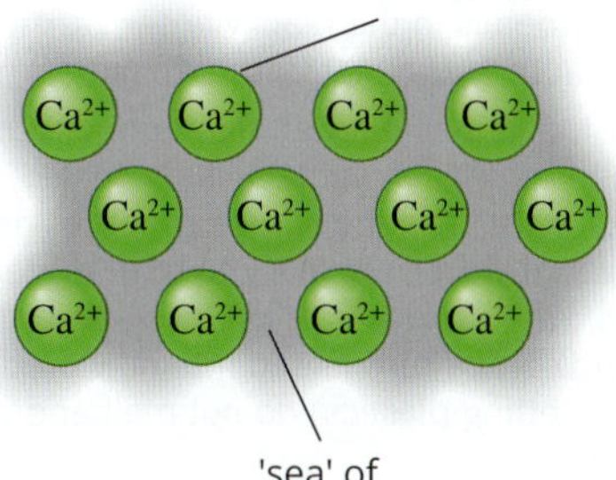

 b Strong electrostatic forces of attraction are present between Ca^{2+} ions and the delocalised valence electrons.

5 Barium has a high melting point because there are strong attractive forces between the positive ions and the delocalised electrons. Barium conducts electricity because the delocalised electrons from the outer shell are free to move through the entire metal, enabling the transfer of a charge throughout the metal.

6 **a** Both potassium and gold have good thermal and electrical conductivity. However, gold has a higher density, and higher melting and boiling points than potassium.
 b sodium
 c silver
 d Sodium and potassium are in group 1. Gold and silver are transition metals.

7 **a** 1+ **b** 2+
 c 3+ **d** 2+

8 The smaller the first ionisation energy of an element, the greater the metallic character of that element.

4.2 Reactivity of metals

1 **a** calcium oxide
 b sodium hydroxide and hydrogen

2 B 3 calcium 4 Zn > Fe > Au

5 **a i** In general, the reactivity of main group metals increases going down a group in the periodic table.
 ii In general, the reactivity of main group metals decreases across a period.
 b Transition metals tend to be less reactive than the elements in groups 1 and 2
 c The trend in metal reactivity corresponds with the periodic trends in first ionisation energy.

6 Potassium is more reactive than sodium because it is lower in group 1 (its valence electrons are further from the protons in the nucleus and less strongly attracted to it, so they are more easily lost in a reaction). Potassium is on the left of calcium in period 4 (so its nucleus has a lower effective nuclear charge and its valence electron is less strongly held within the atom) and so potassium is more reactive than calcium.

4.3 Producing and recycling metals

CSA: E-waste

1 Mercury is a neurotoxin. Symptoms of prolonged exposure include tremors, headaches, short-term memory loss, incoordination, weakness, loss of appetite, altered sense of taste and smell, numbness and tingling in the hands and feet, insomnia, and excessive sweating.
Lead is a cumulative toxin which can affect multiple body systems. Lead exposure can permanently damage the brain and impair intellectual development.
Cadmium poisoning symptoms include anaemia and kidney failure. Exposure increases the chance of developing cancer.

2 **a** In a circular economy the ideal is to eliminate waste so that the products in use today are used as raw materials tomorrow. This creates a closed loop.
 b While the recycling process described for mobile phones is arguably an improvement on current practice, it does not create closed loops. The process will create waste, including CO_2, and all components of the e-waste, such as plastics, are not being recycled.

3 The aims of a circular economy could be better realised if mobile phones were used for longer before being replaced and if they were designed so that they could be more readily repaired and, ultimately, more readily recycled. Improved rates of recycling of the different components in the phones, perhaps encouraged by legislation, as well as improved recycling techniques would also be desirable.

Key questions

1 In a *linear economy* raw materials are used to make a product, and after its use the product is thrown away. A *circular economy* is based on a model of production and consumption that aims to design out waste and pollution, keep products and materials in use, and regenerate natural systems.

2 Metal recycling meets some of the aims of a circular economy by:
- returning materials to the production cycle
- avoiding using metals as landfill
- saving energy (less energy is used than if the metals are directly extracted from their ores)
- reducing greenhouse emissions compared to mining
- minimising the impact of ore extraction on the environment.

3 Not all products containing metals are recycled; some products are discarded in landfill. It can be uneconomic to recover all the metallic material from some items that contain many different components and only small concentrations of metals. Some waste items are also hard to disassemble. The market price of the metal is a factor, with higher recovery rates for precious metals such as gold and platinum. Furthermore, at the national level, if mineral ores or manufactured products are exported to other countries these materials are not recycled within the country of origin.

4 The market price of gold and platinum is higher than for copper and aluminium, so the economic incentive for their recovery is greater.

5 Use of high-temperature open furnaces can allow lead to enter the immediate environment. In fact, a study has shown that people in developing countries who are engaged in recycling spent batteries in this manner have developed neurological disorders, with high blood lead levels and intellectual disabilities.

Chapter 4 review

1 a aluminium Al, copper Cu, gold Au, iron Fe, silver Ag
  b aluminium: period 3, group 13
    copper: period 4, group 11
    gold: period 6, group 11
    iron: period 4, group 8
    silver: period 5, group 11
  c gold and silver
  d copper, gold, iron and silver
  e gold

2 a low density
  b high electrical conductivity
  c high tensile strength

3 electrical conductivity

4 C

5 The diagram below shows a two-dimensional view of a metallic lattice. The metal atoms are arranged in an ordered manner. The metal atoms lose their valence electrons and form an 'electron sea'. Electrostatic attraction between the electrons and the metal cations hold the lattice together.

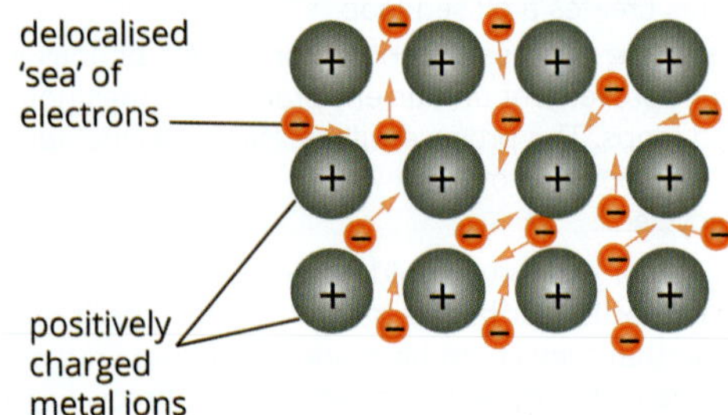

6 a i valence electrons that are not restricted to a region between two atoms
    ii a regular three-dimensional arrangement of a very large number of positive ions (cations)
    iii the electrostatic attraction between a lattice of cations and delocalised electrons
  b valence (outer-shell) electrons

7 a When a current is applied to the copper wire, the free-moving, delocalised electrons move from one end to the other and so the copper wire conducts electricity.
  b The delocalised electrons in the metal spoon obtain energy from the boiling mixture and move more quickly. These electrons move freely throughout the spoon, colliding with other electrons and metal ions, transferring energy so that the spoon becomes warmer and, eventually, too hot to hold.
  c A lot of energy is required to overcome the strong forces of attraction between the iron ions and the delocalised electrons in the metal lattice, in order for the iron to change from a solid to a liquid.
  d As the copper is drawn out, the copper ions are forced apart and the delocalised electrons rearrange themselves around these ions and re-establish strong forces of attraction.

8 Reactive metals, such as sodium and potassium, react with water to form the metal hydroxide and hydrogen gas. The hydrogen gas is observed as bubbles.

9 a false b true
  c false d true

10 B 11 D 12 Student answers will vary.

13 a Na: group 1, period 3
    K: group 1, period 4
    Ca: group 2, period 4
  b Na: $1s^22s^22p^63s^1$
    K: $1s^22s^22p^63s^23p^64s^1$
    Ca: $1s^22s^22p^63s^23p^64s^2$
  c i The atoms of Na are smaller than those of K, so the delocalised valence electrons of Na are closer to the positive nuclear charge than those of K. The electrostatic forces of attraction between delocalised electrons and cations are stronger in Na, so Na requires more energy to overcome the metallic bonding to boil the metal.
    ii Valence electrons are in the fourth shell in the atoms of both Ca and K. However, there are twice as many valence electrons in the atoms of Ca. Also, the charge on a calcium cation is 2+ as opposed to 1+ on the potassium cation. Therefore the electrostatic forces of attraction between delocalised electrons and cations are stronger in Ca and so it requires more energy to overcome the metallic bonding to boil the metal.

14 D 15 B

16 The reaction on the left is more vigorous, so the metal must be more reactive. Iron is a more reactive metal than silver and so iron must be on the left. Silver is less reactive than iron and so silver must be on the right.

17 Based on the order of metals in the reactivity series, metal A is copper, metal B is sodium and metal C is aluminium.

18 When metals are heated in a smelter at high temperatures they melt without decomposing. When subsequently cooled, they reform a metallic lattice that has the same properties as the original metal. Plastics, however, are likely to burn or decompose at high temperatures. While some plastics (thermoplastics) will melt when heated and can be remoulded into new products, other plastics (thermosetting plastics) do not melt and cannot be recycled in this way.

19 a A circular economy aims to reduce waste, pollution (including greenhouse gas emissions), use of natural resources, environmental damage and energy use.
  b Recycling metals reduces the cost of raw materials for manufacturers. Furthermore, with a reduced need for mining natural resources there is less environmental damage, and greenhouse gas and pollutant emissions are reduced.
  c Some metals are less valuable so there is less financial incentive to recycle products made of these metals. It can be difficult to extract metals that are used in products that are complex and use many other components, such as smartphones. Other metals, such as lithium, are not easily extracted from wastes and considerable energy is required.

20 Metals at the top of the reactivity series, such as sodium, are rarely used in their pure form. They are mainly used in the form of their compounds, e.g. as sodium chloride (salt). Many of these metals are also found as compounds in large quantities in nature. So there is little financial incentive to extract the pure elements. Furthermore, the extraction of the highly reactive metals requires larger quantities of energy and is more expensive.
On the other hand, metals at the bottom of the series, such as platinum and gold, tend to be valuable and require less energy to recycle. They are almost completely recovered after use and then reused.

Chapter 5 Ionic compounds

5.1 Properties of ionic compounds

CSA: How fluoride ions make tooth enamel harder

1 a Ionic compounds are hard.
 b Ionic compounds are hard.

2 The overall charge on fluoroapatite can be calculated by multiplying the charge on each ion by how many of that ion there are, then finding a total charge for all ions. That is:
Ca^{2+}: 10 × 2+ = 20+
PO_4^{3-}: 6 × 3– = 18–
F^-: 2 × 1– = 2–
Total charge = 0

3 SiF_6^{2-}

Key questions

1 Ionic compounds are formed when non-metal atoms react with metal atoms. In this process, non-metal atoms **gain** electrons to form **negatively** charged ions called **anions** and metal atoms **lose** electrons to form **positively** charged ions called **cations**. The ions formed pack together in a three-dimensional lattice held strongly together by **electrostatic** forces of attraction.

2 In an ionic compound like sodium chloride, positively and negatively charged ions are held together in a three-dimensional lattice by strong electrostatic forces of attraction. This gives the compound a high melting point, but because the ions in the lattice cannot move, the compound will not conduct electricity in the solid state. If the compound is melted or dissolved in water, however, the lattice breaks down, the ions can now move freely and so the compound can conduct an electric current.

3 a Ionic compounds have high melting points and are hard.
 b Ionic compounds conduct electricity in the molten state or in solution.
 c Ionic compounds are brittle.

4 When an ionic compound is hit with a hammer, the layers of ions within the ionic lattice move relative to each other. This causes ions with like charges to be adjacent to each other and they repel. This causes the lattice to shatter, as seen in the diagrams in Figure 5.1.7 on page 164.

5 a ionic compounds only
 b metals and ionic compounds
 c metals only
 d metals and ionic compounds
 e metals only

6 Substance A is an ionic compound because it has a high melting point and will not conduct electricity at 100°C because it is a solid at that temperature. It will conduct electricity at 1000°C because that is above its melting point and it will be in a liquid state. Substance C is also an ionic compound for similar reasons. Substance B is a metal because it will conduct electricity both in the solid and molten state. The fact that it is insoluble in water also indicates that it could be a metal.
Some ionic compounds are soluble in water and some are not, so solubility alone cannot be used to decide whether substances are ionic compounds.

5.2 Formation of ionic compounds

TY 5.2.1 3Ca (2,8,8,2) + 2P (2,8,5) → $3Ca^{2+}$ (2,8,8) + $2P^{3-}$ (2,8,8)

TY 5.2.2 BaF_2

1 a KBr(s) → K^+(aq) + Br^-(aq)
 b $Ca(NO_3)_2$(s) → Ca^{2+}(aq) + $2NO_3^-$(aq)
 c Na_2S(s) → $2Na^+$(aq) + S^{2-}(aq)
 d $FeCl_3$(s) → Fe^{3+}(aq) + $3Cl^-$(aq)
 e $Al_2(SO_4)_3$(s) → $2Al^{3+}$(aq) + $3SO_4^{2-}$(aq)

2 A, D and E

3 Potassium and sulfur will react together to form a compound, potassium sulfide. During this process each sulfur atom will **gain two** electron(s) to form a **negatively** charged sulfide ion with the symbol S^{2-}. Each sulfide ion will have the same stable electron configuration as an atom of **argon**, which is the **noble gas** element nearest to it on the periodic table. Also during the reaction, each potassium atom will **lose one** electron(s) to achieve the same stable electron configuration as an atom of **argon**.

4 a $ZnCl_2$ b K_2O c Sr_3N_2 d Na_2CO_3
 e $Al_2(SO_4)_3$ f $Zn_3(PO_4)_2$ g CuCl h Fe_2O_3
 i $Cr_2(SO_4)_2$

5 a magnesium sulfide b potassium oxide
 c iron(II) sulfate d barium nitrate
 e copper(I) sulfate f iron(III)cyanide
 g gold(III) dichromate h lead(IV) phosphate

6 Cations: calcium, Ca^{2+}, aluminium, Al^{3+}.
Anions: nitrogen (nitride ion, N^{3-}), fluorine (fluoride ion, F^-) and phosphorus (phosphide ion, P^{3-}).

7 a

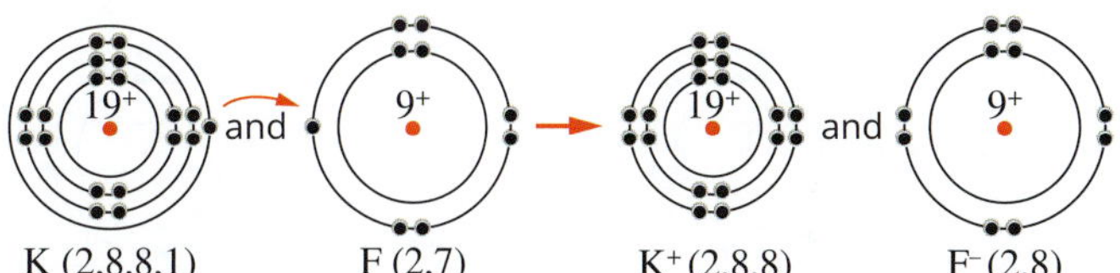

b

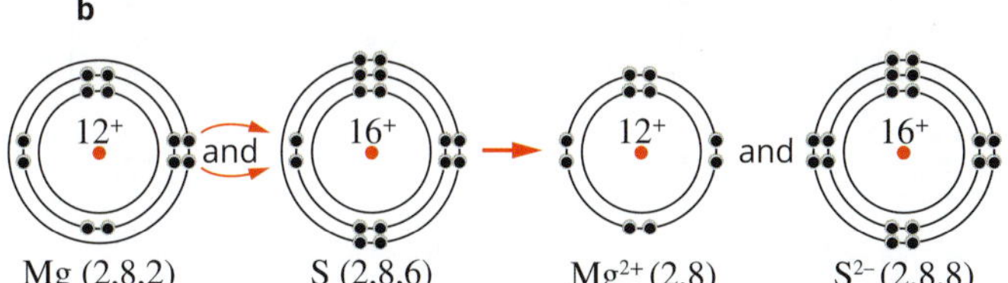

c

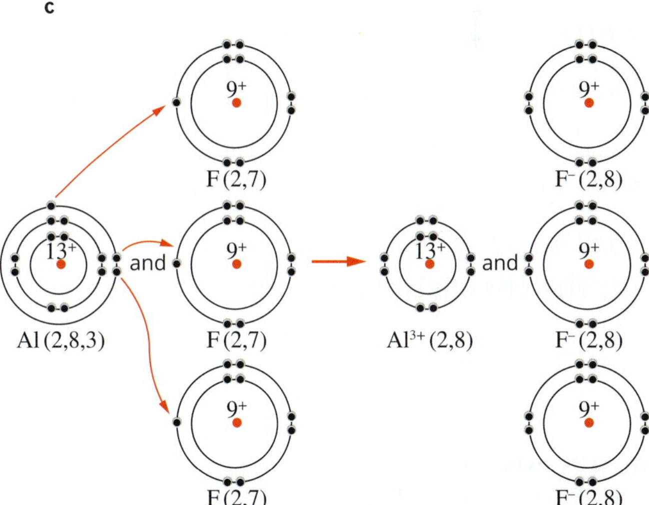

d

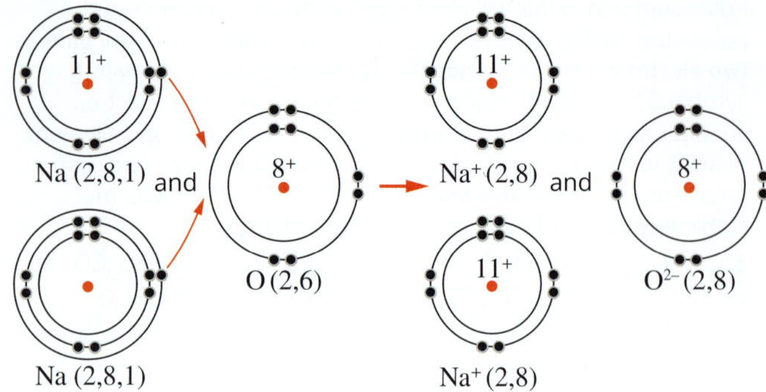

e

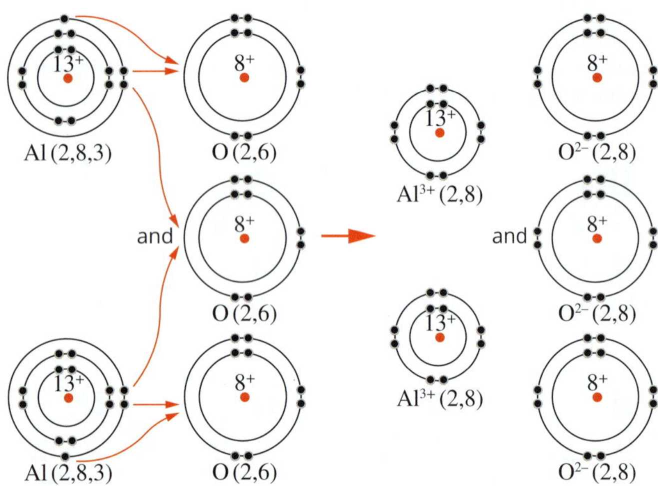

8 a Na (2,8,1) + Cl (2,8,7) → Na^+ (2,8) + Cl^- (2,8,8)
b Mg (2,8,2) + O (2,6) → Mg^{2+} (2,8) + O^{2-} (2,8)
c 2Al (2,8,3) + 3S (2,8,6) → $2Al^{3+}$ (2,8) + $3S^{2-}$ (2,8,8)

5.3 Precipitation reactions

TY 5.3.1 Compounds containing sodium ions or nitrate ions are usually soluble, so sodium nitrate will not form a precipitate. Compounds containing sulfide ions are usually insoluble, so copper(II) sulfide will form as a precipitate.

TY 5.3.2 $CuSO_4(aq) + 2NaOH(aq) \rightarrow Cu(OH)_2(s) + Na_2SO_4(aq)$
$Na^+(aq)$ and $SO_4^{2-}(aq)$ are spectator ions.

TY 5.3.3 $Ba^{2+}(aq) + 2OH^-(aq) \rightarrow Ba(OH)_2(s)$

CSA: The chemistry of colour

1 Sn^{4+}
2 $Na_2CrO_4(aq) + Pb(NO_3)_2(aq) \rightarrow PbCrO_4(s) + 2NaNO_3(aq)$
3 Any soluble cadmium compound, for example: $CdCl_2$, $Cd(NO_3)_2$, $CdSO_4$, $Cd(CH_3COO)_2$

Key questions

1 a $KBr(s) \rightarrow K^+(aq) + Br^-(aq)$
b $Ca(NO_3)_2(s) \rightarrow Ca^{2+}(aq)\ 2NO_3^-(aq)$
c $Na_2S(s) \rightarrow 2Na^+(aq) + S^{2-}(aq)$
d $FeCl_3(s) \rightarrow Fe^{3+}(aq) + 3Cl^-(aq)$
e $Al_2(SO_4)_3 \rightarrow 2Al^{3+}(aq) + 3SO_4^{2-}(aq)$

2 a i $CaCO_3$ ii no precipitate iii MgS
iv $Fe(OH)_2$ v Ag_3PO_4
b i $NO_3^-(aq)$ and $K^+(aq)$ ii no spectator ions
iii $Na^+(aq)$ and $SO_4^{2-}(aq)$ iv $Cl^-(aq)$ and $NH_4^+(aq)$
v $Na^+(aq)$ and $NO_3^-(aq)$

3 a i magnesium sulfide
ii silver chloride
iii aluminium hydroxide
iv magnesium hydroxide
b i $K_2S(aq) + MgCl_2(aq) \rightarrow MgS(s) + 2KCl(aq)$
ii $CuCl_2(aq) + 2AgNO_3(aq) \rightarrow 2AgCl(s) + Cu(NO_3)_2(aq)$
iii $AlCl_3(aq) + 3KOH(aq) \rightarrow Al(OH)_3(s) + 3KCl(aq)$
iv $MgSO_4(aq) + 2NaOH(aq) \rightarrow Mg(OH)_2(s) + Na_2SO_4(aq)$

4 a i $NH_4Cl(aq) + AgNO_3(aq) \rightarrow AgCl(s) + NH_4NO_3(aq)$
ii $Ag^+(aq) + Cl^-(aq) \rightarrow AgCl(s)$
b i $Cu(NO_3)_2(aq) + K_2CO_3(aq) \rightarrow CuCO_3(s) + 2KNO_3(aq)$
ii $Cu^{2+}(aq) + CO_3^{2-}(aq) \rightarrow CuCO_3(s)$
c i $2K_3PO_4(aq) + 3MgSO_4(aq) \rightarrow Mg_3(PO_4)_2(s) + 3K_2SO_4(aq)$
ii $3Mg^{2+}(aq) + 2PO_4^{3-}(aq) \rightarrow Mg_3(PO_4)_2(s)$
d i $Ca(OH)_2(aq) + FeCl_2(aq) \rightarrow Fe(OH)_2(s) + CaCl_2(aq)$
ii $Fe^{2+}(aq) + 2OH^-(aq) \rightarrow Fe(OH)_2(s)$
e i $Ba(NO_3)_2(aq) + (NH_4)_2SO_4(aq) \rightarrow BaSO_4(s) + 2NH_4NO_3(aq)$
ii $Ba^{2+}(aq) + SO_4^{2}(aq)^- \rightarrow BaSO_4(s)$
f i $Pb(CH_3COO)_2(aq) + Na_2SO_4(aq) \rightarrow PbSO_4(s) + 2NaCH_3COO(aq)$
ii $Pb^{2+}(aq) + SO_4^{2-}(aq) \rightarrow PbSO_4(s)$

5 a NH_4^+, NO_3^- b K^+, NO_3^- c K^+, SO_4^{2-}
d Ca^{2+}, Cl^- e NH_4^+, NO_3^- f Na^+, CH_3COO^-

6 a barium sulfate
b $Ba(CH_3COO)_2(aq) + K_2SO_4(aq) \rightarrow BaSO_4(s) + 2KCH_3COO(aq)$
c $Ba^{2+}(aq) + SO_4^{2-}(aq) \rightarrow BaSO_4(s)$

Chapter 5 review

1 D 2 A 3 D

4 a The electrostatic forces of attraction between the positive and negative ions are strong and will be overcome only at high temperatures.
b The strong electrostatic forces of attraction between the ions mean that a strong force is needed to break up the lattice, giving the ionic crystals the property of hardness. However, the crystal lattice will shatter when a strong force is applied, suddenly causing ions of like charge to become adjacent to each other and be repelled.
c In the solid state, the ions are not free to move. However, when the solid melts or dissolves in water, the ions are free to move and conduct electricity.

5 Statement 1: Ionic compounds are able to conduct electricity in the molten state.
Statement 2: Ionic compounds are not able to conduct electricity in the solid state.
Statement 3: Ionic compounds are hard and have high melting points..
Statement 4: Ionic compounds are brittle.

6 a 2,8 b 2,8
c 2,8 d 2,8

7 a KBr b MgI_2 c CaO
d AlF_3 e Ca_3N_2

8 Fe^{2+} would react in the presence of PO_4^{3-} and S^{2-} to produce the insoluble compounds $Fe_3(PO_4)_2$ and FeS.

9 a $CuNO_3$ b CrF_2 c K_2CO_3
d $Mg(HCO_3)_2$ e $Ni_3(PO_4)_2$

10 a ammonium carbonate
b copper(II) nitrate
c copper(II) nitrite
d chromium(III) bromide
e tin(II) dihydrogen phosphate
f lead(IV) hydrogen sulfite

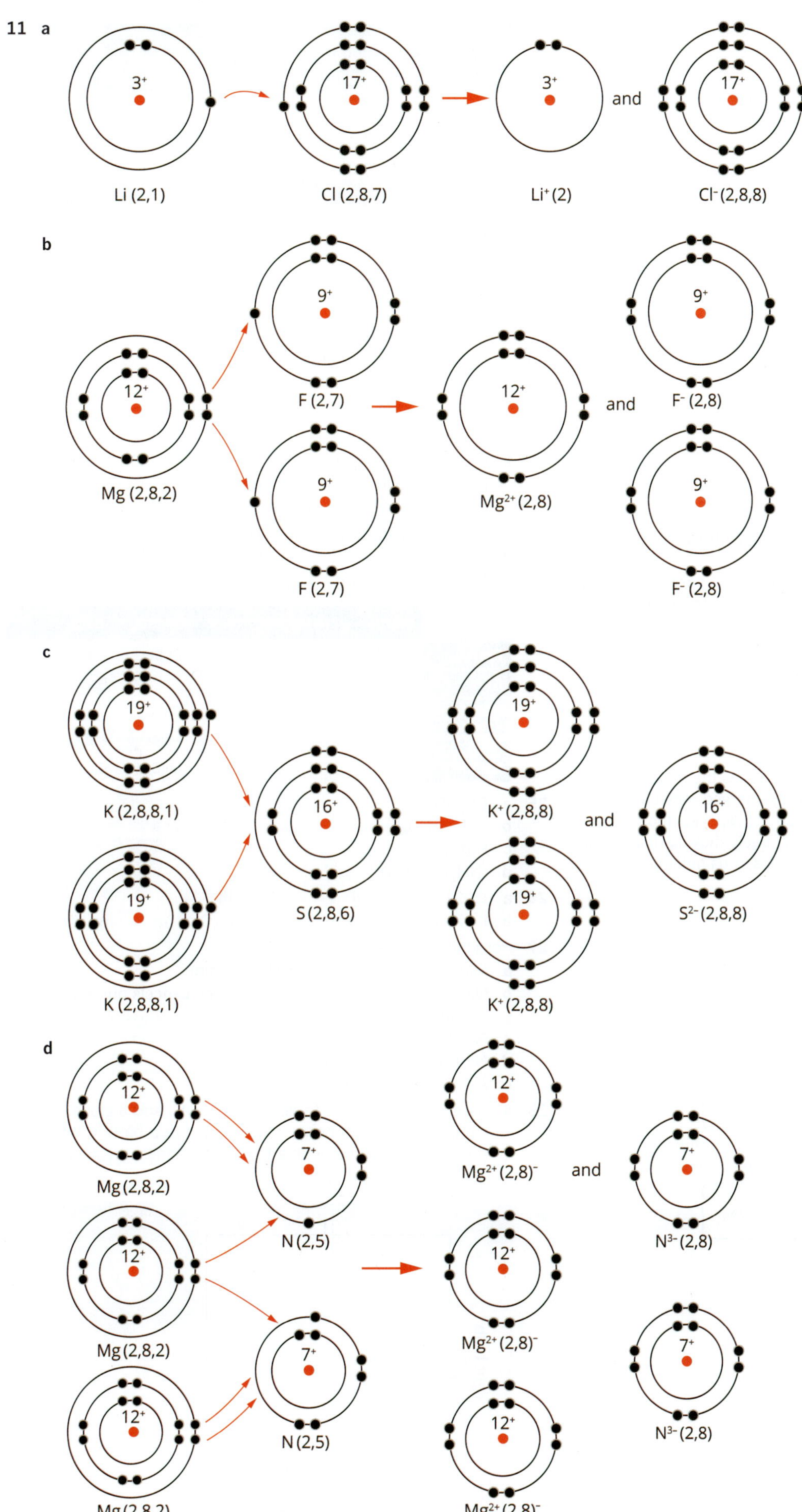
11 a
3^+
17^+
3^+
and
17^+
Li (2,1)
Cl (2,8,7)
Li^+ (2)
Cl^- (2,8,8)
b
9^+
9^+
12^+
F (2,7)
12^+
and
F^- (2,8)
9^+
9^+
Mg (2,8,2)
Mg^{2+} (2,8)
F (2,7)
F^- (2,8)
c
19^+
19^+
K (2,8,8,1)
16^+
K^+ (2,8,8)
and
16^+
19^+
19^+
S (2,8,6)
S^{2-} (2,8,8)
K (2,8,8,1)
K^+ (2,8,8)
d
12^+
12^+
7^+
Mg (2,8,2)
Mg^{2+} (2,8)$^-$
and
7^+
12^+
N (2,5)
N^{3-} (2,8)
12^+
Mg (2,8,2)
Mg^{2+} (2,8)$^-$
7^+
7^+
12^+
12^+
N (2,5)
N^{3-} (2,8)
Mg (2,8,2)
Mg^{2+} (2,8)$^-$

12 **a** Mg (2,8,2) + 2Cl (2,8,7) ⟶ Mg^{2+} (2,8) + $2Cl^-$ (2,8,8)
b 2Al (2,8,3) + 3O (2,6) ⟶ $2Al^{3+}$ (2,8) + $3O^{2-}$ (2,8)
c 3Na (2,8,1) + P (2,8,5) ⟶ $3Na^+$(2,8) + P^{3-}(2,8,8)

13 The subscripts represent the ratio of metal to non-metal ions in the ionic compound.

14 Elements in group 17 of the periodic table have seven electrons in their outer shell, so only need to gain one electron to satisfy the octet rule.

15 **a** barium sulfate **b** none
c lead(II) sulfate **d** none

16 B **17** B

18 **a** K^+, F^-, Ca^{2+} and O^{2-}
b Both ions in potassium fluoride are singly charged. Both ions in calcium oxide have double charges. The forces of attraction between the two double charged ions in calcium oxide will be much stronger than that between the single-charged ions of potassium fluoride. The melting point of calcium oxide (2572°C) will therefore be higher than that of potassium fluoride (858°C)

19 **a** Agree. In metals the outermost shell electrons are delocalised throughout the metallic lattice. In ionic compounds the metal atom loses its outermost electron(s) to the non-metal atoms.
b Disagree. Particles of opposite charge will attract, not repel each other.
c Disagree. In ionic compounds the negatively charged ions are held in fixed positions between positively charged ions. In metals, however, the negatively charged outer shell electrons are free to move, not held in fixed positions.
d Disagree. In metals it is the freely moving delocalised electrons that conduct the electric current. In molten ionic compounds, however, both positively and negatively charged ions move, therefore conducting the electric current.
e Agree. When metals are drawn into a wire layers of positively charged metal ions slide over each other as the metal 'stretches'. This means that positively charged metal ions are moving past each other. They do not repel each other, however, because electrons from the freely moving delocalised electron 'sea' move in between the ions, thus holding them together.

20 A possible answer is shown.

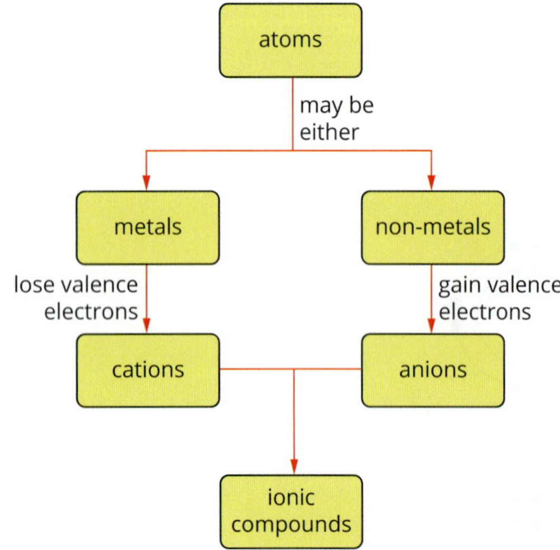

21

Element	Electrons lost or gained when forming an ion?	Noble gas with same electron configuration as ion formed
phosphorus	gained	argon
lithium	lost	helium
oxygen	gained	neon
aluminium	lost	neon
potassium	lost	argon
bromine	gained	krypton
sulfur	gained	argon

22 **a** CD_3 **b** EF
c G_3H **d** KL

23 **a** $NH_4Cl(aq) + AgNO_3(aq) \longrightarrow NH_4NO_3(aq) + AgCl(s)$
$Ag^+(aq) + Cl^-(aq) \longrightarrow AgCl(s)$
b $FeCl_2(aq) + Na_2S(aq) \longrightarrow FeS(s) + 2NaCl(aq)$
$Fe^{2+}(aq) + S^{2-}(aq) \longrightarrow FeS(s)$
c $Fe(NO_3)_3(aq) + 3KOH(aq) \longrightarrow 3KNO_3(aq) + Fe(OH)_3(s)$
$Fe^{3+}(aq) + 3OH^-(aq) \longrightarrow Fe(OH)_3(s)$
d $CuSO_4(aq) + 2NaOH(aq) \longrightarrow Cu(OH)_2(s) + Na_2SO_4(aq)$
$Cu^{2+}(aq) + 2OH^-(aq) \longrightarrow Cu(OH)_2(s)$
e $Ba(NO_3)_2(aq) + Na_2SO_4(aq) \longrightarrow BaSO_4(s) + 2NaNO_3(aq)$
$Ba^{2+}(aq) + SO_4^{2-}(aq) \longrightarrow BaSO_4(s)$

24

	$NaOH$	$(NH_4)_3PO_4$	NaI	$MgSO_4$	$BaCl_2$
$Pb(NO_3)_2$	$Pb(OH)_2$	$Pb_3(PO_4)_2$	PbI_2	$PbSO_4$	$PbCl_2$
KI					BaI_2
$CaCl_2$	$Ca(OH)_2$	$Ca_3(PO_4)_2$		$CaSO_4$	
Na_2CO_3				$MgCO_3$	$BaCO_3$
Na_2S				MgS	BaS

25 **a** $Ca(ClO_4)_2$
b $Al_4[Fe(CN)_6]_3$
c $Fe(ClO_4)_3$
d $(NH_4)_4[Fe(CN)_6]$

26 **a** **i** 2+ **ii** 2– **iii** 3+
b **i** X_3N_2 **ii** PbY_2 **iii** $Z_2(Cr_2O_7)_3$ **iv** Z_2Y_3

27 Possible combinations are:
a $MgCl_2$ or MgF_2 (2+ cation with 1– anion)
b NaCl or CaS (cation and anion charge cancels out; e.g. 1+ cation with a 1– anion)
c Na_2O or K_2S (1+ cation with 2– anion)
d Na_3N or Li_3P (1+ cation with 3– anion)
e $AlCl_3$ or AlF_3 (3+ cation with 1– anion)

28 **a** Assemble equipment to test conductivity. Add a globe to a circuit containing a power source, such as a battery. When the electrodes are touching the solid magnesium chloride, the globe will not light up.

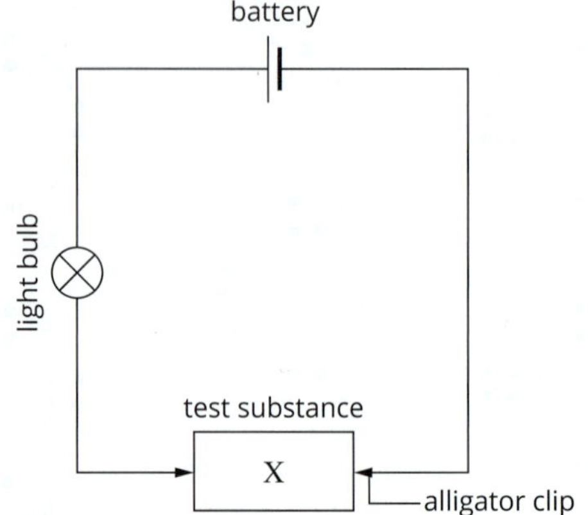

b Dissolve some solid sodium chloride (about 1 teaspoon per 200 mL) in deionised water. Using the same equipment, place the two electrodes in the solution, but don't allow them to touch; the globe will glow.

c If a crystal of sodium chloride was hit firmly with a hammer, it would shatter. Care is needed—safety glasses must be worn.

Chapter 6 Separation and identification of components of mixtures

6.1 How substances dissolve

CSA: Vitamin C and vitamin A: Similar but different

1 Vitamin D molecules would be largely non-polar, whereas those of vitamin B would be polar.

2 Whereas vitamin C is excreted in urine on a daily basis, vitamin A is not excreted in urine, but is stored in body fat. The concentration of vitamin A will therefore increase over time if relatively large quantities of it are being consumed and hypervitaminosis can develop.

3 Since vitamin E molecules have a long non-polar hydrocarbon chain and only one polar –OH group, the vitamin is likely to be relatively non-polar overall and fat soluble.

Key questions

1 **a** solute **b** solution
c solvent **d** solute

2 Nitrogen gas (**b**), ethane (**e**) and tetrachloromethane (**f**) molecules are all symmetrical and non-polar and so will be insoluble in water. Molecules in canola oil (**d**) have long, non-polar hydrocarbon chains so it will also be insoluble in water. Glucose (**c**) and ethanoic acid (**g**) are all polar molecules and will be soluble in water. Nitric acid (**a**) is a strong acid that ionises in water.

3 D and F

4 Methanol dissolving: $CH_3OH(l) \xrightarrow{H_2O(l)} CH_3OH(aq)$
Nitric acid dissolving: $HNO_3(l) + H_2O(l) \rightarrow H_3O^+(aq) + NO_3^-(aq)$

5 Sodium chloride is an ionic compound consisting of sodium and chloride **ions**. In solid sodium chloride, the two different ions form a **lattice**, which is held together by an **electrostatic** force of attraction called an **ionic bond**. When water is added to solid sodium chloride, water molecules attach themselves to ions in the solid by forces of **ion–dipole attraction**. When the lattice breaks up and the solid dissolves, the sodium and chloride ions are now surrounded by water molecules and are said to be **hydrated**.

6 **a** $MgSO_4(s) \xrightarrow{H_2O} Mg^{2+}(aq) + SO_4^{2-}(aq)$
b $Cu(NO_3)_2(s) \xrightarrow{H_2O} Cu^{2+}(aq) + 2NO_3^-(aq)$
c $(NH_4)_2S(s) \xrightarrow{H_2O} 2NH_4^+(aq) + S^{2-}(aq)$
d $Al_2(SO_4)_3(s) \xrightarrow{H_2O} 2Al^{3+}(aq) + 3SO_4^{2-}(aq)$
e $Na_3PO_4(s) \xrightarrow{H_2O} 3Na^+(aq) + PO_4^{3-}(aq)$

7

	Ionising molecular compounds	Ionic compounds	Non-ionising molecular compounds
Examples	sulfuric acid, H_2SO_4	$Ca(OH)_2$	propanol, C_3H_7OH
Type of particles present before dissolving occurs	molecules	ions	molecules
Type of particles present after dissolving occurs	ions	ions	molecules
Equation for dissolving process	$H_2SO_4(l) + 2H_2O(l) \rightarrow 2H_3O^+(aq) + SO_4^{2-}(aq)$	$Ca(OH)_2(s) \xrightarrow{H_2O} Ca^{2+}(aq) + 2OH^-(aq)$	$C_3H_7OH(l) \xrightarrow{H_2O} C_3H_7OH(aq)$

8 **a** Hydrogen chloride only
b Neither methanol nor hydrogen chloride
c Both methanol and hydrogen chloride
d Hydrogen chloride only.

6.2 Principles of chromatography

TY 6.2.1 $R_f = 0.3$

CSA: Investigating the ingredients of whipped cream

1 **a** The stationary phase is silica and the mobile phase is a mixture of liquids, pentane, hexane and diethyl ether.
b Ultraviolet light is used to detect the spots on the TLC plates because the components are colourless and invisible. They fluoresce and become visible under ultraviolet light.
c It is probable that the scientists found they obtained better separation of the components on the TLC plates using a mixture of solvents.

2

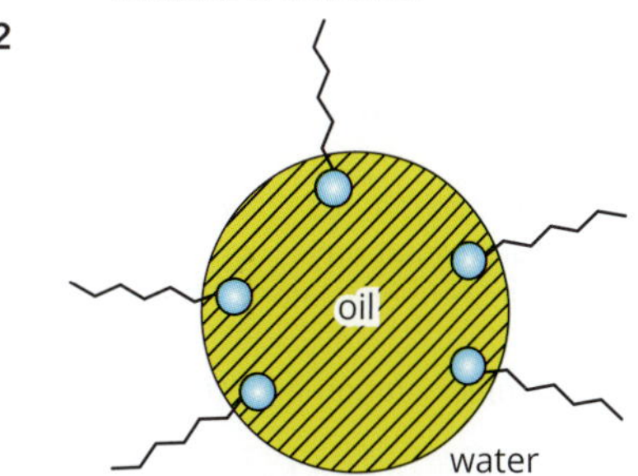

Key questions

1 Ethanol is the mobile phase in this example. The pigments contain the components to be separated and the paper is the stationary phase.

2 components: the different compounds in the mixture, which can be separated by chromatography
mobile phase: the solvent that moves over the stationary phase in chromatography
stationary phase: the components of a mixture undergo adsorption to this phase
adsorption: the attraction of one substance to the surface of another
desorption: the breaking of the attraction between a substance and the surface to which the substance is adsorbed

3 For parts **a**, **b**, **c**, it would be useful to set your answers out in a table. Measurements may vary slightly depending on the screen or book you are measuring from, so this is a model answer only. Yours may vary slightly.

Band	a Distance from origin (mm)	b R_f	c Compound
Light green	20	0.33	chlorophyll b
Dark green	27	0.45	chlorophyll a
Orange	40	0.67	xanthophyll
Yellow	50	0.83	ß-carotene
Solvent front	60	–	–

d The chromatogram would probably be different because separation of components depends on their solubility in the mobile phase (as well as strength of adsorption to the stationary phase). The polarity of the solvent used in TLC and paper chromatography will affect the R_f of the sample components. A polar solvent will dissolve polar samples readily; a non-polar solvent will dissolve non-polar samples readily.

4 (1) Dissolve a sample of pure phenacetin in a volume of chloroform. This is the standard solution.
(2) Dissolve a tablet of the analgesic in chloroform. This is the sample solution.
(3) Place a small spot of the sample solution near the bottom of a thin-layer plate.
Place a spot of the standard solution next to it, at the same distance from the bottom of the plate.
(4) When the spots are dry, place the plate in a container with a small volume of solvent, such as chloroform. The lower edge of the plate, but not the spots, should be immersed.
(5) Allow the solvent to rise until it almost reaches the top of the plate and then remove the plate from the container.
(6) Let the plate dry and examine it under ultraviolet light (phenacetin is colourless but fluoresces under ultraviolet light). If a spot from the sample appears at the same distance from the origin as the spot from the standard solution, the tablet probably contains phenacetin.

Chapter 6 review

1 A solution is most likely to form when the polarity of the solute is similar to that of the solvent. The bonds formed between solute and solvent are then similar to those that existed between solute particles and between solvent particles. Water, being polar, is therefore a good solvent for ionic and polar substances.

2 Propanol is a polar molecule with a hydroxyl group. The hydroxyl group is able to form hydrogen bonds with water molecules and will therefore dissolve in water. Propane is a non-polar hydrocarbon and so can only interact with other molecules by weak dispersion forces. It is not able to form hydrogen bonds with other molecules and so will not dissolve in water.

3 **a** Dissociation
b **i** $Cu^{2+}(aq)$, $NO_3^-(aq)$
ii $Zn^{2+}(aq)$, $SO_4^{2-}(aq)$
iii $NH_4^+(aq)$, $PO_4^{3-}(aq)$

4 **a** $CH_3OH(l) \xrightarrow{H_2O(l)} CH_3OH(aq)$
b $C_{12}H_{22}O_{11}(l) \xrightarrow{H_2O(l)} C_{12}H_{22}O_{11}(aq)$

5 **a** $MgSO_4(s) \xrightarrow{H_2O(l)} Mg^{2+}(aq) + SO_4^{2-}(aq)$
b $Na_2S(s) \xrightarrow{H_2O(l)} 2Na^+(aq) + S^{2-}(aq)$
c $KOH(s) \xrightarrow{H_2O(l)} K^+(aq) + OH^-(aq)$
d $(CH_3COO)_2Cu(s) \xrightarrow{H_2O(l)} 2CH_3COO^-(aq) + Cu^{2+}(aq)$
e $Li_2SO_4(s) \xrightarrow{H_2O(l)} 2Li^+(aq) + SO_4^{2-}(aq)$

6 Ion–dipole interaction. A potassium ion has a positive charge and so will attract the negative part (the oxygen atom) of a polar water molecule. Several water molecules will orient themselves around the potassium ion so that their oxygen atoms, which carry a partial negative charge, are closer to the potassium ion than the hydrogen atoms

7 C

8

i	Ionic compound	**e**	Dissolves in water by dissociating, then forming ion–dipole bonds with water
ii	Compound composed of polar molecules with –OH groups	**d**	Dissolves in water by forming hydrogen bonds with water
iii	Compound composed of small polar molecules in which a hydrogen atom is covalently bonded to an atom of a group 17 element	**b**	Dissolves in water by ionising then forming ion–dipole bonds with water
iv	Non-polar molecular compound	**c**	Does not dissolve in water
v	Compound composed of covalent molecules with a large non-polar end and one –OH group.	**a**	Does not dissolve in water due to the size of the molecule

9 Propan-1-ol has a polar hydroxyl group which will form hydrogen bonds to water molecules. These hydrogen bonds are strong enough to overcome the intermolecular bonds holding propan-1-ol molecules together, so the two liquids will mix. Once this happens, each propan-1-ol molecule will be surrounded by water molecules.

10 Thin layer chromatography is a technique that allows you to determine the **composition** and **purity** of different types of substances. In this technique, a thin layer of a solid **stationary phase** is applied to a plate. The **components** of the sample are carried over the surface of the stationary phase by the solvent, or **mobile phase**. The components separate, depending on the relative attractions of compounds towards the two phases. The individual components are seen as **spots** on the plate, which can be identified by calculating their $\boldsymbol{R_f}$ **values**.

11 **a** Water was **absorbed** by the towel as the wet swimmer dried himself.
A thin layer of grease **adsorbed** onto the cup when it was washed in the dirty water.
b Absorb: Atoms or molecules are taken *into* the material.
Adsorb: Atoms or molecules accumulate and bond weakly to the *surface* of a solid or liquid.

12 **a** I_2, CH_4 and C_2H_4
b $C_6H_{12}O_6$ and C_3H_7OH
c HI and HNO_3

13 Hexane. Since benzene is a non-polar solvent, it will dissolve non-polar solutes best. KCl is an ionic soid which therefore would not dissolve well. Glycerol and ethanol are relatively small molecules with polar –OH groups, so they would also not dissolve well in benzene.

14 **a** Methanol, butanol, pentane
b Pentane, butanol, methanol
c Methanol is a small molecule with a hydroxyl group. Methanol can form hydrogen bonds to other molecules and so will be readily soluble in water. It will not dissolve in non-polar solvents such as hexane, because the forces of attraction between methanol molecules will be too strong for the hexane molecules to break them. Pentane is a non-polar molecule and is unable to form hydrogen bonds to other molecules, so it is not soluble in water. It is, however, soluble in compounds that are non-polar like itself, and so will be readily soluble in hexane. Butanol has a hydroxyl group and so can form

hydrogen bonds to water, but it also has quite a long non-polar hydrocarbon chain, which reduces its solubility in water but enhances its solubility in non-polar solvents such as hexane. Butanol will be partially soluble in both water and hexane.

15 **a** Ammonia is a highly polar molecule and forms hydrogen bonds with water. It is therefore very soluble in water. Methane, however, is non-polar. Weak (dispersion) forces would occur between methane and water, but these are unable to disrupt the stronger hydrogen bonds between water molecules. Therefore, methane does not dissolve in water.

b Glucose dissolves in water because it has very polar—OH groups that can form hydrogen bonds with water molecules. Sodium chloride is ionic; hence there are ion–dipole attractions between the ions and water. These attractions are strong enough to overcome the attraction between the sodium ions and chloride ions in the solid NaCl lattice.

16 The R_f value is the ratio of the distance a component has moved from the origin to the distance from the origin to the solvent front.

$R_f(\text{blue}) = \frac{8}{10} = 0.8$

$R_f(\text{purple}) = \frac{6}{10} = 0.6$

$R_f(\text{yellow}) = \frac{2}{10} = 0.2$

17 R_f = distance dye has moved/distance solvent front has moved.

Blue dye $R_f = \frac{7.5}{9.0} = 0.83$

Red dye $R_f = \frac{5.2}{9.0} = 0.58$

18 **a** If the solvent were above the level of the origin, the compounds under test would dissolve and disperse throughout the solvent.

b Components in a mixture undergoing chromatography cannot move faster than the solvent that is carrying them over the stationary phase. R_f values must therefore be less than one.

c 2

d B: blue; C: green. They can be identified on the basis of their colour and R_f values.

e purple

f 0.63; 0.13

19 A component that appears near the top of a paper chromatogram is likely to be more soluble in the mobile phase and adsorb less strongly to the stationary phase than a component near the bottom of the chromatogram.

20 **a** 3.2 cm **b** 13 cm

c

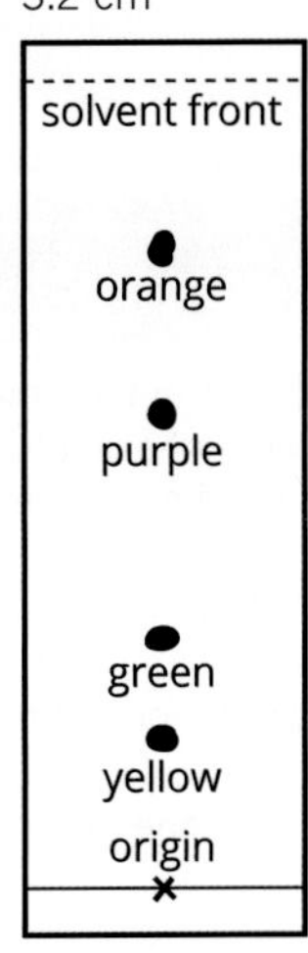

21 **a** Taurine, glycine and an unknown. Two of the three spots produced by the medicine match those produced by taurine and glycine. The third spot does not match any of the amino acid standards and represents an unknown substance.

b Two ways of visualising the spots are:
- viewing the chromatogram under UV light
- spraying the finished chromatogram with a compound that causes the amino acids to fluoresce.

c $R_f = 0.12$

d leucine

Unit 1 Area of Study 1 Review

1 C	**2** B	**3** C	**4** D	**5** C
6 C	**7** B	**8** B	**9** D	**10** A
11 C	**12** A	**13** A	**14** D	**15** B

16 **a** $1s^22s^22p^63s^2$

b **i** 0.160 nm

ii Na atom would have a larger radius as its effective nuclear charge is lower, pulling its electrons less strongly toward the nucleus.

c **i** It is a lattice of positively charged magnesium ions surrounded by a 'sea' of valence electrons. The lattice is held together by the electrostatic attraction between the valence electrons and cations.

ii The electrons are not localised; they are free to move and so can conduct an electric current.

d **i** Observations: bubbles of gas evolved and increase in temperature

$Mg(s) + 2HCl(aq) \longrightarrow MgCl_2(aq) + H_2(g)$

or $Mg(s) + 2H^+(aq) \longrightarrow Mg^{2+}(aq) + H_2(g)$

ii Any suitable example such as K, Na, Ca.

17 **a** **i** Ca **ii** Ar **iii** C **iv** Na or Mg

v Li **vi** N **vii** F

b Elements in group 1 are metals, which react by giving away electrons. Chemical reactivity increases because the outer-shell electrons are further from the nucleus as one moves down the group, and so are more readily released in a reaction, which makes the metals more reactive.

18 **a** Chlorine is on the right side of the periodic table and sodium is on the left. Atomic radius decreases across a period because the increasing effective nuclear charge pulls the outer-shell electrons more tightly to the nucleus, causing the radius of the atom to decrease.

b Fluorine is further to the right on the periodic table than lithium and effective nuclear charge increases from left to right across the periodic table. As effective nuclear charge increases, the electrons are held more tightly to the nucleus and more energy is required to remove the first one.

c Ba and Be are in the same group, with Be higher than Ba. Going down a group, the atom size is increasing, meaning the outer-shell electrons are further from the nucleus. The outer electrons of Be are, therefore, held more tightly and are less readily released.

d The *s*-block elements have an *s*-subshell as their outer occupied electron subshell. The *s*-subshell can take 1 or 2 electrons, so the block is only 2 groups wide.

19 **a** $Pb(NO_3)_2(aq) + 2KCl(aq) \longrightarrow PbCl_2(s) + 2KNO_3(aq)$

b $Pb^{2+}(aq) + 2Cl^-(aq) \longrightarrow PbCl_2(s)$

c **i** lead(II) chloride

ii potassium ions and nitrate ions

20 a The forces between the molecules of ice are intermolecular hydrogen bonds. The bonds between the H and O atoms within the water molecules are covalent bonds. Covalent bonds are much stronger than hydrogen bonds and so require much more energy, and thus a higher temperature, to break.

b Ethyne has the structure CHCH. The C atoms have a triple bond between them, each using three of their four valence electrons to form the triple bond. The fourth valence electron of each carbon atom forms a covalent bond with a hydrogen atom. There are no lone pairs and only two electron groups on the carbons. So these adopt a linear arrangement.

Hydrogen peroxide has the following structure:

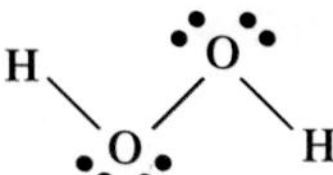

Each oxygen has six valence electrons. Two are involved in bonding leaving two pairs of non-bonding electrons. So there are four electron groups around each oxygen atom. These assume a tetrahedral arrangement to minimise repulsion. So the molecule is not linear.

21 a H:N:H (with H below N; electron-dot structure of ammonia)

b The four electron pairs form a tetrahedral arrangement around the atom due to the repulsion of the electron pairs. The result is a triangular pyramid shaped molecule.

covalent bond

hydrogen bond

c i :N≡N: **ii** :O=C=O:

d i N_2 has no polar bonds, so the intermolecular bonds are weak dispersion forces.

Although the bonds within CO_2 are polar, the molecule overall is symmetrical, so the dipoles cancel and the overall molecule is non-polar. Therefore, the only intermolecular forces are dispersion forces.

ii Nitrogen is a highly electronegative element; consequently, the bonds between the atoms of nitrogen and hydrogen are highly polarised. The ammonia molecule is a dipole because its shape is not symmetrical. There is an electrostatic attraction between the nitrogen atom of one ammonia molecule and the hydrogen atom of a nearby ammonia molecule. These attractions between these dipoles are known as hydrogen bonds.

22 a :O=O:

b The intramolecular bonds in oxygen are stronger because they are double covalent bonds, as opposed to the single covalent bonds in hydrogen peroxide.

c i Dispersion forces

ii Hydrogen bonds, which are significantly stronger than the dispersion forces between oxygen molecules. Dispersion forces also exist between molecules of hydrogen peroxide.

Because the hydrogen peroxide molecule is larger than that of oxygen, the dispersion forces between hydrogen peroxide molecules are stronger than those between oxygen molecules.

d Oxygen has six valence electrons, so achieves stability when it forms two covalent bonds. Nitrogen only has five valence electrons, so forms three covalent bonds to achieve stability.

23 a Allotropes are different physical forms of the same element.

b Both consist of carbon atoms covalently bonded to other carbon atoms.

c Diamond is a three-dimensional lattice in which each carbon atom is covalently bonded to four other carbon atoms in a tetrahedral configuration. So strong bonding extends throughout the lattice.

Graphite consists of layers of carbon atoms in which each atom is covalently bonded to three other carbon atoms, making strong layers. There are weaker dispersion forces between the layers. The one electron of each carbon atom not involved in bonding is delocalised.

d The delocalised electrons in graphite are free to move and conduct electricity. In diamond, each carbon atom is bonded with four other carbon atoms so there are no free electrons.

e Because of the weak bonding between the layers of graphite, the layers can slide over one another and thus can slide onto a page.

24 a i $MgCl_2(s) \xrightarrow{H_2O(l)} Mg^{2+}(aq) + 2Cl^-(aq)$

ii ionic bonds

iii ion–dipole interactions

b i $C_2H_5OH(l) \xrightarrow{H_2O(l)} C_2H_5OH(aq)$

ii hydrogen bonds and dispersion forces

iii hydrogen bonds and dispersion forces

c The magnesium chloride solution would be the better conductor as many charged particles in the form of $Mg^{2+}(aq)$ and $Cl^-(aq)$ ions are present in the solution after the $MgCl_2$ has dissolved; the dissolved ethanol does not contain charged particles.

25 a i no change

ii A white precipitate, magnesium carbonate, is formed.

iii A green precipitate, copper (II) carbonate, is produced.

iv no change

v no change

vi no change

b ii $Mg^{2+}(aq) + CO_3^{2-}(aq) \rightarrow MgCO_3(s)$

iii $Cu^{2+}(aq) + CO_3^{2-}(aq) \rightarrow CuCO_3(s)$

c

	Sodium carbonate	Potassium nitrate	Magnesium nitrate	Copper(II) nitrate
Sodium carbonate	x	x	white precipitate	green precipitate
Potassium nitrate	x	x	x	x
Magnesium nitrate	white precipitate	x	x	x
Copper(II) nitrate	green precipitate	x	x	x

Sodium carbonate solution will react with two of the other solutions. It will form a white precipitate with one solution, and a green precipitate with another. No reaction will be observed with the third solution.

Potassium nitrate solution will not react with any of the other three solutions.

Magnesium nitrate solution will only react with one of the other solutions, forming a white precipitate.

Copper(II) nitrate solution will only react with one of the other solutions to form a green precipitate.

26 The Rf value for the blue colour is:

$$\frac{\text{distance travelled by component}}{\text{distance travelled by solvent front}} = \frac{9}{12} = 0.75$$

When the blue spot has moved 15 cm, the solvent will have moved $\frac{15.0}{0.75} = 20$ cm

27 **a**

```
   ••
H ×• P •• H
   •×
   H
```

b PH_3 is a polar molecule. It is asymmetrical, so it has an overall molecular dipole. (The P–H bond is not particularly polar, but the overall asymmetry of the molecule results in an overall molecular dipole.)

c In a PH_3 molecule, there are four pairs of electrons around the central P atom. These electron pairs adopt a tetrahedral geometry. Since there is one lone pair, the molecular shape is a triagonal pyramid.

d $M_r(PH_3) = 34.0$
$M_r(NH_3) = 17.0$

e Since H is bonded to N in ammonia, NH_3 molecules are able to form hydrogen bonds between molecules. Between PH_3 molecules there are dipole–dipole forces. Hydrogen bonding is stronger than dipole–dipole forces, so NH_3 has the higher melting point.

f $^{32}_{15}P$

g $1s^2 2s^2 2p^6 3s^2 3p^3$

28 **a** A: 0.60
B: 0.48
C: 0.38
D: 0.20
E: 0.10

b A: leucine and/or isoleucine
B: β-phenylalanine
C: proline and/or valine and/or tyrosine
D: threonine and/or hydroxyproline and/or serine and/or glycine
E: lysine and/or arginine and/or taurine

c A: leucine and/or isoleucine
B: β-phenylalanine
C: proline
D: serine
E: arginine

d In the second run, the two solvents could not clarify the identity of A, but did help identify C, D and E. A two-way chromatogram produces better separation of components of complex mixtures, permitting easier isolation and identification.

29 **a** Sugar ($C_{12}H_{22}O_{11}$) and ethanol (CH_3CH_2OH) dissolve by forming hydrogen bonds with water. Salt (NaCl) dissolves by dissociating into ions, and forming ion–dipole interactions with water. Hydrochloric acid dissolves by ionising to form H_3O^+ ions and Cl^- ions.

b The conductivity of the solution.

c The solute used.

d Hydrochloric acid is a strong acid, and ionises completely. Thus, when in solution it will produce many ions, which will lead to a higher conductivity. Vinegar is a weak acid that partially ionises. Fewer ions in solution will lead to a lower conductivity.

e Ethanol and sugar both dissolve by forming hydrogen bonds with water. No ions are formed in this process, therefore conductivity gives no indication of solubility for these substances.

f The design of the experiment does not investigate the stated aim, therefore the conclusion is not valid. The aim of the experiment is to investigate the different *ways* substances dissolve, yet the conclusion is about the relationship between solubility and conductivity.

g For example:
- Change the aim and hypothesis of the investigation.
- Use the same amount of each solute.
- Use all ionic substances, as these all dissolve in the same way.

30 **a** $1s^2$

b Group 18, period 1. Helium is the first of the noble gases.

c Helium would be expected to be highly unreactive, due to its position at the top of the noble gases group.

d Helium-3 and helium-4 both contain 2 protons and 2 electrons. However, helium-3 contains 1 neutron, whereas helium-4 contains 2 neutrons.

e Helium exists as single atoms. The only forces between these atoms are dispersion forces. Furthermore, as helium is very small, these dispersion forces are very weak. As the forces between helium atoms are so weak, it has both a low melting point and a low boiling point.

f Low density makes it good for buoyancy; Low reactivity means it provides a safe atmosphere and it can be a carrier gas without interfering with the substance being analysed; liquid helium exists at temperatures below −269°C, making it useful in medical and scientific research that requires extremely low temperatures.

g **i** A linear economy operates on a 'take-make-dispose' model. There is little attempt to recapture, reuse or recycle materials. The circular economy model focuses on the optimal use and reuse of resources from the extraction of raw materials to production then consumption.

ii Low reactivity means that it is not possible to capture the used helium in a compound; its low density means that once released, helium continues to rise until it leaves Earth's atmosphere. These properties make it difficult to use helium in a way described by the circular economy model.

h **i** Hydrogen is the only gas with a density lower than helium, thus it would be a good replacement in balloons and airships (blimps); hydrogen can be produced from other substances, making it more sustainable.

ii Hydrogen is highly reactive and explosive. It would need to be kept away from oxygen and sources of ignition.

Chapter 7 Quantifying atoms and compounds

7.1 Relative mass

TY 7.1.1 $A_r(B) = 10.81$

TY 7.1.1 Abundance of 62.95 isotope = 70.50%
Abundance of 64.95 isotope = 29.50%

1 B

2 ^{39}K would be most abundant. The relative atomic mass of 39.1 is closest to the mass number of 39. There needs to be a majority of isotopes with the mass number 39 for the weighted average to be just above 39.

3 B

4 $A_r(H) = 1.008$

5 8.00%.

6 **a** % abundance ^{90}Zr = 51%
% abundance ^{91}Zr = 11%
% abundance ^{92}Zr = 17%
% abundance ^{94}Zr = 17%
% abundance ^{96}Zr = 4.1%

b $A_r(Zr) = 90$

7.2 Avogadro's constant

TY 7.2.1 $N(CO_2) = 9.6 \times 10^{23}$ molecules

TY 7.2.2 $n(H) = 1.5$ mol

TY 7.2.3 $N(H) = 8.4 \times 10^{23}$ atoms

TY 7.2.4 $N(H) = 1.5 \times 10^{24}$ atoms

TY 7.2.5 $n(Mg) = \frac{N}{N_A} = 0.0013$ mol

1 $6.022\,140\,76 \times 10^{23}$

2 **a** amount of substance counted in moles
  b actual number of particles
  c Avogadro's constant, i.e. 6.02×10^{23}

3 It could mean one mole of hydrogen atoms (H) or one mole of hydrogen molecules (H_2).

4 **a** $N(Na) = 1.8 \times 10^{24}$ atoms
  b $N(Fe) = 9.0 \times 10^{21}$ atoms
  c $N(CO_2) = 1.72 \times 10^{19}$ molecules

5 **a** $n(H_2O$ molecules$) = 1.0$ mol
  b n(Ne atoms) = 0.42 mol
  c n(ethanol molecules) = 53 mol

6 **a** n(Cl atoms) = 0.80 mol
  b n(H atoms) = 4.8 mol
  c n(O atoms) = 6.0 mol

7 **a** 1.3 mol **b** 0.23 mol
  c 3.3×10^{-4} mol **d** 0.66 mol

7.3 Molar mass

TY 7.3.1 $M_r = 63.0$ g mol^{-1}

TY 7.3.2 $n(CH_4) = 6.25$ mol

TY 7.3.3 $m(Na_2CO_3) = 496$ g

CSA: The sting of a bee

1 130.0 g mol^{-1} **2** 7.7×10^{-9} mol

3 4.6×10^{15} molecules

TY 7.3.4 $N(C_{12}H_{22}O_{11}) = 7.4 \times 10^{21}$ molecules

Key questions

1 The molar mass of a compound can be calculated by adding the relative atomic masses for each atom present in the compound's formula. This value is then expressed in grams per mol (g mol^{-1}).

2 $m = n \times M$

3 **a** 71.0 g mol^{-1} **b** 176.0 g mol^{-1} **c** 249.6 g mol^{-1}

4 **a** 96.0 g **b** 24.0 g **c** 255 g

5 **a** 2.5 mol **b** 0.37 mol **c** 0.28 mol

6 **a** 6.0×10^{23} atoms
  b 6.0×10^{22} atoms
  c 6.0×10^{21} atoms

7 **a** 5.0 mol **b** 1.35 mol **c** 3.9×10^{-5} mol

8 **a** **i** 9.00×10^{23} molecules
  ii 1.1×10^{24} molecules
  b 6.02×10^{22} atoms
  c 2.4×10^{25} atoms

7.4 Percentage composition, empirical and molecular formulas

TY 7.4.1 35.0%

TY 7.4.2 CH_2O

CSA: Analysing a life-saving rat poison

1 The mass of oxygen in the organic compound can be determined by subtracting the mass of carbon plus the mass of hydrogen from the original mass of the compound:
$m(O) = m(\text{compound}) - m(C) - m(H)$

2

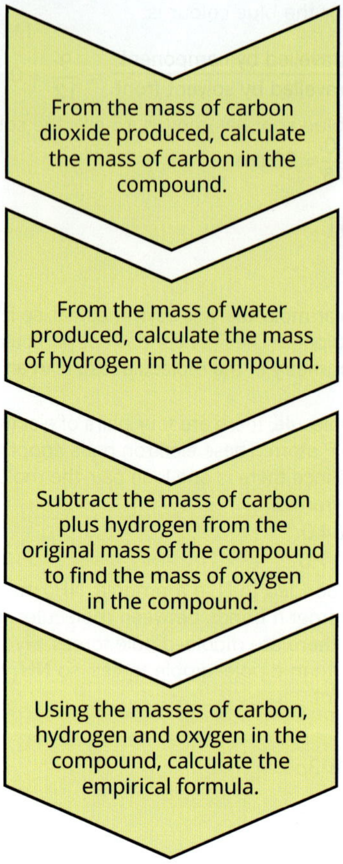

From the mass of carbon dioxide produced, calculate the mass of carbon in the compound.

From the mass of water produced, calculate the mass of hydrogen in the compound.

Subtract the mass of carbon plus hydrogen from the original mass of the compound to find the mass of oxygen in the compound.

Using the masses of carbon, hydrogen and oxygen in the compound, calculate the empirical formula.

3 $C_{19}H_{16}O_4$.

TY 7.4.3 C_4H_{10}

Key questions

1 The percentage composition of a given compound tells you the proportion by mass of the different elements in that compound.

2 $C_6H_{12}O_6$, because the ratio of the atoms could be simplified to CH_2O while still using whole numbers. The molecular formula of a molecule with this molecular formula would be CH_2O.

3 **a** 26.2% **b** 52.16% **c** 46.8%

4 **a** HCl **b** CO **c** $C_5H_{10}O_2$ **d** CH_4

5 CH_2

6 **a** C_6H_6 **b** H_2O_2 **c** $C_3H_6O_3$ **d** NO_2 **e** $C_{11}H_{22}$

7 **a** CH_2O **b** $C_6H_{12}O_6$

Chapter 7 review

1 C

2 The relative atomic mass of carbon is the weighted average of the isotopic masses of isotopes of an element. Relative isotopic masses are for the mass of individual isotopes. For example, isotopes of carbon have different relative isotopic masses (i.e. ^{12}C, ^{13}C and ^{14}C). Small amounts of ^{13}C and ^{14}C make the relative atomic mass of carbon slightly greater than 12.

3 g mol^{-1}

4 **a** The relative isotopic mass of an isotope is the mass of an atom of that isotope relative to the mass of an atom of ^{12}C, taken as 12 units exactly. For example, the relative isotopic mass of the lighter of the two chlorine isotopes is 34.969.
  b The relative atomic mass of an element is the weighted average of the relative masses of the isotopes of the element on the ^{12}C scale. For example, the relative atomic mass of boron is 10.81.
  c The molar mass of an element is the mass of 1 mol of the element. It is equal to the relative atomic mass of the element expressed in grams. For example, the molar mass of magnesium is 24.31 g mol^{-1}. Note that relative atomic

mass and molar mass of an element are numerically equal. However, relative atomic mass has no units because it is the mass of one atom of the element compared with the mass of one atom of the carbon-12 isotope.

5 The empirical formula of a compound gives the simplest whole-number ratio of elements in that compound. The molecular formula gives the actual number of each type of element in one molecule of the compound.

6 $A_r(Pd) = 106.4$

7 **a** $A_r(Ar) = 39.96$
$A_r(K) = 39.11$
b Although potassium atoms have one more proton than argon atoms, the most abundant isotope of argon has 22 neutrons, giving it a relative atomic mass close to 40. The most abundant isotope of potassium has only 20 neutrons, giving it a relative atomic mass close to 39.

8 **a** Peak heights: $^{50}Cr = 0.3$ units, $^{52}Cr = 12$ units, $^{53}Cr = 1$ unit, $^{54}Cr = 0.2$ units, Total height = 13.5 units
Percentages: $^{50}Cr = 2.2\%$, $^{52}Cr = 88.9\%$, $^{53}Cr = 7.4\%$, $^{54}Cr = 1.5\%$
b $A_r(Cr) = 52$

9 **a** 0.748 mol **b** 15.0 mol
c 3.8×10^4 mol **d** 1.7×10^{-24} mol

10 B

11 **a** **i** 8.73×10^{23} molecules **ii** 3.49×10^{24} atoms
b **i** 3.47×10^{23} molecules **ii** 1.04×10^{24} atoms
c **i** 9.21×10^{21} molecules **ii** 4.61×10^{22} atoms
d **i** 1.5×10^{24} molecules **ii** 6.8×10^{25} atoms

12 **a** $M(Na) = 23.0\ g\ mol^{-1}$
b $M(HNO_3) = 63.0\ g\ mol^{-1}$
c $M(Mg(NO_3)_2) = 148.3\ g\ mol^{-1}$
d $M(FeCl_3{\cdot}6H_2O) = 270.3\ g\ mol^{-1}$

13 **a** 2.4 g **b** 81 g **c** 0.21 g **d** 340 g

14 **a** 1.0 mol **b** 0.389 mol
c 0.002 78 mol **d** 2.7×10^4 mol

15 **a** 93.8% **b** 20% **c** 60.0%

16 **a** CO **b** CO_2 **c** C_2H_4O **d** C_6H_5Cl

17 $C_2H_5NO_2$

18 **a** $C_4H_5N_2O$ **b** $M = 194\ g\ mol^{-1}$
c The molar mass of one empirical formula unit
$(C_2H_5NO_2) = 4 \times 12.0 + 5 \times 1.0 + 2 \times 14.0 + 16.0$
$= 97.0\ g\ mol^{-1}$

The number of empirical formula units in the molecular formula $= \frac{194}{97.0} = 2$

The molecular formula is $C_8H_{10}N_4O_2$.
d 0.005 15 mol

19 Proportions of the isotopes are 48.0% and 52.0%.

20 **a** 6.67×10^{-23} g **b** 3.0×10^{-23} g **c** 7.3×10^{-23} g

21 $m(Fe)$ needed = 62.0 g

22 **a** **i** $n(NaCl) = 0.100$ mol **ii** $n(Na^+) = 0.100$ mol
$n(Cl^-) = 0.100$ mol
b **i** $n(CaCl_2) = 0.405$ mol **ii** $n(Ca^{2+}) = 0.405$ mol
$n(Cl^-) = 0.810$ mol
c **i** $n(Fe_2(SO_4)_3) = 0.00420$ mol **ii** $n(Fe^{3+}) = 0.00840$ mol
$n(SO_4^{2-}) = 0.0126$ mol

23 **a** $M(\text{substance}) = \frac{72}{0.5} = 144\ g\ mol^{-1}$
b $M(\text{substance}) = \frac{10}{0.1} = 100\ g\ mol^{-1}$

24 C

25 **a** $M(\text{antibiotic}) = 1.25 \times 10^4\ g\ mol^{-1}$ **c** 9.6×10^{16} molecules
b $n(\text{antibiotic}) = 1.6 \times 10^{-7}$ mol

26 $M(Ni) = \frac{0.370}{0.0573} = 58.9\ g\ mol^{-1}$

27 **a** C_2H_6O **b** C_2H_6O

28 **a** D, F, E, A, C, B
b

	Metal	Oxygen
Mass (g)	0.542	0.216
Relative atomic mass	40.1	16.0
Moles	0.0135 mol	0.0135 mol
Ratio	1	1

c calcium (relative atomic mass of 40.1)

Chapter 8 Organic compounds

8.1 Organic materials

1 Compounds such as petrol, polymers and cosmetics are **carbon**-based compounds. Many of these compounds are currently produced from **crude oil**, which is a mixture made up of the remains of marine microorganisms, such as **bacteria** and plankton that died millions of years ago. The great age of these deposits explains why petrol is called a **fossil** fuel. Crude oil is a **non-renewable** resource because no more carbon is being added to the environment.

2 Organic compounds are carbon-based compounds which typically also have hydrogen and oxygen in them. They may also contain nitrogen, sulfur, phosphorus and halogens (fluorine, chlorine, bromine, iodine).

3 Photosynthesis is the name of the process by which carbon dioxide is changed by plants into glucose. This involves the following reaction:

$$6CO_2(g) + 6H_2O(l) \xrightarrow{\text{UV light, chlorophyll}} C_6H_{12}O_6(aq) + 6O_2(g)$$

Water and carbon dioxide are converted by a series of complex reactions in the presence of UV light and the molecule chlorophyll into glucose and oxygen.

4 Organic chemicals that are made from crude oil are non-renewable because crude oil is a fossil fuel and is not being made at a rate that is anywhere close to the rate at which we use it.

5 The glucose comes from photosynthesis in plants. This is a renewable process as the glucose can be replaced at a rate that is similar to the rate at which we use it, so the polyethene made from ethanol derived from glucose is also renewable.

6 **a** If the polymers are recycled, rather than thrown away, the organic compounds are reused, rather than new hydrocarbons being refined from crude oil.
b If the government insists by legislation that petrol and diesel-fuelled cars are replaced by electric cars, then it will have to make this possible by providing infrastructure, while the non-renewable resource, crude oil, will be saved for uses where it cannot be replaced.
c Since crude oil is a non-renewable resource, it is important to find other ways to make the organic chemicals that we require, so replacement materials, such as plant-based biomass, need to be found.

8.2 Hydrocarbons

TY 8.2.1 3-methylpentane

TY 8.2.2 2,3-dimethylbut-1-ene

1 **a** propane **b**

```
    H   H   H
    |   |   |
H — C — C — C — H
    |   |   |
    H   H   H
```

c $CH_3CH_2CH_3$ **d** $C_3H_8(g) + 5O_2(g) \rightarrow 3CO_2(g) + 4H_2O(l)$

2 a alkanes: C_nH_{2n+2}; alkenes: C_nH_{2n}

b Alkanes have only C–C single bonds, so have the largest ratio of hydrogen atoms to carbon atoms; alkenes have one C=C double bond, so have fewer hydrogens per carbon than do alkanes.

3 a Structural isomers of a compound, such as C_5H_{12}, have the same molecular formula, but they differ from each other by having a different arrangement of the carbon atoms in the molecule, and hence a different structure.

b

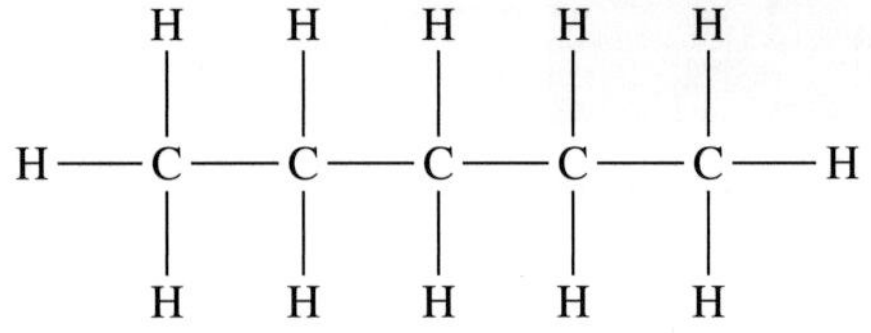

pentane

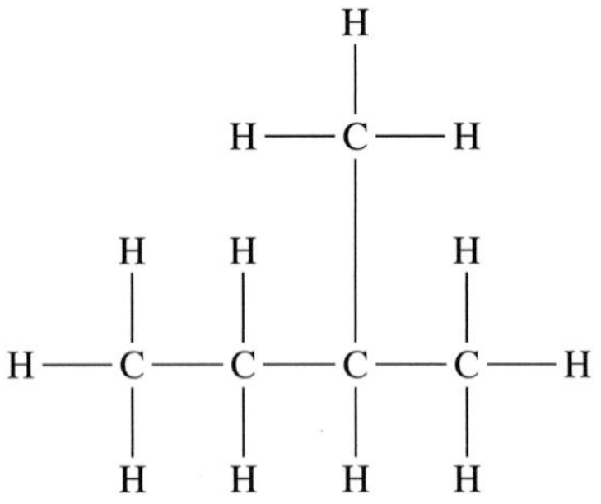

2-methylbutane

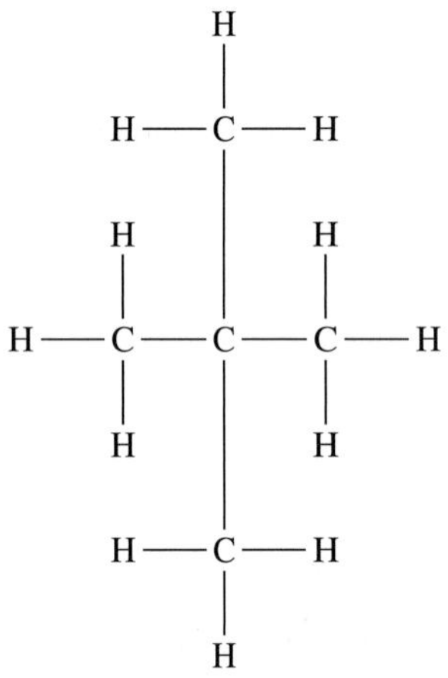

2,2-dimethylpropane

4 a pent-1-ene

b $CH_2{=}CHCH_2CH_2CH_3$ or $CH_3CH_2CH_2CH{=}CH_2$ or $CH_3CH_2CH_2CHCH_2$ or $CH_2CHCH_2CH_2CH_3$

5 a but-2-ene

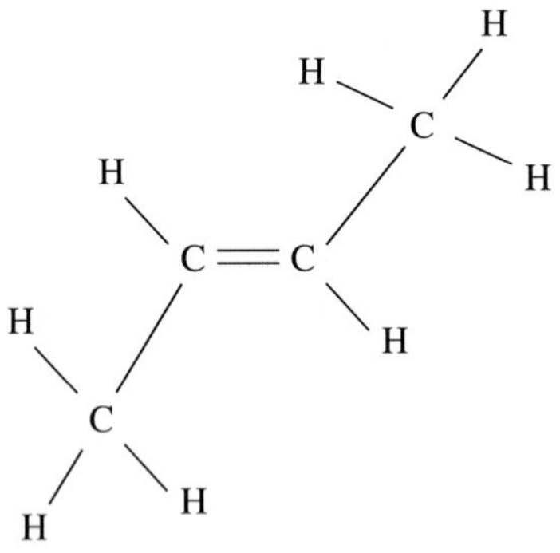

b 4-methylpent-1-ene

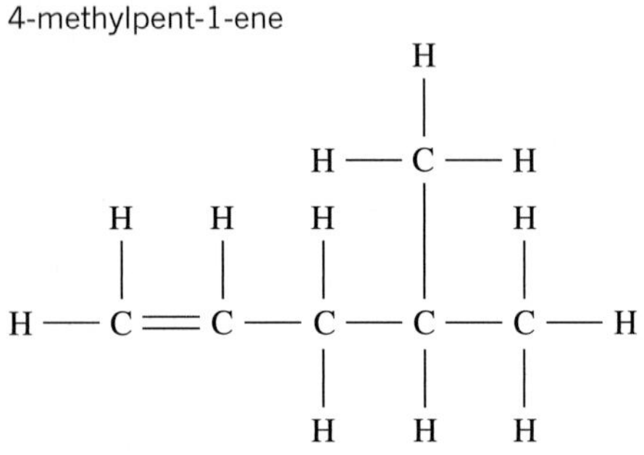

6 a hexane.
b 3-methylhexane
c 2,4-dimethylhexane
d 2,2-dimethylbutane

7 a hexane

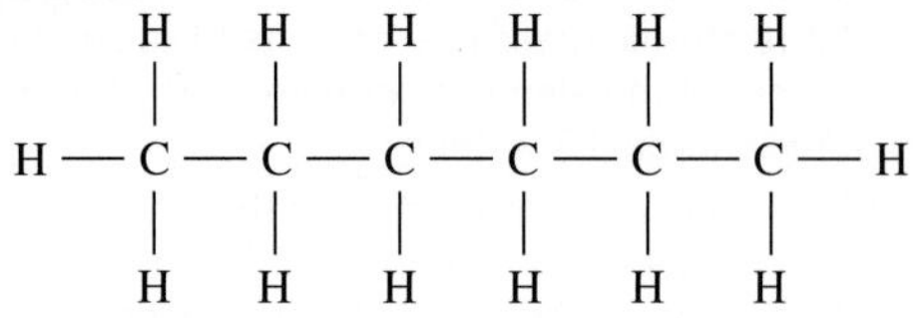

b 3-methylhexane

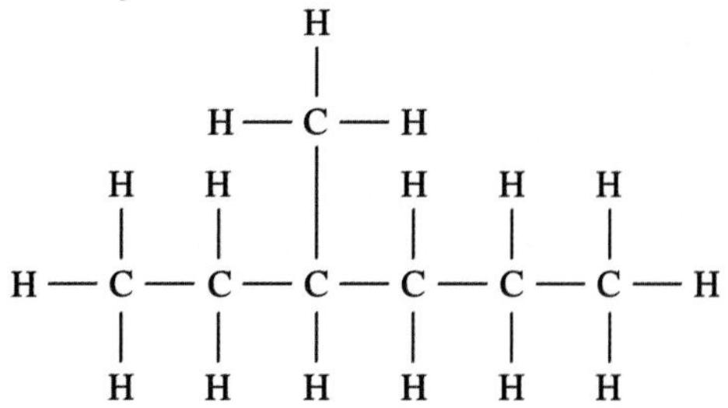

c 3,3-dimethylpentane

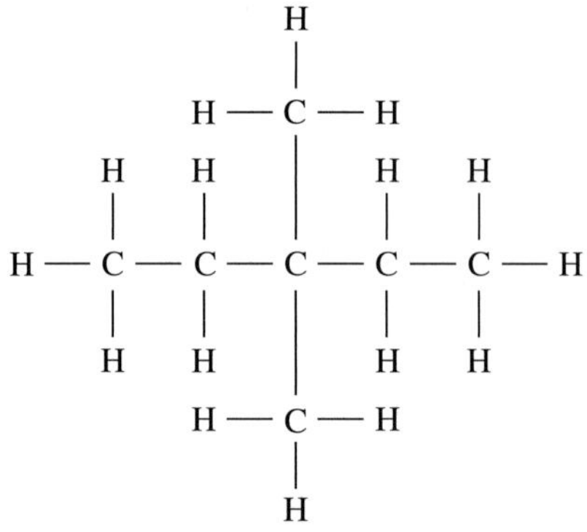

d 3-ethyl-2-methylpentane

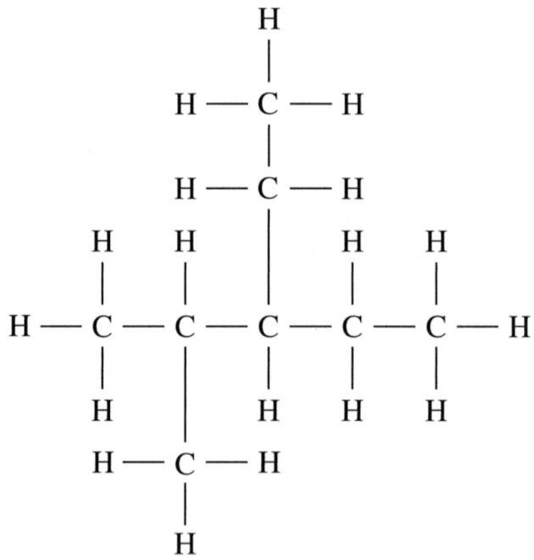

8 a

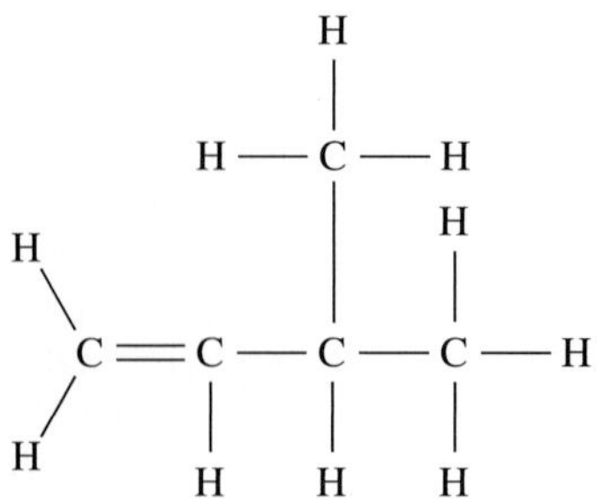

b C_5H_{10}

c

```
 H     H   H   H   H
  \    |   |   |   |
   C = C — C — C — C — H
  /        |   |   |
 H         H   H   H
```

pent-1-ene

```
    H   H   H       H
    |   |   |       |
H — C — C = C — C — C — H
    |           |   |
    H           H   H
```

pent-2-ene

```
        H
        |
    H — C — H
        |   H   H
 H      |   |   |
  \     |   |   |
   C =  C — C — C — H
  /         |   |
 H          H   H
```

2-methylbut-1-ene

```
            H
            |
        H — C — H
    H       |       H
    |       |       |
H — C — C = C — C — H
    |       |   |
    H       H   H
```

2-methylbut-2-ene

d 3-methylbut-2-ene is actually the same molecule as 2-methylbut-2-ene, but is just being viewed from the other end of the molecule. IUPAC rules specify the the carbon atoms in the chain are numbered from the end of the chain that will give the smallest numbers to double-bonded carbon atoms.

e The molecular formula of this compound is C_5H_{10}, so all isomers must have the same molecular formula. The proposed isomer, dimethylpropene, would have the molecular formula C_5H_{11}, and it would have 5 bonds around the 2nd carbon atom in the chain, which is not possible.

9 a i

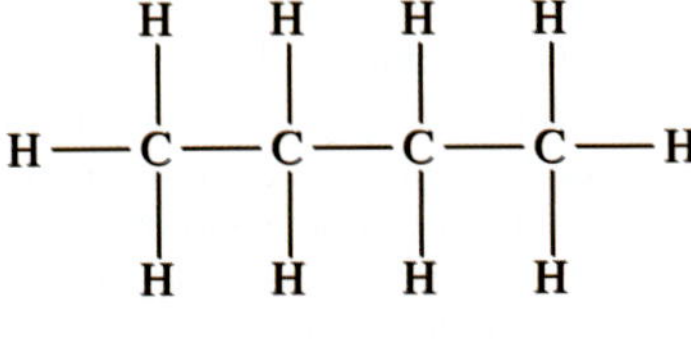

$CH_3(CH_2)_2CH_3$

ii It is not possible to have a methyl branch (or any branch) on the first carbon, so 1-methyl is not a branch, it is the continuation of the main carbon chain.

iii butane

b i

```
      CH3   CH3
      |     |
CH3 — CH —  CH — CH3
```

2,3-dimethylbutane

And

```
                  CH3
                  |
CH3 — CH2 —  C — CH3
                  |
                  CH3
```

2,2-dimethylbutane

ii The methyl branches are not numbered. These could both be on carbon number 2, or one could be on carbon number 2 and the other on carbon number 3.

iii 2,2-dimethylbutane or 2,3-dimethylbutane

c i 

```
        H
        |
    H — C — H
    H   |   H   H
    |   |   |   |
H — C — C — C — C — H
    |   |   |   |
    H   H   H   H
```

ii The smallest possible number should always be selected for numbering a branch. Carbon number 3 from one end of the chain is actually carbon number 2 from the other end.

iii 2-methylbutane

10

Alkene	Incorrect semi-structural formula	Mistake	Correct semi-structural formula
but-2-ene	$CH_2CHCH_2CH_3$	Double bond is in the wrong place. Formula given is for but-1-ene	$CH_3CHCHCH_3$
2-methylprop-1-ene	$CH_2CH(CH_3)_2$	Five bonds around carbon number 2. Extra H on carbon number 2.	$CH_2C(CH_3)_2$
2,3-dimethylpent-2-ene	$CH_3C(CH_3)CCH_2CH_3$	Missing methyl group on carbon number 3.	$CH_3C(CH_3)C(CH_3)CH_2CH_3$

8.3 Haloalkanes

TY 8.3.1 2-bromo-3-methylpentane

CSA: Haloalkanes and the ozone layer

1 A free radical has an unpaired electron. Covalent bonds are formed when two unpaired electrons are shared between two atoms, so a free radical is very reactive because it has an electron already available to form a bond, whereas before new bonds can be formed in other reactions the existing bonds have to be broken.

2 Because higher energy light is needed to break the bonds in O_2, compared to O_3, this suggests that the bonds in ozone are weaker than the covalent double bond in an oxygen molecule.

3 The preferential breaking of the carbon–chlorine bond suggests that this bond is not as strong as the carbon–fluorine bond, which does not break.

4 The four isomers are 1-chlorobutane, 2-chlorobutane, 1-chloro2methylpropane and 2-chloro-2-methylpropane.

Key questions

1 **a** CH_3CH_2Cl
b CH_3Br
c CH_3CHICH_3

2 **a** 1-bromopropane

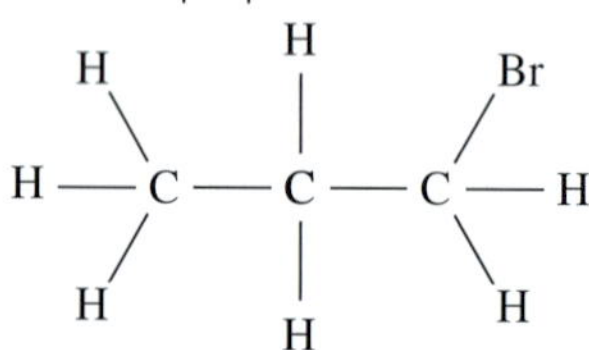

b 2-chlorobutane

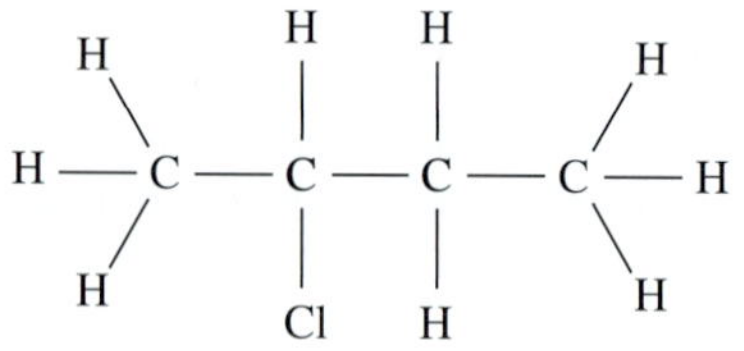

c 1,2-dichloroethane

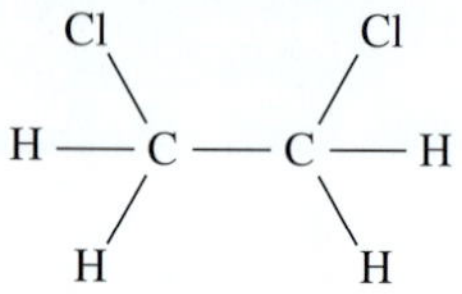

d 1,1-dibromoethane

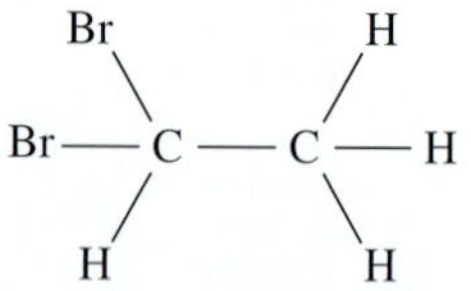

3 **a** 2-chloro-2-iodopropane
b 1,1,2-tribromoethane
c tetrachloromethane

4 1-chlorobutane, 2-chlorobutane, 1-chloro-2-methylbutane, and 2-chloro-2-methylbutane

5 Because the bromine atom could be bonded to carbon number 1 or carbon number 2, a number must be included in the name of bromobutane. So, it must be either 1-bromobutane or 2-bromobutane.

6 Having two chlorine atoms in the molecule increases the possibilities for isomers significantly. The chlorine atoms may be on the same carbon, on adjacent carbons, or even on the two carbons at either end of the molecule. When there is just one chlorine atom in the molecule, there are only two isomers.
For C_3H_7Cl the isomers are: $CH_3CH_2CH_2Cl$ (1-chloropropane) and $CH_3CHClCH_3$ (2-chloropropane).
For $C_3H_6Cl_2$ the isomers are: $CHCl_2CH_2CH_3$ (1,1-dichloropropane), $CH_2ClCHClCH_3$ (1,2-dichloropropane), $CH_2ClCH_2CH_2Cl$ (1,3-dichloropropane), $CH_3CCl_2CH_3$ (2,2-dichloropropane).

7 **a** The molecule is polar because there is one C–Br bond, which is polar, and so the molecule is asymmetrical.
b The molecule is polar because there is one C–Cl bond, which is polar, and so the molecule is asymmetrical.
c The molecule is polar because there are two C–Br bonds, which are polar, on the same carbon, making the molecule asymmetrical.
d The molecule is polar because the C–Cl bond is more polar than the C–I bond, (chlorine is more electronegative than iodine), so even though there is a polar bond on each carbon, these are not equal, so do not cancel out. The molecule is asymmetrical.

8.4 Alcohols and carboxylic acids

TY 8.4.1 propan-2-ol
TY 8.4.2 2,4-dimethylpentan-3-ol
TY 8.4.3 5-methylhexanoic acid

1 The hydroxyl, –OH, functional group forms hydrogen bonds with water molecules. Because methanol and ethanol have only one and two carbon atoms respectively, the non-polar (alkyl) part of the molecule is not large enough to overcome this attraction.

2 **a** The presence of the hydroxyl functional group enables stronger intermolecular forces to be formed between alcohol molecules. This attractive force is stronger than the dispersion forces between alkane molecules, so more energy is required to separate alcohols than alkanes with the same number of carbon atoms, hence the boiling points of alcohols are higher.
b As the length of the carbon chain in alcohols increases, the strength of the dispersion forces between the molecules increases (number of electrons in the molecules increases), so more energy is required to separate the molecules and so the boiling points of the alcohols increase.

3 **a** $C_nH_{2n+1}OH$ or $C_nH_{2n+2}O$
b The molecule $C_5H_{12}OH$ is not a member of the alcohol homologous series because it has too many hydrogen atoms in its formula. With 5 carbon atoms, it should have the formula $C_5H_{11}OH$ to belong to the alcohol homologous series.

4

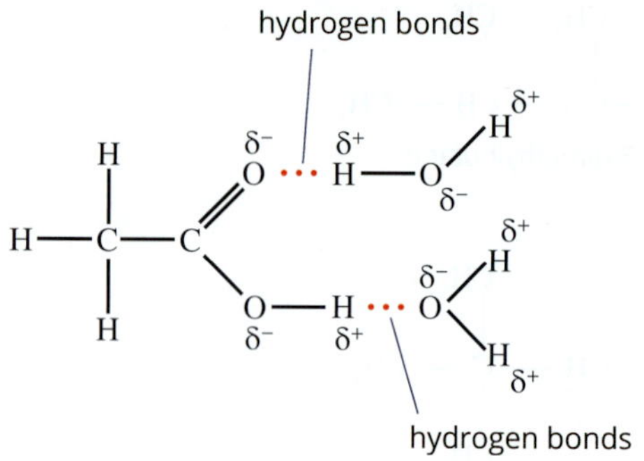

hydrogen bonds

hydrogen bonds

5 **a** $CH_3CH_2CH_2COOH$
b $CH_3CH(CH_3)COOH$
c $CH_3CH_2CH(CH_3)CH(CH_3)COOH$

6 **a** propan-2-ol
b 2-methylbutan-2-ol
c 2-methylbutanoic acid
d 3,4-dimethylpentanoic acid

7 **a** pentan-3-ol

```
      H    H    H    H    H
      |    |    |    |    |
H —  C —  C —  C —  C —  C — H
      |    |    |    |    |
      H    H    O    H    H
                |
                H
```

b heptanoic acid

```
      H     H     H     H     H     H         O
      |     |     |     |     |     |        //
H —  C  —  C  —  C  —  C  —  C  —  C  —  C
      |     |     |     |     |     |        \
      H     H     H     H     H     H         O — H
```

c 3-methylpentan-1-ol

```
             H    H    H    H    H
             |    |    |    |    |
H — O —  C —  C —  C —  C —  C — H
             |    |    |    |    |
             H    H    |    H    H
                  H —  C — H
                       |
                       H
```

d 2,5-dimethylhexan-3-ol

```
                               H
                    H          |
                    |     H —  C — H
      H    H        O    H     |    H
      |    |        |    |     |    |
H —  C —  C —  C —  C —  C —  C — H
      |    |        |    |     |    |
      H    |        H    H     H    H
      H —  C — H
           |
           H
```

e 3-methylbutanoic acid

```
      H     H     H     O
      |     |     |     ||
H —  C  —  C  —  C  —  C — O — H
      |     |     |
      H     |     H
      H  —  C — H
            |
            H
```

8 **a** The correct name is butan-1-ol.

```
      H    H    H    H
      |    |    |    |
H —  C —  C —  C —  C — O — H
      4    3    2    1
      |    |    |    |
      H    H    H    H
```

b The correct name is 2-methylheptan-3-ol.

```
      H
      |
H —  C — H
      1
      |
      |    OH   H    H    H    H
      |    |    |    |    |    |
H —  C —  C —  C —  C —  C —  C — H
      2    3    4    5    6    7
      |    |    |    |    |    |
      |    H    H    H    H    H
H —  C — H
      |
      H
```

c The correct name is 3-methylbutanoic acid.

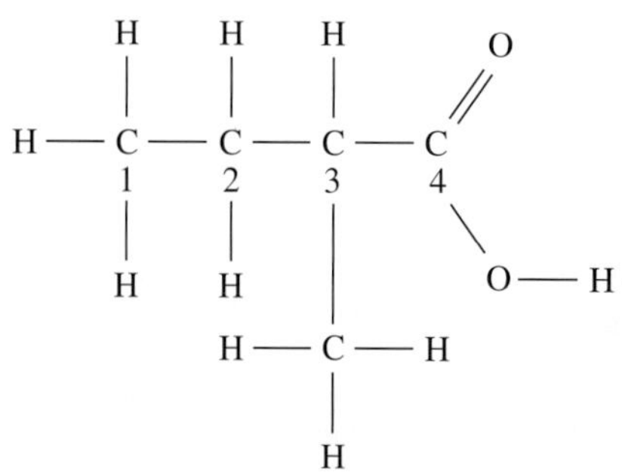

d The correct name is 2,3-dimethylbutan-2-ol.

```
                         H
                         |
                    H —  C — H
      H                  |     H
      |                  |     |
H —  C —  CH  —  C  —  C — H
      4    3       2     1
      |    |       |     |
      H    |       OH    H
      H —  C — H
           |
           H
```

Chapter 8 review

1 C **2** D **3** B **4** D

5 **a** ethane

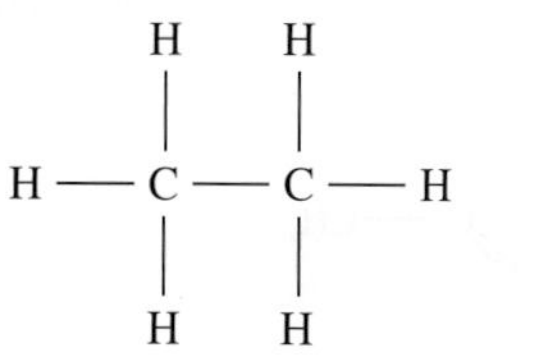

b 2-methylpropane

```
      H     H     H
      |     |     |
H —  C  —  C  —  C — H
      |     |     |
      H     |     H
      H  —  C — H
            |
            H
```

c 3-methylpentane

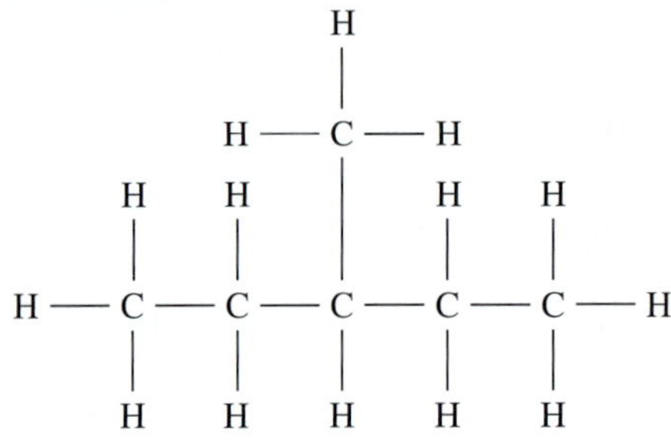

d pentane

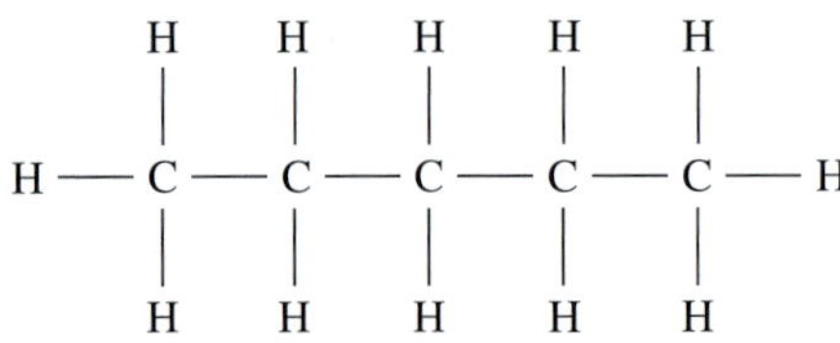

6 **a** $C_{16}H_{34}$ is an alkane. **b** $C_{17}H_{36}$
c $C_{15}H_{32}$ **d** $C_{16}H_{32}$

7 **a** Alkene **b** Alkane **c** Alkene
d Alkane **e** Alkane

8 **a** An alkene contains one double carbon–carbon bond, which requires two carbon atoms. The first alkene is therefore ethene.
b The carbon atom has four electrons in the outer shell, which are available for sharing with other atoms to produce four covalent bonds.

9 1,1-dibromopropane, 1,2-dibromopropane, 2,2-dibromopropane and 1,3-dibromopropane

10 Student answers will vary.

11 **a** $CH_4(g) + 2O_2(g) \rightarrow CO_2(g) + 2H_2O(l)$
b $2C_6H_{14}(l) + 19O_2(g) \rightarrow 12CO_2(g) + 14H_2O(l)$
c C_4H_{10}: $2C_4H_{10}(g) + 13O_2(g) \rightarrow 8CO_2(g) + 10H_2O(l)$

12 Renewable means that there is a continuous source of the fuel, so as they are used up they can be replenished. There is only a limited supply of fossil fuels present in the Earth's crust as crude oil, because these were formed from the remains of prehistoric marine microorganisms, such as bacteria and plankton, which have been converted into hydrocarbons over millions of years. Because this process is no longer occurring, our supply of fossil fuels is not renewable.

13 While alcohols have hydrogen bonding between the molecules, carboxylic acid molecules pair up due to the hydrogen bonding between the carboxyl groups, forming dimers. As a result, the dispersion forces between the dimers are equivalent to a molecule with double the molecular mass, so the intermolecular forces are stronger than those between the equivalent alcohols.

Hydrogen bonding between the two carboxyl groups holds the two molecules together in a dimer.

14 C

15 Using polyethene (polyethylene) made from sugarcane instead of from crude oil products means that the non-renewable crude oil products can be saved to for use in making other products. Polyethene made from sugarcane is considered renewable and thus sustainable, because the sugarcane can be regrown quickly after it had been used.

16 **a** hex-2-ene
b 2-methylbutane
c 2,3-dimethylbut-1-ene
d 2-methylpentan-3-ol
e 3-methylbutanoic acid

17 This compound is 3-methylpent-2-ene. The numbering of carbon chain must start at the end nearest the functional group (the carbon-carbon double bond), so the molecule is 3-methylpen-2-ene, rather than 3-methyl pent-3-ene.

18

	Alkanes	Alkenes	Haloalkanes	Alcohols
Solubility in water	does not dissolve	does not dissolve	small haloalkanes dissolve slightly	small alcohols dissolve well in water, due to formation of hydrogen bonds between the hydroxyl group and water molecules
Boiling point	low, increases with size of carbon chain	low, increases with size of carbon chain	higher than alkanes and alkenes, increases with size of carbon chain and with size of halogen atom or number of halogen atoms	higher than haloalkanes due to the strength of the hydrogen bonding between molecules
Bonding between molecules	dispersion forces	dispersion forces	dipole–dipole attraction and dispersion forces	hydrogen bonding and dispersion forces

19 The terms in bold have been corrected.
a **Alkanes** have two more hydrogen atoms per carbon atom than **alkenes**. / Alkenes have two **less** hydrogen atoms per carbon atom than alkanes.
b A haloalkane with five carbon atoms and one chlorine atom bonded to the end carbon could be called **1-chloropentane**. / A haloalkane with **three** carbon atoms and one chlorine atom bonded to the end carbon could be called 1-chloropropane.
c Pentane has **3** structural isomers.
d Alkanes are **saturated** hydrocarbons. / **Alkenes** are unsaturated hydrocarbons.
e The carboxylic acid with seven carbon atoms is called **heptanoic acid**.
f The **hydroxyl** functional group is found in alcohols and has the formula –OH.
g Compounds with the same molecular formula can **have molecules that are structural isomers of each other**.

20 a As the number of carbon atoms increases, the number of electrons in the molecules also increases, the attraction between temporary dipoles formed by the random movement of the electrons increases, and so the strength of the dispersion forces between the molecules increases. Because this intermolecular force of attraction has increased, more energy is needed to separate the molecules, and so the boiling point increases.

b Haloalkanes, alcohols or carboxylic acids would all have a higher boiling point than their corresponding alkane. The functional groups in these compounds enable the intermolecular forces to be stronger, so more energy is needed to separate the molecules than would be the case for alkanes, and so the boiling point is higher.

21 a 1-chloropropane is a molecule with three carbon atoms. While the two end (terminal) carbon atoms are equivalent to each other, the second carbon in the chain is not, so the number is required to distinguish whether the chlorine atom is bonded to an end carbon (designated as carbon number 1) or the second carbon in the chain. Chloroethane has only two carbon atoms which are equivalent to each other, so no distinction between the atoms is needed.

b 2,2-dibromopropane has a higher molecular mass, and more electrons than 2-bromopropane, so it has stronger dispersion forces between its molecules. As a result, more energy is needed to separate the molecules when the state changes from liquid to gas, and so it has a higher boiling point.

c While the hydroxyl group in ethanol is able to make hydrogen bonds with water molecules and enable the ethanol to dissolve in water, octan-1-ol has a much larger non-polar chain than ethanol (8 carbon atoms, compared to 2 carbon atoms), so the polar hydroxyl group has less influence over its solubility and so octan-1-ol cannot dissolve in water.

d In a carboxylic acid, the carboxyl group is always on carbon number 1 (by convention), so no number is needed for the carboxyl group. In comparison, the hydroxyl group can be anywhere along the carbon chain, so needs to be numbered in molecules larger than ethanol.

22 a

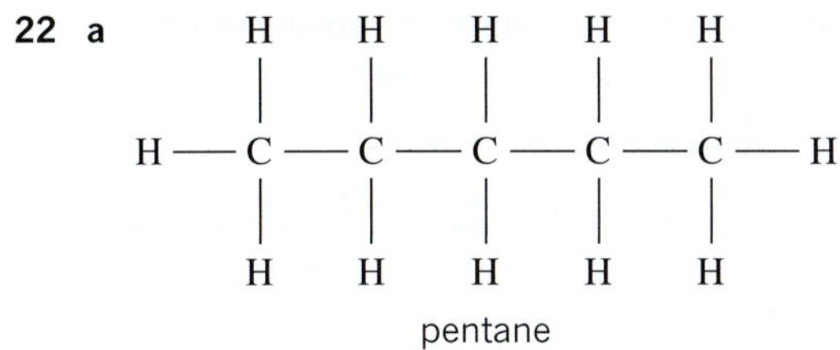

pentane

b

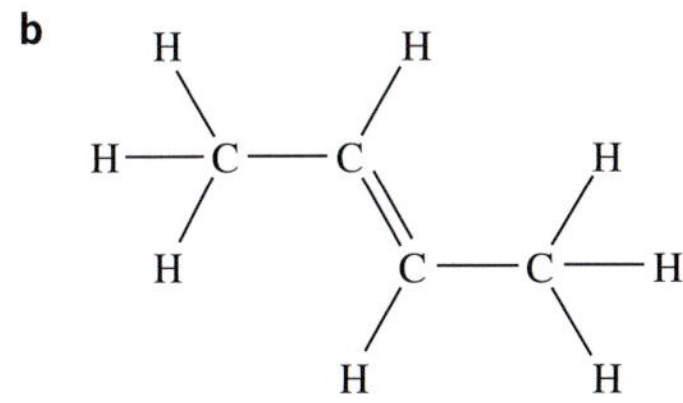

but-2-ene

c

2,3-dimethylbutane

d

iodoethane

Chapter 9 Polymers and society

9.1 Polymer formation

1 a An organic reaction where two or more molecules combine to form a larger molecule, without another product being formed.

b Monomers join when a covalent double bond (usually C=C) breaks. A very long molecule forms without the loss of another smaller molecule.

2 a chloroethane

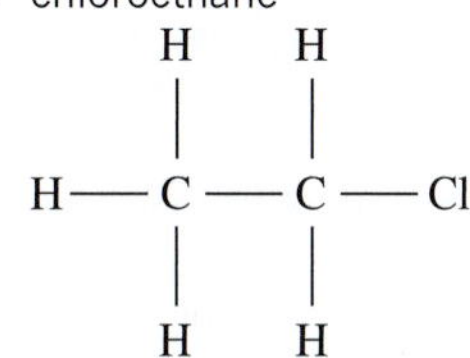

chloroethene

b Chloroethene can form a polymer due to its C=C double bond.

3 Polymers usually consist of thousands of monomer repeating units, so this would not be considered a polymer.

4

5 a Condensation reaction: Type of reaction in which two molecules combine to form a single molecule, usually with the loss of a small molecule such as water.

b Condensation polymerisation occurs between the functional groups on monomers, which react to form larger structural units while releasing smaller molecules (e.g. water) as a by-product.

6

7 4 repeating units

8 a High-density polyethene (HDPE) is made of relatively unbranched chains of polyethene, which can pack together more closely than the chains of low-density polyethene (LDPE). LDPE contains branched chains of polyethene that cannot pack together as closely. HDPE is therefore stronger and slightly less flexible than LDPE. Both HDPE and LDPE are chemically unreactive, waterproof, non-conductors and only slightly permeable to gases.

b i LDPE ii HDPE iii LDPE

9 No. The melting point depends upon the length of the molecules and the degree of branching, which are related to manufacturing conditions These two factors are likely to vary between different manufacturers.

9.2 Thermoplastic and thermosetting polymers

1 a thermosetting
b thermoplastic
c thermosetting
d thermoplastic

2 a covalent bonds within the chains and dispersion forces between chains
b dispersion forces between chains

3 Thermosetting items are difficult to recycle as they do not melt. They are often disposed of in landfill. Thermoplastics that can be recycled offer better sustainability.

4 All the atoms in a thermosetting polymer are connected by strong covalent bonds, so it has a higher resistance to heat than a thermoplastic polymer. A thermoplastic polymer is made up of many individual polymer molecules held together by dispersion forces. The dispersion forces can be broken by heat, causing the polymer to melt.

5 A thermosetting polymer is more likely to produce toxic gases when heated. If it is heated strongly, the covalent bonds will break and smaller, often toxic, compounds will form. Thermoplastic polymers initially melt when heated.

9.3 Designing polymers for a purpose

1 a polypropene

$$n\ (CH_3)(H)C{=}C(F)(H) \longrightarrow \left[-C(CH_3)(H)-C(H)(H)-\right]_n$$

b Teflon

$$n\ (F)(F)C{=}C(F)(F) \xrightarrow{\text{catalyst}} \left[-C(F)(F)-C(F)(F)-\right]_n$$

c polypropenenitrile

$$n\ (H)(H)C{=}C(CH)(H) \longrightarrow \left[-C(H)(H)-C(CN)(H)-\right]_n$$

2

$$-CH_2-C(H)(CH_3)-CH_2-C(H)(Cl)-CH_2-C(H)(CH_3)-CH_2-C(H)(Cl)-$$

3 a 4 b 1,1-difluoroethene

4 –H (in polyethene)
–F (in Teflon)
–Cl (in polyvinyl chloride)
$-C_6H_5$ (in polystyrene)
$-NC_{12}H_8$ (in polyvinyl carbazole)

5 a dispersion forces
b dipole–dipole attractions
c dispersion forces
d dispersion forces
e dipole–dipole attractions

6 a lowest to highest melting point: polypropene, polychloroethene, polytetrafluoroethene
b The melting point of a polymer increases with the length of the molecules. A fair comparison of the impact of the monomer would require polymer chains of equivalent length, degree of branching and arrangement of the chains relative to each other.

7 a Pentane is added to polystyrene to make polystyrene foam. As the polystyrene is heated, the pentane turns to a gas, so the trapped gas in the polymer makes it a foam. Regular polystyrene does not contain any gas.
b insulating, lightweight, tough and inert

8 a mass of polystyrene = 950 g
mass of polystyrene foam = 50 g
b The low density of a foam means that a small amount of polymer can be used to produce a large object.

9.4 Recycling plastics

CSA: Some choices are smarter than others

1 coffee cup: convenient to use, but adds to landfill
starch-based container: can be made from waste and composts easily
biopolyethene cup: made from renewable waste, but adds to landfill
compostable spoon: drains resources, but does not add to landfill
PET drink bottle: resources used in manufacture, but does not contribute to waste or landfill
fork made from crude oil: uses resources in its manufacture and contributes to landfill

2 a Factors to consider include: whether the source material is renewable or non-renewable, the energy required to manufacture the product, whether the product can be recycled or whether it is compostable. You would also need the packaging to serve its purpose of containing the food without having an impact on its taste.
b A ranking from most sustainable to least: starch-based container, compostable spoon, PET drink container, biopolyethene cup, biodegradable coffee cup, non-recyclable fork.

3 Student answers will vary.

4 The recycling codes from 1 to 7 are used all across Australia. Container deposit schemes vary between states and territories.

Key questions

1 a Plastics need to have a code number from 1 to 7 on them, and their recycling status checked before they are added to the recycle bin. The numbers have the recycling triangle around them.
  b thermoplastic

2 suitable for recycle bins: 2 L milk cartons, 600 mL plastic water bottles not suitable for recycle bins: plastic wrap, polystyrene foam, Teflon tape

3 Student answers will vary.

4 PET drink bottles are easy to separate from waste and they are easy to remould into useful products. The majority of plastic water bottles are PET, so it is an obvious market to target.

5 Sorting machines are expensive, but they can do the job of many humans and often do the job more effectively. As the quality of the sorting technology improves, the potential gains increase. The sorting of garbage is not usually considered to be a pleasant job for humans. To progress to a circular economy, the product should be made from a renewable resource and the product needs to either be remoulded to a new product or compostable to turn into non-toxic substances.

6 Many recycling projects produce products of a lower grade, such as outside furniture and mats. The weaving of PET into clothing is producing a higher quality, more desirable product. The impact on scarce raw material supplies is alleviated through the use of recycled material.

9.5 Innovations in polymer manufacture

1 a bio-plastic: a polymer produced from living organisms or biomass
  b bio-monomer: a monomer produced from living organisms or biomass
  c compostable: product that will degrade within 90 days
  d microbial: a process or reaction caused by microorganisms

2 Biodegradable has proved to be a misleading term. Many biodegradable items do not actually degrade in landfill. They require elevated temperatures or added bacteria to degrade within a reasonable time period. The term compostable has been introduced to apply to plastics that generally degrade in normal landfill conditions in a reasonable time.

3 a sugar cane waste
  b dairy waste
  c potato-processing waste
  d any plastic waste

4 a For a polymer to be soluble in water, its side chain needs to have significant dipoles. A repeating –OH group is an example of the dipole necessary.
  b casing around capsules, soluble stitches

5 Composting leads to breakdown of a molecule. Sometimes the products of breakdown are toxic and damaging to the environment. Composting might also produce significant levels of greenhouse gases such as methane and CO_2.

6 a The action of yeast on biomass can cause the carbohydrates to ferment to ethanol and CO_2.
  b If polyethene uses ethanol made from biomass, it means less crude oil is being used. This preserves scarce resources and it uses up waste biomass.

7 a A: example of a linear economy—drains scarce resources and creates landfill.
  B: potato starch is a waste material so it is good to find a use for it. However, producing landfill is not helpful.
  C: the addition of some lactic acid will lower the environmental impact of the polymer, especially if it leads to a compostable product. Less crude oil is used.
  D: crude oil is used, but the final product does not end up in landfill.
  b Perhaps option C offers the most benefit. The lactic acid used for the copolymer can be made from biomass waste so it does not drain resources and it composts to non-toxic substances.

8 a mechanical recycling: PET is heated and remoulded into a new shape. The polymer molecules have not been altered. New PET drink bottles or faux fur jackets are examples of mechanical recycling.
  b organic recycling: microorganisms are able to break the ester bonds in PET to produce smaller and less toxic molecules. PETase is the organism used.
  c chemical recycling: heat and catalysts are used to break the long polymer molecules and to form smaller molecules that are a form of synthetic oil. It is possible to re-form the same monomers to make further PET, but the synthetic oil can be used to make other monomers.

Chapter 9 review

1 B 2 D 3 C 4 all except B

5 a H(Cl)C=CH$_2$ (structure: C=C with H, Cl on one carbon and H, H on the other)
  b 6
  c dipole–dipole attractions.

6 a Monomers are small molecules that are able to react to form long chains of repeating units, called polymers. They often contain a carbon–carbon double bond.
  b When a thermoplastic is heated, the bonds between molecules are broken and the molecules become free to move, so the plastic melts.
  c When a thermosetting is heated it does not melt, but at high temperatures covalent bonds are broken and the material decomposes or burns. It cannot be moulded into a different shape.
  d A cross-link is one or more covalent bonds that connect neighbouring polymer chains.

7 a false b false c false d true

8 a The ethene molecule has a carbon-to-carbon double bond.
  b $H_2C=CH_2$ (structure: C=C with two H on each carbon)
  c Ethane cannot undergo addition polymerisation because it is a saturated compound.

9 a —C(CH_3)(H)—C(H)(H)—C(CH_3)(H)—C(H)(H)—C(CH_3)(H)—C(H)(H)—
  b —C(Cl)(H)—C(H)(Cl)—C(Cl)(H)—C(H)(Cl)—C(Cl)(H)—C(H)(Cl)—
  c —C(CH_3)($OCOCH_3$)—C(H)(H)—C(CH_3)($OCOCH_3$)—C(H)(H)—C(CH_3)($OCOCH_3$)—C(H)(Cl)—

10 a i

ii

iii

b

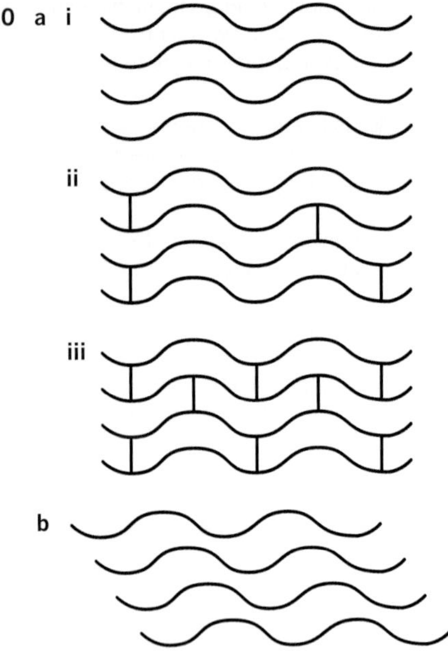

polymer easily stretched

Some stretching but polymer tends to return to its original shape.

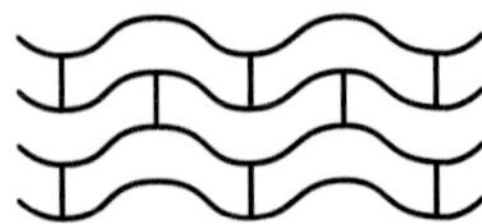

no stretching possible

11 a Thermosetting polymers have much stronger inter-chain bonds than thermoplastic polymers.

b The inter-chain bonds in thermosetting polymers are about the same strength as the covalent bonds within chains. When heated strongly both the inter-chain and within-chain bonds break.

12 a thermosetting

b thermoplastic

c thermosetting

d thermoplastic

e thermoplastic

13 a the polymer used in saucepan handles

b Saucepan handles are much harder and more resistant to the effect of heat than elastic bands. The polymer on the outside of the golf ball is also very hard, but, unlike the polymer in saucepan handles, is quite elastic.

14 Look for 1 to 7 recycle codes on the product itself, not just the packaging. Expanded polystyrene foam should not be recycled.

15 a A polymer formed when more than one type of monomer is used.

b A copolymer can display a blend of the best properties of each monomer. Sometimes only a small amount of a second monomer significantly improves the performance of the polymer.

16 Thermoplastic, as it can be melted and remoulded.

17 a biobased—a material made from substances derived from living organisms

b biodegradable—potentially will degrade under the right conditions

c compostable—can degrade in a useful time frame, usually less than 90 days.

18 a

CH_3 H
C=C
H   H

propene

b 3

c The melting point is likely to be higher than polyethene due to the larger side group (assuming similar chain lengths and degrees of branching).

d This polymer is recyclable—separate the items made from this polymer and remould them.

19 a

H   H
C=C
H   O—$COCH_3$

b PVA contains oxygen atoms in its side-group. They add a significant dipole to the molecule and allow hydrogen bonds to form with water.

c Being water soluble means the polymer might fail in a moist environment. It will degrade more easily, however, than most polymers.

20

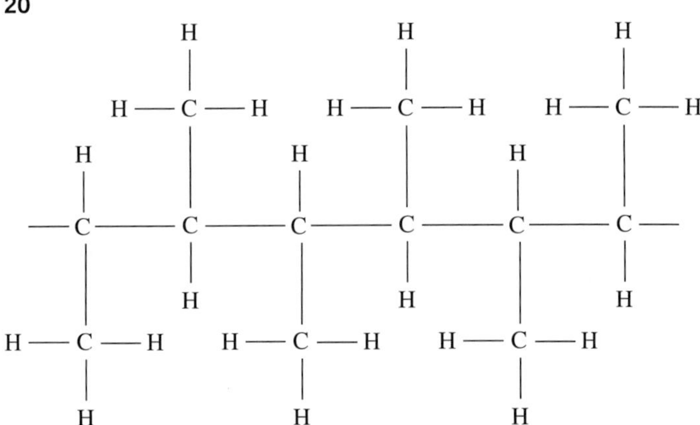

21 Left: C_8H_8 (styrene); middle: C_3H_3N (propenenitrile); right: C_4H_6 (buta-1,3-diene)

22 a The melting point and the toughness will increase with longer polymer molecules

b Incorporation of a large side group will make the polymer tougher and harder.

c A foaming agent will expand the polymer material significantly. It will have a much larger volume and a lower density.

23 The relative molecular mass would have increased. The melting point of the polymer would have increased. The overall strength of inter-chain forces would have increased. The electrical conductivity of the polymer would have remained the same.

24 a Answer to include—polyethene: crude oil fractional distillation to isolate ethane. Ethane cracked to ethene. Ethene polymerised to polyethene. Biopolyethene—carbohydrate ferments to ethanol. Ethanol cracked to ethene. Ethene polymerised.

b World consumption of polyethene is extremely high—there is not enough waste biomass to support the large-scale, rapid production of bioethanol. Fermentation is a slow reaction.

25 a

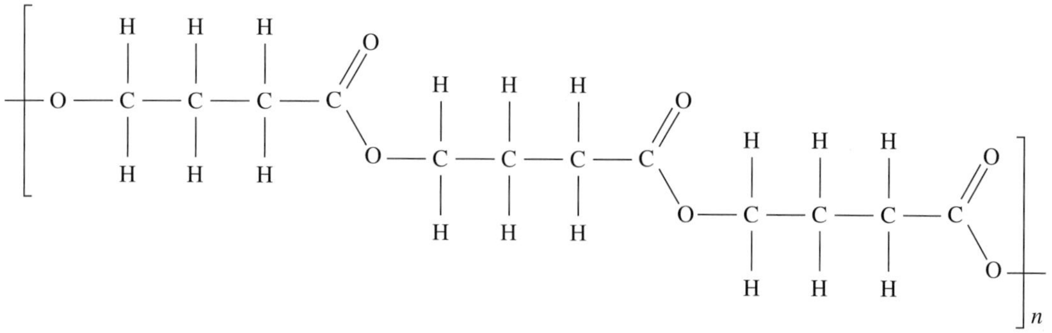

b Condensation and addition polymers both have long molecules. The characteristic structure of the chain, however, will differ. Condensation polymers are likely to have some oxygen or nitrogen atoms in the chain, while the chains of addition polymers contain carbon atoms only. Addition polymers form when unsaturated monomers react, while condensation polymers form when the functional groups on each end of the monomers react.

c The chain of a condensation polymer often contains oxygen and/or nitrogen atoms. These atoms are susceptible to attack from microorganisms.

26 PLA can be manufactured from biomass and it is fully compostable to non-toxic substances. Its production is not draining scarce resources and it is not adding to landfill. Its use leads to little impact on the environment.

27 a Most polyethene is made from non-renewable crude oil. Once used, it is added to landfill. Therefore, it is a drain on resources and a problem in landfill. This is an example of the linear economy in action, often referred to as 'take-make-dispose'.

b A biomonomer is made from biomass of some form. The most common source is bioethanol made by fermentation. This removes the problem of polyethene being a drain on non-renewable resources.

c In mechanical recycling, the polyethene is shredded and remoulded into a new product. The fraction of polyethene recycled has been increased since supermarkets have started collecting soft plastic wrap.

d Licella heats plastic material under pressure. The addition of a catalyst leads to the polymer being broken down to smaller, synthetic oil. This is an example of chemical recycling, where the polymer structure is changed in the recycling process.

e Addition polymers have only carbon atoms in their chains. Microorganisms cannot break the bonds in the chains, so the polymer cannot undergo organic recycling.

Unit 1 Area of Study 2 Review

1	C	**2**	A	**3**	D	**4**	B
5	C	**6**	D	**7**	C	**8**	A
9	B	**10**	A	**11**	C	**12**	B

13 a Crude oil is fossilised organic material, mostly of plant and microbial origin.

b i It is a mixture of hydrocarbons, mostly alkanes from C1 to about C70.

ii Any correct formulas of two alkanes with general formula C_nH_{2n+2}; e.g. C_5H_{12}, C_6H_{14}.

14 a Magnesium has three different isotopes, ^{24}Mg, ^{25}Mg and ^{26}Mg. The most abundant of these isotopes is ^{12}Mg.

b $A_r = 24.3$

15 a 20 mol **b** 70 mol **c** 4.21×10^{25}

16 a but-2-ene

b heptane

c 2,2,3-trimethylpentane

d butanoic acid

e 2-methylpropan-1-ol

17 a

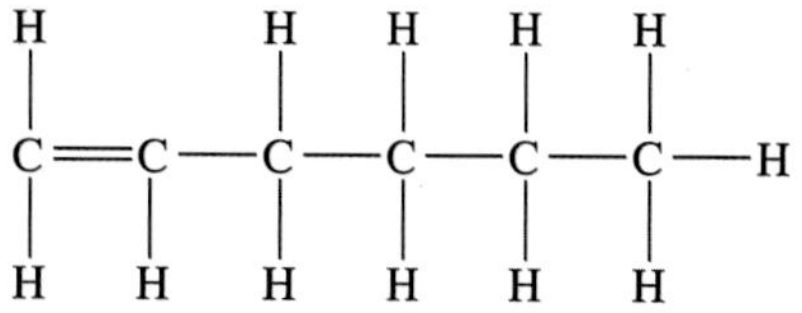

b

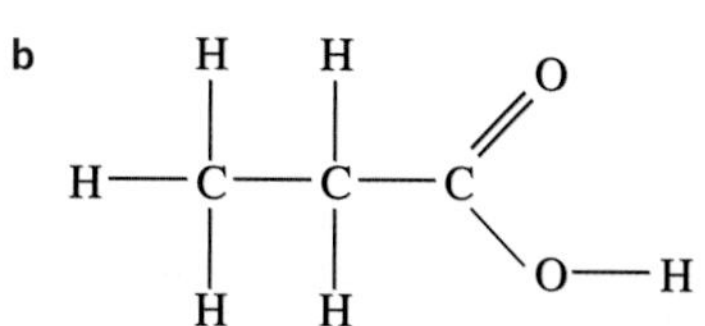

c

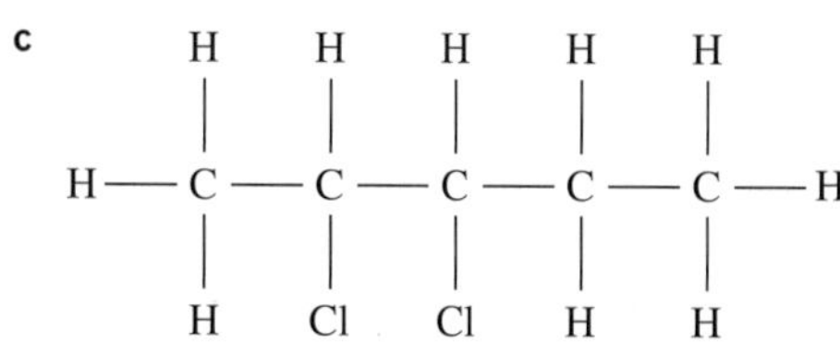

d

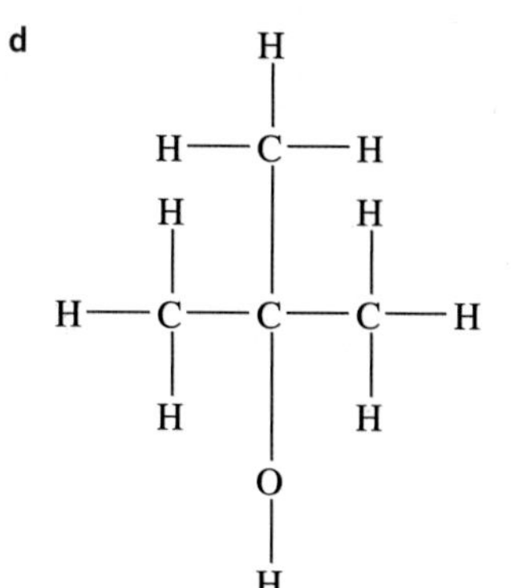

18 a $CH_3(CH_2)_4COOH$ or $CH_3CH_2CH_2CH_2CH_2COOH$

b $CH_3C(CH_3)_2CH_2CH(C_2H_5)\ CH_2CH_2CH_3$ or $(CH_3)_3CCH_2CH(C_2H_5)CH_2CH_2CH_3$

c $CH_3CH(OH)CH_2CH_3$

d $CH_3C(CH_3)CH_2CH(CH_3)CH_2CH_3$

19 a

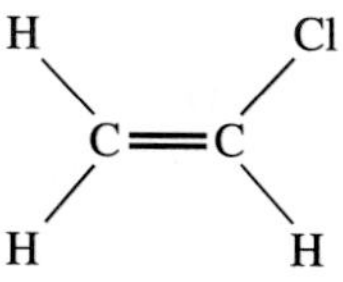

b i The plastic will char and blacken.

ii Thermosetting polymers have strong cross-links between polymer chains so the layers cannot slide past each other and melt.

20 **a** A monomer is a small molecule that is able to bond with other monomers to form a long chain molecule called a polymer; e.g. ethene molecules are the monomers that join together to form the polyethene polymer.

b A thermoplastic polymer is one that softens on heating, but becomes hard again when cooled, e.g. polyethene. A thermosetting polymer is one that doesn't soften on gentle heating, but if heated sufficiently it will char, e.g. urea-formaldehyde.

c A branched polymer is a linear polymer that has some of the atoms forming branches attached to the polymer backbone, e.g. low-density polyethene. Relatively weak intermolecular forces exist between chains.

A cross-linked polymer has covalent bonds linking polymer chains, resulting in a rigid polymer that does not soften on heating, e.g. urea-formaldehyde.

d Addition polymer: long molecule formed when the C=C double bonds in the monomer break, allowing the monomers to join to each other. Polyethene is an example.

Condensation polymer: long molecule formed when the functional groups on the ends of the monomers react, allowing the monomers to join and a small molecule such as water is also formed. Polyester is an example.

21 M(X) = 35.5 g mol^{-1}

The element must be Cl.

22 **a** N_2O **b** NO_2 and N_2O_4 **c** 4.21×10^{24} atoms

23 **a** Hexane molecules have no significant dipoles as the structure only contains carbon and hydrogen atoms. The very small dipoles will cancel each other out as well.

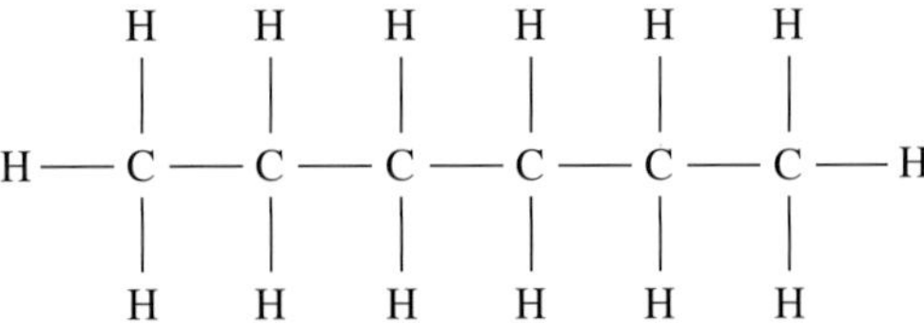

b To be used in the food industry, the molecule must have no long-term health impact on humans. Some hydrocarbons are toxic to humans. Also, hexane is a liquid; some smaller alkanes would also be non-polar, but not suitable as a solvent in the gas state.

24 **a** C_2H_4O **b** 88 g mol^{-1} **c** $C_4H_8O_2$

d (Students should be able to draw the two structures shown under the heading Compound 1. Some students might also be aware of another group of molecules shown under the heading Compound 2, which are not studied until Unit 4.)

Compound 1

butanoic acid

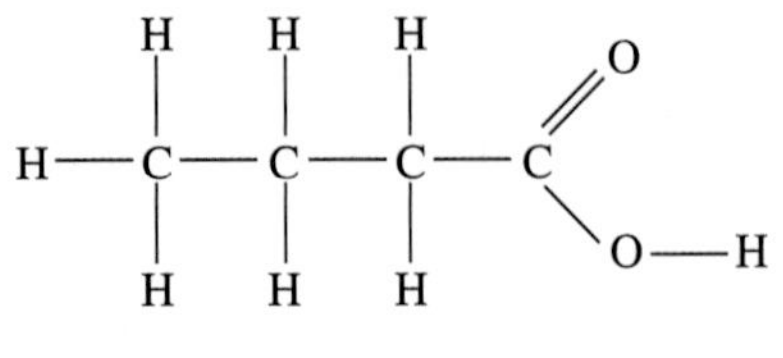

or

2-methylpropanoic acid

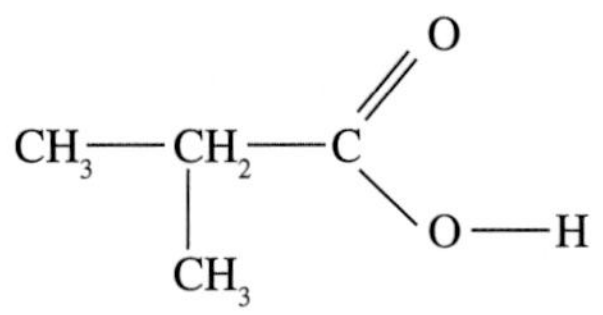

Compound 2

methyl propanoate

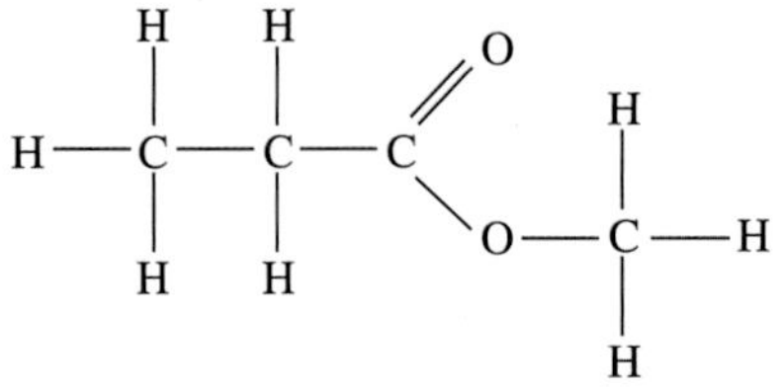

or

ethyl ethanoate

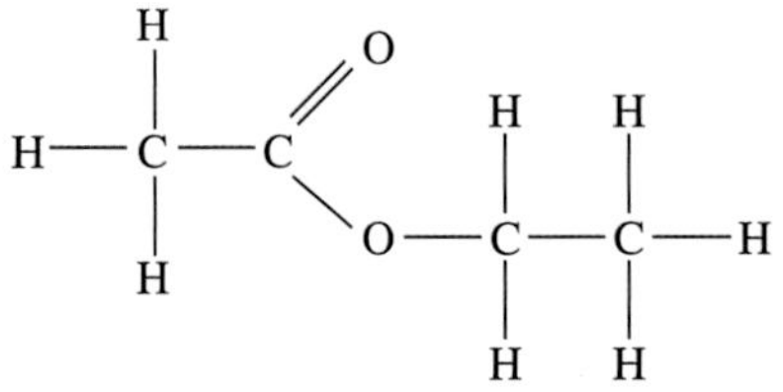

or

propyl methanoate

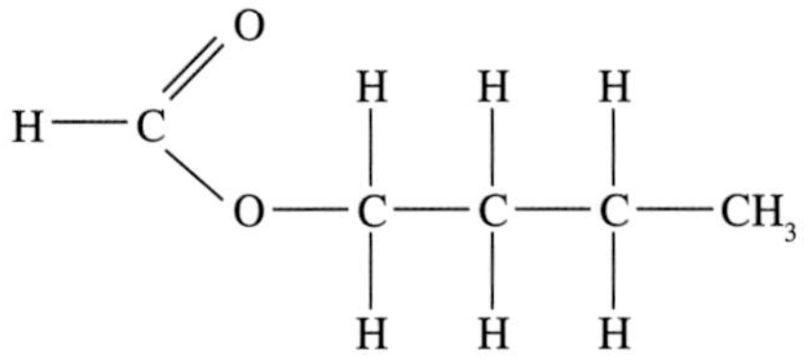

25 **a** $CHCl_3$

b Formula mass of $CHCl_3$ is 119.5 g mol^{-1}, so the molecular formula is the same as the empirical formula.

c

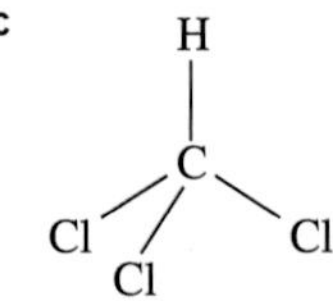

d There are three polar C–Cl bonds in the molecule; the molecule is not symmetrical, so it is polar overall. The interactions between the molecules are dipole–dipole interactions (as well as dispersion forces).

e 48 g

26 **a** 35% **b** 4.25 kg **c** 1.49 kg **d** 106 mol

27 **a** (below)

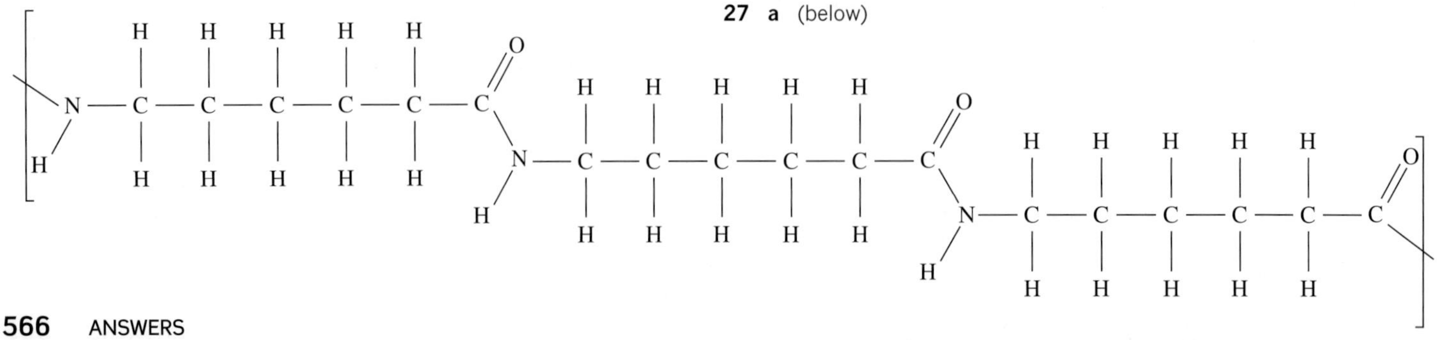

b 2999

c The polymer will have nitrogen and oxygen atoms in the polymer chains. These atoms are susceptible to the action of microorganisms. Addition polymers like polyethene have only carbon atoms in the chains.

28 a Casein will be a condensation polymer. Natural polymers form from reactions between functional groups.

b Casein is a good example of a circular economy. It can be made from waste milk, so non-renewable crude oil is not required. As it degrades readily in water, it composts easily to form harmless products. Casein is therefore a far more sustainable option, both in source material and waste created.

29 a

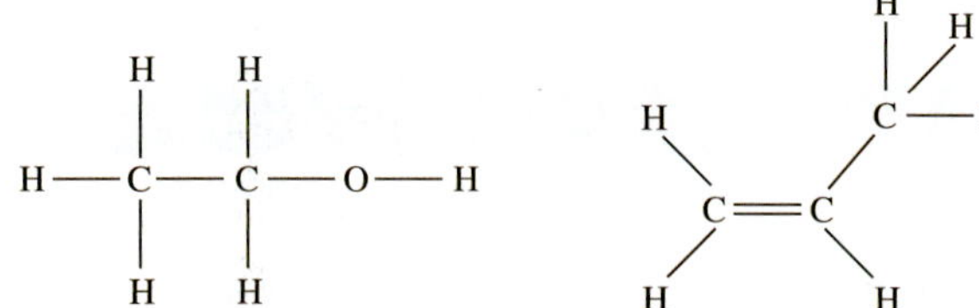

b i 34.8 %   ii 53.3%

c Ethanoic acid has a higher boiling point. Both molecules can form hydrogen bonds, but two hydrogen bonds can form between each pair of ethanoic acid molecules, whereas each pair of ethanol molecules have only one hydrogen bond between them.

d Many microorganisms function best around 35 °C. If the temperature is too high, the organism might be killed. Reactions with chemicals derived from crude oil are not limited by the impact on living organisms.

e The production of either ethanol or ethanoic acid from biomass offers many advantages. Biomass is often a waste material, so it is good to find a use for the waste. If biomass is not used, scarce crude oil resources have to be used instead.

30 a i mass of magnesium

ii mass of ash

b The precision of most data, except the data of student D, is high. The results of the other six students have produced a relatively linear graph. Student D's results can be viewed as an outlier, where the mass obtained is lower than expected.

c i Very close to 0.800 g

ii MgO.

iii The mole ratio is not close enough to give 1 : 1. It is often found in this experiment that some magnesium oxide escapes, leaving the mass of oxide a little lower than it should be. This means the mass of oxygen will be low and the ratio drifts towards 1.1 : 1 or 1.2 : 1

d MgO, as magnesium has an electrovalency of 2+ (2 electrons in the outer shell of its atoms), while oxygen has an electrovalency of 2– (6 electrons in the outer shell), and forms O^{2-} ions.

Chapter 10 Water as a unique chemical

10.1 Essential water

CSA: The importance of floating ice

1 4°C.

2 between 3 and 5°C

3 A solid will float in a liquid when it has a lower density than the liquid. The density of liquid water from 0°C to 12°C is always above 0.999 50. This means solid ice, with a density of 0.9168 g cm^{-3}, is lower at all of these temperatures and will float in water.

4 a 5000 g

b 4999 g

5 At temperatures at or below 0°C, ice forms and floats on top of the liquid water. If ice was denser than liquid water, lakes and oceans would freeze from the bottom up, killing all life in the body of water.

Key questions

1 Surface water has the highest risk of contamination since it can be easily polluted. It is not a protected water source like mains water, where the water is tested and treated if needed.

2 a Physical properties that are unusual to water include:

- relatively high melting and boiling points for its molecular size
- decrease in density on freezing
- high heat capacity
- high latent heat of fusion and evaporation for a substance of its molecular size.

b The bond between H and O atoms in water is highly polar. As a result, hydrogen bonds exist between water molecules. Hydrogen bonds are stronger than other intermolecular bonds (although still weaker than the covalent intramolecular bonds), and so require more energy to break. Thus, water has a relatively high melting and boiling points. Hydrogen bonding between water molecules in ice results in a very open arrangement of molecules, so ice is less dense than liquid water.

3

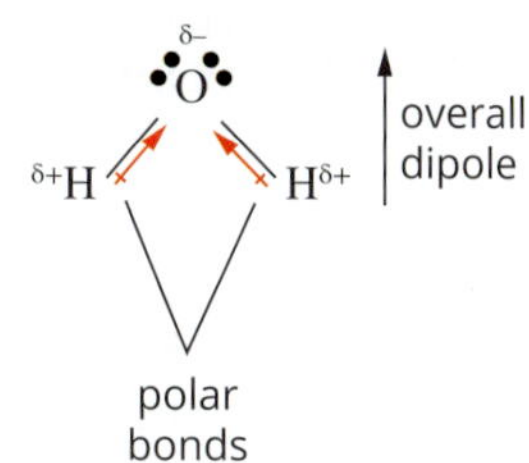

4 Only 2.5% of the water on Earth is freshwater. Most of the freshwater on Earth is locked in ice caps, glaciers or the soil.

5 Australia is the driest inhabited continent. Rainfall in Australia is extremely variable. Most of the rain that falls evaporates before it can enter rivers and reservoirs.

6 Each water molecule has two hydrogen atoms and one oxygen atom. The oxygen atom has two pairs of non-bonding electrons, each of which can form one hydrogen bond. This means the maximum number of water molecules with which one water molecule can form hydrogen bonds is four; up to two hydrogen bonds involving the two hydrogen atoms and up to two hydrogen bonds involving the two pairs of non-bonding electrons on the oxygen atom.

7 a H_2S, H_2Se, H_2Te, H_2Po, H_2O

b With the exception of water, the boiling points of the group 16 hydrides increase going down the group. This is due to increasing strength of dispersion forces as the molecules increase in mass. Water has a significantly higher boiling point than the other compounds due to the hydrogen bonds between its molecules. It is the only group 16 hydride that has intermolecular hydrogen bonds.

10.2 Heat capacity

TY 10.2.1 70.5 kJ

1 Specific heat capacity is the amount of energy required to raise the temperature 1 gram of a substance by 1 degree Celsius.

2 copper, iron, chlorofluorocarbon, concrete, glass, aluminium, wood, ethanediol, water

3 29 J **4** 94 kJ

5 **a** 2.0×10^2 kJ **b** 6.2 kJ

6 31.6°C

7 $c = 0.39\ J\ g^{-1}\ °C^{-1}$
From Figure 10.2.3: the substance is copper.

10.3 Latent heat

TY 10.3.1 170 kJ

1 At the melting point, the flat region of the graph represents the substance changing from a solid to a liquid. The energy change per mole is equal to the latent heat of fusion. At the boiling point, the flat region of the graph represents the substance changing from a liquid to a gas. The energy change per mole is equal to the latent heat of vaporisation.

2 hydrogen bonding

3 The latent heat of fusion is the energy required to change the state of water from a solid to a liquid. This requires the disruption of only a smaller number of the hydrogen bonds between water molecules. The latent heat of vaporisation is much higher as the phase change from liquid to gas requires all of the hydrogen bonds between water molecules to be disrupted.

4 102 kJ **5** 257.4 kJ **6** 100 kJ

Chapter 10 review

1 Water is a **polar** molecule. Within a single molecule, hydrogen and oxygen atoms are held together by strong **covalent bonds**. Between different molecules, the most significant forces are **hydrogen bonds**. It is the relatively **high** strength of the intermolecular forces that give water its unique properties of:
- relatively **high** boiling point, 100°C
- relatively **high** latent heat values 6.0 kJ mol^{-1} and 40.7 kJ mol^{-1}
- relatively **high** specific heat capacity 4.18 J g^{-1} $°C^{-1}$.

2 **a** water molecule
b hydrogen atom
c hydrogen bond (and dispersion forces)
d covalent bond

3 A

4 Most Australians live in the capital cities and their drinking water is supplied from reservoirs, as cities in Australia obtain their water from protected sources.

5 oceans, ice caps and glaciers, groundwater, ground ice and permafrost, lakes, soil moisture, atmosphere as water vapour, rivers

6 Intermolecular forces are those between one molecule and other molecules. For water, these are hydrogen bonds. Intramolecular forces are those holding the atoms together within a molecule. For water, these are covalent bonds. The breaking of covalent bonds requires a chemical reaction and new substances are formed. Covalent bonds are stronger. Evidence for this is the high temperatures required to disrupt the bonds between the oxygen and hydrogen atoms inside the water molecule and so decompose it into its constituent gases. Changing liquid water into gaseous water involves breaking hydrogen bonds to separate one molecule from another. The lower temperatures needed to do so indicate that hydrogen bonds are weaker.

7 Water has a significantly higher boiling point than hydrogen sulfide due to the hydrogen bonds between water molecules. Hydrogen sulfide cannot form hydrogen bonds.

8 It is the high polarity of the water molecule that allows relatively strong hydrogen bonding to occur between molecules. As a consequence, a relatively large quantity of energy is required to break the hydrogen bonds between water molecules when water changes from a liquid to a gas. This gives water a relatively high boiling point.

9 The type of bonds present in the substance determine its temperature increase. It is the type of intermolecular bonds present in the substance that determine the heat capacity of the substance and therefore the temperature change that occurs when heat energy is applied.

10

Definition	Property
a the temperature at which a liquid evaporates to form a gas	boiling point
b the heat energy required to melt a solid to a liquid at its melting point	latent heat of fusion
c the amount of heat energy required to increase a specific mass of a substance by a certain amount, e.g. 1 g by 1°C	specific heat capacity
d the heat energy required to evaporate a liquid to a gas at its boiling point	latent heat of vaporisation

11 Water has a high latent heat of vaporisation. Water is effective as a cooler because it absorbs a relatively large amount of energy when it evaporates, giving it a high latent heat of vaporisation.

12 85.7 kJ **13** 4.2×10^3 kJ **14** 566 kJ

15 9.6 kJ **16** 442 g

17 Difference in heat energies = 964 kJ

18 **a** The crystal lattice of ice is disrupted and molecules have greater freedom of movement. During this time, all solid ice is being converted to liquid water.
b The added energy is taken up in overcoming the hydrogen bonds between molecules, separating the molecules to form a gas.

19 68°C **20** 0.42 J g^{-1} $°C^{-1}$

21 **a** The specific heat capacity of ethanol is almost half that of water. For the same amount of energy the change in temperature of ethanol will be almost twice that of water.
b The change in temperature in each substance will depend on this respective specific heat capacities. The specific heat capacity of each substance is affected by the intermolecular forces present. In both ethanol and water, the most significant intermolecular force is hydrogen bonding, between the H attached to the O on one molecule, and the non-bonding pairs on the O of another molecule. The overall effect of the hydrogen bonding in the water is stronger because each molecule can form up to four hydrogen bonds, whereas each ethanol molecule can only form two. Water is also a smaller molecule, so the hydrogen bonding has a greater effect 'per molecule'. Therefore, water can absorb more heat energy, has a greater SHC and will not increase in temperature as much as the ethanol. Heating 100 g of water from 0°C to 20°C will require more energy than the energy needed to heat 100 g of ethanol over the same temperature range.

22 0.80 J g^{-1} $°C^{-1}$

Chapter 11 Acid-base reactions

11.1 Acids and bases

1 Using the Brønsted–Lowry theory of acids and bases, an acid–base reaction is a reaction in which a **proton** transfer occurs. This theory states that:
- Acids are proton **donors**.
- Bases are proton **acceptors**.
- Hence acids and bases must always act together.
- Acids and bases react with each other and form their respective conjugates. Acids **donate** a proton and form their conjugate **base**. Bases **accept** a proton and form their conjugate **acid**.
- The formulas of conjugate pairs differ by a $\mathbf{H^+}$.
- An **amphiprotic** substance can act as an acid or a base, depending on what the other reactant is.

2 **a** $HCl(g) + H_2O(l) \rightarrow H_3O^+(aq) + Cl^-(aq)$
b HCl/Cl^- and H_3O^+/H_2O

3 H_2SO_4/HSO_4^- and $H_2NO_3^+/HNO_3$

4 **a i** HF – acid; OH^- – base
ii HF/F^- and H_2O/OH^-
b i HCOOH – acid; H_2O – base
ii $HCOOH/HCOO^-$ and H_3O^+/H_2O
c i CH_3NH_2 – base; HCl – acid
ii $CH_3NH_3^+/CH_3NH_2$ and HCl/Cl^-

5 **a** NH_4^+ **b** CH_3COOH **c** $H_2PO_4^-$ **d** HCO_3^-

6 Acting as an acid, whereby the reactant donates one proton:
a $HCO_3^- + H_2O(l) \rightleftharpoons CO_3^{2-}(aq) + H_3O^+(aq)$
b $HPO_4^{2-} + H_2O(l) \rightleftharpoons PO_4^{3-}(aq) + H_3O^+(aq)$
c $HSO_4^- + H_2O(l) \rightleftharpoons SO_4^{2-}(aq) + H_3O^+(aq)$
d $H_2O(l) + H_2O(l) \rightleftharpoons OH^-(aq) + H_3O^+(aq)$
Acting as a base, whereby the reactant accepts one proton:
a $HCO_3^- + H_2O(l) \rightleftharpoons H_2CO_3(aq) + OH^-(aq)$
b $HPO_4^{2-} + H_2O(l) \rightleftharpoons H_2PO_4^-(aq) + OH^-(aq)$
c $HSO_4^- + H_2O(l) \rightleftharpoons H_2SO_4(aq) + OH^-(aq)$
d $H_2O + H_2O(l) \rightleftharpoons H_3O^+(aq) + OH^-(aq)$

7 **a** $H_2CO_3(aq) + H_2O(l) \rightleftharpoons HCO_3^-(aq) + H_3O^+(aq)$
$HCO_3^-(aq) + H_2O(l) \rightleftharpoons CO_3^{2-}(aq) + H_3O^+(aq)$
b $H_3AsO_4(aq) + H_2O(l) \rightleftharpoons H_3O^+(aq) + H_2AsO_4^-(aq)$
$H_2AsO_4^-(aq) + H_2O(l) \rightleftharpoons H_3O^+(aq) + HAsO_4^{2-}(aq)$
$HAsO_4^{2-}(aq) + H_2O(l) \rightleftharpoons H_3O^+(aq) + AsO_4^{3-}(aq)$
c Amphiprotic species are: **a** $HCO_3^-(aq)$, **b** $H_2AsO_4^-(aq)$, $HAsO_4^{2-}(aq)$

11.2 Strength of acids and bases

1 A **concentrated** solution has a larger amount of solute dissolved in a specific volume of solvent, whereas a **dilute** solution has a smaller amount of solute dissolved in the same specific volume of solvent. In chemistry, these terms should not be confused with the meaning of strong and weak when applied to acids and bases. All the molecules of a **strong** acid ionise, whereas at any instant, a small proportion of the molecules of a **weak** acid ionise. This is true for bases as well. Concentrated and dilute refer to moles per **litre**. Strong and weak refer to what **extent** the molecules **ionise** in water.

2 **a** $HClO_4(aq) + H_2O(l) \rightarrow H_3O^+(aq) + ClO_4^-(aq)$
b $HCN(aq) + H_2O(l) \rightleftharpoons H_3O^+(aq) + CN^-(aq)$
c $CH_3NH_2(aq) + H_2O(l) \rightleftharpoons CH_3NH_3^+(aq) + OH^-(aq)$

3 **a** strong acid **b** weak acid
c strong base **d** weak base

4 A concentrated solution of a weak acid would have a high concentration per litre of CH_3COOH molecules but a low concentration of CH_3COO^- ions and H_3O^+ ions because the acid only partially ionises in water. A dilute solution of a strong hydrochloric acid will have almost zero concentration of HCl molecules and a concentration of Cl^- ions and H_3O^+ ions equal to the concentration of the HCl solution, because HCl molecules fully ionise.

5 Stronger acids more readily ionise, forming ions in solution. As perchloric acid is stronger, more hydronium ions would be present than in a solution of ethanoic acid, making it a better conductor of electricity.

11.3 Reactions of acids and bases

TY 11.3.1 $H_3O^+(aq) + OH^-(aq) \rightarrow 2H_2O(l)$

TY 11.3.2 Products of this reaction are sodium chloride in solution, water, and carbon dioxide gas.
$H^+(aq) + HCO_3^-(aq) \rightarrow H_2O(l) + CO_2(g)$

TY 11.3.3 $6H^+(aq) + 2Al(s) \rightarrow 2Al^{3+}(aq) + 3H_2(g)$

CSA: Benefits of neutralisation

1 **a** $HCOOH(aq) + NH_3(aq) \rightleftharpoons HCOO^-(aq) + NH_4^+(aq)$
b $HCOOH(aq) + NaHCO_3(aq) \rightarrow HCOO^-(aq) + Na^+(aq) + CO_2(g) + H_2O(l)$

2 They are a weak acid and a weak base and so will not hurt the body, whereas strong acids and strong bases, such as HCl and NaOH, will cause serious burns to the skin.

Key questions

1 **a** $Mg(s) + H_2SO_4(aq) \rightarrow MgSO_4(aq) + H_2(g)$
$Mg(s) + 2H^+(aq) \rightarrow Mg^{2+}(aq) + H_2(g)$
b $Ca(s) + 2HCl(aq) \rightarrow CaCl_2(aq) + H_2(g)$
$Ca(s) + 2H^+(aq) \rightarrow Ca^{2+}(aq) + H_2(g)$
c $Zn(s) + 2CH_3COOH(aq) \rightarrow Zn(CH_3COO)_2(aq) + H_2(g)$
$Zn(s) + 2H^+(aq) \rightarrow Zn^{2+}(aq) + H_2(g)$
d $2Al(s) + 6HNO_3(aq) \rightarrow 2Al(NO_3)_3(aq) + 3H_2(g)$
$2Al(s) + 6H^+(aq) \rightarrow 2Al^{3+}(aq) + 3H_2(g)$

2 **a** magnesium sulfate **b** calcium chloride
c zinc ethanoate **d** aluminium nitrate

3 **a i** $ZnCO_3(s) + H_2SO_4(aq) \rightarrow ZnSO_4(aq) + CO_2(g) + H_2O(l)$
ii $ZnCO_3(s) + 2H^+(aq) \rightarrow Zn^{2+}(aq) + CO_2(g) + H_2O(l)$
b i $Ca(s) + 2HNO_3(aq) \rightarrow Ca(NO_3)_2(aq) + H_2(g)$
ii $Ca(s) + 2H^+(aq) \rightarrow Ca^{2+}(aq) + H_2(g)$
c i $Cu(OH)_2(s) + 2HNO_3(aq) \rightarrow Cu(NO_3)_2(aq) + 2H_2O(l)$
ii $Cu(OH)_2(s) + 2H^+(aq) \rightarrow Cu^{2+}(aq) + 2H_2O(l)$
d i $Mg(HCO_3)_2(s) + 2HCl(aq) \rightarrow MgCl_2(aq) + 2H_2O(l) + 2CO_2(g)$
ii $Mg(HCO_3)_2(s) + 2H^+(aq) \rightarrow Mg^{2+}(aq) + H_2O(l) + CO_2(g)$

4 **a** $2KOH(aq) + H_2SO_4(aq) \rightarrow K_2SO_4(aq) + 2H_2O(l)$
$OH^-(aq) + H^+(aq) \rightarrow H_2O(l)$
b $NaOH(aq) + HNO_3(aq) \rightarrow NaNO_3(aq) + H_2O(l)$
$OH^-(aq) + H^+(aq) \rightarrow H_2O(l)$
c $Mg(OH)_2(s) + 2HCl(aq) \rightarrow MgCl_2(aq) + 2H_2O(l)$
$Mg(OH)_2(s) + 2H^+(aq) \rightarrow Mg^{2+}(aq) + 2H_2O(l)$
d $CuCO_3(s) + H_2SO_4(aq) \rightarrow CuSO_4(aq) + H_2O(l) + CO_2(g)$
$CuCO_3(s) + 2H^+(aq) \rightarrow Cu^{2+}(aq) + H_2O(l) + CO_2(g)$
e $KHCO_3(aq) + HF(aq) \rightarrow KF(aq) + H_2O(l) + CO_2(g)$
$HCO_3^-(aq) + H^+(aq) \rightarrow H_2O(l) + CO_2(g)$
f $Zn(s) + 2HNO_3(aq) \rightarrow Zn(NO_3)_2(aq) + H_2(g)$
$Zn(s) + 2H^+(aq) \rightarrow Zn^{2+}(aq) + H_2(g)$
g $CaCO_3(s) + 2HCl(aq) \rightarrow CaCl_2(aq) + H_2O(l) + CO_2(g)$
$CaCO_3(s) + 2H^+(aq) \rightarrow Ca^{2+}(aq) + H_2O(l) + CO_2(g)$
h $NaHCO_3(s) + CH_3COOH(aq) \rightarrow CH_3COONa(aq) + H_2O(l) + CO_2(g)$
$NaHCO_3(s) + H^+(aq) \rightarrow Na^+(aq) + H_2O(l) + CO_2(g)$

5 a These steps outline a possible experimental design that could qualitatively demonstrate the strengths of the acids.
(1) Ensure strips of zinc are of the same size.
(2) Clean each strip with steel wool and then place one strip into each of four test tubes.
(3) Pour acid A into the first test tube and begin timing how long it takes for the zinc strip to disappear.
(4) Write down your observations and time into the results table.
(5) Repeat steps 3 and 4 for each of the subsequent acids.
b hydrogen gas and a salt, which would exist as ions in aqueous solution
c You could determine the strength of the acids by recording the time it takes for the metal to disappear. The strongest acid would be the one with the shortest time and the weakest acid would be the one with the longest time.
d acid B
e See the middle column in the table below.
f from weakest to strongest: A, D, C, B
g Possible weak acids could be ethanoic acid or carbonic acid, and the stronger acids could be HNO_3, HCl or H_2SO_4.

Acid added to the zinc strip	Observations	Time taken for reaction to go to completion (s)
A	some bubbling and a slow reaction where the metal disappears slowly	342
B	rapid bubbling and a vigorous reaction where the metal disappears almost instantly	22
C	some quite rapid bubbling and a fairly vigorous reaction where the metal disappears quickly	65
D	some bubbling and a slower reaction than with acids B and C; the metal disappears quite slowly	178

11.4 pH: A measure of acidity

TY 11.4.1 $[H_3O^+] = 5.6 \times 10^{-6}$ M
$[OH^-] = 1.8 \times 10^{-9}$ M
TY 11.4.2 pH = 8.2
TY 11.4.3 pH = 12.3
TY 11.4.4 $[H_3O^+] = 4 \times 10^{-11}$ M
TY 11.4.5 pH = 12.25

1 $[OH^-] = 1.00 \times 10^{-12}$ M 2 $[OH^-] = 1.75 \times 10^{-5}$ M
3 $[OH^-] = 1.0 \times 10^{-10}$ M 4 pH = 4
5 pH = 2 6 $[H_3O^+] = 1.0 \times 10^{-6}$ M 7 pH = 1.3
8 a HCl – strong acid, CH_3COOH – weak acid, NaOH – strong base, NH_3 – weak base.
b 0.01 M HCl: $[H_3O^+] = 0.01$ M; $[OH^-] = 1 \times 10^{-12}$ M
0.01 M NaOH: $[H_3O^+] = 1 \times 10^{-12}$ M; $[OH^-] = 0.01$ M
c 0.01 M HCl: pH = 2
0.01 M NaOH: pH = 12
9 If it were a strong acid, the pH for a 0.1 M solution would be 1. For a 0.1 M strong acid, which fully ionises, the $[H_3O^+] = 0.1$ M or 1×10^{-1} M, so the pH = 1.
Because the pH is 4.3, this indicates the acid has not fully ionised and so it must be a weak acid solution.

11.5 Measuring pH

1 a colourless b yellow c greenish yellow
2 Universal indicator is a mixture of different indicators.
3 a yellow
b The solution would change colour from yellow to violet.
4 a pH = 0–1 b pH = 8.5–14 c pH = 5–6
5 a between blue-violet and red—possibly magenta.
b red
c blue or blue-violet
6 a methyl orange b methyl violet
7 a strong base b weak acid c neutral solution
8 a A pH meter was probably used because the true pH of pure water at 25°C is 7.00. These readings are very close to each other and so are precise, and they are very close to the true value and so are accurate.
b Universal indicator is possibly used because these readings are within 1 pH unit and close to each other within 1 pH unit, as expected for universal indicator. This means they are within the range of precision and accuracy of universal indicator.
c These values are within 2 pH units, so red cabbage indicator was probably used. The precision and accuracy is within the expected range for this indicator.

11.6 Acid-base reactions in the environment

CSA: Other impacts of ocean acidity

1 a Krill eggs do not hatch successfully at a lower pH. Therefore, an increase in ocean acidity is predicted to have a harmful effect on the species of plankton and krill.
b Krill and plankton are a major food source for many marine organisms, from small fish, such as sardines, to huge mammals, such as whales. If the number of krill cannot sustain organisms that feed directly on them, such as seals, penguins and small fish, then organisms such as killer whales and larger fish would not survive either. This would lead to an overall decrease in the number of large species that live within specific ecosystems. They are the base of the ocean food web and, ultimately, food for humans.
2 Gradual destruction of coral reefs, which offer protection to coastal communities from storms and erosion. Also destruction of the economies of coastal communities as important tourist attractions are lost.

Key questions

1 a $CO_2(g) \rightleftharpoons CO_2(aq)$
b $CO_2(aq) + H_2O(l) \rightleftharpoons H_2CO_3(aq)$
c Step 1. $H_2CO_3(aq) + H_2O(l) \rightleftharpoons H_3O^+(aq) + HCO_3^-(aq)$
Step 2. $HCO_3^-(aq) + H_2O(l) \rightleftharpoons H_3O^+(aq) + CO_3^{2-}(aq)$
2 a calcium carbonate
b calcification
c $Ca^{2+}(aq) + CO_3^{2-}(aq) \rightleftharpoons CaCO_3(s)$
3 $CO_3^{2-}(aq) + H_3O^+(aq) \rightleftharpoons HCO_3^-(aq) + H_2O(l)$.
4 When equation 1 moves towards the right as CO_2 atmospheric concentrations increase, more CO_2 dissolves in the water in the oceans producing H_2CO_3 (equation 2). The carbonic acid dissociates into HCO_3^- ions and H_3O^+ ions, which increases the $[H_3O^+]$, reducing the pH of the ocean (equation 3).

Chapter 11 review

1 A
2 a NH_4^+ b HCl
c HCO_3^- d $CH_3COOH(aq)$
3 a $PO_4^{3-}(aq) + H_2O(l) \rightarrow HPO_4^{2-}(aq) + OH^-(aq)$
b $H_2PO_4^-$ accepts a proton from water, and acts as a base:
$H_2PO_4^-(aq) + H_2O(l) \rightarrow H_3PO_4(aq) + OH^-(aq)$
$H_2PO_4^-$ donates a proton to the water, and acts as an acid:
$H_2PO_4^-(aq) + H_2O(l) \rightarrow HPO_4^{2-}(aq) + H_3O^+(aq)$
c $H_2S(aq) + H_2O(l) \rightarrow HS^-(aq) + H_3O^+(aq)$
4 A 5 B
6 a Cl^- b OH^-
c O^{2-} d SO_4^{2-}

7 **a** A Brønsted–Lowry acid is a proton donor.
b A strong base is a substance that ionises completely in water.
c The conjugate acid of a base contains one more hydrogen ion (proton) than the base.

8 **a** Sulfuric acid (H_2SO_4) is a diprotic acid because each molecule can donate two protons to a base:
i.e. $H_2SO_4(aq) + H_2O(l) \longrightarrow H_3O^+(aq) + HSO_4^-(aq)$
$HSO_4^-(aq) + H_2O(l) \rightleftharpoons H_3O^+(aq) + SO_4^{2-}(aq)$
The HSO_4^- ion, however, is amphiprotic because it can act as either an acid or a base, depending on the environment. In water it will undergo both acid and base reactions. For example:
As an acid: $HSO_4^-(aq) + H_2O(l) \rightleftharpoons H_3O^+(aq) + SO_4^{2-}(aq)$
As a base: $HSO_4^-(aq) + H_2O(l) \rightleftharpoons OH^-(aq) + H_2SO_4(aq)$
b A strong acid is one that ionises completely in solution (e.g. HCl). A concentrated acid is one in which there is a large amount of acid dissolved in a given volume of solution; for example, 5 M HCl and 5 M CH_3COOH are concentrated acids.

9

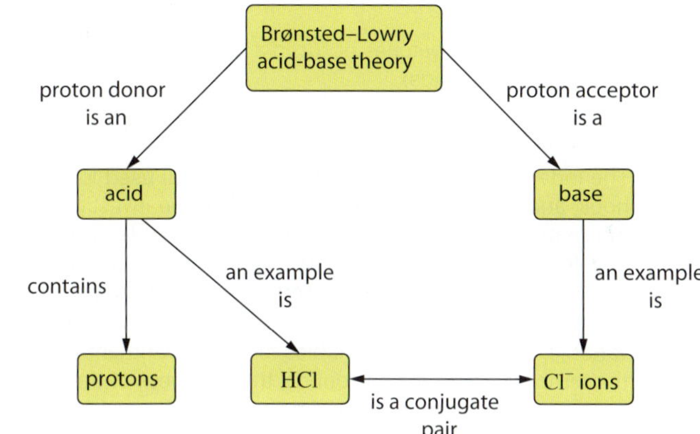

10 A

11 **a** $HClO_4(aq) + H_2O \longrightarrow H_3O^+(aq) + ClO_4^-(aq)$
b $HClO_3(aq) + H_2O(l) \rightleftharpoons H_3O^+(aq) + ClO_3^-(aq)$
c $NH_3(aq) + H_2O(l) \rightleftharpoons NH_4^+(aq) + OH^-(aq)$
d $H_2PO_4^-(aq) + H_2O(l) \rightleftharpoons H_3PO_4\ (aq) + OH^-(aq)$

12 **a** You might use universal indicator in situations where it is more convenient and where the closeness to the true value (accuracy) and exact reproducibility (precision) are not essential. Such cases might be measuring the pH of soil or a swimming pool.
b You would use a pH meter when the pH measurement must be as accurate and precise as possible. Examples might be the measurement of the pH of blood, when determining the exact concentration of an experimental acid solution, where determining the pH would allow the researcher to calculate the [H+].
c Because a pH meter can generally give measurements to within 0.01 of a pH unit, these reading are more accurate (closer to the true value) and can be more readily reproduced within this range (more precise) than universal indicator.

13 **a** Calcification is the process that occurs when aquatic animals, such as such as shellfish, sea stars, coral, sea snails, crabs and lobsters absorb calcium ions and carbonate ions from seawater to build and maintain the calcium carbonate protective structures essential for their survival.
Decalcification is the process where increased H^+ ions in the oceans react with carbonate ions, reducing the concentration of free CO_3^{2-} ions in seawater, and making it more difficult for marine creatures to build or maintain their protective structures.
b Calcification: $Ca^{2+}(aq) + CO_3^{2-}(aq) \rightleftharpoons CaCO_3(s)$
Decalcification: $H_3O^+(aq) + CO_3^{2-}(aq) \rightleftharpoons HCO_3^-(aq) + H_2O(l)$

14 **a** $HNO_3(aq) + KOH(aq) \longrightarrow KNO_3(aq) + H_2O(l)$
b $H_2SO_4(aq) + K_2CO_3(aq) \longrightarrow K_2SO_4(aq) + H_2O(l) + CO_2(g)$
c $2H_3PO_4(aq) + 3Ca(HCO_3)_2(s) \longrightarrow Ca_3(PO_4)_2(s) + 6CO_2(g) + 6H_2O(l)$
d $2HF(aq) + Zn(OH)_2(s) \longrightarrow ZnF_2(aq) + 2H_2O(l)$

15 **a** $2Al(s) + 3H_2SO_4(aq) \longrightarrow Al_2(SO_4)_3(aq) + 3H_2(g)$
b $2Al(s) + 6H^+(aq) \longrightarrow 2Al^{3+}(aq) + 3H_2(g)$

16 **a** $2H^+\ (aq) + MgCO_3(s) \longrightarrow Mg^{2+}(aq) + CO_2(g) + H_2O(l)$
b Antacids are usually liquids or chewable tablets that neutralise the acid in your stomach to relieve the effects of indigestion and heartburn.

17 **a** 10^{-11} M **b** 10^{-9} M **c** 1.8×10^{-6} M
d 2.9×10^{-3} M **e** 1.5×10^{-13} M **f** 4.5×10^{-2} M

18 **a** slightly basic **b** $[H^+] = 10^{-7.4} = 4.0 \times 10^{-8}$

19 D

20 **a** **i** $[H_3O^+] = 0.1$ M **ii** $[OH^-] = 1.0 \times 10^{-13}$ M
b **i** $[H_3O^+] = 1 \times 10^{-3}$ **ii** $[OH^-] = 1 \times 10^{-11}$ M
c **i** $[H_3O^+] = 1 \times 10^{-7}$ **ii** $[OH^-] = 1 \times 10^{-7}$ M
d **i** $[H_3O^+] = 1 \times 10^{-11.7}$ **ii** $[OH^-] = 5 \times 10^{-3}$ M

21 The cola drink is 100 times more acidic than the black coffee.

22 $[OH^-] = 2.5 \times 10^{-9}$ M

23 **a** $[H^+] = 1 \times 10^{-3}$ M
$[OH^-] = 1.0 \times 10^{-11}$ M
b $[H^+] = 1 \times 10^{-10}$ M
$[OH^-] = 1.0 \times 10^{-4}$ M
c $[H^+] = 3.16 \times 10^{-9}$ M
$[OH^-] = 3.16 \times 10^{-6}$ M
d $[H^+] = 1.58 \times 10^{-6}$ M
$[OH^-] = 6.3 \times 10^{-9}$ M
e $[H^+] = 2.5 \times 10^{-10}$ M
$[OH^-] = 4.0 \times 10^{-5}$ M
f $[H^+] = 3.16 \times 10^{-14}$ M
$[OH^-] = 0.316$ M

24 **a** $CO_2(aq)$, $H_2CO_3(aq)$, $H^+(aq)$, $HCO_3^-(aq)$, $CO_3^{2-}(aq)$ and $OH^-(aq)$
b $H_2CO_3(aq) + H_2O(l) \rightleftharpoons HCO_3^-(aq) + H_3O^+(aq)$
$HCO_3^-(aq) + H_2O(l) \rightleftharpoons CO_3^{2-}(aq) + H_3O^+(aq)$

25 Solution A: weaker base, few freely moving charged particles—ammonia
Solution B: neutral, no freely moving charged particles—glucose
Solution C: strong base, many freely moving charged particles—sodium hydroxide
Solution D: strong acid, many freely moving charged particles—hydrochloric acid
Solution E: weaker acid, few freely moving charged particles—ethanoic acid

Chapter 12 Redox reactions

12.1 Introducing redox reactions

TY 12.1.1 Oxidation: $Fe(s) \longrightarrow Fe^{2+}(aq) + 2e^-$
Reduction: $Sn^{2+}(aq) + 2e^- \longrightarrow Sn(s)$

TY 12.1.2 $Ag^+(aq) + e^- \longrightarrow Ag(s)$
The $Ag^+(aq)$ is being reduced.
$Cu(s) \longrightarrow Cu^{2+}(aq) + 2e^-$
The $Ag^+(aq)$ is being reduced.

TY 12.1.3 $2H_2O(l) + 2e^- \longrightarrow H_2(g) + 2OH^-(aq)$
$K(s) \longrightarrow K^+(aq) + e^-$
$2K(s) + 2H_2O(l) \longrightarrow K^+(aq) + H_2(g) + 2OH^-(aq)$

TY 12.1.4 **a** $Cr_2O_7^{2-}(aq) + 14H^+(aq) + 6e^- \longrightarrow 2Cr^{3+}(aq) + 7H_2O(l)$
b $Cr_2O_7^{2-}(aq) + 14H^+(aq) + 6I^-(aq) \longrightarrow 2Cr^{3+}(aq) + 7H_2O(l) + 3I_2(aq)$

1 a reduction b reduction
c oxidation d oxidation

2 a $Fe(s) \longrightarrow Fe^{2+}(aq) + 2e^-$ – oxidation
b $K(s) \longrightarrow K^+(aq) + e^-$ – oxidation
c $F_2(g) + 2e^- \longrightarrow 2F^-(aq)$ – reduction
d $O_2(g) + 4e^- \longrightarrow 2O^{2-}(s)$ – reduction

3 a $SO_4^{2-}(aq) + 4H^+(aq) + 2e^- \longrightarrow SO_2(g) + 2H_2O(l)$
b $H_2O_2(aq) \longrightarrow O_2(g) + 2H^+(aq) + 2e^-$
c $H_2S(g) \longrightarrow S(s) + 2H^+(aq) + 2e^-$
d $MnO_4^-(aq) + 4H^+(aq) + 3e^- \longrightarrow MnO_2(s) + 2H_2O(l)$

4 a Fe(s) has been oxidised to $Fe^{2+}(aq)$.
b $Fe(s) \longrightarrow Fe^{2+}(aq) + 2e^-$
c $H^+(aq)$
d $H^+(aq)$ has been reduced to H_2
e $2H^+(aq) + 2e^- \longrightarrow H_2(g)$
f Fe(s)
g $Fe^{2+}(aq)/Fe(s)$ and $H^+(aq)/H_2(g)$

5 a Magnesium is oxidised, copper ions are reduced.
b $Mg(s) \longrightarrow Mg^{2+}(aq) + 2e^-$
c $Cu^{2+}(aq) + 2e^- \longrightarrow Cu(s)$
d $Mg(s) + Cu^{2+}(aq) \longrightarrow Mg^{2+}(aq) + Cu(s)$
e oxidising agent Cu^{2+}; reducing agent Mg
f The solution loses some of its blue colour due to the loss of $Cu^{2+}(aq)$, which reacts to form Cu(s).

6 a $Sn^{4+}(aq)$ b Sn(s)

7 a K_2O
b K(s)
c $K(s) \longrightarrow K^+(s) + e^-$
d $O_2(g)$
e $O_2(g) + 4e^- \longrightarrow 2O^{2-}(s)$
f $4K(s) + O_2(g) \longrightarrow 2K_2O(s)$
g Potassium has been **oxidised** by **oxygen** to form potassium ions. The **oxygen** has gained electrons from the **potassium**. The oxygen has been **reduced** by **potassium** to form oxide ions. The **potassium** has lost electrons to the **oxygen**.

12.2 Metal displacement reactions

TY 12.2.1 $Cu^{2+}(aq) + Co(s) \longrightarrow Cu(s) + Co^{2+}(aq)$

1 a Al b K c Mg

2 a Ni^{2+} b Cu^{2+} c Ag^+

3 Oxidising agents in order from weakest to strongest: Na^+, Al^{3+}, Cr^{3+}, Fe^{2+}, Pb^{2+}, Au^+

Reducing agents in order from weakest to strongest: Ag, Cu, Ni, Zn, Mn, Mg

4 For reactions to occur spontaneously, the aqueous cation in the solution must be a stronger oxidising agent than the cation of the metal added.
a Yes b No c Yes
d Yes e Yes

5 a $3Cu^{2+}(aq) + 2Al(s) \longrightarrow 3Cu(s) + 2Al^{3+}(aq)$
b no reaction
c $2Ag^+(aq) + Zn(s) \longrightarrow 2Ag(s) + Zn^{2+}(aq)$

6 Use chromium to test the solutions.

The iron(II) nitrate will react with the chromium, but the zinc nitrate will not. This is because $Zn^{2+}(aq)$ is not a strong enough oxidising agent to react with Cr, but $Fe^{2+}(aq)$ is.

12.3 Redox reactions in society

TY 12.3.1 a reducing agent: Ni

oxidising agent: Cu

b oxidation half-equation: $Ni(s) \longrightarrow Ni^{2+}(aq) + 2e^-$
reduction half-equation: $Cu^{2+}(aq) + 2e^- \longrightarrow Cu(s)$

c The electrons will flow from the electrode which is the reducing agent, through the external circuit, to the other electrode.

d, e

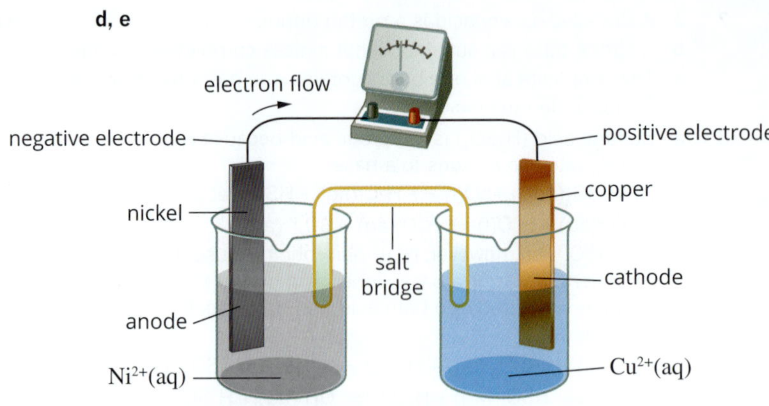

CSA: The earliest batteries

1 Iron is a better reducing agent than copper, so iron would be the anode (where oxidation occurred) and copper would be the cathode.

2 Two possibilities:
- count how many twitches occur in a fixed period of time (perhaps 30 seconds)
- use a ruler to measure how far apart the legs become when they twitch

3 a The independent variable could be the metal used for the hook, and metals other than copper could be investigated. Alternatively, other possible independent variables could be how long the frog had been dead, how long its legs were, and what species of frog is being used in the experiment.

b If the first option in part **a** was used for the independent variable, then controlled variables for the experiment would be: the temperature of the surrounds, the thickness of the copper hook, the size of the frog, the time since the frog had died, the time since the frog had been skinned, the species of frog.

If other options are used for the independent variable, then the controlled variables should be as above, but including the metal used for the hook, and not including the selected independent variable.

Key questions

1 The electrodes in a half-cell need to conduct electricity and to be able to connect to the wires of the external circuit, so they must be metal or graphite.

2 The salt bridge in a galvanic cell completes the circuit by allowing positive and negative ions to move between the two half-cells.

3

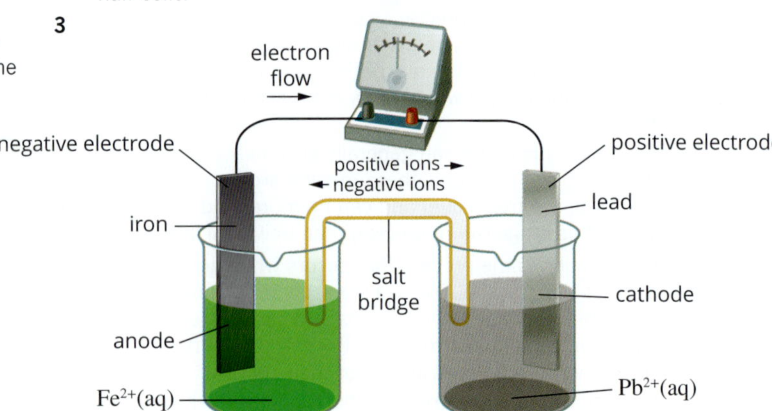

4 A cell is a single pair of electrodes in an electrolyte, typically made up of two half-cells, whereas a battery is made up of two or more cells connected in series.

5 a zinc, Zn
b chromium, Cr
c iron, Fe

6 Primary cells provide a lightweight, portable source of electricity that is useful in wireless devices and children's toys. Because various metals, which are a non-renewable resource, are used in the construction of these cells, they must be recycled, otherwise we will run out of them.

7

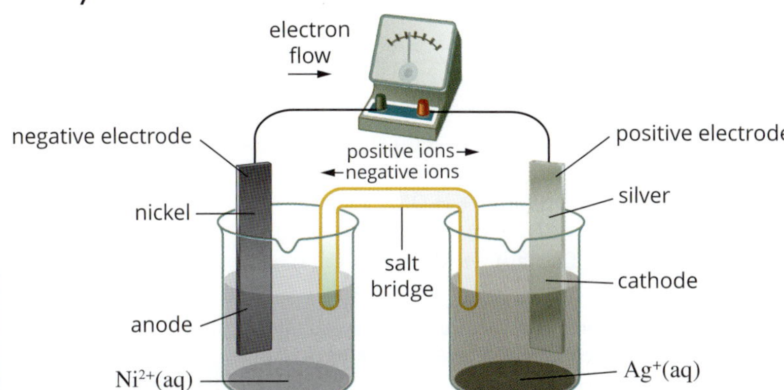

Chapter 12 review

1 a Oxidation is the gain of oxygen atoms; reduction is the loss of oxygen atoms.
b Oxidation is the loss of electrons; reduction is the gain of electrons.

2 The reactivity series arranges metals (and other elements) in order of their ability to react with **oxidising** agents. The most reactive metals are found at the **bottom** of the reactivity series, on the **right**-hand side. For a redox reaction to occur, an oxidising agent must react with a **reducing** agent. In this reaction the oxidising agent must be **higher** in the reactivity series than the conjugate **oxidising** agent of the reducing agent in the reaction. During the reaction, the oxidising agent **gains** electrons and the reducing agent **loses** electrons.

3 a oxidation b reduction
c reduction d oxidation

4 B

5 a $Ag^+(aq) + e^- \rightarrow Ag(s)$
b $Cu(s) \rightarrow Cu^{2+}(aq) + 2e^-$
c $Zn(s) \rightarrow Zn^{2+}(aq) + 2e^-$

6 magnesium, aluminium, zinc, iron, tin, silver

7 C

8 A

9

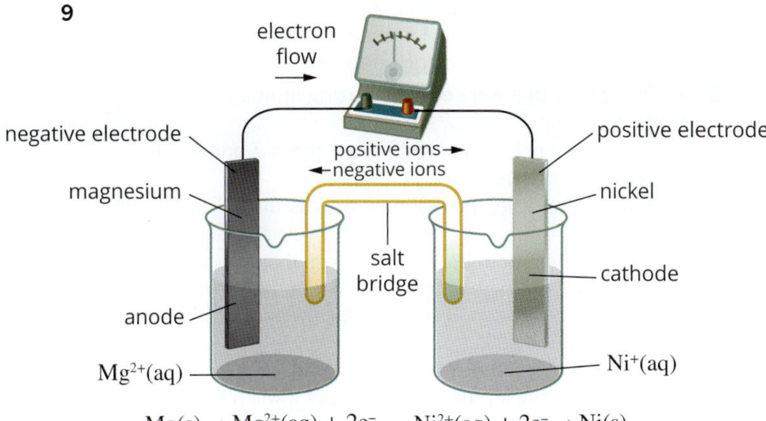

10 D

11 Different answers are possible. The intention here is that the concept map will show how the **anode** is where **oxidation** occurs and the cathode is where **reduction** occurs; a **galvanic cell** is made up of two **half-cells**; the **negative** electrode is the source of **electrons** and the **electrons** move towards the **positive** electrode.

12 a strongest: $Ag^+(aq)$; weakest: $Mg^{2+}(aq)$
b strongest: Mg(s); weakest: Ag(s)
c A coating of silver will form on the lead when it is placed in silver nitrate solution because Ag^+ ions are stronger oxidising agents than Pb^{2+} ions. Neither of the other oxidising agents are strong enough to react.
d zinc and magnesium

13 a spontaneous reaction will occur: $Zn(s) + 2Ag\ NO_3(aq) \rightarrow Zn(NO_3)_2(aq) + 2Ag(s)$
b no reaction
c no reaction
d spontaneous reaction will occur: $Mg(s) + Pb(NO_3)_2(aq) \rightarrow Mg(NO_3)_2(aq) + Pb(s)$
e no reaction
f spontaneous reaction will occur: $Fe(s) + Cu(NO_3)_2(aq) \rightarrow Fe(NO_3)_2(aq) + Cu(s)$

14 Coatings of metals other than iron would be expected on the nails placed in 1 M solutions of $CuSO_4$ and $Pb(NO_3)_2$. These solutions contain oxidising agents that are strong enough to react with iron metal. The iron, in turn, reduces the metal ions in the solution, forming a coating on the nail.

15 The metal could be manganese, zinc or chromium.

16 a Most mobile phones have a single rechargeable cell due to the constrictions of a lack of space for the battery and a desire for the weight of the phone to be kept to a minimum.
b An alkaline cell is most likely to be used for a wireless mouse, as it is cheap and can be replaced quickly when it goes flat, rather than having to wait for it to be recharged.
c Electric cars have rechargeable batteries in them with many cells because the energy requirements of the car are very high.
d An energy storage unit for a home solar power system would also be a rechargeable battery, as there is a need for constant recharging, and a lot of energy needs to be stored.

17 a $Ta_2O_5(s) + 10H^+(aq) + 10e^- \rightarrow 2Ta(s) + 5H_2O(l)$
b $SO_3^{2-}(aq) + H_2O(l) \rightarrow SO_4^{2-}(aq) + 2H^+(aq) + 2e^-$
c $IO_3^-(aq) + 6H^+(aq) + 6e^- \rightarrow I^-(aq) + 3H_2O(l)$

18 a $[H_2O_2(aq) + 2H^+(aq) + 2e^- \rightarrow 2H_2O(l)] \times 4$
$PbS(s) + 4H_2O(l) \rightarrow PbSO_4(s) + 8H^+(aq) + 8e^-$
$4H_2O_2(aq) + PbS(s) \rightarrow PbSO_4(s) + 4H_2O(l)$
b $I_2(aq) + 2e^- \rightarrow 2I^-(aq)$
$H_2S(g) \rightarrow S(s) + 2H^+(aq) + 2e^-$
$I_2(aq) + H_2S(g) \rightarrow 2I^-(aq) + S(s) + 2H^+(aq)$
c $[SO_3^{2-}(aq) + H_2O(l) \rightarrow SO_4^{2-}(aq) + 2H^+(aq) + 2e^-] \times 5$
$[MnO_4^-(aq) + 8H^+(aq) + 5e^- \rightarrow Mn^{2+}(aq) + 4H_2O(l)] \times 2$
$5SO_3^{2-}(aq) + 2MnO_4^-(aq) + 6H^+(aq) \rightarrow 5SO_4^{2-}(aq) + 2Mn^{2+}(aq) + 3H_2O(l)$
d $[NO(g) + 2H_2O(l) \rightarrow NO_3^-(aq) + 4H^+(aq) + 3e^-] \times 2$
$Cr_2O_7^{2-}(aq) + 14H^+(aq) + 6e^- \rightarrow 2Cr^{3+}(aq) + 7H_2O(l)$
$2NO(g) + Cr_2O_7^{2-}(aq) + 6H^+(aq) \rightarrow 2NO_3^-(aq) + 2Cr^{3+}(aq) + 3H_2O(l)$

19 Place a small volume (approx. 10 mL) of each solution into separate clean test tubes. Add a small piece of magnesium ribbon to each solution. The magnesium should be coated by displaced metal in the silver nitrate and tin(II) nitrate solutions, but not in the potassium nitrate. To confirm the identity of the remaining two solutions, take a fresh sample of the two solutions which reacted with the magnesium, in two separate test tubes, and add a small piece of copper. Copper will displace silver from the solution, giving silver deposit and a blue solution. Copper will not displace tin from the solution.

20 The galvanic cell will not generate electricity because the zinc is in the same half-cell as the Cu^{2+} solution, so there will be a direct reaction between these two and no electrons will be transferred to the other half-cell. Also the other half-cell has the wrong electrode in it, lead, but this will not react with the Zn^{2+} solution.

To ensure that electricity is produced, the zinc electrode must be placed in the Zn^{2+} solution (it may need to be cleaned first) and a copper electrode should be placed in the Cu^{2+} solution.

21 in order of increasing reactivity: X, Z, W, Y

22 **a** If as much as possible is recycled, supplies of lithium would be prolonged more than those of cobalt.

b 1.4×10^9 electric vehicle battery packs

c Student answers may vary.

Unit 2 Area of Study 1 Review

1	A	**2**	B	**3**	D	**4**	A
5	B	**6**	A	**7**	D	**8**	D
9	C	**10**	A	**11**	B	**12**	D
13	D	**14**	D	**15**	B		

16 **a** 1.0 M hydrochloric acid; lemon juice; distilled water; 1.0 M sodium hydroxide

b The first in the list is 1.0 M hydrochloric acid, because it is a stronger acid than lemon juice, so will have a lower pH than lemon juice. The estimated pH of 1.0 M HCl is 0.

The last substance in the list will be 1.0 M sodium hydroxide. This is a strong base and will have an estimated pH of 14. There is no other base in the list, so 1.0 M sodium hydroxide will be last.

17 The stronger the intermolecular bonds, the higher the boiling point of a substance.

a Water molecules are highly polar, since hydrogen is bonded to oxygen. So, between water molecules, there are hydrogen bonds as well as weak dispersion forces.

H_2S is much less polar, so between H_2S molecules there are dipole–dipole interactions and weak dispersion forces. As H_2S molecules are larger than water molecules, the dispersion forces between H_2S molecules are more significant than those between water molecules. However, hydrogen bonds are much stronger than dipole–dipole interactions, so, overall, the intermolecular bonds between H_2S molecules are weaker than those between water molecules and the boiling point of H_2S is consequently lower.

b The polarity of H_2Se is about the same as that of H_2S. However, H_2Se consists of much larger molecules, so the dispersion forces between them are more significant. Therefore, overall, the intermolecular forces between H_2S molecules are weaker than those between H_2Se molecules and the boiling point of H_2S is consequently lower.

18 **a** Water has a high latent heat of vaporisation (40.7 kJ mol^{-1}). This means that 40.7 kJ of energy is needed to change 1 mol of water (18 g) from liquid to gas. This is due to the strength of the hydrogen bonds between water molecules. To change the water from liquid to gas, most of these bonds must break and considerable energy is needed to do this.

b Because considerable energy is needed to change liquid water into gaseous water, and also because the volume of the oceans is so massive, the energy available from the Sun is not enough to change the temperature of the oceans by any great amount, so the temperature stays somewhat constant.

19 **a** 8.6×10^4 J

b 8.0×10^3 J

c 13°C

d The high heat capacity of water is due to the relatively strong hydrogen bonds between water molecules.

e Water is used as a coolant in factories and in car radiators

20 An increase in the level of atmospheric carbon dioxide since the Industrial Revolution in the mid-1700s has resulted in an increased amount of dissolved carbon dioxide in the oceans. CO_2 is an acidic gas that forms carbonic acid with water.

$$CO_2(g) + H_2O(l) \rightarrow H_2CO_3(aq)$$

and

$$H_2CO_3(aq) \rightarrow HCO_3^-(aq) + H^+(aq)$$

So with the increased concentration of $H^+(aq)$, the pH of the oceans has decreased.

You may have mentioned sulfur dioxide in your answer. While sulfur dioxide is a gas which dissolves in water to create acid rain, its presence has resulted in localised effects rather than the change in the oceans.

21 **a** **i** $2HCl(aq) + Zn(s) \rightarrow ZnCl_2(aq) + H_2(g)$

ii $2HNO_3(aq) + Ca(OH)_2(aq) \rightarrow Ca(NO_3)_2(aq) + 2H_2O(l)$

iii $H_2SO_4(aq) + Na_2CO_3(s) \rightarrow Na_2SO_4(aq) + CO_2(g) + H_2O(l)$

iv $Cu(s) + 2AgNO_3(aq) \rightarrow Cu(NO_3)_2(aq) + 2Ag(s)$

b **i** $Zn(s) + 2H^+(aq) \rightarrow Zn^{2+}(aq) + H_2(g)$

ii $H^+(aq) + OH^-(aq) \rightarrow H_2O(l)$

iii $2H^+(aq) + CO_3^{2-}(aq) \rightarrow CO_2(g) + H_2O(l)$

iv $Mg(s) + Cu^{2+}(aq) \rightarrow Mg^{2+}(aq) + Cu(s)$

22 **a** 3.00 **b** 11.4 **c** 3.12 **d** 12.5

23 **a** A strong acid is an acid that readily donates protons. Or, an acid that completely ionises in water.

b H_2CO_3 or H_2SO_4 or H_3PO_4 (or other correct polyprotic acids)

c **i** $HCO_3^-(aq) + H_2O(l) \rightarrow H_2CO_3(aq) + OH^-(aq)$

ii $HCO_3^-(aq) + H_2O(l) \rightarrow CO_3^{2-}(aq) + H_3O^+(aq)$

d **i** H_3O^+

ii OH^-

24 $[H_3O^+] = 10^{-pH}$

For seawater with pH = 8.15, $[H_3O^+] = 10^{-8.15} = 7.07 \times 10^{-9}$ M (to two sig figs: 7.1×10^{-9} M)

For seawater with pH = 7.85, $[H_3O^+] = 10^{-5.85} = 1.4 \times 10^{-8}$ M

Since $1.4 \times 10^{-8} \approx 2 \times 7.0 \times 10^{-9}$, so the seawater with pH of 7.85 is 100% more acidic (twice as acidic) as the seawater with pH of 8.14.

25 **a** Statement III is correct because ethanoic acid is a weak acid, while hydrochloric acid is a strong acid. Weak acids react less vigorously with calcium carbonate than strong acids.

Statement IV is correct because both acids are monoprotic acids, so 1 mole of the acid will react with (neutralise) 1 mole of sodium hydroxide, NaOH. Because the two solutions have the same concentration (0.10 mol L^{-1}), the same volume of each acid will produce the same number of mole of H^+ ions, so will neutralise the same amount of NaOH.

b Statement I is incorrect because hydrochloric acid is a strong acid and ethanoic acid is a weak acid. While the two acids are not the same strength, they are the same concentration.

Statement II is incorrect because the concentration of H^+ ions in the 0.10 mol L^{-1} solution of HCl will be greater than that in the 0.10 mol L^{-1} solution of ethanoic acid (because it is a stronger acid), so the pH of the HCl will be lower than that of the ethanoic acid, since $pH = -\log[H_3O^+]$.

26 **a** The results that were obtained using litmus paper are the least precise.

b D. It is the only solution with a pH above 7 and sodium hydroxide is a base.

c B. As it is an acid, the pH must be less than 7. So it cannot be C or D.

A and B must both be acids (pH less than 7 or turns litmus red). B has the higher conductivity and, as they all have the same concentration, B must ionise the most and so must be the stronger acid.

d The pH of A must be greater than 1.5 but less than 7. It is an acid as it turns litmus paper red, but it will have a higher pH than solution B because it has a lower conductivity.

e i 0.03 M
ii 3×10^{-11} M

f The pH of solution C (7.0) and the fact that it does not change the colour of either of the litmus papers suggests that solution C is water. While the pH could point to another neutral solution, such as NaCl(aq), the lack of conductivity demonstrates that there are no ions present. Water has a very low concentration of ions $[H_3O^+] = [OH^-] = 1.0 \times 10^{-7}$ M and this is too small to conduct electricity to any measurable extent.

27

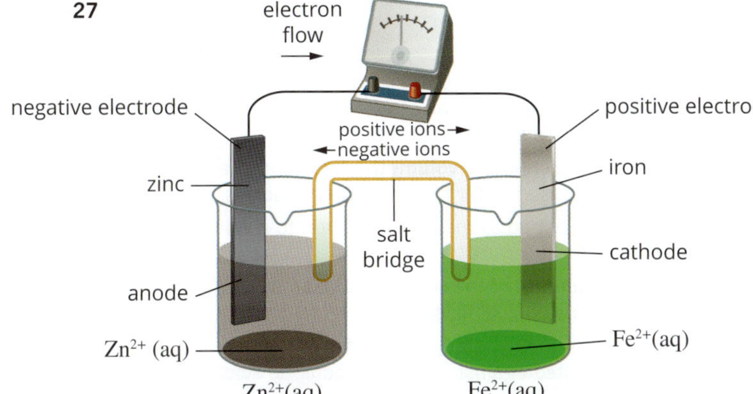

a The reducing agent is zinc and the oxidising agent is Fe^{2+}(aq)

b The reactions occurring at the two half-cells are: Zn(s) $\longrightarrow$ Zn^{2+}(aq) + $2e^-$ and Fe^{2+}(aq) + $2e^-$ $\longrightarrow$ Fe(s)

c The electrons are flowing through the external circuit (the wire) from the Zn(s)/Zn^{2+}(aq) half-cell to the Fe(s)/Fe^{2+}(aq) half-cell (as shown in the diagram).

d The negative electrode is the zinc electrode. The positive electrode is the iron electrode.

e The anode is the zinc electrode and the cathode is the iron electrode.

Chapter 13 Measuring solubility and concentration

13.1 Measuring solubility

TY 13.1.1 140 g

TY 13.1.2 210 g

TY 13.1.3 10 g

CSA: Using precipitation reactions in the purification of Melbourne's water

1 a CaO(s) + H_2O(l) $\longrightarrow$ $Ca(OH)_2$(aq)
b $Ca(OH)_2$(aq) $\longrightarrow$ Ca^{2+}(aq) + $2OH^-$(aq)

2 Fe^{3+}(aq) + $3OH^-$(aq) $\longrightarrow$ $Fe(OH)_3$(s)

Key questions

1 The solubility of a substance can be measured by how much **solute** will dissolve in a given quantity of **solvent** at a given **temperature**. An **unsaturated** solution is one which is able to dissolve more solute at a given temperature. A **supersaturated** solution is one in which there is more solute dissolved at a given temperature than is necessary to make the solution saturated. A **saturated** solution is one which contains the maximum amount of dissolved solute at a given temperature.

2
- Any point below a solubility curve represents **an unsaturated solution**.
- Any point above a solubility curve represents **a supersaturated solution**.
- Any point on a solubility curve represents **a saturated solution**.

3 a 120 g b 50 g c 40 g

4 a 120 g b 24 g c 360 g

5 a 80 g b 32 g c 80 g

6 a unsaturated b saturated
c unsaturated d supersaturated
e unsaturated

7 30 g

8 a Adding sodium sulfate solution would not form a precipitate. Anions from sodium sulfide solution would form precipitates with chromium(IIII) and nickel(II) ions in solution. The precipitated metal compounds could then be removed from the wastewater.
b $2Cr^{3+}$(aq) + $3CO_3^{2-}$(aq) $\longrightarrow$ $Cr_2(CO_3)_3$(s)
c hydroxide, carbonate, phosphate and sulfide compounds of sodium, potassium and ammonium

13.2 Calculating concentration

TY 13.2.1 20.0 g L^{-1}

TY 13.2.2 9.90 ppm

TY 13.2.3 0.48 M

TY 13.2.4 0.667 M

TY 13.2.5 0.0025 mol

TY 13.2.6 0.0250 M

TY 13.2.7 c(NaOH) = 0.400%(m/v)

CSA: Diluting strong acids

1 16 mL 2 2.5 g

Key questions

1 Every concentration unit has two parts—a first part and a second part. The first part provides information about the **quantity of solute**; the second part provides information about the **quantity of solution**. For example, if a solution has a concentration of 1.6%(m/v), this indicates that in the solution there is **1.6 g** of solute dissolved in **100 mL** of solution.

2 a 48.5 g L^{-1} b 57%(v/v) c 13 ppm

3 a The formula for calculating the molar concentration, *c*, of a solution is: $c = \frac{n}{V}$
b The formula for calculating the number of moles, *n*, of solute in a solution is: $n = c \times V$
c The formula for calculating the volume, *V*, of a solution is: $V = \frac{n}{c}$

4 a increases b decreases c does not change
d increases e does not change.

5 a 3.4 M b 0.74 M c 0.035 M
d 0.14 M e 0.038 M

6 a i 0.20 mol ii 0.078 mol iii 5.6×10^{-4} mol
b i 5.3 g ii 2.3 g

7 a 20 mL b 6.4 M c 31 mL d 15 g L^{-1}

8 a i 95 g L^{-1} ii 2.1 mol L^{-1}
b c(ethanol) = 12.0%(v/v)

Chapter 13 review

1 B 2 C 3 A 4 D

5 a 48 g b 200 g c 70 g

6 a 70°C b 10°C c 60°C

7 a c(NaOH) = 0.30 M
b $c(NH_3)$ = 7.12 M
c $c(HNO_3)$ = 0.67 M

8 a i 0.018 mol ii 0.075 mol iii 0.0395 mol
b i 1.1 g ii 22 g iii 3.87 g

9 a 0.11 L b 20 L

10 a 2.4×10^{-3} M b 0.11 M

11 a 190 mL b 9.8 L

12

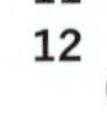

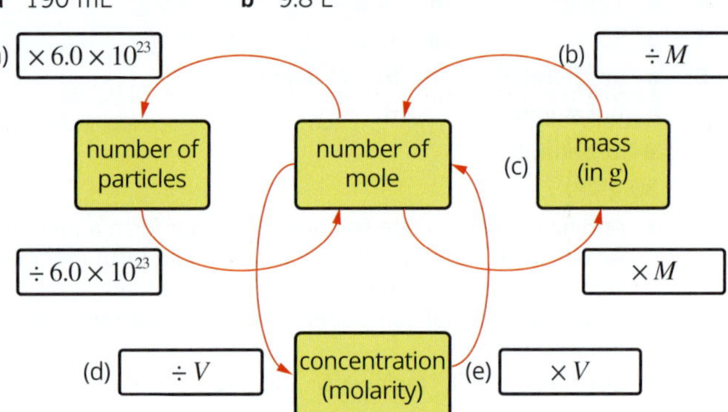

13 False

At 30°C $NaNO_3$ will form a saturated solution when 102 g dissolves in 100 g water. A supersaturated solution of $NaNO_3$ in 200 g water would contain more than 2 × 102 = 204 g.

At 20°C KNO_3 will form a saturated solution when 36 g dissolves in 100 g water. A saturated solution of KNO_3 in 400 g water would contain a maximum of 4 × 36 = 144 g.

14 a For most gases, as temperature increases the amount of gas dissolved in water decreases.

b The bubbles of gas that appear at the bottom of a heated saucepan before boiling point is reached are bubbles of air. The bubbles contain mainly nitrogen and oxygen, which make up the largest proportion of gases in the air. The bubbles form because as the temperature increases, the solubility of the gases in water decreases.

c 5.6 L

15 60 g

16 the 200 litre tank

17 the solubility of potassium nitrate in water at 35°C = 52 g/100 g.

18 a $Al^{3+}(aq) + PO_4^{3-}(aq) \longrightarrow AlPO_4(s)$

b $Cd^{2+}(aq) + 2OH^-(aq) \longrightarrow Cd(OH)_2(s)$

c $Hg^{2+}(aq) + S^{2-}(aq) \longrightarrow HgS(s)$

19 $Pb^{2+}(aq) + SO_4^{2-}(aq) \longrightarrow PbSO_4(s)$

$Pb^{2+}(aq) + 2OH^-(aq) \longrightarrow Pb(OH)_2(s)$

$Cr^{3+}(aq) + 3OH^-(aq) \longrightarrow Cr(OH)_3(s)$

With few exceptions, compounds containing a hydroxide ion are insoluble in water. Lead(II) hydroxide and chromium(III) hydroxide are not included in the exceptions, so these compounds would form precipitates.

Most sulfates are soluble in water, but lead(II) sulfate is an exception and so would form a precipitate.

All nitrate compounds are soluble in water and so would not form a precipitate.

20 3.06 L

21 a $Al_2(SO_4)_3(s) \xrightarrow{H_2O(l)} 2Al^{3+}(aq) + 3SO_4^{2-}(aq)$

b i 0.0777 mol L^{-1}

ii 0.233 mol L^{-1}

iii 0.155 mol L^{-1}

c 2.24%(m/v)

22 a 0.36 mol L^{-1} b 0.72 mol L^{-1}.

23 a 2.4 g b 1.4 g

c 0.039 g d 16 g

24

	Concentration		
compound/ion	%(m/v)	mol L^{-1}	ppm
NaOH	a 0.400	0.100	b 4×10^3
Cd^{2+}	c 3.00×10^{-4}	d 2.67×10^{-5}	3.00

25 a 3.42%(m/v)

b 0.856 mol L^{-1}

c 34.2 g L^{-1}

26 a i 2.99×10^{-5} mol L^{-1} ii 1.9×10^{-4} %(m/v)

b 1.9 ppm

Chapter 14 Analysis for acids and bases

14.1 Principles of volumetric analysis

TY 14.1.1 1.03 M

1 A

2 A sample of anhydrous sodium carbonate of approximately 2 g is weighed accurately. (The solid must be dry if it is to be used as a **primary standard**.)

The solid is tipped into a **volumetric flask** and shaken with about 50 mL of deionised water until the solid dissolves.

More water is added to make the solution to a volume of exactly 100.0 mL.

A 20.00 mL **aliquot** of the solution is taken by using a **pipette** and placed in a conical flask.

A few drops of methyl orange **indicator** are added and the mixture is titrated against dilute hydrochloric acid.

3 a A standard solution is a solution of accurately known concentration. A primary standard is a substance that is readily obtained in a pure form, has a known formula and can be stored without deteriorating or reacting with the atmosphere. It should also be cheap and have a high molar mass.

b The equivalence point in a titration occurs when the reactants have been mixed in the mole ratio shown by the reaction equation. The end point occurs when the indicator changes colour.

c A burette is a piece of equipment capable of delivering different volumes of a liquid accurately (generally up to 50.00 mL). Pipettes usually deliver only a fixed volume of liquid (e.g. 20.00 mL).

d An aliquot is the volume of liquid delivered from a pipette. A titre is delivered by a burette and is the volume needed to reach the end point of a titration.

4 The total volume of solution must be 500 mL. Adding 500 mL to a solid does not produce 500 mL of solution.

5 Average titre of sulfuric acid = 25.42 mL

6 a i methyl orange ii phenolphthalein

iii phenol red iv methyl red

b yellow

7 0.1728 M

8 a 13.8 g b 4.8 g

9 The volumetric flask should be rinsed with deionised water, the pipette with standardised sodium hydroxide solution, the burette with ethanoic acid (CH_3COOH) solution and the conical flask with deionised water.

14.2 Stoichiometry

TY 14.2.1 1.00 mol

TY 14.2.2 10.0 mL

TY 14.2.3 0.108 M

TY 14.2.4 6.45 M

CSA: Determination of the ammonia content of window cleaner

1 The concentrations of commercial supplies of strong acids such as HCl cannot be accurately specified. The stock HCl solution must first be diluted and then standardised by titration to determine the accurate concentration of the solution.

2 No. The average titre is not appropriate as the student has used Trials 3–5 to calculate the average. Trial 4 is not concordant as the difference between the lowest titre (24.16) and the highest (24.29) is 0.13 mL, which is greater than the 0.10 mL range required. The average titre should be calculated using Trials 2, 3 and 5. Average titre = 24.30 mL.

3 c(HCl) in the window cleaner = 2.27 M

4 Step 1: The student added 250 mL of deionised water instead of adding 225 mL of deionised water, so that the bottom of the meniscus is on the calibration line.

5 Examples of systematic errors: The concentration of the HCl solution is not exactly 0.187 M; the 20.00 mL pipette may deliver slightly more or less than 20.00 mL; there may be other components in the window cleaner other than ammonia that react with the HCl.

6 **i** and **iv**
- **i** Rinsing the 20.00 mL pipette with deionised water only will further dilute the diluted window cleaner, resulting in a lower average titre of HCl. The calculated ammonia concentration will be lower than the specified concentration.
- **ii** Rinsing the burette with deionised water only will dilute the HCl, therefore a larger volume of HCl is required to react with the ammonia in the diluted window cleaner aliquot. The average titre will be higher, meaning that the calculated ammonia concentration will be higher than the specified concentration.
- **iii** The volumetric flask should be rinsed with deionised water only, so there is no effect on the calculated concentration.
- **iv** Using phenolphthalein indicator will result in the end point occurring before the equivalence point, resulting in a lower average titre. The calculated ammonia concentration will be lower than the specified concentration.

7 PPE: Ammonia solution causes skin irritation and serious eye irritation therefore gloves and safety glasses should be worn. Ammonia solution may cause respiratory irritation, so work in a fume cupboard or well-ventilated area.

Key questions

1 **a** $\frac{n(HNO_3)}{n(CO_2)} = \frac{2}{1}$

b $\frac{n(HNO_3)}{n(Na_2CO_3)} = \frac{2}{1}$

c $\frac{n(Na_2CO_3)}{n(H_2O)} = \frac{1}{1}$

2 The recommended order for completing solution volume–volume stoichiometry calculations is:

Step 1: Write a balanced equation for the reaction.

Step 2: Identify the known and unknown substances in the question.

Step 3: Calculate the amount, in mol, of the known substance using $n = c \times V$.

Step 4: Use mole ratios from the equation to calculate the amount of the unknown substance.

Step 5: Calculate the volume of the unknown substance using $V = \frac{n}{c}$.

There is no single correct order as steps 1, 2 and 3 can be done in any order.

3 0.685 mol

4 **a** $HNO_3(aq) + KOH(aq) \rightarrow KNO_3(aq) + H_2O(l)$
b 0.001 985 mol
c 0.001 985 mol
d 0.1087 M

5 **a** 32.0 mL **b** 0.229 M

6 0.714 M

7 **a** $Na_2CO_3(aq) + 2HCl(aq) \rightarrow 2NaCl(aq) + CO_2(g) + H2O(l)$
b $n(Na_2CO_3) = 0.0127$ mol
c $n(HCl) = 0.0254$ mol
d c(HCl) in the diluted cleaner = 1.06 M
e c(HCl) in the commercial concrete cleaner = 10.6 M

8 **a** 0.047 06 M
b $Na_2CO_3(aq) + 2HCl(aq) \rightarrow 2NaCl(aq) + CO_2(g) + H_2O(l)$
c yellow to pink
d 20.98 mL
e 0.089 72 M

Chapter 14 review

1 0.2645 M

2 **a** A primary standard has a very high level of purity; has a known formula; is stable (e.g. will not react with atmospheric gases, such as carbon dioxide and water vapour); has a high molar mass to minimise errors in weighing; is readily available; and is relatively inexpensive.

b
- Accurately weigh the primary standard on an analytical balance.
- Transfer the weighed sample to a volumetric flask using a dry glass funnel.
- Rinse any remaining solid particles into the flask using deionised water.
- Half-fill the volumetric flask with water and shake to dissolve the sample.
- When the sample has dissolved, add water to the calibration mark and shake the flask again.
- Determine the concentration of the primary standard.

3 Solid sodium hydroxide reacts with carbon dioxide and absorbs water from the atmosphere.

4 **a** 24.22, 24.20 and 24.16 mL
b 24.19 mL

5 titration 1—phenolphthalein
titration 2—methyl orange

6 **a** A standard solution of potassium hydrogen phthalate is prepared from a primary standard and has an accurately known concentration.
b Students' answers will vary; for example, **iii**, **i**, **vii**, **ii**, **iv**, **ix**, **v**, **viii**, **vi**. Other correct answers are possible, but **i** comes after **iii**, **iv** after **vii** and **ii**, **ix** after **iv**, and **viii** after **ix** and **v**.
c Concordant results are titres that are within a 0.10 mL range, from highest to lowest.
d The 25.70 mL titre is discarded as it is not concordant. Using the concordant titres, 25.12, 25.10, and 25.14 mL, the average titre is 25.12 mL.
e Random errors follow no regular pattern. For example, the differences between the concordant titres could be explained by the difficulty judging the fraction between the two 0.1 mL scale markings on the burette.

7 **a** $HNO_3(aq) + KOH(aq) \rightarrow KNO_3(aq) + H_2O(l)$
b 0.001 971 mol
c 0.001 971 mol
d 0.1018 M

8 **a** $H_2SO_4(aq) + 2KOH(aq) \rightarrow 2K_2SO_4(aq) + 2H_2O(l)$
b 15.1 mL

9 **a** $2HNO_3(aq) + Ba(OH)_2(aq) \rightarrow Ba(NO_3)_2(aq) + 2H_2O(l)$
b 0.116 M

10 0.953 M

11 **a** phenolphthalein
b The pH decreases. The basic solution was in the conical flask; therefore, as acid is added, the concentration of H^+ ions increases and the pH decreases.

c The relative strengths of the acid and base being titrated determine the pH of the solution once the acid and base have been mixed in the correct mole ratios. Therefore, the pH will be greater than 7 as NaOH is a strong base and ethanoic acid is a weak acid.

12 A

13 C

14 a $H_2SO_4(aq) + K_2CO_3(aq) \longrightarrow K_2SO_4(aq) + H_2O(l) + CO_2(g)$

b 0.035 51 M

c 0.0315 M

15 a 0.041 66 M

b 0.007 100 M

16 a potassium hydrogen phthalate

b i vinegar ii deionised water iii NaOH solution

c Rinsing the pipette with deionised water would further dilute the diluted vinegar. This means less sodium hydroxide solution would need to be added to neutralise the ethanoic acid in the diluted vinegar. It would then appear that a smaller amount of sodium hydroxide reacted; hence, it would appear that a smaller amount of ethanoic acid had reacted. The calculated concentration of ethanoic acid in the vinegar would be lower than the actual value.

d colourless to pink.

17 a 0.000 849 mol

b 0.001 70 mol

c 0.0849 M

d 4.76 g L^{-1}

18 a 0.182 M

b i The calculated concentration of the ammonia solution would be lower. Rinsing the pipette with deionised water would slightly dilute the ammonia solution. This means less hydrochloric acid would need to be added to neutralise the ammonia. It would then appear that a smaller amount of hydrochloric acid reacted; hence, it would appear that a smaller amount of ammonia had reacted.

ii The calculated concentration of the ammonia solution would be higher. Rinsing the burette with deionised water would slightly dilute the hydrochloric acid solution. This means more hydrochloric acid solution would need to be added to neutralise the ammonia. It would then appear that a greater amount of hydrochloric acid reacted; hence, it would appear that a greater amount of ammonia had reacted.

iii The calculated concentration of the ammonia solution would not be affected. Rinsing the conical flask with deionised water would have no effect on the titre volume as the amount of ammonia in the conical flask is accurately known.

iv The calculated concentration of the ammonia solution would be higher. The end point using phenolphthalein would occur after the equivalence point. It would then appear that a greater amount of hydrochloric acid reacted; hence, it would appear that a greater amount of ammonia had reacted.

19 a Phenophthalein will change colour from colourless to pink at the end point.

b 12.8%(m/v)

20 a Anhydrous sodium carbonate meets the following criteria for a primary standard:

- is readily obtainable in a pure form
- has a known chemical formula
- is easy to store without deteriorating or reacting with the atmosphere
- has a high molar mass to minimise the effect of errors in weighing
- is inexpensive.

b Use an analytical balance to accurately weigh an amount of Na_2CO_3 in a clean, dry weighing bottle or beaker. Transfer the solid into a volumetric flask using a dry funnel. Wash any solid particles from the bottle or beaker using deionised water. Half fill the volumetric flask with deionised water, stopper and swirl the flask vigorously to dissolve the solid Na_2CO_3. Add deionised water up to the calibration line on the neck of the volumetric flask. The bottom of the meniscus should be level with the calibration mark. Stopper and shake the flask to ensure an even concentration throughout.

c The final rinse of a burette and a pipette must be with the acid or base they are to be filled with to avoid dilution of the solution. The volumetric flask and conical flask must only be rinsed with deionised water.

d 13.7 M

21 0.13 M

22 a 4.01 mol L^{-1}

b 29.7%(m/v)

Chapter 15 Gases

15.1 Greenhouse gases

1 Three of: methane, carbon dioxide, water, nitrous oxide

2 The natural greenhouse effect occurs due to naturally occurring gases in the atmosphere trapping radiation from the Sun near the Earth's surface. The enhanced greenhouse effect has been caused by human activities, increasing the amount of greenhouse gases in the atmosphere. This causes more heat to be trapped and the Earth is warmed beyond what is natural.

3 Methane is the most potent greenhouse gas. Each methane molecule can absorb more than 25 times as much infrared radiation than a carbon dioxide molecule. Even small increases in amounts of methane can lead to much higher absorption of infrared radiation.

4 Many sources of energy for human use involve the combustion of fossil fuels. Combustion of fossil fuels releases carbon dioxide and water into the atmosphere. Carbon dioxide and water are both greenhouse gases.

5 The fuel, methane, and the products of combustion, carbon dioxide and water, are all greenhouse gases. Combustion of methane would increase the carbon dioxide released to the atmosphere. However, methane is a more potent greenhouse gas, so if it were allowed to escape to the atmosphere from the landfill it would make a greater contribution to the greenhouse effect than the carbon dioxide that would be produced. The environment will be less impacted if the methane is combusted to carbon dioxide.

15.2 Introducing properties of gases

TY 15.2.1 0.700 L

TY 15.2.2 90.3 kPa

TY 15.2.3 373 K

TY 15.2.4 87 L

TY 15.2.5 150 L

TY 15.2.6 25.5 L mol^{-1}

TY 15.2.7 44.0 g mol^{-1}

CSA: Experimental determination of the molar mass of a gas

1 a Trial 1: 3.7×10^{-3} mol
Trial 2: 3.4×10^{-3} mol
Trial 3: 3.8×10^{-3} mol
Trial 4: 3.4×10^{-3} mol

b Trial 1: 59 g mol^{-1}
Trial 2: 67 g mol^{-1}
Trial 3: 65 g mol^{-1}
Trial 4: 68 g mol^{-1}

c 65 g mol^{-1}

d The experimentally determined result is significantly higher than expected. This may be a result of unwanted gas displacing the water during the experiment, such as air or water vapour, which results in a higher apparent volume of gas measured. The pressure measured also did not take into account the added pressure of water vapour in the atmosphere. This would contribute to a greater amount of mol of gas measured, thus reducing the determined molar mass.

2 $p(C_4H_{10}) = 2.8 \times 10^2$ kPa

Key questions

1 **a** 14.00 kPa **b** 430 kPa **c** 1.18 atm

2 **a** 393 K **b** 128 K

3 **a** 4.5 L **b** 86.7 L **c** 6.21 L

4 **a** 1.2×10^2 kPa **b** 17 g
c 31°C **d** 34.3 L mol^{-1}

5 There is a greater amount, in mol, of nitrogen.

6 **a** 1.12×10^7 L
b **i** 88.8 kPa **ii** 1.12×10^3 m

7 2.00 g mol^{-1}

15.3 Calculations involving gases

TY 15.3.1 10.9 kg

TY 15.3.2 620 L

TY 15.3.3 83.4 L

TY 15.3.4 200 mL

1 15.0 L **2** 0.75 L

3 **a** 702 g **b** 616 g

4 **a** 0.222 g **b** 2.76 L

5 24.3 g **6** 4.07 L

Chapter 15 review

1 D

2 The natural greenhouse effect occurs due to naturally occurring gases, such as carbon dioxide and water, in the atmosphere trapping radiation from the Sun near the Earth's surface. This is essential for maintaining a stable temperature required for life. The enhanced greenhouse effect has been caused by human activities increasing the amount of greenhouse gases, such as carbon dioxide, methane and nitrous oxide, in the atmosphere. This causes more heat to be trapped and the Earth is warmed beyond the natural effect. The resultant climate change has serious environmental impacts.

3 Climate change is a result of global warming due to the enhanced greenhouse effect. Combustion of fossil fuels causes a large increase in the amount of carbon dioxide in the atmosphere. Carbon dioxide is a greenhouse gas, so contributes to the enhanced greenhouse effect.

4 **a** 19 L
b The volume of CO_2 will be the same at SLC as the volume of O_2. It is a feature of gases that the same amount, in mol, of all gases will occupy the same volume at the same temperature and pressure. It does not vary with the type of gas.

5 **a** $V_m = \frac{V}{n}$

$= \frac{10.0}{0.356}$

$= 28.1$ L mol^{-1}

b The temperature is higher than 25°C. The volume calculated in part **a** is higher than 24.8 L, which is the molar volume at SLC. If the gases are at the same pressure, the greater volume must be due to a higher temperature as volume is directly proportional to temperature, i.e. $V \propto T$.

6 **a** 6.2 L **b** 64.5 g **c** 19 g

7 B **8** 305°C

9 **a** 20.1 g **b** 24.4 g **c** 11.0 g

10 3.7 tonnes

11 8.45×10^3 L

12 **a** 5 L **b** 10 L **c** 5 g

13 **a** The containers have equal numbers of molecules. With pressure, volume and temperature the same, n will be the same.
b carbon dioxide
c carbon dioxide

14 **a** 0.22 mol **b** 44 g mol^{-1}

15 **a** 8.94×10^3 mol **b** 257 kg

16 **a** 44.0 g **b** 24.8 L. **c** 1.77 g L^{-1}

17 **a** 25 L **b** 44.7 g

18 46.2 g **19** 3.1×10^{10} L

Chapter 16 Analysis for salts

16.1 Testing for salts in water

1 **a** Minerals from soil and rocks, marine aerosols and salt deposits from ancient oceans.
b Mining, agriculture, domestic sources, sewage treatment plants, stormwater and industry.

2 Essential heavy metal: zinc, iron, manganese, cobalt, copper, nickel

Non-essential heavy metal: lead, cadmium, arsenic, mercury.

3 Toxic heavy metal levels are increased in our waterways by combustion of fuels and wastes (indirect), improper disposal of batteries, natural deposits in the Earth, mine dust or effluent, stormwater run-off, intentional or accidental spills, leaching from landfill and agriculture.

4 Organometallic substances must have at least one metal–carbon bond. The options (**b**) CH_3Hg^+, (**e**) $[Fe(CN)_6]^{3-}$, (**f**) $Pb(C_2H_5)_4$ and (**g**) $Zn(CH_3)_2$ are organometallic compounds. The other options are salts that do not have a metal-carbon bond.

5 **a** Hard water is water that requires a lot of soap to lather or froth.
b Calcium(II), magnesium(II), iron(II) or manganese(II)

6 **a** Lead sources are lead pipes, batteries, leaded petrol, paints, ammunition and cosmetics.
Nickel sources are power plants, waste incinerators, improper disposal of batteries.
b Precipitation reactions will remove these heavy metals from the water supply. For example, adding NaOH will form precipitates of $Pb(OH)_2(s)$ and $Ni(OH)_2(s)$

7 5100 µS cm^{-1}

16.2 Quantitative analysis of salts

TY 16.2.1 $NiCl_2{\cdot}6H_2O$

TY 16.2.2 0.859 g

1 **a** hydrate **b** anhydrous **c** hydrate
d hydrate **e** anhydrous

2 Weigh out a sample of the $FeSO_4$ crystals ($FeSO_4{\cdot}xH_2O$). Place the crystals in the oven and heat until the sample loses its crystal structure and becomes a powder ($FeSO_4$). Weigh the $FeSO_4$ powder. Calculate $n(FeSO_4)$. Subtract the mass of $FeSO_4{\cdot}xH_2O$ from $FeSO_4$. This result will be the mass of H_2O loss. Calculate $n(H_2O)$. Divide $n(FeSO_4)$ by $n(H_2O)$ and round to the nearest whole number. The final number will be the value x for $FeSO_4{\cdot}xH_2O$.

3 **a** $Cu(OH)_2$
b $Cu^{2+}(aq) + 2OH^-(aq) \longrightarrow Cu(OH)_2(s)$

4 $CoCl_2{\cdot}6H_2O$ **5** 5.28 g

6 **a** PbI_2
b $Pb^{2+}(aq) + 2I^-(aq) \longrightarrow PbI_2(s)$
c 1.71 g

7 **a** $Cu^{2+}(aq) + 2OH^-(aq) \longrightarrow Cu(OH)_2(s)$
b 0.267 g

8 a $AgNO_3(aq) + NaCl(aq) \longrightarrow AgCl(s) + NaNO_3(aq)$
b 7.27 g of NaCl

16.3 Instrumental analysis for salts

TY 16.3.1 8.8 mg L^{-1}

TY 16.3.2 8.92×10^{-3} mol g^{-1}

CSA: Flow injection analysers

1

2

	Concentration NO_3^- (μg L^{-1})
blank	0
standard check 50 μg L^{-1}	50
Site 1	26
Site 2	16
Site 3	87
Site 4	30
Site 5	19

3 No, there is no contamination as the concentrations are 0 μg L^{-1}.

4 The concentration of the 'Standard check 50 μg L^{-1}' was 50 μg L^{-1}. As it is the expected concentration, there are no issues with the chemistry or the instrument.

5 Site 3 had an unusually high nitrate concentration compared to the other four sites. Possible sources of nitrate can be sewage, stormwater or domestic sources. You will not have mining or agriculture in an urban area.

Key questions

1 Copper sulfate is blue because it transmits blue light and absorbs light of other frequencies. Since a colorimeter measures the amount of light absorbed by a sample, light of a colour other than blue must be used when measuring the concentration of a copper sulfate solution.

2 a false b false
c true d true

3 0.11 M

4 A colorimeter uses light filters to measure absorbance whereas UV–visible spectrophotometers use specific wavelengths.

5 The metal complex must be coloured, soluble and stable.

6 The volume of the extract is a key component of the calculations converting the ion concentration to the quantity of ion per mass of soil. An inaccurate volume will cause the final result to be inaccurate.

7 a Plot of absorbance versus manganese(II) concentration:

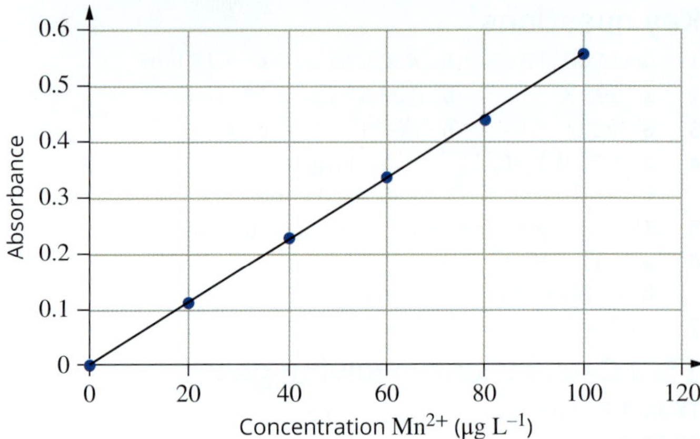

b Concentration = 70 mg L^{-1}

8 a There are two strong absorption peaks at 420 nm and 660 nm.
b Either 660 nm or 420 nm wavelengths could be used as chlorophyll absorbs strongly at both.

9 a Plot of absorbance versus phosphate concentration:

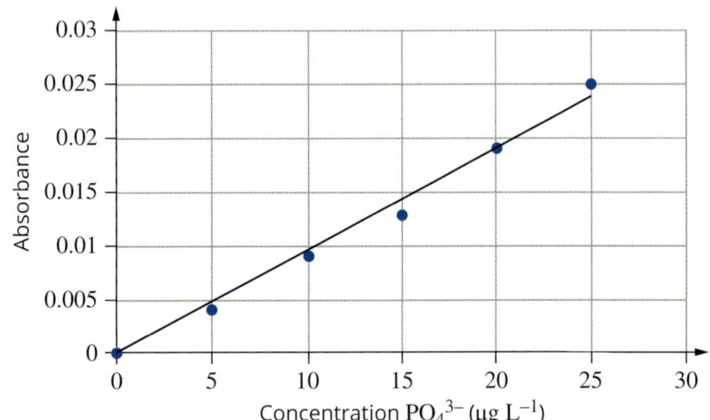

b 11 μg L^{-1}
c 55 μg L^{-1}
d 4.2 μg g^{-1}

Chapter 16 Review

1 B 2 B 3 D 4 C

5 solution B

6 Any three of the following human activities: mining, the burning of fuels, leaching from landfill, run-off from agriculture, discharge from sewage treatment plants, stormwater, industrial wastes and domestic sources

7 a 0.080 M solution
b 0.30 M solution

8 a 580 nm
b Light is absorbed in the yellow–orange region.

9 a any mineral containing zinc
b Zinc can be removed by precipitation reactions. For example, addition of calcium hydroxide will result in the precipitation of insoluble zinc hydroxide.

10 Another example of an organometallic compound is methylmercury. Organometallic compounds are substances with at least one carbon–metal bond.

11 The colorimeter uses filters, whereas UV–visible spectrophotometers use monochromators. The colorimeter uses filters of complementary colours to measure absorption of a coloured sample. UV–visible spectrophotometer uses a selected wavelength to measure absorbance. Colorimeters only measures visible colours whereas the UV–visible spectrophotometer measures wavelengths in both visible and UV region.

12 380 mM **13** 20 202 µS cm^{-1} **14** 123 g

15 $x = 7$

The Epsom salts hydrate is: $MgSO_4{\cdot}7H_2O$

16 **a**

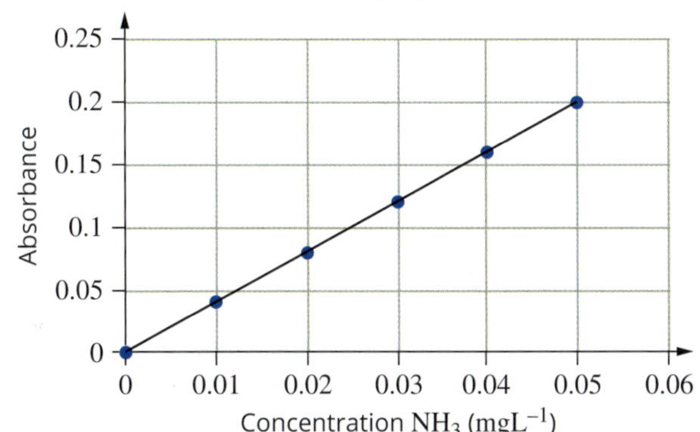

b sample A: 0.015 mg L^{-1}
sample B: 0.033 mg L^{-1}
sample C: 0.025 mg L^{-1}

17 %(P) = 1.2% **18** %(Cd) = 11.9%

19 %(NaCl) = 3.35%

20 **a** 3.2 mg L^{-1}
b 0.32%
c Orange light of wavelength 600 nm is strongly absorbed by a blue solution.

21 **a** **i** 0.535 g **ii** 5350 mg L^{-1}
b **i**

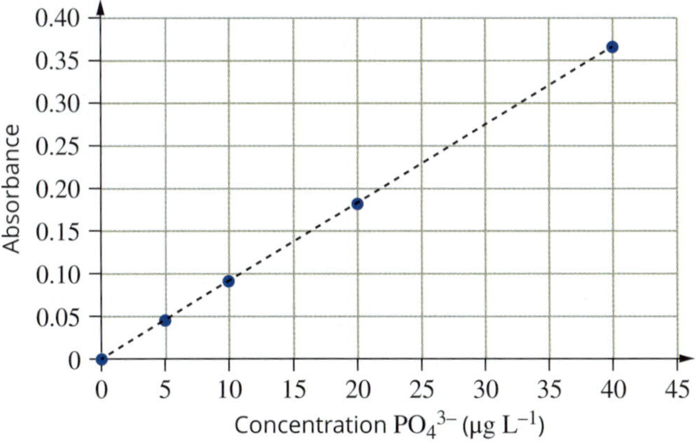

ii 8 µmol L^{-1}
iii High concentrations of phosphate in a waterbody is known as eutrophication.

c **i** Natural sources of these metals would be the minerals in the Yarra River catchment. Man-made sources of these metals would be stormwater, agriculture (the upper Yarra River flows through agricultural land), domestic and industrial accidental or intentional leaks and sewage treatment plants. Specific sources for each metal are:
- lead: lead pipes, incorrect disposal of batteries, paints, and cosmetics as well as residues of leaded petrol
- copper: copper pipes, roofing materials and algicides
- zinc: incorrect disposal of batteries, fertilisers, roofing materials (galvanised iron)

ii Metal cation or organometallic forms.

22 **a**

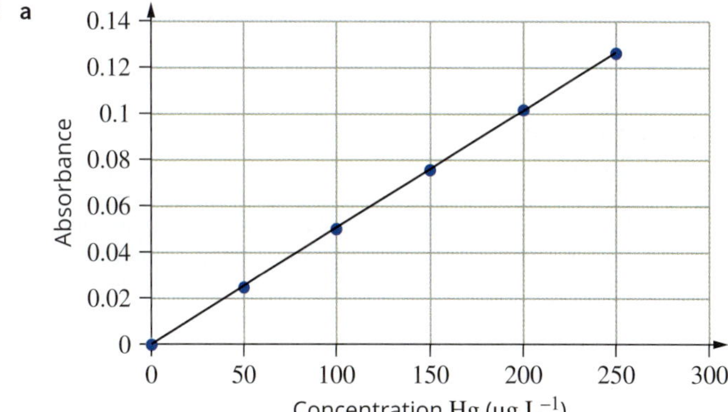

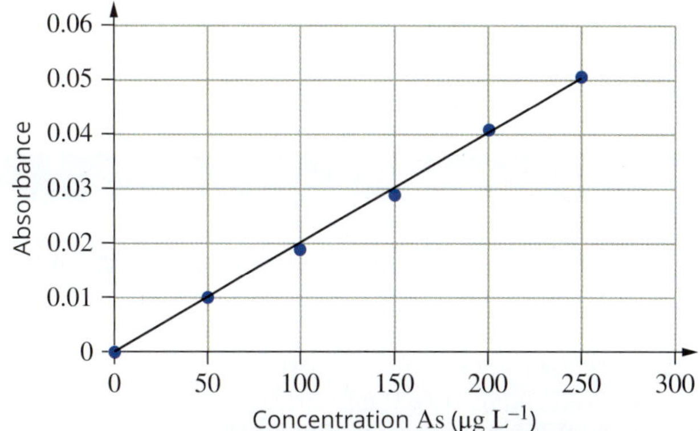

b

	Mercury concentration µg L^{-1}	Arsenic concentration µg L^{-1}
Soil sample 1	215	243
Soil sample 2	60	250
Soil sample 3	156	224

c

	Mercury concentration µg g^{-1}	Arsenic concentration µg g^{-1}
Soil sample 1	65	74
Soil sample 2	14	60
Soil sample 3	43	62

d **i** To convert milligrams to micrograms, you multiply by 1000.
To convert kilograms to grams, you divide by 1000.
To convert mg kg^{-1} to µg kg^{-1}, you multiply by 1000.
To convert µg kg^{-1} to µg g^{-1}, you divide by 1000.
Therefore, mg kg^{-1} = µg g^{-1}.

ii Samples 1 and 3 are contaminated with both arsenic and mercury, as both heavy metals are above the Victorian EPA guidelines. Sample 2 is contaminated with mercury; the arsenic concentration is below 20 mg kg^{-1}.

Unit 2 Area of Study 2 Review

1 A **2** A **3** A **4** B
5 B **6** C **7** B **8** D
9 D **10** D **11** B **12** C
13 B **14** B **15** C
16 250 mL
17 **a** 17.9 g **b** 40.9 g **c** 36.2 g

18 **a** 0.067% **b** 6.7×10^2 ppm.

19 **a** 16 g

b **i** 6 g **ii** 9 g

c Since graphite is insoluble, none will dissolve.

d 0.688 mol L^{-1}

e Add the mixture to 50 g water and heat to 80°C. Both salts are soluble at this temperature, but graphite is insoluble. The graphite can be separated by filtration of the hot solution, leaving both K_2SO_4 and NaCl dissolved in the hot solution. Allow the filtrate to cool to 20°C. Some of the K_2SO_4 will precipitate as the solution is saturated with this salt at 20°C. The precipitated potassium sulfate can be recovered by filtration. The filtrate should be dried in an oven to obtain a pure sample of potassium sulfate.

f Sodium chloride is soluble at both 20°C and 80°C, and the solution is unsaturated at both temperatures, so NaCl will not precipitate out of solution. The filtrate solution contains both sodium chloride and potassium sulfate, so a pure sample could not be obtained, even if it was evaporated to dryness.

20

Equipment	Rinse with:
burette	distilled water, then sodium hydroxide solution
pipette	distilled water, then ammonium chloride solution
conical flask	distilled water

21 **a** $Fe^{3+}(aq) + 3OH^-(aq) \longrightarrow Fe(OH)_3(s)$

b $2Fe(OH)_3(s) \longrightarrow Fe_2O_3(s) + 3H_2O(g)$

c **i** 0.766 g **ii** 0.0106 M

22 **a** 0.592 M

b 0.370 M

c **i** $Pb^{2+}(aq) + 2I^-(aq) \longrightarrow PbI_2(s)$

ii 13.6 g

23 **a** 0.840 g

b **i** $NaHCO_3(aq) + HCl(aq) \longrightarrow NaCl(aq) + CO_2(g) + H_2O(l)$

ii 0.0285 M

24 **a** **i** 1.65×10^7 Pa

ii 163 atm

b 688 g

c 4.27×10^3 L

d 617 balloons

25 **a** A sample of a solution containing Ni^{2+} ions is scanned in a UV–visible spectrophotometer to determine its absorbance across a range of wavelengths. The wavelength that gives the highest absorbance is then chosen.

b Prepare a series of standard solutions containing nickel ions. Measure their absorbance at an appropriate wavelength and plot a calibration curve. Measure the absorbance of the unknown solution and determine its concentration from the calibration curve.

26 **a** 0.0102 mol

b 23.7 L mol^{-1}

c 25.2 L

d 5.95%

e Possible systematic errors (there may be other correct answers):

- It is difficult to accurately measure volumes in a measuring cylinder. An upside-down burette would be more accurate.
- The gas in the measuring cylinder could contain water vapour as well as O_2. It would be useful to dry the O_2 gas before measuring the volume.

27 **a** **i** 110 g

ii Crystals of $NaNO_3$ would precipitate out of the solution.

iii 138 g

b 400 g

c **i** A solution containing 30 g KNO_3 in 100 g water is unsaturated because up to 87 g can dissolve in that quantity of water at 50°C. The solution containing 87 g of KNO_3 is saturated; it contains the maximum amount of KNO_3 that can dissolve at that temperature. The one containing 90 g of KNO_3 is supersaturated; it contains more than the maximum amount of KNO_3 that can dissolve at that temperature and would be an unstable solution.

ii The three solutions can be identified by observing what happens when a small amount of solute is added to each solution. A small amount of solute added to an unsaturated solution will dissolve. A small amount of solute added to a saturated solution will not dissolve. A small amount of solute added to a supersaturated solution or even knocking the solution will cause the excess solute to crystallise out of the solution.

iii On cooling the solutions to 30°C, crystals will precipitate out of the saturated and supersaturated solutions, but not from the unsaturated one, because even at 30°C the solution can hold more than 30 g KNO_3 in 100 g water.

d 194 g

28 Using sodium hydroxide as the primary standard—sodium hydroxide pellets readily absorb water from the atmosphere so its chemical formula would be unknown.

- Washing the volumetric flask with tap water—it should be washed with distilled or deionised water.
- Tapping the pellets from the watch glass into the flask—all of the weighed sodium hydroxide should be washed into the volumetric flask, rather than leaving some on the watch glass (or reweigh the watch glass with the 'remaining sodium hydroxide').
- Using 500 mL of water—the student should add water up to the calibration mark on the volumetric flask, i.e. dissolve the base in 500 mL of solution.
- Using warm water—the student should use water at the temperature used to calibrate the flask. Water expands when heated. Furthermore, heat is released when sodium hydroxide is dissolved in water.
- A stopper should be inserted into the flask before it is shaken. Skin contact with sodium hydroxide should be avoided as it is a corrosive chemical.

29 **a** (Other correct answers may be possible.)

i For the backpack, ensure that:

- all flatulence enters the bag
- the cow is healthy, eating food that is representative of Victorian cattle and in an environment that represents Victorian cattle-growing conditions
- the presence of the bag does not disrupt the cow's regular behaviour.

ii For the capture chamber, ensure that:

- the artificial conditions are set to be similar to a Victorian agricultural environment
- the artificial environment does not disrupt the cow's regular eating and digesting
- the cow's food is representative of normal Victorian fodder.

b 226 L per cow, each day

c 30%

Glossary

A

absolute temperature scale The absolute temperature scale is measured in kelvin (T [K] = T [°C] + 273). The scale starts at absolute zero, the temperature at which atoms and molecules have minimum kinetic energy.

absolute zero A temperature of –273°C or 0 K. Molecules and atoms have minimum kinetic energy at this temperature.

absorbance A measure of the capacity of a substance to absorb light of a specified wavelength.

accepted value A value for a quantity which has been found in published scientific reports, such as journals, or reliable websites, such as those of universities, and values which may be calculated using a correctly balanced equation for the reaction and stoichiometry.

accuracy The value of a measurement is considered to be accurate if it is judged to be close to the true value of the quantity being measured.

accurate If the average of a set of measurements of a quantity is very close to the true or accepted value of the quantity, then the measurement is described as accurate.

acid A substance capable of donating a hydrogen ion (proton).

acid–base indicator A substance that changes colour depending on the concentration of H_3O^+ ions in solution.

acid–base reaction A reaction in which one substance, an acid, donates a hydrogen ion (proton) to another substance, a base.

acidic proton A proton bonded to an electronegative element (oxygen, nitrogen of fluorine) that is donated to a base during an acid–base reaction.

acidic solution An aqueous solution in which the concentration of hydronium ions (H_3O^+) is greater than the concentration of hydroxide ions (OH^-). At 25°C, pH < 7.

acidity The concentration of H_3O^+ ions in an aqueous solution. Acidity is measured using the pH scale.

addition polymer A polymer that is formed by an addition reaction, where many monomers bond together by rearrangement of C=C double bonds without the loss of any atom or molecule. An addition polymer is made from unsaturated monomers.

addition polymerisation The process by which a polymer is formed by an addition reaction, where many unsaturated monomers bond together by rearrangement of C=C double bonds without the loss of any atom or molecule.

addition reaction A reaction in which a molecule binds to an unsaturated hydrocarbon, forming a single carbon–carbon bond. In this process two reactant molecules form one product.

adsorption The attraction of one substance to the surface of another.

affiliation A close connection (between work associates or social groups) which may affect the beliefs of a writer.

aggregation The clumping of suspended particles in water until they are heavy enough to settled to the bottom.

aim A statement, starting with 'To' which describes in detail what will be investigated.

alcohol An organic compound containing the hydroxyl (–OH) functional group; its name ends in '-ol'.

aliquot A volume of liquid measured by a pipette.

alkali A soluble base or a solution of a soluble base.

alkali metal A group 1 metal — Li, Na, K, Rb, Cs and Fr.

alkaline cell A commercial electrochemical cell with an alkaline electrolyte that is a moist paste rather than a solution.

alkane A saturated hydrocarbon; general formula C_nH_{2n+2}.

alkene An unsaturated hydrocarbon containing one carbon–carbon double bond; general formula C_nH_{2n}.

alkyl group A group obtained by removing a hydrogen atom from an alkane; general formula C_nH_{2n+1}, e.g. methyl ($-CH_3$).

alkyl side chain (branch) Also known as alkyl side group. Alkyl side chains have one less hydrogen atom than the corresponding alkane of the same name, so the general formula of an alkyl side chain is $-C_nH_{2n+1}$.

allotrope Different forms of the same element in which the atoms combine in different ways.

alloy A substance formed when other materials (e.g. carbon, other metals) are mixed with a metal.

amorphous A structure that has no consistent arrangement of particles.

amount A measure used by chemists for counting particles; the unit is the mole.

amphiprotic The ability to act as an acid (proton donor) and also as a base (proton acceptor).

analogue Technology that involves measuring infinitely variable values. Analogue devices use a pointer or needle on a scale rather than a number display like a digital device.

anhydrous Containing no water.

anion A negatively charged ion, e.g. a chloride ion, Cl^-.

anode An electrode at which an oxidation reaction occurs.

antacids Substances that neutralise stomach acidity. They are often soluble metal hydroxides or carbonates or a mixture of both.

aqueous Containing water.

aqueous solution When a chemical species has been dissolved in water, the resulting solution is said to be aqueous. This can be shown by writing '(aq)' after the name or symbol of the chemical.

asymmetrical molecule A molecule in which the polar bonds are unevenly (or asymmetrically) distributed. The bond dipoles do not cancel and an overall molecular dipole is created.

atom The basic building block of matter. It is made up of subatomic particles — protons, neutrons and electrons.

atomic number The number of protons in the nucleus of an atom; identical to the charge number of the nucleus; symbol Z.

atomic radius A measurement used for the size of atoms; the distance from the nucleus to the outermost electrons.

atomic theory of matter A theory proposed by John Dalton in 1802 which states that all matter is made up of atoms. He stated that atoms are indivisible, atoms of the same element are identical and compounds are made up of different types of atoms in fixed ratios.

average titre The average of concordant titres.

Avogadro's constant The number of particles in a mole; symbol $N_A = 6.02 \times 10^{23}$ mol^{-1}.

B

bar graph A graph used for organising and displaying discrete data. To construct a column graph, equal width rectangular bars are constructed for each category with height equal to the observed frequency of the category.

base A substance capable of accepting a hydrogen ion (proton).

basic solution An aqueous solution in which $[H_3O^+] < [OH^-]$. For a basic solution at 25°C, pH > 7.

battery A combination of cells connected in series.

bias Measured values consistently in one direction from the actual value; they may be too high or too low.

biomagnification The build-up of heavy metals or other toxins in higher-order predators in a food chain.

biobased Describes a material intentionally made from substances derived from living (or once-living) organisms.

biodegradable Capable of being decomposed by bacteria or other living organisms.

bio-derived Material that is made from plant products.

bioethanol Ethanol that is made by fermenting the sugar and starch component of plants using yeast.

bio-monomer Monomers chemically synthesised from biomass, such as starch, sugar or general food waste.

bio-plastic Polymers formed from renewable biomass such as starch or sugar.

boiling point The temperature at which the state changes from liquid to gas.

bio-polyethene (bio-PE): Polyethene formed from ethanol that has been formed by fermentation.

bio-polyethylene terephthalate (bio-PET): PET formed using ethane-1,2-diol that has been formed from bioethanol.

bio-polypropene (bio-PP): Polypropene formed from propene that has been formed from either ethanol or butan-1-ol from renewable biomass.

blast furnace A type of furnace used for the extraction of metals from their ores, including iron from iron oxides.

block (periodic table) One of four main parts of the periodic table where elements have the same highest energy subshell filled, i.e. *s*, *p*, *d* or *f* subshell.

Bohr model A theory of the atom proposed by Niels Bohr which states that electrons in an atom occupy fixed, circular orbits that correspond to specific energy levels.

boiling point The temperature at which a substance changes state from liquid to gas.

bore water Water collected in aquifers (underground water-bearing rock) below the Earth's surface. Bore water may be accessed by drilling and sinking a bore pipe into the aquifer.

branches The regular or irregular attachment of side chains to a polymer's backbone chain.

brittle Shatters when given a sharp tap.

Brønsted–Lowry theory A theory that defines an acid as a proton (hydrogen ion) donor and a base as a proton acceptor. In the reaction $HCl(g) + H_2O(l) \rightarrow H_3O^+(aq) + Cl^-(aq)$, HCl is the proton donor and is classified as an acid. H_2O is a proton acceptor and is classified as a base.

burette A graduated glass tube with a tap at one end that delivers known volumes of liquid.

C

calcification The building of a deposit of insoluble calcium salts, mainly calcium carbonate. It occurs in the formation of bone and in the development of shells in marine organisms.

calibration The process of determining, checking or rectifying the graduation of any instrument giving quantitative measurements.

calibration curve A plot of data involving two variables that is used to determine the values for one of the variables.

carbon-12 The isotope of carbon that has a mass number of 12. The isotope contains 6 protons and 6 neutrons. One atom of carbon-12 is taken as having a mass of exactly 12 units. This is the standard from which all other relative masses are calculated.

carboxyl group A functional group made up of a carbon atom, two oxygen atoms and a hydrogen atom. One oxygen atom forms a double bond to the carbon atom and the other oxygen atom forms single bonds to both the carbon and the hydrogen atoms. Written as –COOH in condensed structural formulas.

carboxylate ion A negative ion created by the loss of an H+ ion from a carboxylic acid.

carboxylic acid An organic compound containing the carboxyl (–COOH) functional group; its name ends in '-oic acid'.

catalyst A substance that increases the rate of a reaction but is not consumed in the reaction. The catalyst provides a new reaction pathway with a lower activation energy.

cathode An electrode at which a reduction reaction occurs.

cation A positively charged ion.

ceramic Material that is produced by the firing (heating followed by cooling) of clay.

chemical formula A representation of a substance using symbols for its constituent elements. It shows the ratio of atoms present in the substance.

chemical property A property which can only be measured by reacting the substance chemically, i.e. with another substance, or by heating to decompose.

chemical recycling Any process by which a polymer is chemically reduced to its original monomer.

chemical symbol A symbolic representation of an element, usually one or two letters, where the first letter is capitalised and the second letter is lower case, e.g. carbon's symbol is C and sodium's symbol is Na.

chromatogram The output of a chromatography procedure. In TLC and paper chromatography, it is the pattern of bands or spots formed on a plate or on the paper. In HPLC, it is the graph produced.

chromatography A technique for separating the components of a mixture. The components are carried by a mobile phase over the adsorbent surface of the stationary phase.

circular economy An economy based on a model of production and consumption that aims to design out waste and pollution, keep products and materials in use, and regenerate natural systems.

coefficient In a chemical equation, a whole number placed in front of a formula to balance the equation.

coke A solid that contains 80–90% carbon. It is produced by strongly heating coal in the absence of air.

colorimeter An instrument that measures the absorbance of a selected colour of light by a sample solution.

column graph *See bar graph.*

combustion A rapid reaction with oxygen accompanied by the release of large amounts of heat; also called burning.

complementary colours Pairs of colours that are often described as 'opposites'. When combined with each other, these colours 'cancel' each other out to form a greyscale colour such as white or black.

complete combustion A hydrocarbon undergoes complete combustion with oxygen at high temperatures when the only products are carbon dioxide and water.

components The chemicals in a mixture. The components can be separated by chromatography.

compostable Capable of disintegrating into natural elements within 90 days in a compost environment, leaving no toxicity in the soil.

compound A pure substance made up of different types of atoms combined in a fixed ratio.

concentrated solution A solution that has a relatively high ratio of solute to solvent.

concentration A measure of how much solute is dissolved in a specified volume of solution.

concordant titre A set of titres that vary within a narrow range, e.g. within 0.10 mL from smallest to largest titre.

condensation polymer A polymer formed by a condensation reaction, involving the elimination of a small molecule (often water) when monomers bond together. The monomers have functional groups at both ends of the molecule.

condensed electronic configuration A shortened way of writing electronic configuration by using the noble gas before the element.

conductor An object or type of material that permits the flow of electric charges, e.g. a wire is an electrical conductor that can carry electricity along its length.

conflict element Elements mined in areas of war and conflict, which makes their use unsustainable.

conical flask A laboratory flask with a flat, circular base and a cylindrical neck. Used in titrations in preference to a beaker as it is easier to swirl the flask without spilling the contents.

conjugate acid The conjugate acid of a base contains one more hydrogen ion (proton) than the base, e.g. HCl is the conjugate acid of Cl^-.

conjugate acid–base pair An acid and its conjugate base. The conjugate base contains one less hydrogen ion (proton) than the acid.

conjugate base The conjugate base of an acid contains one less hydrogen ion (proton) than the acid, e.g. Cl^- is the conjugate base of HCl.

conjugate redox pair An oxidising agent and its corresponding reduced form, e.g. Cu^{2+}/Cu.

continuous data Any number value within a given range that can be measured.

controlled experiment An experiment in which only one variable is changed, one is measured and all other variables are controlled.

controlled variable A variable that must be kept constant during a investigation.

copolymer A polymer that is made from two or more monomers.

covalent bond The force of attraction formed when one or more pairs of electrons are shared between two nuclei.

covalent lattice a three-dimensional lattice structure formed from covalently bonded non-metal atoms.

covalent layer lattice An arrangement of atoms in a lattice in which there are strong covalent bonds between the atoms that have formed in a layer.

covalent network lattice An arrangement of atoms in a lattice in which there are strong covalent bonds between the atoms in all three dimensions.

cracking A chemical process during which carbon–carbon bonds in alkanes are broken to form smaller molecules and some unsaturated molecules.

critical element Element heavily relied on by industry and society, which faces some form of supply uncertainty.

critical thinking The objective analysis and evaluation of information in order to form a judgement.

cross-link A covalent bond between different chains of atoms in a polymer or other complex molecule.

crude oil A mixture of hydrocarbons that originates from the remains of prehistoric marine microorganisms. The organisms have been broken down by high temperatures and pressures over millions of years.

crystal A solid made up of atoms or molecules arranged in a repeating three-dimensional pattern.

crystal lattice The symmetrical three-dimensional arrangement of atoms or ions inside a crystal.

crystalline region A region where polymer molecules line up parallel to each other and pack closely together.

crystallise To form solid crystals.

D

decalcification The removal or loss of calcium or calcium compounds.

deionised water Water that has had all ions removed and is used for cleaning apparatus and preparation of solutions.

delocalise Spread out.

delocalised electron An electron that is not restricted to the region between two atoms.

density A measure of the amount of mass per unit volume. It has the SI units of kg m^{-3}, but is commonly quoted in g cm^{-3}.

dependent variable The variable that is measured or observed to determine the effect of changes in the independent variable.

desalinated seawater Fresh water made by removing the salt from seawater.

desorption The breaking of the attraction between a substance and the surface to which the substance is adsorbed.

diamond A form of pure carbon that is the hardest naturally occurring substance.

diatomic molecule A molecule formed from two atoms only, e.g. Cl_2.

dilute solution A solution that has a relatively low ratio of solute to solvent.

dilution The addition of a solvent to a solution to reduce its concentration.

dilution factor The ratio of the final volume to the aliquot volume.

$$\text{dilution factor} = \frac{\text{volume of the diluted solution}}{\text{volume of the undiluted solution}}$$

dimer A molecule composed of two identical subunits that may be molecules in their own right, joined by strong intermolecular forces, such as hydrogen bonds.

dipole The separation of positive and negative charges in a molecule.

dipole–dipole attraction A form of intermolecular force that occurs between polar molecules where the partially positively charged end of one molecule is attracted to the partially negatively charged end of another molecule.

diprotic acid An acid that can ionise in water to form two H_3O^+ ions.

discrete data Values that can be counted or measured, but which can only have certain values.

dispersion force The force of attraction between molecules due to temporary dipoles induced in the molecules. The temporary dipoles are the result of random fluctuations in the electron density.

displace The transfer of electrons from an element to a positive ion which results in the ion leaving the solution as an element, e.g. when zinc is placed in a solution of copper(II) ions the displacement reaction is $Cu^{2+}(aq) + Zn(s) \rightarrow Zn^{2+}(aq) + Cu(s)$.

dissociate To break up.

dissociation A process in which molecules or ionic compounds separate or split into smaller particles such as atoms or ions. Examples of dissociation reactions include the solution of NaCl solid in water, forming $Na^+(aq)$ and $Cl^-(aq)$ ions, and the reaction of HCl gas with water, forming $H^+(aq)$ and $Cl^-(aq)$.

dissolution The process of dissolving a solute in a solvent to form a solution.

dissolve To incorporate a solid or gas into a liquid so as to form a solution.

double covalent bond A covalent bond in which four electrons (two electron pairs) are shared.

dry cell A commercial galvanic cell with an electrolyte that is a moist paste rather than a solution.

ductile Able to be drawn into a wire.

E

effective nuclear charge The net positive charge experienced by an electron in an outer shell of an atom. It indicates the attractive force felt by the valence electrons towards the nucleus.

elastomer A rubbery material composed of long molecules that is capable of recovering its original shape after being stretched.

electrical conductivity (EC) The degree to which a specified material or solution conducts electricity.

electrochemical cell A device that converts chemical energy into electrical energy, or vice versa.

electrode A solid conductor in a half-cell at which oxidation or reduction reactions occur.

electrolysis The production of a reaction by the passage of electrical energy from a power supply through a conducting liquid.

electrolyte A solution or molten substance that conducts electricity by means of the movement of ions, e.g. a solution of sodium chloride (table salt).

electromagnetic radiation A form of energy that moves through space. Visible light, radio waves and X-rays are forms of electromagnetic radiation.

electromagnetic spectrum All possible frequencies of electromagnetic radiation shown in order of their wavelengths or frequencies.

electron A negatively charged, subatomic particle that occupies the region around the nucleus of an atom.

electron density The concentration of electrons that usually refer to the regions around an atom or molecule.

electron group A region of negative charge around an atom, which could be either different types of covalent bonds or a non-bonding pair.

electron shell In the shell model of an atom, an electron shell is the fixed energy level that corresponds to a circular orbit of the electrons. In the Schrödinger model, a shell contains subshells and orbitals of equal or similar energy.

electron transfer diagram A diagram that shows how electrons move from a metal atom to a non-metal to form ions.

electronegativity The ability of an atom to attract electrons in a covalent bond towards itself.

electronic configuration In the shell model of an atom, the electronic configuration is a means of representing the number of electrons in each shell.

electrostatic attraction The force of attraction between a positively charged particle and a negatively charged particle.

electrovalency The charge on an ion.

element A substance made up of atoms with the same atomic number.

emission spectrum A spectrum produced when an element is excited by heat or radiation. It appears as distinct lines characteristic of the element.

empirical formula A formula that shows the simplest whole number ratio of the elements in a compound, e.g. CH_2 is the empirical formula of propene (C_3H_6).

end point A point in a titration at which the indicator changes colour, usually marking the completion of the reaction.

endangered element Element at risk of depletion on Earth, as natural deposits are used up.

energy level One of the different shells of an atom in which an electron can be found.

enhanced greenhouse effect An increase in the temperature of Earth's surface due to an increased concentration of greenhouse gases, as a result of human activities.

equivalence point The point in a titration at which the reactants have reacted in their correct mole ratios.

ester A compound produced by the reaction between a carboxylic acid and an alcohol with the elimination of a molecule of water.

ethics Moral beliefs and rules about right and wrong.

eutrophication A process by which pollution from sources such as chemical fertiliser or sewage cause the over-enrichment of water by nutrients. This causes the overgrowth and decay of plants, de-oxygenation of water and the death of organisms.

e-waste A popular term for electronic waste, i.e. electronic products that are unwanted, not working, and at the end of their life cycle.

excited state Describes an atom in which electrons occupy higher energy levels than the lowest possible energy levels.

exponential relationship A mathematical relationship between two variables, in which one rapidly becomes larger as the other increases by equal amounts. The graph of an exponential function looks like a curve that starts off with a very flat slope but starts getting steeper and steeper over time.

external circuit The section of an electrochemical cell in which electrons move. This section of the circuit will include the wires attached to the electrodes.

F

ferrous A material containing iron.

first ionisation energy The energy required to remove one electron from an atom of an element in the gas phase.

flame test A test for the presence of certain metals or metal ions. When burnt in a flame, certain metals and metal ions produce characteristic colours.

flocculation A a compound, most commonly a polymer, is added to drinking water to bind suspended particles together to aid in aggregation.

fractional distillation A separation method based on the different boiling points of the components of a mixture, such as crude oil. The fractionating tower contains a series of trays holding condensed liquid, which vapour rising up the tower must bubble through to provide better separation.

full equation A representation of a reaction that uses formulas.

functional group An atom or group of atoms in an organic molecule that largely determine the molecule's properties and reactions, e.g. –OH, –COOH.

G

galvanic cells A type of electrochemical cell also known as a voltaic cell; a device that converts chemical energy into electrical energy.

galvanometer An instrument for detecting electric current.

gas constant The constant, *R*, in the ideal gas equation $pV = nRT$. $R = 8.31\ J\ K^{-1}\ mol^{-1}$ when pressure is measured in kPa, volume in L and temperature in K.

gemstone A precious stone that is usually ionic in structure.

general formula A formula for an organic compound which replaces the number of carbons with *n*, and relates the number of hydrogens to the value of *n*. Can be applied consistently throughout a homologous series to predict chemical formulas.

graphite A form of carbon in which the carbon atoms are arranged in layers.

gravimetric analysis A technique used for the quantitative determination of the amount of solute in a solution based on the mass of a solid.

green chemistry An approach to chemistry that aims to design products and processes that efficiently use renewable raw materials, and minimise hazardous effects on human health and the environment.

greenhouse effect The warming of the Earth's atmosphere due to the absorption of infrared radiation by gases such as carbon dioxide, water and methane.

greenhouse gas A gas that is able to absorb and re-radiate heat radiation. These gases contribute to the greenhouse effect. Examples are carbon dioxide, methane and water vapour.

ground state A term used to describe an atom in which the electrons occupy the lowest possible energy levels.

group (periodic table) A vertical column of elements in the periodic table.

H

half-cell Half an electrochemical cell, which contains the oxidant and conjugate reductant. When two half-cells are combined, a galvanic cell is formed.

half-equation A balanced chemical equation that shows the loss or gain of electrons by a species during oxidation or reduction, e.g. the oxidation of magnesium is written as the half-equation $Mg(s) \rightarrow Mg^{2+}(aq) + 2e^-$.

haloalkane A molecule derived from alkanes that contain at least the one halogen functional group.

halogen A group 17 element, e.g. F, Cl, Br, I, At.

hard water Water that requires a lot of soap to obtain a lather or froth.

hardness Water hardness is a measure of the amount of metal ions (mainly calcium and magnesium) in the water; the more minerals in the water, the harder the water. This 'hardness' means it is hard to get soapsuds from soap or detergents in this particular water. This happens because the metal ions react strongly with the negatively charged ions in soap molecules to form insoluble compounds. This effectively removes soap from the solution, so more soap is needed to achieve a lather.

heat capacity A measure of a substance's capacity to absorb and store heat energy. The heat capacity of 1 g of water is $4.18\ J\ °C^{-1}$. This tells you that 1 g of water will absorb 4.18 J of heat energy to heat up by 1°C.

heavy metal A metal with high density; usually used to describe a metal that poses a threat to health.

high-density polyethene (HDPE) A form of the polymer polyethene formed from polymer chains with very few, short branches. This means the polymer chains are packed together closely, making the polymer dense. HDPE can have a percentage crystallinity as high as 95% and has excellent mechanical properties. HDPE is used to make pipes, buckets and food containers, such as milk bottles.

homogeneous Uniform. The components of a homogeneous substance are uniformly distributed throughout the substance, e.g. a solution is homogeneous because the solute and the solvent cannot be distinguished from each other.

homologous series A series of compounds with similar properties and the same general formula, in which each member contains one CH_2 unit more than the previous member.

hydrated An ion surrounded by water molecules. Hydrated ions can be found in aqueous solutions or crystalline solids.

hydride A compound in which hydrogen is bonded to another element. HF, HCl and HI are hydrides of group 17 elements.

hydrocarbon A compound that contains carbon and hydrogen only, e.g. the alkanes, alkenes and alkynes.

hydrogen bond A type of intermolecular, dipole–dipole force where a hydrogen atom is covalently bonded to a highly electronegative atom such as oxygen, nitrogen or fluorine. Due to the disparity of electronegativity values between the atoms involved, the hydrogen develops a partial positive charge and bonds to lone pairs of electrons on neighbouring atoms of oxygen, nitrogen or fluorine.

hydrolysis Any chemical reaction in which a molecule of water breaks one or more chemical bonds.

hydronium ion The $H_3O^+(aq)$ ion.

hydroxide ion The $OH^-(aq)$ ion.

hydroxyl group A functional group made up of an oxygen atom with a hydrogen atom bonded to it. Written as –OH in condensed structural formulas.

hypothesis A prediction based on previous knowledge; a possible outcome of the experiment.

I

ideal gas A gas that obeys the gas equations at all temperatures and pressures.

ideal gas constant A constant value denoted by the symbol *R* and equal to a value of $8.31\ J\ K^{-1}\ mol^{-1}$. Used in the formula $pV = nRT$ where pressure is measured in kPa, volume in L, amount in mol and temperature in K.

ideal gas equation The equation that describes the behaviour of an ideal gas $pV = nRT$.

immiscible Two liquids are immiscible if they cannot be mixed without separating from each other.

incomplete combustion A combustion reaction that takes place when oxygen is limited. Incomplete combustion of hydrocarbons produces carbon and carbon monoxide and water.

independent variable A variable that is changed by the researcher.

index (plural: indices) A number written as a superscript after the 10 in scientific notation which shows how many times the 10 is to be multiplied to give the required value.

indicator A substance that is different colours in its acid and base forms.

inert Not chemically reactive.

in-situ leaching Also known as solution mining. The process involves leaving the ore where it is in the ground, and recovering the minerals from it by dissolving them and pumping the solution to the surface where the minerals can be recovered.

instantaneous dipole A net dipole formed in a molecule due to temporary fluctuations in the electron density in the molecule.

intermolecular force An electrostatic force of attraction between molecules, including dipole–dipole forces, hydrogen bonds and dispersion forces.

intramolecular bond A force that holds the atoms within a molecule together.

inverse relationship A relationship in which one variable increases as the other one decreases.

ion A positively or negatively charged atom or group of atoms.

ion–dipole attraction The attraction that forms between dissociated ions and polar water molecules when an ionic solid dissolves in water.

ionic bonding A type of chemical bonding that involves the electrostatic attraction between oppositely charged ions.

ionic bonding model A description of the three-dimensional lattice arrangement of positive and negative ions for an ionic compound in the solid state. The model helps to explain the properties of ionic compounds.

ionic compound A type of chemical compound that involves the electrostatic attraction between oppositely charged ions.

ionic equation An equation for a reaction that only includes the ions that are involved in the reaction, e.g. $Cu^{2+}(aq) + 2OH^-(aq) \rightarrow Cu(OH)_2(s)$.

ionic lattice A regularly repeating three-dimensional arrangement of positively and negatively charged ions.

ionic product of water See *ionisation constant of water*.

ionisation (i) The removal of one or more electrons from an atom or ion; (ii) the reaction of a molecular substance with a solvent to form ions in solution.

ionisation constant of water The equilibrium constant K_w, where $K_w = [H_3O^+][OH^-]$. At 25°C, $K_w = 1.00 \times 10^{-14}$ M^2.

ionisation energy The energy required to remove one electron from an atom of an element in the gas phase.

ionisation reaction The reaction of an acid or base with water.

ionise The reaction of a molecular substance with a solvent to form ions in solution. When some polar molecules dissolve in water they ionise to form a hydronium ion, e.g. $HCl(g) + H_2O(l) \rightarrow H_3O^+(aq) + Cl^-(aq)$.

isotope Each of two or more forms of the same element that contain equal numbers of protons but different numbers of neutrons in their nuclei, e.g. ^{12}C and ^{13}C are isotopes of carbon.

K

kelvin scale The absolute temperature scale is measured in kelvin (T (K) = T (°C) + 273).

kinetic energy The energy that a particle or body has due to its motion ($KE = \frac{1}{2}mv^2$).

kinetic molecular theory A theory that aims to explain the behaviour of gases by assuming gases are composed of a large number of particles in random motion; these particles move in straight lines and have elastic collisions. The gas particles are very small and there is no attraction between the particles. The average kinetic energy of the gas particles is related to the temperature of the gas.

L

latent heat The heat energy required to change the state of a substance without changing the temperature.

latent heat of fusion The energy required to change a fixed amount of solid to liquid at its melting temperature. The latent heat of fusion of water is 6.0 kJ mol^{-1}, meaning 6.0 kJ of energy is needed to change 1 mole of water from a solid to a liquid at 0°C.

latent heat of vaporisation The energy required to change a fixed amount of liquid to a gas at its boiling temperature. The latent heat of vaporisation of water is 40.7 kJ mol^{-1}, meaning 40.7 kJ is needed to change 1 mole of water from a liquid to a gas at 100°C.

lattice A regular arrangement of large numbers of atoms, ions or molecules.

Lewis structure See *electron dot structure*.

limestone A mineral composed of calcium carbonate ($CaCO_3$).

limewater test A test for carbon dioxide gas. The presence of carbon dioxide is detected by bubbling the gas through a calcium hydroxide solution ($Ca(OH)_2(aq)$). The limewater reacts with the carbon dioxide and turns milky.

line graph A graph which shows a linear trend in data.

line of best fit A line which generally represents the relationship represented by the data plotted in a scatter graph.

linear economy An economy in which raw materials are used to make a product, and after its use the product is thrown away.

linear low-density polyethene (LLDPE) An intermediate form of the polymer polyethene that has many short branches.

linear trend A linear trend is represented on a graph by a straight line in which the y-values have equal differences as the x values increase.

literature value A value for a quantity which has been found in published scientific reports, such as journals, or reliable websites, such as those of universities.

locant Number indicating the location of a functional group.

logbook A bound book in which you record every detail of your research.

lone pair See *lone electron pair*.

low-density polyethene (LDPE) A form of the polymer polyethene which has a high degree of short and long chain branching. This means the polymer chains do not pack together closely in the crystal structure. It has therefore weaker intermolecular forces, resulting in a lower tensile strength and increased ductility.

lustrous Having the quality of reflecting light in a glossy and shiny way.

M

main group element An element in groups 1, 2 or 13–18 in the periodic table.

malleable Able to be bent or beaten into sheets.

mass number The number of protons and neutrons in the nucleus of an atom.

mass spectrometer An instrument that measures the mass-to-charge ratio of particles.

mass spectrum A graph of data produced from a mass spectrometer which shows the abundance or relative intensity of each particle, and their mass-to-charge, m/z, ratios.

mass–mass stoichiometry Calculates the mass of an unknown reactant or product using the mass of the known reactant or product, as well as the stoichiometry of that reaction.

matter Anything that has mass and occupies space.

mean The sum of all the values in a data set divided by the number of values in the data set. It is commonly known as the average of a set of numbers.

measurement error The difference between the true value and a measured value.

mechanical recycling Refers to operations that aim to recover plastic waste using mechanical process such as grinding and separating.

melting point The temperature at which the state changes from solid to liquid.

meniscus The curved upper surface of a liquid, as seen in a glass measuring tube.

metal complex A central metal atom surrounded by molecules or ions.

metal displacement reaction A reaction in which a metal causes the ions of another metal in solution to gain electrons and so precipitate out as the solid metal. The metal to be displaced must be less reactive (higher on the electrochemical series) than the metal that is added, e.g. Zn metal will displace Cu from a solution of Cu^{2+} ions.

metallic bonding The electrostatic attractive forces between delocalised valence electrons and positively charged metal ions.

metallic bonding model A model that explains the properties and behaviour of metals in terms of the particles in metals.

metalloid An element that displays both metallic and non-metallic properties, e.g. germanium, silicon, arsenic, tellurium.

method The set of specific steps that are taken to collect data during an investigation. Also known as the procedure.

methodology The general approach taken to investigate the research question.

microbial Relating to microbes or microorganisms, often bacteria.

microplastic Extremely small pieces of plastic debris in the environment resulting from the disposal and breakdown of consumer products and industrial waste.

mineral A naturally occurring inorganic substance that is solid and can be represented by a chemical formula, e.g. quartz.

miscible Liquids that can be mixed in any ratio to form a homogeneous solution.

mistake An avoidable error.

mobile phase The phase that moves over the stationary phase in a chromatographic separation.

model A description that scientists use to represent the important features of what they are trying to describe.

molar mass The mass of 1 mole of a substance measured in g mol^{-1}; symbol M.

molar volume The volume occupied by 1 mole of gas at a specified set of conditions. At standard temperature (0°C or 273 K) and pressure (100 kPa), the molar volume of a gas is 22.7 L mol^{-1}.

molarity The amount of solute, in moles, dissolved in 1 litre of solution (mol L^{-1}).

mole The amount of substance that contains the same number of fundamental particles as there are atoms in 12 g of carbon-12; symbol n; unit mol.

mole ratio The ratio of species involved in a chemical reaction, based on the ratio of their coefficients in the reaction equation.

molecular formula A formula of a compound that gives the actual number and type of atoms present in a molecule. It may be the same as or different from the empirical formula.

molecule A group of two or more atoms covalently bonded together, and representing the smallest fundamental unit of a chemical compound.

molten Materials that are normally found as solids but are liquid, melted, due to elevated temperature.

monochromator An instrument that transmits a narrow band of wavelengths of light or other radiation. The name is from the Greek words *mono* ('single') and *chroma* ('colour').

monomer A small molecule that can react to form long chains of repeating units, called polymers.

monoprotic acid An acid molecule that generates only one hydronium ion when ionised in water.

N

natural indicator An indicator that can be found in plants, such as red cabbage, turmeric, cherries, beetroot and grape juice. Like commercial indicators, natural indicators can be used to determine whether a substance is acidic or basic.

natural polymer Long chains of repeated units which are obtained from nature, meaning they can be observed in and extracted from living things.

neutral solution A solution in which the concentrations of H_3O^+ ions equals the concentration of OH^- ions; is neither acidic nor basic. At 25°C, a neutral solution has a pH of 7.

neutralisation reaction An acid reacts with a base in stoichiometric proportions to form a salt plus water.

neutron An uncharged subatomic particle found in the nucleus of an atom.

noble gas An unreactive gaseous element in group 18 of the periodic table. With the exception of helium, noble gases have eight electrons in their outer shells.

noble gas notation See *condensed electronic configuration.*

non-bonding electron An outer-shell electron that is not shared between atoms.

non-bonding pair Pair of electrons in the outershell of an atom that are not part of a covalent bond.

non-ferrous A metal other than iron or steel.

non-polar Bonds or molecules that do not have a permanent dipole. They have an even distribution of charge.

non-renewable Resources which are used up at a faster rate than they can be replaced.

nucleon A particle that makes up the nucleus of an atom, i.e. protons and neutrons.

nucleus The positively charged core at the centre of an atom, consisting of protons and neutrons.

nuclide notation A way of representing an atom using the element symbol, atomic number (Z) and mass number (A).

O

objective Information or a viewpoint based on facts and free of bias.

observation The action or process of carefully noticing details. In science, we expect to use all senses except taste to make observations.

ocean acidity The reduction in the pH of the ocean caused mainly by absorption of atmospheric carbon dioxide.

octet rule A rule used as part of the explanation for electron configuration and in bonding. The rule is that during a chemical reaction, atoms tend to lose, gain or share their valence electrons, so that there are eight electrons in the outer shell.

orbital In the Schrödinger model, an orbital is a component of a subshell. It is a region of space in which electrons move. Each orbital holds two electrons.

ore A mineral or an aggregate of minerals that contains a valuable constituent, such as a metal, which is mined or extracted.

organic chemistry The study of compounds that have a hydrocarbon backbone, their properties and reactions.

organic compound A compound composed of molecules based on a carbon backbone.

organic recycling The use of enzymes or microorganisms to compost a polymer back to its monomer or to non-toxic smaller molecules.

organometallic compound A substance that contains at least one carbon–metal bond.

origin The point at which a small spot of a mixture is placed so that it can be separated by paper or thin-layer chromatography.

outlier A value that lies outside most of the other values in a set of data.

oxidant A reactant that causes another reactant to lose electrons during a redox reaction. This reactant is, itself, reduced and gains electrons, e.g. in the reaction between magnesium and oxygen, the oxygen is the oxidising agent, as it causes magnesium to lose electrons and form Mg^{2+}: $2Mg(s) + O_2(g) \rightarrow 2MgO(s)$.

oxidation The process by which a chemical species such as a metal atom or a non-metal ion loses electrons. An oxidation half-equation will show the electrons as products (on the right-hand side of the arrow).

oxidised The state of having lost electrons. When a substance is oxidised, the electrons are written on the right-hand side of the arrow.

oxidising agent See *oxidant.*

P

paper chromatography An analytical technique for separating and identifying mixtures that uses paper as the stationary phase.

parallax error The perceived shift in an object's position as it is viewed from different angles.

partial pressure The pressure exerted by an individual gas in a mixture of gases. The total pressure is the sum of the partial pressures.

pascal Unit of pressure equal to 1 newton per square metre (1 N m^{-2}),

where $\text{pressure} = \frac{\text{force}}{\text{area}}$.

parts per million (ppm) A unit of concentration which states the number of grams of solute in 1 million grams of solution. It is equivalent to the number of mg of solute per kg of solution.

peer reviewed A scholarly work of high academic quality in which the author's research or ideas have been scrutinised by others who are experts in the same field (peers).

percentage change A calculation which determines how results have changed over the duration of an experiment (for example). The equation for this is:

$$\text{percentage change} = \frac{(\text{final value} - \text{initial value})}{\text{initial value}} \times 100$$

percentage composition The proportion by mass of the different elements in a compound. Percentage by mass of an element in a compound.

$$\text{percentage composition} = \frac{\text{mass of the element present}}{\text{total mass of the compound}} \times 100$$

percentage difference A calculation which compares experimental results to accepted or theoretical results and expresses them as a percentage.

$$\text{percentage difference} = \frac{(\text{experimental value} - \text{theoretical value})}{\text{theoretical value}} \times 100$$

period (periodic table) A horizontal row of elements in the periodic table. The start of a new period corresponds to the outer electron of that element beginning a new shell.

periodic law The way properties of elements vary periodically with their atomic number.

periodic table A table that organises the elements by grouping them according to their electronic configurations.

periodicity The periodic pattern of properties of the elements.

permanent dipole A net dipole formed in a molecule to permanent differences in electron density due to electronegativity differences between atoms and overall asymmetry in the molecule.

pH A measure of acidity and the concentration of hydronium ions, in solution. Acidic solutions have a pH value less than 7 at 25°C, and bases have a pH value greater than 7 (at 25°C). Mathematically, pH is defined as $pH = -\log_{10}[H_3O^+]$.

pH curve A graph of pH against volume of titrant.

pH scale See *pH.*

photosynthesis A reaction that occurs in the leaves of plants between carbon dioxide and water, in the presence of sunlight and chlorophyll, to form glucose and oxygen. The photosynthesis process can be represented by the equation: $6CO_2(g) + 6H_2O(l) \rightarrow C_6H_{12}O_6(aq) + 6O_2(g)$.

physical property A property which can be measured without changing the substance into another substance.

pie chart A graph or chart which represents percentages of a whole sample set as slices of a circular 'pie'.

pipette A calibrated glass tube used to transfer known volumes of liquid.

plant-sourced biomass Organic matter that has come from plants.

plastic A property of a material that can be reshaped by application of heat and pressure. In society, polymers are often referred to as plastics.

plasticiser Small molecules that soften a plastic by weakening intermolecular attractions between polymer chains.

polar Bonds or molecules with a permanent dipole. They have an uneven distribution of charge.

polarity The measure of how polar a molecule or bond is. The difference in charge between the positive and negative ends of an electric dipole. The difference in charge between the positive and negative ends of a polar molecule or covalent bond.

polyatomic ion An ion that is made up of more than one element, e.g. the carbonate ion (CO_3^{2-}).

polyatomic molecule A molecule that consists of more than two types of atoms, e.g. H_2O.

polymer A long-chain molecule that is formed by the reaction of large numbers of repeating units (monomers).

polymerisation The process of synthesising a polymer.

polyprotic acid An acid molecule that generates more than one hydronium ion when ionised in water.

potable water Water that is suitable for drinking.

precipitate The solid formed during a reaction in which two or more solutions are mixed.

precipitation reaction A reaction between substances in solution in which one of the products is insoluble.

precise When repeated measurements of the same quantity give values that are in close agreement.

precision A measure of the repeatability or reproducibility of scientific measurements and refers to how close two or more measurements are to each other.

pressure A measure of a force exerted on a surface per unit area of the surface. Common units for gas pressure are Pa, kPa and atm.

primary cell A galvanic cell that is non-rechargeable because the products of the reaction migrate away from the electrodes.

primary data Measurements or observations that you collect during your investigation.

primary smelter An installation or factory that is used to extract metals from their ores using high temperatures. Scrap metal is also processed in some primary smelters.

primary source A source that is a first-hand account.

primary standard A substance of known high purity which may be dissolved in a known volume of solvent.

principle A general scientific law which explains how something happens or works.

processed data Data which has been manipulated by calculations or graphical processes to enable conclusions about the data to be reached.

proton A positively charged, subatomic particle bound to neutrons in the nucleus of an atom.

purity In chemical terms, a material is pure if it only contains a single substance, without any other substances present.

pyramidal A molecular shape formed when there are four electron groups and one non-bonding pair around a central atom.

Q

qualitative Relating to quality and not measured values.

qualitative analysis An analysis to determine the identity of the chemical(s) present in a substance.

qualitative data Data which can be observed and relates to a type or category, such as colour, or states (such as gas, liquid or solid).

quantised In specific quantities or chunks.

quantitative Relating to measured values rather than quality.

quantitative data Measured numeric values which are usually accompanied by a relevant unit.

quantitative analysis An analysis to determine the concentration of the chemicals present in a mixture.

quantum mechanics A branch of science that describes the behaviour of extremely small particles such as electrons.

R

radioactive Spontaneously undergoing nuclear decay to produce radiation such as beta particles, alpha particles and gamma rays.

random error An error that follows no regular pattern. (The effects of random errors can be reduced by taking the average of many observations.)

raw data The information and results collected and recorded during an experiment.

reactivity The ease with which a chemical can undergo reactions.

reactivity series of metals A ranking of metals in increasing order of their reactivity (ability to be oxidised), with the half-equations written as reduction equations of the corresponding ion. Least reactive metals are at the top and most reactive metals are at the bottom.

recycled water Water recovered by the purification of wastewater.

redox reaction A reaction in which electron transfer occurs from the reducing agent to the oxidising agent. In a redox reaction, both oxidation and reduction occur.

reduced The state of having gained electrons. When a substance is reduced, the electrons are written on the left-hand side of the arrow (as reactants) in the half-equation.

reducing agent A reactant that causes another reactant to gain electrons during a redox reaction. This reactant is oxidised and loses electrons; e.g. in the reaction between magnesium and oxygen, the magnesium is the reducing agent, as it causes oxygen to gain electrons and form O^{2-} ions: $2Mg(s) + O_2(g) \rightarrow 2MgO(s)$.

reductant See *reducing agent*.

reduction The process by which a chemical species gains electrons. A reduction half-equation will show the electrons on the reactant side (left-hand side) of the equation.

relative atomic mass The weighted average of the relative isotopic masses of an element on the scale where ^{12}C is 12 units exactly; symbol A_r.

relative isotopic abundance The percentage abundance of a particular isotope in a sample of an element.

relative isotopic mass The mass of an atom of the isotope relative to the mass of an atom of carbon-12 taken as 12 units exactly.

relative mass The mass of an atom, molecule or compound relative to the mass of an atom of ^{12}C, taken 12 units exactly.

renewable Not finite; that is, can be replenished within a reasonable time (e.g. a human lifespan).

repeatability The consistency of results when an experiment is repeated many times.

replication The process of repeating the steps of a method.

reproducibility The closeness of the agreement between measurements of the same quantity, carried out under changed conditions of measurements.

research question A statement defining what is being investigated.

resolution The smallest change in the measured quantity that causes a perceptible change in the value shown by the measuring instrument.

retardation factor (R_f) The ratio of the distance a component has moved from the origin to the distance the solvent has moved from the origin.

risk assessment A formal way of identifying risks and assessing potential harm from hazards in an experiment.

roasting A process of heating a sulfide ore to a high temperature in air.

S

safety data sheet (SDS) A summary of the risks of using a particular chemical, including measures to be followed to reduce risk.

salinity The presence of salt in water and soil that can damage plants or inhibit their growth.

salt A substance formed from a metal or ammonium cation and an anion. Salts are the products of reactions between acids and bases, metal oxides, carbonates and reactive metals.

salt bridge An electrical connection between the two half-cells in a galvanic cell; it is usually made from a material saturated in electrolyte solution.

saturated (i) A hydrocarbon that is composed of molecules with only carbon–carbon single bonds. (ii) Combined with or containing all the solute that can normally be dissolved at a particular temperature.

saturated solution A solution that cannot dissolve any more solute at the given temperature.

scatter graph A graph in which dots are used to represent values for two different numeric variables. The position of each dot on the horizontal and vertical axis indicates values for an individual data point.

Schrödinger model A model for the behaviour of electrons in atoms. It describes electrons as having wave-like properties.

scientific method A method of investigation in which a research question is investigated by forming a hypothesis and then testing it.

scientific notation The standard way to express a number as the product of a real number and power of 10. Also known as standard form.

scrap metal Discarded metal that is suitable for reprocessing.

secondary cell An accumulator or rechargeable cell. Recharging can occur because the products formed in the cell during discharge remain in contact with the electrodes in a convertible form.

secondary smelter An installation or factory that is used to extract metals from scrap metal using high temperatures.

secondary source A source derived from the original data or account.

seed crystal A small crystal from which a large crystal of the same material can typically be grown.

self-ionisation An ionisation reaction of pure water in which water behaves as both an acid and a base.

semi-structural formula A condensed formula that summarises the structural formula of a compound in a single line of text.

side group A group of atoms attached to a backbone chain of a long molecule.

significant figures Digits in a number that are reliable and are necessary to indicate a quantity. These are limited in chemistry by the equipment used, or, in a calculation, by the number of significant figures in the least accurate piece of data used.

single covalent bond A covalent bond in which two electrons are shared between two nuclei. It is depicted in a valence structure as a line between the two atoms involved.

slag A mixture of waste materials left over after a desired metal has been removed from its ore.

smelting A process of applying heat to ore to extract a metal.

soil matrix The term used to describe all the components of soil, including sand, silt, clay, salts, organic matter and living organisms.

soil texture The classification of soil based the percentage of sand, silt and clay particles.

solubility A measure of the amount of solute dissolved in a given amount of solvent at a given temperature.

solubility curve A graph of solubility versus temperature for a particular solute dissolved in a particular solvent.

solubility table A reference table that can be used to predict the solubility of ionic compounds.

solute A substance that dissolves in a solvent, e.g. sugar is the solute when it dissolves in water.

solution A homogeneous mixture of a solute dissolved in a solvent.

solution volume–volume stoichiometry Calculations using titration data to determine the number of moles of the reactants in a chemical reaction. These calculations can be used to determine the concentration of acids and bases in aqueous solutions.

solvent A substance, usually a liquid, which is able to dissolve a solute to form a solution. Water is a very good solvent.

specific heat capacity The amount of energy required to raise the temperature of an amount of a substance, usually 1 gram, by 1°C. The unit for specific heat capacity is usually $J\ g^{-1}\ °C^{-1}$, e.g. the specific heat capacity of water is $4.18\ J\ g^{-1}\ °C^{-1}$.

spectator ion An ion that remains in solution and is unchanged in the course of a reaction. Spectator ions are not included in ionic equations.

spectroscopy The study of the way that radiation, such as light and radio waves, interacts with matter.

spontaneous redox reaction A redox reaction that occurs naturally.

standard atmosphere One standard atmosphere (1 atm) is the pressure required to support 760 millilitres of mercury (760 mmHg) in a mercury barometer at 25°C. One atmosphere pressure is approximately the pressure experienced at sea level.

standard form The standard way to express a number as the product of a real number and power of 10. Also known as scientific notation.

standard laboratory conditions (SLC) Conditions of temperature and pressure relevant to a gas, where the temperature is 298 K (25°C) and the pressure is 100 kPa.

standard solution A solution that has an accurately known concentration.

standardised The process by which the concentration of a solution is accurately determined, often through titration with a standard solution.

standards Standards are materials with an identity and concentration that are precisely known. A standard can provide a reference that can be used to calibrate analytical instruments or determine unknown concentrations.

stationary phase A solid, or a solid that is coated in a viscous liquid, used in chromatography. The components of a mixture undergo adsorption to this phase as they are carried along by the mobile phase.

steel A generally hard, strong, durable and not malleable alloy of iron and carbon, usually containing 0.2–1.5% carbon, often with other constituents such as manganese, chromium, nickel, molybdenum, copper, tungsten, cobalt or silicon, depending on the desired alloy properties.

stem name The name that corresponds to the prefix for the longest chain of carbons in the molecule.

stock solution A large volume of a common reagent with a known concentration.

stoichiometry The calculation of relative amounts of reactants and products in a chemical reaction. Chemical equations give the ratios of the amounts (moles) of the reactants and products.

strong acid An acid that readily donates a hydrogen ion (proton) to a base.

strong base A base that readily accepts a hydrogen ion (proton) from an acid.

structural formula A formula that represents the three-dimensional arrangement of atoms in a molecule and shows all bonds, as well as all atoms.

structural isomer A compound with the same molecular formula, but different structures.

subatomic particle A particle that makes up an atom—protons, neutrons and electrons.

sublimation The process by which a substance goes directly from the solid phase to the gaseous phase, without passing through a liquid phase.

subjective A viewpoint or information which is based on personal opinion and feelings, rather than facts.

subshell A component of a shell in the Schrödinger model, made up of orbitals. Each subshell can be regarded as an energy level that electrons can occupy.

substitution reaction A reaction that involves the replacement of an atom or group of atoms with another atom or group of atoms.

supersaturated solution An unstable solution that has more solute dissolved at a given temperature than a saturated solution.

sustainability Using resources, such as minerals and energy, to meet our own needs without compromising the ability of future generations to meet their needs.

symmetrical molecule A molecule in which the polar bonds are evenly (or symmetrically) distributed. The bond dipoles cancel out and do not create an overall molecular dipole.

systematic error An error that produces a constant bias in measurement. (Systematic errors are eliminated or minimised through calibration of apparatus and the careful design of a procedure.)

T

temporary dipole A net dipole formed in a molecule due to temporary fluctuations in the electron density in the molecule.

tensile strength The maximum resistance of a material to a force which is pulling it apart before breaking, measured as the maximum stress the material can withstand without tearing.

terminal carbon A carbon atom that is on the end of a carbon chain.

tetrahedral The shape of a molecule with a central atom surrounded by four other atoms. The bond angle between two outer atoms and the central atom is 109.5°.

theory A plausible hypothesis that is supported by a considerable amount of evidence.

thermoplastic A thermoplastic polymer will soften and melt when heated, allowing it to be remoulded or recycled. When heated sufficiently, the intermolecular bonds break, allowing the molecules to become free to move and be remoulded.

thermosetting When a thermosetting polymer is heated, it does not melt, but at high temperatures, covalent bonds are broken and it decomposes or burns. It cannot be moulded into a different shape.

thin-layer chromatography An analytical technique for separating and identifying mixtures; it uses a thin layer of fine powder spread on a glass or plastic plate as the stationary phase.

titration The process used to determine the concentration of a reactant where one solution is added from a burette to a known volume of another solution.

titration curve A plot of the pH versus the equivalents of acid or base added during the titration of the other.

titre The volume of liquid, measured by a burette, used in a titration.

total dissolved solids (TDS) The total amount of mobile charged ions, including minerals, salts or metal ions dissolved in a given quantity of water.

transition metal An element in groups 3–12 of the periodic table.

trend An observed pattern of data in a particular direction.

trigonal planar A molecular shape formed when there are three electron groups and no non-bonding pairs around a central atom.

triple covalent bond A covalent bond in which six electrons are shared between two nuclei. It is depicted in a valence structure as three lines between the two atoms involved.

triprotic acid An acid molecule that generates three hydronium ions when ionised in water.

true value The value that would be obtained under perfect conditions.

U

uncertainty An error associated with measurements made during experimental work.

unsaturated A hydrocarbon composed of molecules with one or more carbon–carbon double or triple bonds.

unsaturated solution A solution that contains less solute dissolved at a given temperature than a saturated solution.

UV–visible spectrophotometer An analytical device that measures the absorbance of a solution in the UV–visible region of the spectrum.

V

valence electron An electron found in the valence shell; an outermost electron in an atom or ion.

valence shell The highest energy shell (outer shell) of an atom that contains electrons.

valence shell electron pair repulsion (VSEPR) theory A model used to predict the shape of molecules. The basis of VSEPR is that the valence electron pairs surrounding an atom mutually repel each other, and therefore adopt an arrangement that minimises this repulsion, thus determining the molecular shape.

valid When an experiment or investigation does test the stated aims and hypothesis.

validity Whether an experiment or investigation is testing the stated hypothesis and aims.

variable Any factor that can be controlled, changed and measured in an experiment.

voltaic cell A type of electrochemical cell that is also known as a galvanic cell; a device that converts chemical energy into electrical energy.

volume The amount of space that a substance occupies. It can be calculated by multiplying length by width by depth of a regular solid. Otherwise, it can be determined by finding the volume of water displaced by the substance.

volumetric analysis Analysis using measurement of volumes, e.g. titration.

volumetric flask A laboratory flask calibrated to contain a precise volume.

W

water of hydration Water molecules that form an essential part of a crystal arrangement, but that are not directly bonded to the cations and anions in the crystal structure.

weak acid An acid that is partly ionised in water.

weak base A base that accepts hydrogen ions (protons) from acids to a limited extent.

index

G

H

I

K

L

M

N

O

P

W

Y

Z

ttributions

Cover and back cover: Gaff, Karl/Science Photo Library.

123RF: alex_star, p. 340cl; alicenerr, p. 150rb; bigy00, p. 449rb; blueneo, p. 316 (bottle); cloud7days, p. 316 (cutlery); dgool, p. 340tc; Elnur, p. 473l; foxterrier2005, p. 192tl; Lebedev, Anton, p. 194b; maridav, p. 303r; Minishev, Dmytrii, p. 292bl; Monteil, Fabien, p. 308lt; New Africa, p. 316 (casein waste); porselen, p. 340bl; purpleanvil, p. 32b; sablin, p. 316 (landfill); Tihonovs, Aleksandrs, p. 150lb; Tsartsianidis, George, p. 430r; tykhyi, p. 150lt.

agefotostock: Arnold Mondadori Editore S.P., p.202bl.

Alamy Stock Photo: TMI, pp. 56-7, 210-14, 322-5; Addictive Stock, pp. 132-3; Allen, Tim, p. 425t; Arif, Alisha, p. 444r; B. Murton/Southampton Oceanography Centre/Science Photo Library, p. 177; British Antarctic Survey/Science Photo Library, p. 473r; Bulgar, Wladimir/Science Photo Library, p. 251b; Chillmaid, Martyn F., pp. 146br, 445tr; Chillmaid, Martyn F./Science Photo Library, p. 450t; Chronicle, p. 73; Clarivan, Carlos, p.216cl; Cooper, Ashley/Science Photo Library, p. 471cc; Degginger, Phil, pp. 392, 393; Dunkley, Dan/Science Photo Library, p. 48t; Dunn, Andrew, p. 60br; Fermariello, Mauro/Science Photo Library, p. 281r; Graham, Tim, p.458; Hauser,Steffen/botanikfoto, p. 375rc; insidefoto srl, p. 87tl; Jenkins, Carolyn, p. 431t; Jensen, Mikkel Juul/Science Photo Library, pp. 86tl, 216b, 471rb; John Birdsall Social Issues Photo Library/Science Photo Library, p. 409br; Joseph Stroscio, Robert Celotta/nAtional Institute of Standard and Technology/Science Photo Library, p. 60bl; Lenk, Piter, pp. 159, 331t; Leslie Garland Picture Library, p. 400c; Levine, Richard, p. 462; Lourens, Smak, p. 61br; McLean, Robert, p. 444l; Microgen Images/Science Photo Library, p. 22l; nagelestock.com, pp. 246-7; Newscom, p. 291b; Pictorial Press Ltd, p. 79br; PNL, p. 180c; Price, Simon, p. 480b; Prisma by Dukas Presseagentur GmbH, pp. x, 1-3; R. E. Litchfield/Science Photo Library, p. 295rt; Schneider, Karsten/Science Photo Library, p. 345t; Sci-Comm Studios/Science Photo Library, p. 74bl; Science History Images, p. 408t; Science Photo Library, pp. 4l; 224tl, 231, 387, 408b; Sciencephotos, pp. 350t, 389b; Shields, Martin, p. 4r; TEK IMAGE/Science Photo Library, p. 215; The Natural History Museum, p. 349; Trenchard, Tommy E, p. 86tr; Trevor Clifford Photography/Science Photo Library, p. 365tr.

Art Gallery of New South Wales: *Five bells 1963* by John Olsen. Purchased with funds provided by the Art Gallery Society of New South Wales 1999. Photo: Mim Stirling, AGNSW © John Olsen/Copyright Agency. Licensed by Viscopy, 2017. 133.1999, p. 138tl; *Untitled 2010* by Mavis Ngallametta. Gift of the artist 2015. Donated through the Australian Government's Cultural Gifts Program. © Mavis Ngallametta. Licensed by Martin Browne Contemporary. 156.2015, p. 138tr.

Auscape: Woerle, Frank, p. 453 (Kakadu plum).

Bank of France: Bank of France, Public domain, via Wikimedia Commons.

Book, Natalie: p. 305 (tape, cling wrap, fibres, foam balls, superglue, and Perspex).

Chemistry Education Association (CEA): front cover (logo), p. iiib.

Chillmaid, Martyn F.: p. 366b.

Corbis: 145/William Andrew, p. 106lt; Egmont Strigl/Westend61, p. 127bl; Lowell, Georgia, p. 124t; Lucas, Ken, p. 145b; Murray, Louise, p. 517t; Yamashita, Michael S., p. 499t.

Cross, Malcolm: p. 202t.

CSIRO: scienceimage, p. 126b; © Copyright CSIRO Australia. Extract from "Tackling plastic waste". Accessed on August 11, 2022. https://www.csiro.au/en/research/environmental-impacts/recycling/plastics, p. 310.

Derby Museum and Art Gallery: *The Alchymist, In Search of the Philosopher's Stone, Discovers Phosphorus, and prays for the successful Conclusion of his operation, as was the custom of the Ancient Chymical Astrologers 1171* by Joseph Wright of Derby, now in Derby Museum and Art Gallery, Derby, U.K. 1883-152, p. 87bc.

Derry, Lanna: pp. 24tr; 24trb; 39t.

DK Images: Dunning, Mike, p. 190.

Faber-Castell: p. 126cl.

Fotolia: ALCE, p. 494t; ASK-Fotografie, p. 426lb; ASP Inc, p. 434; Attila, Barabas, p. 230; Biebel, Harald, p. 150lc; bit24, p. 216cr; blauviolette, p. 236bl; Copri, p. 127rt; Dirima, p. 343tr; Disto89, p. 288tl; Dudarev, Mikhail, p. 345b; EcoView, p. 218cr; ermess, p. 340cr; EuToch, p. 293tl; Freer, p. 288bt; haveseen, p. 339t; Jeayesy, p. 288tc; Kilala, p. 350bl; Lustil, p. 127cl; Matoska, Jan, p. 364t; Megasquib, p. 14b; Mirco251280, p. 135cl; molekuul.be, p. 289b; Nomad_Soul, p. 288tr; Olson, Tyler, p. 431b; Pabijan, p. 223t; Piccillo, Nicholas, p. 223br; piotr_roae, p. 164br; Popova, Olga, p. 288bb; pumpuija, p. 241t; rekmp, p. 474b; scottchan, p. 340cc; shaiith, p. 473c; Toome, Maksim, p. 340tl; van Hateren, Ina, p. 199.

Friends of Sherwood Arboretum Association: p. 453 (Davidson plum).

Getty Images: Ardian, Dimas/Stringer, p. 233t; Auscape/UIG, p. 494b; BanksPhotos, p. 517b; Berg, Alistair p. 287; Bourseiller, Philippe/Getty Images Reportage Archive, p. 493; Chillmaid, Martyn F., p. 445rb; desmon jiag, p. 98-9; Husmo, Arnulf, p. 388b; jacek_kadaj, p. 443; Jacobs Stock Photography Ltd, p. 469; MacGregor, p. 291t; merrymoonmary, pp. 326-27, 414-16, 526-31; Miller, Jeff, p. 328-9; New York Daily News Archive, p. 101tr; Richards, Paul J./AFP via Getty Images, p. 430l; Science Photo Library, pp. 363bl, 363tr, 426lt, 432; Terry, Sheila/Science Photo Library, p. 499b; Unangst, Andrew, p. 122br; Universal History Archive, p. 183b; UniversalImagesGroup, p. 477br; Winters, Charles D., pp. 178, 401, 419t, 500tl; Yu, James, p. 418.

Hogendoorn, Bob: Chemistry Education Association, p. 449bl.

images-of-elements.com: p. 136b.

Ixom Operations Pty Ltd: p. 14t.

LEGO®: © Lego, pp. 250r, 285.

Lochman Transparencies: p. 426cr.

Manilda Group: Dean Holland via Manildra Group, p. 250l.

materialstoday.com: 'Biodegradability of polylactide bottles in real and simulated composting conditions' by Kale, Gaurav, Auras, Rafael, Singh, Sher Paul, Narayan, Ramani. Polymer Testing, vol. 26, issue 8, 2007, pp. 1049-1061, ISSN 0142-9418, p. 315br.

McBroom, Alice: p. 351.

Mohamed Sapi'ee, Amirah Fatin: p. 163.

Moylan, Mick: pp. 335l and 335c.

Museums Victoria: *Basket,* bathi marrin, *Milingimbi, Eastern Arnhem Land, Northern Territory, Australia, 1930* from the Rev T.T. (Ted) Webb Collection. Photographer: Caroline Harding/ Copyright Museums Victoria/All Rights Reserved. Reproduced with consent from Milingimbi Art and Culture., p. 118cl.

Nanollose: © Nanollose, pp. 316 (nanollose fibre - cotton/ rayon heading), 317t.

NASA: MSFC/David Higginbotham/Emmett Given, p. 137; NASA, ESA, J. Hester and A. Loll (Arizona State University), p. 248; Luthi, D., et al.. 2008; Etheridge, D.M., et al. 2010; Vostok ice core data/J.R. Petit et al.; NOAA Mauna Loa CO2 record, p. 379t; Courtesy NASA/JPL-Caltech, p. 398l.

National Health and Medical Research Council: 'Australian Drinking Water Guidelines Paper 6 National Water Quality Management Strategy' by the National Health and Medical Research Council, National Resource Management Ministerial Council, Commonwealth of Australia, Canberra, p. 501tl.

Nestlé: © Nestlé, p. 314b.

Newspix: Currie, Ian, p. 61bl.

Nine: Armao, Joe/*The Age* (Melbourne), p. 399b; Kilponen, Dallas/The Sydney Morning Herald, p. 500b.

Pasco Scientific: p. 30.

Patagonia: © Patagonia, p. 316 (jacket).

Pearson Education Ltd.: pp. 44l, 44r, 61tr; Prentice Hall Inc, p. 474t.

Powerhouse Museum: Myers, Michael, p. 164t.

Prendergast, Sarah: Sarah Prendergast from Beaconhills College, p. 426br.

Replas Plastics Australia Pty Ltd: pp. 308br, 309bl, 309br, 316 (bench seat), 316 (polyrok), 316 (footpath).

Reuters: Brindicci, Marcos, p. 531t; Karpukhin, Sergei, p. 124br.

Royal Australian Chemical Institute (RACI): p. 79t.

Rural Industries Research and Development Corporation: 'Food Safety of Australian Plant Bushfoods' 2001 by Hegarty, M.P. and Hegarty, E.E. Plantchem Pty Ltd and R.B.H. Wills Centre for Advancement of Food Technology and Nutrition, University of Newcastle. Publication No. 01/28 Project No. AGP-1A, p. 15.

Science Photo Library: pp. 27, 67b, 68c, 70t, 143, 366c, 388t, 470tr; Andrew Lambert Photography, pp. 24tl, 144, 146tl, 357, 374tl, 375lb, 445lb, 510tl; British Antarctic Survey, pp. 63b, 383b; Brookes, Andrew, National Physical Laboratory, p. 218l; Camm, Martin/Carwadine/Nature Picture Library, p. 218cl; Chemical Heritage Foundation, p. 60cl; Chillmaid, Martyn F., pp. 82t, 192br, 305bt, 389t, 445tl; Clarivan, Carlos, pp. 217, 333; Custom Medical Stock Photo, p. 135l; David Gallan/Nature Picture Library, p. 471cl; Dr. Burgess, Jeremy, pp. 303l, 344, 495; GIPHOTOSTOCK, p. 423; Hinsch, Jan, p. 383t; King-Holmes, James, p. 308bl; Klemm, Horst/Greatstock, p. 330t; Landmann, Patrick, p. 277bt; Mettelart, Tatiana/Look at Sciences, p. 275br; NASA's Goddard Space Flight Center, p. 271; National Oceanic and Atmospheric Administration (NOAA), pp. 381l, 381r; Natural History Museum, London, p. 363bc; Pasieka, Alfred, p. 335r; Phillips, David M./Science Source, p. 417; Saurer, Friedrich, pp. 277bbr, 279br; Schroeder, Monica, p. 470bl; Stammers, Sinclair, p. 202bc; Steger, Volker, p. 183t; TEK Image, p. 122bl; Trevor Clifford Photography, p. 158; Turtle Rock Scientific/ Science Source, p. 29; West, Jim, p. 337l; Yilmaz, I. Noyan, p. 471rt.

Science Source: Winters, Charles D., p. 58-9.

Shutterstock.com: 2728747, p. 225; aerophoto, p. 308lb; Africa Studio, p. 293cl; Alybaba, p. 453t; asharkyu, p. 153; AtlasStudio, p. 310; Auruskevicius, Marijus, p. 162t; Berman, Eugene, p. 135r; Blanda, Eva, p. 365br; Boltaca, Nadezda, p. 150rt; Boykung, p. 138cr; Bro Studio, p. 22br; Chaay_Tee, p. 428; CP DC Press, p. 81; De Armas, Esteban, p. 84; Designua, p. 114bl; dio5050, p. 336l; fifg, p. 155; Finesell, p. 125r; Fox, Ryan R, p. 164bl; FromMyEyes, p. 296b; Garrett, Mia, p. 503; Genarik, p. 375rb; goffkein.pro, p. 293cr; Gorpenyuk, Maksym, p. 67t; Grozetskaya, Tatiana, p. 124bl; haryigit, p. 110t; Idea tank, p. 515bl; jannoon028, p. 182; joshimerbin, p. 471 (Mars); JP Lavoie, p. 165cr; Karamitros, Dimitrios, p. 413b; Keni, p. 515br; Kryvenok, Anastasiia, p. 340tr; Lopatin, Viacheslav, p. 135c; loraks, p. 152; Magnetix, p. 368; Makarov, Alexandr, p. 151bl; Mastersky, p. 453 (quandong); MikeDotta, p. 315tl; Mr Privacy, p. 125l; Mueller, Ulrich, p. 316 (compost); nikcoa, p. 316 (starch); O2creationz, p. 32t; OSweetNature, p. 92bl; Othnay, Alon, p. 160; oticki, p. 87cr; Pengjan, Sitthipong, p. 501bl; petrmalinak, p. 409bl; pixelrain, p. 197t; PomInOz, pp. 249l, p. 316 (combine harvester); Provasilich, p. 262bl; Rawpixel.com, p. 188-89; rktz, p. 437; Russ, Albert, p. 145t; Salman, Vladimir, p. 340br; Schulze, Ken, p. 340bc; sergeevspb, p. 127rb; Shuttertum, p. 515bc; siamionau pavel, p. 223bl; Singkham, p. 102tr; sspopov, p. 316 (crude oil drop); Standard Studio, p. 295rb; SteAck, p. 205tl; Stiop, Alexy, p. 452; Tham KC, p. 295rc; Tinti, Stefano, p. 306; Tooykrub, p. 429; Torrenta Y, p. 163l; Urban by Habit, p. 16; Vecton, p. 378; Whiteaster, p. 193l; Zaccherini, Claudio, p. 305bb; Zaiats, Oleksandr, p. 312l.

Springer Link: 'Determination of mono- and diacylglycerols from E 471 food emulsifiers in aerosol whipping cream by high-performance thin-layer chromatography–fluorescence detection' by Oellig, C., Blankart, M., Hinrichs, J. et al.. Anal Bioanal Chem 412, 7441–7451 (2020). https://doi.org/10.1007/s00216-020-02876-2. Licensed under Attribution 4.0 International (CC BY 4.0), link to licence: https://creativecommons.org/licenses/by/4.0/.

Trützschler Switzerland AG: p. 316 (spools of fibre).

United Nations Commission for Europe (UNECE): pp. 13, 20.

Vanderkruk, Kellie: p. 518b.

Victorian Curriculum and Assessment Authority (VCAA): Selected extracts from the VCE Chemistry Study Design (2023–2027) are copyright Victorian Curriculum and Assessment Authority (VCAA), reproduced by permission. VCE® is a registered trademark of the VCAA. The VCAA does not endorse this product and makes no warranties regarding the correctness or accuracy of its content. To the extent permitted by law, the VCAA excludes all liability for any loss or damage suffered or incurred as a result of accessing, using or relying on the content. Current VCE Study Designs and related content can be accessed directly at www.vcaa.vic.edu.aupp. 1-3, 59, 99, 133, 159, 189, 215, 247, 287, 329, 349, 387, 417, 443, 469, 493, 533.

Wikimedia Commons: Benjah-bmm27, p. 279bl.

Wikipedia: Lindsay, Dan © 2009 Porsche Ceramic Composite Brakes on Porsche Carrera S (997)., p. 162b.

Groups

Key (12 Mg 24.3 / 1090 / 650 / 1.3 / [Ne]$3s^2$ / Magnesium):
- ATOMIC NUMBER — 12
- SYMBOL — Mg
- ELECTRON STRUCTURE — [Ne]$3s^2$
- RELATIVE ATOMIC MASS ( ) indicates most stable isotope — 24.3
- BOILING POINT °C — 1090
- MELTING POINT °C — 650
- ELECTRONEGATIVITY — 1.3

s BLOCK | d BLOCK | p BLOCK

Periods	1	2	3	4	5	6	7	8	9	10	11	12	13	14	15	16	17	18
1									1 **H** 1.0 -252.9 -259 2.2 $1s^1$ Hydrogen									2 **He** 4.0 -268.9 -272 – $1s^2$ Helium
2	3 **Li** 6.9 1342 180 1.0 $1s^22s^1$ Lithium	4 **Be** 9.0 2468 1287 1.6 $1s^22s^2$ Beryllium											5 **B** 10.8 4000 2077 2.0 $1s^22s^22p^1$ Boron	6 **C** 12.0 4827 3500 2.6 $1s^22s^22p^2$ Carbon	7 **N** 14.0 -195.8 -210 3.0 $1s^22s^22p^3$ Nitrogen	8 **O** 16.0 -183 -219 3.4 $1s^22s^22p^4$ Oxygen	9 **F** 19.0 -188.1 -220 4.0 $1s^22s^22p^5$ Fluorine	10 **Ne** 20.2 -246 -249 – $1s^22s^22p^6$ Neon
3	11 **Na** 23.0 882.9 97.8 0.9 [Ne]$3s^1$ Sodium	12 **Mg** 24.3 1090 650 1.3 [Ne]$3s^2$ Magnesium											13 **Al** 27.0 2519 660.3 1.6 [Ne]$3s^23p^1$ Aluminium	14 **Si** 28.1 3265 1414 1.9 [Ne]$3s^23p^2$ Silicon	15 **P** 31.0 280.5 44.2 2.2 [Ne]$3s^23p^3$ Phosphorus	16 **S** 32.1 444.6 115.2 2.6 [Ne]$3s^23p^4$ Sulfur	17 **Cl** 35.5 -34 -102 3.2 [Ne]$3s^23p^5$ Chlorine	18 **Ar** 39.9 -185.8 -189 – [Ne]$3s^23p^6$ Argon
4	19 **K** 39.1 758.8 63.4 0.8 [Ar]$4s^1$ Potassium	20 **Ca** 40.1 1484 842 1.0 [Ar]$4s^2$ Calcium	21 **Sc** 45.0 2836 1541 1.4 [Ar]$3d^14s^2$ Scandium	22 **Ti** 47.9 3287 1670 1.5 [Ar]$3d^24s^2$ Titanium	23 **V** 50.9 3407 1910 1.6 [Ar]$3d^34s^2$ Vanadium	24 **Cr** 52.0 2671 1907 1.7 [Ar]$3d^54s^1$ Chromium	25 **Mn** 54.9 2061 1246 1.6 [Ar]$3d^54s^2$ Manganese	26 **Fe** 55.8 2861 1538 1.8 [Ar]$3d^64s^2$ Iron	27 **Co** 58.9 2927 1495 1.9 [Ar]$3d^74s^2$ Cobalt	28 **Ni** 58.7 2913 1455 1.9 [Ar]$3d^84s^2$ Nickel	29 **Cu** 63.5 2560 1085 1.9 [Ar]$3d^{10}4s^1$ Copper	30 **Zn** 65.4 907 419.5 1.6 [Ar]$3d^{10}4s^2$ Zinc	31 **Ga** 69.7 2229 27.8 1.8 [Ar]$3d^{10}4s^24p^1$ Gallium	32 **Ge** 72.6 2833 938.2 2.0 [Ar]$3d^{10}4s^24p^2$ Germanium	33 **As** 74.9 613 816.8 2.2 [Ar]$3d^{10}4s^24p^3$ Arsenic	34 **Se** 79.0 684.8 220.8 2.6 [Ar]$3d^{10}4s^24p^4$ Selenium	35 **Br** 79.9 58.8 -7.1 3.0 [Ar]$3d^{10}4s^24p^5$ Bromine	36 **Kr** 83.8 -153.4 -157 – [Ar]$3d^{10}4s^24p^6$ Krypton
5	37 **Rb** 85.5 687.8 39 0.8 [Kr]$5s^1$ Rubidium	38 **Sr** 87.6 1377 769 1.0 [Kr]$5s^2$ Strontium	39 **Y** 88.9 3345 1522 1.2 [Kr]$4d^15s^2$ Yttrium	40 **Zr** 91.2 4406 1854 1.3 [Kr]$4d^25s^2$ Zirconium	41 **Nb** 92.9 4741 2477 1.6 [Kr]$4d^45s^1$ Niobium	42 **Mo** 96.0 4639 2622 2.2 [Kr]$4d^55s^1$ Molybdenum	43 **Tc** (98) 4262 2157 2.1 [Kr]$4d^55s^2$ Technetium	44 **Ru** 101.1 4147 2333 2.2 [Kr]$4d^75s^1$ Ruthenium	45 **Rh** 102.9 3695 1963 2.3 [Kr]$4d^85s^1$ Rhodium	46 **Pd** 106.4 2963 1555 2.2 [Kr]$4d^{10}5s^0$ Palladium	47 **Ag** 107.9 2162 961.8 1.9 [Kr]$4d^{10}5s^1$ Silver	48 **Cd** 112.4 766.8 321.1 1.7 [Kr]$4d^{10}5s^2$ Cadmium	49 **In** 114.8 2072 156.6 1.8 [Kr]$4d^{10}5s^25p^1$ Indium	50 **Sn** 118.7 2602 231.9 2.0 [Kr]$4d^{10}5s^25p^2$ Tin	51 **Sb** 121.8 1587 630.6 2.0 [Kr]$4d^{10}5s^25p^3$ Antimony	52 **Te** 127.6 987.8 449.5 2.1 [Kr]$4d^{10}5s^25p^4$ Tellurium	53 **I** 126.9 184.4 113.7 2.7 [Kr]$4d^{10}5s^25p^5$ Iodine	54 **Xe** 131.3 -108.1 -112 2.6 [Kr]$4d^{10}5s^25p^6$ Xenon
6	55 **Cs** 132.9 670.8 28.5 0.8 [Xe]$6s^1$ Caesium	56 **Ba** 137.3 1845 727 0.9 [Xe]$6s^2$ Barium	57 **La** 138.9 2464 920 1.1 [Xe]$5d^16s^2$ Lanthanum	72 **Hf** 178.5 4600 2233 1.3 [Xe]$4f^{14}5d^26s^2$ Hafnium	73 **Ta** 180.9 5455 3017 1.5 [Xe]$4f^{14}5d^36s^2$ Tantalum	74 **W** 183.9 5555 3414 1.7 [Xe]$4f^{14}5d^46s^2$ Tungsten	75 **Re** 186.2 5596 3454 1.9 [Xe]$4f^{14}5d^56s^2$ Rhenium	76 **Os** 190.2 5008 3033 2.2 [Xe]$4f^{14}5d^66s^2$ Osmium	77 **Ir** 192.2 4428 2446 2.2 [Xe]$4f^{14}5d^76s^2$ Iridium	78 **Pt** 195.1 3825 1768 2.2 [Xe]$4f^{14}5d^96s^1$ Platinum	79 **Au** 197.0 2836 1064 2.4 [Xe]$4f^{14}5d^{10}6s^1$ Gold	80 **Hg** 200.6 356.6 -38.8 1.9 [Xe]$4f^{14}5d^{10}6s^2$ Mercury	81 **Tl** 204.4 1473 303.8 1.8 [Xe]$4f^{14}5d^{10}6s^26p^1$ Thallium	82 **Pb** 207.2 1749 327.5 1.8 [Xe]$4f^{14}5d^{10}6s^26p^2$ Lead	83 **Bi** 209.0 1564 271.4 1.9 [Xe]$4f^{14}5d^{10}6s^26p^3$ Bismuth	84 **Po** (210) 962 253.8 2.0 [Xe]$4f^{14}5d^{10}6s^26p^4$ Polonium	85 **At** (210) 366.8 301.8 2.2 [Xe]$4f^{14}5d^{10}6s^26p^5$ Astatine	86 **Rn** (222) -61.8 -71.2 – [Xe]$4f^{14}5d^{10}6s^26p^6$ Radon
7	87 **Fr** (223) 676.8 27 0.7 [Rn]$7s^1$ Francium	88 **Ra** (226) 1140 699.8 0.9 [Rn]$7s^2$ Radium	89 **Ac** (227) 3200 1050 1.1 [Rn]$6d^17s^2$ Actinium	104 **Rf** (261) – – – [Rn]$5f^{14}6d^27s^2$ Rutherfordium	105 **Db** (262) – – – [Rn]$5f^{14}6d^37s^2$ Dubnium	106 **Sg** (266) – – – [Rn]$5f^{14}6d^47s^2$ Seaborgium	107 **Bh** (264) – – – [Rn]$5f^{14}6d^57s^2$ Bohrium	108 **Hs** (267) – – – [Rn]$5f^{14}6d^67s^2$ Hassium	109 **Mt** (268) – – – [Rn]$5f^{14}6d^77s^2$ Meitnerium	110 **Ds** (271) – – – [Rn]$5f^{14}6d^87s^2$ Darmstadtium	111 **Rg** (272) – – – [Rn]$5f^{14}6d^97s^2$ Roentgenium	112 **Cn** (285) – – – [Rn]$5f^{14}6d^{10}7s^2$ Copernicium	113 **Nh** (280) – – – [Rn]$5f^{14}6d^{10}7s^27p^1$ Nihonium	114 **Fl** (289) – – – [Rn]$5f^{14}6d^{10}7s^27p^2$ Flerovium	115 **Mc** (289) – – – [Rn]$5f^{14}6d^{10}7s^27p^3$ Moscovium	116 **Lv** (292) – – – [Rn]$5f^{14}6d^{10}7s^27p^4$ Livermorium	117 **Ts** (294) – – – [Rn]$5f^{14}6d^{10}7s^27p^5$ Tennessine	118 **Og** (294) – – – [Rn]$5f^{14}6d^{10}7s^27p^6$ Oganesson

f BLOCK

58 **Ce** 140.1 3443 799 1.1 [Xe]$4f^25d^06s^2$ Cerium	59 **Pr** 140.9 3520 931 1.1 [Xe]$4f^35d^06s^2$ Praseodymium	60 **Nd** 144.2 3074 1016 1.1 [Xe]$4f^45d^06s^2$ Neodymium	61 **Pm** (145) 3000 1042 – [Xe]$4f^55d^06s^2$ Promethium	62 **Sm** 150.4 1794 1072 1.1 [Xe]$4f^65d^06s^2$ Samarium	63 **Eu** 152.0 1794 1072 – [Xe]$4f^75d^06s^2$ Europium	64 **Gd** 157.3 3273 1313 1.2 [Xe]$4f^75d^16s^2$ Gadolinium	65 **Tb** 158.9 3230 1359 – [Xe]$4f^95d^06s^2$ Terbium	66 **Dy** 162.5 2567 1412 1.2 [Xe]$4f^{10}5d^06s^2$ Dysprosium	67 **Ho** 164.9 2700 1472 1.2 [Xe]$4f^{11}5d^06s^2$ Holmium	68 **Er** 167.3 2868 1529 1.2 [Xe]$4f^{12}5d^06s^2$ Erbium	69 **Tm** 168.9 1950 1545 1.3 [Xe]$4f^{13}5d^06s^2$ Thulium	70 **Yb** 173.1 1196 824 – [Xe]$4f^{14}5d^06s^2$ Ytterbium	71 **Lu** 175.0 3402 1663 1.0 [Xe]$4f^{14}5d^16s^2$ Lutetium
90 **Th** 232.0 4875 1750 1.3 [Rn]$5f^06d^27s^2$ Thorium	91 **Pa** 231.0 – 1572 1.5 [Rn]$5f^26d^17s^2$ Protactinium	92 **U** 238.0 4131 1135 1.7 [Rn]$5f^36d^17s^2$ Uranium	93 **Np** (237) 3902 644 1.3 [Rn]$5f^46d^17s^2$ Neptunium	94 **Pu** (244) 3228 640 1.3 [Rn]$5f^66d^07s^2$ Plutonium	95 **Am** (243) 2011 1176 – [Rn]$5f^76d^07s^2$ Americium	96 **Cm** (247) 3110 1340 – [Rn]$5f^76d^17s^2$ Curium	97 **Bk** (247) – 986 – [Rn]$5f^96d^07s^2$ Berkelium	98 **Cf** (251) – 900 – [Rn]$5f^{10}6d^07s^2$ Californium	99 **Es** (252) – – – [Rn]$5f^{11}6d^07s^2$ Einsteinium	100 **Fm** (257) – – – [Rn]$5f^{12}6d^07s^2$ Fermium	101 **Md** (258) – 827 – [Rn]$5f^{13}6d^07s^2$ Mendelevium	102 **No** (259) – – – [Rn]$5f^{14}6d^07s^2$ Nobelium	103 **Lr** (262) – – – [Rn]$5f^{14}6d^17s^2$ Lawrencium